Table for flow conversion

| Unit | $m^3$/sec | $m^3$/day | ℓ/sec | $ft^3$/sec | $ft^3$/day | ac-ft/day | gal/min | gal/day | mgd |
|---|---|---|---|---|---|---|---|---|---|
| 1 cubic meter/ second | 1 | $8.64 \times 10^4$ | $10^3$ | 35.31 | $3.051 \times 10^6$ | 70.05 | $1.58 \times 10^4$ | $2.282 \times 10^7$ | 22.824 |
| 1 cubic meter/ day | $1.157 \times 10^{-5}$ | 1 | 0.0116 | $4.09 \times 10^{-4}$ | 35.31 | $8.1 \times 10^{-4}$ | 0.1835 | 264.17 | $2.64 \times 10^{-4}$ |
| 1 liter/ second | 0.001 | 86.4 | 1 | 0.0353 | 3051.2 | 0.070 | 15.85 | $2.28 \times 10^4$ | $2.28 \times 10^{-2}$ |
| 1 cu. foot/ second | 0.0283 | 2446.6 | 28.32 | 1 | $8.64 \times 10^4$ | 1.984 | 448.8 | $6.46 \times 10^5$ | 0.646 |
| 1 cu. foot/ day | $3.28 \times 10^{-7}$ | 0.02832 | $3.28 \times 10^{-4}$ | $1.16 \times 10^{-5}$ | 1 | $2.3 \times 10^{-5}$ | $5.19 \times 10^{-3}$ | 7.48 | $7.48 \times 10^{-6}$ |
| 1 acre-foot/ day | 0.0143 | 1233.5 | 14.276 | 0.5042 | 43,560 | 1 | 226.28 | $3.259 \times 10^5$ | 0.3258 |
| 1 gallon/ minute | $6.3 \times 10^{-5}$ | 5.451 | 0.0631 | $2.23 \times 10^{-3}$ | 192.5 | $4.42 \times 10^{-3}$ | 1 | 1440 | $1.44 \times 10^{-3}$ |
| 1 gallon/ day | $4.3 \times 10^{-8}$ | $3.79 \times 10^{-3}$ | $4.382 \times 10^{-5}$ | $1.55 \times 10^{-6}$ | 0.11337 | $3.07 \times 10^{-6}$ | $6.94 \times 10^{-4}$ | 1 | $10^{-6}$ |
| 1 million gallons/ day | $4.38 \times 10^{-2}$ | 3785 | 43.82 | 1.55 | $1.337 \times 10^5$ | 3.07 | 694 | $10^6$ | 1 |

SECOND EDITION

# APPLIED HYDROGEOLOGY

C. W. FETTER

University of Wisconsin—Oshkosh

Merrill Publishing Company
A Bell & Howell Information Company
Columbus Toronto London Melbourne

*This book is dedicated to my wife,*
*Nancy, and my children: Bill, Rob, and Elizabeth.*

Cover Art: Michael Linley

Published by Merrill Publishing Company
A Bell & Howell Information Company
Columbus, Ohio 43216

This book was set in Times Roman and Optima.

Administrative Editor: David Gordon
Production Coordinator: Molly Kyle
Art Coordinator: Mark Garrett
Cover Designer: Cathy Watterson
Text Designer: Cynthia Brunk



Library of Congress Catalog Card Number: 87-62679
International Standard Book Number: 0-675-20887-4
Printed in the United States of America
2 3 4 5 6 7—91 90 89 88

# Preface

Since the publication of the first edition of *Applied Hydrogeology,* there has been a tremendous growth in need for persons trained in hydrogeology as well as a greater desire for basic hydrogeological training among people who are expert primarily in other fields. This demand results from greater understanding of the fragile nature of ground-water quality and the realization that ground-water contamination is more prevalent than previously known. Consequently, this second edition emphasizes methods of site characterization, ground-water monitoring, contaminant hydrogeology, and computer modeling. I have drawn on the consulting I have done in recent years for new knowledge about contaminant hydrogeology, additional practical methods of site characterization, and material for new case studies.

*Applied Hydrogeology* is intended as an introductory textbook for courses in hydrogeology at either advanced undergraduate or dual graduate/undergraduate levels and as a reference book for the working professional. The reader is expected to have a working knowledge of college algebra; calculus is helpful, but not necessary for practical understanding of the material. A background in college chemistry is necessary to understand the chapter on Water Chemistry. The book stresses application of mathematics to problem solving rather than derivation of theory. To this end, you will find many example problems with step-by-step solutions. Case studies in many chapters enhance understanding of the occurrence and movement of ground water in a variety of geologic settings. The expanded appendices, which are tables of various functions and tables for unit conversions, provide additional data for solving problems in well hydraulics, water chemistry, and contaminant transport. A glossary of hydrogeological terms makes the book a valuable reference.

I am grateful to all who helped me in this project. The draft manuscript was carefully reviewed by Dr. Jay Lehr of the National Water Well Association, Professor Max L. Anderson of the University of Wisconsin—Platteville, Professor June A. Oberdorfer of San Jose State University, and Professor David Huntley of San Diego State University. Dr. Richard Parizek of Pennsylvania State

University furnished the photographs that appear in Chapter 12. Dr. Robert A. Griffin of the Illinois State Geological Survey served as a sounding board for many ideas on contaminant transport and furnished some original data. In addition, Professor Ernst Kastning of Radford University reviewed the manuscript for accuracy. My thanks also to Charles Joslin, Sue Birch, and Mark Garrett, whose efforts resulted in an excellent art program, and to Gail Zaucha for compiling the index. In the past eight years I have had many contacts with professional hydrogeologists, each of whom has shared insights into the application of hydrogeology to solve practical problems. Finally, I wish to thank all the students in my classes who used the first edition and made many helpful comments on possible improvements.

The National Water Well Association has granted permission to use figures from *Ground Water* and the *Ground Water Monitoring Review;* the American Geophysical Union has permitted me to use figures from *Water Resources Research, Journal of Geophysical Research* and *Transactions (EOS).*

C. W. Fetter

# Foreword

In 1953, I enrolled as a chemical engineering student at Princeton University, only to find that my sensitivity to the malodorousness of the laboratory drove me out-of-doors and into the arms of the Geologic Engineering faculty, who promised me a bright future with lots of fresh air. Once there, I cast about for a specialty field and an undergraduate thesis topic. In a roundabout way, I found it one day on the cover of *U.S. News and World Report,* which asked the rhetorical question, "Is the U.S. Running Out of Water?" This article proved to be the first national news story on a subject that the media now deal with daily. It opened up for me an insatiable fascination with a liquid once joked about as "a great drink, but one that will never sell." How wrong we were. Back at the Geological Engineering Department, the faculty focused my interest in water under the ground; the field of ground-water geology was barely crawling out of its crib. My studies thus began in 1954 with the only American textbook on the subject, *Ground Water,* written by C. F. Tolman way back in 1937. Dr. Tolman held this educational monopoly for over two decades, before David K. Todd brought matters a bit more up-to-date with his 1959 text, *Ground Water Hydrology*. Todd's book also stood alone for a few years, until Stan Davis and Roger Deweist published their 1966 book, *Hydrogeology*. But it was Allan Freeze and John Cherry who opened the floodgates of ground-water literature with their 1979 book, *Ground Water*. Since that year, every major publisher and quite a few minor ones have enlisted a ground-water scientist to pen his version of the state of our art. The preponderance of the texts are good and useful, but tend to be unbalanced because of the authors' natural biases toward their specialties, which run the gamut from well hydraulics to contaminant transport to mathematical modeling. An exception to this rule is the book now in your hands or on your desk. Bill Fetter has approached ground-water hydrology with a balanced viewpoint and a desire to arm the practitioner with a full array of tools and a detailed perception of how each facet of our field of study integrates with the others to yield a clear perspective of our conceptual physical model of how ground water occurs, moves, affects, and is affected by our planet earth.

Bill Fetter is an academic with a pragmatic bent. He is as at home in the classroom as on a consulting assignment. He began his career as a staff hydrogeologist for a consulting firm and now divides his time between teaching and consulting. In preparing this revision, he mined the recently published hydrogeologic literature, attended numerous professional meetings and short courses, and then tested this newly gained knowledge in field applications. For example, Bill has been an expert consultant in hydrogeology to the United States Environmental Protection Agency, in which role he designed and directed the hydrogeological study and ground-water monitoring program at a Superfund hazardous waste site in Seymour, Indiana. This project was a state-of-the-art investigation of organic chemical contamination of ground water, described as a case study in Chapter 10 of this edition.

Only a decade ago, our ranks were still dominated by geologists. Their interests beyond ground-water geology usually focused on well hydraulics. Today, we have been joined by chemists, biologists, mathematicians, and civil and environmental engineers whose primary interests are contaminant transport. Though, collectively, all our colleagues show a thoughtful concern for water management and the legislative framework within which it must operate, Bill has dealt with it all in a particularly practical way.

Since the first edition of *Applied Hydrogeology* appeared in 1980, our understanding of many of the factors that control contaminant transport has advanced dramatically. Our ability to obtain representative ground-water samples from monitoring wells has improved markedly, and the nearly universal appreciation and utilization of numerical ground-water models has permanently altered the practice of ground-water hydrology. Bill deals with these issues in this revised text in new or significantly expanded chapters.

Thus, as an introductory college text in ground-water hydrology, this book offers superior opportunities to teacher and student alike. For the new immigrant to ground water from a related discipline, few books can promote self-learning as well as this as a result of the self-effacing nature of the narrative which happily lacks the pompous arrogance too many of us exhibit in our erudite tomes.

Finally, as a ready reference on nearly all subjects relevant to ground-water science and as a practical glossary of significant terms, *Applied Hydrogeology* belongs on every practitioner's bookshelf.

These are exciting times in ground-water science. In a relatively short time, we have moved from nearly total anonymity to frequent celebrity in the scientific community. Our ranks have swelled from a relatively exclusive club of reticent, yet often smug specialists, to an effusive population of highly trained, enthusiastic, and often optimistic men and women enjoying the spotlight of the nation's environmental concerns. When our field first began to achieve public recognition and when the government first seriously established goals to protect our valuable natural resource, Charles L. McGuinness, a former Chief of the Ground Water Branch of the U.S. Geological Survey, said, with valid concern:

> Of all the things that might be said about ground water in today's world, one that seems highly appropriate to me is an expression of amazement. After years—in fact, decades—in which students of ground water felt that we were just voices crying in the wilderness, the world has suddenly discovered our subject. Now we don't know whether to laugh or to cry, because the world suddenly wants from us more than we have to give—more knowledge than was ever demanded before, and more than we ever dreamed would be needed.

In the nearly two decades since those words were spoken, history is proving that we have been equal to the task and are able to provide for society's needs in developing and protecting our vast ground-water resources. Bill Fetter's contribution is one reason for my confidence.

Jay Lehr

# Contents

## THREE
## Runoff and Streamflow 37



## FOUR
## Soil Moisture and Ground Water 63

## FIVE
## Principles of Ground-Water Flow 115



## SIX
## Ground-Water Flow to Wells 161

## ELEVEN
## Ground-Water Development and Management 443



## TWELVE
## Field Methods 479

## THIRTEEN
## Ground-Water Models 525



## Case Studies

# ONE

# Water

If anyone be too lazy to keep his dam in proper condition, and does not keep it so; if then the dam breaks and all the fields are flooded, then shall he in whose dam the break occurred be sold for money and the money shall replace the corn which he has caused to be ruined.

Code of Hammurabi, Section 53 (1760 B.C.)

## 1.1 WATER

Water is the elixir of life; without it life is not possible. Although many environmental factors determine the density and distribution of vegetation, one of the most important is the amount of precipitation. Agriculture can flourish in some deserts, but only with water either pumped from the ground or imported from other areas. Civilizations have flourished with the development of reliable water supplies—and then collapsed as the water supply failed. This is a book about the occurrence of water, both at the surface and in the ground.

A person requires about 3 quarts (3 liters) of potable water per day to maintain the essential fluids of the body. Primitive people in arid lands exist with this amount as their total consumption. A single cycle of a flush toilet may use 6 gallons (23 liters) of water. In New York City the per capita water usage exceeds 260 gallons (1000 liters) daily. Much of this use is for industrial, municipal, and commercial purposes; for personal purposes, the typical American uses 50 to 80 gallons (200 to 300 liters) per day. Even greater quantities of water are required for energy and food production.

In 1985 the total offstream water use in the United States has been estimated to be 404 billion gallons (1530 liters) per day of fresh and saline water. This does not include water used for hydroelectric power generation and other instream uses, but does include water used for thermoelectric power plant cooling. Fresh-water use in 1985 included 75 billion gallons (284 billion liters) per day of ground water and 270 billion gallons (1022 billion liters) per day of surface water. Per capita fresh-water use was 1440 gallons (5450 liters) per day. Consumptive use of water—that is, water evaporated during use—was about 110 billion gallons (416 billion liters) per day (1).

A common goal of all countries is to increase economic production. It is generally thought this will result in a better lifestyle for the citizens. While the validity of this assumption has been questioned by some individuals in the heavily industrialized countries, the goal of governments at all levels still remains to promote economic expansion. This will naturally increase per capita water usage, although water and energy conservation may help. Inasmuch as the populations of most countries are growing, it is likely the total use of water will increase, even with conservation measures.

The United States has had a history of increasing water use. Figure 1.1 illustrates the withdrawal of fresh water in the United States for offstream uses such as municipal supply, industrial, domestic, irrigation, and electrical power plant cooling. Between 1950 and 1980 total fresh-water use increased by 117 percent while per capita fresh-water use increased by 42 percent (Figure 1.2).

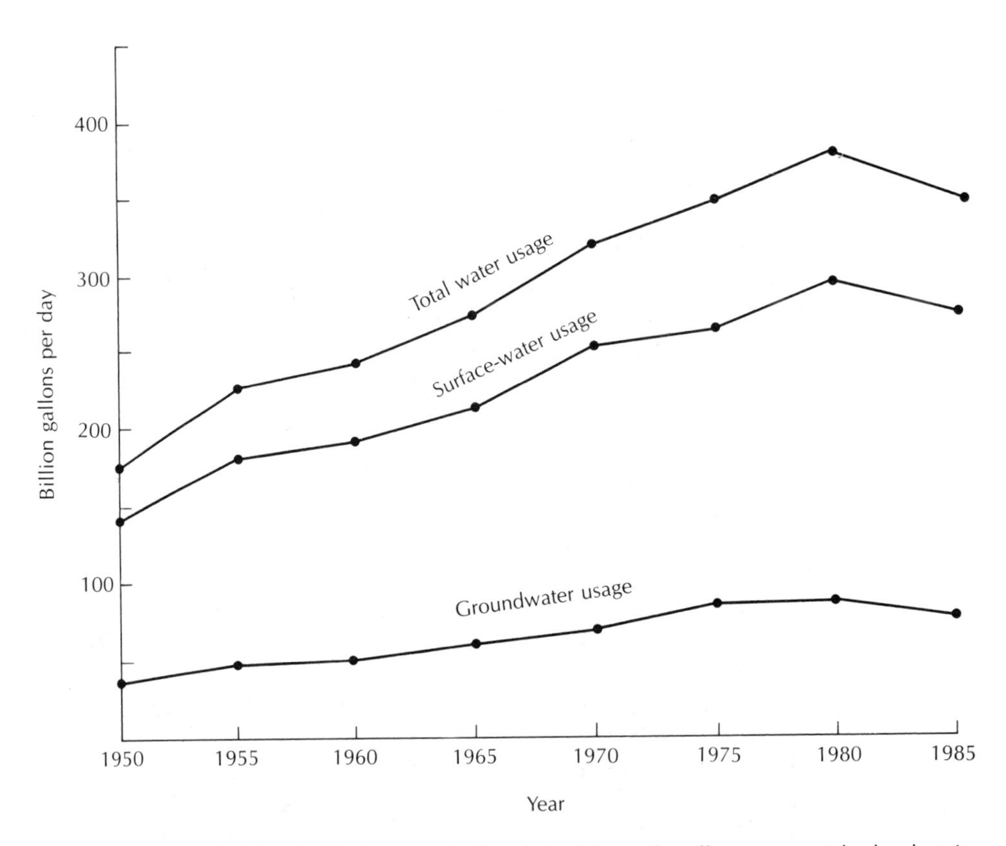


**FIGURE 1.1** Withdrawal of fresh water in the United States for all uses except hydroelectric power generation. Source: U.S. Geological Survey (1–6).

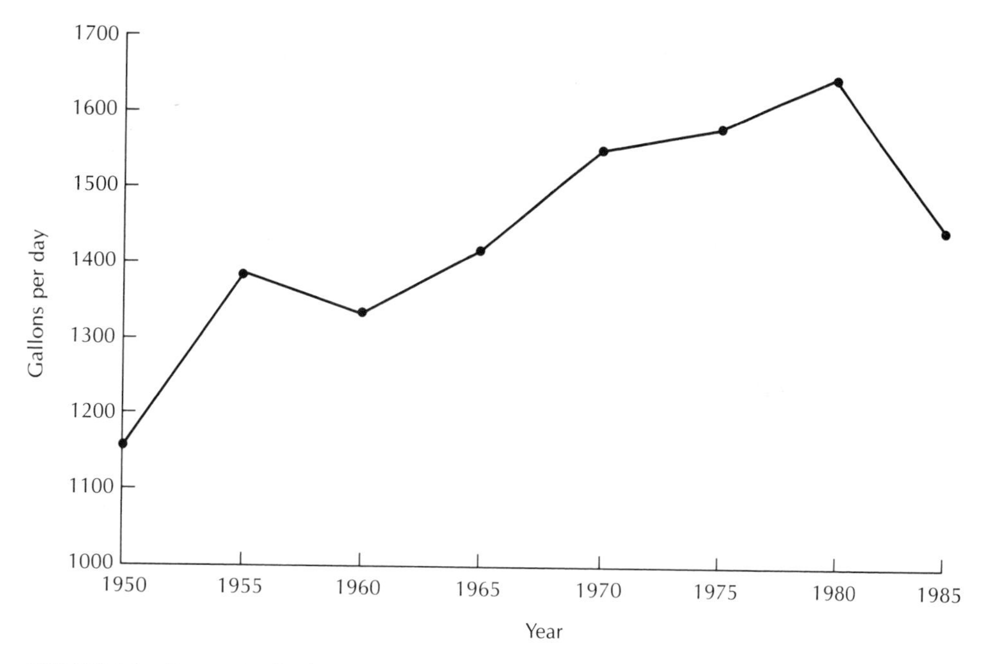


**FIGURE 1.2** Per capita fresh-water usage in the United States for all uses except hydroelectric power generation. Source: U.S. Geological Survey (1, 6).

Between 1980 and 1985 fresh-water use declined for the first time, from 373 billion gallons (1412 billion liters) per day to 345 billion gallons (1306 billion liters) per day. The 1985 fresh ground-water use of 75 billion gallons (284 billion liters) was less than in either 1980 (83 billion gallons per day) or 1975 (82 billion gallons per day) (1, 6). This may indicate that the trend of increasing water use has been broken.

## 1.2 HYDROLOGY AND HYDROGEOLOGY

As viewed from a spacecraft, the earth appears to have a blue-green cast owing to the vast quantities of water covering the globe. The oceans may be obscured by billowing swirls of clouds. These vast quantities of water distinguish Earth from the other planets in the solar system.

**Hydrology** is the study of water. In the broadest sense, hydrology addresses the occurrence, distribution, movement, and chemistry of all waters of the earth. **Hydrogeology** encompasses the interrelationships of geologic materials and processes with water. (A similar term, **geohydrology,** is sometimes used as a synonym for hydrogeology, although it more properly describes an engineering field dealing with subsurface fluid hydrology.)

The physiography, surficial geology, and topography of a drainage basin, together with the vegetation, influence the relationship between precipitation over the basin and water draining from it. The creation and distribution of precipitation is heavily influenced by the presence of mountain ranges and other topographic features. Running water and ground water are geologic agents that help shape the land. The movement and chemistry of ground water is heavily dependent upon geology.

Hydrogeology is both a descriptive and an analytic science. The development and management of water resources are important parts of hydrogeology as well.

## 1.3 THE HYDROLOGIC CYCLE

An account of the water supply of the world would reveal that saline water in the oceans accounts for 97.2 percent of the total. Land areas hold 2.8 percent of the total. Ice caps and glaciers hold 2.14 percent; ground water to a depth of 13,000 feet (4000 meters) accounts for 0.61 percent of the total; soil moisture 0.005 percent; fresh-water lakes 0.009 percent; rivers 0.0001 percent; and saline lakes 0.008 percent (5). Over 75 percent of the water in land areas is locked in glacial ice or is saline (Figure 1.3).

Only a small percentage of the world's total water supply is available to humans as fresh water. More than 98 percent of the available fresh water is ground water, which far exceeds the volume of surface water. At any given time, only 0.001 percent of the total supply of water is in the atmosphere. However, atmospheric water circulates very rapidly, so that each year enough water falls to cover the conterminous United States to a depth of 30 inches (75 centimeters) (7). Of this amount, 22 inches (55 centimeters) are returned to the atmosphere through evaporation and transpiration by growing plants, while 8 inches (20 centimeters) flow into the oceans as rivers (7). Although the previous sentence implies that the **hydrologic cycle** begins with water from the oceans, the cycle actually has no beginning and no end. As most of the water is in the oceans, it is convenient to describe the hydrologic cycle as starting with the oceans.

Water evaporates from the surface of the oceans. The amount of evaporated water varies, being greatest near the equator, where solar radiation is more intense. Evaporated water is pure, since when it is carried into the atmosphere the salts of the sea are left behind. Water vapor moves through the atmosphere as an integral part of the phenomena we term "the weather." When atmospheric conditions are suitable, water vapor condenses and forms droplets. These droplets may fall to the sea or onto land or may revaporize while still aloft.

Precipitation that falls on the land surface enters into a number of different pathways of the hydrologic cycle. Some of the water will drain across the land into a stream channel. This is termed **overland flow.** If the surface soil is porous, some water will seep into the ground by a process termed **infiltration.** The water clings to soil particles, and this soil moisture may be drawn into the

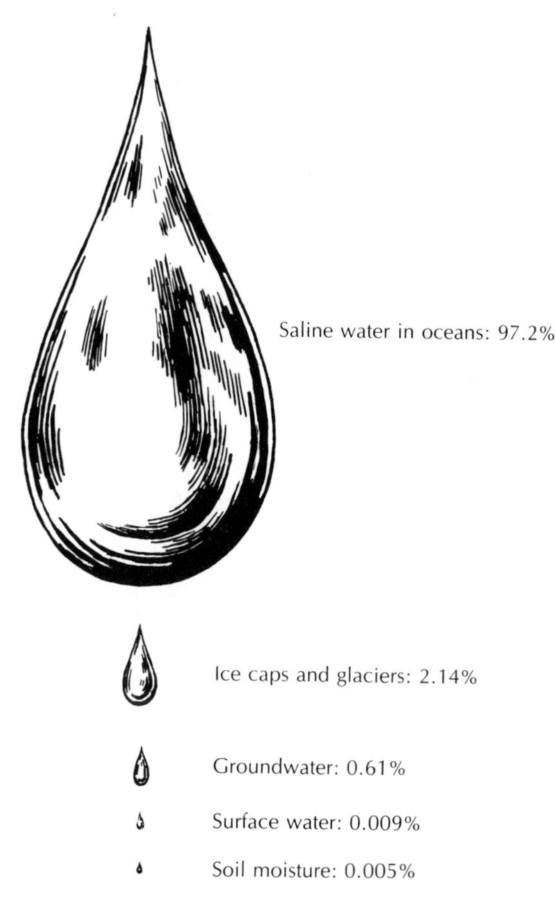


**FIGURE 1.3** Distribution of world's water supply.

rootlets of growing plants. After the plant uses the water, it is transpired as vapor into the atmosphere.

Excess soil moisture is pulled downward by gravity. At some depth, the soil or rock is saturated with water. The top of the saturated zone is the **water table,** and below the water table is **ground water.** Ground water flows through the rock and soil layers of the earth until it discharges as a spring or as seepage into a stream, lake, or ocean (Figure 1.4).

Water flowing in a stream can come from overland flow or from ground water that has seeped into the stream bed. The ground-water contribution to a stream is termed **baseflow,** while the total flow in a stream is **runoff.**

Evaporation is not restricted to open water bodies, such as the ocean, lakes, streams, and reservoirs. Precipitation intercepted by leaves and other vegetative surfaces can also evaporate, as can water detained in land-surface

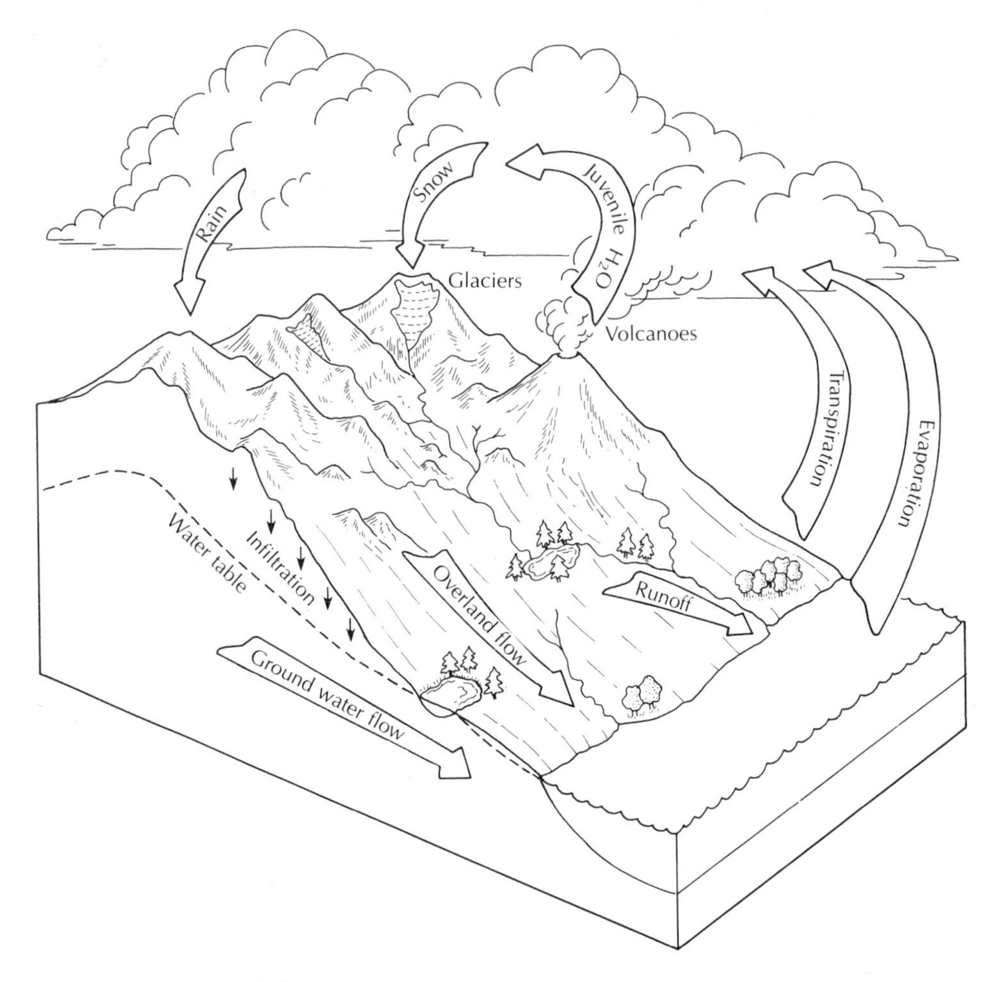


FIGURE 1.4 The hydrologic cycle.

depressions or soil moisture in the upper layers of the soil. Direct evaporation of ground water can take place when the saturated zone is at or near the land surface.

## 1.4 ENERGY TRANSFORMATIONS

The hydrologic cycle is an open system in which solar radiation serves as a source of constant energy. This is most evident in the evaporation and atmospheric circulation of water. The energy of a flowing river is due to the work done by solar energy evaporating water from the ocean surface and lifting it to higher elevations, where it falls to earth.

When water changes from one state to another (liquid, vapor, or solid), there is an accompanying change in the heat energy of the water. The **heat energy** is the amount of thermal energy contained by a substance. A **calorie** of heat is defined as the energy necessary to raise the temperature of one gram of pure water from 14.5° C to 15.5° C. The evaporation of 1 gram of water at 15° C requires an input of 590 calories of energy (the **latent heat of vaporization**). The water vapor retains this added heat energy. When water vapor condenses to the liquid form, 590 calories per gram are released (the **latent heat of condensation**).

In order to melt one gram of ice at 0° C, 80 calories of heat, the **latent heat of fusion,** must be added. The resultant temperature of the water is still 0° C, although a gram of water at 0° C has more heat energy than an equivalent weight of ice. Water can pass from the solid state to the vapor state (sublimation) with the addition of 700 calories per gram. Formation of frost and freezing of water result in the release of heat energy.

The transportation of water through the hydrologic cycle and the accompanying heat transfers are vital to the heat balance of the earth. At the equator, the amount of solar radiation is fairly constant through the year, while at the poles it varies from zero during the polar winter to amounts greater than those at the equator during the polar summer. During polar winters the land is in shadow, as the sun does not strike the ground. On the other hand, during the summers, the sun shines continuously. Over the year, the Northern Hemisphere northward of 38° latitude has a net heat loss, as the outgoing terrestrial radiation to space exceeds the incoming solar radiation that is absorbed. Between the equator and 38° N, there is more solar radiation absorbed than terrestrial radiation lost to space. In order to balance these anomalies, heat is transferred by currents in the oceans and through the atmosphere as movement of air masses and water vapor. This creates climatic conditions and changing weather patterns that profoundly affect the hydrologic cycle.

## 1.5 THE HYDROLOGIC EQUATION

The hydrologic cycle is a useful concept, but is quantitatively rather vague. The **hydrologic equation** provides a quantitative means of evaluating the hydrologic cycle. This fundamental equation is a simple statement of the **law of mass conservation.** It may be expressed as

$$\text{Inflow} = \text{Outflow} \pm \text{Changes in Storage}$$

If we consider any hydrologic system—for instance, a lake—it has a certain volume of water at a given time. There are a number of inflows that add water: precipitation that falls on the lake surface, streams that flow into the lake, ground water that seeps into the lake, and overland flow from nearby land surfaces. Water also leaves the lake through evaporation, transpiration by emergent aquatic vegetation, outlet streams, and ground-water seepage from the lake bottom. If, over a given period of time, the total inflows are greater than the total outflows, the lake level will rise as more water accumulates. If the outflows exceed the inflows over a time period, the volume of water in the lake will decrease. Any differences between rates of inflow and outflow in a hydrologic system will result in a change in the volume of water stored in the system.

The hydrologic equation can be applied to systems of any size. It is as useful for a small reservoir as it is for an entire continent. The equation is time-dependent. The elements of inflow must be measured over the same time periods as the outflows.

The application of the hydrologic equation to a watershed is illustrated in the following case study.

---

## CASE STUDY: MONO LAKE

*Half a dozen little mountain brooks flow into Mono Lake, but not a stream of any kind flows out of it. What it does with its surplus water is a dark and bloody mystery.*

Mark Twain

Mono Lake lies on the eastern slope of the Sierra Nevada near the east entrance to Yosemite National Park. Mono Lake is a terminal lake, which means that although water enters the lake by precipitation and by streams and ground water flowing into the lake, it can leave only by evaporation. The lake level fluctuates with climatic changes. The volume of water that leaves the lake by evaporation is the product of the surface area times the depth of evaporation. If the volume that leaves by evaporation is exactly balanced by the inflow, the lake level will not change. If the inflow exceeds evaporation, the water level will rise. If the inflow is less than evaporation, the lake level will fall.

When it was first surveyed in 1856, the elevation of Mono Lake was 6407 feet above sea level. Climatic effects of moister and drier periods caused the lake level to rise to as much as 6428 feet in 1919 and then to fall to 6410 feet by 1941. In that year water was first diverted from four of the five major streams feeding Mono Lake into the Los Angeles Aqueduct and thence to southern California.

The lake had a surface area of 53,500 acres when the surface was at an elevation of 6410 feet. Because of diversions averaging 100,000 acre-feet per year, by 1981 the lake elevation was reduced to 6371.5 feet. At this elevation the lake has a surface area of about 40,000 acres. The annual diversion of 100,000 acre-feet would cover the 40,000-acre lake stage to a depth of 2.5 feet. The water level is falling because the amount of diversions from the Mono Lake basin plus the natural evaporation is far in excess of the amount of precipitation plus surface- and ground-water drainage into the basin. If the 100,000 acre-feet per year diversion is continued, it is estimated that the lake will stabilize at an elevation of 6323 feet, with a corresponding surface area of 22,500 acres. At that size Mono Lake will be small enough that evaporation from the lake will be reduced to the amount of direct precipitation into the lake plus the volume of inflow from the ground water and undiverted streamflow so that the hydrologic equation will again balance. Should the diversions be reduced or eliminated, the level of Mono Lake would rise.

One consequence of the reduction of the volume of Mono Lake has been an increase in the salinity. In its original, natural condition Mono Lake had a salinity of 5.2 percent. At the 1981 elevation it was 9.5 percent and with continued diversions could reach 27 percent. The increasing salinity has resulted in a reduction of the brine shrimp and brine fly populations of the lake. Ultimately this may disrupt an important part of the food chain for the Mono Lake basin as many nesting and migratory birds feast on the tiny brine shrimp and brine flies (8).

## 1.6 HYDROGEOLOGISTS

The professional hydrogeologist has a wide variety of occupations from which to choose. Employment may be found with federal agencies, United Nations groups, state agencies, and local government. Energy and mining companies may call upon the services of hydrogeologists to help provide water where it is needed or perhaps remove it where it is unwanted. Private consulting organizations also employ many individuals trained in hydrogeology. Water resource management districts and planning agencies often include hydrogeologists on their staffs.

Hydrogeology is an interdisciplinary field. The hydrogeologist usually has training in geology, hydrology, chemistry, mathematics, and physics. Hydrogeologists are also being trained in such areas of engineering as fluid mechanics and flow through porous media, as well as in computer science. Such training is necessary, as hydrogeologists must be able to communicate effectively with engineers, planners, ecologists, resource managers, and other professionals. By the same token, an understanding of the basic principles of hydrogeology is useful to soil scientists, engineers, planners, foresters, and others in similar fields. Modeling of hydrologic systems is another area requiring knowledge of a number of disciplines.

## 1.7 APPLIED HYDROGEOLOGY

The argument has been made that "hydrogeology is more than a classical science" (9). Traditional studies in hydrogeology have focused on either the mathematical treatment of flow through porous media or on a general geologic description of the distribution of rock formations in which ground water occurs. One occasionally even finds a paper describing the theoretical flow of fluids through an idealized porous medium that probably does not occur in nature. Likewise, many reports on the ground-water geology of an area make no attempt to evaluate how much water is available for use. Neither type of study has much practical value in and of itself.

Hydrogeologists are being employed as problem solvers and decision makers (9). They need to identify a problem, define the data needs, design a field program for collection of data, propose alternative solutions to the problem, and implement the preferred solution.

Hydrogeology is also being recognized by both hydrogeologists (10) and planners (11) as an important part of environmental planning. As population and economic growth expand, human use of the natural environment is becoming more intense. Incidents of environmental degradation are less likely to occur when the planner and the hydrogeologist cooperate.

The need for hydrogeological services has never been greater. The traditional role of the hydrogeologist has been the location and development of ground-water supplies. In many parts of the world there is an urgent need for the development of new ground-water supplies. Expanding population in many developing countries has put pressure on available water supplies. Surface-water supplies are especially susceptible to bacterial and viral contamination, especially where sanitation is primitive. Drought in some areas of Africa has now persisted for a generation. Inexpensive water wells in such areas can be virtual lifesavers.

In many developed countries disposal of industrial and municipal waste has resulted in the contamination of ground water with such things as hydrocarbons (12), pesticides (13, 14), radionuclides (15), organic solvents (16, 17), and metals (18). Agricultural practices have also resulted in degradation of ground and surface water (19), as have septic tanks (20), road salting (21), municipal landfills (22), and mining (23).

While there is great cause for concern over the amount of ground-water contamination, it has been pointed out that perhaps only one or two percent of the available ground-water resource has been contaminated (24). Yet, the areas where contamination has ocurred are frequently the areas of more concentrated population, where the water is in greatest need (25). This brings us to the two areas where hydrogeologists can make the most significant impact and greatest contribution to society in the industrialized nations: proper siting, design, and construction of land disposal facilities so as to avoid future ground-water contamination (26) and restoration of water quality in aquifiers that have been contaminated (27).

Hydrogeological studies are an important part of the overall analysis of sites proposed for major construction projects. Coal and nuclear power plants, dams, tunnels, pumped storage reservoirs, and land-treatment systems for wastewater are typical examples of such projects. Solution mining of water-soluble uranium ore is another application of hydrogeology. Hydrogeology is also fundamental to environmental-impact analysis.

A glance toward the future suggests there will be fewer natural resources, more intensive use of the land, and a greater desire to promote environmental conservation along with economic growth. The challenges to the hydrogeologist will grow, as will the opportunities.

## 1.8 SOURCES OF HYDROGEOLOGIC INFORMATION

Hydrogeologic information is available from a wide range of sources. In terms of sheer volume, the Water Resources Division of the U.S. Geological Survey is the leading source in the United States. This agency collects basic data on streamflow, surface-water quality, ground-water levels, and ground-water quality. The USGS also conducts water resources investigations and basic research. USGS publications are available in libraries that are designated depositories of federal documents; these publications are also available from the U.S. Government Printing Office.

The U.S. Geological Survey maintains a computerized central storage facility for water resources data called the National Water Data Storage and Retrieval System, generally known by the acronym WATSTORE (28). Ground-water data in WATSTORE can be accessed by authorized users by means of an on-line computer retrieval system known as the Ground-Water Site Inventory (GWSI) file (29). The ability to access the tens of thousands of data files on both ground and surface water held by the USGS is a powerful tool. The GWSI system has the ability to reduce the data to make *x-y* plots, graphs, and tables.

The National Water Well Association maintains Ground Water On-Line, a computerized data base of bibliographic information (30). As of 1986 there were more than 50,000 documents indexed by more than 700 hydrogeological descriptors. New documents were being indexed at a rate of 500 per month. A bibliographic search can be conducted by computer which will seek out all of the indexed documents that correspond to the selected descriptors. The search can be conducted on-line by using a personal computer and a modem. An inter-library loan service is also available to obtain copies of the articles the data base search found. There are many other bibliographic data base search services as well that might be helpful to the hydrogeologist.

The National Oceanic and Atmospheric Administration is the parent organization of the Weather Bureau. *The Climatic Record of the United States* is published for each state and contains precipitation, temperature, evaporation, and other climatic data. Other U.S. federal agencies that may conduct studies related to hydrogeology include the Corps of Engineers, Bureau of Land Management, Bureau of Reclamation, Soil Conservation Service, Environmental Protection Agency, Nuclear Regulatory Agency, and Department of Energy.

In most states, there are one or more agencies responsible for water-oriented research and other activities. The functions, responsibilities, and organizational formats of state agencies in water resources activities vary from state to state. Typical agency designations include State Department of Water Resources or Water Survey, State Geological Survey, Department of Conservation or Natural Resources, and State Department or Board of Health. In many states, various responsibilities are allocated among several agencies. In addition, Congress has established provisions for a water resources research center or institute in each state and Puerto Rico. These are associated with a major university in each state.

Reports of current research and recent developments in hydrogeology and ground water are included in the following journals:

- ☐ *Bulletin, International Association of Scientific Hydrology*
- ☐ *Ground Water*

- ☐ *Ground Water Monitoring Review*
- ☐ *Journal American Water Works Association*
- ☐ *Journal of Hydrology*
- ☐ *Memoirs, International Association of Hydrogeologists*
- ☐ *Transactions, American Society of Civil Engineers*
- ☐ *Water Resources Bulletin*
- ☐ *Water Resources Research*

A number of professional organizations sponsor symposia and meetings where technical sessions on hydrogeology or ground water are held. These include the following:

- ☐ American Geophysical Union
- ☐ American Institute of Hydrology
- ☐ American Society of Civil Engineers
- ☐ American Water Resources Association
- ☐ Geological Society of America
- ☐ Geological Society of Canada
- ☐ International Association of Hydrogeologists
- ☐ International Association of Scientific Hydrology
- ☐ International Water Resources Association
- ☐ National Water Well Association

## REFERENCES


1. SOLLEY, W. B. U.S. Geological Survey, personal communication of preliminary data, 1987.
2. MURRAY, C. R., and E. B. REEVES. *Estimated Use of Water in the United States, 1970*. U.S. Geological Survey Circular 676, 1972, 37 pp.
3. MAC KICHAN, K. A., and J. C. KAMMERER. *Estimated Use of Water in the United States, 1960*. U.S. Geological Survey Circular 456, 1961, 26 pp.
4. MURRAY, C. R. *Estimated Use of Water in the United States, 1965*. U.S. Geological Survey Circular 556, 1968, 53 pp.
5. MURRAY, C. R., and E. B. REEVES. *Estimated Use of Water in the United States, 1975*. U.S. Geological Survey Circular 765, 1977, 39 pp.
6. SOLLEY, W. B., E. B. CHASE, and W. B. MANN IV. *Estimated Use of Water in the United States in 1980*. U.S. Geological Survey Circular 1001, 1983, 56 pp.
7. Federal Council for Science and Technology. *Scientific Hydrology*. Washington, D.C., U.S. Government Printing Office, 1962.
8. ANON. "Mono Lake: Endangered Oasis." Position Paper of the Mono Lake Committee. Lee Vining, Calif., 1985, 45 pp.
9. STEPHENSON, D. "Hydrogeology Is More than a Classical Science." *Ground Water,* 12, no. 3 (1974):148–51.

**10.** SOMMERS, D. A. "Put Hydrogeology into Planning" *Ground Water,* 8, no. 6 (1970):2–7.

**11.** MC HARG, I.L. *Design with Nature*. Garden City, N.Y.: The Natural History Press, 1969, 197 pp.

**12.** KRAMER, W. H. "Ground-Water Pollution from Gasoline." *Ground Water Monitoring Review,* 2 (Spring 1982):18–22.

**13.** ROTHSCHILD, E. R., R. J. MANSER, and M. P. ANDERSON. "Investigation of Aldicarb in Ground Water in Selected Areas of the Central Sand Plain of Wisconsin." *Ground Water,* 20 (1982):437–45.

**14.** DIERBERG, F. E., and C. J. GIVEN. "Aldicarb Studies in Ground Water from Florida Citrus Groves and Their Relationship to Ground-Water Protection." *Ground Water,* 24 (1986):16–22.

**15.** WHITE, R. B., and R. B. GAINER. "Control of Ground Water Contamination at an Active Uranium Mill." *Ground Water Monitoring Review,* 5 (Spring 1985):75–82.

**16.** GRAY, W. J., and J. L. HOFFMAN. "Numerical Model Study of Ground-Water Contamination from Price's Landfill, New Jersey—Data Base and Flow Simulation." *Ground Water,* 21 (1983):7–14.

**17.** GOODENKAUF, O., and J. C. ATKINSON. "Occurrence of Volatile Organic Chemicals in Nebraska Ground Water." *Ground Water,* 24 (1986):231–33.

**18.** NOVAKOVIC, B. "Impact and Recovery of Chromium Waste Leaked beneath an Industrial Plant." Proceedings of the Fourth National Symposium and Exposition on Aquifer Restoration and Ground Water Monitoring, National Water Well Association (1984):394–400.

**19.** BECK, B. F., L. ASMUSSEN, and R. LEONARD. "Relationship of Geology, Physiography, Agricultural Land Use, and Ground-Water Quality in Southwest Georgia." *Ground Water,* 23 (1985):627–34.

**20.** SMITH, S. D., and D. H. MYOTT. "Effect of Cesspool Discharge on Ground-Water Quality on Long Island, New York." *Journal American Water Works Association,* 67 (1975):456–58.

**21.** OBERTS, G. L. "Pollutants Associated with Sand and Salt Applied to Roads in Minnesota." *Water Resources Bulletin*. 22 (1986):479–83.

**22.** NOSS, R. R., and E. T. JOHNSON. "Field Monitoring of the Adams, Massachusetts, Landfill Leachate Plume." Proceedings of the Fourth National Symposium and Exposition on Aquifer Restoration and Ground Water Monitoring. National Water Well Association (1984):356–62.

**23.** DAVIS, P. R., and W. C. WALTON. "Factors Involved in Evaluating Ground Water Impacts of Deep Coal Mine Drainage." *Water Resources Bulletin,* 18 (1982): 841–48.

**24.** LEHR, J. H. "A Problem Yes—A Disaster No." *Ground Water,* 19 (1981):2–3.

**25.** HALL, C. W. "Keynote Address—The Fifth National Symposium and Exposition on Aquifer Restoration and Ground Water Monitoring." *Ground Water Monitoring Review,* 5 (Summer 1985):4–7.

**26.** LEHR, J. H. "A View Toward Substantial Elimination of Future Ground Water Contamination." *Ground Water Monitoring Review,* 4 (Spring 1984):4.

**27.** LEHR, J. H., and D. M. NIELSEN. "Aquifer Restoration and Ground-Water Rehabilitation—A Light at the End of the Tunnel." *Ground Water,* 20 (1982):650–56..

**28.** BAKER, C. H., JR., and D. G. FOULK. "WATSTORE User's Guide, Volume 2, Ground-Water File." U.S. Geological Survey Open-File Report 75–589 (1980 revision), 159 pp.

**29.** MERCER M. W., and C. O. MORGAN. "Storage and Retrieval of Ground-Water Data at the U.S. Geological Survey." U.S. Geological Survey Circular 856 (1982):9.

**30.** NATIONAL WATER WELL ASSOCIATION. *Newsletter of the Association of Ground Water Scientists and Engineers,* 2 (February 1986):1.

# TWO

# Evaporation and Precipitation

Rivers depend for their existence on the rains and on the waters within the earth, as the earth is hollow, and has water in its cavities.

Anaxagoras of Clazomenae (500–428 B.C.)

## 2.1 EVAPORATION

Water molecules are continually being exchanged between a liquid and atmospheric water vapor. If the number passing to the vapor state exceeds the number joining the liquid, the result is **evaporation.** When water passes from the liquid to the vapor state, it will absorb 590 calories of heat from the evaporative surface for every gram of water evaporated. The vapor pressure of the liquid is directly proportional to the temperature. Evaporation will proceed until the air becomes saturated with moisture. The **absolute humidity** of a given air mass is the number of grams of water per cubic meter of air.

At any given temperature, air can hold a maximum amount of moisture: the **saturation humidity.** This is directly proportional to the temperature of the air. Table 2.1 gives the saturation humidity for several environmental temperatures. The **relative humidity** for an air mass is the percent ratio of the absolute humidity to the saturation humidity for the temperature of the air mass. As the relative humidity approaches 100 percent, evaporation ceases.

Condensation occurs when the air mass can no longer hold all of its humidity. This happens when an air mass is cooled and the saturation humidity value drops. If the absolute humidity remains constant, the relative humidity will rise. When it reaches 100 percent, any further cooling will result in condensation. The **dew point** for an air mass is the temperature at which condensation will begin. As condensation is the reverse of evaporation, the process of condensation releases 590 calories of heat to the surroundings per gram of water: the latent heat of condensation.

Evaporation of water takes place from free-water surfaces—lakes, res-

**TABLE 2.1** Saturation humidity of air (grams per cubic meter)

| Temperature (°C) | Humidity |
|---|---|
| −25 | 0.705 |
| −20 | 1.074 |
| −15 | 1.605 |
| −10 | 2.358 |
| − 5 | 3.407 |
| 0 | 4.874 |
| 5 | 6.797 |
| 10 | 9.399 |
| 15 | 12.83 |
| 20 | 17.30 |
| 25 | 23.05 |
| 30 | 30.38 |

Source: *Handbook of Chemistry and Physics* (Cleveland, Ohio: CRC Publishing Company, 1976).

ervoirs, puddles, dew droplets, etc. The rate is dependent upon factors such as the water temperature and the temperature and absolute humidity of the layer of air just above the free-water surface. Solar radiation is the driving energy force behind evaporation, as it warms both the water and the air. The rate of evaporation is also related to the wind—especially over land. The wind carries vapor away from the free-water surface and keeps absolute humidity low. By disturbing the water surface, the wind may also increase the rate of molecular diffusion from it.

Evaporation from lakes and reservoirs is an important consideration in water-budget studies. It can be computed for a lake or reservoir if all of the inflows (precipitation over the surface, surface-water inflow, and ground-water inflow) and the outflows (ground-water outseepage, spillway discharge, and pumpage) and change in storage are known. The hydrologic equation (inflow = outflow ± changes in storage) is used. All of these factors, with the exception of the ground-water flux, can be measured with an error of perhaps ±10 percent. In a carefully prepared water-budget study for Lake Hefner, Oklahoma, daily evaporation was computed to an accuracy of 5 to 10 percent (1). For many reservoirs, monthly or annual evaporation can be computed fairly easily. The most difficult factor to measure is the ground-water flux.

Free-water evaporation is measured quite simply by using shallow pans. The most commonly used is the **land pan.** The U.S. Weather Bureau maintains about 450 evaporation stations using Class A land pans. Similar pans are used in Canada. They are 4 feet (122 centimeters) in diameter and 10 inches (25.4 centimeters) deep, made of unpainted galvanized metal. Land pans are placed on supports so that air can circulate all around. Water depths from 7 to 8 inches

**TABLE 2.2** Class A land pan coefficients for midwestern United States

| | | | |
|---|---|---|---|
| January | 0.62 | July | 0.76 |
| February | 0.72 | August | 0.75 |
| March | 0.77 | September | 0.73 |
| April | 0.77 | October | 0.69 |
| May | 0.78 | November | 0.63 |
| June | 0.77 | December | 0.58 |
| | Annual 0.75 | | |

Source: W. J. Roberts and J. B. Stall, Illinois State Water Survey Report of Investigation 57, 1967.

(18 to 20 centimeters) are maintained. Records are kept of the daily depth of water, the volume of water added to replace evaporated water, and the daily precipitation into the pan. Using the hydrologic budget, the daily evaporation can be computed. Errors may result from splash caused by heavy rainfall and drinking by birds. The wind movement is also measured and expressed in units of miles per day. (A steady wind blowing at a velocity of 10 miles per hour would have a 24-hour wind movement of 240 miles per day.)

The water in a Class A land pan will be warmed much more readily by solar radiation than the surface waters of a lake or reservoir. The chief reason is the difference between the water depth in the pan and the depth of the surface layer of reservoir water. The pan may also gain or lose heat through the sides and bottom, a process that does not occur in reservoirs. For these reasons, observed pan evaporation is multiplied by a factor with a value less than 1.0, the pan coefficient, to estimate reservoir evaporation during the period of observation. Detailed studies in the United States Midwest have yielded monthly pan coefficients ranging from 0.58 in December to 0.78 in May, with an annual value of 0.75 (2) (see Table 2.2).

The U.S. Weather Bureau has developed a lake evaporation nomograph (3). From this diagram, daily lake evaporation can be determined using mean daily temperature, solar radiation in langleys* per day, mean daily dew point temperature, and wind movement in miles per day. The graph in Figure 2.1 is entered from the left side at the mean daily air temperature. As an example, this is 75° F. A horizontal line is drawn across the chart along the 75° F axis. Perpendicular lines are dropped at the intersections of the values of solar radiation and mean daily dew point temperature. In the example, these are 500 langleys per day and 50° F. The right-hand perpendicular extends from the mean daily dew point temperature to the total daily wind movement. The example value is 200 miles per day. From this intersection, a horizontal line is drawn toward the

*A langley is a measure of solar radiation equal to one calorie per square centimeter of surface. In the SI system (International System of Units based on the meter, kilogram, second, and ampere), the unit is the joule per square meter, which is equal to $4.184 \times 10^4$ langleys.

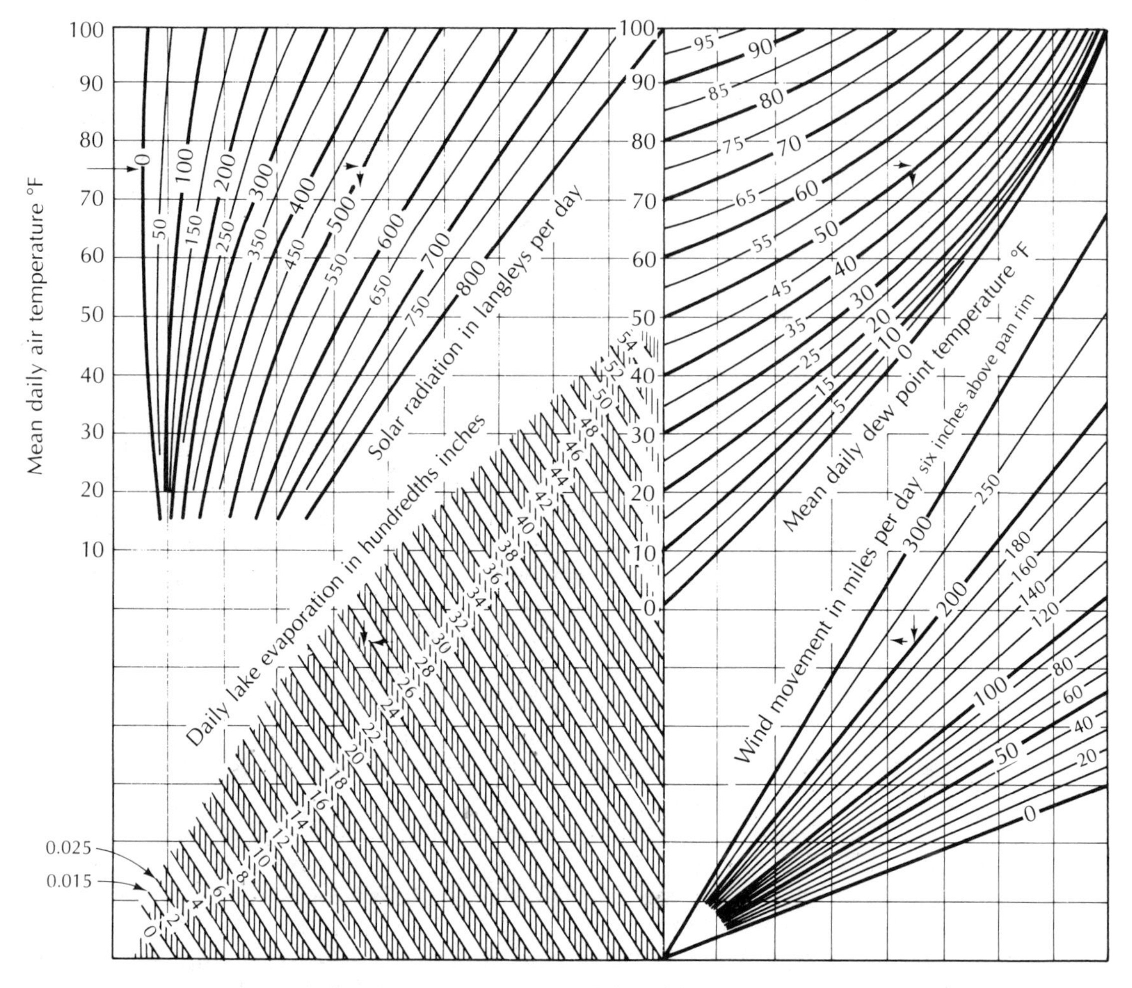


**FIGURE 2.1** Nomograph used to determine the value of daily lake evaporation for shallow lakes if solar radiation, mean daily air temperature, mean daily dew point temperature, and wind movement are known. Source: United States National Weather Service (3).

left. This horizontal line and the left-hand perpendicular will intersect in a field indicating the mean daily lake evaporation. For the example in Figure 2.1, this is 0.25 inch per day.

In some instances, it may be necessary to estimate evaporation without the availability of evaporation pan data. Such estimates are possible via methods based on heat budgets (4). The energy budget for a reservoir may be used to find the amount of energy used for evaporation, which in turn can yield the amount of evaporation. Other methods use aerodynamic data and vapor pressures of the water and air in empirical formulas (5). Some investigators have been able to combine these two approaches (6).

## 2.2 TRANSPIRATION

Free-water evaporation is only part of the mechanism for mass transfer of water to the atmosphere. Growing plants are continuously pumping water from the ground into the atmosphere through a process called **transpiration** (7). Water is drawn into a plant rootlet from the soil moisture owing to osmotic pressure, whereupon it moves through the plant to the leaves. The turgidity of nonwoody vascular plants is caused by the cellular pressures of the contained water. The water is passed as vapor through openings in the surface of the leaves known as **stomata.** Air also passes through these openings. A small portion (less than 1 percent) of the water is used to manufacture plant tissue, but most is transpired to the atmosphere. The process of transpiration accounts for most of the vapor losses from a land-dominated drainage basin.

The amount of transpiration is a function of the density and size of the vegetation. As an example, transpiration from a cornfield in May, when the plants are a few centimeters high, is much less than in August, when they may exceed 7 feet (2 meters) in height. Transpiration is obviously important only during the growing season; about 95 percent takes place during the daylight hours, when photosynthesis is occurring. Transpiration is also limited by available soil moisture. When the soil-moisture content becomes so low that the surface tension of the soil–water interface exceeds the osmotic pressure of the roots, water will no longer enter the roots. This is termed the **wilting point** of the soil.

When available water becomes limited, deep-rooted plants are more resistant to drought wilting than shallow-rooted plants, as the former can draw moisture from deeper layers. Also, some plants have fewer stomata and can close them through the use of special cells to reduce water loss during drought periods. Such drought-resistant species can transpire less water during periods of stress. **Phreatophytes** are plants with a tap root system extending to the water table. They can transpire at a high rate even in the desert, so long as the water table does not drop below the tap root. In areas of low precipitation, the native vegetation is adapted to existing with minimal water. These desert plants are called **xerophytes.** They have a shallow root system that spreads out away from the plant.

Aquatic plants, or **hydrophytes,** are a special case. They exist with their root systems submerged, and the special cells some plants have to close the stomata are lacking. As long as adequate water is available, transpiration proceeds at a high rate. The rate of transpiration is controlled by the amount of solar energy and the heat content of the water. The water loss from a pond is about the same, whether or not emergent aquatic vegetation is present.

Measurement of transpiration can be performed under carefully controlled laboratory conditions. A **phytometer** is a sealed container partially filled with soil. Transpiration by plants rooted in the soil causes an increase in the humidity, which can be measured in the air space around the plant. However, such laboratory studies reveal little about the behavior of plants in natural or agricultural conditions.

## 2.3 EVAPOTRANSPIRATION

Under field conditions it is not possible to separate evaporation from transpiration totally. Indeed, we are generally concerned with the total water loss, or **evapotranspiration,** from a basin. Whether the loss is due to free-water

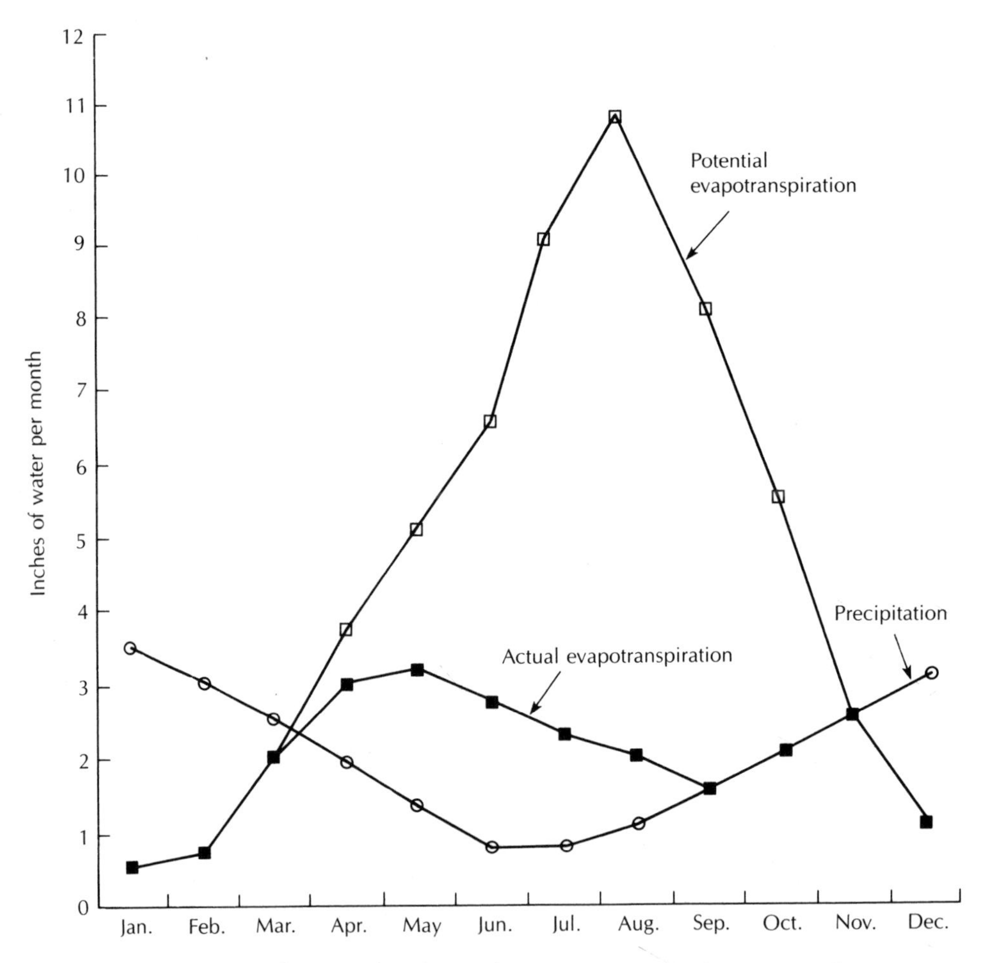


**FIGURE 2.2** Diagram of potential and actual evapotranspiration in an area of warm summers and cool winters. **A.** Dry summers and moist winters with limited soil-moisture storage capacity. **B.** Little seasonal variation in precipitation and ample soil-moisture storage capacity.

evaporation, plant transpiration, or soil-moisture evaporation is of little importance.

The term **potential evapotranspiration** was introduced by Thornthwaite (8) as equal to "the water loss which will occur if at no time there is a deficiency of water in the soil for the use of vegetation." Thornthwaite recognized that there is an upper limit to the amount of water an ecosystem will lose by evapotranspiration. The majority of the water loss due to evapotranspiration takes place during the summer months, with little or no loss during the winter. Because there is often not sufficient water available from soil moisture, the term **actual evapotranspiration** is used to describe the amount of evapotranspiration that occurs under field conditions. Figure 2.2A shows potential evapotranspiration and actual evapotranspiration for a region with a warm, dry summer and a

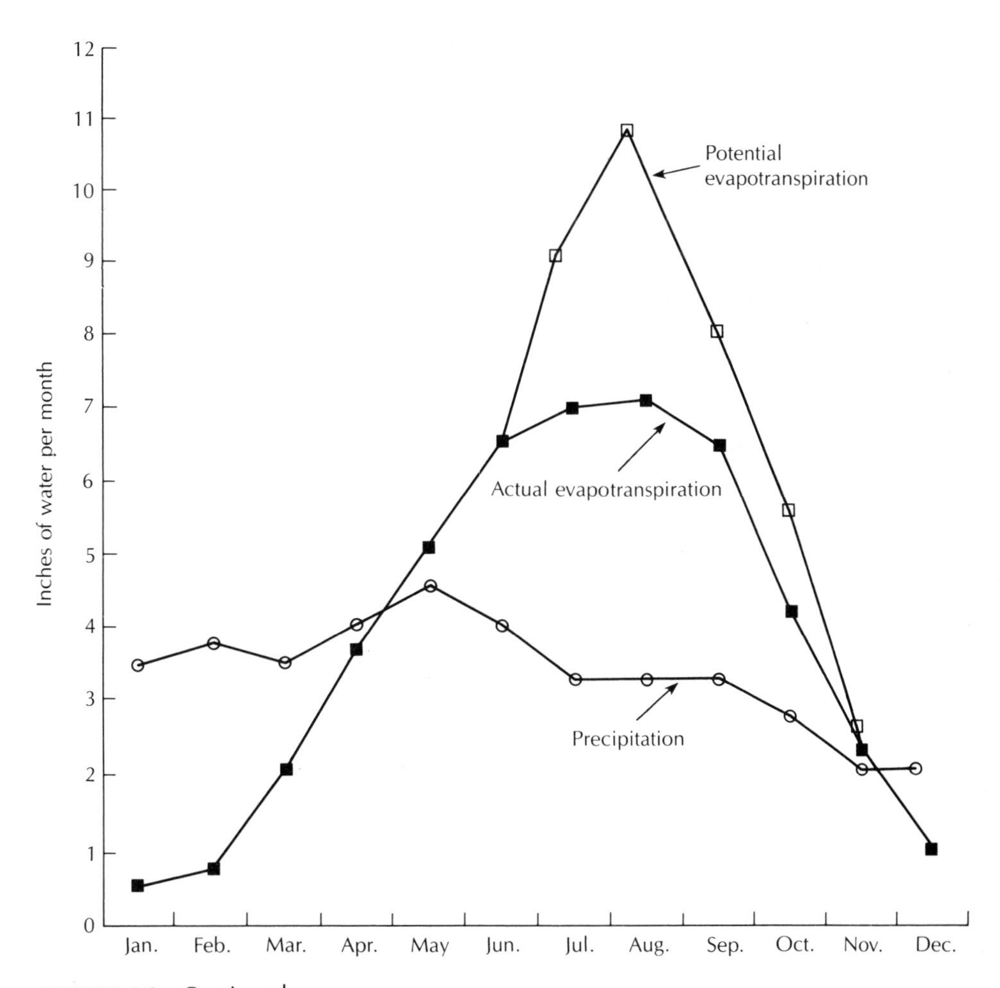


**FIGURE 2.2** Continued

cool, moist fall, winter, and spring. Under these conditions the actual evapotranspiration is much less than the potential, especially if the soil-moisture storage capacity is limited. In months when the potential evapotranspiration is less than the rainfall, some of the demand will be met by drawing upon moisture stored in the soil. When available soil moisture is depleted, the actual evapotranspiration will be limited to the monthly precipitation. Figure 2.2B shows potential and actual evapotranspiration in an area where the precipitation is more or less evenly distributed through the year. This circumstance results in the actual evapotranspiration being closer to the potential value.

Thornthwaite's method is based upon the assumption that potential evapotranspiration was dependent only upon meteorological conditions and ignored the effect of vegetative density and maturity. While this assumption is not correct, the method devised by Thornthwaite to compute potential evapotranspiration is still useful. The only necessary factors to input are mean monthly air temperature, latitude, and month (9, 10). The last two factors yield average monthly sunlight. The Thornthwaite method is reasonably accurate in determining annual values, especially in humid areas. As no factor for vegetative growth is included, values computed for spring and early summer are too high, as the crop is just emerging; midsummer values may be too low.

Another method of estimating potential evapotranspiration was developed by Blaney and Criddle (11). This method introduces a crop factor, which varies as the growing season progresses. Thus, some of the objections to the Thornthwaite method are overcome, but the effects of wind and relative humidity on evapotranspiration remain unaccounted for.

Evapotranspiration can be measured directly using a **lysimeter**—a large container holding soil and plants. The lysimeter is set outdoors, and the initial soil-moisture content is determined. Precipitation into the lysimeter and any irrigation water added are measured. Changes in soil-moisture storage reveal how much of the added water is lost to evapotranspiration. It is necessary to design the lysimeter so that any moisture in excess of that specifically retained by the soil is collected. The following equation can be used with the lysimeter:

$$E_T = S_i + P + I - S_f - D \qquad \textbf{(2–1)}$$

where

$E_T$ is the evapotranspiration for a period
$S_i$ is the volume of initial soil moisture
$S_f$ is the volume of final soil moisture
$P$ is the precipitation into the lysimeter
$I$ is the irrigation water added to the lysimeter
$D$ is the excess moisture drained from the soil

Lysimeters should be designed so that they accurately reproduce the soil type and profile, moisture content, and type and size of vegetation of the surrounding area. They should be buried so that the soil surface is at the same level inside and outside the container. Soil-moisture changes can be determined by sampling the soil, by means of moisture meters, or by weighing the entire mass of soil, water, and plants. Whatever method is employed, operation of a lysimeter is both time-consuming and expensive. If water is applied to the lysimeter at a rate sufficient to keep the soil at, or nearly at, field capacity (see Section 4.8), the lysimeter will measure potential evapotranspiration.

When the soil moisture drops below the amount that the soil can hold against gravity by surface tension (the field capacity), available water may limit evapotranspiration to some value less than the potential evapotranspiration. The plants are required to draw upon soil moisture and, as this diminishes and less water is extracted, actual evapotranspiration falls below the potential. If the soil moisture drops too low, the plants may wither and die. The soil-moisture content below which plants can no longer obtain moisture is the wilting point.

There is some uncertainty about the rate of evapotranspiration when the soil moisture is between the wilting point and the field capacity. Some have suggested that it proceeds at a rate equal to the potential evapotranspiration until the wilting point is reached (12), while others have suggested that the evapotranspiration rate is linearly proportional to the ratio of the remaining available soil moisture to the initial available soil moisture (9). Soil texture and unsaturated soil permeability play major roles in determining the rate of actual evapotranspiration (13).

Evapotranspiration is the major use of water in all but extremely humid, cool climates. If evapotranspiration were reduced, then runoff or ground-water infiltration or both could increase. This would increase the available water supply. Studies have shown that basin runoff from a forested watershed has increased following the timbering of the forest (14). The increase is greatest during the first year, when there is little reforestation. As the forest regrows, the runoff again decreases. Cutting of forests to increase runoff may also result in increased erosion from the uplands and concurrent sedimentation in the lowlands. Conversion of one plant cover to another can also affect the evapotranspiration rate. In arid Arizona, the conversion of a plot of land formerly covered with chaparral to grasses resulted in streamflow increases of several hundred percent. This was due in part to lower evapotranspiration, as the grass was not as deep-rooted as the chaparral (15). However, in Colorado, the conversion of sagebrush to bunchgrass had no appreciable effect on the amount of watershed runoff, although an increase in cattle forage did result (16).

In some areas of the humid eastern United States, which were originally wooded, marginal farms are being abandoned. The old fields are gradually reverting to forest. There has been a concomitant decrease in streamflow from these watersheds. The replacement of deciduous forests with conifers results in an increase in evapotranspiration (17).

Experiments have shown that evaporation from small lakes and reservoirs can be reduced by applying a monolayer of a fatty alcohol to the water surface (18). This has not proven to be practical, however, owing to the cost of a treatment and the rapid rate at which the fatty alcohol dissipates. Likewise, fatty alcohols have been used as antitranspirants in treating plants and soils. However, concentrations high enough to reduce transpiration also reduce crop growth (19). Chemical antievaporants and antitranspirants have not yielded the hoped-for success.

## 2.4 CONDENSATION

When an air mass with a relative humidity lower than 100 percent is cooled without losing moisture, the relative humidity will approach 100 percent as the dew point temperature is approached. When the air mass is saturated, **condensation** may start to occur. Condensation generally requires a surface or nucleus on which to form. The morning dew or frost is the result of condensation taking place on plants or other surfaces. Rain or ice needs nuclei in the range of 0.1 to 10 μm. Particles serving as nuclei include clay minerals, salt, and combustion products.

In the absence of sufficient nuclei, the air mass may become supersaturated without the formation of raindrops or ice crystals. This is the theory behind artificial precipitation augmentation. "Cloud-seeding" procedures involve the addition of artificial nuclei, including silver iodide and dry ice, to the atmosphere. Research has shown that even in severe droughts, atmospheric conditions conducive to successful seeding may sometimes occur during the summer (20).

Once droplets or ice crystals have formed, they initially grow by attraction (diffusion) of water vapor as well as additional condensation. Rising air masses or upward movements of clouds tend to keep newly formed fog and cloud elements aloft. These elements are in the size range of 10 to 50 μm. As cloud elements collide and coalesce, raindrops begin to form. When the raindrops start to fall, further collisions occur, so that some raindrops may grow as large as 0.2 inch (6 millimeters) in diameter. Rain that falls through an unsaturated air mass may evaporate before it reaches the ground. Falling ice crystals grow by diffusion and collision to form snowflakes. The largest snowflakes form when temperatures are close to freezing.

## 2.5 FORMATION OF PRECIPITATION

In order for precipitation to occur, several conditions must be met: (a) a humid air mass must be cooled to the dew point temperature, (b) condensation or freezing nuclei must be present, (c) droplets must coalesce to form raindrops, and (d) the raindrops must be of sufficient size when they leave the clouds to ensure that they will not totally evaporate before they reach the ground.

Air masses are cooled by a process known as **adiabatic expansion,** which occurs when the air mass rises in the atmosphere. Since the atmosphere becomes less dense with altitude, a rising air mass must expand owing to the lower pressure. If there is no exchange of heat between the air mass and its surroundings, the laws of thermodynamics dictate that the temperature will fall.

When the rising air mass is dry—that is, the relative humidity is lower than 100 percent—the rate of cooling is 1° C for every 328-foot (100-meter) rise in height. This is the **dry adiabatic lapse rate.** When the air mass reaches the dew point temperature, further lifting and cooling will cause condensation. The latent heat of vaporization is released; hence, the **wet adiabatic lapse rate** is lower than the dry rate. The exact value depends upon the amount of condensation occurring.

Under normal conditions, air temperature decreases with increasing altitude at a mean rate of 0.7° C for every 328 feet. Owing to uneven or unsteady heating or cooling, the temperature gradient or lapse rate may be more or less than 0.7° C per 328 feet. **Temperature inversions,** or layers of warm air overlying cooler air, exist and are typically caused by warm air masses overriding cold fronts or by conductive cooling of the earth's surface. Solar radiation during the day causes high temperature gradients.

Most rising air masses can be attributed to one of three causal factors: movement of weather fronts, convective processes, and orographic effects. **Frontal precipitation** is caused by the lifting of an air mass by a moving weather

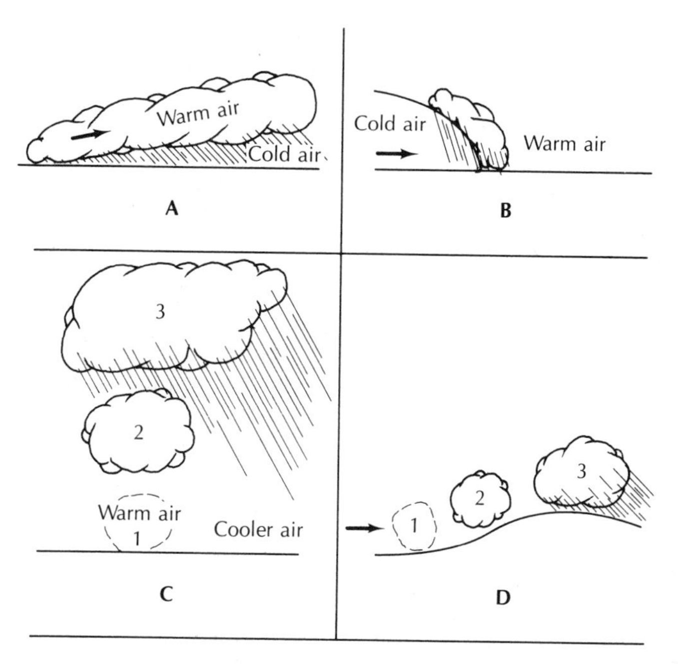


**FIGURE 2.3** Precipitation caused by adiabatic lifting of an air mass may be the result of the following activity: **A.** A warm front pushing over a cold front. **B.** A cold front colliding with a warm front. **C.** Uneven heating near the surface causing a warm air mass to rise convectionally. **D.** Orographic lifting caused by prevailing winds blowing over a topographic high.

front. If a warm front is moving upward over a colder, more dense air mass, precipitation and cloudiness will extend for several hundred miles ahead of the surface front (Figure 2.3A). The slope of a front of this type is small and the rate of ascent of the warm air mass is slow; hence, precipitation is generally light. Should a cold front be moving, it will typically be faster than a warm front. The cold front is steeper, warm air is forced upward more rapidly, and heavier rain amounts may be recorded—especially near the surface front (Figure 2.3B).

Uneven heating of an air mass at the surface, or cooling at the top of an air mass, may cause it to be warmer than the surrounding air. The denser air will flow beneath it, causing the air mass to rise. It will continue to rise, and cool adiabatically, until the temperature is equal to its surroundings. This can cause **convectional** rising. Summer thunderstorms and associated cumulus clouds are a result of this process (Figure 2.3C).

If a moving air mass is forced upward over a mountain range, it must gain altitude. **Orographic** cooling and precipitation may result. The vegetation of the Black Hills of South Dakota is different from the surrounding grassland prairie. There is sufficient precipitation for forest to grow and, from a distance, clouds can be often seen hovering over the hills. This is orographically caused condensation, as air masses are forced upward as they move from west to east over the central Black Hills (Figure 2.3D).

## 2.6 MEASUREMENT OF PRECIPITATION

Any open container can be used to catch and measure rainfall. Experiments have shown that the size of the opening has little effect on the catch, except for very small (less than one-inch-diameter) gages (21). The United States Standard Rain Gage has an opening 8 inches (20.3 centimeters) in diameter, while the Canadian standard gage is 9 centimeters (3.57 inches) in diameter. These are manually read gages; the water is emptied and the gages read once a day. The catch of precipitation gages is affected by high winds. Such gages generally catch less than the true amount of rainfall because of updrafts around the gage opening. The location of the gage is also critical. In one study, two identical 8-inch gages were placed 10 feet (3 meters) apart on a ridge. One gage consistently caught 50 percent more rainfall than the other (22). Gages should be placed as close to the ground as possible in order to avoid wind. They should be in the open, away from trees and buildings. Low bushes and shrubs can provide a windbreak. Level ground is best, with the top of the gage horizontal. On steep slopes, it may be desirable to have the orifice opening parallel to the slope.

The effect of wind is greatest for light rain or snow. Some rain gages are equipped with a shield, or wind deflector, around the opening in order to overcome wind problems. This will improve the catch of snow, but it will still be less than 100 percent effective in substantial winds.

There are a number of different types of recording rain gages available that can automatically measure or weigh the precipitation. The temporal distribution of precipitation through a day can thus be obtained. Such data are necessary for any studies of precipitation intensity. For remote areas, recording rain gages can be used to record daily precipitation for long time periods. In such circumstances, manual gages could provide only a total rainfall for the period between readings.

In the United States there are some 13,500 precipitation stations, for the most part operated by trained volunteers. Daily records from these weather stations are published monthly on a state-by-state basis in *Climatological Data;* data from recording stations are published in *Hourly Precipitation Data.* Both of these are publications of the U.S. Environmental Data Service. Canada has about 2000 precipitation stations, the data from which are published by the Canadian Atmospheric Environment Service in the *Monthly Record of Observations.*

As every viewer of local television news and weather programs knows, radar can be used to detect areas of precipitation. Rain droplets or snow particles reflect part of the directed radar beam back to the originating station. The amount of reflected energy is directly proportional to the intensity of the precipitation. The radar apparatus measures precipitation in the atmosphere. As the beam is at an oblique angle to the ground, the farther the distance from the station, the greater the altitude of the precipitation being measured. Radar measurements of precipitation may not accurately indicate ground precipitation. Evaporation may occur between the point of measurement and the ground, or wind may cause the precipitation to drift so that it falls to earth at some place other than that indicated by the radar (23).

Some special radar equipment can convert the intensity of the radar reflection into precipitation rates. The rates are integrated over time to yield a depth of total precipitation over an area. This yields data about precipitation rates between ground stations. The use of radar in combination with conventional ground-station rain gages can give improved areal measurement of precipitation (24).

## 2.7 SNOW MEASUREMENTS

The measurement of snowfall in standard rain gages is subject to error due to turbulence around the gage. The snow that is caught is melted and the water equivalent reported. If only an approximation is required, a water content of 10 percent of the snow depth can be assumed. However, as anyone who regularly shovels snow knows, the density of newly fallen snow can vary considerably.

In northern and mountainous climates, the accumulation of snow on the ground is an important hydrologic parameter. In some areas, the runoff of melting snow in the spring is a predominant source of water for reservoirs used for water supply, irrigation, and power generation. A thick accumulation of snow can also mean a high flood potential when snowmelt occurs in the spring. Melting snow also recharges soil moisture and the water table.

Snow surveys are made periodically through the winter to measure the thickness and water content of accumulated snow. A thin-walled tube with a sharp leading edge is driven through the snow to the ground. The tube and the snow contained within are weighed, and the weight of the empty tube subtracted to determine the weight of the snow. A snow survey requires that someone make traverses, stopping at predetermined stations to make measurements. The snow courses should sample representative terrain, vegetative cover, and altitude of the catchment area.

The extent of snow cover can be mapped using satellite photography (25). The resulting data, combined with data from snow-course surveys, can be used to determine the total volume of water in the snowpack. Melting of the snowpack can begin only when the temperature of the snow has risen to 0° C. Initial meltwater clings to snow granules by surface tension, so that at least 2 to 8 percent of the snowpack must melt before runoff begins. Energy-balance methods can be used to predict daily snowmelt (26).

## 2.8 EFFECTIVE DEPTH OF PRECIPITATION

In water-budget studies, it is necessary to know the average depth of precipitation over a drainage basin. This may be determined for time periods ranging from the duration of part of a single storm to a year. The data are generally measurements of precipitation and/or equivalent snowfall at a number of points throughout the drainage basin.

A problem is created if data are missing at one or more stations. This can occur as a result of equipment malfunction or operator absence. To solve the problem, three close precipitation stations with full records that are evenly spaced around the station with a missing record are used. The following equation yields an estimate of the missing data at Station Z. The mean annual precipitation *(N)* at Station Z and the three index stations, A, B, and C, as well as the actual precipitation *(P)* at the index stations for the time period over which data are missing are needed:

$$P_Z = \frac{1}{3}\left[\frac{N_Z}{N_A}P_A + \frac{N_Z}{N_B}P_B + \frac{N_Z}{N_C}P_C\right] \quad \textbf{(2–2)}$$

If the rain gage network is of uniform density, then a simple arithmetic average of the point-rainfall data for each station is sufficient to determine the **effective uniform depth (EUD)** of precipitation over the drainage basin (Figure 2.4).

If the rain-gage network is not uniform, then some adjustment is necessary. The most accurate method, excluding use of radar data, is to draw a precipitation contour map with lines of equal rainfall **(isohyets).** In drawing the isohyets, such factors as known influence of topography on precipitation can be taken into account. Simple linear interpolation between precipitation stations

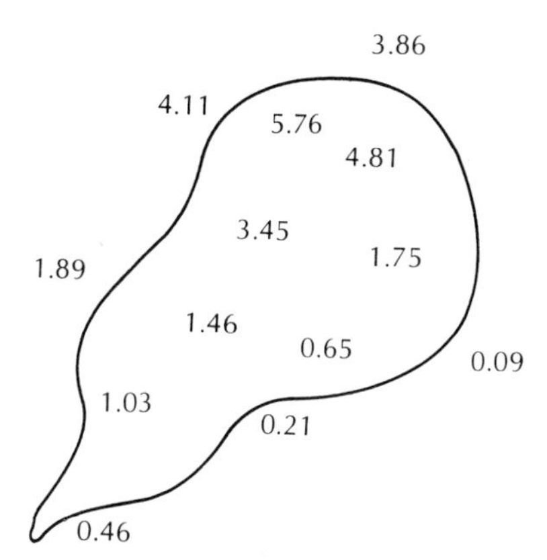


**FIGURE 2.4** Precipitation-gage network over a drainage basin. Precipitation amounts are given in inches. Station locations are at decimal points.

can also be used. The area bounded by adjacent isohyets is measured with a planimeter, and the average depth of precipitation over the area is the mean of the bounding isohyets. The effective uniform depth of precipitation is the weighted average based on the relative size of each isohyetal area (Figure 2.5). The drawback of the isohyetal method is that the isohyets must be redrawn and the areas remeasured for each analysis.

The **Theissen method** to adjust for nonuniform gage distribution uses a weighing factor for each rain gage. The factor is based on the size of the area

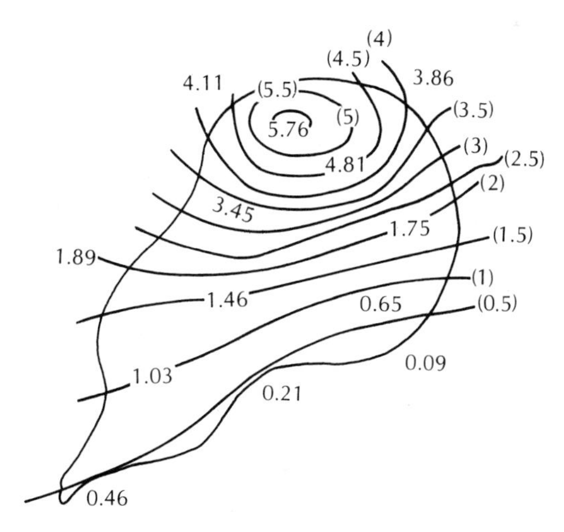


**FIGURE 2.5** Isohyetal lines for the precipitation-gage network of Figure 2.4. The isohyets show contours of equal rainfall depth with a contour interval of 0.5 inch. The contours are based on simple linear interpolation.

within the drainage basin which is closest to a given rain gage. These areas are irregular polygons. The method of constructing them can be described rather easily; however, it takes a bit of practice to master the technique. The rain-gage network is drawn on a map of the drainage basin. Adjacent stations are connected by a network of lines (Figure 2.6A). Should there be doubt as to which stations to connect, lines should be between the closest stations. A perpendicular line is then drawn at the midpoint of each line connecting two stations (Figure 2.6B), and extensions of the perpendicular bisectors are used to draw polygons around each station (Figure 2.6C). It is best to start with a centrally located

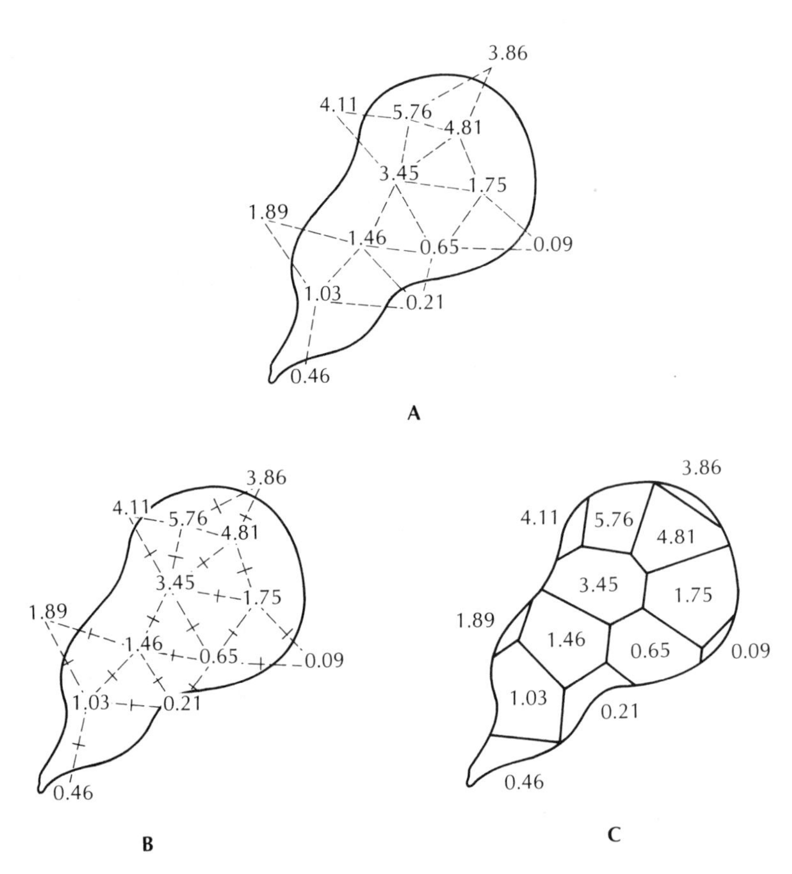


FIGURE 2.6 Theissen polygons based on the rain-gage network of Figure 2.4. **A.** The stations are connected with lines. **B.** The perpendicular bisector of each line is found. **C.** The bisectors are extended to form the polygons around each station.

station and then expand the polygonal network outward. The area of each polygon is measured, and a weighted average for each station's precipitation is used to find the EUD.

---

**EXAMPLE PROBLEM** Determine the effective uniform depth of precipitation using the arithmetic mean, isohyetal, and Theissen methods.

### ARITHMETIC MEAN METHOD

Figure 2.4 shows a drainage basin with seven stations in its boundaries. An additional six stations are located outside the drainage divide. In the arithmetic mean method, only the gages inside the drainage basin boundary are considered.

Arithmetic mean

$$= \frac{1.03 + 0.65 + 1.46 + 1.75 + 4.81 + 3.45 + 5.76}{7}$$

$$= 2.70 \text{ in.}$$

### ISOHYETAL METHOD

The first step is to draw lines of equal precipitation (isohyets) on the drainage basin map. Isohyets are usually whole numbers or decimals (every 0.1 inch; every 0.5 inch; every 1 millimeter, etc.). The following rules apply:

1. Isohyets never cross.
2. Isohyets never split.
3. Isohyets never meet.
4. A station that does not fall on an isohyet will be between two isohyets. The isohyets will both be equal (either larger or smaller than the station value) or one will be larger and one smaller.
5. Adjacent isohyets must be equal or only one contour interval different in value.
6. Isohyets should be scaled between stations using linear interpolation.

Figure 2.5 shows the isohyetal map of the problem area.

The area between adjacent isohyets is determined by use of a planimeter. The equivalent uniform depth of precipitation between isohyets is usually assumed to be equal to the median value of the two isohyets. For example, the EUD between a 1-inch isohyet and a 2-inch isohyet is 1.5 inches. For areas enclosed by a single isohyet, judgment should be used to estimate the equivalent uniform depth. The weighted average precipitation is based on the equivalent uniform depth of precipitation between adjacent isohyets and their areas.

| *A*<br>Isohyet (in.) | *B*<br>Estimated EUD | *C*<br>Net Area (sq mi) | *D*<br>Percent of Total Area | *E*<br>Weighted Precipitation (in.) (*B* × *D*) |
|---|---|---|---|---|
| 5.5+ | 5.6 | 1.1 | 0.8 | 0.045 |
| 5.0–5.5 | 5.25 | 7.6 | 5.3 | 0.278 |
| 4.5–5.0 | 4.75 | 10.6 | 7.4 | 0.352 |
| 4.0–4.5 | 4.25 | 9.5 | 6.7 | 0.285 |
| 3.5–4.0 | 3.75 | 8.6 | 6.0 | 0.225 |
| 3.0–3.5 | 3.25 | 8.3 | 5.8 | 0.189 |
| 2.5–3.0 | 2.75 | 10.7 | 7.5 | 0.206 |
| 2.0–2.5 | 2.25 | 12.3 | 8.6 | 0.194 |
| 1.5–2.0 | 1.75 | 15.1 | 10.6 | 0.186 |
| 1.0–1.5 | 1.25 | 23.8 | 16.7 | 0.209 |
| 0.5–1.0 | 0.75 | 31.2 | 21.8 | 0.164 |
| <0.5 | 0.3 | 4.0 | 2.8 | 0.008 |
| TOTAL | | 142.8 sq mi | | 2.34 in. NET EUD |

### THEISSEN METHOD

This method provides for the nonuniform distribution of gages by determining a weighting factor for each gage. A weighted mean of the precipitation values can then be computed. Theissen polygons for the example problem are shown in Figure 2.6C. The area of each polygon is determined by a planimeter.

| *A*<br>Station Precipitation (in.) | *B*<br>Net Area (sq mi) | *C*<br>Percent of Total Area | *D*<br>Weighted Precipitation (in.) (*A* × *C*) |
|---|---|---|---|
| 5.76 | 16.9 | 11.9 | 0.686 |
| 4.81 | 16.1 | 11.4 | 0.546 |
| 4.11 | 3.4 | 2.4 | 0.099 |
| 3.86 | 1.6 | 1.1 | 0.044 |
| 3.45 | 19.3 | 13.6 | 0.470 |
| 1.89 | 2.5 | 1.8 | 0.033 |
| 1.75 | 12.0 | 8.5 | 0.148 |
| 1.46 | 19.8 | 14.0 | 0.204 |
| 1.03 | 18.0 | 12.7 | 0.131 |
| 0.65 | 17.0 | 12.0 | 0.078 |
| 0.46 | 6.0 | 4.2 | 0.019 |
| 0.21 | 7.2 | 5.1 | 0.011 |
| 0.09 | 2.0 | 1.4 | 0.001 |
| TOTAL | 141.8 sq mi | | 2.47 in. NET EUD |

A weighted mean of the EUD is found, based on the depth of precipitation and the area of the polygon within the basin boundary.

## REFERENCES

1. HARBECK, G. E., and F. W. KENNON. "The Water Budget Control." In *Water-Loss Investigations: Lake Hefner Studies Technical Report.* U.S. Geological Survey Professional Paper 269, 1954.
2. ROBERTS, W. J., and J. B. STALL. *Lake Evaporation in Illinois.* Illinois State Water Survey Report of Investigation 57, 1967, 44 pp.
3. KOHLER, M. A., T. J. NORDENSON, and W. E. FOX. *Evaporation from Ponds and Lakes.* U.S. Weather Bureau Research Paper 38, 1955.
4. PENMAN, H. L. "Natural Evaporation from Open Water, Bare Soil, and Grass." *Proceedings of the Royal Society* (London), ser. A, 193 (1948):120–45.
5. HARBECK, G. E. *A Practical Field Technique for Measuring Reservoir Evaporation Utilizing Mass-Transfer Theory.* U.S. Geological Survey Professional Paper 272-E, 1962, pp. 101–5.
6. KOHLER, M. A., and L. H. PARMELE. "Generalized Estimates of Free-Water Evaporation." *Water Resources Research,* 3 (1967):997–1005.
7. HENDRICKS, D. W., and V. E. HANSEN. "Mechanics of Transpiration." *American Society of Civil Engineers, Journal of Irrigation and Drainage Division,* 88 (June 1962):67–82.
8. THORNTHWAITE, C. W. "Report of the Committee on Transpiration and Evaporation, 1943–1944." *Transactions, American Geophysical Union,* 25 (1944):687.
9. THORNTHWAITE, C. W., and J. R. MATHER. *The Water Balance,* Publication 8. Centerton, N.J.: Laboratory of Climatology, 1955, pp. 1–86.
10. THORNTHWAITE, C. W., and J. R. MATHER. *Instructions and Tables for Computing Potential Evapotranspiration and the Water Balance,* Publication 10. Centerton, N.J.: Laboratory of Climatology, 1957, pp. 185–311.
11. BLANEY, H. F., and W. D. CRIDDLE. *Determining Water Requirements in Irrigation Areas from Climatological and Irrigation Data.* U.S. Department of Agriculture, Soil Conservation Service Technical Paper 96, 1950.
12. VEIHMEYER, F. J., and A. H. HENDIRCKSON. "Does Transpiration Decrease as Soil Moisture Decreases?" *Transactions, American Geophysical Union,* 36 (1955):425–48.
13. MOLZ, F. J., I. REMSON, A. A. FUNGAROLI, and R. L. DRAKE. "Soil Moisture Availability for Transpiration." *Water Resources Research,* 4 (1968):1161–70.
14. HIBBERT, A. R. "Forest Treatment Effects on Water Yield." In *Forest Hydrology,* ed.
W. E. Sopper and H. W. Lull. Oxford, England: Pergamon Press, 1967, pp. 527–43.
15. HIBBERT, A. R. "Increases in Streamflow after Converting Chaparral to Grass." *Water Resources Research,* 7 (1971):71–80.
16. SHOWN, L. M., G. C. LUSBY, and F. A. BRANSON. "Soil Moisture Effects of Conversion of Sagebrush Cover to Bunchgrass Cover." *Water Resources Bulletin,* 8 (1972):1265–72.

17. URIE, D. H. "Influences of Forest Cover on Groundwater Recharge Timing and Use." In *International Symposium on Forest Hydrology,* ed. W. E. Sopper and H. W. Hull. Oxford, England: Pergamon Press, 1967, pp. 313–24.
18. BARTHOLIC, J. F., J. R. RUNKELS, and E. B. STENMARK. "Effects of a Monolayer on Reservoir Temperature and Evaporation." *Water Resources Research,* 3 (1967):173–80.
19. GALE, J., E. B. ROBERTS, and R. M. HAGEN. "High Alcohols as Antitranspirants." *Water Resources Research,* 3 (1967):437–41.
20. HUFF, F. A., and R. G. SEMONIN. "Potential of Precipitation Modification in Severe Droughts." *Journal of Applied Meteorology,* 14 (1975):974–79.
21. HUFF, F. A. "Comparison between Standard and Small Orifice Rain Gauges." *Transactions, American Geophysical Union,* 30 (1955):689–94.
22. COURT, A. "Reliability of Hourly Precipitation Data." *Journal of Geophysical Research,* 65 (1960):4017–24.
23. STOUT, G. E., and E. A. MUELLER. "Survey of Relationships between Rainfall Rate and Radar Reflectivity in the Measurement of Precipitation." *Journal of Applied Meteorology,* 7 (1968):465–74.
24. WILSON, J. W. "Integration of Radar and Rain Gauge Data for Improved Rainfall Measurements." *Journal of Applied Meteorology,* 9 (1970):489–97.
25. BARNES, J. C., and C. J. BOWLEY. "Snow Cover Distribution as Mapped from Satellite Photography." *Water Resources Research,* 4 (1968):257–72.
26. PRICE, A. G., and T. DUNN. "Energy Balance Computations of Snowmelt in a Subarctic Area." *Water Resources Research,* 12 (1976):686–94.

## PROBLEMS

1. A swimming pool has a length of 50 meters and a width of 25 meters. During July the Class A land pan evaporation is 17.0 centimeters. If the pan coefficient is 0.80, what is the monthly water loss from the pool due to evaporation?
2. A reservoir has a surface area of 690 acres. The following table shows the monthly inflow of surface water, outflow as releases from the reservoir via the spillway, direct precipitation into the reservoir, and evaporation from the reservoir. The reservoir elevation was 701 feet on January 1st. Compute the reservoir elevation at the end of each month.

| Month | Inflow (acre-feet) | Outflow (acre-feet) | Precipitation | | Evaporation | | Net Change | | Elevation (feet) |
|---|---|---|---|---|---|---|---|---|---|
| | | | (inches) | (acre-feet) | (inches) | (acre-feet) | (acre-feet) | (feet) | |
| Dec | | | | | | | | | 701 |
| Jan | 1732 | 175 | 2.75 | | 1.05 | | | | |
| Feb | 1755 | 190 | 3.05 | | 1.55 | | | | |
| Mar | 872 | 232 | 3.76 | | 2.05 | | | | |
| Apr | 955 | 375 | 4.11 | | 2.80 | | | | |
| May | 708 | 525 | 2.70 | | 3.75 | | | | |
| Jun | 312 | 955 | 1.05 | | 4.25 | | | | |
| Jul | 102 | 1720 | .75 | | 5.15 | | | | |
| Aug | 37 | 2250 | 1.25 | | 5.76 | | | | |
| Sep | 175 | 1575 | 1.55 | | 4.92 | | | | |
| Oct | 575 | 550 | 3.79 | | 3.02 | | | | |
| Nov | 1250 | 175 | 4.53 | | 1.75 | | | | |
| Dec | 1875 | 125 | 5.01 | | 0.60 | | | | |

3. Figure 2.7 on p. 36 is a map of a drainage basin and the rainfall amounts during a storm at a number of precipitation stations both within and outside the drainage basin. Make a Theissen network drawing for the drainage basin. The exact station location is the decimal point in the rainfall amount. The relative size of the area associated with each Theissen polygon can be measured with a planimeter or estimated by tracing the Theissen network on cross-section paper and counting the number of squares in each polygon. Estimate the effective uniform depth of precipitation over the drainage basin.

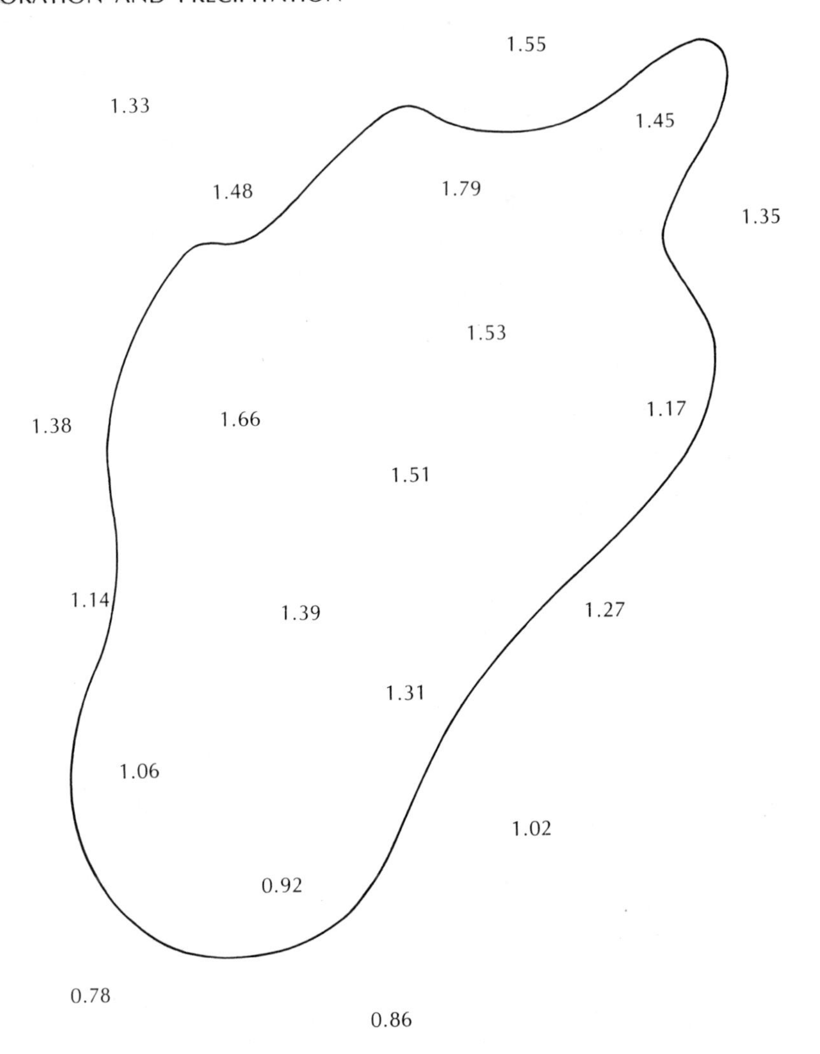


**FIGURE 2.7** Base map for problem 2.3.

# THREE

# Runoff and Streamflow

The land had great depth of soil and gathered the water into itself and stored it up into the soil . . . as though it were a sort of natural water jar; it drew down into the natural hollow the water which it had absorbed from the high ground and so afforded in all districts of the country liberal sources of springs and rivers. . . .

*Critias,* Plato (427–347 B.C.)

## 3.1 EVENTS DURING PRECIPITATION

During a precipitation event, some of the rainfall is intercepted by vegetation before it reaches the ground **(interception).** This may later fall to the ground or evaporate. In a heavily forested area, most of the precipitation is caught by leaves and twigs. For a period at the start of a summer thunderstorm, no raindrops reach the forest floor, although drops can be heard striking the leaves overhead. When the storage capacity of the leaf surfaces is exhausted, water will run down tree trunks and drip downward **(stem flow)** (1–3). The amount of water intercepted by dense forests ranges from 8 to 35 percent of total annual precipitation (4). In a mixed hardwood forest in the northeastern United States, it averaged 20 percent in the summer and winter seasons (5). While evaporation of intercepted water reduces the net transpiration by the plants, in some cases most of the evaporated water is simply lost. One study concluded that only about 10 percent of the intercepted water actually reduced evapotranspiration (6).

The water reaching the ground can infiltrate into the soil, form puddles, or flow as a thin sheet of water across the land surface. Hydrologists refer to the water trapped in puddles as **depression storage.** It ultimately evaporates or infiltrates.

The overland flow process, sometimes called **Horton overland flow** after Robert Horton (7, 8), occurs only when the precipitation rate exceeds the infiltration capacity. In areas in which soils have a high infiltration capacity, this process may occur only during very intense storms or when the soil is saturated or frozen. In order for overland flow to occur, the infiltration capacity of the soil must first be exceeded; then the depression storage must be filled (Figure 3.1).

If the unsaturated zone is uniformly permeable, most of the infiltrated water percolates vertically. Should layers of soil with a lower vertical hydraulic conductivity occur beneath the surface, then infiltrated water may move horizontally in the unsaturated zone. This **interflow** may be substantial in some drainage basins and contribute significantly to total streamflow. Thin permeable soil overlying fractured bedrock of low permeability would provide a geologic condition contributing to significant interflow (Figure 3.2).

Water will fall directly onto the surfaces of lakes and reservoirs during the period of precipitation. This amount might not be considerable for streams, but for lakes and reservoirs it could be. Lake Michigan and its associated water bodies have a surface area of 22,300 square miles. The land area of the surround-

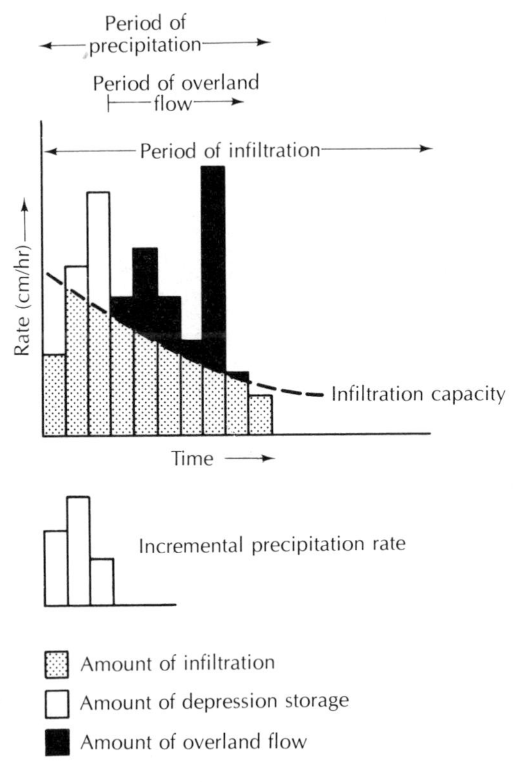


**FIGURE 3.1** Incremental precipitation rate and its dissociation into amounts of infiltration, depression storage, and overland flow. Infiltration begins when the precipitation does. Overland flow does not begin until the depression storage is exhausted. Overland flow continues past the termination of precipitation. Infiltration will continue as long as there is any water in depression storage—usually past the period of overland flow.

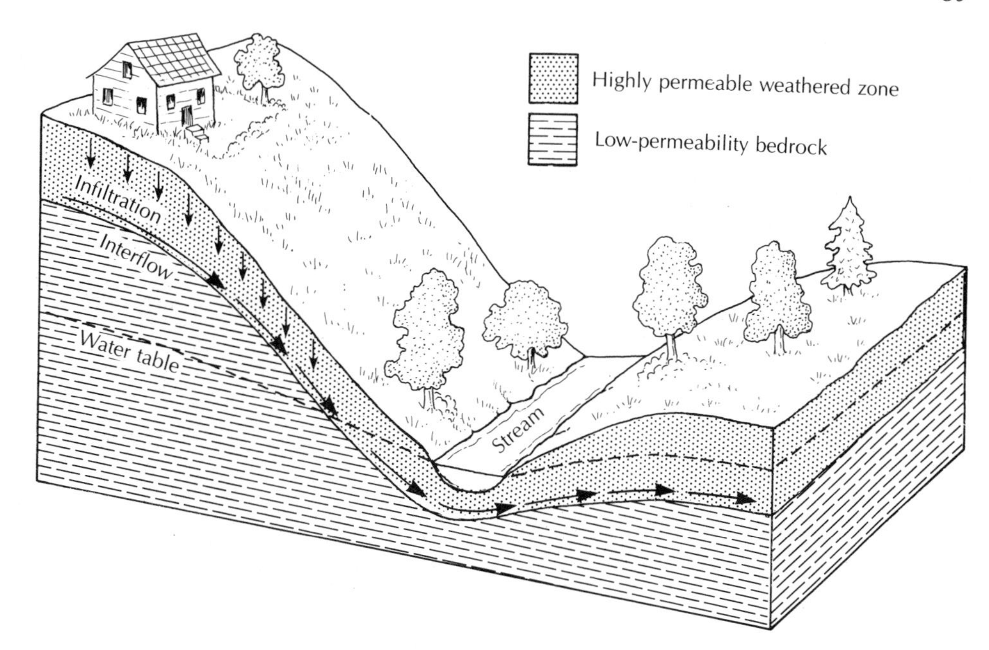


FIGURE 3.2 Interflow developing where a highly permeable but thin layer of weathered rock overlies a bedrock unit of lower permeability.

ing drainage basin is 45,000 square miles (9). Assuming equal distribution of precipitation over the entire Lake Michigan basin, about one-third falls as **direct precipitation** on a water body.

Infiltrated water that reaches the water table becomes stored in the ground-water reservoir. This is not static storage, as ground water is in constant movement. While freshly infiltrated precipitation is entering the ground-water reservoir, other ground water, known as **baseflow,** is discharging into a stream. If infiltration causes the water table to rise, ground-water discharge into nearby streams will also increase. For baseflow streams, the amount of ground-water discharge is directly proportional to the hydraulic gradient toward the stream (Figure 3.3).

The runoff cycle in which so much emphasis is placed on Horton overland flow has been criticized on several fronts (10). Horton overland flow is rarely observed in the field, except after very heavy precipitation events. This is especially true if the ground is covered with vegetation or humus, such as leaf litter (11). Horton overland flow appears to be more common in arid regions or areas in which the soil has been compacted by vehicles, animals, etc. (12). Overland runoff can also occur when precipitation falls on soils that are saturated.

Water that infiltrates into the soil on a slope can move downslope as lateral unsaturated flow in the soil zone. This has been called **throughflow** (11). The difference between throughflow and interflow is that throughflow emerges as seepage at the foot of the slope rather than entering a stream, as does inter-

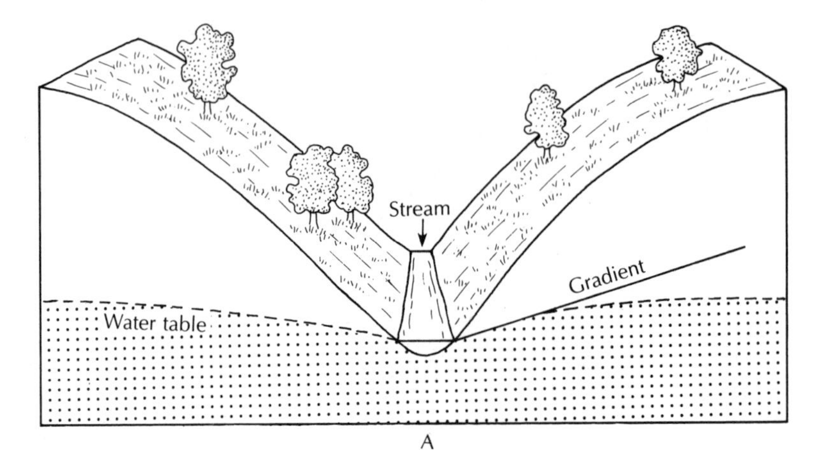


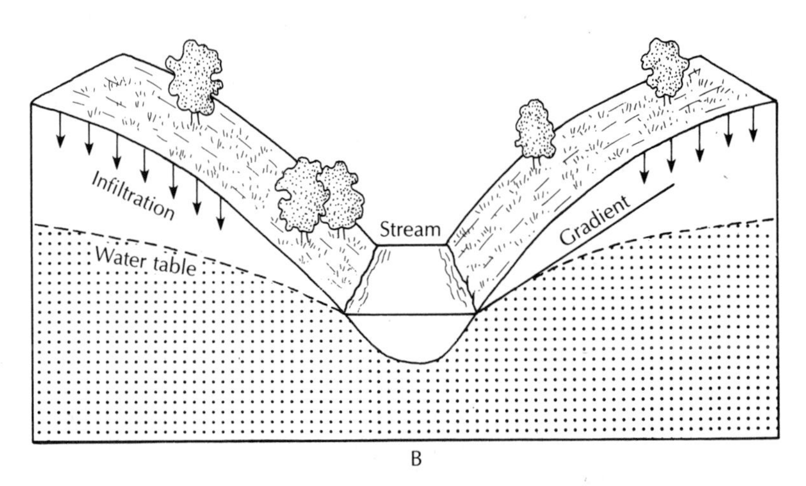


**FIGURE 3.3** Influence of the water-table gradient on baseflow. The stream in Part A is being fed by ground water with a low hydraulic gradient. A gentle rain does not produce overland flow, but infiltration raises the water table. The increased hydraulic gradient of Part B causes more baseflow to the stream, which is now deeper and has a greater discharge.

flow. Thus, the throughflow appears as overland flow before entering a stream channel. This overland flow is called **return flow** (13) to distinguish it from Horton overland flow.

A more comprehensive concept of the hydrologic cycle has been proposed by Dunne (12). In arid to subhumid climates where there is thin vegetation, Horton overland flow is the main contributor to the storm peak and comprises most of the streamflow. In humid climates, Horton overland flow is not significant, but interflow, return flow, and direct precipitation on the channel are important. Where there are thin soils and gentle, concave slopes, direct precipitation and return flow are more important than interflow. On steep, straight slopes, interflow becomes much more important, although return flow and direct precipitation still cause the peaks.

## 3.2 HYDROGRAPH SEPARATION

A stream **hydrograph** shows the discharge of a river at a single location as a function of time. While the total streamflow shown on the hydrograph gives no indication of its origin, it is possible to break down the hydrograph into components such as overland flow, baseflow, interflow, and direct precipitation. The model presented in this section is based on the Horton runoff cycle; it would be most useful for arid-zone hydrology.

### 3.2.1 BASEFLOW RECESSIONS

The hydrograph of a stream during a period with no excess precipitation will decay, following an exponential curve. The discharge is composed entirely of ground-water contributions. As the stream drains water from the ground-water reservoir, the water table falls, leaving less and less ground water to feed the stream. If there were no replenishment of the ground-water reservoir, baseflow to the stream would become zero. Figure 3.4 shows a baseflow recession curve for a stream in a climate with a dry summer season.

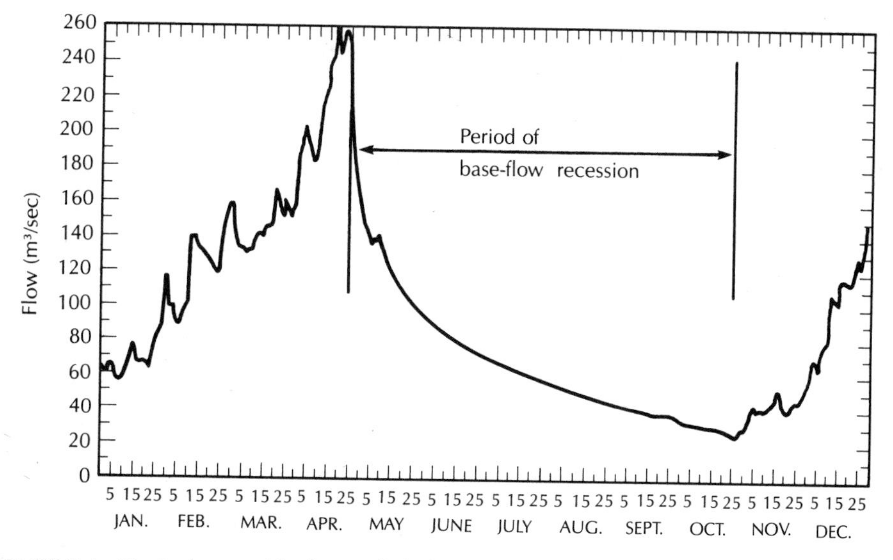


**FIGURE 3.4** Typical annual hydrograph for a river with a long dry summer season: Lualaba River, Central Africa. Source: C. O. Wisler and E. F. Brater, eds., *Hydrology*, 2nd ed. (New York: John Wiley & Sons, 1959). Used with permission.

The **baseflow recession** for a drainage basin is a hydromorphic characteristic. It is a function of the overall topography, drainage pattern, soils, and geology of the watershed. Figure 3.5 illustrates this by showing the annual summer recession of a river for six consecutive years. The start of the baseflow recession was considered to be the day when the annual discharge dropped below 3500 cubic feet per second. The recession is similar from year to year. The baseflow of the stream decreases during a dry period because as ground water drains into the stream, the water table falls. A lower water table means that the rate at which ground water seeps into the stream declines. Picture in your mind a bucket with a hole near the bottom. As the water drains from the bucket, the water level (water table) falls and the stream of water draining from the bucket (baseflow to streams) declines in volume. The stream of water draining (baseflow) will not increase until the water in the bucket is replenished (recharge) and the water level (water table) rises.

The baseflow recession equation is

$$Q = Q_0 e^{-at} \tag{3-1}$$

where

- $Q$ is the flow at some time $t$ after the recession started
- $Q_0$ is the flow at the start of the recession
- $a$ is a recession constant for the basin
- $t$ is the time since the recession began

---

**EXAMPLE PROBLEM**

**Part A:** Find the recession constant for the basin of Figure 3.5.

If

$$Q = Q_0 e^{-at}$$

then

$$\begin{aligned} e^{-at} &= Q/Q_0 \\ -at &= \ln Q/Q_0 \\ a &= -1/t \ln Q/Q_0 \end{aligned}$$

From Figure 3.5, $Q_0 = 3500$ cubic feet per second. After 100 days, $Q = 1500$ cubic feet per second.

$$\begin{aligned} a &= -1/t \ln Q/Q_0 = -1/100 \ln 1500/3500 \\ &= -0.01 \ln 0.4286 \\ &= -0.01 \times (-0.847) \end{aligned}$$

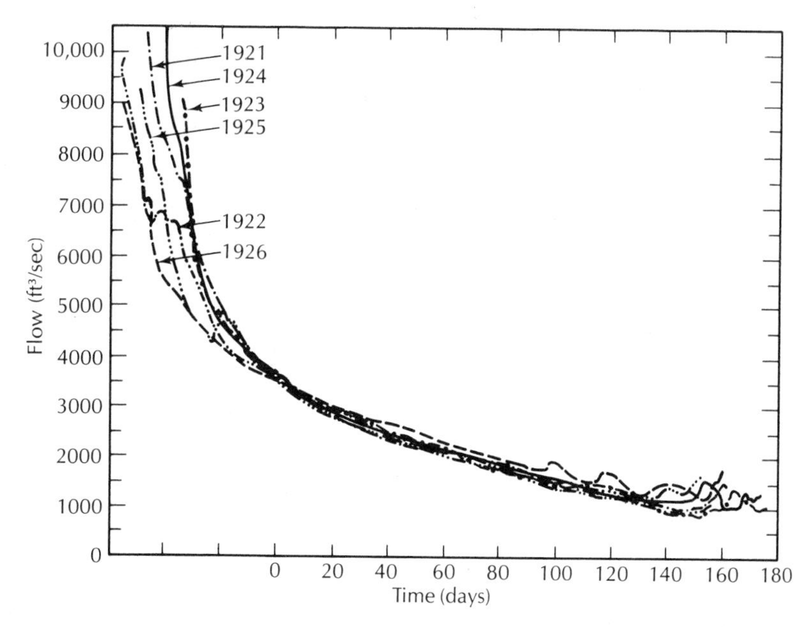


**FIGURE 3.5** Annual baseflow recessions for six consecutive years for the Lualaba River, Central Africa. Source: C. O. Wisler and E. F. Brater, eds., *Hydrology*, 2nd ed. (New York: John Wiley & Sons, 1959). Used with permission.

Therefore,

$$a = 8.47 \times 10^{-3}$$

if $Q$ is in cubic feet and $t$ is in days.

**Part B:** What would the baseflow be after 40 days of recession?

$$\begin{aligned} Q &= Q_0 e^{-at} \\ &= 3500 \exp(-8.47 \times 10^{-3} \times 40) \\ &= 3500 \times 0.713 \end{aligned}$$

Therefore,

$$Q = 2494 \text{ ft}^3/\text{sec}$$

---

### 3.2.2 STORM HYDROGRAPH

While the baseflow component of a stream is somewhat constant, the total discharge of the stream may fluctuate greatly through the year. The difference is due to the episodic nature of precipitation events that contribute over-

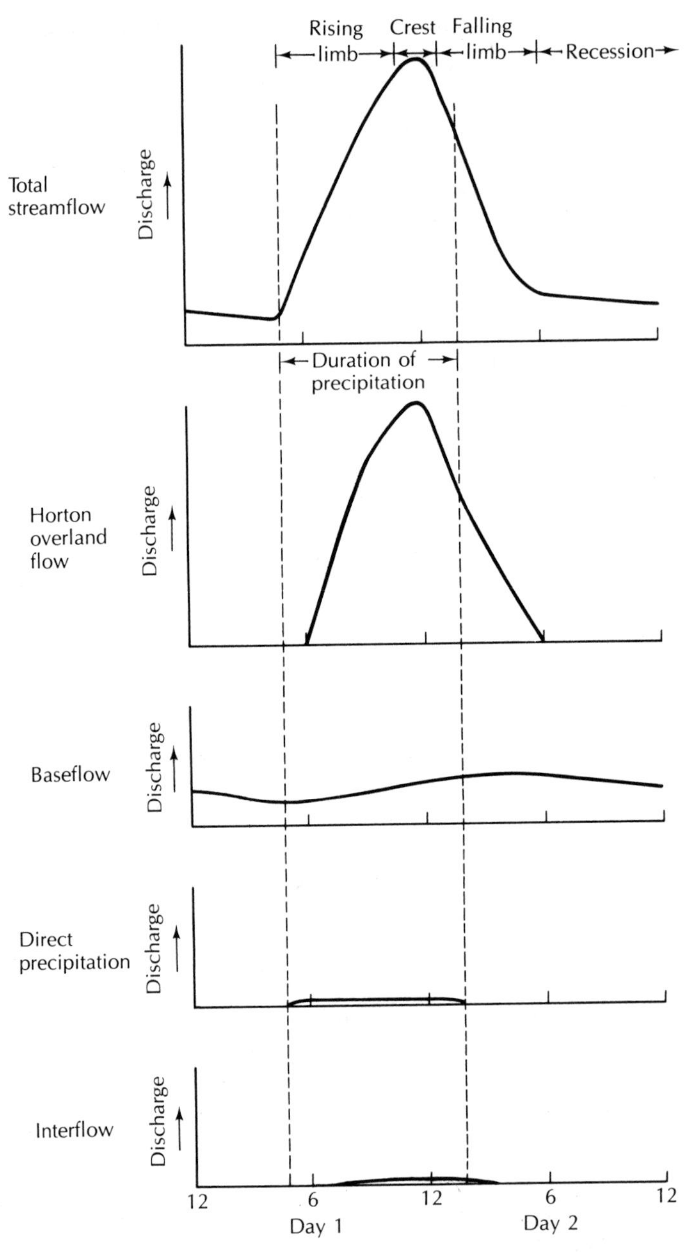


**FIGURE 3.6** Hypothetical storm hydrograph for a period of evenly distributed precipitation, separated into Horton overland flow, direct precipitation, and interflow.

land flow, interflow, and direct precipitation. For most drainage basins, direct precipitation adds only a modest amount of water to the stream. Interflow is a factor that can be highly variable, depending upon the geology of the drainage basin. A deep, sandy soil might not induce any interflow; on the other hand, a lava landscape covered by loose rubble might have no overland flow but great amounts of interflow at the base of the rubble where it overlies a hard, low-permeability lava flow. Steeply sloping land also promotes interflow. The most consistent factor in the storm hydrograph is overland flow. Figure 3.6 shows a hypothetical storm hydrograph broken down into overland flow, interflow, direct precipitation, and baseflow recession. The baseflow component is given for a stream that continues to receive ground-water discharge through the duration of the overland-flow peak.

One of the tasks in analyzing a storm hydrograph is to separate the overland-flow component from the baseflow. Generally, it is first assumed that both the direct precipitation and the interflow components are inconsequential; however, the hydrogeologist should be aware of the general geology and surface slope of the drainage area before assuming the inconsequence of the latter component. The overland flow is assumed to end some fixed time after the storm peak. As a general rule of thumb, this can be approximated by the formula (14)

$$D = A^{0.2} \tag{3-2}^*$$

where

$D$ is the number of days between the storm peak and the end of overland flow

$A$ is the drainage basin area in square miles

The exponential constant of 0.2 is somewhat arbitrary; thus, blind use of the preceding formula could result in error. The value will depend upon many drainage basin characteristics, such as mean slope, vegetation, drainage density, roughness, etc.

The baseflow recession that existed prior to the storm peak is extended until it is approximately under the storm peak. It is then drawn so as to rise to meet the stream hydrograph at a point $D$ days after the peak. In Figure 3.7, a storm hydrograph for a drainage basin of 2100 square miles has been separated. For the given basin, $D$ is equal to 4.6 days.

### 3.2.3 GAINING AND LOSING STREAMS

The typical stream of a humid region receives ground-water discharge; therefore, as one goes downstream the baseflow increases, even if no tributaries enter. This is a **gaining,** or **effluent,** stream. The water table slopes toward the stream, so that the hydraulic gradient of the aquifer is toward the stream. Figure 3.8B shows a cross section through a gaining reach of a stream.

---

*In the metric system, Equation 3-2 is $D$ (days) $= 0.827\ A^{0.2}$ (square kilometers).

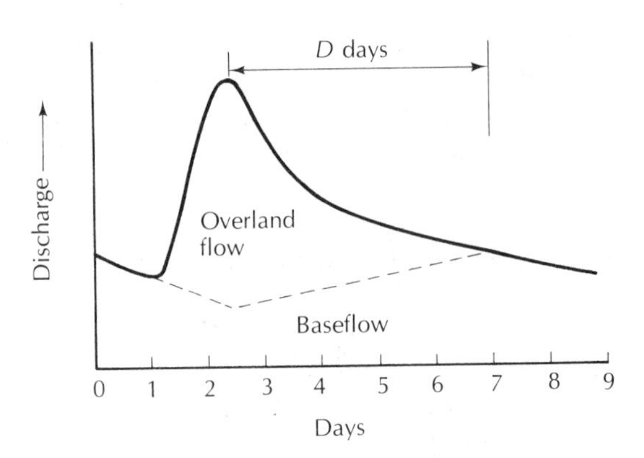


**FIGURE 3.7** Hydrograph separation into overland-flow component and baseflow component for a stream receiving Horton overland flow.

In arid regions, many rivers are fed by overland flow, interflow, and baseflow at high altitudes. As they wind their way to lower elevation, the local precipitation amounts decrease; consequently, there is less infiltration and a lower water table. There may also be a dramatic change in the depth to ground water when a stream draining a high-altitude basin of lower-permeability material flows out onto coarse alluvial materials. For whatever reason, if the bottom of the stream channel is higher than the local water table, water may drain from the stream into the ground (Figure 3.8A). As one goes downstream, less and less water will be found in the channel. The stream is **losing,** or **influent.** The rate of water loss is a function of the depth of water and the hydraulic conductivity of the underlying alluvium. Fine-grained deposits on the channel bottom will retard the rate of loss to the ground water.

A stream that is normally a gaining stream during baseflow recessions may temporarily become a losing stream during floods. If the flood-crest depth in the channel is greater than the local water-table elevation, the hydraulic gradient in the aquifer next to the stream is reversed. Water flows from the stream into the ground (Figure 3.9). The result is a temporary storage of flood water in the aquifer next to the stream. When the flood crest passes, the hydraulic gradient again reverses, and the stream is once again gaining (Figure 3.10).

Heavy ground-water pumping near a stream can lower the water table to an elevation below the level of the stream bottom. The reach of the stream affected by the lowered water table will become a losing stream, while upstream and downstream reaches can still be gaining. Figure 3.11 is an idealization of this phenomenon, based on the behavior of the well field along the Fenton River of the University of Connecticut at Storrs.

Conjunctive use of ground and surface water is discussed in Section 11.10. Legal issues have arisen when ground-water pumpage has depleted streamflow in states where the water in the stream has been appropriated to other uses (Section 11.6).

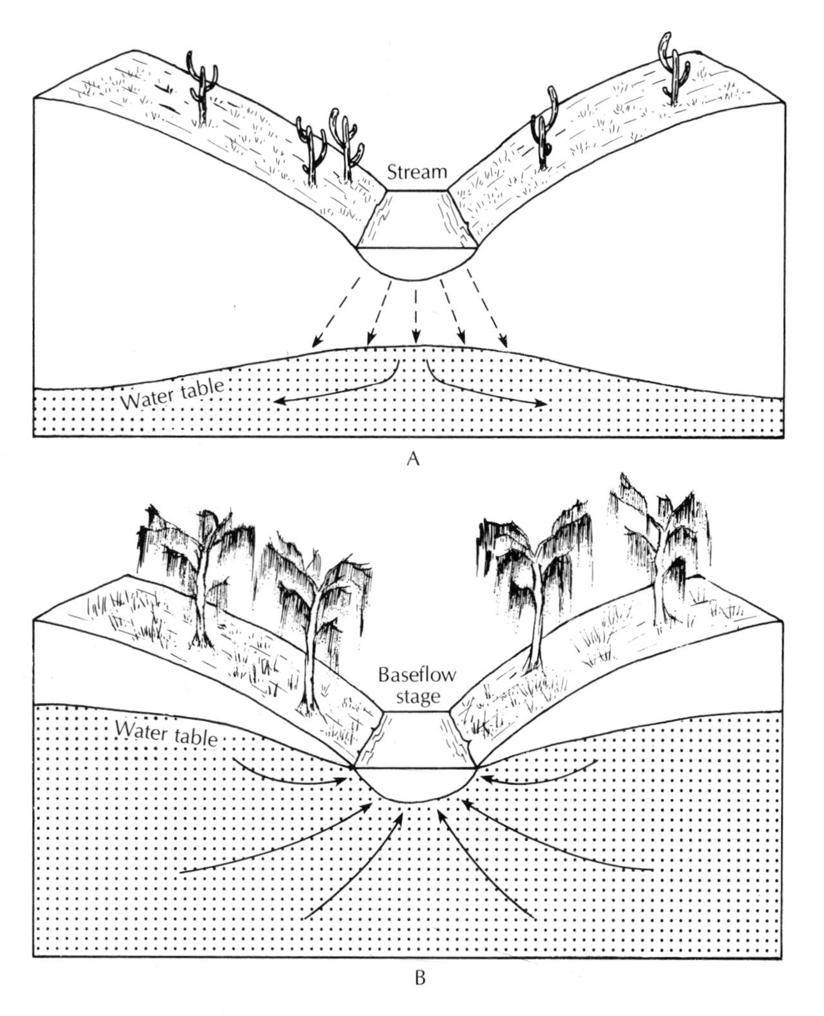


**FIGURE 3.8** **A.** Cross section of a losing stream, which is typical of arid regions, where streams can recharge ground water. **B.** Cross section of a gaining stream, which is typical of humid regions, where ground water recharges streams.

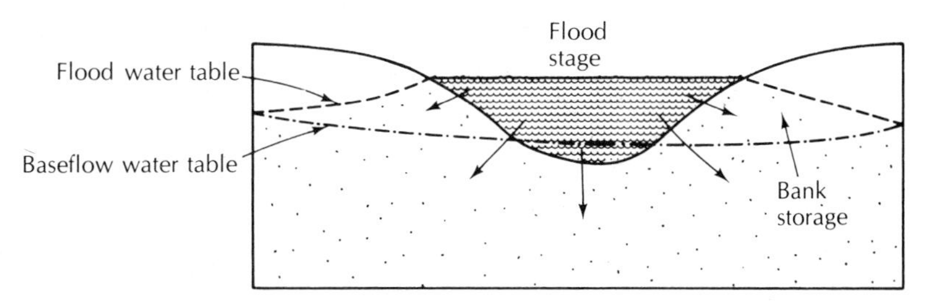


**FIGURE 3.9** A stream that is gaining during low-flow periods can temporarily become a losing stream during flood stage.

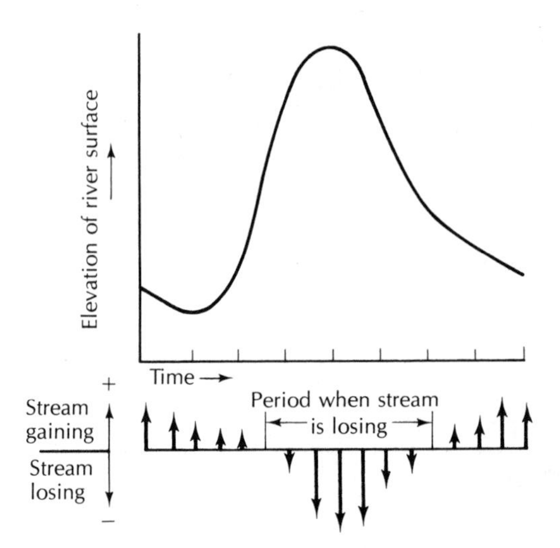


**FIGURE 3.10** Effect of flood state on the ground-water regime adjacent to the river. As the flood peak passes, the normal direction of ground-water flow into the stream is reversed.

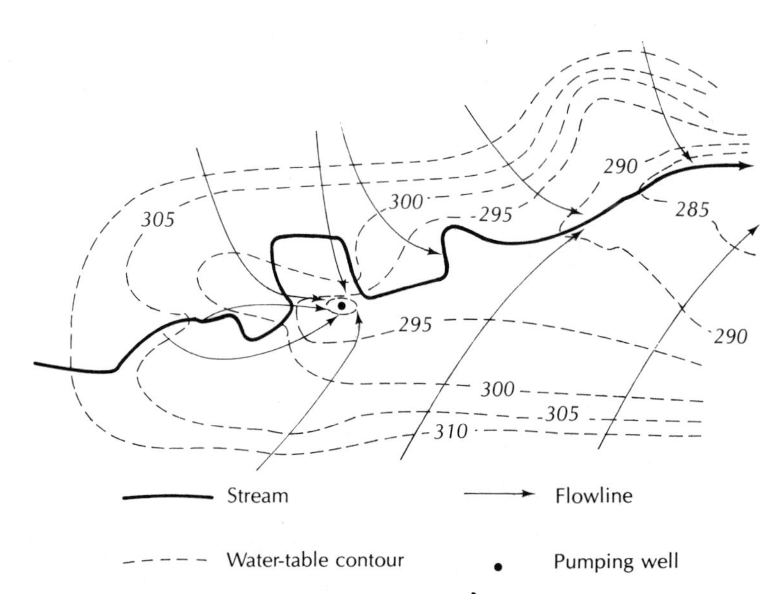


**FIGURE 3.11** Induced stream-bed infiltration caused by a pumping well. Source: P. Rahn, *Ground Water*, 6, no. 3 (1968): 21–32.

## 3.3 RATIONAL EQUATION

A relatively simple method of computing the rainfall-runoff relationship is known as the **rational method.** It is used to predict peak runoff rates from data on the rainfall intensity and a knowledge of land use in the drainage basin. The rational method is of greatest validity when used in analysis of small drainage basins of 200 acres (100 hectares) or less. The rational equation is

$$Q = CIA \tag{3-3}$$ *

where

$Q$ is the peak runoff rate (cubic feet per second)

$I$ is the average rainfall intensity (inches per hour)

$A$ is the drainage area (acres)

$C$ is the runoff coefficient from Table 3.1

The rational equation assumes that the rainfall event lasts long enough for the maximum discharge of the drainage basin to occur. In order for such a simple relationship to hold, the rate of infiltration must also be constant during the storm. In Table 3.1 (p. 50), a range of $C$ values is given for many land uses. The lower values are used for storms of low intensity; storms of greater intensity will have proportionally more runoff, justifying the use of higher $C$ factors.

## 3.4 DURATION CURVES

For design or regulatory purposes, it may be necessary to know how often the discharge of a stream may be less than or greater than a given value. As an example, if a river is considered for a water-supply source, it is necessary to know how much water can be obtained. The average flow is not a particularly useful value, in that possibly 50 percent of the time the river would carry less than the average discharge. Depending upon the available storage and other sources of supply, some flow duration is selected as the reliable flow. For example, the 90 percent duration is the flow that will be equaled or exceeded 90 percent of the time.

**Duration curves** are generally constructed for either daily flow or annual flow, although other time periods could also be considered. Data are ranked from greatest to least flow values. They are then assigned a serial rank, $m$, starting with 1 for the greatest flow and going to $n$, the number of data values. If two or more data values are equal, each should receive a different serial rank. The probability, $P$, in percentage, that a given flow will be equaled or exceeded may be found by the equation

---

*In the metric system, Equation 3-3 is $Q$ (cubic meters/sec) = 0.0028 $CI$ (millimeters/hour) $A$ (hectares).

**TABLE 3.1**

| Description of Area | C |
|---|---|
| Business | |
| Downtown | 0.70–0.95 |
| Neighborhood | 0.50–0.70 |
| Residential | |
| Single-family | 0.30–0.50 |
| Multiunits, detached | 0.40–0.60 |
| Multiunits, attached | 0.60–0.75 |
| Residential suburban | 0.25–0.40 |
| Apartment | 0.50–0.70 |
| Industrial | |
| Light | 0.50–0.80 |
| Heavy | 0.60–0.90 |
| Parks, cemeteries | 0.10–0.25 |
| Playgrounds | 0.20–0.35 |
| Railroad yard | 0.20–0.35 |
| Unimproved | 0.10–0.30 |
| *Character of surface* | |
| Pavement | |
| Asphalt and concrete | 0.70–0.95 |
| Brick | 0.70–0.85 |
| Roofs | 0.75–0.95 |
| Lawns, sandy soil | |
| Flat, up to 2% grade | 0.05–0.10 |
| Average, 2%–7% grade | 0.10–0.15 |
| Steep, over 7% | 0.15–0.20 |
| Lawns, heavy soil | |
| Flat, up to 2% grade | 0.13–0.17 |
| Average, 2%–7% grade | 0.18–0.22 |
| Steep, over 7% | 0.25–0.35 |

Source: American Society of Civil Engineers, "Design and Construction of Sanitary and Storm Sewers," *Manuals and Reports of Engineering Practice No. 37,* 1970.

$$P = 100 \frac{m}{n + 1} \tag{3-4}$$

A plot of $P$ as a function of flow will yield a duration curve showing the percentage of time a given flow is equaled or exceeded. The curve can be plotted on a type of graph paper known as probability paper. This paper is constructed with a special abscissa and ordinates that may be either arithmetic or logarithmic.

Figure 3.12 shows duration curves of daily flow for three rivers in Wisconsin. In order to compare the three directly, the discharge has been computed as cubic feet per second per square mile (cfs/mi$^2$) of drainage basin. This makes

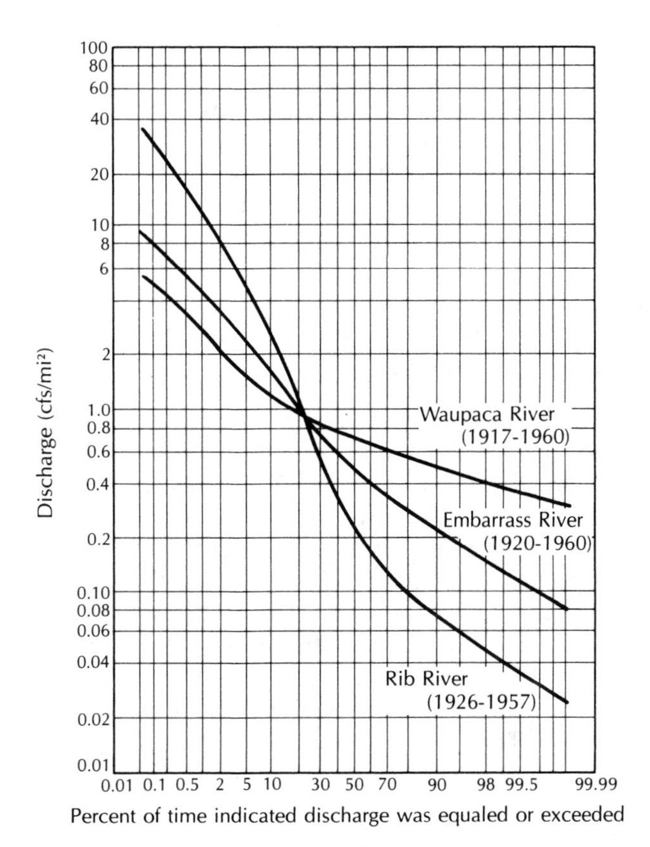


**FIGURE 3.12** Daily duration curves for three streams having different runoff characteristics owing to the differing geology of the drainage basins. Source: U.S. Geological Survey.

the flow independent of the size of the drainage basin. The three streams are all in central Wisconsin and the annual runoff (precipitation less evaporation) is about 11 inches (28 centimeters) per year for all three. Examination of Figure 3.12 reveals a great variability in the distribution of this annual runoff.

The Rib River has high flood values; 1.0 percent of the time flow equals or exceeds 12 cfs/mi$^2$. On the other hand, the 1 percent value for the Waupaca River is 2.7 cfs/mi$^2$, with the Embarrass River intermediate. All three rivers have the same 20 percent flow value: 0.9 cfs/mi$^2$. Whereas the Rib River had the greatest flood flows, it has the smallest low flows. One percent of the time the Rib River discharge is less than 0.04 cfs/mi$^2$. (The graph shows this as 99 percent of the time the flow equals or exceeds 0.04 cfs/mi$^2$.) The Waupaca River has low flows an entire magnitude greater; the flow is less than 0.39 cfs/mi$^2$ for 1 percent of the time. Again, the Embarrass River falls about evenly between the two.

This distribution of runoff is caused by the geology of the drainage basins. The Rib River is located in an area of crystalline bedrock, which has a very low hydraulic conductivity. Part of the drainage basin is in the driftless area, where surficial glacial deposits are lacking, overland flow and return flow are high, and baseflow is scant. The soils are thin, with little water-retaining capacity. The drainage basin of the Waupaca River has thick deposits of unconsolidated sand. Most of the potential overland flow is absorbed by the sand; hence, there are small flood peaks. This water can drain slowly and provide high baseflows. The Embarrass River has thick deposits of glacial drift—but it is till and lake clay—so the hydraulic response of the watershed is intermediate.

## 3.5 DETERMINING GROUND-WATER RECHARGE FROM BASEFLOW

A simple method of estimating ground-water recharge in a basin has been developed. It utilizes stream hydrographs from two or more consecutive years. The baseflow-recession equation (3-1) indicates that $Q_0$ varies logarithmically with time, $t$. A plot of a stream hydrograph with time on an arithmetic scale and discharge on a logarithmic scale will therefore yield a straight line for the baseflow recession. Figure 3.13 shows hypothetical stream hydrographs. The baseflow recessions are shown as dashed lines; they were considered to start when the summer stream level dropped below the adjacent water table and to end when the first spring flood occurred. The total potential ground-water discharge, $V_{tp}$, is the amount that would be discharged by a complete ground-water recession (15). The value of $V_{tp}$ can be found from

$$V_{tp} = \frac{Q_0 t_1}{2.3} \tag{3-5}$$

where

$Q_0$ is the baseflow at the start of the recession

$t_1$ is the time it takes for the baseflow to go from $Q_0$ to $0.1Q_0$.

If one determines the remaining potential ground-water discharge at the end of a recession and then the total potential ground-water discharge at the beginning of the next recession, the difference between the two is the ground-water recharge that has taken place between recessions. The amount of potential baseflow, $V_t$, remaining some time, $t$, after the start of a baseflow recession is given by

$$V_t = \frac{V_{tp}}{10^{(t/t_1)}} \tag{3-6}$$

or

$$V_t = \frac{(Q_0 t_1)/2.3}{10^{(t/t_1)}} \tag{3-7}$$

This analysis assumes that there are no consumptive uses of ground water in the basin so that all ground-water discharge is by means of baseflow to streams. If there are such uses as pumpage or evapotranspiration of ground-

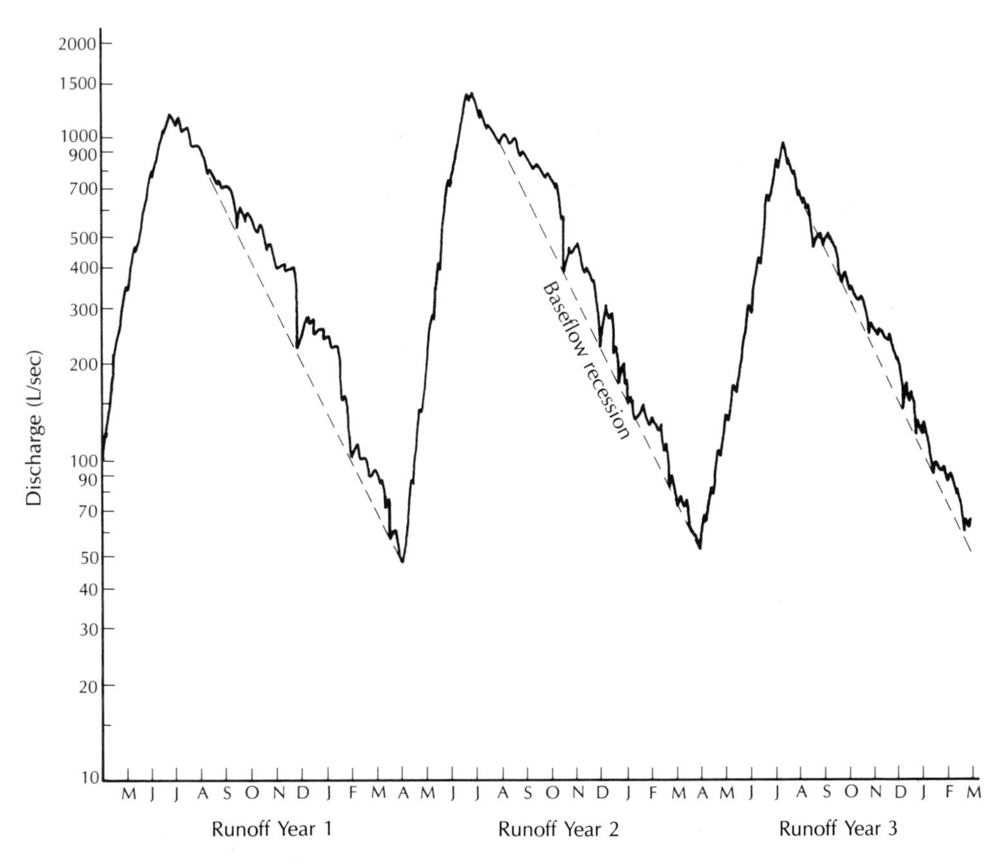


**FIGURE 3.13** Semilogarithmic stream hydrographs showing baseflow recessions.

water by phreatophytes, this use must be added to the amount determined by the baseflow recession method to get total recharge to the ground-water reservoir.

---

**EXAMPLE PROBLEM** Refer to Figure 3.13. Determine the amount of ground-water recharge that takes place from the end of the baseflow recession of Runoff Year 1 to the start of the baseflow recession of Runoff Year 2.

The value of $Q_0$ for the first recession is 760 liters per second and it takes 6.3 months for the discharge to reach $0.1Q_0$:

$$V_{tp} = \frac{Q_0 t_1}{2.3}$$

$$V_{tp} = \frac{760 \text{ L/sec} \times 6.3 \text{ mon} \times 30 \text{ days/mon} \times 1440 \text{ min/day} \times 60 \text{ sec/min}}{2.3}$$

$$V_{tp} = 5.4 \times 10^9 \text{ L}$$

The value of $V_t$ at the end of the recession, which lasts 7.5 months, is

$$V_t = \frac{V_{tp}}{10^{(t/t_1)}} = \frac{5.4 \times 10^9}{10^{(7.5/6.3)}} = \frac{5.4 \times 10^9}{15.5} = 3.5 \times 10^8 \text{ L}$$

For the next year's recession, the value of $Q_0$ is 1000 liters per second, and $t$ is again 6.3 months. Therefore,

$$V_{tp} = \frac{1000 \text{ L/sec} \times 6.3 \times 30 \times 1440 \times 60 \text{ sec}}{2.3}$$
$$= 7.1 \times 10^9 \text{ L}$$

The amount of recharge is equal to the total potential baseflow remaining at the end of the first baseflow recession subtracted from $V_{tp}$ for the beginning of the next recession:

$$\text{Recharge} = 7.1 \times 10^9 \text{ L} - 3.5 \times 10^8 \text{ L}$$
$$= 6.75 \times 10^9 \text{ L}$$

---

## 3.6 MEASUREMENT OF STREAMFLOW

### 3.6.1 STREAM GAGING

Water flowing in an open channel is subject to friction as it comes in contact with the channel bottom and sides. As a result, the fastest current is at the surface in the center of the channel. If a series of careful measurements of flow velocity from the surface downward are made, a parabolic profile will emerge (Figure 3.14). Field studies have shown that the velocity at a depth equal to 0.6 times the total depth is very close to the average velocity for the entire section. The average of measurements made at 0.2 times depth and 0.8 times depth is also used to represent the average velocity of the entire profile.

The flow, $q$, in an open channel with a cross-sectional area, $a$, and average velocity, $v$, can be found from the equation

$$q = va \qquad \textbf{(3-8)}$$

The velocity of flow can be measured by using a **current meter.** The U.S. Geological Survey has standard specifications for two types of meters. Each has a horizontal wheel with sets of small, cone-shaped cups attached. The wheel turns in the current and a cam attached to the spindle of the wheel makes an electrical contact once every revolution. The meter wheel is wired either to a set of headphones and a battery or to a direct readout meter. The operator with a headphone meter counts the clicks over a measured time period, usually thirty to sixty seconds, and uses a calibration curve furnished by the manufacturer to find the velocity.

The **Price-type meter** has a wheel about 5 inches (13 centimeters) in di-

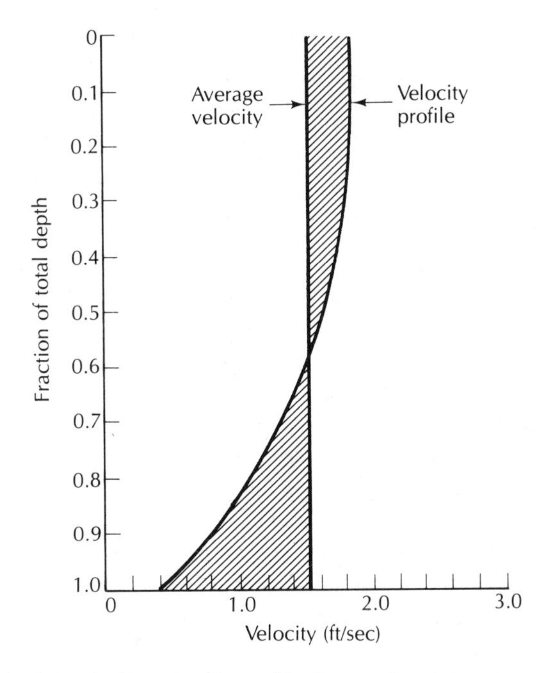


**FIGURE 3.14** Typical parabolic velocity profile for a natural stream.

ameter; it is equipped with a vane to orient the meter perpendicular to the flow. The Price-type meter is usually suspended on a cable and lowered into a river, with a streamlined weight to pull it down. A **pygmy-Price meter** is smaller and is usually mounted on a graduated wading rod. The operator takes to the river with rubber boots and places the rod on the bottom of a stream in order to make the measurements.

In a typical stream, velocity will vary from bank to bank, necessitating a number of measurements. A straight reach of stream with a smooth shoreline, no brush hanging in the water, and no weeds or large rocks should be chosen. Places with back-eddies should be avoided; they will overestimate the total discharge, as the current meter will not distinguish the direction of flow. If a wading rod is to be used, the water must not be too deep or too swift for a person to wade. This is especially important to check when measuring peak flows.

A tape is stretched perpendicularly across the stream or along the bridge. The channel is subdivided into fifteen to thirty segments. At the midpoint of each segment, the depth, $d_i$, is measured and recorded. The meter is then raised to 0.6 times the depth, and the average velocity, $v_i$, for that segment is measured. If the water is deep, the average velocity at 0.8 depth and 0.2 depth should be used. The discharge, $q_i$, for a segment of width, $w_i$, is given by

$$q_i = v_i d_i w_i \tag{3-9}$$

The process is repeated for each segment of the cross section. The total discharge, $Q$, for the river is the sum of the discharge for each segment. For a measurement with $m$ segments,

$$Q = \sum_{i=1}^{m} q_i \tag{3-10}$$

If measurements are made from a bridge using a cable-suspended meter, a swift current may draw the meter downstream. The amount of line let out to measure the depth is thus too great. A correction must be made based on the angle of the cable from the vertical (16).

Current measurements may be made through ice. A series of holes are cut in the ice across the river and the current measured by the preceding method. Because of friction between the ice and the underlying water, the velocity should be measured at 0.2 depth, 0.6 depth, and 0.8 depth and the results averaged.

It is possible to develop an empirical relationship between stream stages (elevation of the water surface above a datum) and discharge. As discharge measurements are slow and costly to make, knowledge of the preceding relationship is useful. A **rating curve** for a stream is made by simultaneously measuring the discharge of a stream and its stage and then repeating the measurements for a number of different stage heights. Stage versus discharge is then plotted as coordinates on graph paper to produce the rating curve (Figure 3.15). If the stream channel does not scour during flooding, and if the stage is not affected by such factors as tributary flow, a simple rating curve is sufficient. Otherwise, a rating

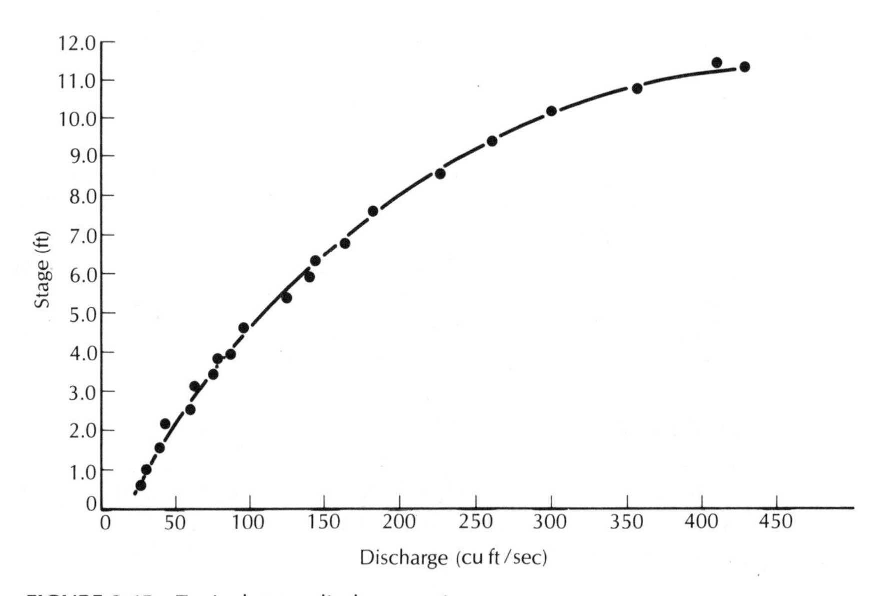


**FIGURE 3.15** Typical stage-discharge rating curve.

curve must also include a factor for water-surface slope (17). Automated stream-gaging stations employ a float device that measures the stage of a river by means of a stilling well connected to the stream. The stage data are transformed into discharge data by using either the rating curve or a rating table based on the curve. In the United States, most stream-stage measurements are recorded in digital form for automatic data processing.

### 3.6.2 WEIRS

The discharge of small streams can be conveniently measured by use of a **weir.** This is a small dam with a spillway opening of specified shape. There are a number of standard shapes for sharp-crested weirs, the most common being a 90-degree V-notch or a rectangular cut-out. A small earthen or concrete dam is built and the weir set into it. The dam will impound a small amount of water that should free fall over the weir crest, or lowest point of the spillway. The elevation of the backwater above the weir crest, $H$, is measured. The discharge over the weir can be found from the following formulas:

Rectangular weir:

$$Q = \tfrac{1}{3}(L - 0.2H)\, H^{3/2} \qquad \textbf{(3-11)}$$

Ninety-degree V-notch weir:

$$Q = 2.5H^{5/2} \qquad \textbf{(3-12)}$$

where

$Q$ is the discharge (cubic feet per second)

$L$ is the length of the weir crest (feet)

$H$ is the head of the backwater above the weir crest (feet)

## 3.7 MANNING EQUATION

In open-channel hydraulics, the average velocity of flow of water may be found from the **Manning equation:**

$$V = \frac{1.49R^{2/3}S^{1/2}}{n} \qquad \textbf{(3-13)*}$$

where

$V$ is the average velocity in feet per second

$R$ is the hydraulic radius, or the ratio of the cross-sectional area of flow in square feet to the wetted perimeter (see Figure 3.16) in feet

*In the metric system, Equation 3-13 is $V$ (meters/sec) $= (1/n)\, R^{2/3}\, S^{1/2}$.

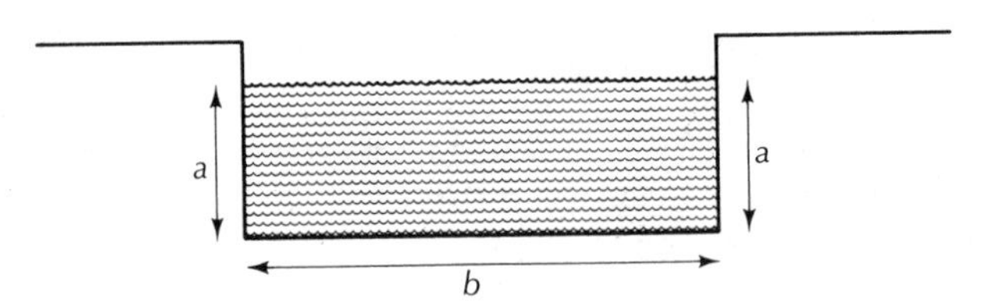


FIGURE 3.16 The cross-sectional area of flow is $a \times b$. The wetted perimeter is $a + b + a$.

$S$ is the energy gradient, which is the slope of the water surface

$n$ is the Manning roughness coefficient

The velocity of flow is dependent upon the amount of friction between the water and the stream channel. Smoother channels will have less friction and, hence, faster flow. Channel roughness contributes to turbulence, which dissipates energy and reduces flow velocity. The following values for $n$ are typical:

| | |
|---|---|
| mountain streams with rocky beds: | 0.04–0.05 |
| winding natural streams with weeds: | 0.035 |
| natural streams with little vegetation: | 0.025 |
| straight, unlined earth canals: | 0.020 |
| smoothed concrete: | 0.012 |

The U.S. Geological Survey has published a series of photographs of rivers for which the value of the Manning roughness coefficient has been computed (18). Field measurements of velocity, slope, area, and wetted perimeter were made and the value of $n$ computed from Equation 3-13. Careful study of the photographs can be used to obtain an estimate for the value of $n$ for a given river under study.

The Manning equation can be used to determine flow in situations that preclude direct measurements. For example, if the current is changing rapidly (during a rising or falling flood peak, for example), conventional streamflow measurements would take too long. It might take the better part of an hour to make a discharge measurement, and flow velocity and discharge could change substantially during the period of measurement. Under these conditions, an instant computation can be made using river cross sections and a measured slope.

## REFERENCES

1. BROWN, J. H., and A. C. BARKER. "An Analysis of Throughflow and Stemflow in Mixed Oak Stands." *Water Resources Research,* 6 (1970):316–23.
2. ROGERSON, T. L., and W. R. BYRNES. "Net Rainfall under Hardwoods and Red Pine in Central Pennsylvania." *Water Resources Research,* 4 (1968):55–58.

3. HELVEY, J. D. "Interception by Eastern White Pine." *Water Resources Research,* 3 (1967):723–30.
4. DUNNE, T., and L. B. LEOPOLD. *Water in Environmental Planning.* San Francisco: W. H. Freeman and Company, 1978, p. 88.
5. TRIMBLE, G. R., JR., and S. WEITZMAN. "Effect of a Hardwood Forest Canopy on Rainfall Intensities." *Transactions, American Geophysical Union,* 35 (1954): 226–34.
6. THORUD, D. B. "The Effect of Applied Interception on Transpiration Rates of Potted Ponderosa Pine." *Water Resources Research,* 3 (1967):443–50.
7. HORTON, R. E. "The Role of Infiltration in the Hydrologic Cycle." *Transactions, American Geophysical Union,* 14 (1933):446–60.
8. HORTON, R. E. "An Approach toward a Physical Interpretation of Infiltration Capacity." *Soil Science Society of America, Proceedings,* 4 (1940):399–417.
9. INTERNATIONAL GREAT LAKES LEVELS BOARD. "Regulation of Great Lakes Water Levels," Appendix A, *Hydrology and Hydraulics.* Report to the International Joint Commission, December 7, 1973.
10. CHORLEY, R. J. "The Hillslope Hydrological Cycle." In *Hillslope Hydrology,* ed. M. J. Kirkby. Chichester, Sussex, England: John Wiley & Sons, 1978, pp. 1–42.
11. KIRKBY, M. J., and R. J. CHORLEY. "Throughflow, Overland Flow, and Erosion." *Bulletin, International Association Scientific Hydrology,* 12 (1967):5–21.
12. DUNNE, T. "Field Studies of Hillslope Flow Processes." In *Hillslope Hydrology,* ed. M. J. Kirkby. Chichester, Sussex, England: John Wiley & Sons, 1978, pp. 227–94.
13. DUNNE, T., and R. D. BLACK. "An Experimental Investigation of Runoff Production in Permeable Soils." *Water Resources Research,* 6 (1970):478–90.
14. LINSLEY, R. K., JR., M. A. KOHLER, and J. L. H. PAULHUS. *Hydrology for Engineers.* New York: McGraw-Hill Book Company, 1975, 230 pp.
15. MEYBOOM, P. "Estimating Groundwater Recharge from Stream Hydrographs." *Journal of Geophysical Research,* 66 (1961):1203–14.
16. CORBETT, D. M., et al. *Stream-Gauging Procedure.* U.S. Geological Survey Water-Supply Paper 888, 1945, pp. 43–51.
17. MITCHELL, W. D. *Stage-Fall-Discharge Relations for Steady Flow in Prismatic Channels.* U.S. Geological Survey Water-Supply Paper 1164, 1954, 112 pp.
18. BARNS, H. H. *Roughness Characteristics of Natural Channels.* U.S. Geological Survey Water-Supply Paper 1849, 1967, 213 pp.

## PROBLEMS

1. Analysis of baseflow recession curves from a drainage basin has yielded a recession constant of $1.2 \times 10^{-2}$ when discharge is in cubic feet per second and time is in days.
   a. If a recession begins with a discharge of 2975 cubic feet per second and $t$ is in days, what will be the flow after 35 days and 70 days?

   **b.** If the recession begins with a discharge of 1165 cubic feet what would the flow be in 40 days?

2. The hydrograph in Figure 3.7 has a drainage basin area of 225 square miles. Use Equation 3-2 to separate the overland-flow component from the baseflow component.
3. An industrial park with flat-roofed buildings, large parking lots, and very little open area has a drainage basin area of 90 acres. The 25-year rainfall event (the amount that would on an average occur once in 25 years) has a precipitation intensity of 2 inches per hour.
   **a.** Use Equation 3-3 to compute the runoff rate.
   **b.** The industrial park is drained by a drainage canal that is 12 feet wide, has vertical walls, and has a bottom slope of 0.005. Use Equation 3-13 to compute the depth of water that would flow in the canal.
4. A rectangular weir is placed in a small stream to measure flow. The value of $L$ is 2 feet and $H$ is 0.23 feet. Use Equation 3-11 to compute the discharge of the stream.
5. A V-notch weir is placed in a road culvert to measure the flow of a stream passing through the culvert. The value of $H$ is 1.12 feet. Use Equation 3-12 to compute the discharge of the stream.
6. The annual flow of the Colorado River at Lees Ferry for the period 1896 to 1956 is given in Table 3.2. Use Equation 3-4 to construct a table of probability values. Use standard probability paper (e.g., Keuffel & Esser Co. 46 8000) to plot a duration curve showing the percent of the time an indicated discharge was equaled or exceeded.

**TABLE 3.2** Annual flow of the Colorado River at Lees Ferry (in millions of acre-feet)

| Year | Flow | Year | Flow | Year | Flow |
|---|---|---|---|---|---|
| 1896 | 10.089 | 1917 | 24.037 | 1938 | 17.545 |
| 1897 | 18.009 | 1918 | 15.364 | 1939 | 11.075 |
| 1898 | 13.815 | 1919 | 12.462 | 1940 | 8.601 |
| 1899 | 15.874 | 1920 | 21.951 | 1941 | 18.148 |
| 1900 | 13.228 | 1921 | 23.015 | 1942 | 19.125 |
| 1901 | 13.582 | 1922 | 18.305 | 1943 | 13.103 |
| 1902 | 9.393 | 1923 | 18.269 | 1944 | 15.154 |
| 1903 | 14.807 | 1924 | 14.201 | 1945 | 13.410 |
| 1904 | 15.645 | 1925 | 13.033 | 1946 | 10.426 |
| 1905 | 16.027 | 1926 | 15.853 | 1947 | 15.473 |
| 1906 | 19.124 | 1927 | 18.616 | 1948 | 15.613 |
| 1907 | 23.402 | 1928 | 17.279 | 1949 | 16.376 |
| 1908 | 12.856 | 1929 | 21.428 | 1950 | 12.894 |
| 1909 | 23.275 | 1930 | 14.888 | 1951 | 11.647 |
| 1910 | 14.248 | 1931 | 7.769 | 1952 | 20.290 |
| 1911 | 16.028 | 1932 | 17.243 | 1953 | 10.670 |
| 1912 | 20.520 | 1933 | 11.356 | 1954 | 7.900 |
| 1913 | 14.473 | 1934 | 5.640 | 1955 | 9.150 |
| 1914 | 21.222 | 1935 | 11.549 | 1956 | 10.720 |
| 1915 | 14.027 | 1936 | 13.800 | | |
| 1916 | 19.201 | 1937 | 13.740 | | |

7. An aqueduct has smooth earthen sides and bottom. The slope of the water surface is 5 feet per mile. The channel is trapezoidal in shape with a 45-degree angle to the sides of the trapezoid and a bottom segment that is 50 feet wide. The water in the aqueduct is 8 feet deep in the center. Use Equation 3-13 to compute the flow in the aqueduct.

# FOUR

# Soil Moisture and Ground Water

Some of the vapour that is formed by day does not rise high because the ratio of the fire that is raising it to the water that is being raised is small. When this cools and descends at night, it is called dew and hoar-frost. When the vapour is frozen before it has condensed to water again it is hoar-frost . . . . It is dew when the vapour has condensed into water and the heat is not so great as to dry up the moisture that has been raised, nor the cold sufficient (owing to the warmth of the climate or season) for the vapour itself to freeze.

*Meteorologica*, Aristotle (384–322 B.C.)

## 4.1 POROSITY OF EARTH MATERIALS

At the time they are formed, some rocks contain void spaces while others are solid. Those rocks occurring near the surface of the earth are not totally solid. The physical and chemical weathering processes there continually decompose and disaggregate rock, thus creating voids. Slight movements of rock masses near the surface can cause rocks to crack or fracture. This also results in openings between rocks.

**Sediments** are assemblages of individual grains that were deposited by water, wind, ice, or gravity. There are openings called **pore spaces** between the sediment grains, so that sediments are not solid.

The cracks, voids, and pore spaces in earth materials are of great importance to hydrogeology. Ground water and soil moisture occur in the voids in otherwise solid earth materials.

### 4.1.1 DEFINITION OF POROSITY

The **porosity** of earth materials is the percentage of the rock or soil that is void of material. It is defined mathematically by the equation

$$n = \frac{100V_v}{V} \quad \textbf{(4-1)}$$

where

$n$ is the porosity (percentage)

$V_v$ is the volume of void space in a unit volume of earth material

$V$ is the unit volume of earth material, including both voids and solids

Laboratory porosity is determined by taking a sample of known volume ($V$). The sample is dried in an oven at 105° C until it reaches a constant weight. This expels moisture clinging to surfaces in the sample, but not water that is hydrated as a part of certain minerals. The dried sample is then submerged in a known volume of water and allowed to remain in a sealed chamber until it is saturated. The volume of the voids ($V_v$) is equal to the original water volume less the volume in the chamber after the saturated sample is removed. This method excludes pores not large enough to contain water molecules and those which are not interconnected.

The total porosity can be computed from the relationship

$$n = 100\,[1 - (P_b/P_d)] \qquad \textbf{(4-2)}$$

where

$n$ is the total porosity as a percentage

$P_b$ is the bulk density of the aquifer material (g/cm$^3$)

$P_d$ is the particle density of the aquifer material (g/cm$^3$)

The bulk density of the aquifer material is the mass of the sample after oven drying divided by the original sample volume (the sample can change volume upon oven drying). The particle density is the oven-dried mass divided by the volume of the mineral matter in the sample as determined by a water-displacement test. For most rock and soil the particle density is about 2.65 g/cm$^3$.

### 4.1.2 POROSITY AND CLASSIFICATION OF SEDIMENTS

The porosity of sediments consists of the void spaces between solid fragments. If the fragments are solid spheres of equal diameters, they can be put together in such a manner that each sphere sits directly on the crest of the underlying sphere (Figure 4.1). This is called **cubic packing,** with an associated porosity of 47.65 percent (1). If the spheres lie in the hollows formed by four adjacent spheres of the underlying layer, the result is **rhombohedral packing,** with a porosity of 25.95 percent (1).

These two configurations represent the extremes of porosity for arrangements of equidimensional spheres with each sphere touching all neighboring spheres. The diameter of the sphere does not influence the porosity. Thus, a room full of bowling balls in cubic packing would have the same porosity as a room full of 1-millimeter ball bearings. The volume of an individual pore would

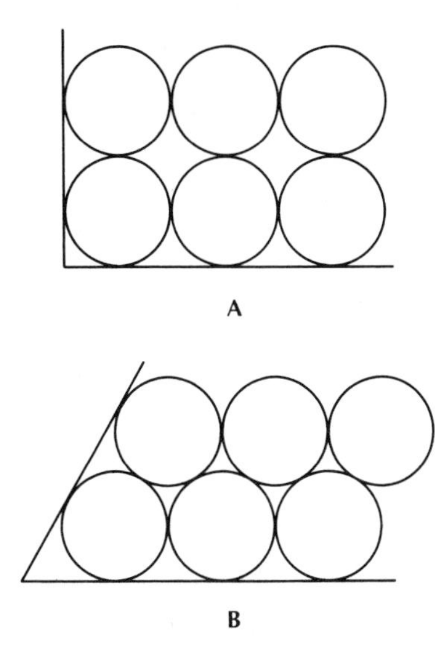


**FIGURE 4.1** **A.** Cubic packing of spheres with a porosity of 47.65 percent. **B.** Rhombohedral packing of spheres with a porosity of 25.95 percent.

be much larger for the bowling balls. The porosity of well-rounded sediments, which have been sorted so that they are all about the same size, is independent of the particle size and falls in the range of about 26 to 48 percent, depending upon the packing.

If a sediment contains a mixture of grain sizes, the porosity will be lowered. The smaller particles can fill the void spaces between the larger ones. The wider the range of grain sizes, the lower the resulting porosity (Figure 4.2). Geologic agents can sort sediments into layers of similar sizes. Wind, running water, and wave action tend to create well-sorted sediments. Other processes, such as glacial action and landslides, result in sediments with a wide range of grain sizes. These poorly sorted sediments have low porosities.

In addition to grain-size sorting, the porosity of sediments is affected by the shape of the grains. Well-rounded grains may be almost perfect spheres, but many grains are very irregular. They can be shaped like rods, disks, or books. Sphere-shaped grains will pack more tightly and have less porosity than particles of other shapes. The fabric or orientation of the particles, if they are not spheres, also influences porosity.

Sediments are classified on the basis of the size (diameter) of the individual grains. There are many classification systems in use. Figure 4.3 shows a common system frequently used by sedimentologists.

The engineering classification of sediments is somewhat different than the geological classification. The American Society of Testing Materials defines sediments on the basis of the grain-size distribution shown in Table 4.1.

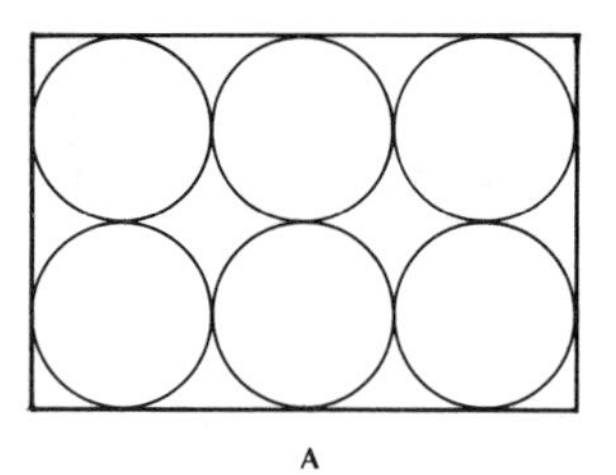


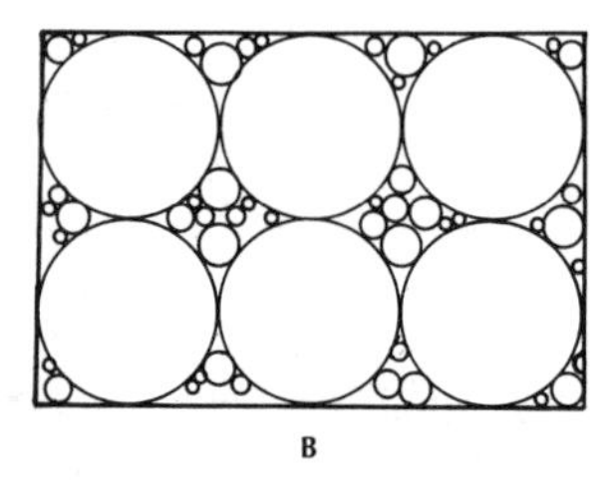


**FIGURE 4.2** **A.** Cubic packing of spheres of equal diameter with a porosity of 47.65 percent. **B.** Cubic packing of spheres with void spaces occupied by grains of smaller diameter, resulting in a much lower overall porosity.

The grain-size distribution of a sediment may be conveniently plotted on semilogarithmic paper. The cumulative percent finer by weight is plotted on the arithmetic scale and the grain size is plotted on the logarithmic scale. The grain size of the sand fraction is determined by shaking the sand through a series of sieves with decreasing mesh openings. The 200 mesh screen, with an opening of 0.075 millimeter, separates the sand fraction from the fines. The gradation of the fines is determined by a hydrometer test, which is based on the rate that the sediment settles in water. Figure 4.4 is a grain-size distribution curve for a silty fine to coarse sand. This sample is somewhat poorly sorted as there is a wide range of grain sizes present. Figure 4.5 is the grain-size distribution curve for a well-sorted fine sand. Less than 5 percent of the sample consisted of fines which

**TABLE 4.1** Engineering grain-size classification

| Name | Size Range (mm) | Example |
|---|---|---|
| Boulder | >305 | Basketball |
| Cobbles | 76–305 | Grapefruit |
| Coarse gravel | 19–76 | Lemon |
| Fine gravel | 4.75–19 | Pea |
| Coarse sand | 2–4.75 | Water softener salt |
| Medium sand | 0.42–2 | Table salt |
| Fine sand | 0.075–0.42 | Powdered sugar |
| Fines | <0.075 | Talcum powder |

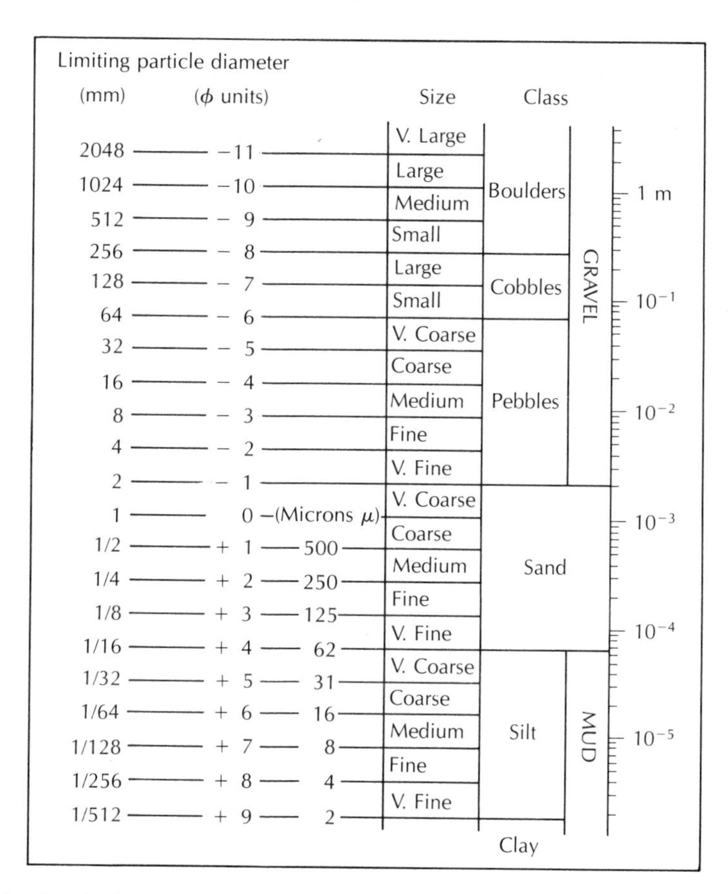


**FIGURE 4.3** Standard sizes of sediments with limiting particle diameters and the $\phi$ scale of sediment size, in which $\phi$ is equal to $\log_2 s$ (the particle diameter). Source: G. M. Friedman and J. E. Sanders, *Principles of Sedimentology* (New York: John Wiley & Sons, 1978). Used with permission.

pass the 200 mesh sieve. A hydrometer test was not performed on this sample because of the lack of fines.

The **uniformity coefficient** of a sediment is a measure of how well or poorly sorted it is. The uniformity coefficient, $C_u$, is the ratio of the grain size that is 60 percent finer by weight, $D_{60}$, to the grain size that is 10 percent finer by weight, $D_{10}$:

$$C_u = D_{60}/D_{10} \tag{4-3}$$

A sample with a $C_u$ less than 4 is well sorted; if the $C_u$ is more than 6 it is poorly sorted. The poorly sorted silty sand in Figure 4.4 has a $C_u$ of 8.3, whereas the well-sorted sand of Figure 4.5 has a $C_u$ of 1.4.

The **effective grain size,** $D_{10}$, is the size corresponding to the 10 percent line on the grain-size curve.

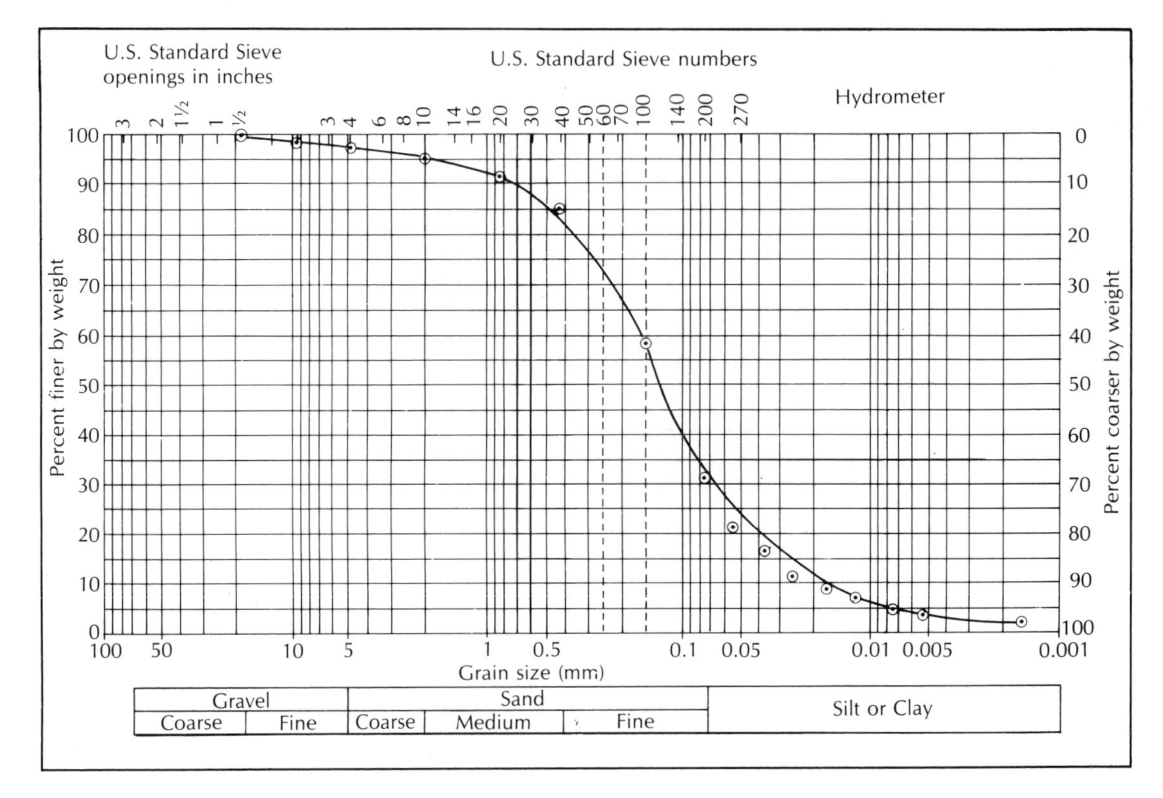


**FIGURE 4.4** Grain-size distribution curve of a silty fine to medium sand.

Clays and some clay-rich or organic soils can have very high porosities. Organic materials do not pack very closely because of their irregular shapes. The dispersive effect of the electrostatic charge present on the surfaces of certain book-shaped clay minerals causes clay particles to be repelled by each other. The result is a relatively large proportion of void space.

The general range of porosity that can be expected for some typical sediments is listed in Table 4.2.

**TABLE 4.2** Porosity ranges for sediments (1–4)

| | |
|---|---|
| Well sorted sand or gravel | 25–50% |
| Sand and gravel, mixed | 20–35% |
| Glacial till | 10–20% |
| Silt | 35–50% |
| Clay | 33–60% |

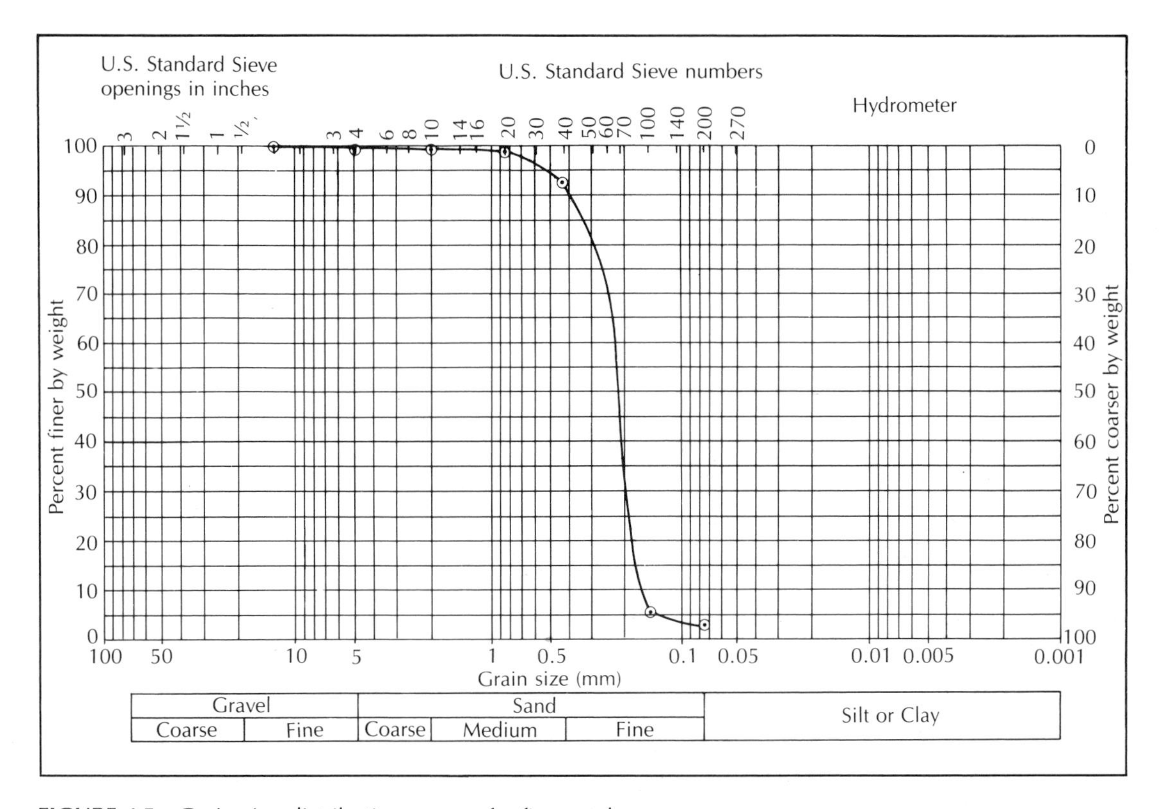


**FIGURE 4.5** Grain-size distribution curve of a fine sand.

### 4.1.3 POROSITY OF SEDIMENTARY ROCKS

**Sedimentary rocks** are formed from sediments through a process known as **diagenesis.** A sediment, which may be either a product of weathering or a chemically precipitated material, is buried. The weight of overlying materials and physicochemical reactions with fluids in the pore spaces induce changes in the sediment. This includes compaction, removal of material, addition of material, and transformation of minerals by replacement or change in mineral phase. Compaction reduces pore volume by rearranging the grains and reshaping them. The deposition of cementing materials such as calcite, dolomite, or silica will reduce porosity, although the dissolution of material that is dissolved by the pore fluid will increase porosity. The primary structures of the sediment may be preserved in the sedimentary rock. The porosity of a sandstone, for instance, will be influenced by the grain size, size sorting, grain shape, and fabric of the original sediment. Diagenesis is a complex process, but in general the primary porosity of a sedimentary rock will be less than that of the original sediment. This is especially true of fine-grained sediments (silts and clays) (Figure 4.6).

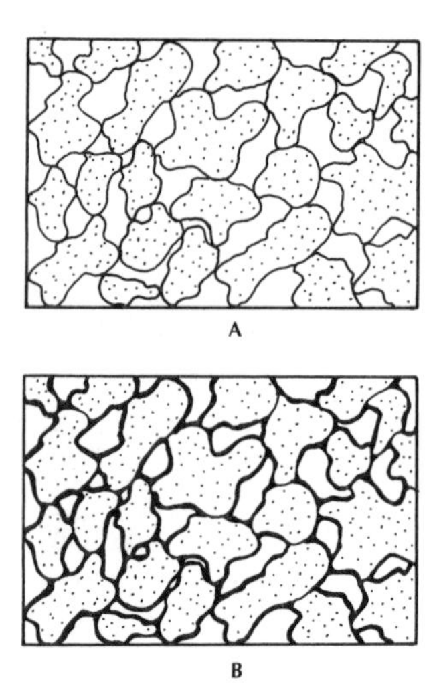


FIGURE 4.6 **A.** A clastic sediment with porosity between the grains. **B.** Reduction in porosity of the clastic sediment due to deposition of cementing materials in the pore spaces.

Rocks at the earth's surface are usually fractured to some degree. The fracturing may be mild, resulting in widely spaced joints. At the other extreme, violent fracturing may completely shatter the rock, resulting in fault breccias. Fractures create secondary porosity in the rock. Ground water can be found in fractured sedimentary rocks in the pores between grains **(primary porosity)** as well as in fractures **(secondary porosity).** Ground water flowing through fractures may enlarge them by solution of material. Bedding planes in the sedimentary rocks may have primary porosity formed during deposition of the sediments and secondary porosity if the rock has moved along a bedding plane.

Some cohesive sediments (those rich in silt and/or clay) are also subject to fracturing. In some cases, this is merely from shrinkage cracks that develop when the sediment dries. However, slumping, loading, or tectonic activity can also cause fracturing in nonplastic cohesive sediments. This fracturing can be a significant source of secondary porosity in such deposits.

Limestones and dolomites are well-known and widespread examples of sedimentary rocks of chemical or biochemical origin. They are formed of calcium carbonate and calcium-magnesium carbonate, respectively. Gypsum, a calcium sulfate, and halite or rock salt (sodium chloride) are also widely distributed common examples of chemical precipitates.

The materials that formed these rocks were originally part of an aqueous solution. Inasmuch as the precipitation process is reversible, the rock can be redissolved. When these rock types are in a zone of circulating ground water, the rock may be removed by solution. Ground water moves initially through pore spaces, as well as along fractures, joints, and bedding planes. As more water moves through the bedding planes, they are preferentially dissolved and enlarged, causing the rock to become very porous. Some limestone formations have openings large enough to permit thousands of tourists a day to pass through. The caverns at Carlsbad, New Mexico, and Ljubljana, Yugoslavia, exemplify such massive porisity. Gypsum and salt may also be cavernous (1).

The percent porosity of sedimentary rocks is highly variable. In clastic rocks, it can range from 3 to 30 percent (2, 5–7). Reported values for limestones and dolomites range from less than 1 to 30 percent (2, 7–9).

### 4.1.4 POROSITY OF PLUTONIC AND METAMORPHIC ROCKS

**Plutonic rocks** (those formed by intrusive igneous processes) and **metamorphic rocks** (those formed by applying heat and pressure to preexisting rocks) are typically thought to have a very low porosity as they are formed of interlocking crystals (2). However, in a deep granite test hole in northern Illinois, porosity was measured at 1.42 percent at a depth of 5248 feet (1600 meters). At this depth there were few fractures in the rock so some of the porosity was possibly primary. In the same borehole the porosity was as great as 2.15 percent where there were fractures present (10).

Two geologic processes, weathering and fracturing, increase overall rock porosity. Rock at depth is under pressure due to the weight of overlying materials. This rock may be fractured by expansion as the overlying weight is removed by erosion. Tectonic stresses in the earth can cause folding and faulting. Rock in a fault shear zone may be extensively fractured. Expansion cracks can form at the crest of a fold. Joint sets in crystalline rock are usually found in three mutually perpendicular directions (11). Fracturing increases porosity of crystalline rocks by about 2 to 5 percent (2, 12). Weathering due to chemical decomposition and physical disintegration operates with greater efficacy with increasing rock porosity. Weathered plutonic and metamorphic rocks can have porosities in the range of 30 to 60 percent (13). Owing to the sheetlike structure of some weathering minerals, such as the micas, porosities can exceed that of loosely packed spheres.

Porosity due to fracturing is concentrated in the rock along the sets of joints and is a function of the width of the openings in the joints. Weathered rock has the pore spaces distributed throughout the rock, although weathering may be more intense along joint or weathering planes.

Figure 4.7 contains histograms of the number of fractures as a function of depth in two deep core holes drilled into 1.47-billion-year-old biotite granite in northern Illinois. The Precambrian/Cambrian unconformity is at a depth of about 1970 feet (600 meters), which is shown as the top of the histograms. Fractures can be seen to be concentrated near the top of the Precambrian basement rock, indicating that weathering occurred during the time the basement rock was at the surface. However, fracturing was also detected near the bottom of the test hole at a depth of 5248 feet (14). This indicates that unloading can create fractures to significant depths in some instances.

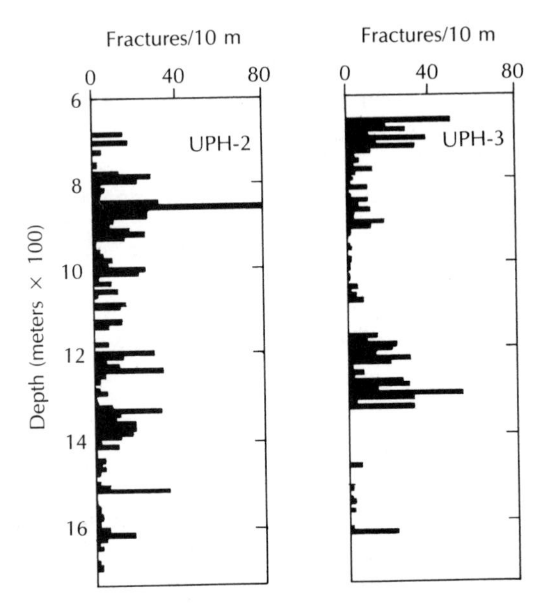


**FIGURE 4.7** Histograms showing number of fractures as a function of depth at two deep core holes into Precambrian-age granite in northern Illinois. Source: B. C. Haimson and T. W. Doe, "State of Stress, Permeability, and Fractures in the Precambrian Granite of Northern Illinois," *Journal of Geophysical Research*, 88, B9 (1983):7355–71.

### 4.1.5 POROSITY OF VOLCANIC ROCKS

**Volcanic rocks** (those formed by extrusive igneous activity) are similar in chemical composition to plutonic rocks because both are formed by the cooling of molten rock (magma). However, extrusive rocks cool and solidify quickly because they are formed in a surface environment; this gives them radically different porosity characteristics. Volcanic rocks include lava flows; deposits of ash and cinders, which can occur in loose, unconsolidated piles; and such rocks as welded tuff.

Lava cooling rapidly at the surface will trap degassing products, resulting in holes in the rock (vesicular texture). The holes create porosity, although they may not be interconnected. Shrinkage cracks that develop in the lava as it cools create joints. Flowing lava can form a crust, which then breaks apart to form a rubbly structure. The broken surface of buried lava flows, the remains of natural lava tubes and tunnels through which molten lava once poured, and stream gravels trapped between lava flows all produce a high porosity in some extrusive rocks. Porosity of basalt, a crystalline extrusive rock that is formed from magma with a low gas content, generally ranges from 1 to 12 percent (15). Pumice, a glassy rock that is formed from a magma with a very high gas content, can have a porosity of as high as 87 percent (2), although the vesicles are not well interconnected.

Pyroclastic deposits are formed by volcanic material thrown into the air when molten. They can have high porosities. Values of porosity of tuff ranging from 14 to 40 percent have been reported (16). Recent volcanic ash may have a porosity of 50 percent. Weathering of volcanic deposits can increase the porosity to values in excess of 60 percent (2).

## 4.2 SPECIFIC YIELD

**Specific yield** ($S_y$) is the ratio of the volume of water that drains from a saturated rock owing to the attraction of gravity to the total volume of the rock (17) (Figure 4.8).

Water molecules cling to surfaces because of surface tension of the water (Figure 4.9). If gravity exerts a stress on a film of water surrounding a mineral grain, some of the film will pull away and drip downward. The remaining film will be thinner, with a greater surface tension so that, eventually, the stress of gravity will be exactly balanced by the surface tension. **Hygroscopic water** is the moisture clinging to the soil particles because of surface tension. At the moisture content of the specific yield, gravity drainage will cease.

If two samples are equivalent with regard to porosity, but the average grain size of one is much smaller than the other, the surface area of the finer sample will be larger. As a result, more water can be held as hygroscopic moisture by the finer grains.

The **specific retention** ($S_r$) of a rock or soil is the ratio of the volume of water a rock can retain against gravity drainage to the total volume of the rock

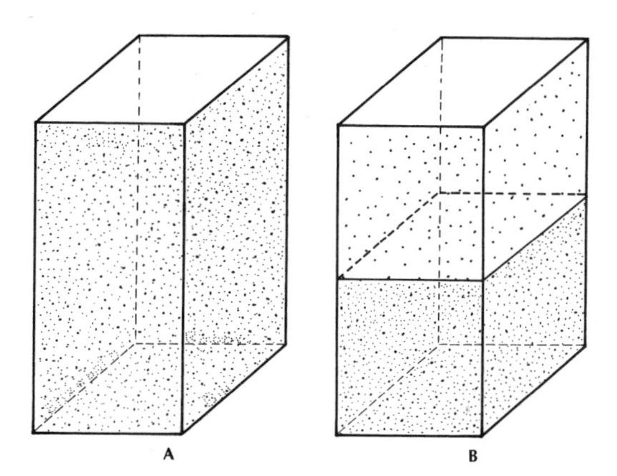


FIGURE 4.8 **A.** A volume of rock saturated with water. **B.** After gravity drainage, 1 unit volume of the rock has been dewatered with a corresponding lowering of the level of saturation. Specific yield is the ratio of the volume of water that drained from the rock owing to gravity to the total rock volume.

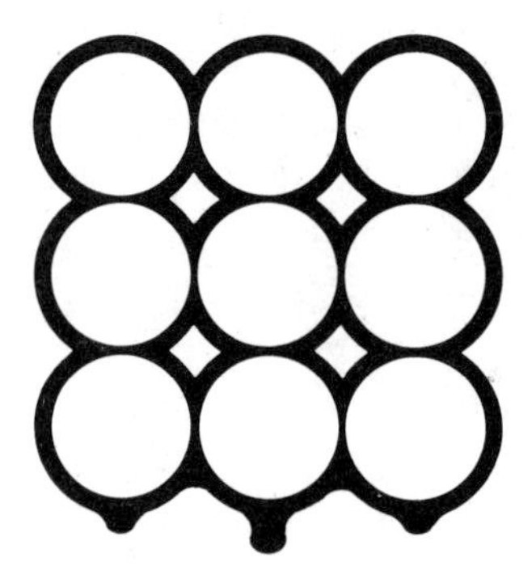

**FIGURE 4.9** Hygroscopic water clinging to spheres owing to surface tension. Gravity attraction is pulling the water downward.

(17). Since the specific yield represents the volume of water that a rock will yield by gravity drainage, with specific retention the remainder, the sum of the two is equal to porosity:

$$n = S_y + S_r \qquad \textbf{(4-4)}$$

The specific retention increases with decreasing grain size, so that a clay may have a porosity of 50 percent with a specific retention of 48 percent.

Table 4.3 lists the specific yield, in percent, for a number of sediment textures. The data for this table were compiled from a large number of samples in various geographic locations. Maximum specific yield occurs in sediments in the medium-to-coarse sand size range (0.5 to 1.0 millimeter). This is shown graphically in Figure 4.10, which plots specific yield as a function of grain size for several hundred samples from the Humboldt River Valley of Nevada.

Both soil formed by weathering processes at the surface and sediments that are depositional generally contain a mixture of clay, silt, and sand. Figure 4.11 is a soil classification triangle showing lines of equal specific yield (18). It is

**TABLE 4.3** Specific yields in percent (18)

| Material | Specific Yield<br>Maximum | <br>Minimum | <br>Average |
|---|---|---|---|
| Clay | 5 | 0 | 2 |
| Sandy clay | 12 | 3 | 7 |
| Silt | 19 | 3 | 18 |
| Fine sand | 28 | 10 | 21 |
| Medium sand | 32 | 15 | 26 |
| Coarse sand | 35 | 20 | 27 |
| Gravelly sand | 35 | 20 | 25 |
| Fine gravel | 35 | 21 | 25 |
| Medium gravel | 26 | 13 | 23 |
| Coarse gravel | 26 | 12 | 22 |

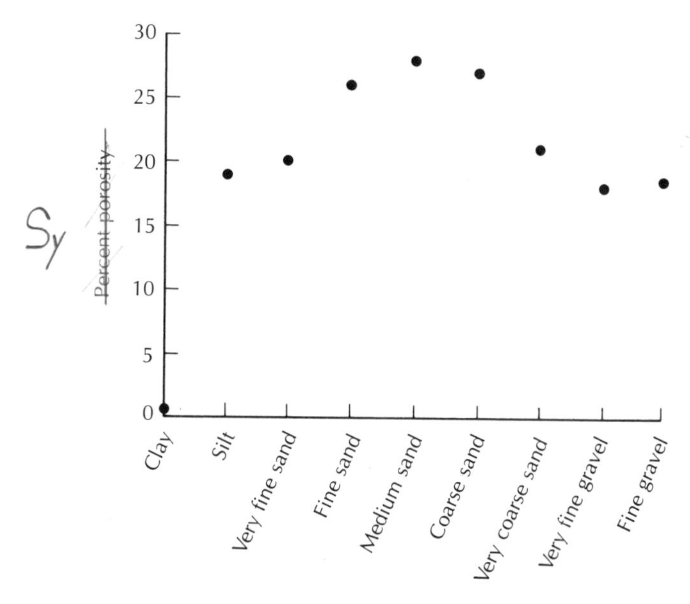


**FIGURE 4.10** Specific yield of sediments from the Humbolt River Valley of Nevada as a function of the median grain size. Source: Data from P. Cohen, U.S. Geological Survey Water-Supply Paper 1975, 1965.

apparent that the specific yield increases rapidly as the percentage of sand increases and as the percentages of silt, and especially clay, decrease.

Specific yield may be determined by laboratory methods. A sample of sediment of known volume is fully saturated. This is usually done in a soil column that is flooded slowly from the bottom, allowing air to escape upward. Water is then allowed to drain from the column (19). Care must be taken to avoid evaporation losses; even for sand-sized grains, columns must be allowed to drain for very long time periods (months) before equilibrium is reached (20). The ratio of the volume of water drained to the volume of the soil column is the specific yield (multiplied by 100 to express the value as a percentage).

The specific yield of sediment and rock can also be determined in the field. Water wells are pumped, and the rate at which the water level falls in nearby wells is measured (21–23). Chapter 6 includes a discussion of such pumping-test methods.

## 4.3 HYDRAULIC CONDUCTIVITY OF EARTH MATERIALS

We have seen that earth materials near the surface generally contain some void space and thus exhibit porosity. Moreover, in most cases, these voids are interconnected to some degree. Water contained in the voids is capable of moving

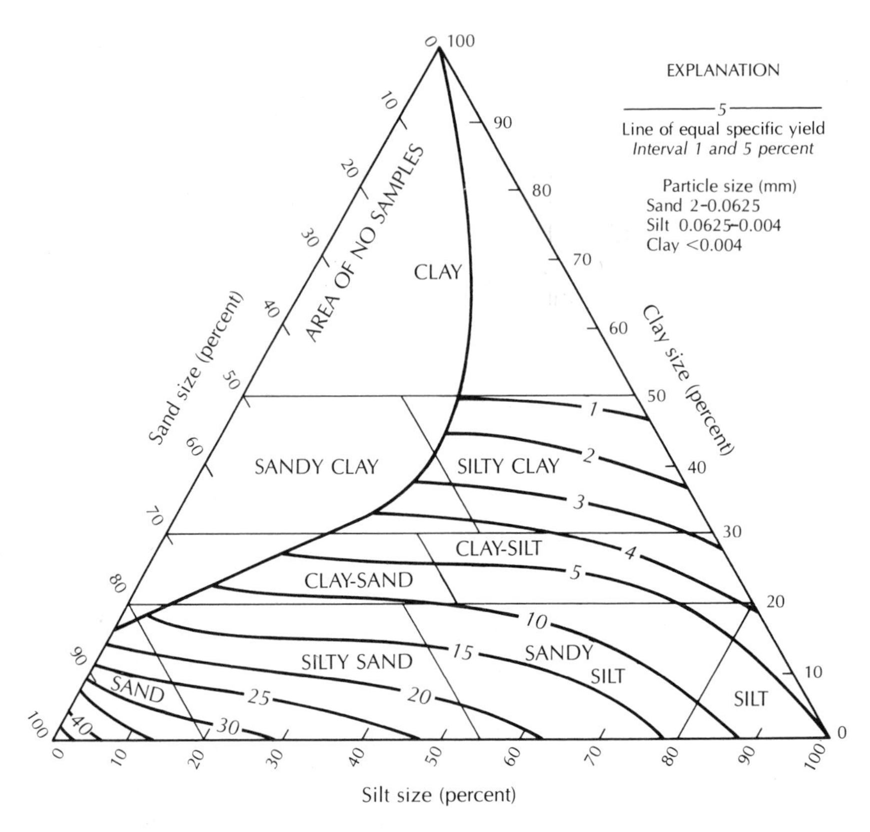


**FIGURE 4.11** Textural classification triangle for unconsolidated materials showing the relation between particle size and specific yield. Source: A.I. Johnson, U.S. Geological Survey Water-Supply Paper 1662-D, 1967.

from one void to another, thus circulating through the soil, sediment, and rock. It is the ability of a rock to transmit water which, together with its ability to hold water, constitute the most significant hydrologic properties. There are some rocks that exhibit porosity but lack interconnected voids, e.g., vesicular basalt. These rocks cannot convey water from one void to another. Some sediments and rocks have porosity, but the pores are so small that water flows through the rock with difficulty. Clay and shale are examples.

### 4.3.1 DARCY'S EXPERIMENT

In the mid-nineteenth century, a French engineer, Henry Darcy, made the first systematic study of the movement of water through a porous medium (24). He studied the movement of water through beds of sand used for water filtration. Darcy found that the rate of water flow through a bed of a "given nature" is proportional to the difference in the height of the water between the two ends of the filter beds and inversely proportional to the length of the flow path. He determined also that the quantity of flow is proportional to a coefficient, $K$, which is dependent upon the nature of the porous medium.

Figure 4.12 illustrates a horizontal pipe filled with sand. Water is applied under pressure through end $A$. The pressure can be measured and observed by means of a thin vertical pipe open in the sand at point $A$. Water flows through the pipe and discharges at point $B$. Another vertical pipe or piezometer is present to measure the pressure at $B$.

Darcy found experimentally that the discharge, $Q$, is proportional to the difference in the height of the water, $h$ (hydraulic head), between the ends and inversely proportional to the flow length, $L$:

$$Q \propto h_A - h_B \text{ and } Q \propto 1/L$$

The flow is also obviously proportional to the cross-sectional area of the pipe, $A$. When combined with the proportionality constant, $K$, the result is the expression known as **Darcy's law:**

$$Q = KA\left(\frac{h_A - h_B}{L}\right) \tag{4-5}$$

This may be expressed in more general terms as

$$Q = -KA\left(\frac{dh}{dl}\right) \tag{4-6}$$

where $dh/dl$ is known as the **hydraulic gradient.** The quantity $dh$ represents the change in head between two points that are very close together, and $dl$ is the small distance between these two points. The negative sign indicates that flow is in the direction of decreasing hydraulic head. The use of the negative sign ne-

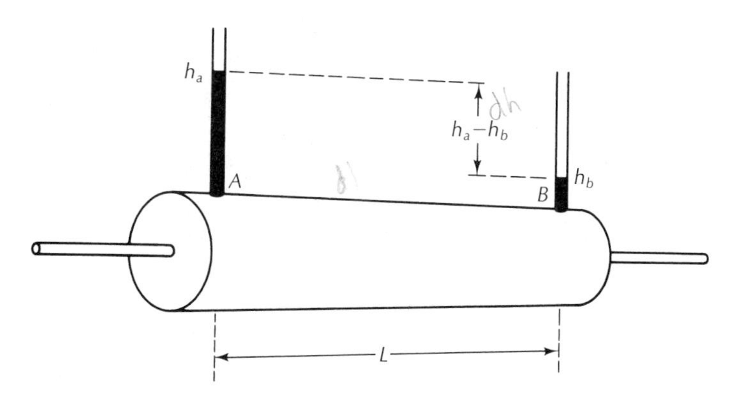


**FIGURE 4.12** Horizontal pipe filled with sand to demonstrate Darcy's experiment. (Darcy's original equipment was actually vertically oriented.)

cessitates careful determination of the sign of the gradient. If the value of $h_2$ at point $X_2$ is greater than $h_1$ at point $X_1$, then flow is from point $X_2$ to $X_1$. If $h_1 > h_2$, then flow is from $X_1$ to $X_2$.

### 4.3.2 HYDRAULIC CONDUCTIVITY

Equation 4-6 can be rearranged to show that the coefficient $K$ has the dimensions of length/time $(L/T)$, or velocity. This coefficient has been termed the **hydraulic conductivity.** In other works, it may be referred to as the coefficient of permeability:

$$K = \frac{Q}{A(dh/dL)} \tag{4-7}$$

Discharge has the dimensions volume/time $(L^3/T)$, area $(L^2)$, and gradient $(L/L)$. Substituting these dimensions into Equation 4-5, the dimensions of $K$ are determined:

$$K = \frac{(L^3/T)}{(L^2)(L/L)} = (L/T)$$

Darcy did not address the fact that the value of $K$ is a function of properties of both the porous medium and the fluid passing through it. It is intuitively obvious that a viscous fluid (one which is thick), such as crude oil, will move at a slower rate than water, which is thinner and has a lower viscosity. The hydraulic conductivity is directly proportional to the **specific weight,** $\gamma$, of the fluid. The specific weight is the force exerted by gravity on a unit volume of the fluid. This represents the driving force of the fluid. Hydraulic conductivity is also inversely proportional to the **dynamic viscosity** of the fluid, $\mu$, which is a measure of the resistance of the fluid to the shearing that is necessary for fluid flow. A proportionality expression for $K$ can be written as

$$K = K_i\left(\frac{\gamma}{\mu}\right) = K_i\left(\frac{\rho g}{\mu}\right) \tag{4-8}$$

where $g$ is the acceleration of gravity and $\rho$ is the density.

The new constant $K_i$, is representative of the properties of the porous medium alone. It is termed the **intrinsic permeability.** This is basically a function of the size of the openings through which the fluid moves. The larger the square of the mean pore diameter, $d$, the lower the flow resistance. The cross-sectional area of a pore is also a function of the shape of the opening. A constant can be used to describe the overall effect of the shape of the pore spaces. Using this dimensionless constant, called the **shape factor,** $C$, the intrinsic permeability is given by the expression

$$K_i = Cd^2 \tag{4-9}$$

The dimensions of $K_i$ are $(L^2)$, or area.

Units for $K_i$ can be in square feet or square centimeters. In the petroleum industry, the **darcy** is used as a unit of intrinsic permeabiliby. (The petroleum engineer is similarly concerned with the occurrence and movement of fluids through porous media.) The darcy is defined as

$$1 \text{ darcy} = \frac{\dfrac{1 \text{ centipoise} \times 1 \text{ cm}^3/\text{sec}}{1 \text{ cm}^2}}{1 \text{ atmosphere}/1 \text{ cm}}$$

This expression can be converted to square centimeters, since

$$1 \text{ centipoise} = 0.01 \text{ dyne-sec/cm}^2$$

and

$$1 \text{ atmosphere} = 1.0132 \times 10^6 \text{ dynes/cm}^2$$

Substituting into the definition of the darcy, it may be seen that

$$1 \text{ darcy} = 9.87 \times 10^{-9} \text{ cm}^2$$

Both the viscosity and the density of a fluid are functions of its temperature. The colder the fluid, the more viscous it is. There is also a more complex relationship between temperature and density, as the density of water decreases with temperature to 4° C, at which temperature it is at a maximum. The hydraulic conductivity of a rock or sediment will vary with the temperature of the water. As solutions become saline, this may also affect the values of specific gravity and viscosity, which will also cause the hydraulic conductivity to vary. The laboratory, or standard, value of hydraulic conductivity is defined for pure water at a temperature of 15.6° C. The units are defined in terms of length/time (Table 4.4). In the United States the unit feet per day is used in hydrogeological practice and centimeters per second is used in soils engineering practice. In the

**TABLE 4.4** Conversion values for hydraulic conductivity

|  | |  |
|---|---|---|
| 1 gal/day/ft$^2$ | = | 0.0408 m/day |
| 1 gal/day/ft$^2$ | = | 0.134 ft/day |
| 1 gal/day/ft$^2$ | = | $4.72 \times 10^{-5}$ cm/sec |
| 1 ft/day | = | 0.305 m/day |
| 1 ft/day | = | 7.48 gal/day/ft$^2$ |
| 1 ft/day | = | $3.53 \times 10^{-4}$ cm/sec |
| 1 cm/sec | = | 864 m/day |
| 1 cm/sec | = | 2835 ft/day |
| 1 cm/sec | = | 21,200 gal/day/ft$^2$ |
| 1 m/day | = | 24.5 gal/day/ft$^2$ |
| 1 m/day | = | 3.28 ft/day |
| 1 m/day | = | 0.00116 cm/sec |

SI system of units hydraulic conductivity is in meters per day. A derived unit of gallons per day per square foot was used in the past. It can be converted to feet per day by dividing by 7.48.

### 4.3.3 PERMEABILITY OF SEDIMENTS

Unconsolidated coarse-grained sediments represent some of the most prolific producers of ground water. Likewise, clays are often used for engineering purposes, such as lining solid-waste disposal sites, because of their extremely low intrinsic permeability. There is obviously a wide-ranging continuum of permeability values for unconsolidated sediments (Table 4.5).

The intrinsic permeability is a function of the size of the pore opening. The smaller the size of the sediment grains, the larger the surface area the water contacts (Figure 4.13). This increases the frictional resistance to flow, which reduces the intrinsic permeability. For well-sorted sediments, the intrinsic permeability is proportional to the grain size of the sediment (25).

For sand-sized alluvial deposits, several factors relating intrinsic permeability to grain size have been noted (26). These observations would hold true for all sedimentary deposits, regardless of origin of deposition.

1. As the median grain size increases, so does permeability. This is due to larger pore openings.
2. Permeability will decrease for a given median diameter as the standard deviation of particle size increases. The increase in standard deviation indicates a more poorly sorted sample, so that the finer material can fill the voids between larger fragments.
3. Coarser samples show a greater decrease in permeability with an increase in standard deviation than do fine samples.
4. Unimodal (one dominant size) samples have a greater permeability

**TABLE 4.5** Ranges of intrinsic permeabilities and hydraulic conductivities for unconsolidated sediments

| Material | Intrinsic Permeability (darcys) | Hydraulic Conductivity (cm/sec) |
|---|---|---|
| Clay | $10^{-6}-10^{-3}$ | $10^{-9}-10^{-6}$ |
| Silt, sandy silts, clayey sands, till | $10^{-3}-10^{-1}$ | $10^{-6}-10^{-4}$ |
| Silty sands, fine sands | $10^{-2}-1$ | $10^{-5}-10^{-3}$ |
| Well-sorted sands, glacial outwash | $1-10^{2}$ | $10^{-3}-10^{-1}$ |
| Well-sorted gravel | $10-10^{3}$ | $10^{-2}-1$ |

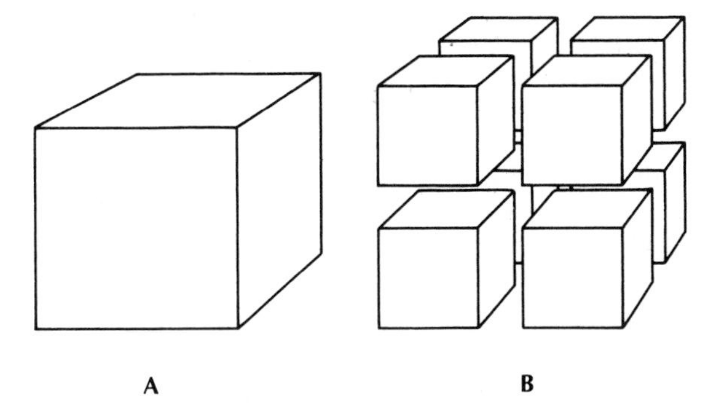


**FIGURE 4.13** Relationship of sediment grain size to surface area of pore spaces. **A.** A cube of sediment with a surface area of six square units. **B.** The cube has been broken into eight pieces, each with a diameter of one-half of the cube in Part A. The surface area has increased to twelve square units—an increase of 100 percent.

than bimodal (two dominant sizes) samples. This is again a result of poorer sorting of the sediment sizes, as the bimodal distribution indicates.

The hydraulic conductivity of sandy sediments can be estimated from the grain-size distribution curve by the **Hazen method** (27). The method is applicable to sands where the effective grain size ($D_{10}$) is between approximately 0.1 and 3.0 millimeters. The Hazen approximation is

$$K = C(D_{10})^2 \quad \textbf{(4-10)}$$

where

$K$ is hydraulic conductivity in cm/sec
$D_{10}$ is the effective grain size in cm
$C$ is a coefficient based on the following table

| | |
|---|---|
| Very fine sand, poorly sorted | 40–80 |
| Fine sand with appreciable fines | 40–80 |
| Medium sand, well sorted | 80–120 |
| Coarse sand, poorly sorted | 80–120 |
| Coarse sand, well sorted, clean | 120–150 |

The Hazen method was developed on the basis of empirical studies done for the design of sand filters for drinking water. Your author has shown its hydrogeological applicability in a study of the permeability of well-sorted medium sand deposited by alluvial processes, as shown by the following case study (28). (The Hazen method may not always work as well as it did in this particular application.)

## CASE STUDY: HYDRAULIC CONDUCTIVITY ESTIMATES IN GLACIAL OUTWASH

A hazardous-waste–processing site was located on a level glacial outwash plain in southern Indiana. There were two aquifers present in the unconsolidated glacial deposits. The upper aquifer consisted of a well-sorted fine to medium sand. There were 27 ground-water monitoring wells installed in this aquifer. The lower aquifer was a poorly sorted fine to coarse sand. There were 9 monitoring wells installed in this aquifer. The grain-size analyses of the sand samples from the screen zones of the wells in each aquifer are summarized in the following table:

| | Upper Aquifer | | Lower Aquifer | |
|---|---|---|---|---|
| | Mean | Range | Mean | Range |
| $D_{10}$ | 0.14 mm | 0.08–0.20 mm | 0.16 mm | 0.09–0.26 mm |
| $D_{60}$ | 0.31 mm | 0.19–0.45 mm | 2.04 mm | 0.35–6.70 mm |
| $C_u$ | 2.29 | 1.50–3.89 | 11.01 | 3.89–33.50 |

The hydraulic conductivities of the sediments at each monitoring well were estimated by the Hazen method, using a coefficient of 100. The hydraulic conductivities of the sediments at each monitoring well were measured by means of a Hvorslev slug test performed on the well (see Section 6.7). The following table compares the results in cm/sec.

| | Geometric Mean (cm/sec) | Range (cm/sec) |
|---|---|---|
| | Upper Aquifer | |
| Hazen method | $1.9 \times 10^{-2}$ | $4.0 \times 10^{-2} - 6.4 \times 10^{-3}$ |
| Hvorslev test | $1.9 \times 10^{-2}$ | $8.9 \times 10^{-2} - 4.2 \times 10^{-3}$ |
| | Lower Aquifer | |
| Hazen method | $1.2 \times 10^{-2}$ | $2.6 \times 10^{-2} - 8.1 \times 10^{-3}$ |
| Hvorslev test | $1.4 \times 10^{-2}$ | $1.7 \times 10^{-1} - 2.6 \times 10^{-3}$ |

The geometric means of the data sets were used to compare the Hvorslev test results with the Hazen method results in the above case study. Hydraulic condictivity values frequently vary by more than two orders of magnitude within the same hydrogeologic unit. An arithmetic mean of such a sample population tends to give more weight to the more permeable values. Some hydrogeologists believe that a more representative description of the average hydraulic conductivity of a hydrologic unit is the **geometric mean.** This is determined by taking the natural log of each value, finding the mean of the natural logs and then obtaining the exponential ($e^x$) of that value to arrive at the geometric mean.

**EXAMPLE PROBLEM** Find the geometric mean of the following set of hydraulic conductivity values and compare it to the arithmetic mean:

| Hydraulic conductivity *(K)* | ln *(K)* |
|---|---|
| $2.17 \times 10^{-2}$ cm/sec | $-3.83$ |
| $2.58 \times 10^{-2}$ cm/sec | $-3.66$ |
| $2.55 \times 10^{-3}$ cm/sec | $-5.97$ |
| $1.67 \times 10^{-1}$ cm/sec | $-1.79$ |
| $9.50 \times 10^{-4}$ cm/sec | $-6.96$ |
| Sum: $2.18 \times 10^{-1}$ cm/sec | $-22.21$ |

Geometric mean: mean ln $(K)$: $-22.21/5 = -4.44$

exp [mean ln $(K)$]: $e^{-4.44} = 1.18 \times 10^{-2}$ cm/sec

Arithmetic mean: $(2.18 \times 10^{-1})/5 = 4.36 \times 10^{-2}$ cm/sec

Field studies of hydraulic permeability based on grain-size analysis of sediments from test borings and slug tests in monitoring wells yield information on the distribution of hydraulic conductivity across the site. Aquifer pumping tests, discussed in Chapter 6, are another way to determine the hydraulic conductivity of rock and soil in the field. Pumping tests integrate the distributed permeability and give an average permeability over a large area.

### 4.3.4 PERMEABILITY OF ROCKS

The intrinsic permeability of rocks is due to primary openings formed with the rock and secondary openings created after the rock was formed. The size of openings, the degree of interconnection, and the amount of open space are all significant.

Clastic sedimentary rocks have primary permeability characteristics similar to those of unconsolidated sediments. However, diagenesis can reduce the size of the throats which connect adjacent pores through cementation and compaction. This could reduce permeability substantially without a large impact on primary porosity. Primary permeability may also be due to sedimentary structures, such as bedding planes.

Crystalline rocks, whether of igneous, metamorphic, or chemical origin, typically have a low primary permeability, in addition to a low porosity. The intergrown crystal structure contains very few openings, so fluids cannot pass through as readily. The exceptions to this are volcanic rocks, which can have a high primary porosity. If the openings are large and well connected, then high permeability may also be present.

Secondary permeability can develop in rocks through fracturing. The increase in permeability is initially due to the number and size of the fracture openings. As water moves through the fractures, minerals may be dissolved from the rock and the fracture enlarged. This increases the permeability. Chemically precipitated rocks (limestone, dolomite, gypsum, halite) are most susceptible to solution enlargement, although even igneous rocks may be so affected. Bedding-plane openings of sedimentary rocks may also be enlarged by solution.

Weathering is another process which can result in an increase in permeability. As the rock is decomposed or disintegrated, the number and size of pore spaces, cracks, and joints can increase.

---

**EXAMPLE PROBLEM** The intrinsic permeability of a consolidated rock is $2.7 \times 10^{-3}$ darcy. What is the conductivity for water at 15° C?

At 15° C for water,

$$\rho = 0.999099 \text{ g/cm}^3$$
$$\mu = 0.011404 \text{ poise}$$

The acceleration of gravity is given as

$$g = 980 \text{ cm/sec}^2$$

As 1 darcy $= 9.87 \times 10^{-9}$ square centimeter, the intrinsic permeability is $2.66 \times 10^{-11}$ square centimeter:

$$K = K_i\left(\frac{\rho g}{\mu}\right) = 2.66 \times 10^{-11} \text{ cm}^2 \times \frac{0.999099 \text{ g/cm}^3 \times 980 \text{ cm/sec}^2}{0.011404 \text{ poise}}$$

$$1 \text{ poise} = \frac{\text{dyne-sec}}{\text{cm}^2}$$

$$1 \text{ dyne} = \frac{\text{g-cm}}{\text{sec}^2}$$

$$1 \text{ poise} = \frac{\text{g}}{\text{sec-cm}}$$

$$K = 2.28 \times 10^{-6} \frac{\text{g/cm}^3 \times \text{cm/sec}^2}{\text{g/sec-cm}}$$
$$= 2.28 \times 10^{-6} \text{ cm/sec}$$

---

## 4.4 EFFECTIVE POROSITY

Fluids flowing through the pores in rock and sediment may not move through the entire volume of pore space. There may be both noninterconnected pores as well as dead-end pores, which have no outlet so that water cannot flow through them. The effective porosity, $n_e$, is the porosity available for fluid flow. The effective pore fraction, *epf*, is the ratio of the porosity available to flow to the total porosity. The effective porosity can be defined by the following equation:

$$n_e = n \times epf \tag{4-11}$$

It has been only recently that much research has been done on determining the effective porosity of earth materials (e.g., 45–48). A recent research effort has studied the effective porosity of fine-grained earth materials (29). One conclusion of that study was that the effective pore fraction is a function of the size of the molecules relative to the size of the openings between pores. In fine-grained materials the size of the openings between pores could be smaller than the diameter of a particular molecule, thus limiting the passageways through which that molecule could move. The effective pore fraction for that molecule would be less than the effective pore fraction for a smaller molecule. In the same study it was found that when water tagged with tritium (a radioactive isotope) flowed through a lacustrine clay in a permeameter, the effective pore fraction was essentially 1.0. Thus the water molecule was small enough to pass through the openings between all of the pores. This suggests that for sediments the pores are interconnected and the effective pore fraction for water molecules should be 1.0.

## 4.5 FORCES ACTING ON GROUND WATER

There are three outside forces acting on the water contained in the ground. The most obvious of these is **gravity,** which pulls water downward. The second force is **external pressure.** Above the zone of saturation, atmospheric pressure is acting. The combination of atmospheric pressure and the weight of overlying water creates pressures in the zone of saturation. The third force is **molecular attraction,** which causes water to adhere to solid surfaces. It also creates surface tension in water when the water is exposed to air. The combination of these two processes is responsible for the phenomenon of capillarity.

When water in the ground is flowing through a porous medium, there are forces resisting the fluid movement. These consist of the **shear stresses** acting tangentially to the surface of the solid and **normal stresses** acting perpendicularly to the surface (30). We can think of these forces collectively as "friction." The internal molecular attraction of the fluid itself resists the movement of fluid molecules past each other. This shearing resistance is known as the viscosity of the fluid.

## 4.6 WATER TABLE

Water may be present beneath the earth's surface as a liquid, solid, or vapor. Other gases may also be present, either in vapor phase or dissolved in water. In the lower zone of porosity, generally all that is present is mineral matter and liquid water. The rock is saturated with water, and the water may also contain dissolved gas. The fluid pressure is greater than atmospheric pressure owing to the weight of overlying water. As the surface is approached, the fluid pressure decreases as the thickness of fluid above it decreases. At some depth, which

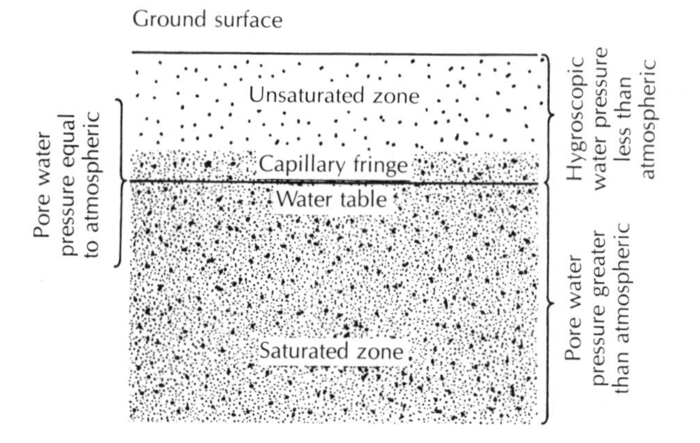


**FIGURE 4.14** Distribution of fluid pressures in the ground with respect to the water table.

varies from place to place, the pressure of the fluid in the pores is equal to atmospheric pressure. The undulating surface at which pore water pressure is equal to atmospheric pressure is called the **water table** (Figure 4.14).

Water in a shallow well (a few feet or so below the water table) will rise to the elevation of the water table at that location. The position of the water table often follows the general shape of the topography, although the water-table relief is not as great as the topographic relief. At all depths below the water table, the rock is generally saturated with water.*

A hypothetical experiment can serve to illustrate the formation of the water table. A box made of clear plastic is filled with sand. A notch is cut in one side of the plastic, and the surface of the sand is smoothed to model a valley draining toward the notch. A fine mist of water is then spread evenly over the surface of the sand, simulating rainfall. The precipitation rate is sufficiently low to preclude any overland flow. The water will move downward through the sand, so that, eventually, a zone of saturation will develop at the bottom. As shown in Figure 4.15A, this zone will have a level surface. As more rainfall is simulated, the water table will rise, continuing to be perfectly flat. It will follow this pattern until the water table reaches the lowest point in the valley.

Continuing rainfall will cause further increases in the height of the water table. In the valley, the water level will be above the surface, so that water will now flow through the notch. Elsewhere, the water table will be higher than the elevation of the notch, and ground water will begin to flow laterally because of

*There are exceptions. The rocks may contain trapped liquid and gaseous hydrocarbons, for example. Or there may be isolated voids, which cannot fill with any fluid.

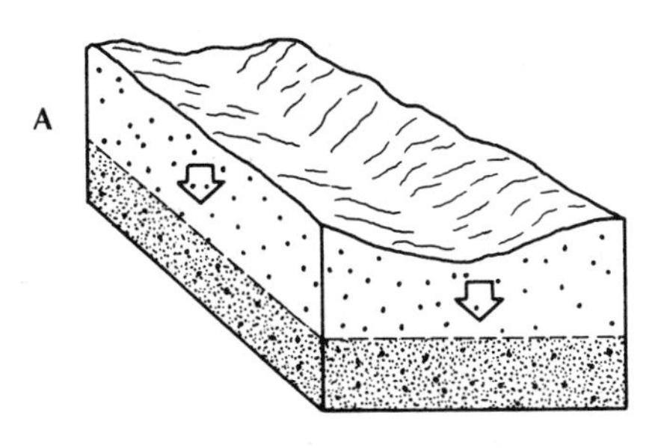


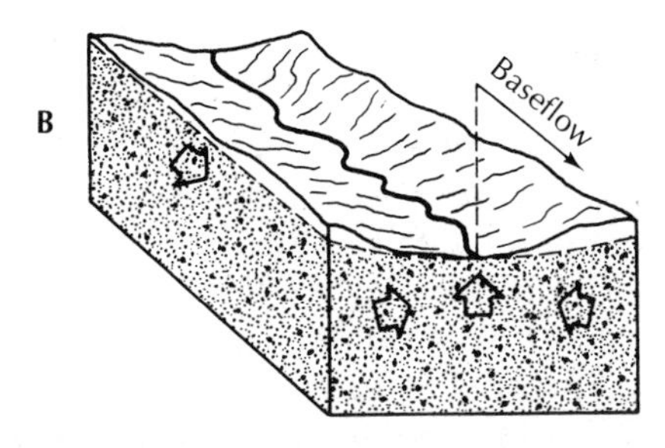


**FIGURE 4.15** **A.** Diagram of a flat-lying water table in an aquifer where there is downward movement of water through the unsaturated zone but no lateral ground-water movement. **B.** Diagram of the water table in a region where water is moving downward through the unsaturated zone to the water table and moving as ground-water flow through the zone of saturation toward a discharge zone along the stream. Net discharge from the aquifer is occurring as baseflow from the stream.

the hydraulic gradient. Now, water will flow through the saturated zone toward the point of discharge (Figure 4.15B).

We can make the following observations, of which items 4 and 5 pertain primarily to humid regions:

1. In the absence of ground-water flow, the water table will be flat.
2. A sloping water table indicates the ground water is flowing.
3. Ground-water discharge zones are in topographical low spots.
4. The water table has the same general shape as the surface topography.
5. Ground water generally flows away from topographical high spots and toward topographic lows.

## 4.7 INFILTRATION

Rainfall reaching the earth's surface can either form puddles, run across the surface, or infiltrate into the ground. Soil has a finite capacity to absorb water.

The infiltration capacity varies not only from soil to soil, but is also different for dry versus moist conditions in the same soil.

If a soil is initially dry, the infiltration capacity is high. Surface effects between the soil particles and the water exert a tension that draws the moisture downward into the soil through labyrinthine capillary passages. As the capillary forces diminish with increased soil-moisture content, the infiltration capacity drops (Figure 4.16). In addition, colloidal particles in the soil swell as the moisture content increases. Eventually, the infiltration capacity reaches a more or less constant value determined by the unsaturated permeability of the soil.

If the precipitation rate is lower than the initial infiltration capacity of the soil, there will be no overland flow (Figure 4.17A). But should the infiltration capacity decrease to a rate lower than the precipitation rate, depression storage will fill, and then overland flow can occur. Part B of the figure shows an idealized version of a rainfall chart: initially there is no overland flow, but depression storage fills, and then an overland-flow component develops as the infiltration capacity decreases with time. If the uniform precipitation rate is initially higher than the infiltration capacity, depression storage starts to fill immediately, with overland flow starting when depression storage is filled (Part C).

Conditions that encourage a high infiltration rate include coarse soils, well-vegetated land, low soil moisture, and a topsoil layer made porous by insects and other burrowing animals, in addition to land-use practices that avoid

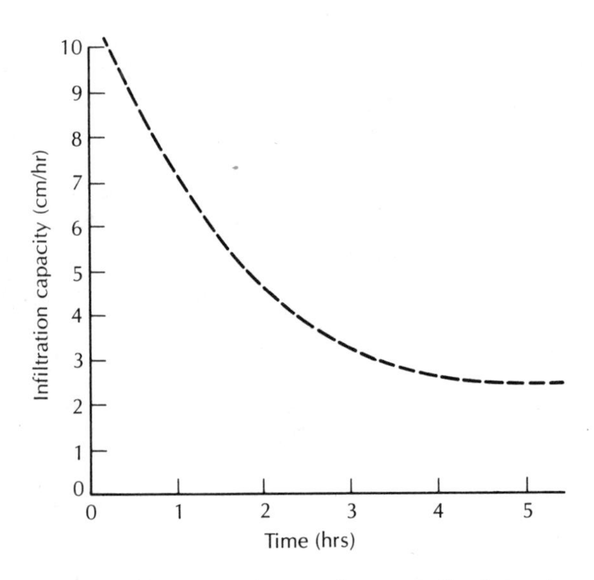


**FIGURE 4.16** Decreasing infiltration capacity of an initially dry soil as the soil-moisture content of the surface layer increases.

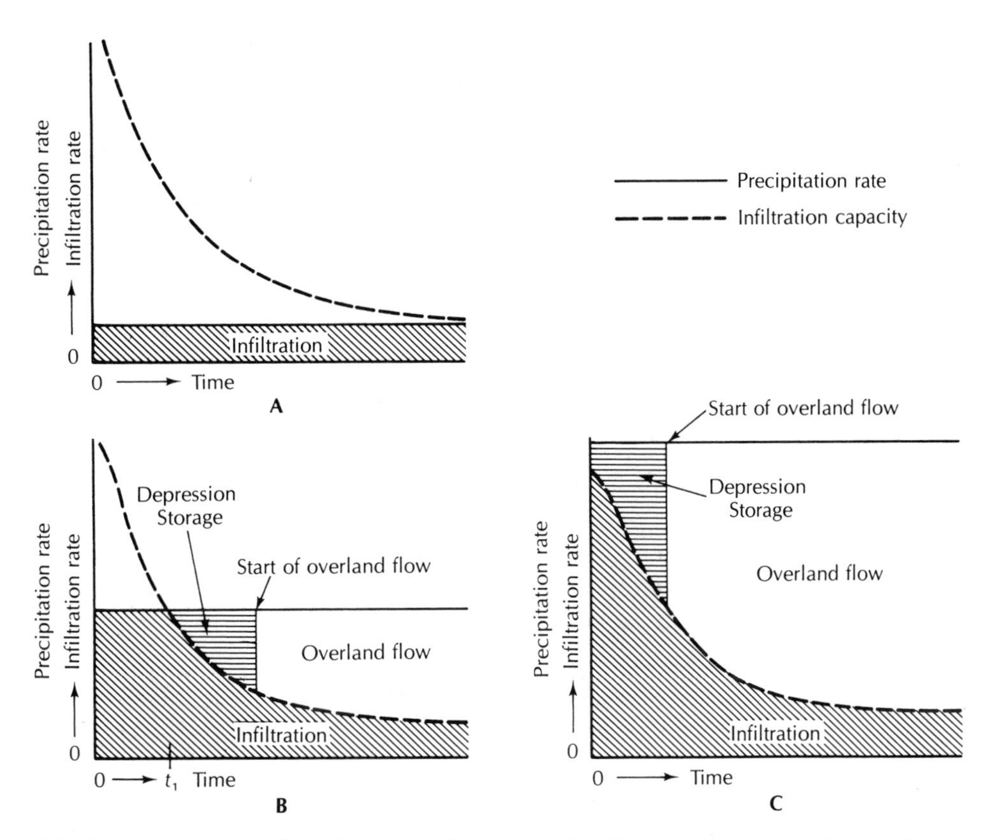


**FIGURE 4.17** Relationship of precipitation rate and infiltration capacity in three different situations: **A.** No depression storage or overland flow. **B.** Filling of depression storage after time, *t*, has elapsed. **C.** Filling of depression storage starting immediately.

soil compaction. Once the final infiltration rate is reached, the depth of ponded water also promotes high infiltration.

Moisture flowing downward through the unsaturated zone from a rainstorm moves as a front. Waves of water pass through a given soil layer, each front displaced by one from a more recent rain (31). Radioactive tracer studies have shown that water infiltrated from a single storm can be identified for its entire journey through the unsaturated zone to the water table (32).

## 4.8 SOIL MOISTURE

The uppermost soil layers may have a three-phase system of soil, air, and water. The water is in both the liquid and vapor phases. Larger pore spaces contain a film of water coating the mineral grains, while the smaller and capillary pores may be saturated. Water movement takes place through the saturated pores, while water vapor passes through the air-filled pores (33).

The soil layers that have the three-phase system of soil/water/air have been termed the **unsaturated zone, zone of aeration,** or **vadose zone.**

The **water content,** $w$, of a soil, expressed as a percentage, is the weight of the contained water, $W_w$, divided by the total weight of the soil mass, $W_s$:

$$w = 100\ (W_w/W_s) \tag{4-12}$$

The **saturation ratio,** $R_s$, is the ratio of the volume of the contained water, $V_w$, to the volume of the voids, $V_v$:

$$R_s = V_w/V_v \tag{4-13}$$

The **volumetric water content,** $\theta$ of a soil is the volume of the contained water, $V_w$, divided by the total volume of the soil sample, $V$:

$$\theta = V_w/V \tag{4-14}$$

If fluid pressures are measured above the water table, they will be found to be negative with respect to local atmospheric pressure. This is called a **tension.** Air

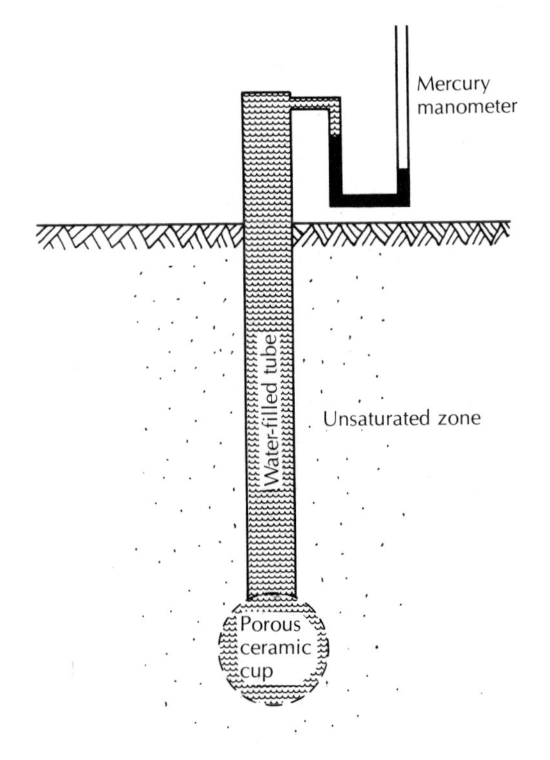


FIGURE 4.18 Porous-cup tensiometer with mercury manometer to measure soil-moisture tension.

may also be present in the voids above the water table. Air pressure is equal to atmospheric pressure above the water table. Water vapor is also present in the voids above the water table.

The pressure in the unsaturated zone is negative, owing to tension of the soil-surface-water contact. The negative pressure head, $\psi$, is measured in the field with a **tensiometer.** This device consists of a tube that is closed at the top, with a ceramic cup at the bottom to provide a porous membrane (Figure 4.18). When the tensiometer is inserted into soil, water within it is in contact with the soil moisture through the porous membrane. The suction exerted on the water in the tensiometer can be measured with a mercury manometer, vacuum gage, or pressure transducer (34).

Water molecules at the water table are subject to an upward attraction due to surface tension of the air–water interface and the molecular attraction of the liquid and solid phases. This is known as **capillarity.**

Capillary pores in the **zone of aeration** can draw water up from the zone of saturation beneath the water table. In very fine-grained soils, this **capillary fringe** can saturate the soil above the water table. However, tensiometer readings will reveal that the head is negative, indicating that the capillary fringe is part of the vadose zone. The zone of aeration is best defined as the zone where the soil moisture is under tension.

Because of irregularities in the size of the openings, capillary water does not rise to an even height above the water table; rather, it forms an irregular fringe. The capillary fringe is higher in fine-grained soils than in coarse-grained ones because of the greater tensions created by the smaller pore openings. It can rise to a height of several feet in silts and clays. The capillary fringe can provide a means of direct evaporation of ground water if the water table is close enough to the surface that the capillary fringe reaches the ground surface. As water evaporates from the soil surface, it can be replaced by water from the zone of saturation drawn upward by capillarity.

Above the capillary fringe, there is moisture coating the solid surfaces of the fragment or rock particles. If the liquid coating becomes too thick to be held by surface tension, a droplet will pull away and be drawn downward by gravity. The fluid can also evaporate and move through the air space in the pores as water vapor.

The amount of vapor movement in the unsaturated zone is much less important than transport in the liquid form (35). However, this might not hold true if the water content of the soil is very low or if there is a strong temperature gradient. The movement of vapor through the unsaturated zone is a function of the temperature and humidity gradients in the soil and molecular diffusion coefficients for water vapor in the soil (36).

The water content of a soil sample can be measured directly by excavation of a soil sample. The volume, $V$, of a moist soil sample is measured and the moist weight, $W_m$, determined. The sample is oven-dried at 105° C until a constant weight is obtained and then the weight of the soil, $W_s$, is determined. The weight of the water is $W_m - W_s$. Water content is hence $(W_m - W_s)/W_s$.

To find the volumetric water content the volume of the water must first be determined by dividing the weight of the water by the specific weight of water, $\gamma_w$. Volumetric water content is then determined as $[(W_m - W_s)/\gamma_w]/V$.

Soil moisture can also be measured indirectly by nondestructive means. One method involves burying small blocks, called resistance cells, in which electrodes are embedded, and then passing an electrical current through the wire. The electrical resistance of the block is proportional to the moisture it contains, which, in turn, is depenent upon the soil moisture. The meter can be calibrated for the soil type and, once calibrated, can be used for repeated measurements at the same location. This type of apparatus is relatively inexpensive, but the buried resistance blocks must be left in the soil.

A more expensive device uses a source of fast neutrons enclosed in a probe. The probe is lowered into a tube in the soil. When the fast neutrons encounter hydrogen atoms in water, they become slow neutrons. The density of slow neutrons produced is a function of the amount of soil moisture. A slow neutron counter is also a part of the probe. The neutron meter must be calibrated against the known water content of one part of the soil profile. The method can be used for repeated soil-moisture measurements in the same access tube. The water content in a spherical volume of 6-inch (15-centimeter) radius is measured by this method.

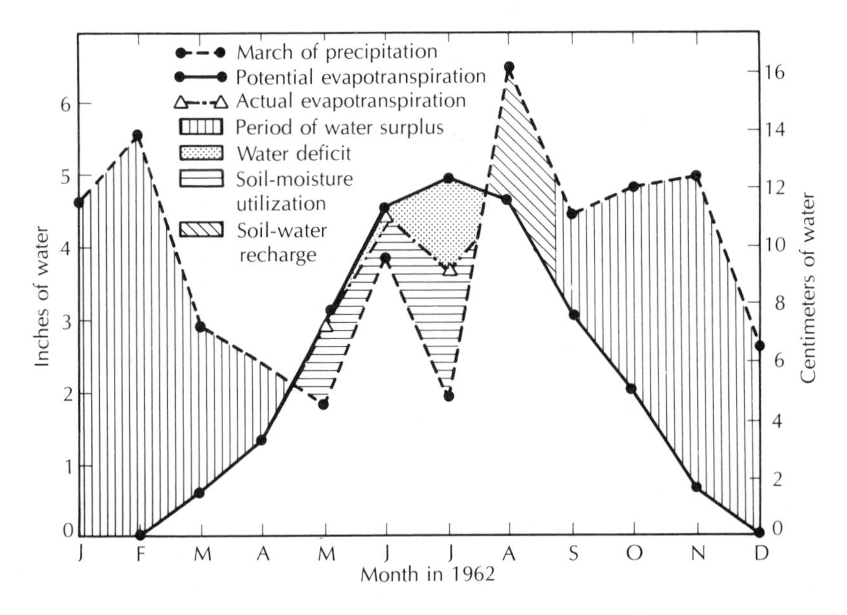


**FIGURE 4.19** Soil-moisture budget for Bridgehampton, New York. The diagram is based on measured precipitation and computed potential and actual evapotranspiration. The Thornthwaite method was used for evapotranspiration computations. Source: C. W. Fetter, Jr. *Bulletin, Geological Society of America,* 87 (1976):401–6.

Soil moisture at a location varies with changes in the amount of precipitation and evapotranspiration. Figure 4.19 shows a soil-moisture budget for a humid area based on measured precipitation and computed potential and actual evapotranspiration. During the period of water surplus, there is moisture available for ground-water recharge and runoff. Major fluctuations are seasonal in nature. In the spring, soil moisture is high, as snowmelt and spring rains have created large amounts of water available for infiltration. At times, the top layer of soil may be completely saturated, even though lower layers are unsaturated. During these periods of very high soil moisture, ground-water recharge can occur. Moisture moves downward by gravity flow. As water is withdrawn from the soil by evapotranspiration, the soil-moisture content drops. When the soil-moisture content of a layer reaches the point at which the force of gravity acting on the water equals the surface tension, gravity drainage ceases. This soil-moisture content is the **field capacity** of the soil. Field capacity is related to specific retention but has different units. It depends upon specific retention, evaporation depth, and the unsaturated permeability characteristic curve of the soil.

The concept of field capacity is somewhat vague. Gravity drainage may take a long period of time to occur. The amount of moisture retained for a few days is much more than that retained for a long period. Table 4.6 shows the soil moisture of a silt-loam (fine-textured) soil as a function of time (37). It can be seen that field capacity is not a single value, but a time-dependent parameter.

Soil moisture becomes lower than the field capacity as evapotranspiration removes still more water. During summer periods, the soil often dries. Occasional rainstorms may cause short-term rises in soil-moisture content, but there is generally no ground-water recharge. Exceptionally heavy summer rains, which replenish the depleted soil moisture and raise the water content above the field capacity, can create a wave of infiltrated water that courses downward through the soil-moisture zone and past the roots of thirsty plants, recharging the ground-water reservoir and thus causing the water table to rise. After fall frosts kill plants and cause deciduous trees to lose their leaves, the rate of evapotranspiration slows greatly, and soil moisture increases if rainfall and infiltration continue. Figure 4.20 illustrates a hypothetical annual cycle of soil moisture for a moderately humid area (20 to 30 inches, or 50 to 75 centimeters per year precipitation).

**TABLE 4.6.** Moisture content of a silt loam as a function of time since saturation (37)

| Time | $\theta$ (%) |
|---|---|
| 1 day | 20.2 |
| 7 days | 17.5 |
| 30 days | 15.9 |
| 60 days | 14.7 |
| 156 days | 13.6 |

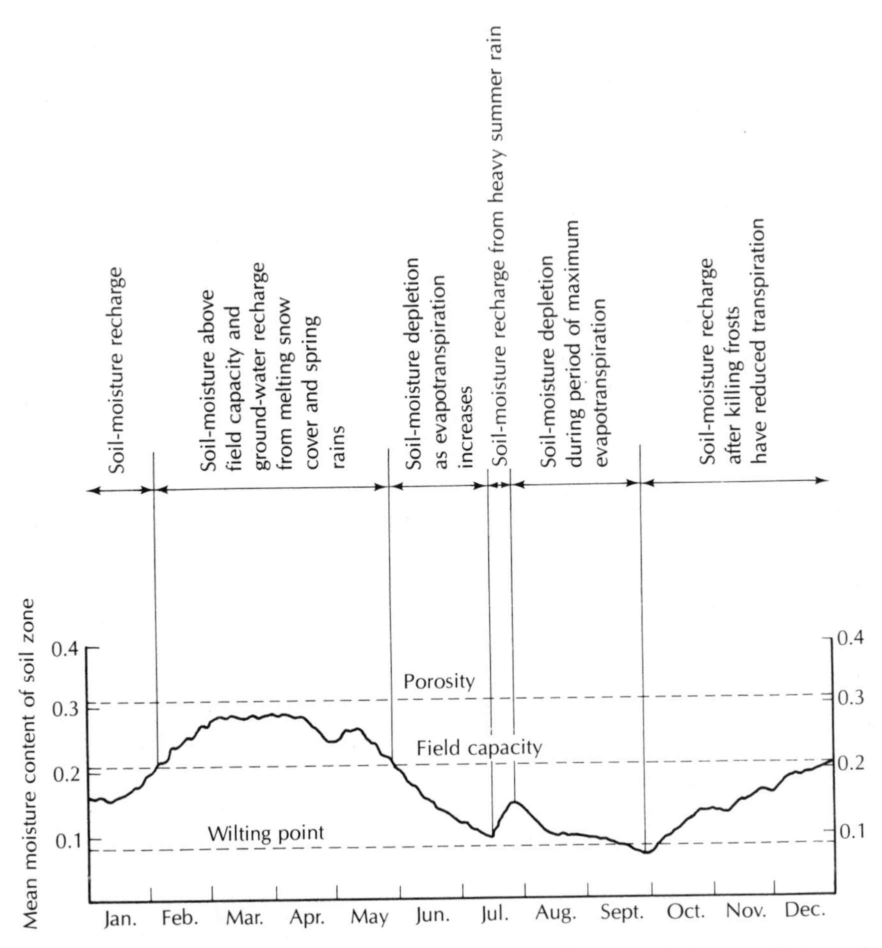


**FIGURE 4.20** Hypothetical fluctuation of soil moisture for a sandy loam soil through an annual cycle in a region with a moderate amount of rainfall (20 to 30 inches [50 to 75 centimeters] per year) and heavy rains in the spring.

If the soil moisture drops too low, the remaining moisture is too tightly bound to the soil particles for the plant roots to withdraw it. The soil-moisture content at which this first occurs is the **wilting point.** Plants wilt and may die for lack of moisture. The wilting-point moisture content is greater for fine-textured soils owing to their greater surface area. Figure 4.21 shows typical wilting-point moisture content values for various soil types. Also shown are generalized field capacity values. The available water capacity of a soil is the difference between the field capacity and the wilting point. Water in a soil above the field will drain downward if the soil is permeable or will waterlog a fine-grained, slowly permeable soil. Brief study of Figure 4.21 shows that the available water capacity is greatest for soils of intermediate texture, e.g., loams and silt loams.

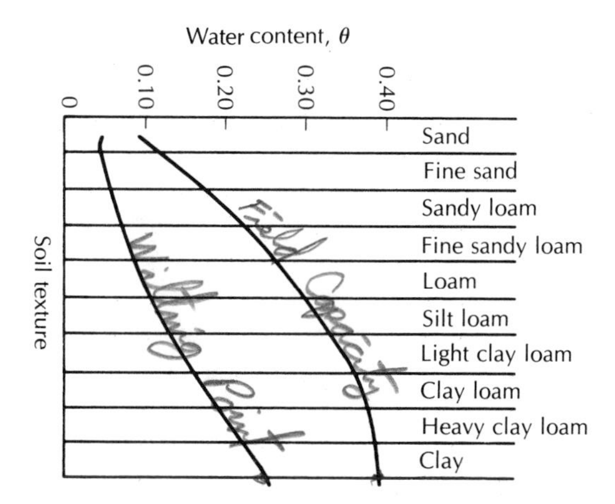


**FIGURE 4.21** Water-holding properties of soils based on texture. The available water supply for a soil is the difference between field capacity and wilting point. Source: U.S. Department of Agriculture, *Yearbook*, 1955.

It should be stressed that the soil-moisture zone extends only to the depth of plant roots. In agricultural soils, this might be a few feet or less. Some trees have tap roots extending tens of feet downward. The unsaturated zone may extend beyond the zone of soil moisture. In this case, there is an **intermediate zone** between the soil-moisture zone and the capillary fringe. In other cases, the soil-moisture zone may extend to the water table—an especially likely occurrence if the water table is within a few feet of the surface.

## 4.9 THEORY OF UNSATURATED FLOW

The infiltration of water into the unsaturated zone can be considered mathematically in terms of both a **gravity potential,** $Z$, and a **moisture potential,** $\psi$ (33). The moisture potential is a negative pressure due to the soil-water attraction. The moisture potential increases with decreasing amounts of soil moisture. Depending upon the soil-moisture content, either the moisture potential or the gravity potential may predominate. At moisture contents close to the specific retention, the gravity potential is greater; but when the soil is very dry, the moisture potential may be several orders of magnitude greater than the gravity potential.

When the soil is not saturated, soil moisture flows downward by gravity flow through interconnected pores that are filled with water. With increasing water content, more pores fill, and the rate of downward water movement increases. Darcy's law is valid for flow in the unsaturated zone, although the **unsaturated hydraulic conductivity,** $K(\theta)$, is not a constant. The unsaturated hydraulic conductivity is a function of the volumetric water content, $\theta$. As $\theta$ increases, so does $K(\theta)$. Figure 4.22 shows the relationship between $K(\theta)$ and $\theta$ for a clay soil. The value of the moisture potential, $\psi$, is also a function of $\theta$,

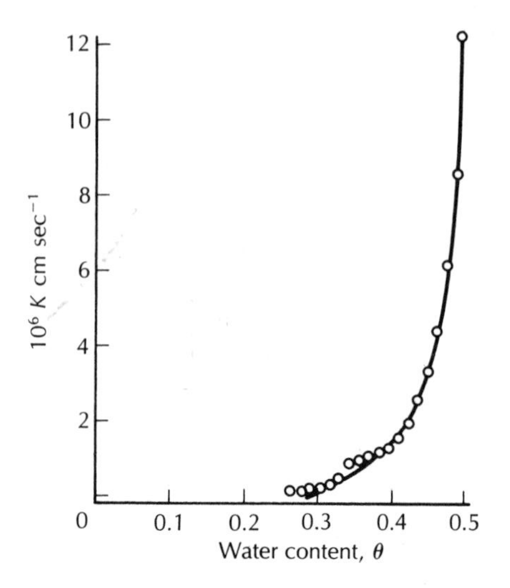


**FIGURE 4.22** The relationship between hydraulic conductivity, $K$, and volumetric water content, $\theta$, for a clay. Source: J. R. Philip, "Theory of Infiltration," in *Advances in Hydroscience*, vol. 5, ed. V. T. Chow (New York: Academic Press, 1969). Used with permission.

often ranging over many orders of magnitude (Figure 4.23). The **total potential,** $\phi$, in unsaturated flow is the sum of the moisture potential, $\psi(\theta)$, and the elevation head, $Z$:

$$\phi = \psi(\theta) + Z \tag{4-15}$$

The relationship of unsaturated hydraulic conductivity and volumetric water content is determined experimentally. A sample of the rock or sediment is placed in a container. The water content is kept constant, and the rate at which water moves through the soil is measured. The value of $K(\theta)$ can be determined by Darcy's law. Figure 4.22 is based on a number of different measurements at differing values of $\theta$.

The moisture potential, $\psi$, is measured by a suction applied to a soil. The curve of $\psi$ versus volumetric water content is a plot of the volumetric water content, starting with a saturated sample to which increased suction is gradually applied. A rather substantial problem occurs because of hysteresis in the volumetric water content–moisture potential relationship. The curve of $\psi$ as a function of $\theta$ is different if it is determined for decreasing as opposed to increasing values of $\theta$. Thus, two curves such as that of Figure 4.23 are needed: one if the sediment has a particular $\theta$ as a result of an increase in soil-water content, and another if the soil is drying. Figure 4.24 shows the impact of hysteresis for an idealized sandy soil. Therefore, to use a value of $\psi$, one must know the prior moisture history of the sample.

Flow in the unsaturated zone is further complicated by the fact that both $K$ and $\psi$ may change as $\theta$ varies, and, by its very nature, unsaturated flow in-

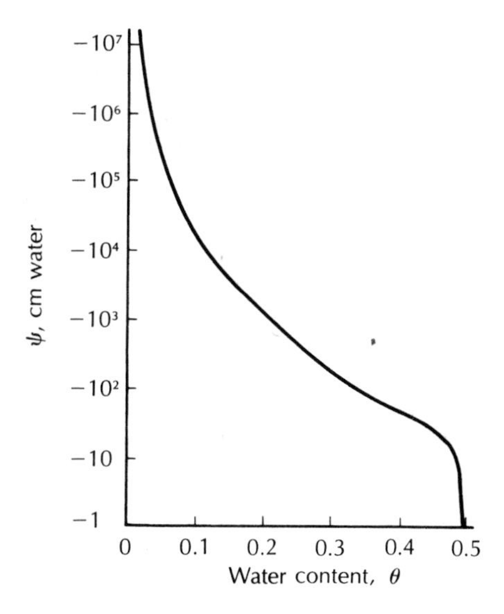


**FIGURE 4.23** The relationship between moisture potential, $\psi$, and volumetric water content, $\theta$, for the clay soil of Figure 4.22. Source: J. R. Philip, "Theory of Infiltration," in *Advances in Hydroscience,* vol. 5, ed. V. T. Chow (New York: Academic Press, 1969). Used with permission.

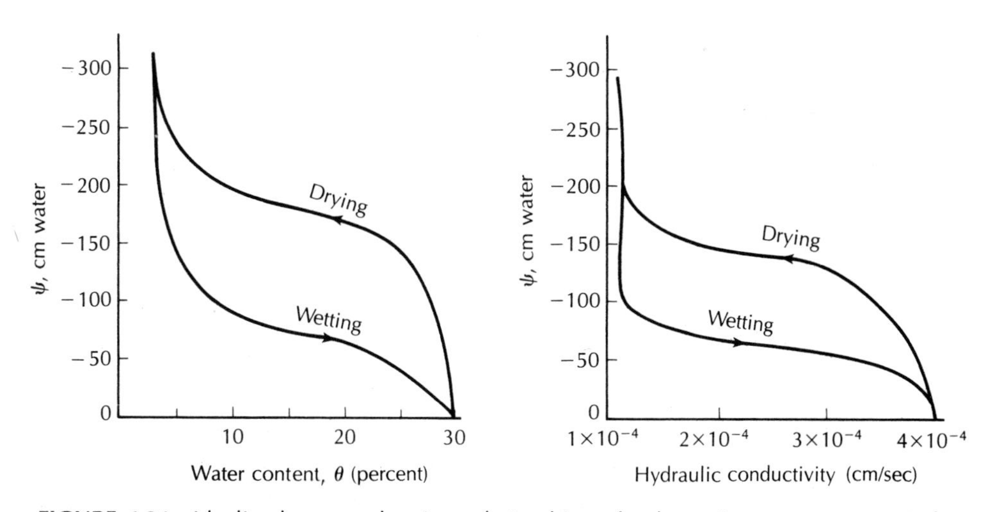


**FIGURE 4.24** Idealized curves showing relationships of volumetric water content, $\theta$, hydraulic conductivity, $K$, and soil-moisture tension head, $\psi$. The effect of hysteresis is included for wetting and drying cycles.

volves many changes in volumetric moisture content as waves of infiltrated water pass. The flow equations are nonlinear and not subject to easy solutions (35, 38).

The downward movement of a slug of infiltrated water is shown in Figure 4.25. During infiltration (Part A), the topmost soil layer is brought up to the saturated water content of almost 0.3. The rainfall rate is exceeding the infiltration capacity, so the vertical hydraulic conductivity of the soil controls the rate of downward movement. After infiltration ceases (Part B), the slug of infiltrated water begins to move downward, although only a small layer remains at the saturated moisture content. Eventually, all of the downward-moving slug is at an unsaturated state, and $K(\theta)$ controls the downward movement (Part C). Finally, the slug reaches the water table and raises it (Part D). By this time, gravity drainage has reduced the upper soil layer to field capacity.

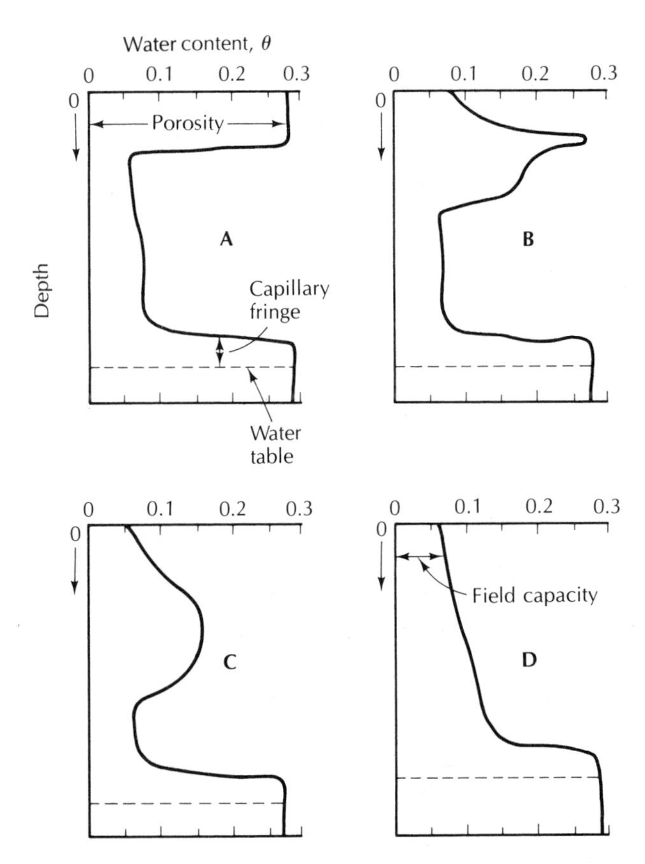


**FIGURE 4.25** Moisture profiles showing the downward passage of a wave of infiltrated water. The soil is saturated at a water content of 0.29 and has a field capacity water content of 0.06. Source: Modified from Agronomy Monograph 17, "Drainage for Agriculture," 1974, pp. 359–405. Used with permission of the American Society of Agronomy.

One important consideration in unsaturated flow is that at low volumetric water contents, the relations that hold true in saturated flow may be invalid. The best example is the fact that for coarse materials, such as sand and gravel, the pores are large and drain quickly. At lower volumetric moisture contents, there may be very few saturated pores. On the other hand, finer-grained soils may have most of the pores still saturated. Thus, at lower values of $\theta$, the unsaturated hydraulic conductivity of a clay may be greater than that of a sand. A layer of sand in a fine-textured, unsaturated soil may retard downward movement of infiltrating water owing to its low unsaturated hydraulic conductivity.

## 4.10 WATER-TABLE RECHARGE

When the front of infiltrating water reaches the capillary fringe, it displaces air in the pore spaces and causes the water table to rise. The capillary fringe is also higher, and the latest-arriving recharge is actually found at the top of the capillary fringe. The time of movement of infiltrating water is a function of the thickness of the unsaturated zone and the vertical unsaturated hydraulic conductivity. The presence of layers of low-permeability material, such as silts and clays, can retard the rate of recharge, even if the layers are thin. The time lag may be only a few hours in the humid regions for very coarse soils with a water table close to the surface. In arid environments, with very infrequent recharge and great depths to the water table, water may take years to pass through the unsaturated zone.

In the arid Hualapai Plateau area of northwestern Arizona, the annual recharge rate is only 0.1 inch (0.25 centimeter) per year. Although rainfall averages 9 to 13 inches (23 to 33 centimeters) per year, potential evapotranspiration is in the range of 72 to 76 inches (183 to 193 centimeters) per year (39). Grassland regions have more precipitation and, consequently, more recharge. In the Sand Hills region of South Dakota and Nebraska, the mean annual precipitation is 18.2 inches (46.3 centimeters), and ground-water recharge amounts to 2.7 inches (6.9 centimeters) per year (40). On eastern Long Island, New York, precipitation averages 46 inches (116.8 centimeters) per year, and evapotranspiration, as computed by the Thornthwaite method, averages 22.6 inches (57.4 centimeters) per year. The soils are permeable, and much of the remaining 23.4 inches (59.4 centimeters) per year of water recharges the water table (41).

The rate at which water-table recharge occurs is variable, depending upon, among other things, the thickness of the unsaturated zone. Where the unsaturated zone is thinner, recharge can reach the water table first, resulting in a localized ground-water mound. The unsaturated zone is thinner in topographically low places; for example, near a lake shore or in a lowland. Soil moisture percolating through the unsaturated zone beneath upland areas takes longer so that if the water table is initially somewhat level, localized high spots can develop on the water table. Localized flow systems can develop which move water laterally from the temporary ground-water mounds toward the water table beneath the uplands, where the infiltration has not reached the water table. Eventually the water table beneath the upland area will rise owing not only to the vertical percolation of infiltrating water but also to the lateral movement from the temporary ground-water mounds. This results in the development of very complex and transitory local flow systems. In very permeable materials, the local ground-water mounds beneath the topographic lows dissipate quickly. However, in less permeable materials the local ground-water mounds could take a rather long time to dissipate. Localized flow reversals could take place with the ground water first flowing from beneath the lowland toward the upland in response to the formation of localized recharge mounds and then moving in the opposite direction as the water table beneath the upland finally is recharged by the percolating soil moisture. It would be anticipated that the flow reversals would affect only the water moving in the very top of an unsaturated flow system (42).

On eastern Long Island, the response of the water table to recharge is fairly rapid. Figure 4.26 is a hydrograph for 1962 of a water-table well with an unsaturated zone on the order of 10 to 13 feet (3 to 4 meters). This well is a few miles from the precipitation station that yielded the data for Figure 4.19. As the water table is being drained by stream and spring discharge, it will decline (a) if there is no recharge or (b) if the amount of recharge is less than ground-water outflow. The hydrograph peaks in March and then begins a recession which lasts until October. Periods of surplus water are indicated in Figure 4.19 for January through April, and then September through December. The peak in March corresponds to snowmelt and spring rains, and the upturn in water levels by No-

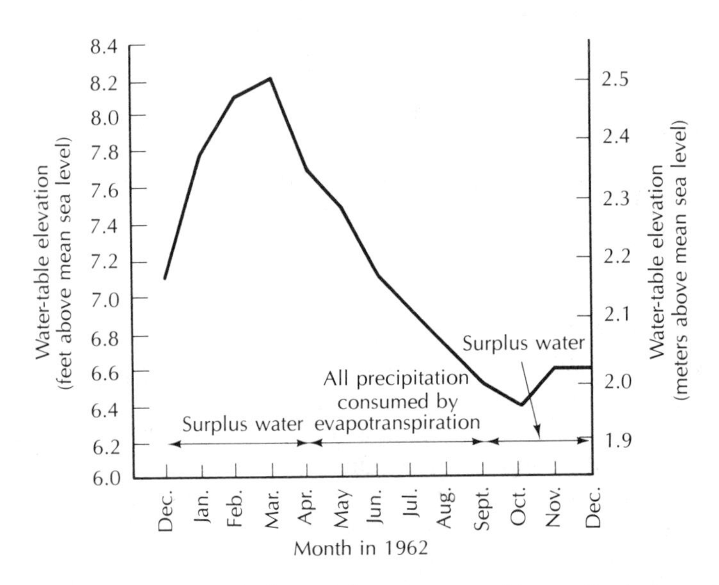


**FIGURE 4.26** Hydrograph of a shallow well in a water-table aquifer in Long Island. The period of record is the same as for the soil-moisture budget diagram of Figure 4.19.

vember is due to fall recharge. The response time of this water-table aquifer is more rapid than for those cases in which the unsaturated zone is many tens of feet thick.

## 4.11 AQUIFERS

Natural earth materials have a very wide range of hydraulic conductivities. Near the earth's surface there are very few, if any, geologic formations that are absolutely impermeable. Weathering, fracturing, and solution have affected most rocks to some degree. However, the rate of ground-water movement can be exceedingly slow in units of low hydraulic conductivity.

An **aquifer** is a geologic unit that can store and transmit water at rates fast enough to supply reasonable amounts to wells. The intrinsic permeability of aquifers would range from about $10^{-2}$ darcy upward. Unconsolidated sands and gravels, sandstones, limestones and dolomites, basalt flows, and fractured plutonic and metamorphic rocks are examples of rock units known to be aquifers.

A **confining layer** is a geologic unit having little or no intrinsic permeability—less than about $10^{-2}$ darcy. This is a somewhat arbitrary limit and depends upon local conditions. In areas of clay, with intrinsic permeabilities of $10^{-4}$ darcy, a silt of $10^{-2}$ darcy might be used to supply water to a small well. On the other hand, the same silt might be considered a confining layer if it were found in an area of coarse gravels with intrinsic permeabilities of 100 darcys. Ground water moves through most confining layers, although the rate of movement is very slow.

Confining layers are sometimes subdivided into aquitards, aquicludes, and aquifuges. An **aquifuge** is an absolutely impermeable unit that will not transmit any water. An **aquitard** is a layer of low permeability that can store ground water and also transmit it slowly from one aquifer to another. The term **leaky confining layer** is also applied to such a unit. An **aquiclude** is also a unit of low permeability, but is located so that it forms an upper or lower boundary to a ground-water flow system. Most authors now use the terms confining layer and leaky confining layer.

Confining layers can be important elements of regional flow systems, and leaky confining layers can transmit significant amounts of water if the cross-sectional area is large. On Long Island, New York, there is a deep aquifer, the Lloyd Sand Member of the Raritan Formation. The recharge to this aquifer passes downward across a confining layer 200 to 300 feet (60 to 100 meters) thick, the Clay Member of the Raritan Formation. Over an areal extent measured in hundreds of square miles, a very low rate of vertical seepage provides recharge to the underlying aquifer. Some of the recharge to the principal artesian aquifer of the southeastern United States takes place by downward leakage through overlying confining layers, although much of the recharge is carried through sinkholes.

Aquifers can be close to the land surface, with continuous layers of materials of high intrinsic permeability extending from the land surface to the base of the aquifer. Such an aquifer is termed a **water-table aquifer** or **unconfined aquifer.** Recharge to the aquifer can be from downward seepage through the unsaturated zone (Figure 4.27). Recharge can also occur through lateral ground-water flow or upward seepage from underlying strata.

Some aquifers, called **confined** or **artesian aquifers,** are overlain by a confining layer. Recharge to them can occur either in a recharge area, where the aquifer crops out, or by slow downward leakage through a leaky confining layer (Figure 4.28). If a tightly cased well is placed through the confining layer, water from the aquifer may rise considerable distances above the top of the aquifer (Figure 4.29). This indicates that the water in the aquifer is under pressure. The **potentiometric surface** for a confined aquifer is the surface representative of the level to which water will rise in a well cased to the aquifer. (The term **piezometric** was used in the past, but it has now been replaced by potentiometric.) If the potentiometric surface of an aquifer is above the land surface, a flowing artesian well may occur. Water will flow from the well casing without need for a pump. Of course, if a pump were installed, the amount of water obtained from the well could be increased.

In some cases, a layer of low-permeability material will be found as a lens in more permeable materials. Water moving downward through the unsaturated zone will be intercepted by this layer and will accumulate on top of the lens. A layer of saturated soil will form above the main water table. This is termed a **perched aquifer** (Figure 4.30). Water moves laterally above the low-

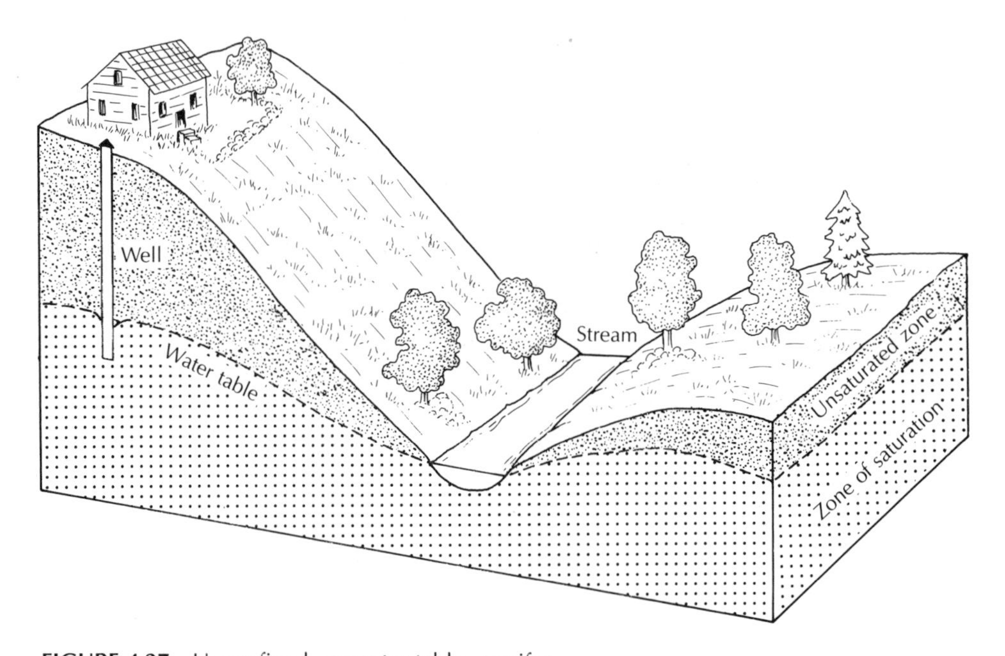


FIGURE 4.27 Unconfined, or water-table, aquifer.

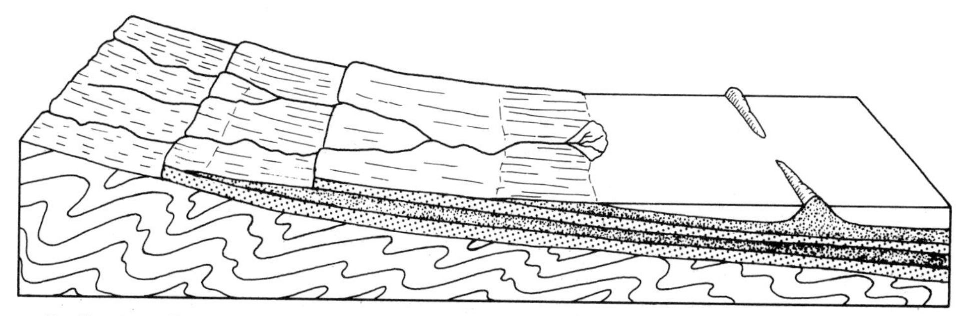

Confined aquifers created by alternating aquifers and confining units deposited on a regional dip.

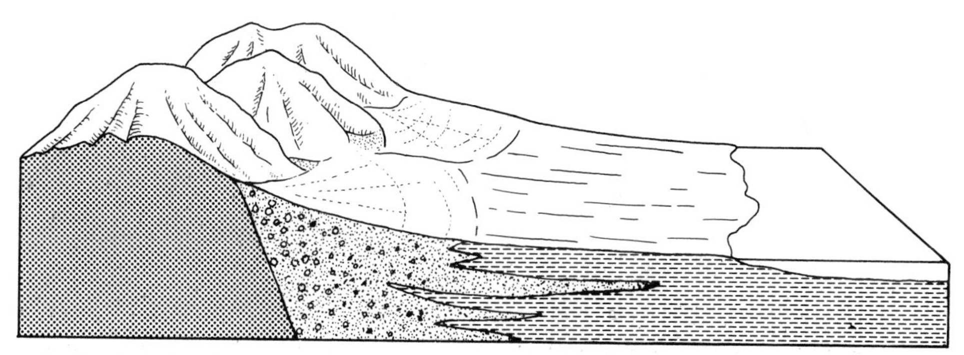

Confined aquifers created by deposition of alternating layers of permeable sand and gravel and impermeable silts and clays deposited in intermontane basins.

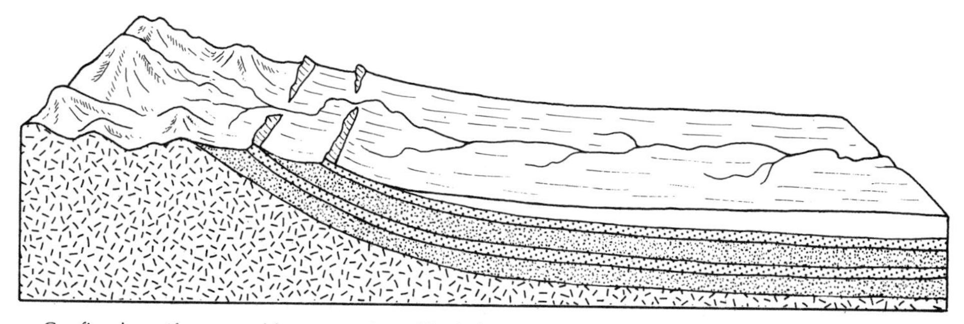

Confined aquifer created by upwarping of beds by intrusions.

**FIGURE 4.28** Confined aquifers created when aquifers are overlain by confining beds.

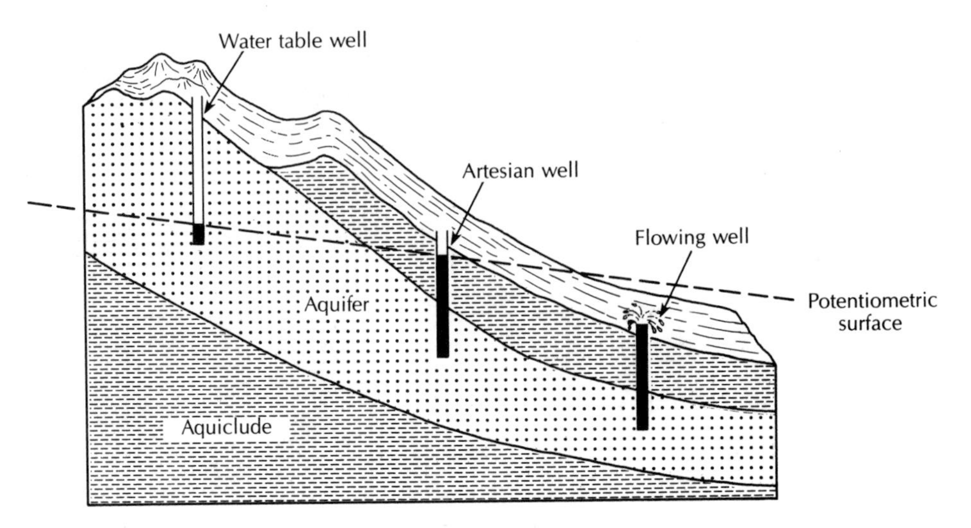


FIGURE 4.29 Artesian and flowing well in confined aquifer.

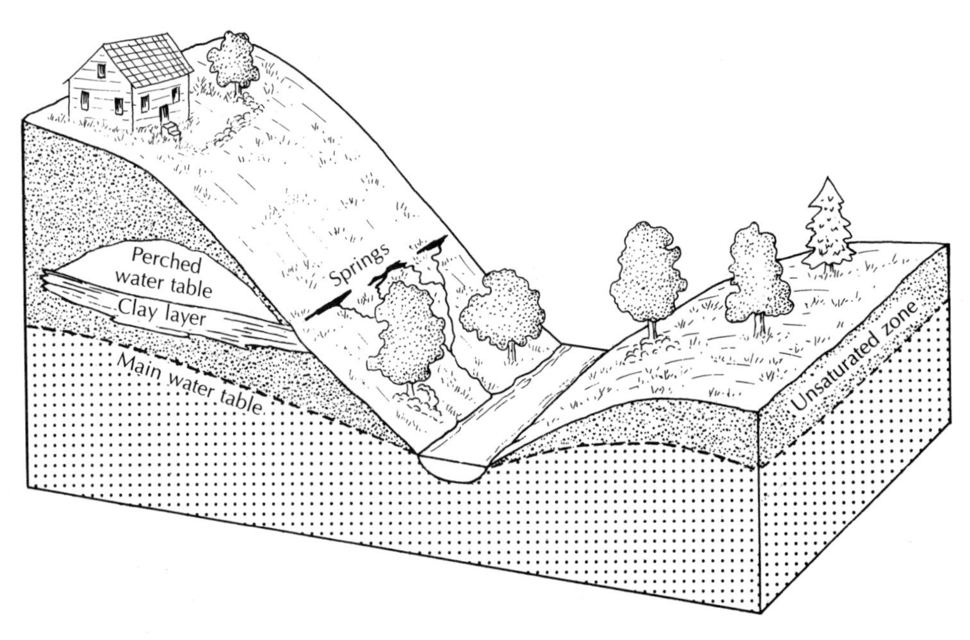


FIGURE 4.30 Perched aquifer formed above the main water table on a low-permeability layer in the unsaturated zone.

permeability layer up to the edge and then seeps downward toward the main water table or forms a spring. Perched aquifers are common in glacial outwash, where lenses of clay formed in small glacial ponds are present. They are also often present in volcanic terranes, where weathered ash zones of low permeability can occur sandwiched between high-permeability basalt layers.

Perched aquifers are usually not very large; most would supply only enough water for household use. Some lakes are perched on low-permeability sediments. Such ponds are especially vulnerable to widely fluctuating lake-stage levels with changes in the amount of rainfall.

## 4.12 AQUIFER CHARACTERISTICS

We have thus far considered the intrinsic permeability of earth materials and their hydraulic conductivity when transmitting water. A useful concept in many studies is aquifer **transmissivity.** This is a measure of the amount of water that can be transmitted horizontally by the full saturated thickness of the aquifer under a hydraulic gradient of 1. The transmissivity, $T$, is the product of the hydraulic conductivity and the saturated thickness of the aquifer, $b$:

$$T = bK \tag{4-16}$$

For a multilayer aquifer, the total transmissivity is the sum of the transmissivity of each of the layers:

$$T = \sum_{i=1}^{n} T_i \tag{4-17}$$

The dimensions of transmissivity are $(L^2/T)$. Common units are square meters per day or square feet per day. Aquifer transmissivity is a concept that assumes flow through the aquifer to be horizontal. In some cases, this is a valid assumption; in others, it is not.

When the head in a saturated aquifer or confining unit changes, water will be either stored or expelled. The **storage coefficient,** or **storativity** $(S)$, is the volume of water that a permeable unit will absorb or expel from storage per unit surface area per unit change in head. It is a dimensionless quantity.

In the saturated zone, the head creates pressure, affecting the arrangement of mineral grains as well as the density of the water in the voids. If the pressure increases, the mineral skeleton will expand; if it drops, the mineral skeleton will contract. This is known as **elasticity.** Likewise, water will contract with an increase in pressure and expand if the pressure drops. When the head in an aquifer or confining bed declines, the aquifer skeleton compresses, which reduces the effective porosity and expels water. Additional water is released as the pore water expands due to lower pressure.

The **specific storage** $(S_s)$ is the amount of water per unit volume of a saturated formation that is stored or expelled from storage owing to compressibility of the mineral skeleton and the pore water per unit change in head. This is also called the **elastic storage coefficient.** The concept can be applied to both aquifers and confining units.

The specific storage is given by the expression (43–45)

$$S_s = \rho_w g(\alpha + n\beta) \tag{4-18}$$

where

$\rho_w$ is the density of the water
$g$ is the acceleration of gravity
$\alpha$ is the compressibility of the aquifer skeleton
$n$ is the porosity
$\beta$ is the compressibility of the water

Specific storage has dimensions of $1/L$. The value of specific storage is very small, generally 0.0001 foot or less.

In a confined aquifer, the head may decline—yet the potentiometric surface remains above the unit (Figure 4.31). Although water is released from stor-

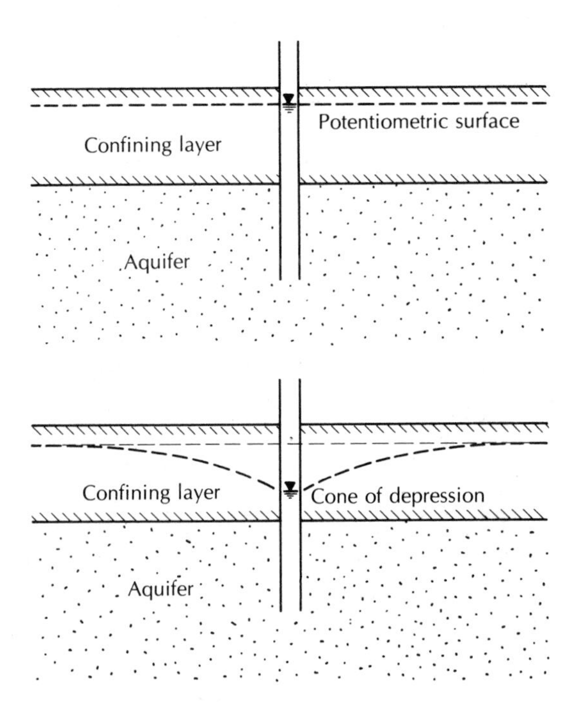


FIGURE 4.31 Diagram showing lowering of the potentiometric surface in a confined aquifer with the resultant water level still above the aquifer materials. In this circumstance, the aquifer remains saturated.

age, the aquifer remains saturated. The storativity *(S)* of a confined aquifer is the product of the specific storage $(S_s)$ and the aquifer thickness *(b)*:

$$S = bS_s \tag{4-19}$$

All of the water released is accounted for by the compressibility of the mineral skeleton and the pore water. The water comes from the entire thickness of the aquifer. The value of the storativity of confined aquifers is on the order of 0.005 or less.

In an unconfined unit, the level of saturation rises or falls with changes in the amount of water in storage. As the water level falls, water drains from the pore spaces. This storage or release is due to the **specific yield** $(S_y)$ of the unit. Water is also stored or expelled depending on the specific storage of the unit. For an unconfined unit, the storativity is found by the formula

$$S = S_y + hS_s \tag{4-20}$$

where $h$ is the thickness of the saturated zone.

The value of $S_y$ is several orders of magnitude greater than $hS_s$ for an unconfined aquifer, and the storativity is usually taken to be equal to the specific yield. For a fine-grained unit, the specific yield may be very small, approaching the same order of magnitude as $hS_s$. Storativity of unconfined aquifers ranges from 0.02 to 0.30.

The volume of water drained from an aquifer, $V_w$, may be found from the equation

$$V_w = SA\Delta h \tag{4-21}$$

where $A$ is the horizontal area and $\Delta h$ is the decline in head.

---

**EXAMPLE PROBLEM**

An unconfined aquifer with a storativity of 0.13 has an area of 123 square miles. The water table drops 5.23 feet during a drought. How much water was lost from storage?

$$\begin{aligned} V_w &= SA\Delta h \\ &= 0.13 \times 123 \text{ mi}^2 \times 2.7878 \times 10^7 \text{ ft}^2/\text{mi}^2 \times 5.23 \text{ ft} \\ &= 2.33 \times 10^9 \text{ ft}^3 \end{aligned}$$

If the same aquifer had been confined with a storativity of 0.0005, what change in the amount of water in storage would have resulted?

$$\begin{aligned} V_w &= 0.0005 \times 123 \text{ mi}^2 \times 2.7878 \times 10^7 \text{ ft}^2/\text{mi}^2 \times 5.23 \text{ ft} \\ &= 8.97 \times 10^6 \text{ ft}^3 \end{aligned}$$

---

## 4.13 HOMOGENEITY AND ISOTROPY

Hydrogeologists are interested in two key properties of geologic formations: hydraulic conductivity and specific storage or specific yield. A third property, the thickness, is also important, since the overall hydrogeologic response of a unit is a function of the product of the hydraulic parameters and the thickness.

A **homogeneous** unit is one that has the same properties at all locations. For a sandstone, this would indicate that the grain-size distribution, porosity, degree of cementation, and thickness are variable only within small limits. The values of the transmissivity and storativity of the unit would be about the same wherever present. A plutonic or metamorphic rock would have the same amount of fracturing everywhere, including the strike and dip of the joint sets. A limestone would have the same amount of jointing and solution openings at all locations.

In **heterogeneous** formations, hydraulic properties change spatially. One example would be a change in thickness. A sandstone that thickens as a wedge is nonhomogeneous, even if porosity, hydraulic conductivity, and specific storage remain constant. The change in thickness results in a change in the hydraulic properties of the unit. Layered units may also be nonhomogeneous. Most sedimentary units were deposited as successive layers of sediments with intervals of nondeposition. These layers are known to vary in thickness, from microscopic layers to those measured in meters. If the hydraulic properties of the layers are different, the entire unit is heterogeneous. The individual beds may be homogeneous. The third type of heterogeneity in a sedimentary unit occurs when a

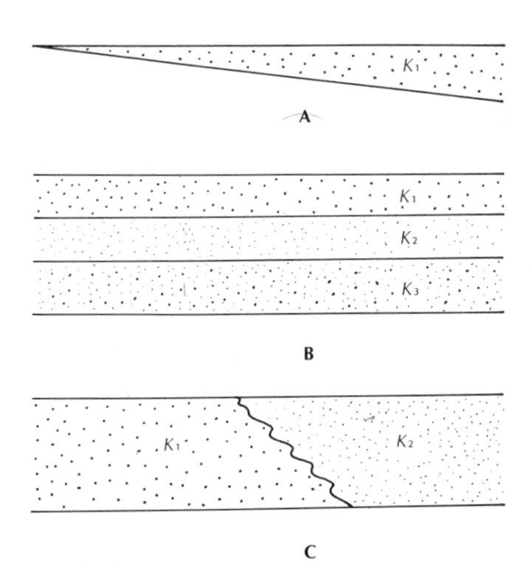


FIGURE 4.32 A. Heterogeneous formation consisting of a sediment which thickens in a wedge. B. Heterogeneous formation consisting of three layers of sediments of differing hydraulic conductivity. C. Heterogeneous formation consisting of sediments with different hydraulic conductivities lying next to each other.

facies change in the unit involves a transformation of hydraulic characteristics as well as lithologic features. Figure 4.32 illustrates these examples of heterogeneity in clastic sedimentary units.

Carbonate units may be heterogeneous (a) if there is a change in thickness or (b) if the degree of solution openings of fractures varies. The formation of solution passageways by moving ground water is typically concentrated along preferred fractures or bedding planes; thus, limestone formations are often heterogeneous. Plutonic rocks may have uneven fracturing or sporadic shear zones which render them heterogeneous. Basalt flows are virtually always heterogeneous by the very nature of the way they are formed. As might be expected, it is a very unusual geologic formation that is perfectly homogeneous. Geologic processes operate at varying rates and over uneven terrain, resulting in heterogeneity.

In a porous medium made of spheres of the same diameter packed uniformly, the geometry of the voids is the same in all directions. Thus, the intrinsic permeability of the unit is the same in all directions, and the unit is said to be **isotropic.** On the other hand, if the geometry of the voids is not uniform, there may be a direction in which the intrinsic permeability is greater. The medium is thus **anisotropic.** For example, a porous medium composed of book-shaped grains arranged in a subparallel manner would have a greater permeability parallel to the grains than crossing the grain orientation (Figure 4.33).

In fractured rock units, the direction of ground-water flow is completely constrained by the direction of the fractures. There may be zero intrinsic perme-

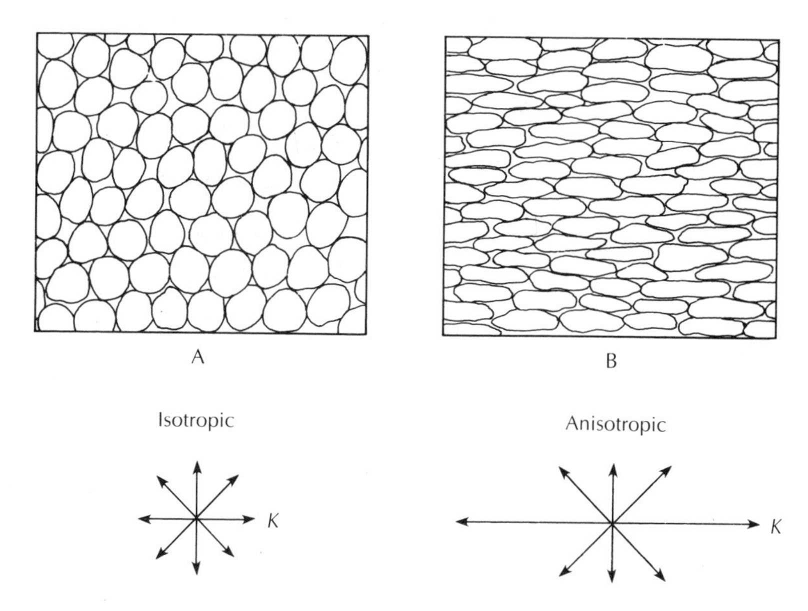


FIGURE 4.33 Grain shape and orientation can affect the isotropy or anisotropy of a sediment.

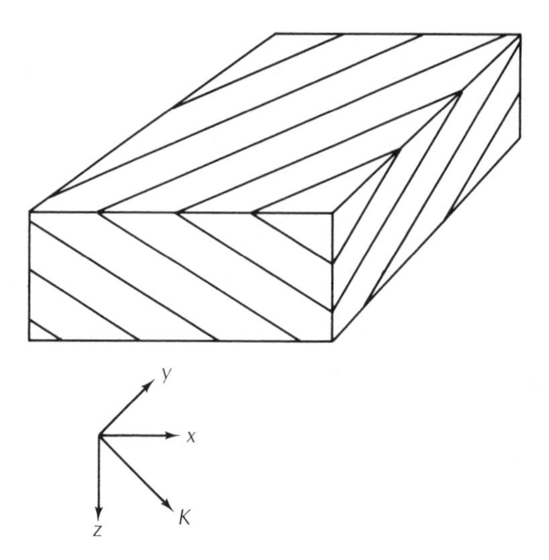


FIGURE 4.34 Anisotropy of fractured rock units due to directional nature of fracturing.

ability in directions not parallel to a set of fractures (Figure 4.34). Basalt flows are highly anisotropic, as flow parallels the dip of the flow as it moves in the interflow zones. Shrinkage cracks in the basalt are vertical, yielding some vertical permeability.

In sedimentary units, there may be several layers, each of which is homogeneous. The equivalent vertical and horizontal hydraulic conductivity can be easily computed. Figure 4.35 shows a three-layered unit, each unit having a different horizontal and vertical hydraulic conductivity ($K_h$ and $K_v$).

The average horizontal conductivity is found from the summation

$$K_h \text{ avg} = \sum_{m=1}^{n} \frac{K_{h_m} d_m}{d} \tag{4-22}$$

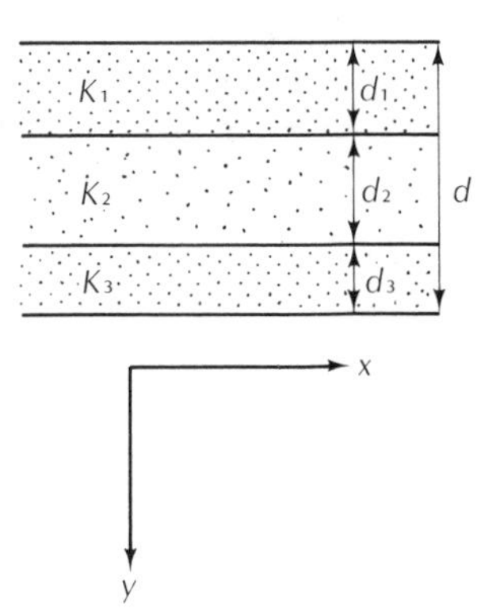


FIGURE 4.35 Heterogeneous formation consisting of three layers of differing hydraulic conductivity.

where $d$ is the total thickness and $d_m$ is the thickness of each layer. The overall vertical hydraulic conductivity is given by

$$K_v \text{ avg} = \frac{d}{\sum_{m=1}^{n} \frac{d_m}{K_{vm}}} \quad \textbf{(4-23)}$$

## REFERENCES


1. MEINZER, O. E. *The Occurrence of Ground Water in the United States.* U.S. Geological Survey Water-Supply Paper 489, 1923, 321 pp.
2. DAVIS, S. N. "Porosity and Permeability in Natural Materials." In *Flow through Porous Media,* ed. R. J. M. DeWiest. New York: Academic Press, 1969, pp. 53–89.
3. COHEN, P. *Water Resources of the Humboldt River Valley near Winnemucca, Nevada.* U.S. Geological Survey Water-Supply Paper 1975, 1965.
4. MAC CARY, L. M., and T. W. LAMBERT. *Reconnaissance of Ground-Water Resources of the Jackson Purchase Region, Kentucky.* U.S. Geological Survey Hydrologic Atlas HA-13, 1962, 9 pp.
5. WINSAUER, W. O., et al. "Resistivity of Brine-Saturated Sands in Relation to Pore Geometry." *American Association of Petroleum Geologists Bulletin,* 36 (1952):253–77.
6. WYLLIE, M. R. J., and M. B. SPANGLER. "Application of Electrical Resistivity Measurements to Problems of Fluid Flow in Porous Media." *American Association of Petroleum Geologists Bulletin,* 36 (1952):359–403.
7. MANGER, G. E. *Porosity and Bulk Density of Sedimentary Rocks.* U. S. Geological Survey Bulletin 1144-E, 1963.
8. ARCHIE, G. E. "Classification of Carbonate Reservoir Rocks and Petro-physical Considerations." *American Association of Petroleum Geologists Bulletin,* 36 (1950):943–61.
9. MURRAY, R. C. "Origin of Porosity in Carbonate Rocks." *Journal of Sedimentary Petrology,* 30 (1960):59–84.
10. DANIELS. J. J., G. R. OLHOEFT, and J. H. SCOTT. "Interpretation of Core and Well Log Physical Property Data from Drill Hole UPH-3, Stephenson County, Illinois." *Journal of Geophysical Research,* 88, B9 (1983):7346–54.
11. KRYNINE, D. P., and W. R. JUDD. *Principles of Engineering Geology and Geotechnics.* New York: McGraw-Hill Book Company, 1957, 730 pp.
12. BRACE, W. F., B. W. PAULDING, JR., and C. SCHOLZ. "Dilatancy in the Fracture of Crystalline Rock." *Journal of Geophysical Research,* 71 (1966):3939–53.
13. STEWART, J. W. *Infiltration and Permeability of Weathered Crystalline Rocks, Georgia Nuclear Laboratory, Dawson County, Georgia.* U.S. Geological Survey Bulletin 1133-D, 1964.
14. HAIMSON, B. C., and T. W. DOE. "State of Stress, Permeability, and Fractures in the Precambrian Granite of Northern Illinois." *Journal of Geophysical Research,* 88, B9 (1983):7355–71.

**15.** SCHOELLER, H. *Les Eaux souterraines*. Paris: Mason et Cie, 1962.

**16.** KELLER, G. V. *Physical Properties of Tuffs in the Oak Spring Formation, Nevada*. U.S. Geological Survey Professional Paper 400-B, 1960.

**17.** MEINZER, O.E. *Outline of Groundwater Hydrology, with Definitions*. U.S. Geological Survey Water-Supply Paper 494, 1923, 71 pp.

**18.** JOHNSON, A. I. *Specific Yield—Compilation of Specific Yields for Various Materials*. U.S. Geological Survey Water-Supply Paper 1662-D, 1967, 74 pp.

**19.** JOHNSON, A. I., R. C. PRILL, and D. A. MORRIS. *Specific Yield—Column Drainage and Centrifuge Moisture Content*. U.S. Geological Survey Water-Supply Paper 1662-A, 1963, 60 pp.

**20.** PRILL, R. C., A. I. JOHNSON, and D. A. MORRIS. *Specific Yield—Laboratory Experiments Showing the Effect of Time on Column Drainage*. U.S. Geological Survey Water-Supply Paper 1662-B, 1965, 55 pp.

**21.** WENZEL, L. K. *Methods of Determining Permeability of Water-Bearing Materials*. U.S. Geological Survey Water-Supply Paper 887, 1942, 192 pp.

**22.** FERRIS, J. G., D. B. KNOWLES, R. H. BROWN, and R. W. STALLMAN. *Theory of Aquifer Tests*. U.S. Geological Survey Water-Supply Paper 1536-E, 1962, 174 pp.

**23.** PRICKETT, T. A. "Type-Curve Solution to Aquifer Tests under Water-Table Conditions." *Ground Water,* 3, no. 5 (1965).

**24.** DARCY, H. *Les Fontaines publiques de la ville de Dijon*. Paris: Victor Dalmont, 1856, 647 pp.

**25.** NORRIS, S. E., and R. E. FIDLER. *Relation of Permeability to Grain Size in a Glacial-Outwash Aquifer at Piketown, Ohio*. U.S. Geological Survey Professional Paper 525-D, 1965, pp. 203–6.

**26.** MASCH, F. E., and K. J. DENNY. "Grain-Size Distribution and Its Effect on the Permeability of Unconsolidated Sands." *Water Resources Research,* 2 (1966):665–77.

**27.** HAZEN, A. "Discussion: Dams on Sand Foundations." *Transactions, American Society of Civil Engineers,* 73 (1911):199

**28.** FETTER, C. W., JR. *Final Hydrogeologic Report, Seymour Recycling Corporation Hazardous Waste Site, Seymour, Indiana*. Submitted to United States Environmental Protection Agency, 1985, 171 pp. and appendices

**29.** PEYTON, G. R., and others. "Effective Porosity of Geologic Materials." Proceedings of the Twelfth Annual Research Symposium, U.S. Environmental Protection Agency, EPA/600/9-86/022 (1986):21–8.

**30.** RUMER, R. R., JR. "Resistance to Flow through Porous Media." In *Flow through Porous Media,* ed. R. J. M. DeWiest. New York: Academic Press, 1969, pp. 91–108.

**31.** SMITH, W. O. "Infiltration in Sands and Its Relation to Groundwater Recharge." *Water Resources Research,* 3 (1967):539–55.

**32.** ZIMMERMAN, U., et al. "Tracers Determine Movement of Soil Moisture and Evapotranspiration." *Science,* 152 (1960):346–47.

**33.** CHILDS, E. C. "Soil Moisture Theory." In *Advances in Hydroscience,* vol. 4, ed. V. T. Chow. New York: Academic Press, 1967, pp. 73–117.

34. WATSON, K. K. "A Recording Field Tensiometer with Rapid Response Characteristics." *Journal of Hydrology,* 5 (1967):33–39.
35. SWARTZENDRUBER, D. "The Flow of Water in Unsaturated Soils." In *Flow through Porous Media,* ed. R. J. M. DeWiest. New York: Academic Press, 1969, pp. 215–92.
36. RIPPLE, C. D., J. RUBIN, and T. E. A. VAN HYLCKAMA. *Estimating Steady-State Evaporation Rates from Bare Soils under Conditions of High Water Table.* U.S. Geological Survey Water-Supply Paper 2019-A, 1972, 39 pp.
37. HILLEL, D. *Soil and Water.* New York: Academic Press, 1971, 288 pp.
38. PHILIP, J. R. "Theory of Infiltration." In *Advances in Hydroscience,* vol. 5, ed. V. T. Chow. New York: Academic Press, 1969, pp. 215–96.
39. HUNTOON, P. "Cambrian Stratigraphic Nomenclature and Ground-Water Prospecting Failures on the Hualapai Plateau, Arizona." *Ground Water,* 15 (1977):426–33.
40. RAHN, P. H., and H. A. PAUL. "Hydrogeology of a Portion of the Sand Hills and Ogallala Aquifer, South Dakota and Nebraska." *Ground Water,* 13 (1975):428–37.
41. FETTER, C. W., JR. "Hydrogeology of the South Fork of Long Island, New York." *Bulletin, Geological Society of America,* 87 (1976):401–6.
42. WINTER, T. C. "The Interaction of Lakes with Variably Saturated Porous Media." *Water Resources Research,* 19 (1983):1203–18.
43. JACOB, C. E. "On the Flow of Water in an Elastic Artesian Aquifer." *Transactions, American Geophysical Union,* 21 (1940):574–86.
44. JACOB, C. E. "Flow of Groundwater." In *Engineering Hydraulics,* ed. H. Rouse. New York: John Wiley & Sons, 1950, pp. 321–86.
45. COOPER, H. H. "The Equation of Groundwater Flow in Fixed and Deforming Coordinates." *Journal of Geophysical Research,* 71 (1966):4785–90.

## PROBLEMS

1. A sample of silty sand has a volume of 70 $cm^3$. At the natural moisture content it weighs 150.7 grams. The sample was then saturated with water and reweighed to a weight of 169.7 grams. The sample was drained by gravity until it reached a constant weight of 146.2 grams. The sample was then oven-dried at 105° C until it reached a constant weight of 139.7 grams. Assume the unit weight of water is 1 gram/$cm^3$. Compute the following:
   - **a.** water content under natural conditions
   - **b.** volumetric water content under natural conditions
   - **c.** saturation ratio under natural conditions
   - **d.** porosity
   - **e.** specific yield
   - **f.** specific retention
   - **g.** water content at saturation
2. The hydraulic conductivity of a silty sand is $1.36 \times 10^{-5}$ cm/sec at 15° C. What is the intrinsic permeability in $cm^2$? At 15° C, density is 0.9991 gm/$cm^3$ and viscosity is 0.0114 poise. Assume $g$ is 980 cm/sec.

3. An aquifer has a specific yield of 0.19. During a drought period the following declines in the water table were noted:

| Area | Size | Decline |
|---|---|---|
| A | 14 mi$^2$ | 2.75 feet |
| B | 7 mi$^2$ | 3.56 feet |
| C | 28 mi$^2$ | 5.42 feet |
| D | 33 mi$^2$ | 7.78 feet |

What was the total volume of water represented by the decline in the water table?

4. An aquifer has three different formations. Formation A has a thickness of 30 feet and a hydraulic conductivity of 7 feet per day. Formation B has a thickness of 15 feet and a conductivity of 78 feet per day. Formation C has a thickness of 22 feet and a conductivity of 17 feet per day. Assume that each individual formation is isotropic and homogeneous. Compute both the overall horizontal and vertical conductivity.
5. Determine the effective grain size and uniformity coefficient of the sediments graphed on Figure 4.36.
6. Use the Hazen method to estimate the hydraulic conductivity of the sediments graphed on Figures 4.4, 4.5, and 4.36.

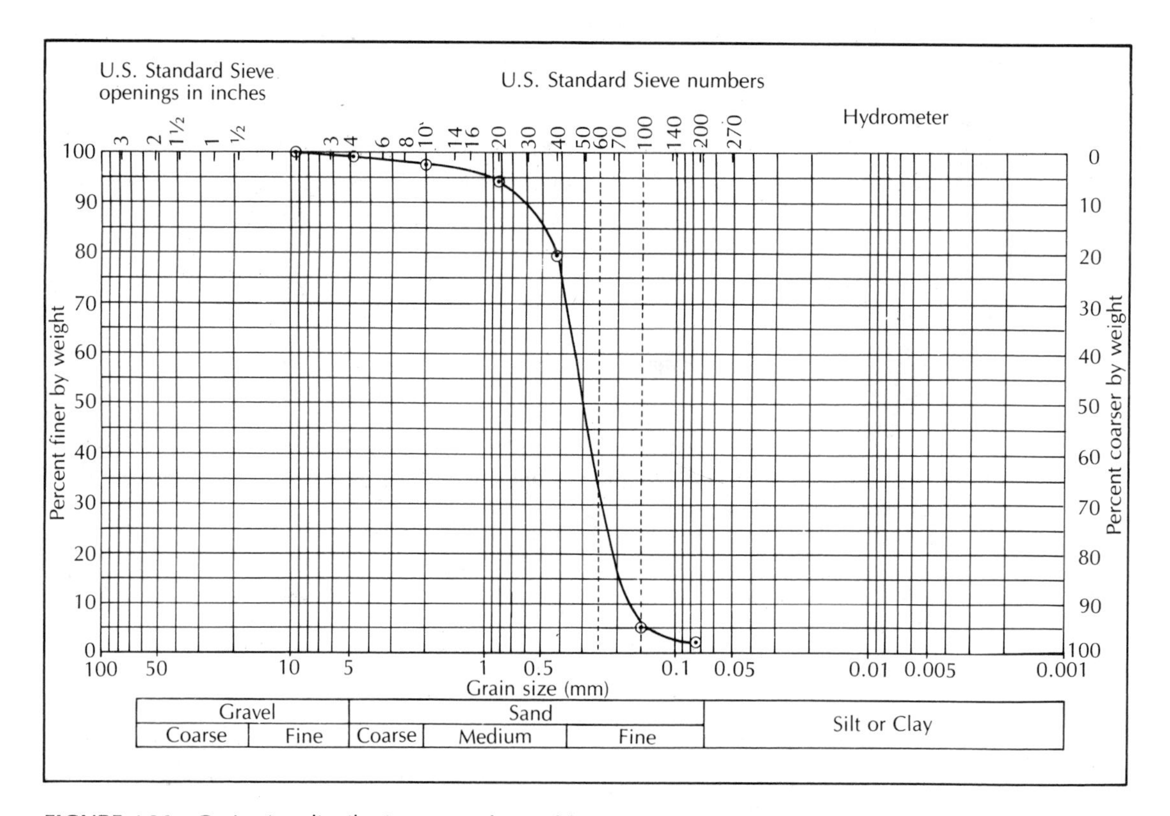


FIGURE 4.36 Grain-size distribution curve for problems 5 and 6.

# FIVE

# Principles of Ground-Water Flow

Just as above the earth, small drops form and these join others, till finally water descends in a body as rain, so too we must suppose that in the earth the water at first trickles together little by little and that the sources of rivers drip, as it were, out of the earth and then unite.

*Meteorologica*, Aristotle (384–322 B.C.)

## 5.1 INTRODUCTION

Ground water possesses energy in mechanical, thermal, and chemical forms. Because the amounts of energy vary spatially, ground water is forced to move from one region to another in nature's attempt to eliminate these energy differentials. The flow of ground water is thus controlled by the laws of physics and thermodynamics. To enable a separate examination of mechanical energy, we will make the simplifying assumption that the water is of nearly constant temperature. Thermal energy must be considered, however, in such applications as geothermal flow systems and burial of radioactive heat sources.

**Energy** is the capacity to do work, which implies that some resistance to change in movement must be overcome. **Work** is done when a force is applied to a fluid while the fluid is moving. Work is equal to the product of the net force exerted and the distance through which the force moves:

$$W = Fd \tag{5-1}$$

where

$W$ is the work
$F$ is the force
$d$ is the distance

The **force** acting on a body is equal to the product of mass of the body and its acceleration (Newton's second law of motion):

$$F = ma \tag{5-2}$$

where

$F$ is the force
$m$ is the mass
$a$ is the acceleration

The unit of **mass** is a kilogram, and if acceleration is expressed in meters per second per second (m/sec$^2$), then the unit of force is the newton (N), or kg·m/sec$^2$.

The **weight** of a body is the gravitational force exerted on it by the earth. The gravitational acceleration, $g$, varies from place to place, but is approximately 9.8 m/sec$^2$. The weight of a body is given by

$$w = mg \tag{5-3}$$

where

$w$ is the weight
$g$ is the acceleration of gravity

Weight has the same units as force. The mass of a body, the weight of which is 1 newton, at a place where $g = 9.80$ m/sec$^2$, is

$$m = \frac{w}{g} = \frac{1 \text{ N}}{9.80 \text{ m/sec}^2} = 0.102 \text{ kg}$$

In the English system, the pound is a unit force; hence, a unit of weight. The unit of mass is the slug. These definitions may seem somewhat confusing, since the pound as a unit of weight is often compared to the kilogram—a unit of mass. Balances and scales are calibrated in kilograms and grams. If we say that a sample weighs 0.13 kilogram, we are saying that the sample has a mass of 0.13 kilogram in the earth's gravitational field. The weight would be 1.27 newtons (0.29 pound). A kilogram has a mass of 1000 grams.

The **density** of a fluid is its mass per unit volume. Density is usually referred to by the Greek letter rho, $\rho$. It has units of g/cm$^3$ or kg/m$^3$:

$$\rho = m/V \tag{5-4}$$

where $V$ is the volume.

The **specific weight** of a substance is its weight per unit volume, indicated by the Greek letter gamma, $\gamma$. The units are newtons/m$^3$ or newtons/cm$^3$. Specific weight changes with location on the earth, since the value of $g$ changes:

$$\gamma = w/V \tag{5-5}$$

**EXAMPLE PROBLEM** A fluid has a density of 1.085 g/cm$^3$. If the acceleration of gravity is 9.81 m/sec$^2$, what is the specific weight of the fluid?

$$\gamma = w/V \qquad w = mg \qquad \rho = m/V$$
$$\gamma = \rho g$$
$$\rho = 1.085 \text{ g/cm}^3 \times 1/1000 \text{ kg/g} \times 10^6 \text{ cm}^3/\text{m}^3$$
$$= 1.085 \times 10^3 \text{ kg/m}^3$$
$$\gamma = 1.085 \times 10^3 \text{ kg/m}^3 \times 9.81 \text{ m/sec}^2$$
$$= 1.064 \times 10^4 \text{ N/m}^3$$

## 5.2 MECHANICAL ENERGY

There are a number of different types of mechanical energy recognized in classical physics. Of these, we will consider kinetic energy, gravitational potential energy, and energy of fluid pressures.

A moving body or fluid tends to remain in motion, according to Newtonian physics. This is because it possesses energy due to its motion—**kinetic energy.** This energy is equal to one-half the product of its mass and the square of the magnitude of the velocity:

$$E_k = \tfrac{1}{2}mv^2 \tag{5-6}$$

where

$E_k$ is the kinetic energy
$v$ is the velocity

If $m$ is in kilograms and $v$ in m/sec, then $E_k$ has the units of kg·m$^2$/sec$^2$ or newton-meters. The unit of energy is the joule, which is one newton-meter. The joule is also the unit of work.

Imagine that a weightless container filled with water of mass $m$ is moved vertically upward a distance, $z$, from some reference surface (a datum). Work is done in moving the mass of water upward. This work is equal to

$$W = Fd = (mg)z \tag{5-7}$$

where

$z$ is the elevation of the center of gravity of the fluid above the reference elevation
$m$ is the mass
$g$ is the acceleration of gravity

The units are kg × m/sec$^2$ × m, or newton-meters, with dimensions of $[ML^2T^{-2}]$.

The mass of water has now acquired energy equal to the work done in lifting the mass. This is a potential energy, due to the position of the fluid mass with respect to the datum. $E_g$ is **gravitational potential energy:**

$$W = E_g = mgz \tag{5-8}$$

A fluid mass has another source of potential energy owing to the **pressure** of the surrounding fluid acting upon it. Pressure is the force per unit area acting on a body:

$$P = F/A \tag{5-9}$$

where

$P$ is the pressure

$A$ is the cross-sectional area perpendicular to the direction of the force

The units of pressure are pascals, or newtons/m$^2$. A newton/m$^2$ is equal to a N-m/m$^3$, or joule/m$^3$. Pressure may thus be thought of as potential energy per unit volume of fluid.

For a unit volume of fluid, the mass, $m$, is numerically equal to the density, $\rho$, since density is defined as mass per unit volume. The total energy of the unit volume of fluid is the sum of the three components—kinetic, gravitational, and fluid-pressure energy:

$$E_{tv} = \tfrac{1}{2}\rho v^2 + \rho gz + P \tag{5-10}$$

where $E_{tv}$ is the total energy per unit volume.

If Equation 5-10 is divided by $\rho$, the result is total energy per unit mass, $E_{tm}$:

$$E_{tm} = \frac{v^2}{2} + gz + \frac{P}{\rho} \tag{5-11}$$

which is known as the **Bernoulli equation.** The derivation of the Bernoulli equation may be found in textbooks on fluid mechanics (1).

For steady flow of a frictionless, incompressible fluid along a smooth line of flow, the sum of the three components is a constant. Each term of Equation 5-11 has the units of $(L/T)^2$:

$$\frac{v^2}{2} + gz + \frac{P}{\rho} = \text{constant} \tag{5-12}$$

Steady flow indicates that the conditions do not change with time. The density of an incompressible fluid would not change with changes in pressure. A frictionless fluid would not require energy to overcome resistance to flow. An ideal fluid would have both of these characteristics; real fluids have neither one. Real fluids are compressible and do suffer frictional flow losses; however, Equation 5-12 is useful for purposes of comparing the components of mechanical energy.

If each term of Equation 5-12 is divided by $g$, the following expression results:

$$\frac{v^2}{2g} + z + \frac{P}{\rho g} = \text{constant} \tag{5-13}$$

This equation expresses all terms in units of energy per unit weight. These are joules/newtons, or m. Thus, Equation 5-13 has the advantage of having all units in length dimensions $(L)$. The first term of $v^2/2g$ is $(\text{m/sec})^2/(\text{m/sec}^2)$, or m; the second term, $z$, is already in m; and the third term, $P/\rho g$, is pascals/$(\text{kg/m}^3)(\text{m/sec}^2)$, or $(\text{N/m}^2)/(\text{kg/m}^3)(\text{m/sec}^2)$, which reduces to m. The sum of these three factors is the total mechanical energy per unit weight, known as the **hydraulic head,** $h$. This is usually measured in the field or laboratory in units of length.

## 5.3 HYDRAULIC HEAD

A **piezometer*** is used to measure the total energy of the fluid flowing through a pipe packed with sand, as shown in Figure 5.1. The piezometer is open at the top and bottom, and water rises in it in direct proportion to the total fluid energy at the point at which the bottom of the piezometer is open in the sand. At point A, which is at an elevation, $z$, above a datum, there is a fluid pressure, $P$. The fluid is flowing at a velocity, $v$. The total energy per unit mass can be found from Equation 5-11.

**EXAMPLE PROBLEM**

At a place where $g = 9.80 \text{ m/sec}^2$ the fluid pressure is $1500 \text{ N/m}^2$; the distance above a reference elevation is 0.75 m; and the fluid density is $1.02 \times 10^3 \text{ kg/m}^3$. The fluid is moving at a velocity of $10^{-6}$ m/sec. Find $E_{tm}$.

$$\begin{aligned} E_{tm} &= gz + \frac{P}{\rho} + \frac{v^2}{2} \\ &= 9.80 \text{ m/sec}^2 \times 0.75 \text{ m} + \frac{1500 \text{ N/m}^2}{1.02 \times 10^3 \text{ kg/m}^3} + \frac{(10^{-6})^2}{2}\frac{\text{m}^2}{\text{sec}^2} \\ &= 7.35 \text{ m}^2/\text{sec}^2 + 1.47 \text{ m}^2/\text{sec}^2 + 5.0 \times 10^{-13} \text{ m}^2/\text{sec}^2 \end{aligned}$$

The total energy per unit mass is $8.82 \text{ m}^2/\text{sec}^2$. The energy is almost exclusively in the pressure and gravitational potential energy terms, which are thirteen orders of magnitude greater than the value of kinetic energy.

*A piezometer is a small-diameter well with a very short well screen or section of slotted pipe at the end. It measures the hydraulic head at a point in the aquifer.

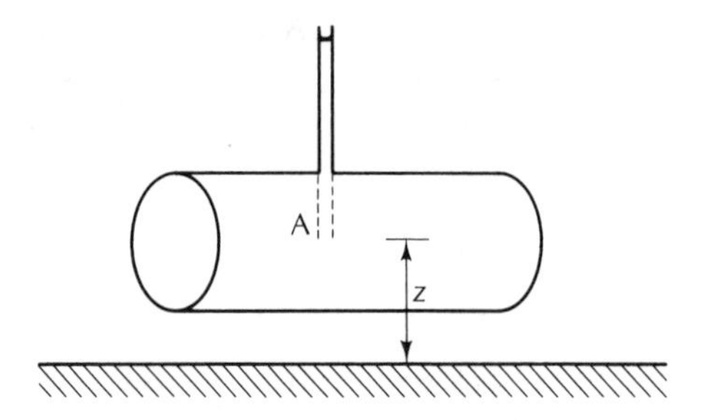


**FIGURE 5.1** Piezometer measuring fluid pressure and the elevation of water.

The preceding problem shows that the amount of energy developed as kinetic energy by flowing ground water is small. The velocity of ground water flowing in porous media under natural hydraulic gradients is very low. The example velocity of $10^{-6}$ meter per second results in a movement of 30 meters per year, which is typical for ground water.

Velocity components of energy may be safely ignored in ground-water flow because they are so much smaller than the other two terms. By dropping $v^2/2g$ from Equation 5-13, the total hydraulic head, $h$, is given by the formula

$$h = z + \frac{P}{\rho g} \tag{5-14}$$

Figure 5.2 shows the components of head. The head is the total mechanical energy per unit weight of water. For a fluid at rest, the pressure at a point is equal to the weight of the overlying water per unit cross-sectional area:

$$P = \rho g h_p \tag{5-15}$$

where $h_p$ is the height of the water column. Substituting into Equation 5-14, we see that

$$h = z + h_p \tag{5-16}$$

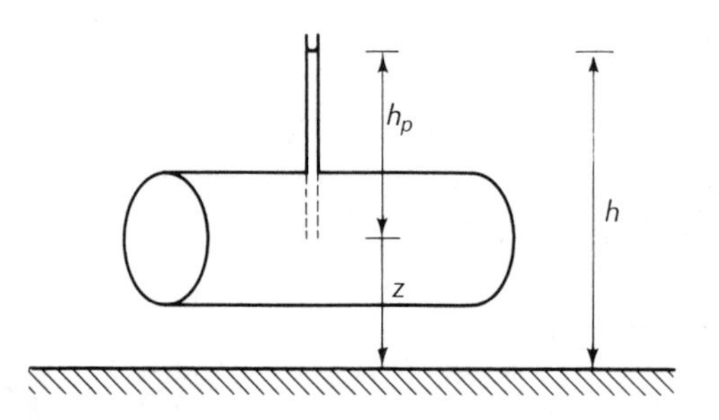


**FIGURE 5.2** Total head, $h$, elevation head, $z$, and pressure head, $h_p$.

The total hydraulic head is equal to the sum of the elevation head and the pressure head. The elevation and pressure heads, when used in the form of Equation 5-16, correlate with energy per unit weight of water with dimensions $L$.

**EXAMPLE PROBLEM** An aquifer has a total saturated depth of 50 meters. At some point in the aquifer, there are no vertical components of flow. What are the total elevation and pressure heads at the top and bottom of the aquifer?

The bottom of the aquifer is assumed to be the reference elevation. At the bottom of the aquifer, the elevation head is zero and the pressure head is 50 meters. At the top of the aquifer, the elevation head is 50 meters, but the pressure head is zero. (We are using gage pressure relating atmospheric pressure to an assumed zero.) The total head in both cases is 50 meters. At an elevation of 25 meters above the bottom, the pressure head is 25 meters, the elevation head 25 meters, and the total head also 50 meters.

## 5.4 FORCE POTENTIAL AND HYDRAULIC HEAD

In Equation 5-11 we showed the total mechanical energy per unit mass to be equal to the sum of the kinetic energy, elevation energy, and pressure. This total potential energy has been termed the **force potential,** and is indicated by the capital Greek letter phi, Φ (2):

$$\Phi = gz + \frac{P}{\rho} = gz + \frac{\rho g h_p}{\rho} = g(z + h_p) \qquad \textbf{(5-17)}$$

Since $z + h_p = h$, the hydraulic head

$$\Phi = gh \qquad \textbf{(5-18)}$$

The force potential is the driving impetus behind ground-water flow and is equal to the product of hydraulic head and the acceleration of gravity. Both force potential and hydraulic head are potentials. Hydraulic head is energy per unit weight and force potential is energy per unit mass.

Figure 5.3 shows a pipe filled with sand with water flowing through it from left to right. The pipe can be rotated to any inclination, with the discharge of water remaining constant. In Part A of the figure, the water flows from point 1 (of elevation $z_1$) to point 2 (of elevation $z_2$), $z_2$ being somewhat greater than $z_1$. In Part B, the slope has been reversed: rather than flowing uphill, the water now flows downhill. However, the fluid pressure head at point 2 ($h_{p2}$) is greater than at point 1 ($h_{p1}$). The fluid is thus moving from a region of low pressure to one of higher pressure. Clearly, neither elevation head nor pressure head alone controls ground-water motion. Part C of Figure 5.3 shows equal elevation heads, with

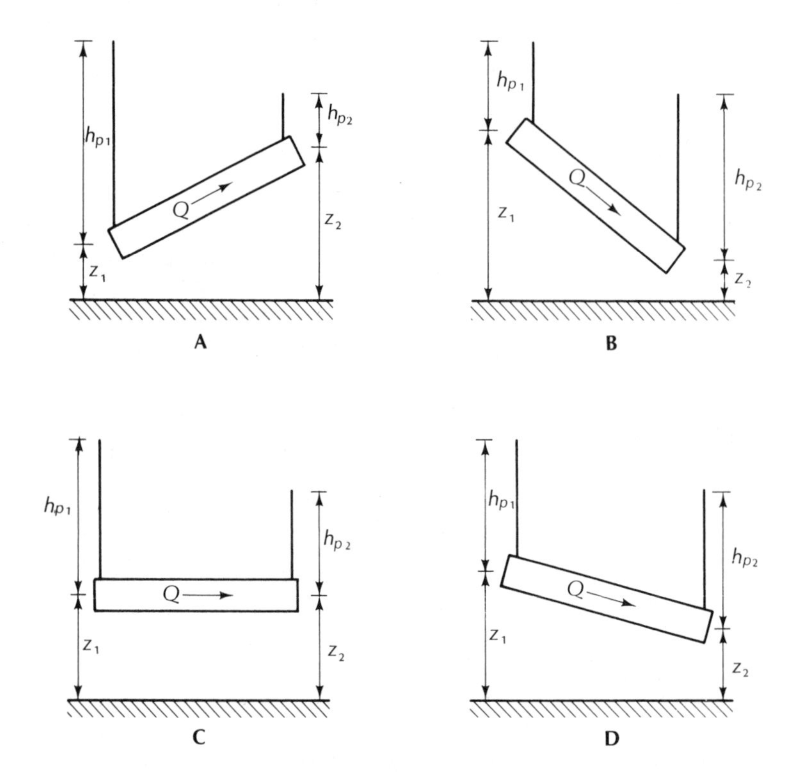


**FIGURE 5.3** Apparatus to demonstrate how changing the slope of a pipe packed with sand will change the components of elevation, $z$, and pressure, $h_p$, heads. The direction of flow, $Q$, is indicated by the arrow.

pressure head declining in the direction of flow. Part D has equal pressure heads, but elevation head declines in the direction of flow.

In this example the total hydraulic head showed the same decrease in the direction of flow. This would be true no matter what the inclination of the pipe, so long as other factors remained constant. Since the force potential is the controlling force in ground-water flow, this demonstrates that the proportion of pressure and elevation head is not a factor.

From Figure 5.3, we see that the force potential and, hence, hydraulic head, decrease in the direction of flow. As ground water moves, it encounters frictional resistance between the fluid and the porous media. The smaller the openings through which the fluid moves, the greater the friction. In overcoming the frictional resistance, some of the force potential is lost. It is transformed into heat (a lower form of energy). Thus, ground water is warmed slightly as it flows, and mechanical energy is converted to thermal energy. Under most circumstances, the resulting change in temperature is not measurable.

## 5.5 DARCY'S LAW

### 5.5.1 DARCY'S LAW IN TERMS OF FORCE AND POTENTIAL

In Section 4.3 it was shown that flow through a pipe filled with sand is proportional to the decrease in hydraulic head divided by the length of the pipe. This ratio is called the **hydraulic gradient.** It should now be apparent that the hydraulic head is the sum of the pressure head and the elevation head. Expressed in terms of hydraulic head, Darcy's law is

$$Q = -KA \frac{dh}{dl} \tag{5-19}$$

Since the fluid potential, $\Phi$, is equal to $gh$, Darcy's law can also be expressed in terms of fluid potential as (2)

$$Q = -\frac{KA}{g}\frac{d\Phi}{dl} \tag{5-20}$$

As expressed above, Darcy's law is in a one-dimensional form, as water flows through the pipe in only one direction. In later sections, we will examine various forms of Darcy's law for two and three directions.

### 5.5.2 THE APPLICABILITY OF DARCY'S LAW

When a fluid at rest starts to move, it must overcome resistance to flow due to the viscosity of the fluid. Slowly moving fluids are dominated by viscous forces. There is a low energy level and the resulting fluid flow is **laminar.** In laminar flow, molecules of water follow smooth lines, called **streamlines** (Figure 5.4A).

As the velocity of flow increases, the moving fluid gains kinetic energy. Eventually, the inertial forces due to movement are more influential than the viscous forces, and the fluid particles begin to rush past each other in an erratic fashion. The result is **turbulent flow,** in which the water molecules no longer move along parallel streamlines (Figure 5.4B).

The **Reynolds number** relates the four factors that determine whether the flow will be laminar or turbulent (1):

$$R = \frac{\rho v d}{\mu} \tag{5-21}$$

where

$R$ is the Reynolds number
$\rho$ is the fluid density

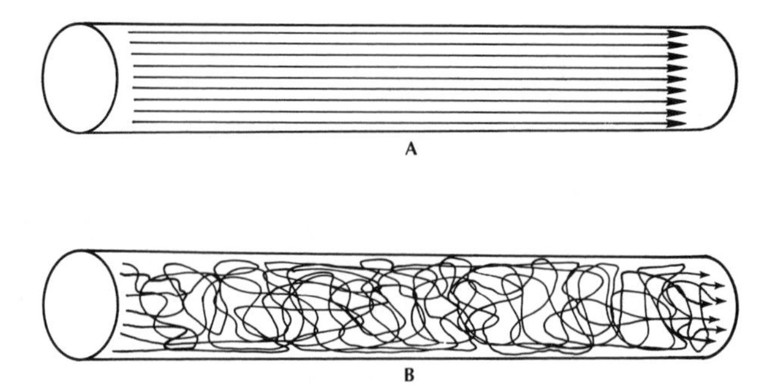


**FIGURE 5.4** **A.** Flow paths of molecules of water in laminar flow. **B.** Flow paths of molecules of water in turbulent flow.

$v$ is the fluid velocity

$d$ is the diameter of the passageway through which the fluid moves

$\mu$ is the viscosity

For open-channel or pipe flow, $d$ is simply the channel width or pipe diameter. In such cases, the transition from laminar to turbulent fluid flow occurs when the average velocity is such that $R$ exceeds a value of 2000 (1). For a porous medium, however, it is not easy to determine the value of $d$. Rather than an average or characteristic pore diameter, the average grain diameter is often used.

Turbulence in ground-water flow is difficult to detect. The inception of fluid turbulent flow in ground water has been reported at a Reynolds number ranging from 60 (3) to 600 (4). However, experimentation has shown that Darcy's law is valid only when conditions are such that the resistive forces of viscosity predominate. These conditions prevail when the Reynolds number is in the range of 1 to 10 (5–7). This means that Darcy's law applies only to very slowly moving ground waters. It is possible to have laminar ground-water flow, but under conditions such that the Reynolds number is so great as to invalidate Darcy's law. Under most natural ground-water conditions, the velocity is sufficiently low for Darcy's law to be valid. Exceptions might be areas of rock with large openings, such as solution openings and basalt flows. Likewise, areas of steep hydraulic gradients, such as the vicinity of a pumping well, might result in high velocities with a correspondingly high Reynolds number.

**EXAMPLE PROBLEM** A sand aquifer has a median grain diameter of 0.5 millimeter. For pure water at 15° C, what is the greatest velocity for which Darcy's law is valid?

At 15° C,

$$\rho = 0.999 \times 10^3 \text{ kg/m}^3$$
$$\mu = 1.15 \times 10^{-3} \text{ Pa-sec}$$

$$R = \frac{\rho v d}{\mu}$$

$$v = \frac{R\mu}{\rho \mathrm{d}}$$

A pascal-second (Pa-sec) is a N-sec/m$^2$. If $R$ cannot exceed 10,

$$v = \frac{10 \times 0.00115 \text{ N-sec/m}^2}{0.999 \times 10^3 \text{ kg/m}^3 \times 0.0005 \text{ m}}$$

$$= 0.023 \text{ m/sec}$$

Darcy's law will be valid for velocities equal to or less than 0.023 meter per second.

---

### 5.5.3 SPECIFIC DISCHARGE AND AVERAGE LINEAR VELOCITY

When water flows through an open channel or a pipe, the discharge, $Q$, is equal to the product of the velocity, $v$, and the cross-sectional area of flow, $A$:

$$Q = vA \quad \textbf{(5-22)}$$

Rearrangement of Equation 5-22 yields an expression for velocity,

$$v = Q/A \quad \textbf{(5-23)}$$

One can apply the same reasoning to Equation 5-19, Darcy's law, for flow through a porous medium:

$$v = \frac{Q}{A} = -K\frac{dh}{dl} \quad \textbf{(5-24)}$$

A moment's reflection will reveal that this velocity is not quite the same as the velocity of water flowing through an open pipe. The discharge is measured as water coming from the pipe. In an open pipe, the cross-sectional area of flow inside the pipe is equivalent to the area of the end of the pipe. However, if the pipe is filled with sand, the open area through which water may flow is much smaller than the cross-sectional area of the pipe. The velocity of flow determined by Equation 5-24 is termed the **specific discharge.*** It is an apparent velocity, representing the velocity at which water would move through an aquifer if the aquifer were an open conduit.

The cross-sectional area of flow for a porous medium is actually much smaller than the dimensions of the aquifer. It is equal to the product of the

---

*The terms discharge velocity and Darcian velocity are synonyms for specific discharge. It would be best to avoid these as their use implies that ground water is moving at this velocity.

effective porosity of the aquifer material and the physical dimensions. Water can move only through the pore spaces. Moreover, part of the pore space is occupied by stagnant water, which clings to the rock material. The effective porosity is that portion of the pore space through which saturated flow occurs.

To find the velocity at which water is actually moving, the specific discharge is divided by the effective porosity to account for the actual open space available for the flow. The result is the **average linear velocity***—a true velocity representing the rate at which the water actually moves through the pore spaces:

$$V_x = \frac{Q}{n_e A} = -\frac{K dh}{n_e dl} \tag{5-25A}$$

where

$V_x$ is the average linear velocity
$n_e$ is the effective porosity

Equation 5-25A does not take into account the factors that account for **dispersion** in flowing ground water. Dispersion is the phenomenon that results because ground water flows through different pores at different rates and various flow paths vary in length. Dispersion is discussed in detail in Section 10.6.4.

Because Equation 5-25A does not include a dispersion factor, it cannot be used to predict the average linear rate of movement of a solute front that is moving at the same rate as the flowing ground water. This is especially true for fine-grained materials, where the process of **diffusion** may be important in the movement of solute from an area of greater to lesser concentration. Diffusion is discussed in Section 10.6.2.

Equation 5-25B is a modification of Darcy's law that can be used to make an estimate of the average linear velocity of a solute front, $V_s$. An experimentally determined Darcian pore factor, *dpf,* is multiplied by the effective porosity, $n_e$, to determine an effective Darcian porosity, $n_{ed}$. Figure 5.5 is an experimental curve that gives the Darcian pore factor as a function of hydraulic conductivity.

$$V_s = -\frac{K dh}{n_{ed} dl} \tag{5-25B}$$

where

$V_s$ is the average linear solute front velocity
$n_{ed}$ is the effective Darcian porosity, which is equal to the Darcian pore factor, *dpf,* times the effective porosity

Equation 5-25B is a useful means of obtaining an estimate of the rate of movement of a solute in flowing ground water. It is an empirical method with no

*The term seepage velocity is a synonym for average linear velocity.

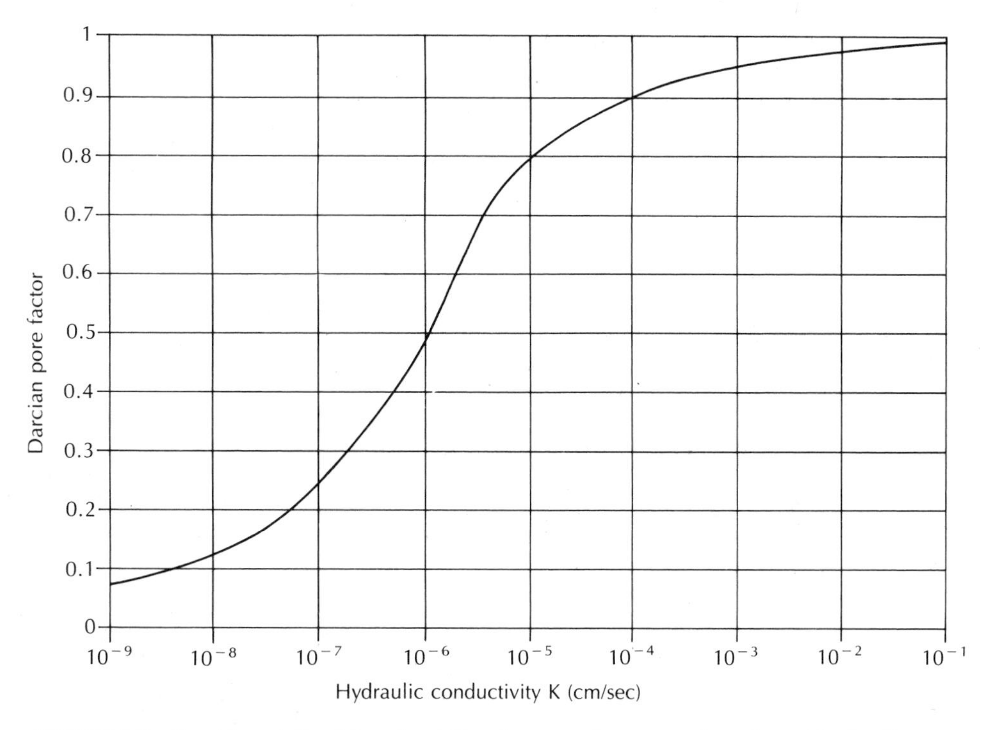


**FIGURE 5.5** Experimentally determined Darcian pore factor as a function of hydraulic conductivity of a sediment. Source: R. A. Griffin, Illinois State Geologic Survey.

theoretical foundation. In Chapter 10 the advection-dispersion equations (Equations 10-7 and 10-8) are given. This is a more accurate method of determining the rate of movement of a solute. In addition, the impact of processes that act to remove the solute from the ground water can be included. However, it is mathematically more complex than Equation 5-25B and requires a knowledge of more hydrodynamic parameters of the formation.

**EXAMPLE PROBLEM** Compute the specific discharge and average linear velocity for water flowing through a pipe filled with sand, with $K = 10^{-4}$ centimeter per second, $dh/dl = 0.01$, $A = 75$ square centimeters, and $n_e = 0.22$.

Specific discharge:

$$v = \frac{Q}{A} = K\frac{dh}{dl}$$

$$= 10^{-4} \text{ cm/sec} \times 0.01$$

$$= 10^{-6} \text{ cm/sec}$$

Average linear velocity:

$$V_x = \frac{Q}{n_e A} = \frac{Kdh}{n_e dl}$$

$$= \frac{1}{0.22} \times 10^{-4} \times 0.01$$

$$= 4.55 \times 10^{-6} \text{ cm/sec}$$

At a rate of $4.55 \times 10^{-6}$ centimeter per second, the water would move 1.43 meters in one year.

## 5.6 PERMEAMETERS

The value of the hydraulic conductivity of earth materials can be measured in the laboratory. Not surprisingly, the devices used to do this are called **permeameters.**

Permeameters all have some type of a chamber to hold a sample of rock or sediment. Rock permeameters hold a core of solid rock, usually cylindrical. Unconsolidated samples may be remolded into the permeameter chamber. It is also possible to make permeability analyses of "undisturbed" samples of unconsolidated materials if they are left in the field-sampling tubes, which become the permeameter sample chambers. If sediments are repacked into the permeameter, they will yield values of hydraulic conductivity that only approximate the value of $K$ for undisturbed material. Recompacted hydraulic conductivities depend upon the density to which the sample is compacted.

The **constant-head permeameter** is used for noncohesive sediments, such as sand and rocks. A chamber with an overflow provides a supply of water at a constant head. Water moves through the sample at a steady rate. The hydraulic conductivity is determined from a variation of Darcy's law:

$$K = \frac{VL}{Ath} \qquad \textbf{(5-26)}$$

where

$V$ is the volume of water discharging in time $t$
$L$ is the length of the sample
$A$ is the cross-sectional area of the sample
$h$ is the hydraulic head
$K$ is the hydraulic conductivity

A constant-head permeameter is illustrated in Figure 5.6. It is critical in such permeameters to have hydraulic gradients approaching those in the field.

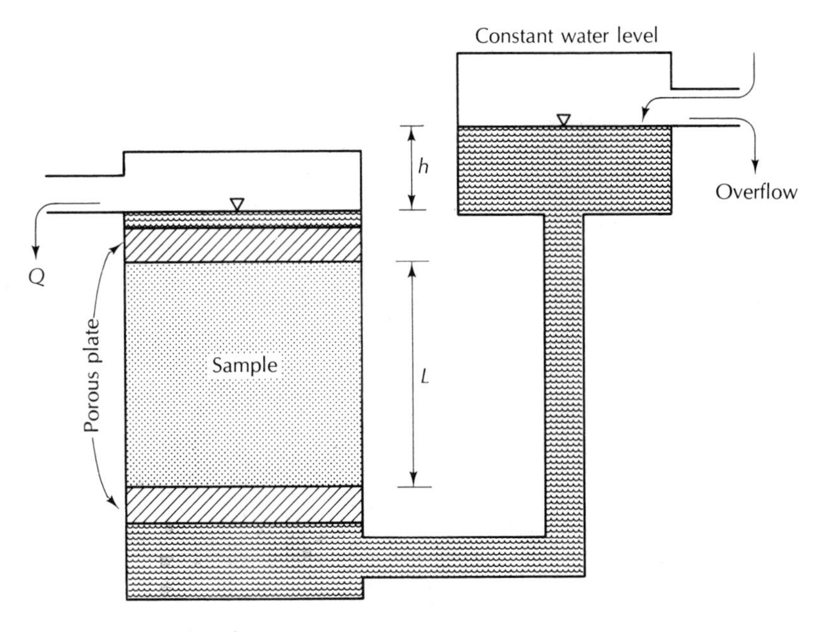


FIGURE 5.6 Constant-head permeameter apparatus.

The head should never be more than about 0.5 of the sample length. Some commercial permeameters permit heads of up to ten times the sample length. Under such conditions, the Reynolds number may become so high that Darcy's law is invalidated. If the permeameter is designed for upward flow, too great an upward-flow velocity may result in quicksand conditions in the permeameter.

For cohesive sediments with low conductivities, a **falling-head permeameter** is used (Figure 5.7). A much smaller volume of water moves through the sample. A falling-head tube is attached to the permeameter. The initial water level above the outlet in the falling-head tube, $h_0$, is noted. After some time period, $t$ (generally several hours), the water level, $h$, is again measured. The inside diameter of the falling-head tube, $d_t$, the length of the sample, $L$, and the diameter of the sample, $d_c$, must also be known. The conductivity, $K$, is found by the formula

$$K = \frac{d_t^2 L}{d_c^2 t} \ln\left(\frac{h_0}{h}\right) \tag{5-27}$$

In using any permeameter, it is critical that the sample be completely saturated. Air bubbles in the sample will reduce the cross-sectional flow area, resulting in lowered measurements of conductivity. The sample must also be tightly pressed against the sidewall of the chamber. If it is not, water may move along the sidewall, avoiding the porous medium. In this case, measurements of conductivity may be too high.

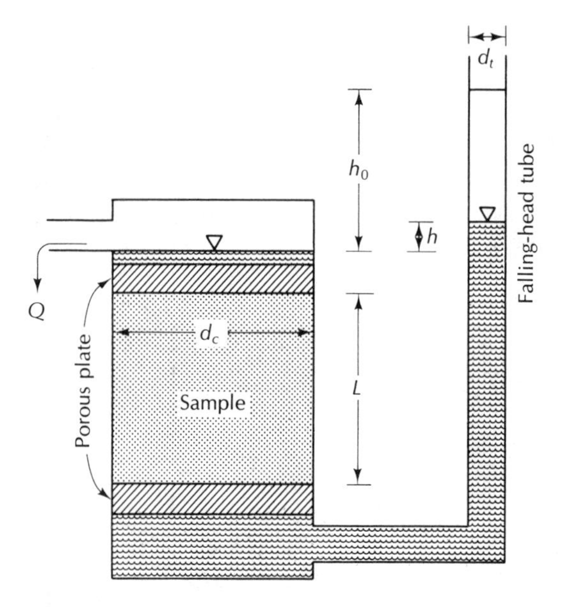


FIGURE 5.7 Falling-head permeameter apparatus.

---

**EXAMPLE PROBLEM** A constant-head permeameter has a sample of medium-grained sand 15 centimeters in length and 25 square centimeters in cross-sectional area. With a head of 5 centimeters, a total of 100 milliliters of water is collected in 12 minutes. Find the hydraulic conductivity.

$$K = \frac{VL}{Ath}$$

$$K = \frac{100 \text{ cm}^3 \times 15 \text{ cm}}{25 \text{ cm}^2 \times 12 \text{ min} \times 60 \text{ sec/min} \times 5 \text{ cm}}$$

$$= 1.67 \times 10^{-2} \text{ cm/sec or } 14.4 \text{ m/day}$$

Note: Units must be such that the resultant answer is in the desired length/time units.

---

**EXAMPLE PROBLEM** A falling-head permeameter containing a silty, fine sand has a falling-head tube diameter of 2 centimeters, a sample diameter of 10 centimeters, and a flow length of 15 centimeters. The initial head is 5 centimeters. It falls to 0.5 centimeter over a period of 528 minutes. Find the hydraulic conductivity.

$$K = \frac{d_t^2 L}{d_c^2 t} \ln\left(\frac{h_0}{h}\right)$$

$$= \frac{2^2 \text{ cm}^2}{10^2 \text{ cm}^2} \times \frac{15 \text{ cm}}{528 \text{ min} \times 60 \text{ sec/min}} \times \ln \frac{5 \text{ cm}}{0.5 \text{ cm}}$$

$$= 4.36 \times 10^{-5} \text{ cm/sec or } 3.77 \times 10^{-2} \text{ m/day}$$

## 5.7 EQUATIONS OF GROUND-WATER FLOW*

### 5.7.1 CONFINED AQUIFERS

The flow of fluids through porous media is governed by the laws of physics. As such, it can be described by differential equations. Since the flow is a function of several variables, it is usually described by partial differential equations in which the spatial coordinates, $x$, $y$, and $z$, and time, $t$, are the independent variables.

In deriving the equations, the laws of conservation for mass and energy are employed. The **law of mass conservation,** or **continuity principle,** states that there can be no net change in the mass of a fluid contained in a small volume of an aquifer. Any change in mass flowing into the small volume of the aquifer must be balanced by a corresponding change in mass flux out of the volume, or a change in the mass stored in the volume, or both. The **law of conservation of energy** is also known as the **first law of thermodynamics.** It states that within any closed system there is a constant amount of energy, which can be neither lost nor increased. It can, however, change form. The **second law of thermodynamics** implies that when energy changes forms, it tends to go from a more useful form, such as mechanical energy, to a less useful form, such as heat. Based upon these principles and Darcy's law, the main equations of ground-water flow have been derived (8–11).

We will consider a very small part of the aquifer, called a **control volume.** The three sides are of lengths $dx$, $dy$, and $dz$, respectively. The area of the faces normal to the $x$-axis is $dydz$; the area of the faces normal to the $z$-axis is $dxdy$ (Figure 5.8).

Assume the aquifer is homogeneous and isotropic. The fluid moves in only one direction through the control volume. However, the actual fluid motion can be subdivided on the basis of the components of flow parallel to the three principal axes. If $q$ is flow per unit cross-sectional area, $\rho_w q_x$ is the portion parallel to the $x$-axis etc., where $\rho_w$ is the fluid density.

The mass flux into the control volume is $\rho_w q_x \, dydz$ along the $x$-axis. The mass flux† out of the control volume is $\rho_w q_x \, dydz + \frac{\delta}{\delta x}(\rho_w q_x) \, dx \, dydz$. The net

*The main equation of ground-water flow is derived in this section following a method used by Jacob (8, 9) and modified by Domenico (10). Those not familiar with differential calculus can skip this material without compromising practical understanding.

†Flux is a rate of flow.

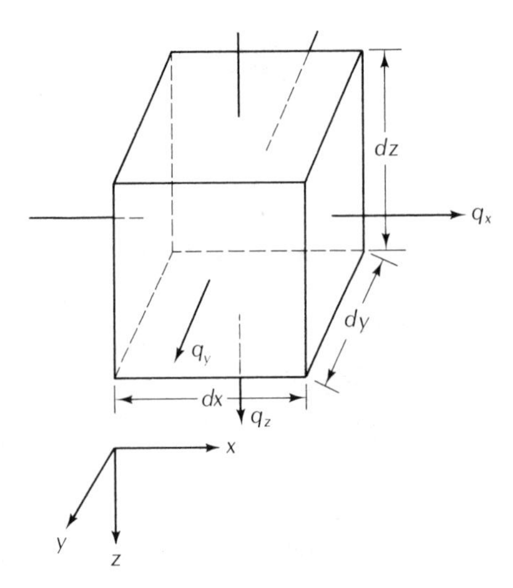


**FIGURE 5.8** Control volume for flow through a confined aquifer.

accumulation in the control volume due to movement parallel to the $x$-axis is equal to the inflow less the outflow, or $-\frac{\delta}{\delta x}(\rho_w q_x)\, dx\, dydz$. Since there are flow components along all three axes, similar terms can be determined for the other two directions: $-\frac{\delta}{\delta y}(\rho_w q_y)\, dy\, dxdz$ and $-\frac{\delta}{\delta z}(\rho_w q_z)\, dz\, dxdy$. Combining these three terms yields the net total accumulation of mass in the control volume:

$$-\left(\frac{\delta}{\delta x}\,\rho_w q_x + \frac{\delta}{\delta y}\,\rho_w q_y + \frac{\delta}{\delta z}\,\rho_w q_z\right) dxdydz \tag{5-28}$$

The volume of water in the control volume is equal to $n\; dxdydz$, where $n$ is the porosity. The initial mass of the water is thus $\rho_w n\; dxdydz$. The volume of solid material is $(1 - n)\; dxdydz$. Any change in the mass of water, $M$, with respect to time is given by

$$\frac{\delta M}{\delta t} = \frac{\delta}{\delta t}(\rho_w n\; dxdydz) \tag{5-29}$$

As the pressure in the control volume changes, the fluid density will change, as will the porosity of the aquifer. The compressibility of water, $\beta$, is defined as the rate of change in density with a change in pressure, $P$:

$$\beta dP = \frac{d\rho_w}{\rho_w} \tag{5-30}$$

The aquifer also changes in volume with a change in pressure. We will assume the only change is vertical. The aquifer compressibility, $\alpha$, is given by

$$\alpha dP = \frac{d(dz)}{dz} \tag{5-31}$$

As the aquifer compresses or expands, $n$ will change, but the volume of solids, $V_s$, will be constant. Likewise, if the only deformation is in the $z$-direction, $d(dx)$ and $d(dy)$ will equal zero:

$$dV_s = 0 = d[(1 - n)\, dxdydz] \tag{5-32}$$

Differentiation of Equation 5-32 yields

$$dzdn = (1 - n)d(dz) \tag{5-33}$$

and

$$dn = \frac{(1 - n)d(dz)}{dz} \tag{5-34}$$

The pressure, $P$, at a point in the aquifer is equal to $P_0 + \rho_w gh$, where $P_0$ is atmospheric pressure, a constant, and $h$ is the height of a column of water above the point. Therefore, $dP = \rho_w g\, dh$, and Equations 5-30 and 5-31 become

$$d\rho_w = \rho_w \beta(\rho_w g\, dh) \tag{5-35}$$

and

$$d(dz) = dz\alpha(\rho_w g\, dh) \tag{5-36}$$

Equation 5-34 can be rearranged if $d(dz)$ is replaced by Equation 5-36:

$$dn = (1 - n)\alpha\rho_w g\, dh \tag{5-37}$$

If $dx$ and $dy$ are constant, the equation for change of mass with time in the control volume, Equation 5-29, can be expressed as

$$\frac{\delta M}{\delta t} = \left[\rho_w n \frac{\delta(dz)}{\delta t} + \rho_w dz \frac{\delta n}{\delta t} + ndz \frac{\delta \rho_w}{\delta t}\right] dxdy \tag{5-38}$$

Substitution of Equations 5-35, 5-36, and 5-37 into Equation 5-38 yields, after minor manipulation,

$$\frac{\delta M}{\delta t} = (\alpha\rho_w g + n\beta\rho_w g)\rho_w\, dxdydz \frac{\delta h}{\delta t} \tag{5-39}$$

The net accumulation of material expressed as Equation 5-28 is equal to Equation 5-39, the change in mass with time:

$$-\left[\frac{\delta(q_x)}{\delta x} + \frac{\delta(q_y)}{\delta y} + \frac{\delta(q_z)}{\delta z}\right] \rho_w\, dxdydz = (\alpha\rho_w g + n\beta\rho_w g)\rho_w\, dxdydz \frac{\delta h}{\delta t} \tag{5-40}$$

From Darcy's law,

$$q_x = -K\frac{\delta h}{\delta x} \quad \textbf{(5-41)}$$

$$q_y = -K\frac{\delta h}{\delta y} \quad \textbf{(5-42)}$$

and

$$q_z = -K\frac{\delta h}{\delta z} \quad \textbf{(5-43)}$$

Substituting these into Equation 5-40 yields the main equation of flow for a confined aquifer:

$$K\left(\frac{\delta^2 h}{\delta x^2} + \frac{\delta^2 h}{\delta y^2} + \frac{\delta^2 h}{\delta z^2}\right) = (\alpha\rho_w g + n\beta\rho_w g)\frac{\delta h}{\delta t} \quad \textbf{(5-44)}$$

which is a general equation for flow in three dimensions for an isotropic, homogeneous porous medium. For two-dimensional flow with no vertical components, the equation can be rearranged and terms introduced from Equations 4-18 and 4-19 for the storativity, $[S = b(\alpha\rho_w g + n\beta\rho_w g)]$, and from Equation 4-16 for the transmissivity, $(T = Kb)$, where $b$ is the aquifer thickness:

$$\frac{\delta^2 h}{\delta x^2} + \frac{\delta^2 h}{\delta y^2} = \frac{S}{T}\frac{\delta h}{\delta t} \quad \textbf{(5-45)}$$

In steady-state flow, there is no change in head with time; for example, in cases in which there is no change in the position or slope of the water table. Under such conditions, time is not one of the independent variables, and steady flow is described by the three-dimensional partial differential equation known as the **Laplace equation:**

$$\frac{\delta^2 h}{\delta x^2} + \frac{\delta^2 h}{\delta y^2} + \frac{\delta^2 h}{\delta z^2} = 0 \quad \textbf{(5-46)}$$

The preceding equations are based on the assumption that all flow comes from water stored in the aquifer. In the field, it is more often than not the case that significant flow is generated from leakage into the aquifer through overlying or underlying confining layers. We will consider the leakage to appear in the control volume as horizontal flow. This assumption is justified on the grounds that the conductivity of the aquifer is usually orders of magnitude greater than that of the confining layer. The law of refraction indicates that, for these conditions, flow in the confining layer will be nearly vertical if flow in the aquifer is horizontal.

The leakage rate, or rate of accumulation, is designated as $w$. The general equation of flow (in two dimensions, since horizontal flow was assumed) is given by

$$\frac{\delta^2 h}{\delta x^2} + \frac{\delta^2 h}{\delta y^2} + \frac{w}{T} = \frac{S}{T}\frac{\delta h}{\delta t} \tag{5-47}$$

The rate of vertical movement, $e$, can be determined from Darcy's law. If the head at the top of the aquitard is $h_0$ and the head in the aquifer just below the aquitard is $h$, the aquitard has a thickness $b'$ and a conductivity (vertical) of $K'$:

$$e = K' \frac{(h_0 - h)}{b'} \tag{5-48}$$

### 5.7.2 UNCONFINED AQUIFERS

Water is derived from storage in water-table aquifers by vertical drainage of water in the pores. This drainage results in a decline in the position of the water table near a pumping well as time progresses. In the case of a confined aquifer, although the potentiometric surface declined, the saturated thickness of the aquifer remained constant. In the case of an unconfined aquifer, the saturated thickness can change with time. Under such conditions, the ability of the aquifer to transmit water—the transmissivity—changes, as it is the product of the conductivity $K$ and the saturated thickness $h$ (assuming that $h$ is measured from the horizontal base of the aquifer).

The general flow equation for two-dimensional unconfined flow is known as the **Boussinesq equation** (12):

$$\frac{\delta}{\delta x}\left(h\frac{\delta h}{\delta x}\right) + \frac{\delta}{\delta y}\left(h\frac{\delta h}{\delta y}\right) = \frac{S_y}{K}\frac{\delta h}{\delta t} \tag{5-49}$$

where $S_y$ is specific yield. This equation is a type of differential equation that cannot be solved using calculus, except in some very specific cases. In mathematical terms, it is nonlinear.

If the drawdown in the aquifer is very small compared with the saturated thickness, the variable thickness, $h$, can be replaced with an average thickness, $b$, that is assumed to be constant over the aquifer. The Boussinesq equation can thus be linearized by this approximation to the form

$$\frac{\delta^2 h}{\delta x^2} + \frac{\delta^2 h}{\delta y^2} = \frac{S_y}{Kb}\frac{\delta h}{\delta t} \tag{5-50}$$

which has the same form as Equation 5-45.

## 5.8 SOLUTION OF FLOW EQUATIONS

The flow of water in an aquifer can be mathematically described by Equation 5-44, 5-45, 5-46, 5-47, or 5-50. These are all partial differential equations in which the head, $h$, is described in terms of the variables $x$, $y$, $z$, and $t$. They are solved by means of a mathematical model consisting of the applicable governing flow equation, equations describing the hydraulic head at each of the boundaries of the aquifer, and equations describing the initial conditions of head in the aquifer.

If the aquifer is homogeneous and isotropic, and the boundaries can be described with algebraic equations, then the mathematical model can be solved by use of an analytical solution based on integral calculus. However, if the aquifer does not correspond to those conditions (e.g., a layered aquifer), then a numerical solution to the mathematical model is needed. Numerical solutions are based on the concept that the partial differential equation can be replaced by a similar equation that can be solved using arithmetic. Likewise, the equations governing initial and boundary conditions are replaced by numerical statements of these conditions. Numerical solutions are typically solved on digital computers. The use of digital-computer models is treated in Chapter 13.

## 5.9 GRADIENT OF HYDRAULIC HEAD

The potential energy, or force potential, $\Phi$, of ground water consists of two parts: elevation and pressure. It is equal to the product of the acceleration of gravity and the total head (2), and represents mechanical energy per unit mass:

$$\Phi = gh \tag{5-51}$$

Force potential is a physical quantity. To obtain it one needs only to measure the heads in an aquifer with piezometers and multiply the results by the acceleration of gravity. If a point in an aquifer has a head of 15.1 meters, and the value of $g$ is 9.81 m/sec$^2$, then $\Phi$ is $15.1 \times 9.81 = 148.1$ m$^2$/sec$^2$. For practical purposes, as $g$ is usually constant throughout an area, most field problems are solved in terms of hydraulic head, $h$.

If the value of $h$ is variable in an aquifer, a contour map may be made showing lines of equal value of $h$ (equipotential lines). Such a map is similar to a topographic map of the land-surface elevation. In three-dimensional cases, one deals with surfaces of equal value of $h$ (equipotential surfaces).

A **potentiometric map** is a contour map of the potentiometric surface (surface of elevation and pressure head summed) of a particular hydrogeologic unit. A **water-table map** is a specific type of a potentiometric surface map for an unconfined aquifer. The water-table map shows the surface where the pressure head is zero (equal to atmospheric) but elevation head varies.

Figure 5.9 shows a family of equipotential lines for a two-dimensional field. Also shown is a vector, known as the gradient of $h$ (grad $h$). Remember, a vector is a directed line segment, so grad $h$ has a magnitude and a direction. It

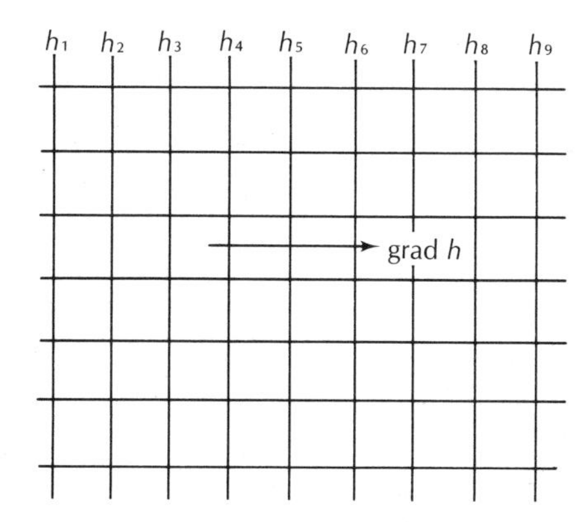


**FIGURE 5.9** Equipotential lines in a two-dimensional flow field and the gradient of $h$.

is roughly analogous to the maximum slope of the potential field. In the notation of differential calculus,

$$\text{grad } h = \frac{dh}{ds} \tag{5-52}$$

Grad $h$ has a direction perpendicular to the equipotential lines.

For isotropic aquifers, the value of $K$ is equal in all directions. In such aquifers, the fluid flow will be parallel to grad $\Phi$, which means that it will also be perpendicular to the equipotential lines. By contrast, anisotropic media have differing values of $K$ in different directions. Under such conditions, the flowlines will not be parallel, but rather oblique to the direction of grad $h$. They will thus cross the equipotential lines at other than a 90-degree angle; however, they cannot be parallel to the equipotential lines.

If the potential is the same everywhere, it will be manifest in a condition such as a flat water table. In this case, grad $h$ equals zero, since there is no slope to $h$. There will be no ground-water flow, since grad $h$ must have a positive value before ground water will move.

## 5.10 FLOW NETS

The Laplace equation (5-46) may be solved for some types of problems by means of the construction of a **flow net** showing equipotential lines and streamlines (13, 14). It is an appropriate method for two-dimensional problems in an isotropic medium; especially adaptable to problems in which the flow boundaries are known before the solution is attempted. Thus, hydrogeological field work is necessary before a solution is undertaken.

Flow nets are bounded by either equipotential lines or streamlines. In Figure 5.10, a flow net is drawn for seepage through an earthen dam resting on

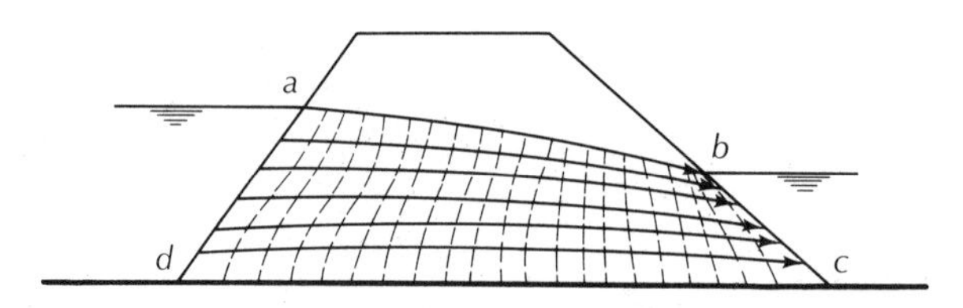


**FIGURE 5.10** A flow net for seepage through an earthen dam resting on an impervious surface.

an impervious surface. The water table in the dam is represented by Line a-b. It is a streamline. Line d-c is also a streamline. Line a-d is an equipotential line, with the head equal to the depth of the impounded water. Line b-c is also an equipotential line, the head being equal to the depth of the tailwater.

A flow net is made by trial-and-error process. The boundaries of the flow region are carefully drawn to scale and a few streamlines sketched on the drawing. They should be evenly spaced across the width of the flow region. These are only a few of the infinite number of streamlines. In many flow nets, at least two of them will be boundaries of the flow region. The streamlines will begin and end at equipotential surfaces. They must intersect these equipotential surfaces at right angles.

The final step is to add the intermediate equipotential lines. They must intersect all streamlines at right angles, including the boundary streamlines. The equipotential lines and streamlines should bound areas that are approximately square. At certain odd angles in the boundaries, these areas may be three- or five-sided. Most beginners at the art of flow-net construction will find an ample supply of paper, pencils, and erasers essential. As a check on the quality of the flow net, the diagonals of the squares can be drawn. These should form smooth curves that intersect each other at right angles. This should, of course, be done on a copy of the final product.

In addition to presenting a graphic display of the ground-water flow directions and potential distribution, the completed flow net can be used to determine the quantity of water flowing by the following formula:

$$Q = \frac{Kph}{f} \tag{5-53}$$

where

$Q$ is the total volume discharge per unit width of aquifer

$K$ is the hydraulic conductivity

$p$ is the number of flow paths bounded by adjacent pairs of streamlines

$h$ is the total head loss over the length of the streamlines

$f$ is the number of squares bounded by any two adjacent streamlines and covering the entire length of flow

**EXAMPLE PROBLEM** In Figure 5.10 there is an earthen dam 13 meters across and 7.5 meters high. The impounded water is 6.2 meters deep, while the tailwater is 2.2 meters deep. The dam is 72 meters long. If the hydraulic conductivity is $6.1 \times 10^{-4}$ centimeter per second, what is the seepage through the dam?

$$K = 6.1 \times 10^{-4} \text{ cm/sec} = 0.527 \text{ m/day}$$

From the flow net, the total head loss, $h$, is $6.2 - 2.2 = 4.0$ meters. There are 6 flow channels $(p)$ and 21 squares of equipotential drop along each flow path $(f)$:

$$\begin{aligned} Q &= \frac{Kph}{f} \times \text{dam length} \\ &= \frac{0.527 \text{ m/day} \times 6 \times 4 \text{ m}}{21} \times \text{dam length} \\ &= 0.60 \text{ m}^3\text{/day for each meter of dam length} \\ &= 0.60 \times 72 \\ &= 43.4 \text{ m}^3\text{/day for the entire 72-meter length of the dam} \end{aligned}$$

The preceding problem illustrates one of the pitfalls of two-dimensional problem solutions. It must be recognized that two-dimensional problems imply a third dimension, with an axis of symmetry perpendicular to the two-dimensional representation. The width of total flow prependicular to this axis must be included to determine the total volume of flow. An alternative method is to state flow in terms of discharge per unit width. For an aquifer, flow might be stated in cubic meters per day per kilometer width of the aquifer (measured orthogonal to the direction of flow).

Further examples of flow nets are given in Figure 5.11. These represent flow beneath a dam with a cutoff and flow beneath a sheet pile. (The cutoff wall beneath the dam and the sheet piling driven into the sediments lengthen the flow path. This decreases the volume of flow beneath the structure.)

## 5.11 REFRACTION OF STREAMLINES

When flow passes from one stratum to another stratum of different hydraulic conductivity, the direction of flow changes (2). The same type of refraction occurs when seismic waves pass through layers of the earth or when light waves pass from air into water. From the continuity principle and Darcy's law, when a fluid moves into a medium with higher permeability, less of the aquifer area is needed to transmit the fluid. Therefore, the flow channels can be narrower. If the flow moves from a region of higher to lower permeability, then the flow channels must be wider to accommodate the same volume of flow.

By analogy to the physical laws governing refraction of light, we know

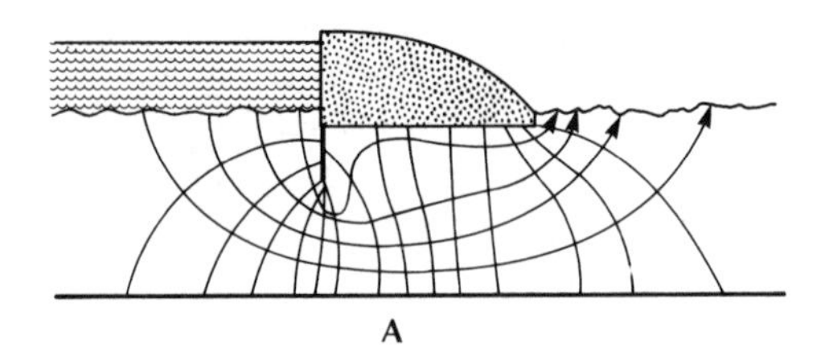


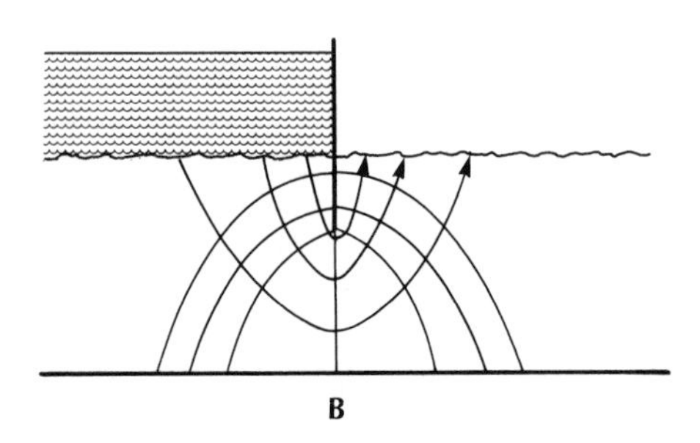


FIGURE 5.11 **A.** Flow net for flow beneath a dam with a cutoff. **B.** Flow beneath a vertical sheet piling.

that the ratio of the hydraulic conductivities is equal to the ratio of the tangent of the angles made by the flow paths with a line perpendicular to the boundary (Figure 5.12A):

$$\frac{K_1}{K_2} = \frac{\tan \sigma_1}{\tan \sigma_2} \tag{5-54}$$

As a consequence, the direction of refraction for flow going from a region of low conductivity to one of high conductivity will be different from that for flow going from high to low conductivity (Figure 5.12B,C). Likewise, if the streamlines are

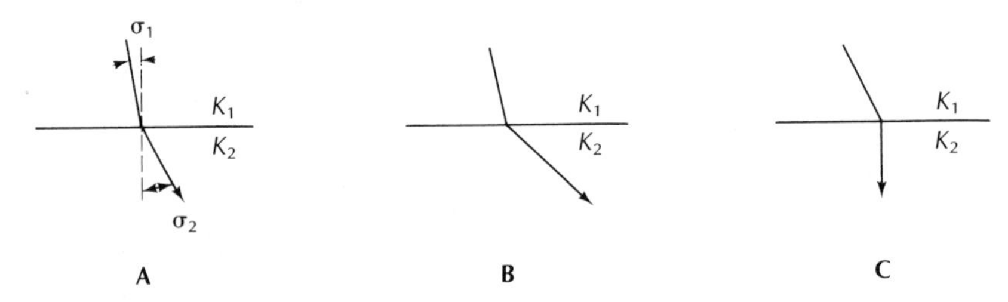


FIGURE 5.12 **A.** Refraction of a streamline crossing a conductivity boundary. **B.** Refracted streamline going from a region of low to high conductivity. **C.** Refracted streamline going from a region of high to low conductivity.

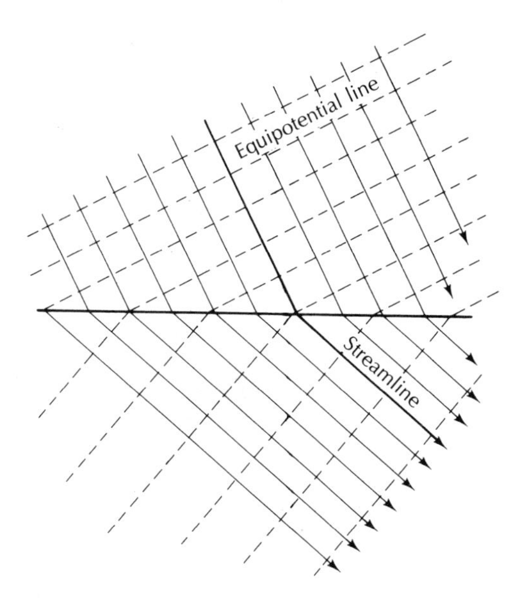


FIGURE 5.13 A flow net with flow crossing a conductivity boundary showing refraction of streamlines and equipotential lines. The hydraulic conductivity above the boundary is less than that below the boundary.

refracted, and they are perpendicular to the equipotential lines, then the equipotential lines must also be refracted. Figure 5.13 shows a portion of a flow net crossing a conductivity boundary.

## 5.12 STEADY FLOW IN A CONFINED AQUIFER

If there is the steady movement of ground water in a confined aquifer, there will be a gradient or slope to the potentiometric surface of the aquifer. Likewise, we know that the water will be moving in the opposite direction of grad $h$. For flow of this type, Darcy's law may be used directly. In Figure 5.14, a portion of a confined aquifer of uniform thickness is shown. The potentiometric surface has a linear gradient; i.e., its two-directional projection is a straight line. There are two observation wells where the hydraulic head can be measured.

The quantity of flow per unit width, $q'$, may be determined from Darcy's law:

$$q' = Kb \frac{dh}{dl} \tag{5-55}$$

where

$K$ is the hydraulic conductivity
$b$ is the aquifer thickness

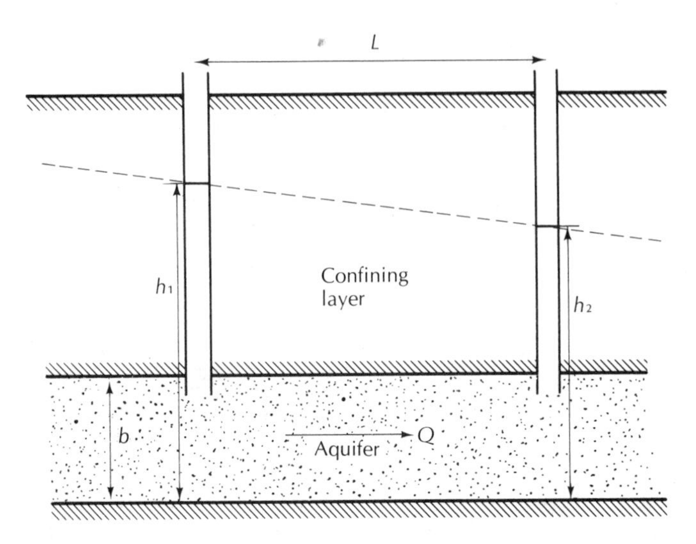


FIGURE 5.14 Steady flow through a confined aquifer of uniform thickness.

$\frac{dh}{dl}$ is the slope of potentiometric surface

$q'$ is the flow per unit width

One may wish to know the head, $h$, at some intermediate distance, $a$, between $h_1$ and $h_2$. This may be found from the equation

$$h = h_1 - \frac{q'}{Kb} a \tag{5-56}$$

**EXAMPLE PROBLEM**

A confined aquifer is 33 meters thick and 7 kilometers wide. Two observation wells are located 1.2 kilometers apart in the direction of flow. The head in Well 1 is 97.5 meters and in Well 2 it is 89.0 meters. The hydraulic conductivity is 1.2 meters per day. What is the total daily flow of water through the aquifer?

$$Q = -Kb \frac{dh}{dl} \text{ width}$$

$$= 1.2 \text{ m/day} \times 33 \text{ m} \times \frac{97.5 \text{ m} - 89.0 \text{ m}}{1200 \text{ m}} \times 7000 \text{ m}$$

$$= 1963.5 \text{ m}^3\text{/day}$$

What is the elevation of the potentiometric surface at a point located 0.3

kilometer from Well $h_1$ and 0.9 kilometer from Well $h_2$? Discharge per unit width is 1963.5 m$^3$/day/7000 m = 0.2805 m$^3$/day:

$$h = h_1 - \frac{q'}{Kb} a$$

$$= 97.5 \text{ m} - \frac{0.2805 \text{ m}^3\text{/day}}{1.2 \text{ m/day} \times 33 \text{ m}} \times 300 \text{ m}$$

$$= 97.5 - 2.125$$

$$= 95.375 \text{ m}$$

## 5.13 STEADY FLOW IN AN UNCONFINED AQUIFER*

In an unconfined aquifer, the fact that the water table is also the upper boundary of the region of flow complicates flow determinations. Figure 5.15 illustrates the problem. On the left side of the figure, the saturated flow region is $h_1$ feet thick. On the right side, it is $h_2$ feet thick, which is $h_1 - h_2$ feet thinner than the left side. If there is no recharge or evaporation as the flow traverses the region, the quantity of water flowing through the left side is equal to that flowing through the right side. From Darcy's law, it is obvious that since the cross-sectional area is smaller on the right side, the hydraulic gradient must be greater. Thus, the gradient of the water table in unconfined flow is not constant; it increases in the direction of flow.

This problem was solved by Dupuit (15), and his assumptions are known as **Dupuit flow.** The assumptions are that (a) the hydraulic gradient is equal to the slope of the water table and (b) for small water-table gradients, the streamlines are horizontal and the equipotential lines are vertical. Solutions based on these assumptions have proved to be very useful in many practical problems. However, the Dupuit assumptions do not allow for a seepage face above the outflow side.

From Darcy's law,

$$q' = -Kh \frac{dh}{dx} \tag{5-57}$$

where $h$ is the saturated thickness of the aquifer. At $x = 0$, $h = h_1$; at $x = L$, $h = h_2$.

*The equations in this section are derived following methods used by Polubarinova-Kochina (16) and Haar (17). Those not familiar with calculus may skip the material from Equation 5-57 to Equation 5-69 without compromising practical understanding.

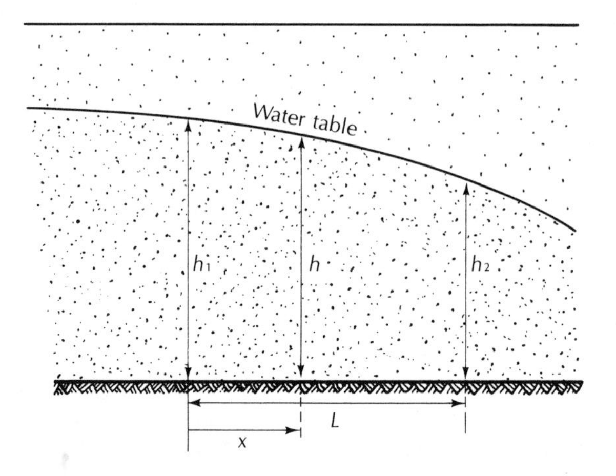


FIGURE 5.15 Steady flow through an unconfined aquifer resting on a horizontal impervious surface.

Equation 5-57 may be set up for integration with the boundary conditions:

$$\int_0^L q'dx = -K\int_{h1}^{h2} hdh$$

Integration of the above yields

$$q'x\Big|_0^L = -K\frac{h^2}{2}\Big|_{h1}^{h2}$$

Substitution of the boundary conditions for $x$ and $h$ yields

$$q'L = -K\left(\frac{h_2^2}{2} - \frac{h_1^2}{2}\right) \tag{5-58}$$

Rearrangement of Equation 5-58 yields the **Dupuit equation:**

$$q' = -\frac{1}{2}K\left(\frac{h_2^2 - h_1^2}{L}\right) \tag{5-59}$$

If we consider a small prism of the unconfined aquifer, it will have the shape of Figure 5.16. On one side it is $h$ units high and slopes in the $x$-direction. Given the Dupuit assumptions, there is no flow in the $z$-direction. The flow in the $x$-direction, per unit width, is $q'_x$. From Darcy's law, the total flow in the $x$-direction through the left face of the prism is

$$q'_x dy = -K\left(h\,\frac{\delta h}{\delta x}\right)_x dy \tag{5-60}$$

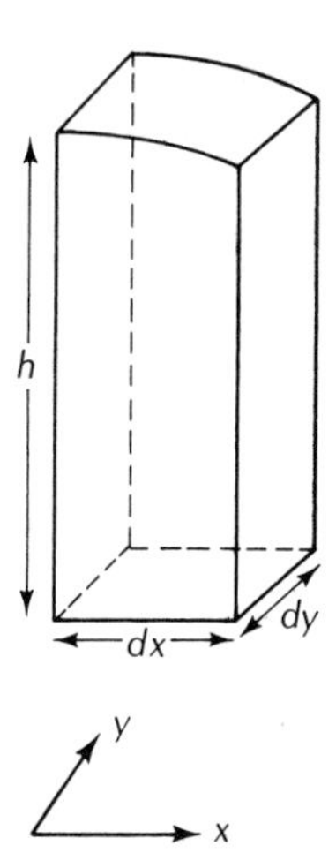


**FIGURE 5.16** Control volume for flow through a prism of an unconfined aquifer with the bottom resting on a horizontal impervious surface and the top coinciding with the water table.

where $dy$ is the width of the face of the prism. The discharge through the right face, $q'_{x+dx'}$ is

$$q'_{x+dx}dy = -K\left(h\frac{\delta h}{\delta x}\right)_{x+dx} dy \tag{5-61}$$

Note that $\left(h\frac{\delta h}{\delta x}\right)$ has different values at each face. The change in flow rate in the $x$-direction between the two faces is given by

$$(q'_{x+dx} - q'_x)dy = -K\frac{\delta}{\delta x}\left(h\frac{\delta h}{\delta x}\right)dxdy \tag{5-62}$$

Through a similar process, it can be shown that the change in the flow rate in the $y$-direction is

$$(q'_{y+dy} - q'_y)dx = -K\frac{\delta}{\delta y}\left(h\frac{\delta h}{\delta y}\right)dydx \tag{5-63}$$

For steady flow, any change in flow through the prism must be equal to a gain or loss of water across the water table. This could be infiltration or evapotranspiration. The net addition or loss is at a rate of $w$, and the volume change within the initial volume is $w\ dxdy$ where $dxdy$ is the area of the surface. If $w$ represents evapotranspiration, it will have a negative value. As the change in flow is equal to the new addition,

$$-K\frac{\delta}{\delta x}\left(h\frac{\delta h}{\delta x}\right)dxdy - K\frac{\delta}{\delta y}\left(h\frac{\delta h}{\delta y}\right)dydx = w\ dxdy \tag{5-64}$$

We can simplify Equation 5-64 by dropping out $dxdy$ and combining the differentials:

$$-K\left(\frac{\delta^2 h^2}{\delta x^2} + \frac{\delta^2 h^2}{\delta y^2}\right) = 2w \tag{5-65}$$

If $w = 0$, then Equation 5-65 reduces to a form of the Laplace equation:

$$\frac{\delta^2 h^2}{\delta x^2} + \frac{\delta^2 h^2}{\delta y^2} = 0 \tag{5-66}$$

If flow is in only one direction, and we align the $x$-axis parallel to the flow, then there is no flow in the $y$-direction, and Equation 5-65 becomes

$$\frac{d^2(h^2)}{dx^2} = -\frac{2w}{K} \tag{5-67}$$

Integration of this equation yields the expression

$$h^2 = -\frac{wx^2}{K} + c_1 x + c_2 \tag{5-68}$$

where $c_1$ and $c_2$ are constants of integration.

The following boundary conditions can be applied: at $x = 0$, $h = h_1$; at $x = L$, $h = h_2$ (Figure 5.17). By substituting these into Equation 5-68, the constants of integration can be evaluated with the following result:

$$h^2 = h_1^2 - \frac{(h_1^2 - h_2^2)x}{L} + \frac{w}{K}(L - x)x \tag{5-69}$$

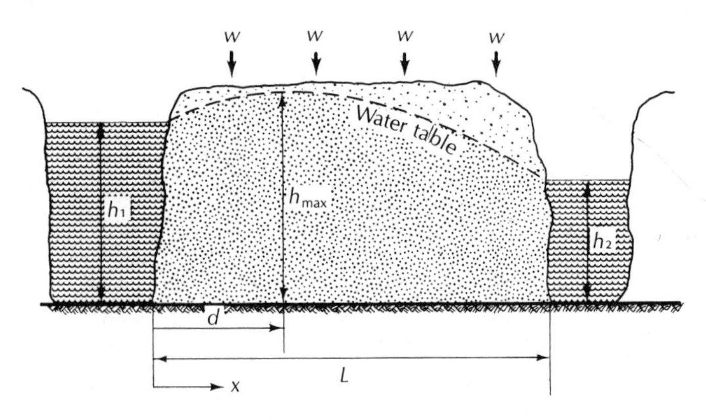


**FIGURE 5.17** Unconfined flow, which is subject to infiltration or evaporation.

This equation can be used to find the elevation of the water table anywhere between two points located $L$ distance apart if the saturated thickness of the aquifer is known at the two end points.

For the case in which there is no infiltration or evaporation, $w = 0$ and Equation 5-69 reduces to

$$h^2 = h_1^2 - \frac{(h_1^2 - h_2^2)x}{L} \quad \textbf{(5-70)}$$

By differentiating Equation 5-69, and because $q'_x = -Kh(dh/dx)$, it may be shown that the discharge per unit width, $q'_x$, at any section $x$ distance from the origin is given by

$$q'_x = \frac{K(h_1^2 - h_2^2)}{2L} - w\left(\frac{L}{2} - x\right) \quad \textbf{(5-71)}$$

If the water table is subject to infiltration, there may be a water divide with a crest in the water table. In this case, $q'_x$ will be zero at the water divide. If $d$ is the distance from the origin to a water divide, then substituting $q'_x = 0$ and $x = d$ into Equation 5-71 yields

$$d = \frac{L}{2} - \frac{K}{w}\frac{(h_1^2 - h_2^2)}{2L} \quad \textbf{(5-72)}$$

Once the distance from the origin to the water divide has been found, then the elevation of the water table at the divide may be determined by substituting $d$ for $x$ in Equation 5-69.

**EXAMPLE PROBLEM**

An unconfined aquifer has a hydraulic conductivity of 0.002 centimeter per second and an effective porosity of 0.27. The aquifer is in a bed of sand with a uniform thickness of 31 meters as measured from the land surface. At Well 1, the water table is 21 meters below land surface. At Well 2, located some 175 meters away, the water table is 23.5 meters from the surface. What is (a) the discharge per unit width, (b) the average linear velocity at Well 1, and (c) the water-table elevation midway between the two wells?

**Part A: From Equation 5-71,**

$$q' = K\frac{(h_1^2 - h_2^2)}{2L}$$

$$h_1 = 31 - 21 = 10 \text{ m}$$

$$h_2 = 31 - 23.5 = 7.5 \text{ m}$$

$$K = 0.002 \text{ cm/sec} = 1.73 \text{ m/day}$$

$$L = 175 \text{ m}$$

$$q' = 1.73 \text{ m/day} \times \frac{10^2 \text{ m}^2 - 7.5^2 \text{ m}^2}{2 \times 175 \text{ m}}$$
$$= 0.22 \text{ m}^3\text{/day per unit width}$$

**Part B: From Equation 5-25A,**

$$v_x = \frac{Q}{n_e A}$$

As $Q = q' \times$ unit width and $A = h_1 \times$ unit width,

$$v_x = \frac{q'}{n_e A}$$
$$= \frac{0.22 \text{ m}^3\text{/day}}{0.27 \times 10 \text{ m} \times 1 \text{ m}} = 0.080 \text{ m/day}$$

**Part C: From Equation 5-50,**

$$h = \sqrt{h_1^2 - (h_1^2 - h_2^2)\frac{x}{L}}$$
$$= \sqrt{10^2 - (10^2 - 7.5^2)\frac{87.5}{175}}$$
$$= 8.84 \text{ m}$$

---

**EXAMPLE PROBLEM** A canal was constructed running parallel to a river 1500 feet away. Both fully penetrate a sand aquifer with a hydraulic conductivity of 1.2 feet per day. The area is subject to rainfall of 1.8 feet per year and evaporation of 1.3 feet per year. The elevation of the water in the river is 31 feet and in the canal it is 27 feet. Determine (a) the water divide, (b) the maximum water-table elevation, (c) the daily discharge per 1000 feet into the river, and (d) the daily discharge per 1000 feet into the canal.

**Part A:**

$$d = \frac{L}{2} - \frac{K}{w}\frac{(h_1^2 - h_2^2)}{2L}$$

$h_1 = 31$ ft

$h_2 = 27$ ft

$L = 1500$ ft

$K = 1.2$ ft/day

$$w = 1.8 \text{ ft/year infiltration} - 1.3 \text{ ft/year evaporation}$$
$$= 0.5 \text{ ft/year accretion}$$
$$= 0.0014 \text{ ft/day}$$
$$d = \frac{1500}{2} - \frac{1.2}{0.0014}\frac{(31^2 - 27^2)}{2 \times 1500}$$
$$= 684 \text{ ft from the river}$$

**Part B:**

$$h = \sqrt{h_1^2 - \frac{(h_1^2 - h_2^2)x}{L} + \frac{w}{K}(L - x)x}$$

$$x = d = 684 \text{ ft}$$
$$h = h_{max}$$
$$h_{max} = \sqrt{31^2 - \frac{(31^2 - 27^2)684}{1500} + \frac{0.0014}{1.2}(1500 - 684)684}$$

$$= 38.8 \text{ ft}$$

**Part C:** $q'_x$ at $x = 0$:

$$q'_x = \frac{K(h_1^2 - h_2^2)}{2L} - w\left(\frac{L}{2} - x\right)$$
$$= \frac{1.2 \times (31^2 - 27^2)}{2 \times 1500} - 0.0014\left(\frac{1500}{2} - 0\right)$$
$$= -0.957 \text{ ft}^3\text{/day/ft and } -957 \text{ ft}^3\text{/day/1000 ft}$$

The negative sign indicates flow is going in a direction opposite to $x$, or into the river.

**Part D:** $q'_x$ at $x = L$:

$$q'_x = \frac{K(h_1^2 - h_2^2)}{2L} - w\left(\frac{L}{2} - x\right)$$
$$= \frac{1.2 \times (31^2 - 27^2)}{2 \times 1500} - 0.0014\left(\frac{1500}{2} - 1500\right)$$
$$= 1.143 \text{ ft}^3\text{/day/ft and } 1143 \text{ ft}^3\text{/day/1000 ft}$$

Flow is in the direction of $x$, or into the canal.

## 5.14 FRESH-WATER–SALINE-WATER RELATIONS

### 5.14.1 COASTAL AQUIFERS

We have assumed to this point that the content of dissolved solids of ground water is so low that it does not affect the physics of flow. However, if fresh ground water is adjacent to saline ground water, the difference in density between the two fluids becomes very important. Due to the difference in dissolved solids, the density of the saline water, $\rho_s$, is greater than the density of fresh water, $\rho_w$. Salt water is found adjacent to fresh water in inland areas, often in the same aquifer, as well as in oceanic coastal areas and oceanic islands. Highly saline water in inland aquifers could be either trapped from the time of formation of the rock unit (connate water) or occur through mineralization due to stagnant flow conditions. At coastal locations, the fresh ground water beneath land is discharging near the coast and mixing with saline ground water beneath the sea floor.

Fresh ground water usually grades into saline water with a steady increase in the content of dissolved solids. In some situations, the contact may be quite sharp; that is, a very thin zone of mixed water. The mixture of fresh water and salt water yields the zone in which there is a salinity gradient. If the aquifer is subject to hydraulic head fluctuations caused by tides, the zone of mixed water will be enlarged. In unconfined coastal aquifers, there is ground-water flow occurring in both the fresh zone and the saline zone (18). Fresh water is flowing upward to discharge near the shoreline, and there is a cyclic flow in the salty water near the interface (Figure 5.18). We will make the simplifying assumption that there is a sharp interface between fresh water and saline water. Although

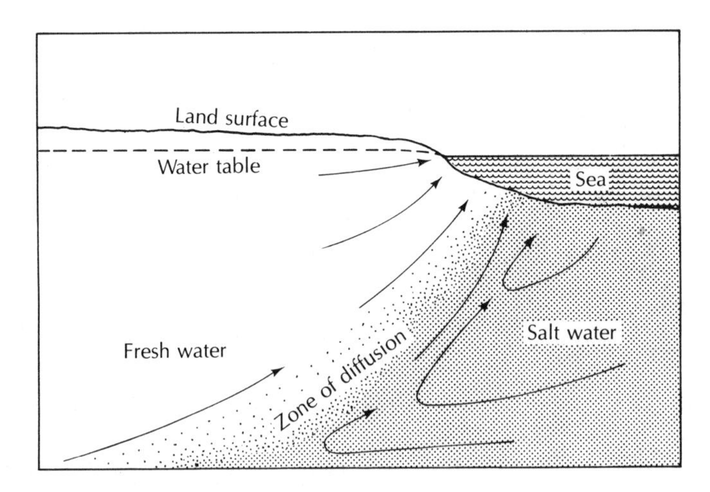


**FIGURE 5.18** Circulation of fresh and saline ground water at a zone of diffusion in a coastal aquifer. Source: H. H. Cooper, Jr., U.S. Geological Survey Circular 1613-C, 1964.

the salt-water interface problem can be studied using dispersion and mass-transport theory (19, 20), the mathematical treatment is beyond the scope of this book. The zone of dispersion is often thin with respect to the overall thickness of the fresh-water lens. Likewise, we will consider only the steady-state case. Solutions have been developed for moving interface problems (19–23), but they are too complex to be considered here.

A number of scientists have made significant contributions to the study of the saline-water–fresh-water interface in coastal aquifers. Studies by W. Baydon-Ghyben (24) and A. Herzberg (25) in the late nineteenth century have been widely cited and have given rise to the Ghyben-Herzberg principle, which we will now discuss. However, their work was antedated by more than half a century by an American, Joseph DuCommun (26), who clearly made the same observations in 1828. Unfortunately, DuCommun is not given just credit in the literature.

These early observers noted that in unconfined coastal aquifers the depth to which fresh water extends below sea level is approximately 40 times the height of the water table above sea level. The (misnamed) **Ghyben-Herzberg principle** states that

$$z_{(x,y)} = \frac{\rho_w}{\rho_s - \rho_w} h_{(x,y)} \quad \textbf{(5-73)}$$

where

$z_{(x,y)}$ is the depth to the salt-water interface below sea level at location $(x,y)$
$h_{(x,y)}$ is the elevation of the water table above sea level at point $(x,y)$
$\rho_w$ is the density of fresh water
$\rho_s$ is the density of salt water

The application of this principle is limited to situations in which both the fresh water and salt water are static.

**EXAMPLE PROBLEM** If $\rho_w$ = 1.000 gram per cubic centimeter and $\rho_s$ = 1.025 gram per cubic centimeter, what is the ratio of $z_{(x,y)}$ to $h_{(x,y)}$?

$$z_{(x,y)} = \frac{\rho_w}{\rho_s - \rho_w} h_{(x,y)}$$

$$= \frac{1.000}{1.025 - 1.000} h_{(x,y)}$$

$$= 40\ h_{(x,y)}$$

Figure 5.19 illustrates the Ghyben-Herzberg principle for an unconfined coastal aquifer. Hubbert (2) pointed out that $h_{(x,y)}$ should actually be the hy-

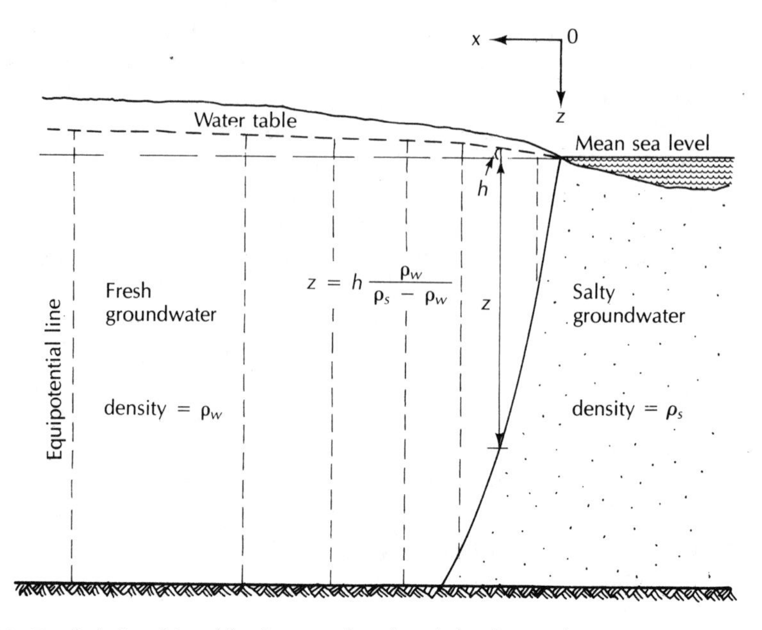


**FIGURE 5.19** Relationship of fresh-water head and depth to salt-water interface.

draulic head at the interface at point $(x,y)$. However, for thin aquifers with a large vertical extent, the Dupuit assumption that equipotential lines are vertical can be made, so that the hydraulic head at the salt-water interface is equal to the elevation of the water table at that location (27). In a study of the salt-water interface on eastern Long Island, it was shown that even at the coastline, where the greatest deviations from the Dupuit assumptions could be expected, there was almost no difference in the position of the interface as computed from potential theory versus Dupuit flow (Figure 5.20).

Flow in coastal aquifers can be described by means of the Dupuit equations in combination with the Ghyben-Herzberg principle. The steady flow of ground water is given by the partial differential equation (27)

$$\frac{\delta^2 h^2}{\delta x^2} + \frac{\delta^2 h^2}{\delta y^2} = \frac{-2w}{K\,(1 + G)} \qquad \textbf{(5-74)}$$

where

$w$ is the recharge to the aquifer

$K$ is the hydraulic conductivity

$G$ is equal to $\dfrac{\rho_w}{\rho_s - \rho_w}$

Should the value of the depth to the salt-water interface, as computed by Equation 5-73, exceed the depth of the aquifer, then the salt-water wedge is

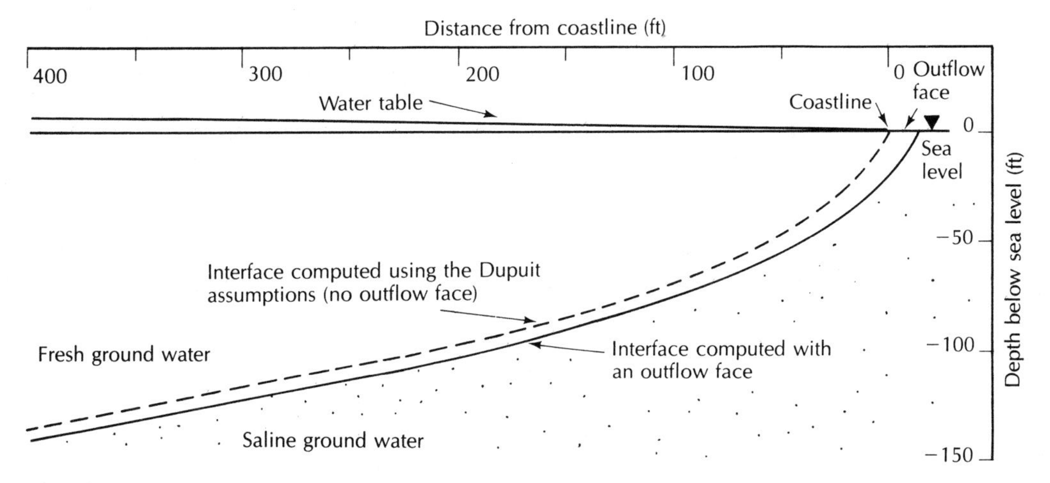


**FIGURE 5.20** Comparison of the position of the salt-water interface on eastern Long Island as computed using the Dupuit assumptions with that computed using an outflow face. Source: C. W. Fetter, Jr., *Water Resources Research,* 8 (1972):1307–14.

missing. This is the case on the left side of Figure 5.19. In this region, the governing equation is (16)

$$\frac{\delta^2 h^2}{\delta x^2} + \frac{\delta^2 h^2}{\delta y^2} = \frac{-w}{K\,(z_m + h)} \tag{5-75}$$

where $z_m$ is the aquifer thickness below sea level. Both Equation 5-74 and Equation 5-75 can be solved for an infinite-strip coastline; that is, one with flow in only one direction. The $x$- and $z$-axes are shown on Figure 5.19.

The **Dupuit-Ghyben-Herzberg model** of one-dimensional flow in coastal aquifers yields the following expression for the $x$- and $z$-coordinates of the interface (28):

$$z = \sqrt{\frac{2q'xG}{K}} \tag{5-76}$$

where $q'$ is the discharge from the aquifer at the coastline, per unit width $[(L^3/T)/L]$.

One of the failings of the Dupuit-Ghyben-Herzberg model of coastal aquifers is that the salt-water interface intercepts the water table at the coastline. This does not allow for vertical components of flow and discharge of the fresh water into the sea floor. Therefore, a simple model has been developed in which the $x$- and $z$-coordinates of the interface are given by the following relation (29):

$$z = \frac{Gq'}{K} + \sqrt{\frac{2Gq'x}{K}} \tag{5-77}$$

Note that Equation 5-77 is identical to 5-76, except that a constant, $Gq'/K$, has been added. Thus, when $x = 0$, $z$ will still have a value. The interface is shown in Figure 5.21.

The width of the outflow face, $x_0$, may be found from the expression

$$x_0 = -\frac{Gq'}{2K} \tag{5-78}$$

and the height of the water table at any distance, $x$, from the coast is given by

$$h = \sqrt{\frac{2q'x}{GK}} \tag{5-79}$$

### 5.14.2 OCEANIC ISLANDS

Oceanic islands are underlain by salty ground water as well as being surrounded by sea water. The fresh water takes the form of a fresh-water lens floating on the more dense salty water. Equation 5-74 also describes the flow of water in an oceanic island. It can be solved for islands of regular shape, such as an infinite strip or circle.

For an infinite-strip island receiving recharge at a rate, $w$, with a width

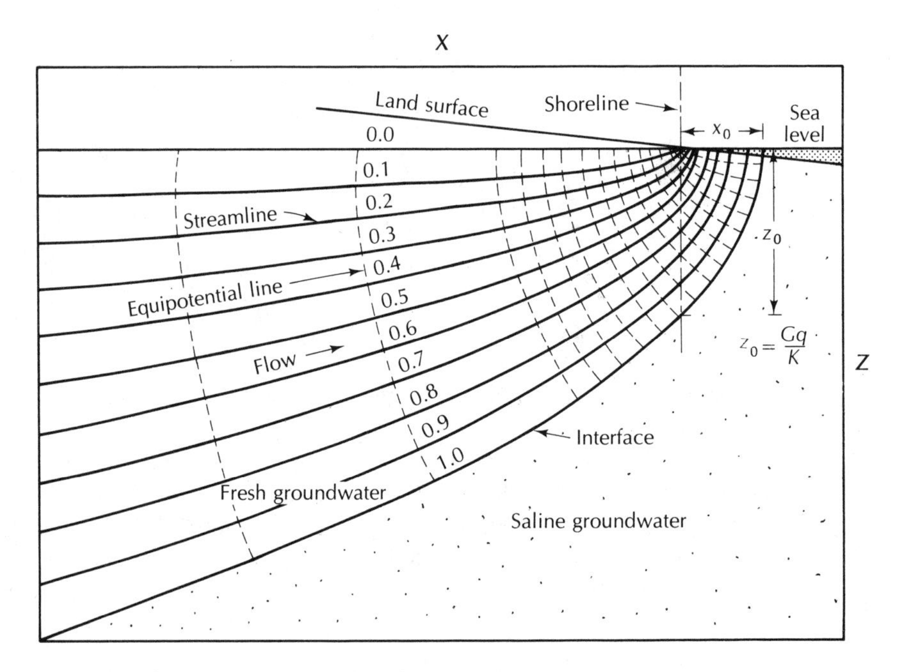


**FIGURE 5.21** Flow pattern near a beach as computed using Equation 5-79. Source: R. E. Glover, U.S. Geological Survey Water-Supply Paper 1613-C, 1964.

equal to $2a$, the head of the water table, $h$, at any distance, $x$, from the shoreline is given by (27):

$$h^2 = \frac{w[a^2 - (a - x)^2]}{K(1 + G)} \tag{5-80}$$

A circular island with a radius of distance $R$ can be evaluated using

$$h^2 = \frac{w(R^2 - r^2)}{2K(1 + G)} \tag{5-81}$$

where $h$ is the head above sea level at some radial distance, $r$, from the center of the island.

**EXAMPLE PROBLEM**

**Part A:** An infinite-strip island has a width of 2 kilometers. The permeability of the sediments is $10^{-2}$ centimeter per second and there is a daily accretion of 0.13 centimeter per day. The density of fresh water is 1.000 and the density of salty ground water is 1.025. Compute a water-table profile across the island using Equation 5-80. Then determine the interface depth using Equation 5-73.

$$G = \frac{1}{1.025 - 1} = 40$$

$$K = 10^{-2} \text{ cm/sec} = 8.64 \text{ m/day}$$

$$w = 0.0013 \text{ m/day}$$

$$a = 1000 \text{ m}$$

$$h^2 = \frac{w[a^2 - (a - x)^2]}{K(1 + G)}$$

$$z = Gh$$

| x (m) | h (m) | z (m) |
|---|---|---|
| 1000 | 1.92 | 76.8 |
| 900 | 1.91 | 76.4 |
| 800 | 1.88 | 75.2 |
| 700 | 1.83 | 73.2 |
| 600 | 1.76 | 70.4 |
| 500 | 1.66 | 66.4 |
| 400 | 1.53 | 61.2 |
| 300 | 1.37 | 54.8 |
| 200 | 1.15 | 46.0 |
| 100 | 0.84 | 33.6 |
| 0 | 0 | 0 |

**Part B:** From the computed profile, it can be seen that the fresh-water lens thins very rapidly in the last 100 meters as the shoreline is approached. As the Dupuit assumptions may not be valid near the coastline, the profile of the interface close to the coast can be computed by use of Equation 5-77. The outflow per unit width, $q'$, is equal to the recharge rate times the half-width.

$$q' = 0.0013\ \text{m}^3/\text{day} \times 1000\ \text{m}$$
$$= 1.3\ \text{m}^3/\text{day/m}$$
$$z = \frac{Gq'}{K} + \sqrt{\frac{2Gq'x}{K}}$$

| x (m) | z (m) |
|---|---|
| 0 | 6.0 |
| 20 | 21.5 |
| 40 | 28.0 |
| 60 | 32.9 |
| 80 | 37.1 |
| 100 | 40.7 |

Find the width of the outflow face:

$$x_0 = -\frac{Gq'}{2K}$$
$$= -3.0\ \text{m}$$

Find the height of the water table at a distance from the coast of 100 meters:

$$h = \sqrt{\frac{2q'x}{GK}}$$
$$= 0.87\ \text{m}$$

Comparison of the position of the salt-water interface in a coastal aquifer as computed by using Equation 5-76, which does not allow for an outflow face, with the position computed by Equation 5–77, which includes an outflow face, will show the results are similar (see Figure 5.20). However, the solution using Equation 5-77 will be more exact at the coastline and in that area will show a greater value of head and a greater depth to the interface. The coastal zone is about 1 to 5 percent of the total width of the island (30). The two equations will yield essentially the same results away from the coastal zone. Use of Equations 5-73, 5-76, 5-80, and 5-81 will result in a slight error near the coastline owing to the lack of allowance for an outflow face.

## 5.15 TIDAL EFFECTS

Aquifers located next to tidal bodies are subjected to short-term fluctuations in the head, $h$, due to the tide. Water-level recorders located in coastal wells show

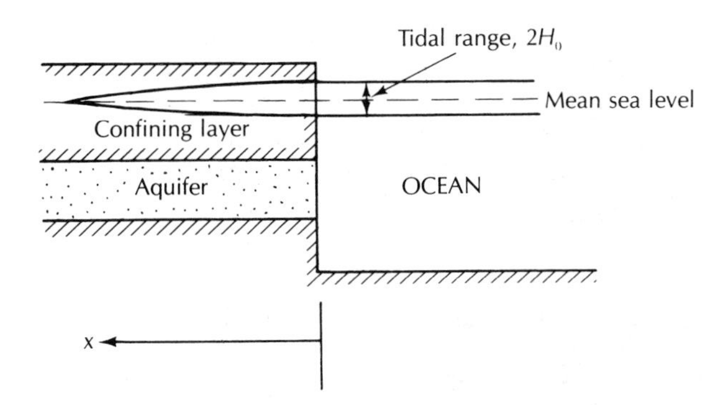


**FIGURE 5.22** Coastal aquifer showing the tidal range, $2H_0$, and the effect of the tide on the potentiometric surface.

a fluctuation in the hydraulic head that parallels the rise and fall of the tide. The amplitude of the fluctuation is greatest at the coast and diminishes as one goes inland.

If we have a confined aquifer, as shown in Figure 5.22, water can enter at the subsurface outcrop. The governing flow equation in one dimension can be used to describe the flow of water into and out of the aquifer as the tide changes.

The amplitude of the tidal change is $H_0$ and the tidal period, or time for the tide to go from one extreme to the other, is $t_0$. At any distance, $x$, inland from the coast, the amplitude of the tidal fluctuation, $H_x$, is given by (9)

$$H_x = H_0 \exp\left(-x\sqrt{\pi S/t_0 T}\right) \quad \textbf{(5-82)}$$

where $S$ and $T$ are the aquifer storativity and transmissivity.

The time lag, $t_\tau$, between the high tide and the peak of the water level (or low tide and the low point in the water level) is given by

$$t_\tau = x\sqrt{t_0 S/4\pi T} \quad \textbf{(5-83)}$$

The preceding equations can also be applied to unconfined flow, as an approximation, if the range of tidal fluctuations is small compared with the saturated aquifer thickness.

## REFERENCES

1. STREETER, V. L. *Fluid Mechanics*. New York: McGraw-Hill Book Company, 1962, 555 pp.
2. HUBBERT, M. K. "The Theory of Ground-Water Motion." *Journal of Geology*, 48, no. 8 (1940): 785–944.
3. SCHNEEBELI, G. "Experiences sur la limite de validité de la loi de Darcy et l'apparition de la turbulence dans un écoulement de filtration." *La Houille Blanche*, 10, no. 2 (1955): 141–49.

4. HUBBERT, M. K. "Darcy's Law and the Field Equations of Flow of Underground Fluids." *Transactions, American Institute of Mining and Metallurgical Engineers,* 207 (1956):222–39.

5. LINDQUIST, E. *On the Flow of Water through Porous Soil.* Premier Congres des grands barrages (Stockholm), 1933, pp. 81–101.

6. ROSE, H. E. "An Investigation into the Laws of Flow of Fluids through Beds of Granular Materials." *Proceedings of the Institute of Mechanical Engineers,* 153 (1945):141–48.

7. ROSE, H. E. "On the Resistance Coefficient—Reynolds Number Relationship for Fluid Flow through a Bed of Granular Materials." *Proceedings of the Institute of Mechanical Engineers,* 153 (1945):154–68.

8. JACOB, C. E. "On the Flow of Water in an Elastic Artesian Aquifer." *Transactions, American Geophysical Union,* 21 (1940):574–86.

9. JACOB, C. E. "Flow of Groundwater." In *Engineering Hydraulics,* ed. H. Rouse. New York: John Wiley & Sons, 1950, pp. 321–86.

10. DOMENICO, P. A. *Concepts and Models in Groundwater Hydrology.* New York: McGraw-Hill Book Company, 1972, 405 pp.

11. COOPER, H. H. "The Equation of Groundwater Flow in Fixed and Deforming Coordinates." *Journal of Geophysical Research,* 71 (1966):4785–90.

12. BOUSSINESQ, J. "Recherches théoriques sur l'écoulement des nappes d'eau infiltrées dans le sol et sur le débit des sources." *Journal de Mathématiques Pures et Appliquées,* 10 (1904):5–78.

13. CASAGRANDE, A. "Seepage through Dams." *Journal of Boston Society of Civil Engineers* (1940):295–337.

14. FORCHHEIMER, P. *Hydraulik.* Leipzig: B. G. Teubner, 1914, 566 pp.

15. DUPUIT, J. *Études théoriques et pratiques sur le mouvement des eaux dans les canaux découverts et à travers les terrains perméables,* 2nd ed. Paris: Dunod, 1863, 300 pp.

16. POLUBARINOVA-KOCHINA, P. Y. *Theory of Ground Water Movement,* trans, R. J. M. DeWiest. Princeton, N.J.: Princeton University Press, 1962, 613 pp.

17. HARR, M. E. *Groundwater and Seepage.* New York: McGraw-Hill Book Company, 1962, 315 pp.

18. COOPER, H. H., JR. A Hypothesis Concerning the Dynamic Balance of Fresh Water and Salt Water in a Coastal Aquifer. *Journal of Geophysical Research,* 64 (1959):461–67.

19. BREDEHOEFT, J. D., and G. F. PINDER. "Mass Transport in Flowing Groundwater." *Water Resources Research,* 9 (1973):194–210.

20. SEGOL, G., G. F. PINDER, and W. G. GRAY. "A Galerkin-Finite Element Technique for Calculating the Transient Position of the Saltwater Front." *Water Resources Research,* 11 (1975):343–47.

21. ANDERSON, M. P. "Unsteady Groundwater Flow beneath Strip Oceanic Islands." *Water Resources Research,* 12 (1976):640–44.

22. COLLINS, M. A., and L. W. GELHAR. "Seawater Intrusion in Layered Aquifers." *Water Resources Research,* 7 (1971):971–79.

23. PINDER, G. F., and H. H. COOPER, JR. "A Numerical Technique for Calculating

the Transient Position of the Saltwater Front." *Water Resources Research,* 6 (1970):875–82.

24. BAYDON-GHYBEN, W., *Nota in verband met de voorgenomen putboring nabij Amsterdam.* Koninklyk Instituut Ingenieurs Tijdschrift (The Hague), 1888–1889, pp. 8–22.
25. HERZBERG, A. "Die Wasserversorgung einiger Nordseebader." *Journal Gasbeleuchtung und Wasserversorgung* (Munich), 44 (1901):815–19, 842–44.
26. DUCOMMUN, J. "On the Cause of Fresh Water Springs, Fountains, and c." *American Journal of Science,* 1st ser., vol. 14 (1828):174–76. As cited in S. N. Davis, "Flotation of Fresh Water on Sea Water, A Historical Note." *Ground Water,* 16, no. 6 (1978):444–45.
27. FETTER, C. W., JR. "Position of the Saline Water Interface beneath Oceanic Islands." *Water Resources Research,* 8 (1972):1307–14.
28. TODD, D. K. "Sea Water Intrusion in Coastal Aquifers." *Transactions, American Geophysical Union,* 34 (1953):749–54.
29. GLOVER, R. E. "The Pattern of Fresh-Water Flow in a Coastal Aquifer." In *Sea Water in Coastal Aquifers,* U. S. Geological Survey Water-Supply Paper 1613-C, 1964, pp. 32–35.
30. VACHER, H. L. Personal communication, 1987.

## PROBLEMS

Note: $g = 9.8$ m/sec$^2$.

1. What is the weight, in newtons, of an object with a mass of 32.1 kg?
2. If the object in problem 1 has a volume of 0.23 cubic meter:
   a. What is its density?
   b. What is its specific weight?
3. A fluid in an aquifer is four meters above a reference datum, the fluid pressure is 2400 newtons/m$^2$ and the flow velocity is $10^{-5}$ m/sec. Fluid density is $1.01 \times 10^3$ kg/m$^3$.
   a. Find total energy per unit mass.
   b. Find total energy per unit weight.
4. A sand aquifer has a median pore diameter of 0.2 mm. The fluid density is $1.003 \times 10^3$ kg/m$^3$ and the fluid viscosity is $1.15 \times 10^{-3}$ pascal-sec. If the flow rate is 0.0016 m/sec, is Darcy's law valid? What is the reason for your answer?
5. The base of a proposed landfill for paper-mill sludge is in a glacial till 30 feet above an aquifer. The vertical hydraulic conductivity of the till is $1 \times 10^{-7}$ cm/sec, the vertical hydraulic gradient is 0.30, and the effective porosity is 0.075. If contaminated fluid (leachate) drains from the paper-mill sludge into the till, how many years would pass until the leachate reached the drinking water aquifer? If the vertical hydraulic conductivity were $3 \times 10^{-6}$ cm/sec, what would the travel time be? If the effective porosity were 0.25, what would the travel time be for each conductivity value?
6. A confined aquifer is 10 feet thick. The potentimetric surface drops 0.54 foot be-

tween two wells that are 792 feet apart. The hydraulic conductivity is 21 feet per day and the effective porosity is 0.17.

**a.** How many cubic feet per day are moving through a strip of the aquifer that is 10 feet wide?

**b.** What is the average linear velocity?

**7.** A constant-head permeameter has cross-sectional area of 156 $cm^2$. The sample is 18 centimeters long. At a head of 5 centimeters, the permeameter discharges 50 $cm^3$ in 193 seconds. What is the hydraulic conductivity in

**a.** cm/sec?

**b.** ft/day?

**8.** An unconfined aquifer has a hydraulic conductivity of $1.7 \times 10^{-3}$ cm/sec. There are two observation wells 328 feet apart. Both penetrate the aquifer to the bottom. In one observation well the water stands 24.6 feet above the bottom and in the other it is 20.0 feet above the bottom.

**a.** What is the discharge per 100-foot-wide strip of the aquifer in $ft^3$/day?

**b.** What is the water-table elevation at a point midway between the two observation wells?

**9.** At a tropical coastal aquifer the ground water is stagnant. The density of fresh water is 0.998 g/$cm^3$ and that of the underlying salt water is 1.024 g/$cm^3$. If the fresh-water head is 7.6 feet above mean sea level, what is the depth to the salt-water interface?

**10.** A falling-head permeameter contains a sample of length 10 centimeters and diameter 4 centimeters. The falling-head tube has an inside diameter of 1 centimeter. The head falls from 5 to 2 centimeters over a period of 70 minutes.

**a.** What is the hydraulic conductivity in cm/sec?

**b.** The water temperature was 21° C. At this temperature the density of water is 0.997 g/$cm^3$ and the viscosity is 0.00988 g/sec-cm. What is the intrinsic permeability of the sample?

**11.** An earthen dam is constructed on an impermeable bedrock layer. It is 750 feet across (i.e., the distance from the water in the reservoir to the tailwaters below the dam is 750 feet). The average hydraulic conductivity of the material used in the dam construction is 0.23 foot per day. The water in the reservoir behind the dam is 75 feet deep and the tailwaters below the dam are 20 feet deep. Compute the volume of water that seeps from the reservoir, through the dam, and into the tailwaters per a 100-foot-wide strip of the dam in $ft^3$/day.

# SIX

# Ground-Water Flow to Wells

One must, however, note that the flow does not always remain the same. Thus when there are rains the flow is increased, for the water on the hills being in excess is more violently squeezed out. But in times of dryness the flow subsides because no additional supply of water comes into the spring. In the case of the best springs, however, the amount of flow does not contract very much. . . . It is also necessary to find the speed of the flow, for the swifter the flow the more water the spring delivers, and the slower it is, the less.

*Dioptra,* Hero of Alexandria (ca. A.D. 65)

## 6.1 INTRODUCTION

Wells are one of the most important aspects of applied hydrogeology. Water wells are used for the extraction of ground water to fill domestic, municipal, industrial, and irrigation needs. Wells have also been used to control salt-water intrusion, remove contaminated water from an aquifer, lower the water table for construction projects, relieve pressures under dams, and drain farmland.

Wells also function to inject fluids into the ground. On Long Island, New York, all ground water pumped for cooling purposes must be returned to the same aquifer by an injection well. Still another function of wells is to dispose of wastewater into isolated aquifers. Finally, as a means of ground-water management, wells are sometimes used to artificially recharge aquifers at rates greater than natural recharge.

The same theoretical considerations apply to wells that extract water and those that inject water. During well pumpage, drawdown of the head in the aquifer around the well occurs; during injection there is an increase in the head in the aquifer. From a mathematical standpoint, injection is handled by using a negative value for the pumping rate.

## 6.2 UNSTEADY RADIAL FLOW

For purposes of analysis, the aquifers that we will consider will be assumed to be isotropic and homogeneous. It will be shown that solutions can be found for cases in which the value of the horizontal conductivity is different from that of the vertical conductivity. However, we will assume that the aquifer has **radial symmetry;** i.e., the value of the horizontal conductivity does not depend on the direction of flow in the aquifer.

Flow toward a well has been termed **radial flow.** It moves as if along the spokes of a wagon wheel toward the hub. We can deal with radial flow by use of a coordinate system called **polar coordinates.** The position of a point in a plane is specified according to its distance and direction from a fixed point or pole. The distance is measured directly from the pole to the point in the plane. The direction is determined by the angle between the line from the pole to the point and a fixed reference line—the **polar axis.** The polar axis is usually drawn horizontally and to the right of the pole, and the angle is measured by a counterclockwise rotation from the polar axis (Figure 6.1). Only the value of the angle ($\theta$) and the radial distance ($r$) need be specified. If the aquifer is isotropic in a horizontal plane, then flow will have radial symmetry. Under these conditions, the equation for confined flow becomes (1)

$$\frac{\delta^2 h}{\delta r^2} + \frac{1 \delta h}{r \delta r} = \frac{S \delta h}{T \delta t} \quad \textbf{(6-1)}$$

where

$h$ is the hydraulic head
$S$ is the storativity
$T$ is the transmissivity
$t$ is the time
$r$ is the distance from the well

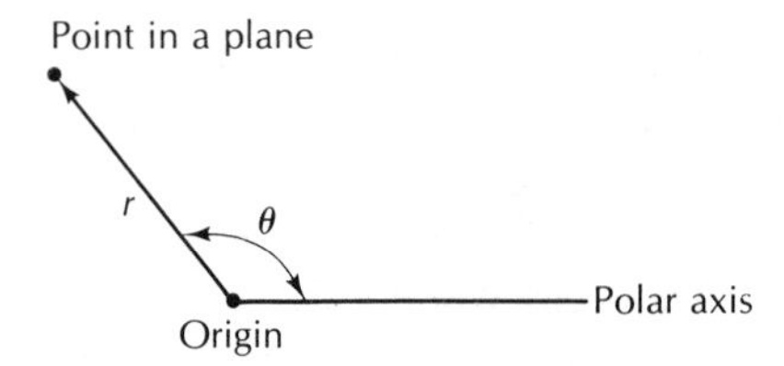


FIGURE 6.1 The use of polar coordinates to describe the position of a point in a plane. It lies a distance ($r$) from the origin, and the angle between the polar axis and a line connecting the point and the origin is $\theta$.

If there is leakage through a confining layer, or recharge to the aquifer, the flow equation is then expressed as

$$\frac{\delta^2 h}{\delta r^2} + \frac{1 \delta h}{r \delta r} + \frac{w}{T} = \frac{S \delta h}{T \delta t} \tag{6-2}$$

where $w$ is the rate of vertical leakage.

Solution of the flow equation for a variety of boundary values for radial flow to wells has yielded a number of extremely useful equations. The mathematics behind the basic solutions involve some rather esoteric mathematical functions. These include Laplace transforms, finite Fourier transforms, Bessel functions, and error functions.* The solutions can be used to determine the drawdown of the potentiometric surface or water table near a pumping well, if the formation characteristics are known. Conversely, if the formation characteristics of an aquifer are unknown, an aquifer test (pumping test) can be made. The well is pumped at a known rate, and the response of the potentiometric surface is measured.

Small-diameter monitoring wells used primarily for sampling ground-water quality or measuring hydraulic head can also be used to determine the hydraulic conductivity of the formation in which the well is screened. A small volume of water is instantaneously added or withdrawn from the well to create an instantaneous head change. The recovery of the water level with time is then recorded. The equations presented in this chapter can then be applied to evaluate the field data and determine the unknown aquifer characteristics.

## 6.3 WELL HYDRAULICS IN A COMPLETELY CONFINED AREALLY EXTENSIVE AQUIFER

When a well is pumped in a completely confined aquifer, the water is obtained from the elastic or specific storage of the aquifer. You will recall that the elastic storage is water that is released from storage by the expansion of the water as pressure in the aquifer is reduced and as the pore space is reduced as the aquifer compacts. The product of the specific storage, $S_s$, and the aquifer thickness is an aquifer parameter called **storativity.** For a confined aquifer, it is generally small (0.005 or less), and pumpage affects a relatively large area of the aquifer. Further, if there is no recharge, the area of drawdown of the potentiometric surface will expand indefinitely as the pumpage continues.

We will assume the following conditions in applying the solution: flow is

*For a review of these functions as they apply to well hydraulics, see M. S. Hantush, "Hydraulics of Wells," in *Advances in Hydroscience,* vol. 1, ed. V. T. Chow (New York, Academic Press, 1964).

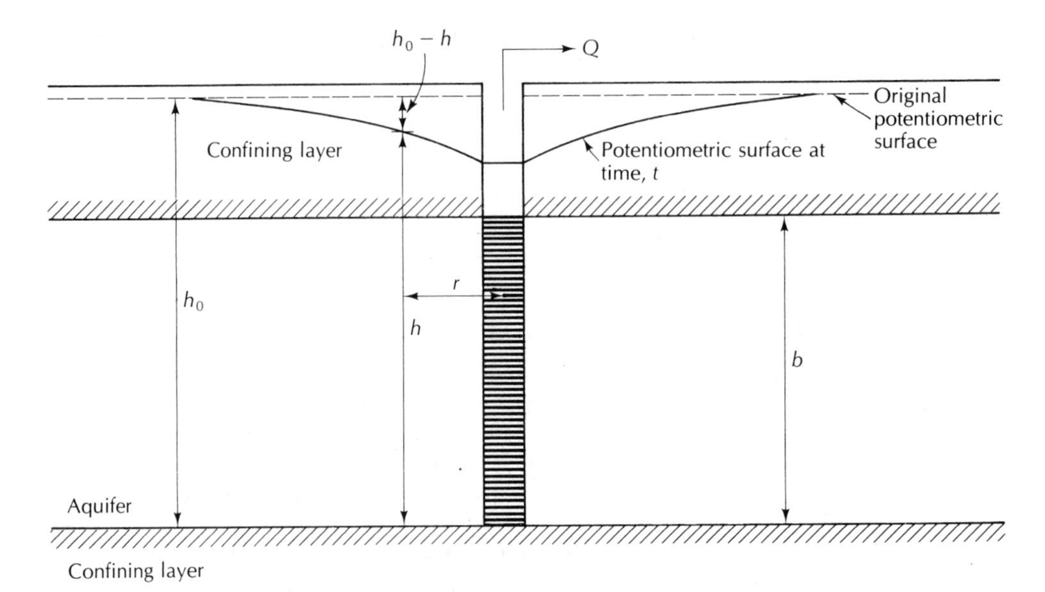


**FIGURE 6.2** Fully penetrating well pumping from a confined aquifer.

in the range of Darcy's law; water is discharged instantaneously from storage; and the aquifer is homogeneous and isotropic, has a constant thickness and a negligible slope, and is of infinite extent. We will also assume that the pumping well and any observation wells fully penetrate the aquifer (i.e., water can enter at any level from the top to the bottom) and that the well diameter is infinitesimal (Figure 6.2).

### 6.3.1 THEIS METHOD

The flow equation for confined aquifers was first solved by C. V. Theis on the basis of the analogy between flow of water in an aquifer and flow of heat in a thermal conductor (2). The original level of the potentiometric surface is $h_0$. The solution that Theis developed is

$$h_0 - h = \frac{Q}{4\pi T}\left[-0.5772 - \ln u + u - \frac{u^2}{2 \cdot 2!} + \frac{u^3}{3 \cdot 3!} - \frac{u^4}{4 \cdot 4!} + \ldots\right] \tag{6-3}$$

where the argument $u$ is given as

$$u = \frac{r^2 S}{4Tt} \tag{6-4}$$

and

$Q$ is the constant pumping rate
$h$ is the hydraulic head at time $t$ since pumping began
$h_0$ is the hydraulic head before the start of pumping
$r$ is the radial distance from the pumping well to the observation well
$T$ is the aquifer transmissivity
$S$ is the aquifer storativity

**Drawdown** is the difference in the elevation of the water level before the well was pumped and the water level at a given time after pumping commenced in either the pumping well or an observation well. It is indicated by the expression $h_0 - h$ in Equation 6-3. The infinite-series term in Equation 6-3 has been called the **well function** and is generally designated as $W(u)$. For a table of the value of $W(u)$ for various values of $u$, see Appendix 1 (3).

**EXAMPLE PROBLEM** A well is located in an aquifer with a conductivity of 15 meters per day and a storativity of 0.005. The aquifer is 20 meters thick and is pumped at a rate of 2725 cubic meters per day. What is the drawdown at a distance of 7 meters from the well after one day of pumping?

$$T = Kb = 15 \text{ m/day} \times 20 \text{ m} = 300 \text{ m}^2/\text{day}$$

$$u = \frac{r^2S}{4Tt} = \frac{(7 \text{ m})^2 \times 0.005}{4 \times 300 \text{ m}^2/\text{day} \times 1 \text{ day}} = 0.0002$$

From the table of $W(u)$ and $u$, if $u = 2 \times 10^{-4}$, $W(u) = 7.94$:

$$h_0 - h = \frac{Q}{4\pi T}W(u) = \frac{2725 \text{ m}^3/\text{day} \times 7.94}{4 \times \pi \times 300 \text{ m}^2/\text{day}} = 5.74 \text{ m}$$

Equations 6-3 and 6-4 are often called the **nonequilibrium** or **Theis equations.** A graph on full-logarithmic paper of $W(u)$ versus $1/u$, as shown in Figure 6.3, has the shape of the cone of depression near the pumping well. This graph is called the **reverse type curve.** This leads to a method for the solution of the nonequilibrium equation when drawdown and time are known but tramsmissivity and storativity are unknown. The procedure for collecting time-drawdown data involves making a pumping test of a well. (Pumping-test procedures are described in Section 6.11.) A pumping test results in a series of drawdown values at different times after the start of pumping for one or more nonpumping observation wells and the extraction well.

The Theis graphical method of solution starts with the construction of the reverse type curve of $W(u)$ plotted against $1/u$ on logarithmic paper (Figure 6.3). Field data for drawdown ($h_0 - h$) at all of the wells are plotted as a function of $t/r^2$ on logarithmic paper of the same scale as used for the type curve

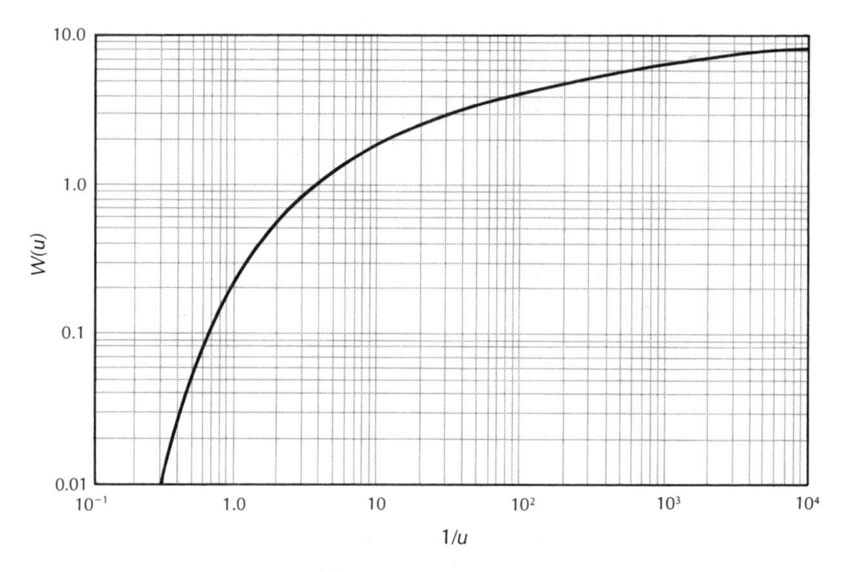


**FIGURE 6.3** The nonequilibrium reverse type curve (Theis curve) for a fully confined aquifer.

(Figure 6.4). By plotting $t/r^2$, data from observation wells located at differing distances from the pumping well can be used. If there is only one observation well, it is sufficient to plot $(h_0 - h)$ as a function of $t$. The graph paper with the data curve is laid on a light table over a type curve drawn to the same scale. The position of the field-data sheet is adjusted, keeping the axes of both sheets parallel, until the data points fall on the type curve (Figure 6.5). Any arbitrary point (not necessarily on the type curve) is selected as a match point. A pin may be pushed through both sheets to mark the match point. From the match point, values of drawdown $(h_0 - h)$, $t/r^2$ or $t$, $W(u)$, and $1/u$ are obtained. A very convenient match point is the intersection of the $W(u) = 1$ and $1/u = 1$ lines on the type curve. This simplifies later arithmetic computations. The value of $u$ is the reciprocal of $1/u$.

If consistent units are used, then Equations 6-3 and 6-4 are used directly. In common usage in the United States, the nonequilibrium equations are often presented in a form that uses inconsistent units (American practical hydrologic units):

$$h_0 - h = \frac{15.3QW(u)}{T} \tag{6-5}$$

$$S = \frac{Ttu}{360r^2} \tag{6-6}$$

where

$h_0 - h$ is the drawdown (feet)

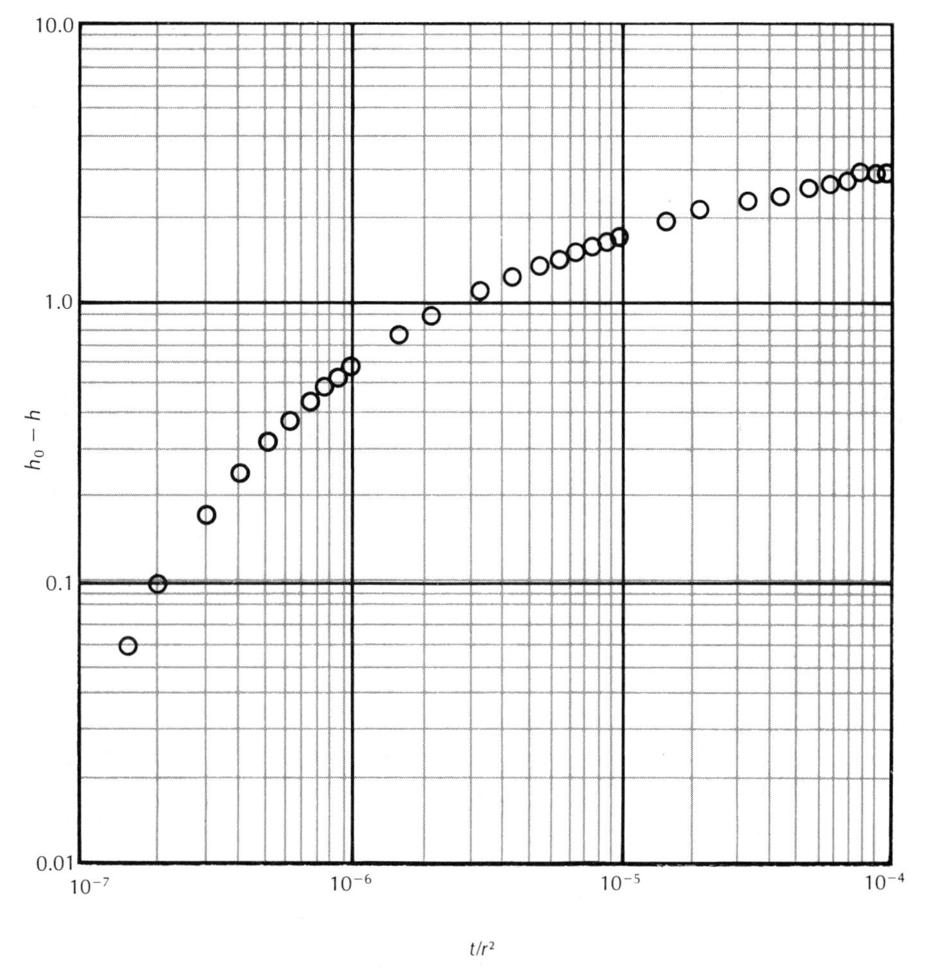


**FIGURE 6.4** Field-data plot of drawdown $(h_0 - h)$ as a function of $t/r^2$ for the Theis graphical method of analysis of pumping-test data.

$Q$ is the well discharge (gallons per minute)
$T$ is the transmissivity (square feet per day)
$r$ is the distance to the observation well (feet)
$S$ is the storativity (dimensionless)
$t$ is the time since pumping began (minutes)

Using the data for $W(u)$ from the match point and the constant pumping rate, $Q$, the value of transmissivity is computed using Equation 6-5.

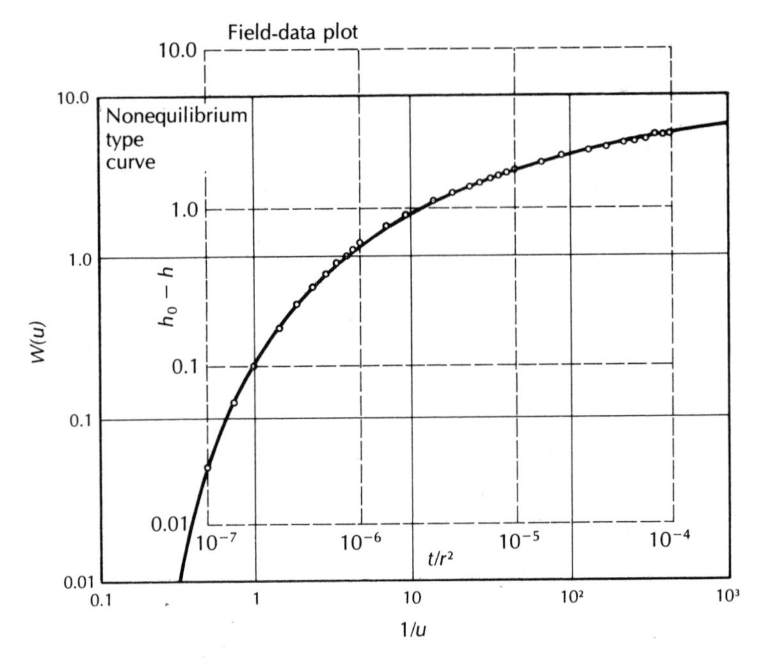


**FIGURE 6.5** Field-data plot of drawdown as a function of $t/r^2$ overlain on a nonequilibrium type curve.

After $T$ has been determined, it is used along with the match-point values for $1/u$ and $t/r^2$ or $t$ to compute the storativity from Equation 6-6.

---

**EXAMPLE PROBLEM** A well in a confined aquifer was pumped at a rate of 220 gallons per minute for about 8 hours. The aquifer was 18 feet thick. Time-drawdown data for an observation well 824 feet away are given in Table 6.1. Find $T$, $K$, and $S$.

In the Theis solution, the data are plotted on logarithmic paper. When overlain on a type curve, the following match point is obtained:

$$W(u) = 1$$
$$1/u = 1$$
$$h_0 - h = 2.4$$
$$t/r^2 = 6.06 \times 10^{-6}$$

Transmissivity:

$$T = \frac{15.3\ QW(u)}{h_0 - h} = \frac{15.3 \times 220 \times 1.0}{2.4}$$
$$= 1403 \text{ ft}^2/\text{day}$$

**TABLE 6.1**

| Time After Pumping Started (min) | $t/r^2$ | Drawdown (ft) |
|---|---|---|
| 3 | $4.46 \times 10^{-6}$ | 0.3 |
| 5 | $7.46 \times 10^{-6}$ | 0.7 |
| 8 | $1.18 \times 10^{-5}$ | 1.3 |
| 12 | $1.77 \times 10^{-5}$ | 2.1 |
| 20 | $2.95 \times 10^{-5}$ | 3.2 |
| 24 | $3.53 \times 10^{-5}$ | 3.6 |
| 30 | $4.42 \times 10^{-5}$ | 4.1 |
| 38 | $5.57 \times 10^{-5}$ | 4.7 |
| 47 | $6.94 \times 10^{-5}$ | 5.1 |
| 50 | $7.41 \times 10^{-5}$ | 5.3 |
| 60 | $8.85 \times 10^{-5}$ | 5.7 |
| 70 | $1.03 \times 10^{-4}$ | 6.1 |
| 80 | $1.18 \times 10^{-4}$ | 6.3 |
| 90 | $1.33 \times 10^{-4}$ | 6.7 |
| 100 | $1.47 \times 10^{-4}$ | 7.0 |
| 130 | $1.92 \times 10^{-4}$ | 7.5 |
| 160 | $2.36 \times 10^{-4}$ | 8.3 |
| 200 | $2.95 \times 10^{-4}$ | 8.5 |
| 260 | $3.83 \times 10^{-4}$ | 9.2 |
| 320 | $4.72 \times 10^{-4}$ | 9.7 |
| 380 | $5.62 \times 10^{-4}$ | 10.2 |
| 500 | $7.35 \times 10^{-4}$ | 10.9 |

Hydraulic conductivity:

$$K = T/b$$

$$= \frac{1403 \text{ ft}^2/\text{day}}{18 \text{ ft}}$$

$$= 77.9 \text{ ft/day}$$

Storativity:

$$S = \frac{Tu}{360} \times t/r^2$$

$$= \frac{1403 \times 1}{360} \times 6.06 \times 10^{-6}$$

$$= 0.00002$$

### 6.3.2 JACOB STRAIGHT-LINE METHOD

C. E. Jacob (4, 5) observed that after the pumping well had been running for some time, higher values of the infinite series became very small, and the nonequilibrium formula could be closely approximated by

$$h_0 - h = (2.3Q/4\pi T) \log_{10} (2.25Tt/Sr^2) \tag{6-7}$$

with consistent units. Equation 6-7 is valid for very small values of $u$. If $(r^2S)/(4Tt)$ is less than 0.05, then all values for Equation 6-3 beyond the first two terms of the infinite series are infinitesimal (6). Truncating Equation 6-3 after the second term of the infinite series yields Equation 6-7. The logarithmic Equation 6-7 will plot as a straight line on semilogarithmic paper if the limiting condition is met. This may be true for large values of $t$ or small values of $r$. Thus, straight-line plots of drawdown versus time can occur after sufficient time has elapsed. In a pumping tests with multiple observation wells, the closer wells will meet the conditions before the more distant ones.

In the **Jacob straight-line method,** a straight line is drawn through the field-data points and extended backward to the zero drawdown axis. It should intercept this axis at some positive value of time (or $t/r^2$ if there are multiple observation wells). This value is designated $t_0$ [or $(t/r^2)_0$]. The value of the drawdown per log cycle $\Delta(h_0 - h)$ is obtained from the slope of the graph. With consistent units for the parameters, the values of transmissivity and storativity may be found from the equations

$$T = \frac{2.3Q}{4\pi\Delta(h_0 - h)} \tag{6-8}$$

and

$$S = \frac{2.25Tt_0}{r^2} \text{ or } 2.25T(t/r^2)_0 \tag{6-9}$$

In American practical units, these equations are expressed as

$$T = \frac{35Q}{\Delta(h_0 - h)} \tag{6-10}$$

$$S = \frac{Tt_0}{640r^2} \tag{6-11}$$

where

$Q$ is the constant rate of pumpage (gallons per minute)

$\Delta(h_0 - h)$ is the drawdown per log cycle of time (feet)

$t$ is the time since pumping began (minutes)

**EXAMPLE PROBLEM**

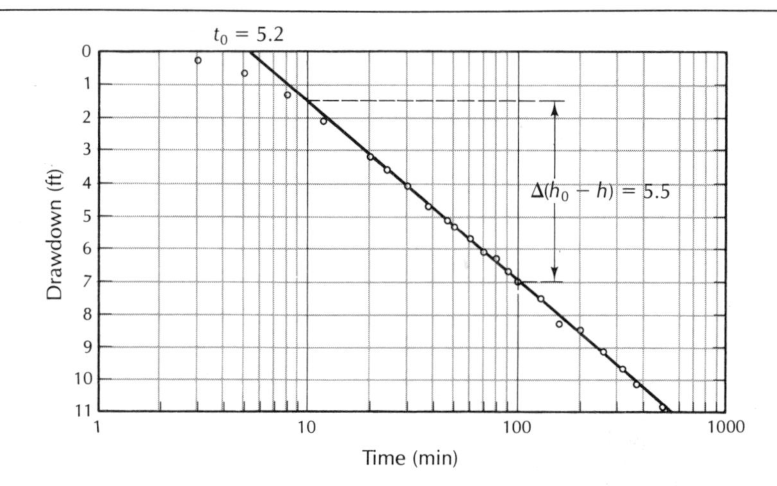


**FIGURE 6.6** Jacob method of solution of pumping-test data for a fully confined aquifer. Drawdown is plotted as a function of time on semilogarithmic paper.

Evaluate the pumping-test data of Table 6.1 by the Jacob method.

The field data are plotted on semilogarithmic paper (Figure 6.6). A straight line is fit to the later time data and extended back to the zero-drawdown axis. The value of $t_0$ is 5.2 minutes, while the drawdown per log cycle is 5.5 feet.

$$T = \frac{35Q}{\Delta(h_0 - h)} = \frac{35 \times 220}{5.5} = 1400 \text{ ft}^2/\text{day}$$

$$S = \frac{Tt_0}{640r^2} = \frac{1400 \times 5.2}{640 \times 824^2} = 0.000017$$

$T$ is the transmissivity (square feet per day)

$r$ is the distance to the pumped well (feet)

In comparing the Jacob solution with the Theis solution in the preceding problems, we see that the resulting answers are almost the same. As these are graphical methods of solution, there will often be a slight variation in the answers, depending upon the accuracy of the graph construction and subjective judgments in matching field data to type curves.

An aquifer test may be made even if there are no observation wells. In this case, drawdown must be measured in the pumping well. There are energy losses as the water rushes into the well, so that the head in the aquifer is higher than the water level in the pumping well. For this reason, aquifer storativity cannot be determined. However, a plot of drawdown versus time for the pumping well can be used to determine aquifer transmissivity.* Either the Theis or Jacob method can be used. It is important that the well be pumped at a constant rate, as any slight fluctuations will immediately affect the water level in the well. Likewise, drawdown data for the start of pumping are affected by the volume of water stored in the well casing. At the start of pumping, the water comes from the well casing rather than from the aquifer, especially when the well diameter is large and/or the pumping rate is small. The measured drawdown data should be adjusted to compensate for this factor (7).

### 6.3.3 DISTANCE-DRAWDOWN METHODS

If simultaneous observations are made of the drawdown in three or more observation wells, a modification of the Jacob straight-line method may be used. If drawdown is measured at the same time in several wells, it is found to vary with the distance from the pumping well in accordance with the Theis equation. Drawdown is plotted on the arithmetic scale as a function of the distance from the pumping well on the logarithmic scale. The drawdown in the wells closest to the pumping well should fall on a straight line.

A line is drawn through the data points for the closest wells. It is extended until it intercepts the zero-drawdown line. This is the distance at which the well is not affecting the water level, and is designated $r_0$. The drawdown per log cycle is $\Delta(h_0 - h)$, as before. With consistent units for the parameters, the values of $T$ and $S$ may be found from

$$T = \frac{2.3Q}{2\pi\Delta(h_0 - h)} \tag{6-12}$$

$$S = \frac{2.25Tt}{r_0^2} \tag{6-13}$$

In American practical hydrologic units,

$$T = \frac{70Q}{\Delta(h_0 - h)} \tag{6-14}$$

$$S = \frac{Tt}{640r_0^2} \tag{6-15}$$

where

$T$ is the transmissivity (square feet per day)

$Q$ is the constant rate of pumpage (gallons per minute)

---

*If, because of turbulent well losses as the water enters the well, the drawdown inside the well is significantly greater than the drawdown in the formation just outside the well, use of time-drawdown data from a single well pump test will understate the formation transmissivity. This can be overcome by measuring the **recovery** of the water level in the well after the pump has been shut down. Time-recovery data can then be plotted and the aquifer transmissivity determined.

$\Delta(h_0 - h)$ is the drawdown per log cycle of distance (feet)

$t$ is the time since pumping began (minutes)

$r_0$ is the intercept of the straight line with the zero-drawdown axis (feet)

---

**EXAMPLE PROBLEM** A well pumping at 400 gallons per minute has observation wells located 10, 40, 150, 300, and 400 feet away. After 200 minutes of pumping, the following drawdowns were observed:

| Distance (ft) | Drawdown (ft) |
|---|---|
| 10 | 15.1 |
| 40 | 9.4 |
| 150 | 4.4 |
| 300 | 1.7 |
| 400 | 0.2 |

The data are plotted in Figure 6.7. The drawdown per log cycle is 8.8 feet and $r_0$ is 460 feet. Find the values of $T$ and $S$.

$$T = \frac{70Q}{\Delta(h_0 - h)}$$

$$= \frac{70 \times 400}{8.8}$$

$$= 3182 \text{ ft}^2/\text{day}$$

$$S = \frac{Tt}{640r_0^2}$$

$$= \frac{3182 \times 200}{640 \times 460^2}$$

$$= 0.0047$$

---

### 6.3.4 COOPER-BREDEHOEFT-PAPADOPULOS METHOD

In some field investigations, the practicing hydrogeologist may be working with low-permeability materials. This is especially likely in site studies for areas of potential waste storage or disposal. The soil materials in these areas may have a conductivity too small to conduct a pumping test. An alternative method of testing involves either injecting into or withdrawing a slug of water of known volume from the well. The rate at which the water level falls or rises is

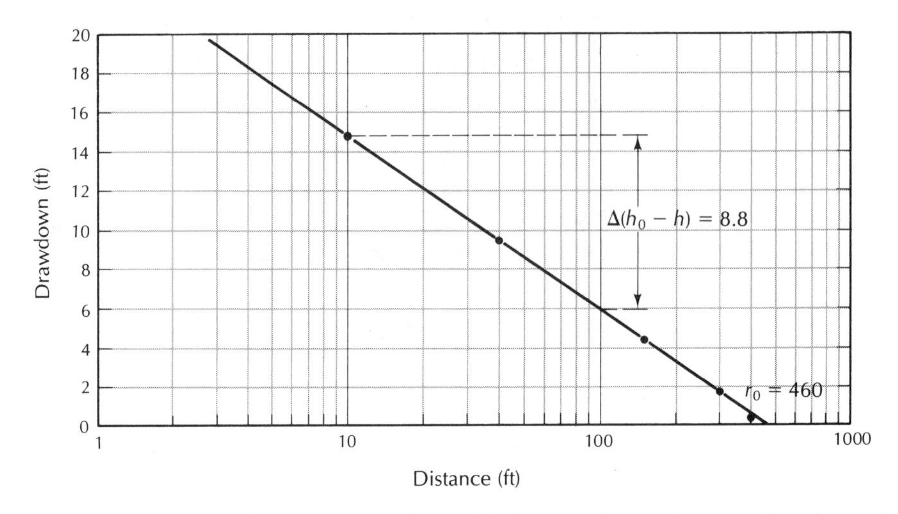


FIGURE 6.7 Variation of the Jacob method of solution of pumping-test data for a fully confined aquifer. Drawdown is plotted as a function of distance to observation well on semilogarithmic paper.

controlled by the formation characteristics. This is known as a **slug test*** (8–10).

A well is drilled and cased above the confined aquifer. A well screen or open hole penetrates the aquifer (Figure 6.8). The well casing has a radius, $r_c$, and the well screen has a radius, $r_s$. Immediately after injection or withdrawal, the water level in the well has an elevation, $H_0$, above or below the initial head. As the water level falls or rises, the difference, $H$, in water-level elevation between that at time $t$ and that at the original head is measured.

A plot of the ratio of the measured head to the head after injection ($H/H_0$) is made as a function of time. The ratio $H/H_0$ is on the arithmetic scale, and time is on the logarithmic scale of semilogarithmic paper:

$$H/H_0 = F(\eta,\mu) \tag{6-16}$$

where

$$\eta = Tt/r_c^2 \tag{6-17}$$

and

$$\mu = r_s^2 S/r_c^2 \tag{6-18}$$

$F(\eta,\mu)$ is a function, the values of which are given in tabulated form in Appendix 2 (10). The tabulated values are plotted as a series of type curves in

---

*The term slug test is sometimes used when the slug is lowered into the well; **bail-down test** is used if a bailer is used to remove a volume of water from the well.

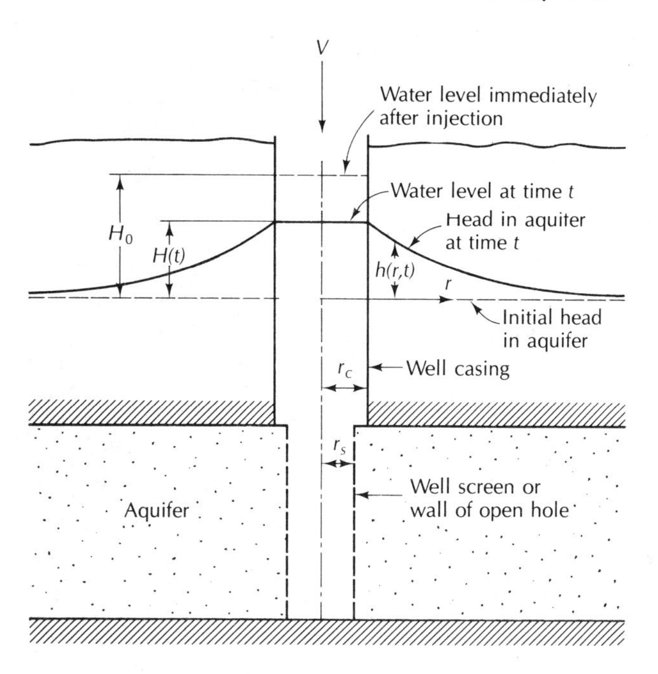


**FIGURE 6.8** Well into which a volume, $V$, of water is suddenly injected for a slug test of a confined aquifer. Source: H. H. Cooper, Jr., J. D. Bredehoeft, and I. S. Papadopulos, *Water Resources Research*, 3 (1967):263–69.

Figure 6.9. The field data should be plotted on the same-scale semilogarithmic paper as the type curves.

The field-data curve is placed over the type curves with the arithmetic axis coincident. That is, the value of $H/H_0 = 1$ for the field data lies on the horizontal axis of 1.0 on the type curve. The data are matched to the type curve ($\mu$), which has the same curvature. The vertical time-axis, $t_1$, which overlays the vertical axis for $Tt/r_c^2 = 1.0$, is selected. The transmissivity is found from

$$T = \frac{1.0 r_c^2}{t_1} \tag{6-19}$$

The value of storativity can be found from the value of the $\mu$-curve for the field data. Since $\mu = (r_s^2/r_c^2)S$,

$$S = (r_c^2 \mu)/r_s^2 \tag{6-20}$$

For small values of $\mu$, however, the curves are often very similar; therefore, in matching the field data, the question of which $\mu$-value to use is often encountered. The use of this method to estimate storativity should be approached with caution. Likewise, the value of $T$ that is determined is representative of the formation only in the immediate vicinity of the test hole.

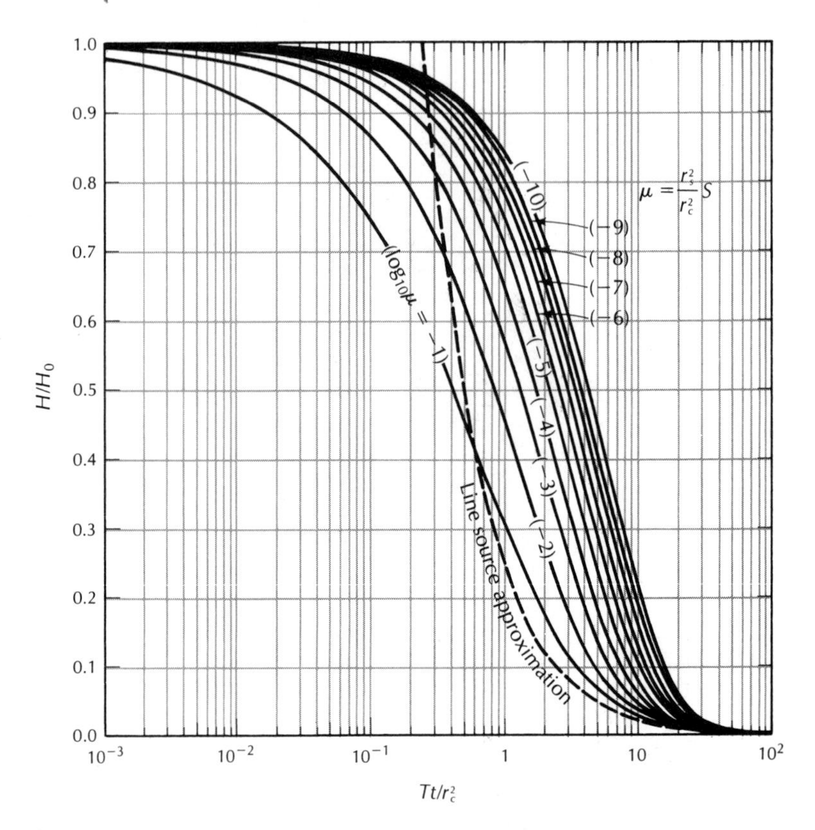


**FIGURE 6.9** Type curves for slug test in a well of finite diameter. Source: S. S. Papadopulos, J. D. Bredehoeft, and H. H. Cooper, Jr., *Water Resources Research*, 9 (1973):1087–89.

**EXAMPLE PROBLEM** A casing with a radius of 7.6 centimeters is installed through a confining layer. A screen with a radius of 5.1 centimeters is installed in a formation with a thickness of 5 meters. A slug of water is injected, raising the water level 0.42 meter. The decline of the head is given in Table 6.2. Find the values of $T$, $K$, and $S$.

**TABLE 6.2**

| Time (sec) | $H$ (m) | $H/H_0$ |
|---|---|---|
| 2 | 0.37 | 0.88 |
| 5 | 0.34 | 0.81 |
| 10 | 0.27 | 0.64 |
| 21 | 0.18 | 0.43 |
| 46 | 0.09 | 0.21 |
| 70 | 0.05 | 0.12 |
| 100 | 0.02 | 0.05 |

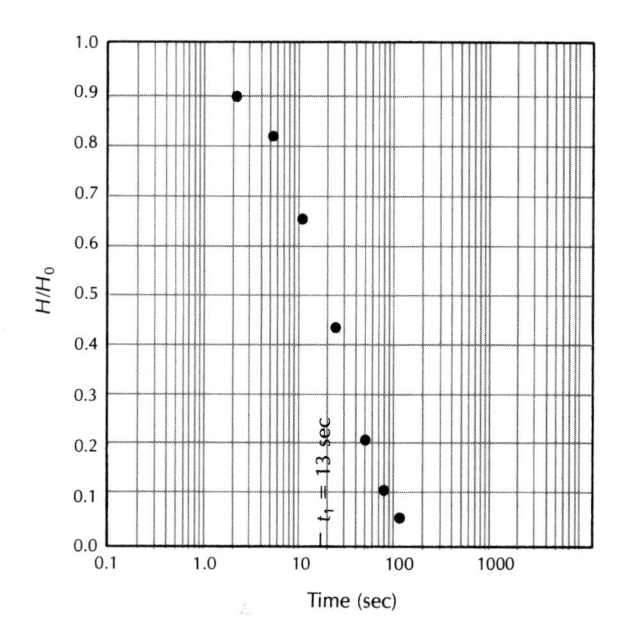


**FIGURE 6.10** Field-data plot of $H/H_0$ as a function of time for a slug-test analysis.

A plot of $H/H_0$ as a function of $t$ is made on semilogarithmic paper (Figure 6.10). It is overlain on the type curve (Figure 6.9). At the axis for $Tt/r^2 = 1.0$, the value $t_1$ is 13 seconds.

$$T = \frac{1.0 r_c^2}{t_1} = \frac{1.0 \times (7.6 \text{ cm})^2}{13 \text{ sec}} = 4.44 \text{ cm}^2/\text{sec}$$

$$\begin{aligned} K &= T/b \\ &= 4.44 \text{ cm}^2/\text{sec}/500 \text{ cm} \\ &= 8.9 \times 10^{-3} \text{ cm/sec} \end{aligned}$$

The $\mu$-curve is $10^{-3}$. With $r_c$ = 7.6 centimeters and $r_s$ = 5.1 centimeters, we find

$$\begin{aligned} S &= (\mu r_c^2)/r_s^2 \\ &= (10^{-3} \times 7.6^2)/(5.1)^2 \\ &= 2 \times 10^{-3} \end{aligned}$$

## 6.4 FLOW IN A SEMICONFINED AQUIFER

Most confined aquifers are not totally isolated from sources of vertical recharge. Semipervious layers, either above or below the aquifer, can leak water into the aquifer if the direction of the hydraulic gradient is favorable (Figure 6.11).

The flow equation for a confined aquifer with recharge is

$$\frac{\delta^2 h}{\delta r^2} + \frac{1\delta h}{r\delta r} - \left(\frac{hK'}{Tb'}\right) = \frac{S}{T}\frac{\delta h}{\delta t} \tag{6-21}$$

where

$K'$ is the vertical hydraulic conductivity of the leaky layer
$b'$ is the thickness of the leaky layer
$h$ is the head
$r$ is the radial distance from the pumping well
$t$ is the time
$S$ is the storativity
$T$ is the transmissivity

### 6.4.1 NO STORAGE IN THE LEAKY CONFINING LAYER

The solution to Equation 6-21 assumes conditions similar to those for the Theis solution, with the following additions: leakage through the confining bed is vertical and proportional to the drawdown; the head in the deposits supplying leakage is constant; and storage in the confining bed is negligible.

The solution (11) to Equation 6-21 is given by

$$h_0 - h = \frac{Q}{4\pi T} W(u, r/B) \tag{6-22}$$

where

$$u = \frac{r^2 S}{4Tt} \tag{6-23}$$

and

$$B = (Tb'/K')^{1/2} \tag{6-24}$$

Equation 6-22 is known as the **Hantush-Jacob formula** for leaky aquifers. The factor $B$ is termed the **leakage factor** (12). The equation is valid for all values of $r_s$ (radius of well screen), providing that

$$t > (30r_s^2 S/T)[1 - (10r_s/b)^2] \text{ and } r_s/B < 0.1$$

The rate of flow from storage in the main aquifer, $q_s$, is found from the formula

$$q_s = Q \exp\left(-Tt/SB^2\right) \tag{6-25}$$

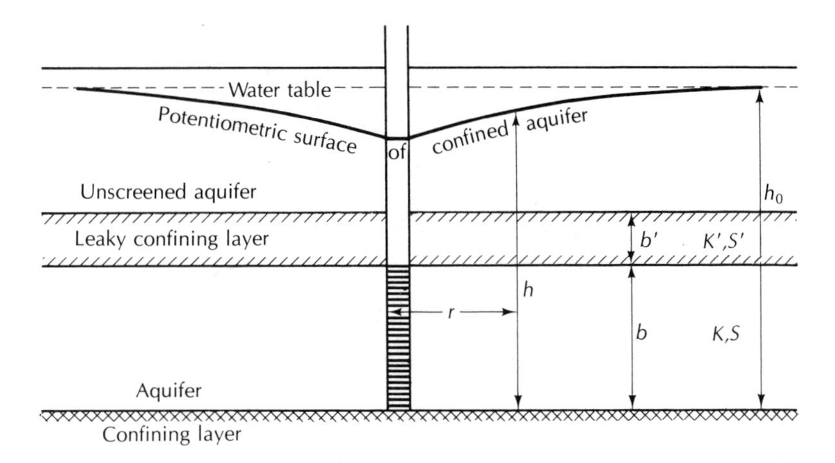


**FIGURE 6.11** Fully penetrating well in an aquifer overlain by a semipermeable confining layer.

The rate of induced leakage, $q_L$, may be found from

$$q_L = Q - q_s \tag{6-26}$$

Values of $W(u,r/B)$ are given in Appendix 3.

### 6.4.2 STORAGE IN THE LEAKY CONFINING LAYER

If there is significant storage in the leaky confining layer, then part of the flow during the initial time period will come from storage in the confining layer (2, 13, 14). It is necessary to use several equations to completely describe the drawdown, depending upon the length of time since pumping began.

The assumptions will be the same as before, except that here the confining layer has finite storativity ($S'$) and there is no drawdown in the unpumped aquifer. The latter assumption may lead to errors for large values of time (14); hence, we will examine only the case in which the time is short (less than $S'b'/10K'$). This is the most useful solution, as hydraulic parameters for both the aquifer and the confining layer can be obtained.

The drawdown equation is

$$h_0 - h = \frac{Q}{4\pi T} H(u,\beta) \tag{6-27}$$

where $H(u,\beta)$ is a function with values tabulated in Appendix 4 and

$$\beta = \frac{r}{4B}(S'/S)^{1/2} \tag{6-28}$$

$$B = [T/(K'/b')]^{1/2} \tag{6-29}$$

where

$K'$ is the vertical conductivity of the confining layer
$b'$ is the thickness of the confining layer
$S'$ is the storativity of the confining layer

and

$$u = \frac{r^2 S}{4Tt}$$

The rate of flow from storage in the main aquifer, $q_s$, is given by

$$q_s = Q \exp(\eta t) \operatorname{erfc}(\sqrt{\eta t}) \tag{6-30}$$

where

$$\eta = (K'/b')(S'/S^2) \tag{6-31}$$

and erfc is the complementary error function, providing that $t < S'b'/10K'$.

The error function erf(*x*) is a function with tabulated values. The com-

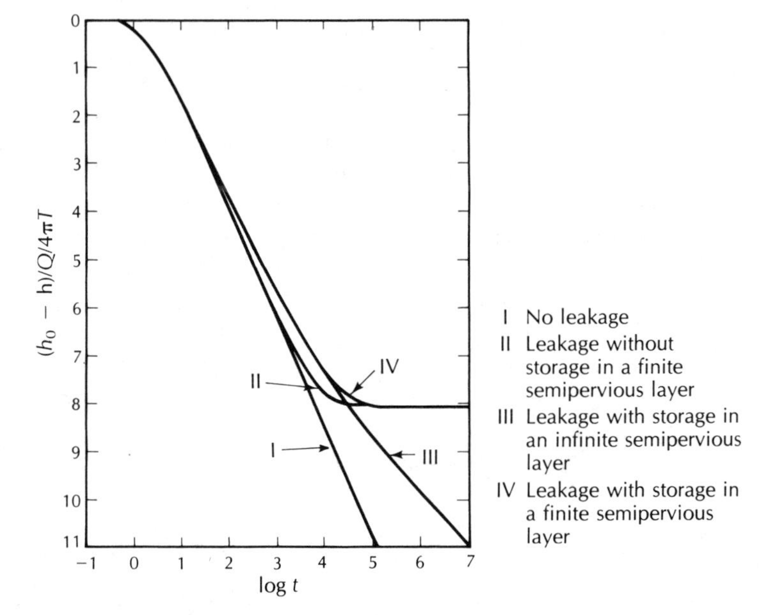


**FIGURE 6.12** Plots of log of dimensionless drawdown as a function of time for an aquifer with various types of overlying confining layer. Source: M. S. Hantush, *Journal of Geophysical Research*, 65 (1960):3713–25.

plementary error function erfc$(x)$ is defined as $1 - \text{erf}(x)$. A table of the error function is given in Appendix 13.

A plot of drawdown as a function of time for several types of confined aquifers is given in Figure 6.12. Other than at very early times, the nonleaky aquifer will decline on a straight line. As there is no recharge, drawdown per log cycle of time will remain constant.

In a leaky aquifer, the drawdown curve will initially follow the nonleaky curve. However, after a finite time interval, the lowered hydraulic head in the aquifer will induce leakage from the confining layer. As part of the well discharge is now coming through the impervious layer, the rate of decline of head will decrease. If there is storage in the confining layer, the rate of drawdown will be lower than if there were no storage. Eventually, the drawdown cone will be large enough so that the pumped water will be coming entirely from leakage through the confining layer and there will be no further drawdown with time. As water is no longer coming from storage in the leaky layer, the two curves coincide.

**EXAMPLE PROBLEM**

An aquifer 10 meters thick is penetrated by a well. It is overlaind by a semipervious layer 1 meter thick with a $K'$ of $10^{-5}$ centimeter per second. There is no storage in the leaky confining layer. The aquifer has a $K$ of $10^{-2}$ centimeter per second and an $S$ of 0.0005. If a well pumps at 500 cubic meters per day, compute values of drawdown at 1, 5, 10, 50, 100, 500, and 1000 meters in order to describe the cone of depression after 1 day.

$$K = 10^{-2} \text{ cm/sec} \times 60 \text{ sec/min} \times 1440 \text{ min/day} \times 10^{-2} \text{ m/cm}$$
$$= 8.64 \text{ m/day}$$
$$K' = 10^{-5} \text{ cm/sec} \times 60 \text{ sec/min} \times 1440 \text{ min/day} \times 10^{-2} \text{ m/cm}$$
$$= 8.64 \times 10^{-3} \text{ m/day}$$
$$b' = 1 \text{ m}$$
$$b = 10 \text{ m}$$
$$T = Kb = 86.4 \text{ m}^2\text{/day}$$
$$B = (Tb'/K')^{1/2}$$
$$= (86.4 \text{ m}^2\text{/day} \times 1 \text{ m}/8.64 \times 10^{-3} \text{ m/day})^{1/2}$$
$$= (10^4)^{1/2} = 100$$
$$u = \frac{r^2S}{4Tt} = \frac{r^2 \times 0.0005}{4 \times 86.4 \times 1} = 1.45 \times 10^{-6}r^2$$
$$r/B = r/100 = 10^{-2}r$$
$$h_0 - h = \frac{Q}{4\pi T} W(u,r/B) = \frac{500}{4 \times \pi \times 86.4} W(u,r/B) = 0.46\ W(u,r/B)$$

As $u = 1.45 \times 10^{-6}r^2$, we can find the value of $u$ for each $r$-value. Then, from the values of $u$ and $r/B$, a value of $W(u,r/B)$ can be found from Appendix 3.

| $r$ | $u$ | $r/B$ | $W(u,r/B)$ |
|---|---|---|---|
| 1 m | $1.45 \times 10^{-6}$ | 0.01 | 9.44 |
| 5 m | $3.63 \times 10^{-5}$ | 0.05 | 6.23 |
| 10 m | $1.45 \times 10^{-4}$ | 0.1 | 4.85 |
| 50 m | $3.63 \times 10^{-3}$ | 0.5 | 1.85 |
| 100 m | $1.45 \times 10^{-2}$ | 1 | 0.842 |
| 500 m | $3.63 \times 10^{-1}$ | 5 | 0.007 |
| 1000 m | 1.45 | 10 | 0.0001 |

From the computed values of $W(u,r/B)$ at each observation point, the drawdown can be computed from $h_0 - h = 0.46\ W(u,r/B)$:

| $r$ (m) | $h_0 - h$ (m) |
|---|---|
| 1 | 4.34 |
| 5 | 2.87 |
| 10 | 2.23 |
| 50 | 0.85 |
| 100 | 0.39 |
| 500 | 0.003 |
| 1000 | 0.000046 |

### 6.4.3 PUMPING TESTS FOR A LEAKY ARTESIAN AQUIFER WITH NO STORAGE IN THE CONFINING LAYER

Pumping tests may be used to determine the formation constants for both an aquifer and an overlying or underlying semipervious layer of finite thickness. Leaky artesian aquifers are also known as **semiconfined aquifers.**

W. C. Walton (15, 16) devised a graphical method based on type curves for $W(u,r/B)$. The type curves are plots on logarithmic paper of $W(u,r/B)$ as a function of $1/u$ for various values of $1/u$ and $r/B$ (Figure 6.13). The type curve for $r/B = 0$ is identical to the Theis nonequilibrium reverse type curve.

Field data are plotted as drawdown versus time, or $t/r^2$ if there are several observation wells. The data curve is placed over the type curve with the axes parallel. The data curve should match one of the type curves for $r/B$; it may have to be matched to an imaginary line interpolated between two $r/B$ lines.

The early drawdown data will tend to fall on the nonequilibrium portion of the type curve. As leakage starts to contribute to flow from the well, the drawdown will follow an $r/B$-type curve.

A match point is picked on the graph. The coordinates on both the data plot and the type curve yield the values of $W(u,r/B)$, $1/u$, $t$ or $t/r^2$, and $h_0 - h$. In addition, the type curve matched yields a value for $r/B$. These values are

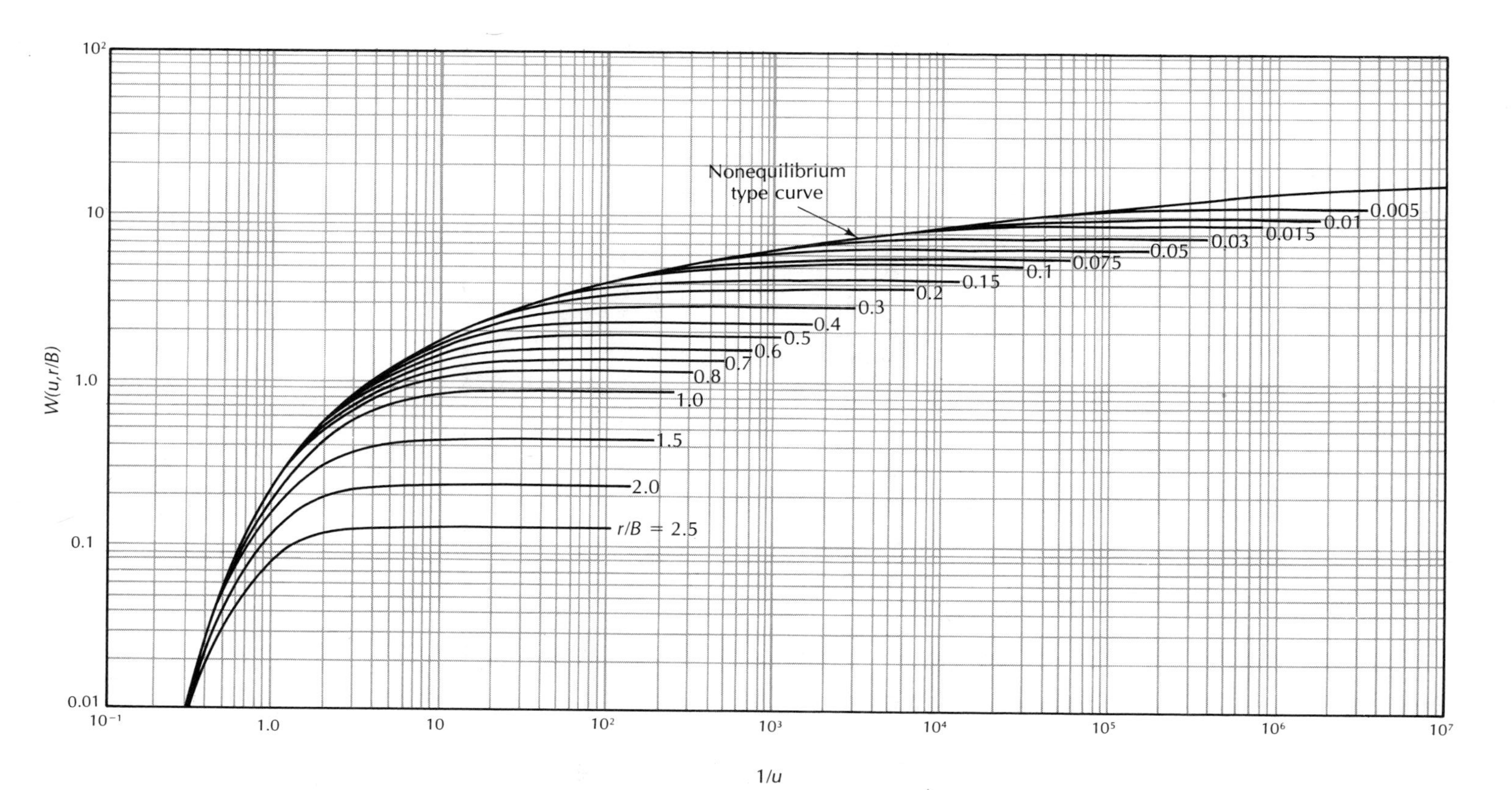


**FIGURE 6.13** Type curves of leaky artesian aquifer in which no water is released from storage in the confining layer. Source: W. C. Walton, Illinois State Water Survey Bulletin 49, 1962.

substituted into the Hantush-Jacob equations to find formation constants for both the aquifer and the semipervious layers.

In American practical hydrologic units, the Hantush-Jacob formulas are

$$h_0 - h = \frac{15.3Q}{T} W(u, r/B) \quad \textbf{(6-32)}$$

$$S = \frac{Ttu}{360r^2} \quad \textbf{(6-33)}$$

$$r/B = r/(Tb'/K')^{1/2} \quad \textbf{(6-34)}$$

$$K' = [Tb'(r/B)^2]/r^2 \quad \textbf{(6-35)}$$

where

$Q$ is the pumping rate (gallons per minute)

$T$ is the transmissivity of the main aquifer (square feet per day)

$t$ is the time since pumping began (minutes)

$S$ is the storativity of the main aquifer

$r$ is the distance from the pumping well to the observation well (feet)

$K'$ is the vertical conductivity of the semipervious layer (feet per day)

$b'$ is the thickness of the semipervious layer (feet)

and

$$B = (Tb'/K')^{1/2}$$

**EXAMPLE PROBLEM** Time drawdown data (16) for a well confined by a stratum of silty fine sand 14 feet thick are given in Table 6.3. Drawdown is measured in an observation well 96 feet away. The well was pumped at 25 gallons per minute. Using the Hantush-Jacob method, find the values of $T$, $S$, and $K$.

**TABLE 6.3**

| Time (min) | Drawdown (ft) |
|---|---|
| 5 | 0.76 |
| 28 | 3.30 |
| 41 | 3.59 |
| 60 | 4.08 |
| 75 | 4.39 |
| 244 | 5.47 |
| 493 | 5.96 |
| 669 | 6.11 |
| 958 | 6.27 |
| 1129 | 6.40 |
| 1185 | 6.42 |

A plot of drawdown versus time must be made. This is shown in Figure 6.14. The match-point values are

$$W(u,r/B) = 1.0$$
$$1/u = 10, \quad u = 0.1$$
$$h_0 - h = 1.9 \text{ ft}$$
$$t = 33 \text{ min}$$

Substituting the match-point values in Equations 6-32, 6-33, 6-34, and 6-35,

$$T = \frac{15.3Q}{h_0 - h} W(u,r/B)$$
$$= \frac{15.3 \times 25 \times 1}{1.9}$$
$$= 201 \text{ ft}^2/\text{day}$$
$$S = \frac{Ttu}{360r^2}$$
$$= \frac{201 \times 33 \times 0.1}{360 \times 96^2}$$
$$= 0.0002$$
$$K' = [Tb'(r/B)^2]/r^2$$
$$= \frac{201 \times 14 \times 0.22^2}{96^2}$$
$$= 0.0148 \text{ ft}^2/\text{day}$$

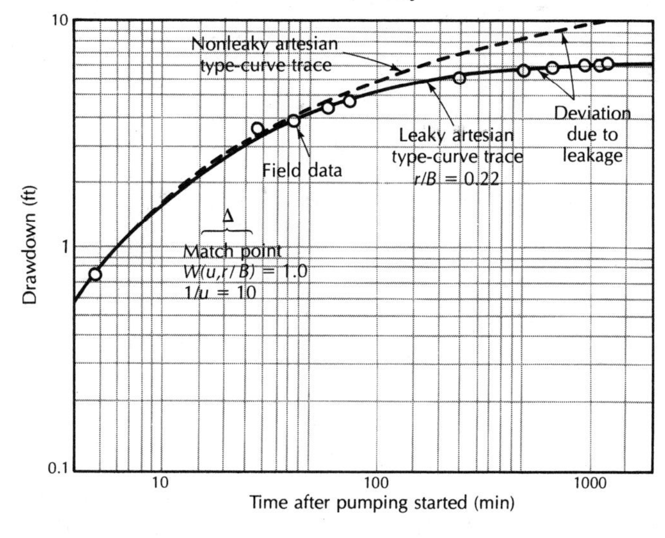


**FIGURE 6.14** Field-data plot of drawdown as a function of time for a leaky confined aquifer. Source: W. C. Walton, Illinois State Water Survey Bulletin 49, 1962.

M. S. Hantush (11) developed an alternative method that does not require the plotting and use of type curves. As can be seen from Figure 6.15, a plot of drawdown versus time on semilogarithmic paper has the shape of an elongated reverse S. At some point, the curve has an inflection. In the **Hantush inflection-point method,** the solution is based on finding this inflection point. For a pumping test with one observation well, the following procedure is used:

1. Plot the drawdown on the arithmetic scale as a function of time since pumping began on the logarithmic scale. If the test has reached equilibrium, then determine the maximum drawdown, $(h_0 - h)_{max}$. If drawdown is still occurring, extrapolate the curve to find $(h_0 - h)_{max}$.
2. The drawdown at the inflection point $(h_0 - h)_i$ is defined as being equal to one-half the maximum drawdown.
3. From the graph, determine the time, $t_i$, when $(h_0 - h)_i$ occurs; also graphically find the slope of the drawdown curve at the inflection point $(m_i)$. This is generally equal to the slope of the straight portion of the drawdown curve. The slope is expressed as drawdown per log-cycle.

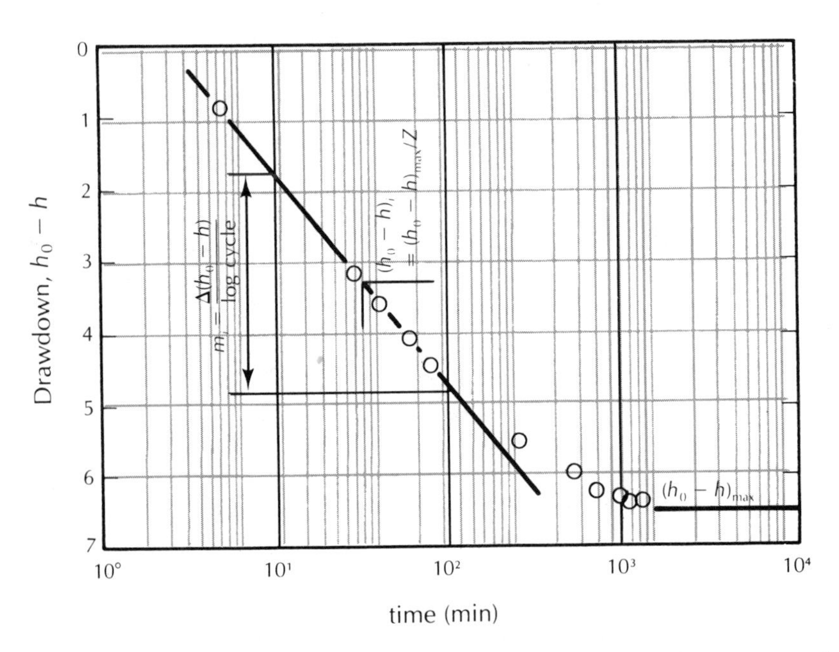


FIGURE 6.15 Plot of drawdown in a confined aquifer as a function of time on semilogarithmic paper for use in the Hantush inflection-point method of analysis.

The following relations hold true for the inflection point:

$$u_i = r^2S/4t_iT = r/2B \tag{6-36}$$

$$m_i = (2.3Q/4\pi T) \exp(-r/B) \tag{6-37}$$

$$(h_0 - h)_i = 0.5 (h_0 - h)_{\max} = \frac{Q}{4\pi T} K_0(r/B) \tag{6-38}$$

$$B = [T/(K'/b')]^{1/2} \tag{6-29}$$

$$f(r/B) = 2.3(h_0 - h)_i/m_i = \exp(r/B)K_0(r/B) \tag{6-39}$$

where $K_0$ is a function with values tabulated in Appendix 5 as $K_0(x)$ and $\exp(x)K_0(x)$.

From the drawdown and slope at the inflection point, the value of $f(r/B)$ may be found:

$$f(r/B) = 2.3(h_0 - h)_i/m_i \tag{6-40}$$

Knowing the value of $f(r/B)$, the function tables may be used to find the value of $r/B$, since $f(x) = \exp(x)K_0x$. Since $r$ is known, the value of $B$ may be easily found.

The transmissivity may be found from the relation

$$T = \frac{QK_0(r/B)}{2\pi(h_0 - h)_{\max}} \tag{6-41}$$

where $K_0$ is a Bessel function, known as a zero-order modified Bessel function of the second kind (see Appendix 5).

The value of the storativity is found from

$$S = 4t_iT/2rB \tag{6-42}$$

The conductivity of the semipervious layer may be determined if its thickness, $b'$, is known:

$$K' = \frac{Tb'}{B^2} \tag{6-43}$$

A computer program for the Hantush inflection-point method has been developed for use on a personal computer (17). Field data are fed into the computer and the program computes $T$, $S$, and $K'/b'$.

### 6.4.4 PUMPING TEST FOR A LEAKY ARTESIAN AQUIFER WITH STORAGE IN THE CONFINING LAYER

A type-curve method can also be used for the leaky confined aquifer with storage in the confining layer. A set of type curves on logarithmic paper are prepared from the tabulated values of $H(u,\beta)$ (Figure 6.16). On logarithmic paper of similar scale, drawdown is plotted against time. The field-data sheet is overlain on the type curves with the $H(u,\beta)$-axis parallel to the drawdown axis, and it is matched to a $\beta$-curve. A match point is selected and values of $H(u,\beta)$, $1/u$, drawdown, and time are obtained. The value of the curve is also noted.

From the match-point value of $H(u,\beta)$ and $h_0 - h$, the value of $T$ is found from Equation 6-27:

$$T = \frac{Q}{4\pi(h_0 - h)} H(u,\beta) \qquad \textbf{(6-44)}$$

and $S$ may be found from Equation 6-23 from the match-point values of $t$, $1/u$, the measured value of $r$, and the computed value of $T$:

$$S = \frac{4Tut}{r^2} \qquad \textbf{(6-45)}$$

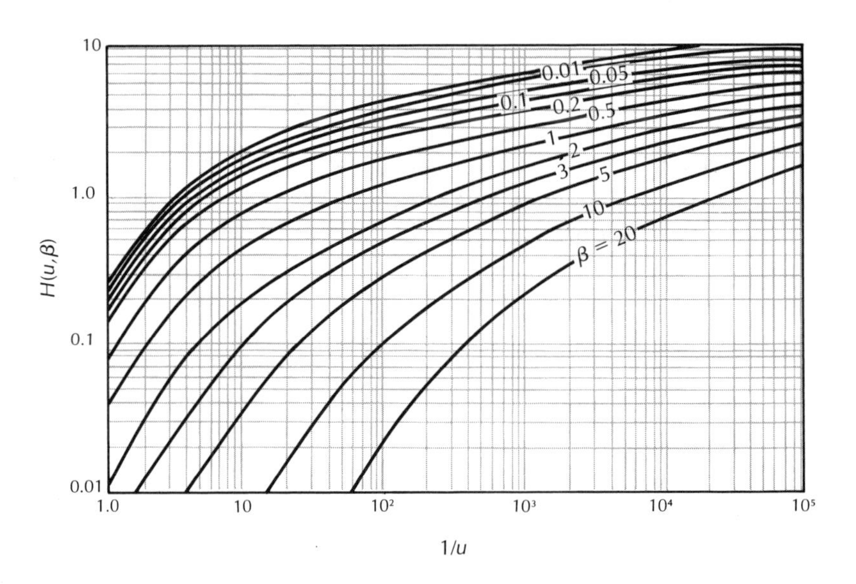


**FIGURE 6.16** Type curves for a well in an aquifer confined by a leaky layer that releases water from storage. Sources: Data from M. S. Hantush, *Journal of Geophysical Research*, 65 (1960):3713–25; type curves from W. C. Walton, *Groundwater Resource Evaluation* (New York: McGraw-Hill Book Company, 1970).

The value of $\beta$ can be used to compute the product $K'S'$ from Equation 6-28:

$$\beta^2 = \frac{r^2 S'}{16B^2 S} \tag{6-46}$$

$$B^2 = T/(K'/b') \tag{6-47}$$

Combining Equations 6-46 and 6-47,

$$K'S' = \frac{16\beta^2 Tb'S}{r^2}$$

If one of the values, either $K'$ or $S'$, is known, the value of the other can be found.

## 6.5 EFFECT OF PARTIAL PENETRATION OF WELLS

In many cases, the open hole or well screen of a pumping well does not extend from the top to the bottom of an aquifer. In all of the cases considered thus far, we assumed the well to penetrate the entire thickness of the aquifer. This caused flow in the aquifer to be essentially horizontal. However, the flow toward a partially penetrating well will be three-dimensional owing to vertical flow components (Figure 6.17). In addition, if the aquifer is anisotropic, the value of the vertical conductivity, $K_v$, and the horizontal conductivity, $K_h$, are important. This will affect both the amount of water pumped from the well and the potential field caused by drawdown.

If an observation well completely penetrates an aquifer, or if a partially penetrating well is located more than $1.5b\sqrt{K_h/K_v}$ distance units from the

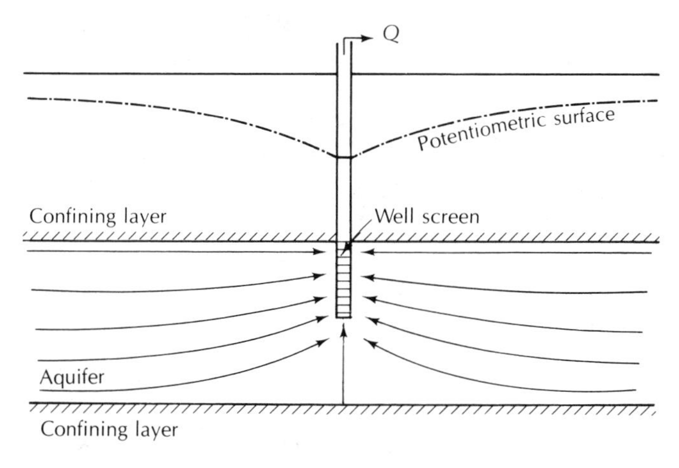


**FIGURE 6.17** Flow lines toward a partially penetrating well in a confined aquifer.

pumped well, the effect of a partially penetrating pumping well is negligible (1). The drawdown is described by either the Hantush-Jacob formula for leaky aquifers or the Theis formula for nonleaky aquifers.

If, however, the pumping well is partially penetrating, and the observation wells are also partially penetrating and located closer to the pumping well than $1.5b\sqrt{K_h/K_v}$ distance units, then the drawdown formula is different and quite complex. Hantush (1) makes the following observations about the drawdown in observation wells in such cases:

1. If two observation wells equidistant from the pumping well are screened in different parts of the aquifer, the time-drawdown curves may be different.
2. Depending upon the length and relative position of observation-well screens, it is possible for a more distant well to have a greater drawdown than a closer well.
3. The effects of partial penetration produce a time-drawdown curve similar in shape to one produced when there is a downward leakage from storage through a thick, semipervious layer.
4. Partial-penetration effects may produce a time-drawdown curve that resembles the effect of a recharge boundary, a fully penetrating well in either a sloping water-table aquifer or an aquifer of nonuniform thickness.

The preceding observations suggest that the hydrogeologist must always use extreme care in collecting and interpreting pumping-test data. Test drilling should be used to delineate the aquifer system before the pumping test. If a pumping test is to be made on a partially penetrating well, fully penetrating observation wells or wells located at a distance of more than $1.5b\sqrt{K_h/K_v}$ would be desirable. The analysis of partial-penetration effects is beyond the scope of this book. The hydrogeologist who must deal with this problem can use the **Hantush partial-penetration method** (18). Using partially penetrating wells, it is also possible to run pumping tests that determine the value of both horizontal and vertical conductivity of the confining layer and the storativity of both the aquifer and the confining layer (19–22).

The pumping-test solutions given in the preceding sections are all based on some simplifying assumptions that may not be valid in certain circumstances. For leaky artesian aquifer flow, the assumption that the storage in the aquifer is negligible may be erroneous (13, 14). The effect of storage in the confining layer occurs early in the pumping test, as leakage water is being furnished from storage in the confining layer. With increasing time, more of the leakage is being contributed by the aquifer above the leaky confining layer, and the amount of water from storage diminishes. If analysis of pumping-test data for a leaky artesian aquifer does not follow an $r/B$-curve for early drawdown data, significant flow from storage should be suspected and an alternative analysis made (13, 14, 22–24).

The Theis solution may overestimate the value of the storativity of shallow elastic aquifers. If the ratio of the average depth of a confined aquifer to its thickness is less than 0.5, the overestimation can be as great as 40 percent owing to three-dimensional consolidation of the aquifer (14). If the average depth-to-thickness ratio is 1.0 or more, then the Theis solution is valid even for deforming aquifers.

## 6.6 WATER-TABLE AQUIFER

A well pumping from a water-table aquifer extracts water by two mechanisms. As with confined aquifers, the decline in pressure in the aquifer yields water because of the elastic storage of the aquifer storativity ($S_s$). The declining water table also yields water as it drains under gravity from the sediments. This is termed specific yield ($S_y$). The flow equation has been solved for radial flow in compressible unconfined aquifers under a number of different conditions and by use of a variety of mathematical gambits (22, 25–36). The many and various equations in these solutions can lead to confusion; however, a qualitative description of the response of a water-table well to pumping may be helpful.

There are three distinct phases of time-drawdown relations in water-table wells. We will examine the response of any typical annular region of the aquifer located a constant distance from the pumping well. Some time after pumping has begun, the pressure in the annular region will drop. As the pressure first drops, the aquifer responds by contributing a small volume of water as a result of the expansion of water and compression of the aquifer. During this time, the aquifer behaves as an artesian aquifer, and the time-drawdown data follow the Theis nonequilibrium curve for $S$ equal to the elastic storativity of the aquifer. Flow is horizontal during this period, as the water is being derived from the entire aquifer thickness.

Following this initial phase, the water table begins to decline. Water is now being derived primarily from the gravity drainage of the aquifer, and there are both horizontal and vertical flow components. The drawdown-time relationship is a function of the ratio of horizontal-to-vertical conductivities of the aquifer, the distance to the pumping well, and the thickness of the aquifer.

As time progresses, the rate of drawdown decreases and the contribution of the particular annular region to the overall well discharge diminishes. Flow is again essentially horizontal, and the time-discharge data again follow a Theis type curve. The Theis curve now corresponds to one with a storativity equal to the specific yield of the aquifer. The importance of the vertical flow component as it affects the average drawdown is directly related to the magnitude of the ratio of the specific yield to the elastic storage coefficient ($S_y/S_s$). As the value of $S_s$ approaches zero, the time duration of the first stage of drawdown also approaches zero. As $S_y$ approaches zero, the length of the first stage increases, so that if $S_y = 0$, the aquifer behaves as an artesian aquifer of storativity $S_s$ (36).

A number of type-curve solutions have been developed (37). The one that we will consider is based on a fully penetrating production well (34, 38). The drawdown response is measured in observation wells that also fully penetrate the aquifer, and the well is pumped at a constant rate.

The flow equation for unconfined aquifers is given by (39)

$$T = \frac{Q}{4\pi(h_0 - h)} W(u_A, u_B, \Gamma) \tag{6-48}$$

where $W(u_A, u_B, \Gamma)$ is the well function for the water-table aquifer, and

$$u_A = \frac{r^2 S}{4Tt} \text{ (for early drawdown data)} \tag{6-49}$$

$$u_B = \frac{r^2 S_y}{4Tt} \text{ (for later drawdown data)} \tag{6-50}$$

$$\Gamma = \frac{r^2 K_v}{b^2 K_h} \tag{6-51}$$

where

s = $h_0 - h$ is the drawdown

$Q$ is the pumping rate

$T$ is the transmissivity

$r$ is the radial distance from the pumping well

$S$ is the storativity

$S_y$ is the specific yield

$t$ is the time

$K_h$ is the horizontal hydraulic conductivity

$K_v$ is the vertical hydraulic conductivity

$b$ is the initial saturated thickness of the aquifer

Two sets of type curves are used. Type-A curves are good for early drawdown data, when instantaneous release of water from storage is occurring. As time elapses, the effects of gravity drainage and vertical flow cause deviations from the nonequilibrium type curve, which is accounted for in the family of Type-A curves. The Type-B curves are used for late drawdown data, when effects of gravity drainage are becoming smaller. The type-B curves end on a Theis curve. Figure 6.18 shows the two sets of type curves for fully penetrating wells. Values of $W(u_A, \Gamma)$ and $W(u_B, \Gamma)$ are found in Appendix 6.

The type curves are then used to evaluate the field data for time and drawdown, which are plotted on logarithmic paper of the same scale as the type curve. The following procedure can be used (36):

1. Superpose the latest time-drawdown data on the Type-B curves. The axes of the graph papers should be parallel and the data matched to the curve with the best fit. At any match point, the values of $W(u_B,\Gamma)$, $u_B$, $t$, and $h_0 - h$ are determined. The value of $\Gamma$ comes from the type curve. The value of $T$ is found using these values and Equation 6-48. The specific yield is found from Equation 6-50.
2. The early drawdown data are then superposed on the Type-A curve for the $\Gamma$-value of the previously matched Type-B curve. A new set of match points is determined. The value of $T$ calculated from Equation 6-48 should be approximately equal to that computed from the Type-B curve. Equation 6-49 can be used to compute the storativity.
3. The value of the horizontal hydraulic conductivity can be determined from

$$K_h = T/b \tag{6-52}$$

4. The value of the vertical hydraulic conductivity can also be computed using the $\Gamma$-value of the matched type curve. Rearrangement of Equation 6-51 yields the following formula:

$$K_v = \frac{\Gamma b^2 K_h}{r^2} \tag{6-53}$$

The preceding analysis is based upon a very low value of drawdown compared with the saturated thickness of the aquifer. If the drawdown is sub-

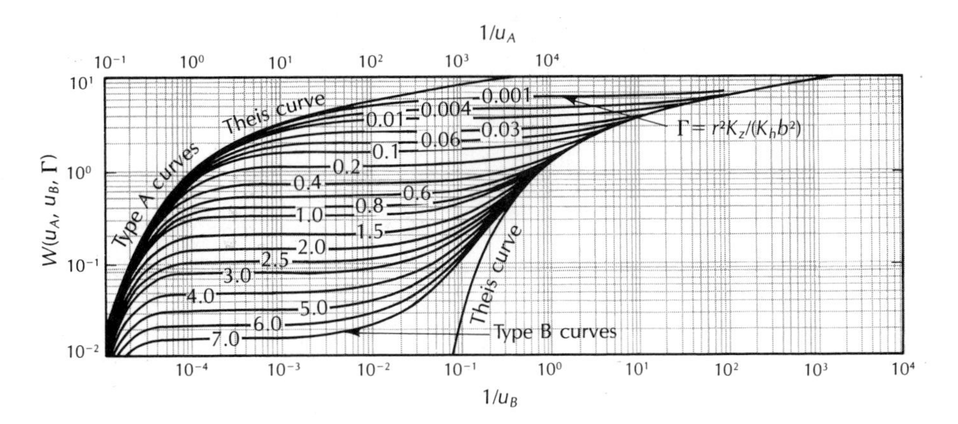


**FIGURE 6.18** Type curves for drawdown data from fully penetrating wells in an unconfined aquifer. Source: S. P. Neuman, *Water Resources Research,* 11 (1975):329–42.

stantial, some authorities (35) suggest that the drawdown data be corrected. The corrected drawdown $(h_0 - h)'$ is found from the relation (11)

$$(h_0 - h)' = (h_0 - h) - [(h_0 - h)^2/2h_0] \tag{6-54}$$

This is normally necessary only for the later time-drawdown data; if $h_0 - h$ is small compared with $h_0$, correction will not be needed.

**EXAMPLE PROBLEM** A well pumping at 1000 gallons per minute fully penetrates an unconfined aquifer with an initial saturated thickness of 100 feet. Time-drawdown data for a well located 200 feet away are plotted on log paper (Figure 6.19). Find $T$, $S_y$, $S$, $K_h$, and $K_v$.

The later time-drawdown data fit best on the $\Gamma = 0.1$ type curve. With the axes of the two sheets of graph paper parallel, the selected match point has values of $W(u_B,\Gamma) = 0.1$, $1/u_B = 10$, $h_0 - h = 0.043$ feet, and $t = 128$ minutes. The value of $u_B$ is 0.1, the pumping rate is equal to $1.9 \times 10^5$ cubic feet per day, and $t = 0.089$ days. From Equation 6-48,

$$\begin{aligned} T &= \frac{Q}{4\pi(h_0 - h)} W(u_B,\Gamma) \\ &= \frac{1.9 \times 10^5 \text{ ft}^3\text{/day} \times 0.1}{4\pi \times 0.043 \text{ ft}} \\ &= 3.52 \times 10^4 \text{ ft}^2\text{/day} \end{aligned}$$

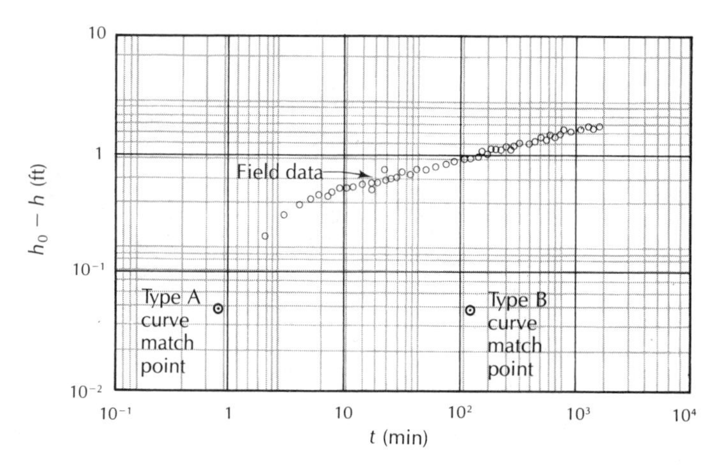


**FIGURE 6.19** Field data for Example Problem of analysis of aquifer test for an unconfined aquifer.

The value of the specific yield can be determined from Equation 6-50:

$$S_y = \frac{4Ttu_B}{r^2}$$

$$= \frac{4 \times 3.52 \times 10^4 \text{ ft}^2/\text{day} \times 0.089 \text{ days} \times 0.1}{(200 \text{ ft})^2}$$

$$= 0.031$$

The early time-drawdown data are now matched to the Type-A curve for $\Gamma = 0.1$. The selected match point is $W(u_A,\Gamma) = 0.1$, $1/u_A = 1.0$, $h_0 - h = 0.041$ foot and $t = 0.9$ minute. The value of $u_A$ is 1.0 and $t = 6.25 \times 10^{-4}$ day. From Equation 6-48,

$$T = \frac{Q}{4\pi(h_0 - h)} W(u_A\Gamma)$$

$$= \frac{1.9 \times 10^5 \text{ ft}^3/\text{day} \times 0.1}{4\pi \times 0.041 \text{ ft}}$$

$$= 3.69 \times 10^4 \text{ ft}^2/\text{day}$$

The storativity value is found from Equation 6-49:

$$S = \frac{4Ttu_A}{r^2}$$

$$= \frac{4 \times 3.69 \times 10^4 \text{ ft}^2/\text{day} \times 6.25 \times 10^{-4} \text{ days} \times 1.0}{(200 \text{ ft})^2}$$

$$= 0.0023$$

The value of the horizontal hydraulic conductivity can be found from Equation 6-52. The average of $T$ is $3.61 \times 10^4$ square feet per day:

$$K_h = T/b$$

$$= (3.61 \times 10^4 \text{ ft}^2/\text{day})/100 \text{ ft}$$

$$= 361 \text{ ft/day}$$

and the value of $K_v$ is determined using Equation 6-53:

$$K_v = \frac{\Gamma b^2 K_h}{r^2}$$

$$= \frac{0.1 \times (100 \text{ ft})^2 \times 361 \text{ ft/day}}{(200 \text{ ft})^2}$$

$$= 9.0 \text{ ft/day}$$

## 6.7 MEASUREMENT OF AQUIFER PARAMETERS BY USE OF PIEZOMETERS

In Section 6.3.4 a method of measuring aquifer parameters by use of a piezometer that fully penetrates a confined aquifer is presented. In many cases piezometers, or auger holes, are installed which do not fully penetrate an aquifer. They are generally installed at a specific depth to monitor head and sample ground-water quality. A very convenient method exists to use these piezometers to determine the hydraulic conductivity of the formation in which the screen is installed. This is the **Hvorslev method** (40).

Figure 6.20 shows the geometry of a piezometer installed in an aquifer. In the case of a piezometer installed into a low-permeability unit, special attention must be paid to the method of construction. In many such cases, a hole is augered into a clay unit, the well and well screen are lowered into the open borehole, and gravel is used to fill the open annulus between the well screen and the wall of the open hole. Under such conditions the radius of the well screen,

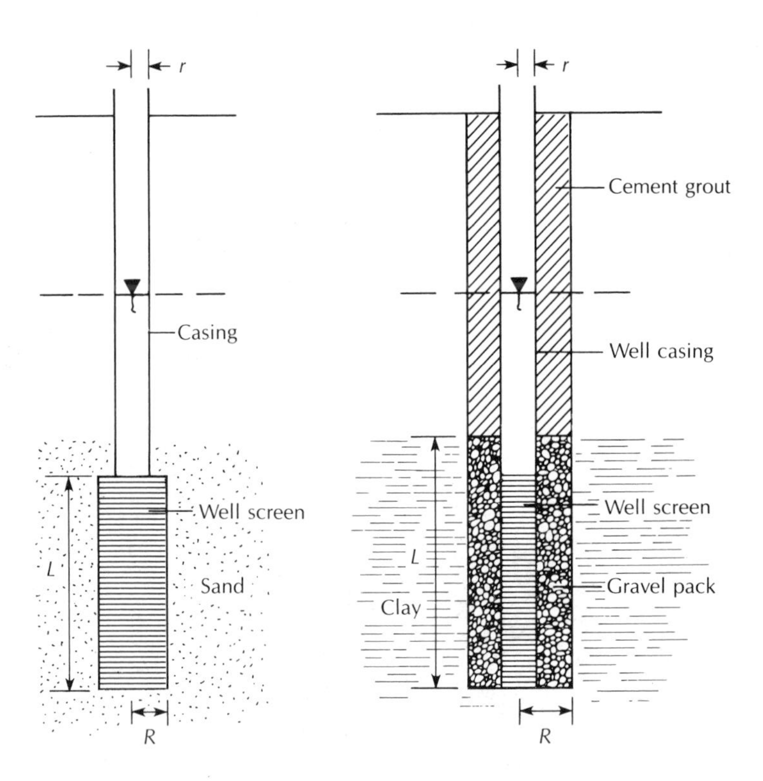


**FIGURE 6.20** Piezometer geometry for Hvorslev method. Note that for a piezometer installed in a low-permeability unit the value $R$ is the radius of the highest permeable zone that includes the gravel pack zone and $L$ is the length of the gravel pack zone.

$R$, is the radius of the bored hole and the length of the well screen, $L$, is the length of the gravel pack. (The gravel pack would typically be extended one to several feet above the well screen and the remainder of the open hole backfilled with some type of grout.)

Naturally the Hvorslev method can be applied only below the water table. Water is either added to the well casing or withdrawn by bailing out the casing with a special tool called a bailer. It is possible to induce the water column in the well casing to rise by rapidly lowering a solid piece of metal, called a slug, into the well and submerging it below the original water surface, or static water level. This is equivalent to adding a volume of water to the well equal to the volume of the slug. The water level in the well is measured prior to the time that the slug is lowered. Immediately after the slug is lowered, the water level in the well is again measured. The water levels are measured at timed intervals as the water level falls back toward the static water level. The height the water level rises above the static water level immediately upon lowering the slug is $h_0$. The height of the water level above the static water level some time, $t$, after the slug is lowered is $h$. The data are plotted by computing the ratio $h/h_0$ and plotting that versus time on semilogarithmic paper, as shown in Figure 6.21. The time-drawdown data should plot on a straight line.

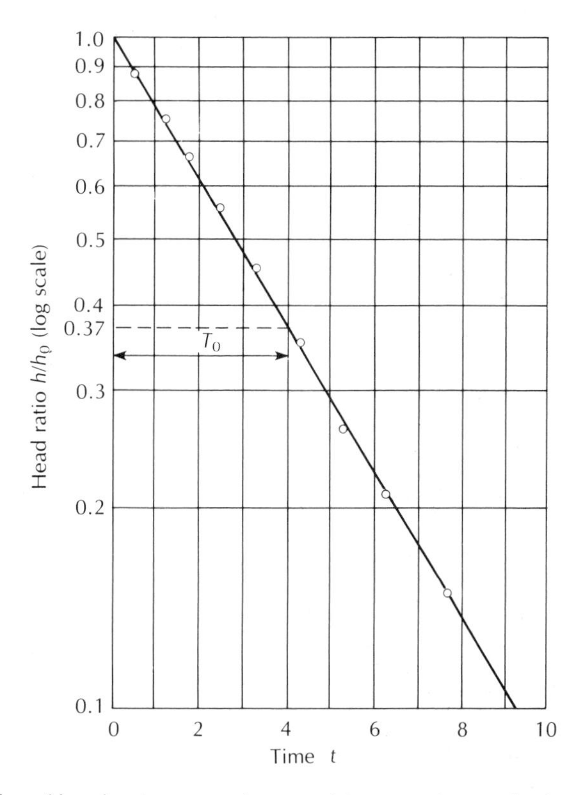


**FIGURE 6.21** Plot of head ratio versus time used for Hvorslev method.

If the length of the piezometer is more than 8 times the radius of the well screen ($L/R>8$), the following formula applies:

$$K = \frac{r^2 \ln (L/R)}{2LT_0} \tag{6-55}$$

where

$K$ is hydraulic permeability
$r$ is the radius of the well casing
$R$ is the radius of the well screen
$L$ is the length of the well screen
$T_0$ is the time it takes for the water level to rise or fall to 37 percent of the initial change (Figure 6.21)

Equation 6-55 is but one of many formulae presented by Hvorslev for differing piezometer geometry and aquifer conditions. However, it is one that is quite useful and could be applied to unconfined conditions for most piezometer designs where the length is typically quite a bit greater than the radius of the well screen. For other conditions, the original paper should be consulted.

There are several other methods described in the literature for analyzing data from slug tests (6, 40, 41). The tests are all performed in the same manner. A piezometer or monitoring well is installed and developed. The initial water level is measured; then the well is either bailed down, or a volume of water is added, or a slug is lowered, or a slug is raised. The water level immediately after the well is perturbated is measured and then the change in water level with time is measured. Piezometers in highly permeable material may recover to the original water level in 30 seconds or less. In such cases, it is necessary to use a pressure transducer to record pressure changes in the well as the water level changes along with automatic electronic signal recording equipment. Commercial equipment of this type can measure the water level every second and automatically record it. Piezometers in low-permeability clays may take hours to days to recover. In these instances the water levels may be easily measured with a steel tape.

Some studies have been performed that compared results of slug-test data analyzed by two or more methods. In one study in highly permeable sands (42), comparable results were obtained by using either the Bouwer and Rice method (41) or the Hvorslev method (40). Another study, performed on a fractured glacial till (43), compared the results of the Cooper-Bredehoeft-Papadopulos method (9, 10) and the Nguyen and Pinder method (44) and found both methods yielded similar results. Methods may vary in the type of wells for which they are applicable and the length of time over which the data may need to be collected.

**EXAMPLE PROBLEM** A slug test is performed by lowering a metal slug into a piezometer that is screened in a coarse sand. The inside diameter of both the well screen and the well casing is 2 inches. The well screen is 10 feet in length. A pressure transducer was used to record the water level every second for the first 10 seconds and less frequently thereafter. The following data were obtained when the slug was rapidly pulled from the piezometer:

| Elapsed time (seconds) | Depth to Water (feet) | Change in Water Level $h$ (feet) | $h/h_0$ |
|---|---|---|---|
| Static Level | 13.99 | | |
| 0 | 14.87 | 0.88 ($h_0$) | 1.000 |
| 1 | 14.59 | 0.60 | 0.682 |
| 2 | 14.37 | 0.38 | 0.432 |
| 3 | 14.20 | 0.21 | 0.239 |
| 4 | 14.11 | 0.12 | 0.136 |
| 5 | 14.05 | 0.06 | 0.068 |
| 6 | 14.03 | 0.04 | 0.045 |
| 7 | 14.01 | 0.02 | 0.023 |
| 8 | 14.00 | 0.01 | 0.011 |
| 9 | 13.99 | 0.00 | 0.000 |

Figure 6.22 contains a plot of the data. The time for the head to rise to 37 percent of the initial change is 1.8 seconds ($T_0$). The following values are obtained from the geometry of the piezometer:

$$r = 0.083 \text{ feet}$$
$$R = 0.083 \text{ feet}$$
$$L = 10 \text{ feet}$$

The ratio $L/R$ is 120.5, which is more than 8, so that Equation 6-55 can be used:

$$K = \frac{r^2 \ln (L/R)}{2LT_0}$$

$$= \frac{(0.083 \text{ ft})^2 \times \ln (10/0.083)}{2 \times 10 \text{ ft} \times 1.8 \text{ sec}}$$

$$= 9.17 \times 10^{-4} \text{ ft/sec} \times 8.64 \times 10^4 \text{ sec/day}$$

$$= 79.2 \text{ ft/day}$$

## 6.8 STEADY-STATE RADIAL FLOW, THEIM EQUATIONS

Prior to the development of the Theis nonequilibrium formula, people doing well analysis assumed a condition of steady flow. That is, the drawdown rate in the pumped well was assumed to be so low as to be essentially zero. This assump-

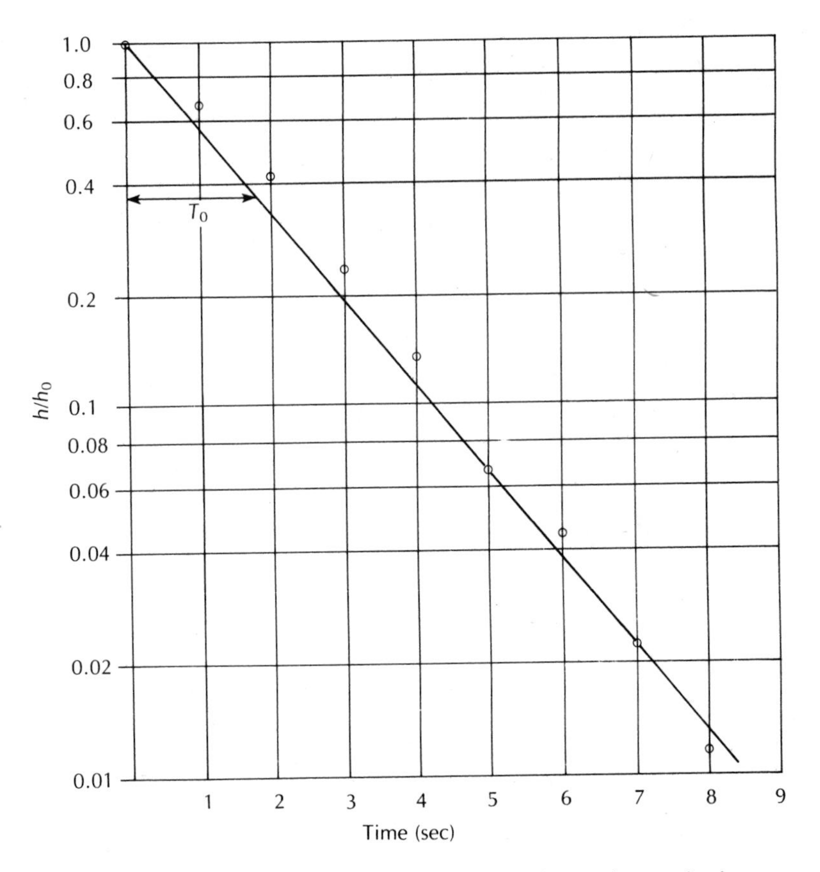


**FIGURE 6.22** Plot of $h/h_0$ versus $t$ for Example Problem of Hvorslev method.

tion is valid for both nonleaky artesian and water-table aquifers if a long period of time has elapsed since the start of pumping (45). For nonleaky artesian aquifers, the appropriate equation is

$$T = \frac{Q}{2\pi(h_2 - h_1)} \ln (r_2/r_1) \tag{6-56}$$

where

$Q$ is the pumping rate
$h_1$ is the head at distance $r_1$ from the pumping well
$h_2$ is the head at distance $r_2$ from the pumping well

For an unconfined aquifer,

$$K = \frac{Q}{\pi(h_2^2 - h_1^2)} \ln (r_2/r_1) \tag{6-57}$$

To find values of $T$ or $K$ from steady-state equations, there must be at least two observation wells at different distances from the pumping well. The well must be pumped long enough for the drawdown to approach a steady-state condition.

The usefulness of steady-state analysis is limited, as values of storativity or specific yield are not obtained. However, transmissivity or hydraulic conductivity values obtained by the above equations are likely to be more accurate than those obtained from a transient analysis.

There may be circumstances in which an existing well that pumps constantly and cannot be shut down must be analyzed. If the well has reached a steady-state condition and discharges at a constant rate, and there are at least two observation wells, the aquifer transmissivity can be found from either Equation 6-56 or 6-57, depending on whether it is a confined or a water-table aquifer.

## 6.9 INTERSECTING PUMPING CONES AND WELL INTERFERENCE

The cases we have considered thus far have involved only one well pumping from an aquifer. However, there are often several wells tapping the same aquifer, resulting in intersecting pumping cones. At any given point in a confined aquifer, the total drawdown is the sum of the individual drawdowns for each well. Because the Laplace equation is linear, the superposition of drawdown effects is found by simple addition. In Figure 6.23, the well interference for a multiple-aquifer well field is presented graphically. Linear superposition is valid only for confined aquifers, in which the value of the transmissivity does not change with drawdown. In water-table aquifers, if the drawdown is significant in relation to the total saturated thickness, the use of linear superposition will result in a predicted composite drawdown that is less than the actual composite drawdown. As a decrease in saturated thickness reduces the transmissivity, the mul-

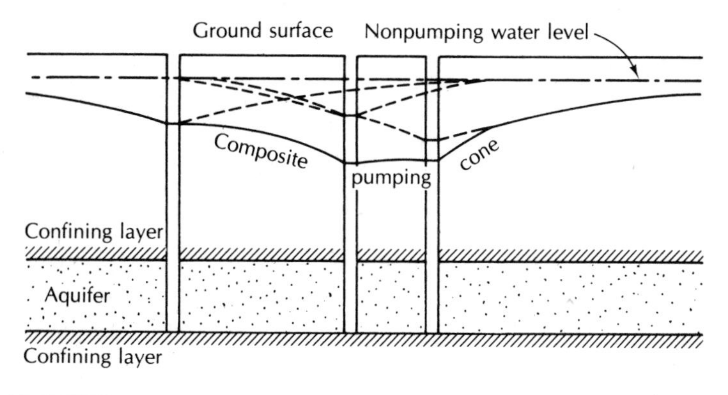


**FIGURE 6.23** Composite pumping cone for three wells tapping the same aquifer. Each well is pumping at a different rate; thus, the pumping level of each is different.

tiple-well system will result in a composite hydraulic gradient greater than that of an equivalent confined system in order to compensate for a reduced value of aquifer transmissivity.

In designing well-field layouts, it is necessary to take into account well interference. The level of the water in the well during pumping determines the length of pipe necessary to carry water to the surface. The characteristics of the well pump and the horsepower requirements of the motor also depend upon the depth to the pumping level. If wells are spaced too closely together, the amount of well interference could be excessive. Aligning wells parallel to a line source of recharge, such as a river, would result in less well interference than would a perpendicular configuration.

## 6.10 EFFECT OF HYDROGEOLOGIC BOUNDARIES

If a well is not located in an aquifer of infinite areal extent, as is the case with all real wells in real aquifers, the drawdown cone will extend until either the well is supplied by vertical recharge or a hydrogeologic boundary is reached. A hy-

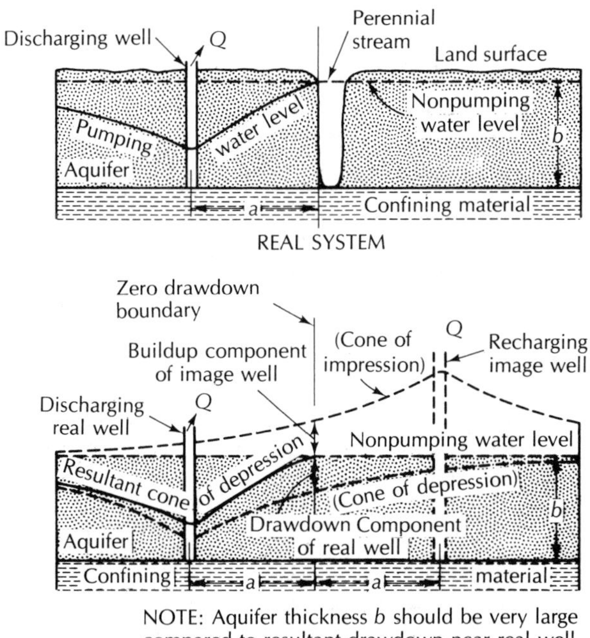


**FIGURE 6.24** Idealized cross section of a well in an aquifer bounded on one side by a stream. Source: J. G. Ferris et al., U.S. Geological Survey Water-Supply Paper 1536-E, 1962.

drogeologic boundary could be the edge of the aquifer, a region of recharge to a fully confined artesian aquifer, or a source of recharge, such as a stream or lake.

Boundaries are considered to be either recharge or barrier boundaries. A **recharge boundary** is a region in which the aquifer is replenished. A **barrier boundary** is an edge of the aquifer, where it terminates, either by thinning or abutting a low-permeability formation, or has been eroded away.

Figure 6.24 shows a well bounded by a recharge boundary. The recharge boundary can be simulated by a recharging **image well** located an equivalent distance away from the recharge boundary but on the opposite side. Figure 6.25 indicates the presence of a barrier boundary. The barrier boundary is simulated by a discharging image well located an equivalent distance away from the boundary but on the opposite side. Boundaries have the most dramatic impact on the drawdown of a pumped well for the aquifer with no source of vertical recharge. As the well withdraws water only from storage in the aquifer, drawdown proceeds as a function of the logarithm of time.

Figure 6.26 shows a theoretical straight-line plot of drawdown as a function of time on semilogarithmic paper. The effect of a recharge boundary is to retard the rate of drawdown. Change in drawdown can become zero if the well

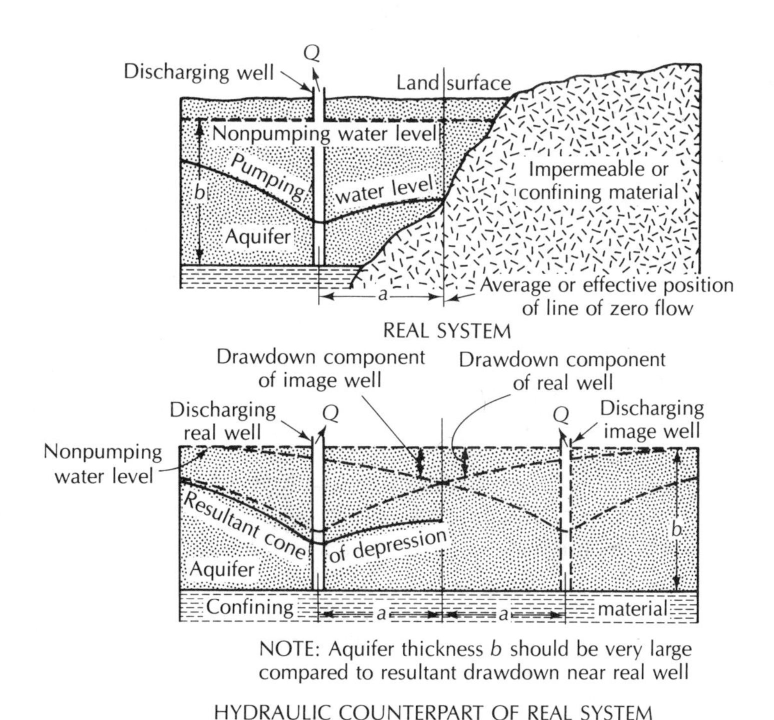


**FIGURE 6.25** Idealized cross section of a well in an aquifer bounded on one side by an impermeable boundary. Source: J. G. Ferris et al., U.S. Geological Survey Water-Supply Paper 1536-E, 1962.

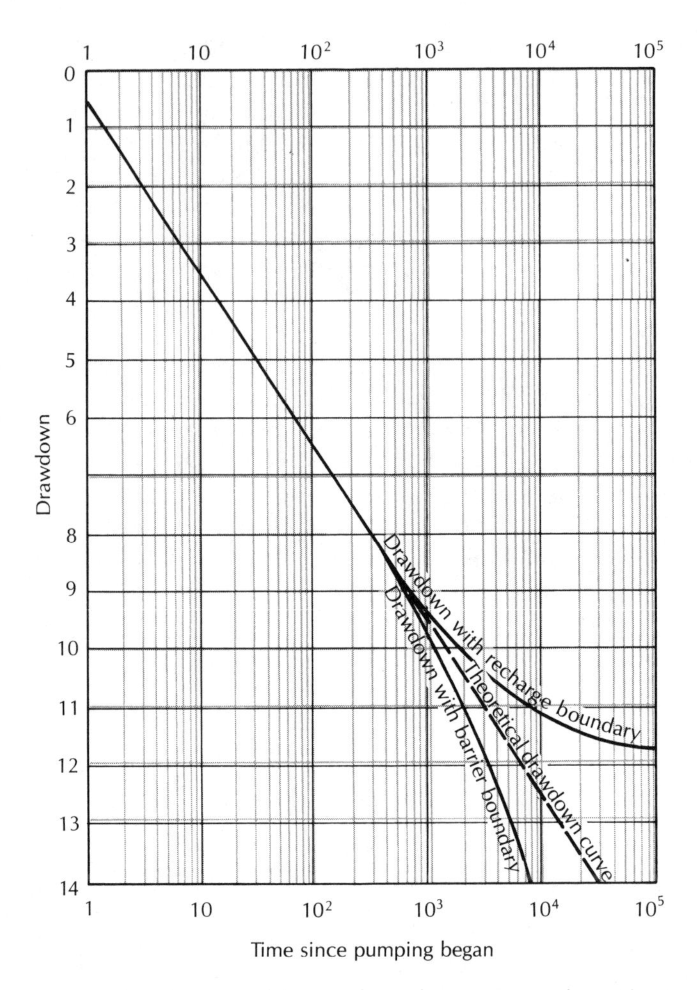


FIGURE 6.26 Impact of recharge and barrier boundaries on semilogarithmic drawdown-time curves.

comes to be supplied entirely with recharged water. The effect of a barrier to flow in some region of the aquifer is to accelerate the drawdown rate. The water level declines faster than the theoretical straight line.

## 6.11 PUMPING-TEST DESIGN

Adequate design and execution of a pumping test involves considerable planning and attention to detail. An understanding of fundamental well hydraulics is necessary, not only for the interpretation of data, but also for the experimental design by which valid and usable data are obtained. The purpose of the pumping test must be established first. Determining the yield of a new well involves simply pumping the well. This type of test, as it is generally conducted, yields only the scantiest information about the aquifer itself. With careful planning, the pumping-well test can yield data to compute the aquifer transmissivity. It can also indicate the general type of aquifer.

If a test well has been drilled prior to the installation of a production well, a reasonable conjecture can be made as to the probability the well will be unconfined, semiconfined, or confined. However, the presence or absence of recharge or barrier boundaries may not be known. Indeed, this is one of the reasons to perform a long-term pumping test. If one makes a semilogarithmic plot of drawdown versus time (e.g., Figure 6.26) one can inspect it to see if the pumping level of the well stabilizes. If this occurs, this means that there is a source of recharge, either vertically by leakage across a semiconfining layer or horizontally from a recharge boundary. If the water level falls faster than the theoretical drawdown curve, then the presence of a barrier boundary must be considered.

The amount of information gained from a pumping test expands greatly if one or more observation wells are involved in addition to the pumping well. Both transmissivity and storativity of the aquifer can be determined, as can the vertical hydraulic conductivity of any overlying semipervious layers. More eloquent tests can be used to determine the value of the vertical anisotropy of the formation. Radial anisotropy and recharge or barrier boundaries can also be detected.

### 6.11.1 SINGLE-WELL PUMPING TESTS

The basics of a single-well pumping test are also applicable to pumping tests involving multiple wells. The first step is to determine the location of the well to be drilled. This is best done on the basis of detailed exploration using geological, geophysical, and perhaps aerial photo techniques. However, the location of the well is often dictated by economic or engineering factors. If economic or engineering factors predominate, the hydrogeologist should determine if there is a reasonable chance of success based on the known hydrogeology of the site.

A test well may be bored as the first step, or the production well may be drilled immediately. The geologist should make a log of the geologic formations encountered. The water level in the drilled hole should be recorded as a function of the depth of the hole; however, this might not be possible if certain drilling techniques such as rotary and reverse rotary are used. Based on the test hole and selected borehole geophysical studies, the hydrogeologist can determine the depth and thickness of potential aquifer zones. An aquifer is selected, and a test or permanent well is installed. If at all feasible, the well should be open throughout the entire thickness of the aquifer. The physical dimensions of the well should be recorded, along with the depth, thickness, and type of aquifer. A description of the aquifer material should be included. An inventory of nearby wells should be made, and it should be determined whether any other wells will be running when an aquifer test is planned. Intersecting cones of depression during an aquifer test should be avoided.

A pump is installed in the completed well. The pump, engine, wiring, piping, and assorted equipment must be reliable. If a pumping test is terminated owing to mechanical failure prior to the planned time, the data may not be sufficient. The discharge pipe from the pump must be equipped with a valve to control the volume of flow and with a means of measuring the flow.

Small pumping rates may be measured by means of a water meter in the line or by filling a container of known volume in a measured time. These methods are generally useful for flows of 100 gallons per minute (6 liters per second) or less. For larger flows, a common method is to use a circular-orifice weir on the discharge pipe (46). Generally, the well-drilling contractor furnishes the equipment and makes the discharge measurements. The hydrogeologist should always check the apparatus and measurements.

There must be a means of conveying the pumped water away from the test site. This is especially true for shallow, unconfined aquifers, where the water could recharge the aquifer and render any pumping-test results useless. If a pumping test runs for several days, the quantity of pumped water that must be conveyed from the site can be considerable.

A means to measure the water level in the well before as well as during pumping must be available. A steel tape, air line, or electrical tape can serve this function (46).

The pumping tests we have studied all have been based on a constant discharge rate. In reality, the water level in the well falls with pumping, the pumping lift increases, and the discharge of the pump tends to decline. To avoid this, the valve on the discharge pipe should be partially closed to restrict the initial discharge. During the course of the test, the valve can be opened as necessary to keep the pumping rate constant. There should be no more than a 10 percent variation in rate during the test, with a smaller pumping-rate variation if possible. After construction, a new well is usually developed by on-and-off pumping, which causes the water to surge back and forth through the well screen or open hole. This increases the yield of the well by washing out fine particles and mud used for drilling. The well should be fully developed before pumping tests are made.

It is necessary to select a measuring point on the well to serve as a fixed reference for water-level measurements. The measuring point should be marked, and its elevation measured and recorded. Prior to the pumping test, the water level should be measured. Other production wells that may be nearby should either be shut down for the duration of the test or pumped at a constant rate. It is difficult to correct pumping-test data for the effect of wells starting and stopping. For the most accurate results, the nonpumping water elevation should be measured several times before pumping begins. Ground-water levels may have a long-term trend of rising or falling. They may also be affected by tides or changes in the barometric pressure. If the static level is found to fluctuate, then detailed pretest measurements must be made for at least twice the expected length of the pumping test. If a long-term linear trend is observed, the drawdown observed during the pumping test must be corrected. The corrected drawdown is the difference between the measured depth to water and the projected static level based on the long-term trend. Tidal fluctuations and fluctuations due to barometric pressure require measurements of the water level in the well and either the tide or air pressure prior to pumping, in order to establish a relationship between them. Measurements of tides or air pressure must then be made during the pumping test to find a corrected static level at the time of each drawdown measurement.

Prior to the test, the discharge of the well should be measured and a planned pumping rate selected. The valve should be preset for this rate. When the pump is started for the test, the valve is adjusted to yield the desired pumping rate. Periodic measurements of discharge and corrections using the valve should be made about every half hour. The time and measurement of pumping rate should be recorded in the field notes.

Water-level readings in the well commence after one minute of pumping. The usual procedure is to record the depth of water below the measuring point and the time of measurement. Computation of drawdown is made later. On the order of ten readings per log cycle of time are made. The first reading could be at one minute, then at one and one-half minutes, and then at every minute up to ten. Between ten and twenty minutes, readings are taken every two minutes; then every five minutes between twenty minutes and one hour; and every ten minutes between one and two hours. After two hours have elapsed, the recommended ten readings per log cycle of time are made. The most work occurs during the first minutes of the pumping test.

There must be some advance planning for the length of time the test is to proceed. For an artesian or leaky artesian aquifer, the test may last twenty-four hours or less. This is often sufficient to delineate values for the formation constants and to determine whether there are any recharge or barrier boundaries. Water-table wells must be pumped for a sufficient duration to preclude any significant effects of vertical flow near the well. The time period is a function of the distance from the pumping well, the conductivity of the formation, and the degree of anisotropy. The time is naturally greater the farther the distance from the pumping well, the lower the horizontal conductivity, and the greater the anisotropy. Pumping tests of unconfined aquifers normally run for several days to several weeks.

Periodically during the pumping test, a sample of the water is collected for chemical analysis. A series of samples will reveal any trend in chemical or bacteriological quality with continued pumping. It is possible to predict final well quality fairly accurately, even on the basis of chemical analysis of water from temporary test wells (47).

Following the collection of time and drawdown data in the pumped well, an analysis is made using the appropriate type curves. It should be remembered that with only one well, the valve of $T$, but not of $S$, can be determined. In another computation usually made at the end of the pumping test, the well yield is divided by the maximum drawdown to obtain the specific capacity, which is widely used as an index of the capacity of the well.

### 6.11.2 PUMPING TESTS WITH OBSERVATION WELLS

If it is feasible, drawdown should be measured in one or more nonpumping wells situated close to the pumping well. These observation wells are often constructed especially for the pumping tests. However, domestic-supply wells, abandoned wells, other wells in a well field, or other wells under construction are sometimes used. The use of one or more observation wells can enable the hydrogeologist to compute the storativity or specific yield of the aquifer. Under some circumstances, the anisotropy of the pumping well and the leakage factor for leaky confined aquifers may be determined. In most cases, the hydrogeologist will be able to employ only one observation well—especially if the aquifer is deep and the area is undeveloped, with no existing wells available.

The selection of the location of the single observation well is critical. It should be located at a radial distance such that the time-drawdown data collected during the planned pumping period will fall on a type curve of unique curvature. If the curvature is too flat, selection of the proper type curve is difficult. After a test boring is made, the hydrogeologist should know the type of aquifer system to be confronted. The formation characteristics are then estimated, and using the correct formula with the planned pumping rate, time-drawdown curves for several hypothetical observation wells at different distances are plotted. On the basis of these curves, the location of the single observation well is selected. As a general rule, it will be closer for a water-table well than for a confined well.

If there are two observation wells, the second should be in a radial line with the first, but at ten times the distance. If there are more than two wells, they should form two or more radial lines from the pumping well. This will indicate any radial anisotropy in the aquifer. A map should be carefully made in the field showing the relative locations of the pumping well and all observation wells. The distance from the pumping well to each observation well should be measured with a steel tape.

Ideally, observation wells should fully penetrate the aquifer, so that they measure the average head in the formation at that location.* Plastic (PVC) pipe with slots cut out in the zone of the aquifer serves as a suitable observation well. The annular space between the drilled hole and the plastic pipe can be backfilled with coarse sand. If the well goes through a confining layer, the hole must be

*Observation wells should also not be screened in aquifers other than the one in which the pumping well is also screened. The author has seen cases where very expensive pumping tests yielded meaningless results because either the pumping well or the observation well was screened in more than one aquifer. Time-drawdown data can be obtained from such misbegotten wells, and misguided hydrogeologists have attempted to evaluate them. The results are, however, are garbage.

backfilled with clay and tamped to prevent leakage around the pipe.

If the observation well has a short well screen, then the screen should be placed such that the head it measures is representative of the average head in the formation at that location. For a confined aquifer, it should be at the middepth of the stratum. For a water-table aquifer, it should be one-third of the distance from the static water table to the bottom of the aquifer.

The observation well should have a rapid response to changes in the water level in the aquifer. One way to test response time is to pour water into the observation well. The induced head should drain away in a fairly short time, usually in a few hours or less in most aquifers. If the observation well does not show a good response, an effort should be made to unclog it. Pouring water in the well may clear it. If the water level is within a few meters of the surface, a plunger on a stick can be used to surge the well to clear it.

Prior to the pumping test, a measuring point should be chosen for each observation well. Usually this is the top of the casing. The elevation of each measuring point should be determined. The depth to the static water level should also be measured prior to pumping. This will be useful in mapping the potentiometric surface.

Depth-to-water measurements are made in the observation wells on the same schedule as the pumping-well measurements. For the first minutes, this probably will necessitate one observer for each well. After twenty minutes, when the readings go to a frequency of every ten minutes, fewer people will be needed. After two and one-half hours, readings are made only every hundred minutes, so that one observer usually can do everything. Also, at the start of the test, one or two additional people should be on hand to measure pump discharge and adjust the flow.

After the end of the scheduled pumping test, recovery measurements can be made in the wells. The water levels will recover at the same rate they fall. In some cases, the drawdown data are affected by uncontrolled variations in the pumping rate. This does not affect the recovery rate. The flow rate for recovery data is equal to the mean discharge for the entire pumping period. In using recovery data, the difference between the water level at the end of pumping and that after a given time since pumping stopped is plotted as a function of the time since pumping stopped. The standard methods of well-test analysis are used. Recovery measurements are a standard part of the aquifer test. In many cases, the recovery data prove to be more useful than the drawdown data; for example, if there was a short period when the well shut down during the test or if the rate of pumping was extremely variable during the drawdown phase of the test.

## REFERENCES

1. HANTUSH, M.S. "Hydraulics of Wells." In *Advances in Hydroscience,* vol. 1, ed. V. T. Chow. New York: Academic Press, 1964, pp. 281–432.

2. THEIS, C. V. "The Lowering of the Piezometric Surface and the Rate and Discharge of a Well Using Ground-Water Storage." *Transactions, American Geophysical Union,* 16 (1935):519–24.
3. WENZEL, L. K. *Methods for Determining Permeability of Water-Bearing Materials with Special Reference to Discharging Well Methods.* U.S. Geological Survey Water-Supply Paper 887, 1942, p. 192
4. COOPER, H. H., JR., and C. E. JACOB. "A Generalized Graphical Method for Evaluating Formation Constants and Summarizing Well-Field History." *Transactions, American Geophysical Union,* 27 (1946):526–34.
5. JACOB, C. E. "Flow of Ground-Water." In *Engineering Hydraulics,* ed. H. Rouse. New York: John Wiley & Sons, 1950, pp. 321–86.
6. HEATH, R. C. *Basic Ground-Water Hydrology.* U.S. Geological Survey Water-Supply Paper 2220, 1983, 85 pp.
7. SCHAFER, D. C. "Casing Storage Can Affect Pumping Test Data." *Johnson Drillers Journal* (January-February 1978):1–5, 10–11.
8. FERRIS, J. G., et al. *Theory of Aquifer Tests.* U.S. Geological Survey Water-Supply Paper 1536-E, 1962.
9. COOPER, H. H., JR., J. D. BREDEHOEFT, and I. S. PAPADOPULOS. "Response of a Finite Diameter Well to an Instantaneous Charge of Water." *Water Resources Research,* 3 (1967):263–69.
10. PAPADOPULOS, I. S., J. D. BREDEHOEFT, and H. H. COOPER, JR. "On the Analysis of 'Slug Test' Data." *Water Resources Research,* 9 (1973):1087–89.
11. HANTUSH, M. S. "Analysis of Data from Pumping Tests in Leaky Aquifers." *Transactions, American Geophysical Union,* 37 (1956):702–14.
12. HANTUSH, M. S., and C. E. JACOB. "Plane Potential Flow of Ground-Water with Linear Leakage." *Transactions, American Geophysical Union,* 35 (1954):917–36.
13. HANTUSH, M. S. "Modification of the Theory of Leaky Aquifers." *Journal of Geophysical Research,* 65 (1960):3713–25.
14. NEUMAN, S. P., and P. A. WITHERSPOON. "Applicability of Current Theories of Flow in Leaky Aquifers." *Water Resources Research,* 5 (1969):817–29.
15. WALTON, W. C. *Selected Analytical Methods for Well and Aquifer Evaluation.* Illinois State Water Survey Bulletin 49, 1962, p. 81.
16. Walton, W. C. *Leaky Artesian Aquifer Conditions in Illinois.* Illinois State Water Survey Report of Investigation 39, 1960.
17. HOLZSCHUH, J. C., III. "A Simple Computer Program for the Determination of Aquifer Characteristics from Pump Test Data." *Ground Water,* 14 (1976): 283–85.
18. HANTUSH, M. S. "Aquifer Test on Partially Penetrating Wells." *Proceedings of the American Society of Civil Engineers,* 87 (1961):171–95.
19. HANTUSH, M. S. "Analysis of Data from Pumping Tests in Anisotropic Aquifers." *Journal of Geophysical Research,* 71 (1960):421–26.
20. HANTUSH, M. S. "Wells in Homogeneous Anisotropic Aquifers." *Water Resources Research,* 2 (1966):273–79.
21. WEEKS, E. P. "Determining the Ratio of Horizontal to Vertical Permeability by Aquifer Test Analysis." *Water Resources Research,* 5 (1969):196–214.

22. BOULTON, N. S., and T. D. STRELTSOVA. "New Equations for Determining the Formation Constants of an Aquifer from Pumping Test Data." *Water Resources Research,* 11 (1975):148–53.

23. STRELTSOVA, T. D. "Analysis of Aquifer-Aquitard Flow." *Water Resources Research,* 12 (1976):415–22.

24. GAMBOLATI, G. "Deviations from the Theis Solution in Aquifers Undergoing Three-Dimensional Consolidation." *Water Resources Research,* 13 (1977):62–68.

25. BOULTON, N. S. "The Drawdown of the Water Table under Non-Steady Conditions Near a Pumped Well in an Unconfined Formation." *Proceedings of the Institute of Civil Engineers* (London), 3, no. 3 (1954):564–79.

26. BOULTON, N. S. *Unsteady Radial Flow to a Pumped Well Allowing for Delayed Yield from Storage.* International Association of Scientific Hydrology Publication 37, 1955, pp. 472–77.

27. BOULTON, N. S. "Analysis of Data from Non-Equilibrium Pumping Tests Allowing for Delayed Yield from Storage." *Proceedings of the Institute of Civil Engineers* (London), 26 (1963):269–82.

28. BOULTON, N. S. "The Influence of the Delayed Drainage on Data from Pumping Tests in Unconfined Aquifers." *Journal of Hydrology,* 19 (1973):157–69.

29. BOULTON, N. S., and J. M. A. PONTIN. "An Extended Theory of Delayed Yield from Storage Applied to Pumping Tests in Unconfined Anisotropic Aquifers." *Journal of Hydrology,* 14 (1971):53–65.

30. STRELTSOVA, T. D. "Unsteady Radial Flow in an Unconfined Aquifer." *Water Resources Research,* 8 (1972):1059–66.

31. STRELTSOVA, T. D. "Unsteady Radial Flow in an Unconfined Aquifer." *Water Resources Research,* 9 (1973):236–42.

32. DAGAN, G. "A Method of Determining the Permeability and Effective Porosity of Unconfined Anisotropic Aquifers." *Water Resources Research,* 3 (1967):1059–71.

33. NEUMAN, S. P. "Theory of Flow in Unconfined Aquifers Considering Delayed Response of the Water Table." *Water Resources Research,* 8 (1972):1031–45.

34. NEUMAN, S. P. "Analysis of Pumping Test Data from Anisotropic Unconfined Aquifers Considering Delayed Gravity Response." *Water Resources Research,* 11 (1975):329–42.

35. NEUMAN, S. P. "Effect of Partial Penetration on Flow in Unconfined Aquifers Considering Delayed Gravity Response." *Water Resources Research,* 10 (1974):303–12.

36. GAMBOLATI, G. "Transient Free Surface Flow to a Well: An Analysis of Theoretical Solutions." *Water Resources Research,* 12 (1976):27–39.

37. WALTON, W. C. "Comprehensive Analysis of Water-Table Aquifer Test Data." *Ground Water,* 16 (1978):311–17.

38. STRELTSOVA, T. D. "Comments on 'Analysis of Pumping Test Data from Anisotropic Unconfined Aquifers Considering Delayed Gravity Response,' by Shlomo P. Neuman." *Water Resources Research,* 12 (1976):113–14.

39. PRICKETT, T. A. "Type Curve Solution to Aquifer Tests under Water Table Conditions." *Ground Water,* 3, no. 3 (1965):5–14.

40. HVORSLEV, M. J. *Time Lag and Soil Permeability in Ground Water Observations.*

U.S. Army Corps of Engineers Waterways Experimentation Station, Bulletin 36, 1951, 50 pp.

41. BOUWER, H., and R. C. RICE. "A Slug Test for Determining Hydraulic Conductivity of Unconfined Aquifers with Completely or Partially Penetrating Wells." *Water Resources Research,* 12 (1976):423–28.
42. FETTER, C. W. *Final Hydrogeologic Report, Seymour Recycling Corporation Hazardous Waste Site, Seymour, Indiana."* Prepared for U.S. Environmental Protection Agency, 1985, 181 pp. and appendices.
43. HERZOG, B. L., and W. J. MORSE. "A Comparison of Laboratory and Field Determined Values of Hydraulic Conductivity at a Hazardous Waste Disposal Site." In *Proceedings, Seventh Madison Waste Conference,* University of Wisconsin (1984):30–52.
44. NGUYEN, V., and G. F. PINDER. "Direct Calculation of Aquifer Parameters in Slug Test Analysis." In J. Rosenshein and G. D. Bennett, eds., *Groundwater Hydraulics,* American Geophysical Union Water Resources Monograph 9 (1984):222–39.
45. THIEM, G. *Hydrologische Methoden.* Leipzig: Gebhardt, 1906, p. 56.
46. Johnson Division, Universal Oil Products Company. *Ground Water and Wells.* St. Paul: Minn., 1966, p. 440.
47. FETTER, C. W., JR. "Use of Test Wells as Water-Quality Predictors." *Journal American Water Works Association,* 67 (1975):516–18.

## PROBLEMS

Type curves will be necessary for the solution of many of these problems. Type curves can be constructed from the data in the Appendices, although this is laborious. Type curves have been published for a number of aquifer tests on confined aquifers. The curves were derived by, among others, the Theis Method, the two methods for leaky artesian aquifers given in this chapter, and the Cooper-Bredehoeft-Papadopulos method. (J. E. Reed, "Type Curves for Selected Problems of Flow to Wells in Confined Aquifers," in *Techniques of Water-Resources Investigations of the United States Geological Survey,* Book 3, Chapter B3, 1980. This is available from the U.S. Government Printing Office, Washington, D.C., or Scientific Publications Company, P.O. Box 23041, Washington, D.C. 20026-3041.)

The published type curves use 3 × 5 cycle logarithmic graph paper with 1.85 inches for each log cycle, such as Keuffel and Esser Co. 46 7522, and semilogarithmic graph paper with 2.00 inches per log cycle, such as Keuffel and Esser Co. 46 6213. The Jacob straight-line methods will require 4-cycle semilogarithmic paper such as Keuffel and Esser Co. 46 6013.

Computer programs, in which the type curves are displayed directly on a video screen, are available from the National Water Well Association, 6375 Riverside Drive, Dublin, Ohio, 43017. The curve-matching is then done right on the screen.

1. A community is installing a new well in a regionally confined aquifer with a transmissivity of 2675 ft$^2$/day and a storativity of 0.0002. The planned pumping rate is 750 gallons per minute. There are several nearby wells tapping the same aquifer, and the hydrogeologist in charge needs to know if the new well will cause significant

interference with these wells. Compute the theoretical drawdown caused by the new well after 30 days of continuous pumping at the following distances: 50, 150, 250, 500, 1000, 3000, 6000, and 10,000 feet.

**2.** If the aquifer in problem 1 is not fully confined, but is overlain by a 10-foot-thick confining layer with a vertical hydraulic conductivity of 0.16 $ft^2$/day and no storativity, what would the drawdown values be after 30 days of pumping at 750 gallons per minute at the indicated distances?

**3.** With reference to the well and aquifer system in problem 1, compute the drawdown at a distance of 250 feet at the following times: 1, 2, 5, 10, 15, 30, and 60 minutes; 2, 5, and 12 hours; and 1, 5, 10, 20, and 30 days.

**4.** With reference to the well and aquifer system in problem 2, compute the drawdown at a distance of 400 feet at the following times: 1, 5, 10, 15, 30, and 60 minutes; 2, 5, and 12 hours; and 1, 5, 10, 20, and 30 days.

**5.** The following data are from a pumping test where a well was pumped at a rate of 200 gallons per minute. Drawdown as shown below was measured in an observation well 250 feet away from the pumped well. The geologist's log of the well is

| | |
|---|---|
| 0–23 feet | Glacial till, brown, clayey |
| 23–77 feet | Dolomite, fractured |
| 77–182 feet | Shale, black, dense |
| 182–217 feet | Sandstone, well-cemented, coarse |
| 217–221 feet | Shale, gray, limy |

A steel well casing was cemented to a depth of 182 feet and the well was extended as an open boring past that point.

| Elapsed Time (minutes) | Drawdown (feet) |
|---|---|
| 0 | 0.00 |
| 1 | 0.66 |
| 1.5 | 0.87 |
| 2.0 | 0.99 |
| 2.5 | 1.11 |
| 3.0 | 1.21 |
| 4.0 | 1.36 |
| 5 | 1.49 |
| 6 | 1.59 |
| 8 | 1.75 |
| 10 | 1.86 |
| 12 | 1.97 |
| 14 | 2.08 |
| 18 | 2.20 |
| 24 | 2.36 |
| 30 | 2.49 |
| 40 | 2.65 |
| 50 | 2.78 |
| 60 | 2.88 |
| 80 | 3.04 |

| Elapsed Time (minutes) | Drawdown (feet) |
|---|---|
| 100 | 3.16 |
| 120 | 3.28 |
| 150 | 3.42 |
| 180 | 3.51 |
| 210 | 3.61 |
| 240 | 3.67 |

**a.** Plot the time-drawdown data on 3 × 5 cycle logarithmic paper. Use the Theis type curve to find the aquifer transmissivity and storativity. Compute the average hydraulic conductivity.

**b.** Replot the data on 4-cycle semilogarithmic paper. Use the Jacob straight-line method to find the aquifer transmissivity and storativity.

**6.** A test well was drilled to a total depth of 117 feet with the following geologist's log:

| | |
|---|---|
| 0–73 feet | Coarse sand |
| 73–82 feet | Clayey sand |
| 82–117 feet | Coarse sand |
| 117 feet | Crystalline bedrock |

The depth to water was 55 feet. The test well was screened from 82 to 117 feet. It was pumped at a rate of 560 gallons per minute. Drawdown was measured in an observation well that was also screened from 82 to 117 feet and was located 82 feet away from the pumping well. The following time-drawdown data were obtained:

| Elapsed Time (minutes) | Drawdown (feet) |
|---|---|
| 0 | 0.00 |
| 1 | 0.90 |
| 2 | 2.15 |
| 3 | 3.05 |
| 4 | 3.64 |
| 5 | 4.07 |
| 6 | 4.52 |
| 7 | 4.74 |
| 8 | 5.02 |
| 9 | 5.21 |
| 10 | 5.53 |
| 15 | 5.72 |
| 20 | 5.97 |
| 30 | 6.12 |
| 40 | 6.20 |
| 50 | 6.25 |
| 60 | 6.27 |
| 90 | 6.29 |
| 120 | 6.29 |

Plot the time-drawdown data on 3 × 5 cycle logarithmic paper. Compute the value of the storativity and transmissivity of the aquifer using the graphical method for leaky aquifers. Find the vertical hydraulic conductivity of the confining layer.

**7.** Compute the value of aquifer storativity and transmissivity of the aquifer in problem 6 using the Hantush inflection-point method.

**8.** A slug test was made with a piezometer that had a casing radius of 2.54 centimeters and a screen of radius 2.54 centimeters. A slug of 4000 $cm^3$ of water was injected; this raised the water level by 197.3 cm. The well completely penetrated a confined stratum that was 2.3 meters thick. The decline in head with time is given in the following chart:

| Elapsed Time (seconds) | Head (feet) |
|---|---|
| 0 | 197.3 |
| 1 | 185.4 |
| 2 | 178.6 |
| 3 | 173.6 |
| 5 | 167.7 |
| 7 | 158.8 |
| 10 | 147.0 |
| 13 | 140.0 |
| 17 | 129.2 |
| 22 | 118.4 |
| 32 | 99.6 |
| 53 | 74.0 |
| 84 | 51.3 |
| 119 | 35.5 |
| 170 | 23.3 |
| 245 | 15.2 |
| 400 | 8.7 |
| 800 | 4.3 |

Plot the data on semilogarithmic paper and find the aquifer transmissivity and storativity using the Cooper-Bredehoeft-Papadopulos method.

**9.** A well in an unconfined aquifer is pumped at a rate of 872 $m^3$/day for 25 hours. The pumping well is fully penetrating. The following drawdown data were recorded in an observation well located 90 meters from the pumping well:

| Time (min) | Drawdown (meters) | Time (min) | Drawdown (meters) |
|---|---|---|---|
| 1.34 | 0.009 | 41 | 0.128 |
| 1.7 | 0.015 | 51 | 0.133 |
| 2.5 | 0.030 | 65 | 0.141 |
| 4 | 0.047 | 85 | 0.146 |
| 5 | 0.054 | 115 | 0.161 |
| 6 | 0.061 | 175 | 0.161 |
| 7.5 | 0.068 | 260 | 0.172 |
| 14 | 0.090 | 300 | 0.173 |

| Time (min) | Drawdown (meters) | Time (min) | Drawdown (meters) |
|---|---|---|---|
| 18 | 0.098 | 370 | 0.173 |
| 21 | 0.103 | 430 | 0.179 |
| 26 | 0.110 | 485 | 0.183 |
| 31 | 0.115 | 665 | 0.186 |
| | | 1320 | 0.200 |
| | | 1520 | 0.204 |

Plot the data on $3 \times 5$ cycle logarithmic paper. Use the water-table aquifer type curves to find the aquifer transmissivity and storativity.

**10.** A well that pumps at a constant rate of 50,000 cubic feet per day has achieved equilibrium so that there is no change in the drawdown with time. (The cone of depression has expanded to include a recharge zone equal to the amount of water being pumped.) The well taps a semiconfined aquifer that is 24 feet thick. An observation well 125 feet away has a head of 273 feet above sea level; another observation well 315 feet away has a head of 291 feet. Compute the value of aquifer transmissivity using the Theim equation.

# SEVEN

# Regional Ground-Water Flow

Again, most springs are in the neighborhood of mountains and high grounds, whereas if we except rivers, water rarely appears in the plains. For mountains and high ground suspended over the country like a saturated sponge, make the water ooze out and trickle together in minute quantities but in many places. They also receive a great deal of water falling as rain.

*Meteorologica,* Aristotle (384–322 B.C.)

## 7.1 INTRODUCTION

In the zone of actively flowing ground water, the water moves through the porous media under the influence of the fluid potential. This movement is a three-dimensional phenomenon, yet we are usually forced to represent it on a two-dimensional medium. In the diagrams in this chapter, the reader will have to imagine the implied third dimension. We will start by examining steady flow through isotropic, homogeneous media and then include the effects of nonhomogeneity and anisotropy.

Flow nets will be used to illustrate the various regional flow patterns. These are a means of portraying the solution to the Laplace equation (5-46), which governs steady flow. The various solutions will represent differing conditions of hydraulic conductivity and aquifer geometry. This type of flow net is constructed by drawing streamlines on a potential field. The potential fields are solutions to a mathematical model of the aquifer systems. Laplace's equation was solved either analytically (1, 2) or numerically (3, 4) with different boundary conditions. One of the most critical boundary conditions is the shape of the water table or potentiometric surface. The streamlines are drawn to illustrate some of the possible flow paths.

# 7.2 STEADY REGIONAL GROUND-WATER FLOW IN UNCONFINED AQUIFERS

## 7.2.1 RECHARGE AND DISCHARGE AREAS

In unconfined aquifers, some characteristics are common to most **recharge areas**; likewise, most **discharge areas** have some common denominators. Recharge areas are usually in topographical high places; discharge areas are located in topographic lows. In the recharge areas, there is often a rather deep unsaturated zone between the water table and the land surface. Conversely, the water table is found either close to or at the land surface in discharge areas.

Streamlines on a flow net tend to diverge from recharge areas and converge toward discharge areas. This convergence will not occur if the discharge zone is large, such as a coastline. A water-table contour map can often be used to locate ground-water recharge and discharge areas. Streamlines can be drawn on the basis of ground-water contours, crossing them at right angles if the aquifer is isotropic.

In the field, vegetation and surface water can sometimes be used to locate discharge areas. There may be some physical manifestation of the discharging ground water, which can take the form of a spring, seep, lake, or stream. The presence of vegetation common to wet soils may be indicative of discharge areas. In arid regions, ground water may be discharged as evaporation or transpiration. In such cases, a thicker-than-normal cover of vegetation or a salt deposit may indicate a discharge area.

The aforementioned physical manifestations sometimes betoken a ground-water discharge area—but not always. In nonhomogeneous materials, a low-permeability layer may, for example, form a perched aquifer, which could result in a wetland or pond. A thorough evaluation of all hydrogeologic information should always be made.

Many field studies conducted in humid regions note that the water table in unconfined aquifers usually has the same general shape as the surface topography. This is not surprising, since recharge taking place in topographical high areas has a greater potential energy than recharge in lower areas. This greater energy is reflected in the higher elevations of the water table in those locations. This generalization may not be true in arid regions.

## 7.2.2 GROUND-WATER FLOW PATTERNS IN HOMOGENEOUS AQUIFERS

A descriptive model of regional steady-state ground-water flow in an unconfined aquifer was first presented by Hubbert (5). He demonstrated that the hydraulic head at a point in a potential field represents the elevation to which water will rise in a **piezometer** that is open only at that point. At the point where an equipotential line intersects the water table, water in the piezometer will rise to the water table. Elsewhere, water in a piezometer intersecting the equipotential line may be above or below the water table, depending upon the relative hydraulic potential.

In Figure 7.1, the water levels in both Piezometer A and Piezometer B are equal, as they both end on the same equipotential line. Piezometer A is located on the water table, while Piezometer B is in an area where the water table has a higher elevation; hence, a greater potential, than A. The water level of Piezometer B is lower than the water table at that location. If a shallow piezometer were installed next to Piezometer B, the water level would be higher in the shallower piezometer. In this area, the value of the hydraulic potential decreases downward, indicating that the direction of flow is downward. Areas with this distribution of potential are recharge areas for a water-table aquifer.

In certain places in the potential field, in adjacent piezometers of different depths, there will be a higher water level in the deeper one. Such is the case with Piezometers C and D. Here, the hydraulic potential increases with depth, indicating upward flow. This is a discharge area.

Having identified the direction of flow at various points in this potential field, we can draw flowlines (Figure 7.2). If the diagram had equal horizontal and vertical scales and the medium were isotropic, the flowlines would cross the equipotential lines at right angles. However, owing to the vertical exaggeration of the drawing, the lines will cross at less than a right angle, even though the

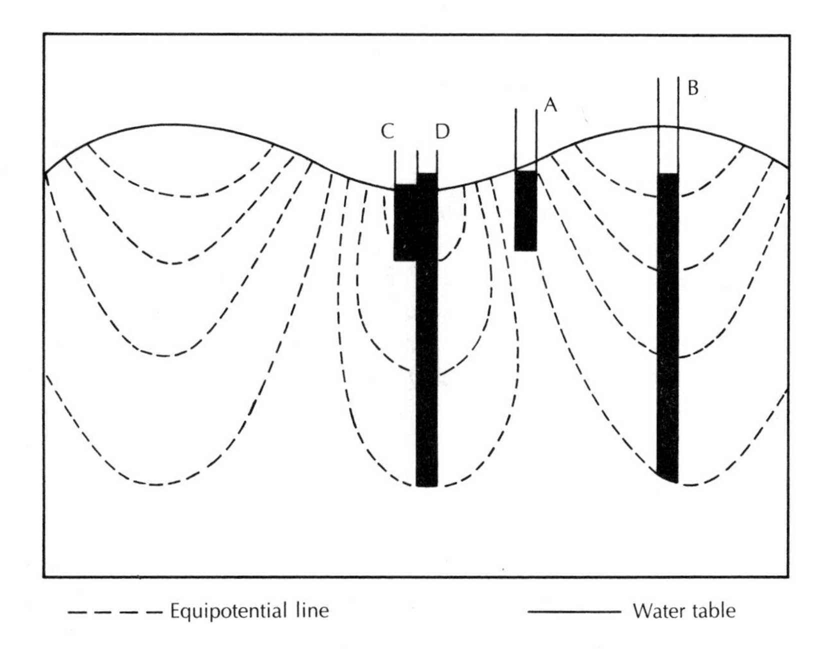


**FIGURE 7.1** Water level in a piezometer will rise to the elevation of the hydraulic head represented by the potential at the open end of the piezometer. The water level of Piezometers A and B is the same, since both terminate on the same equipotential line.

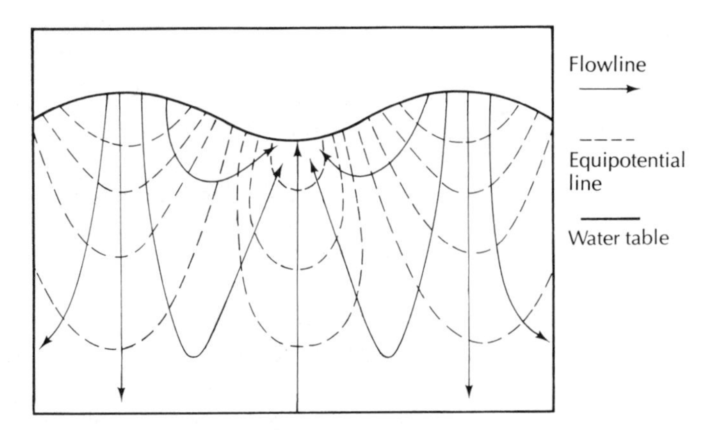


**FIGURE 7.2** Flowlines based on the equipotential field of Figure 7.1. The aquifer is assumed to be very deep. Source: M. K. Hubbert, *Journal of Geology*, 48, no. 8 (1940):795–944. Used with permission of University of Chicago Press.

flow field is in an isotropic medium.* Hubbert's model was for an unconfined aquifer of great depth. The crest of the water table represents a ground-water divide, with flow going in opposite directions on either side. The valley bottoms are areas of concentrated ground-water discharge into streams, with the streamlines converging toward them.

Not all drainage basins have concentrated discharge areas. If ground water is discharged primarily by evapotranspiration, or if there is no major topographic valley, the discharge area may be quite widespread. In such cases, the regional flow pattern will not have converging flowlines. Figure 7.3A shows a flow net for a cross section of a homogeneous drainage basin with a gently sloping water table and a horizontal, impermeable lower boundary (1). The entire upper half of the drainage basin is in the recharge area, while the lower portion forms a discharge area. There is no vertical exaggeration in the drawing, so that streamlines cross equipotential lines at right angles. In recharge areas, the angle between the water table and equipotential lines is oblique and points upstream; at the hinge or midline, it is a right angle; and in ground-water discharge areas, it is oblique and points downstream (4). If a major valley is present in the discharge area, ground-water discharge is concentrated in the valley (4). This changes the potential field, and the entire area above the valley is a recharge area. Figure 7.3B shows a flow net for this situation. There is no vertical exaggeration in the figure.

In Parts A and B of Figure 7.3, there is only one flow system; i.e., a single recharge area and a single discharge area. This occurs because the surface

*For a discussion of the problem of drawing streamlines, see R. O. Von Everdingen, *Groundwater Flow Diagrams in Sections with Exaggerated Vertical Scale*, Geological Survey of Canada Paper 63-27, 1963.

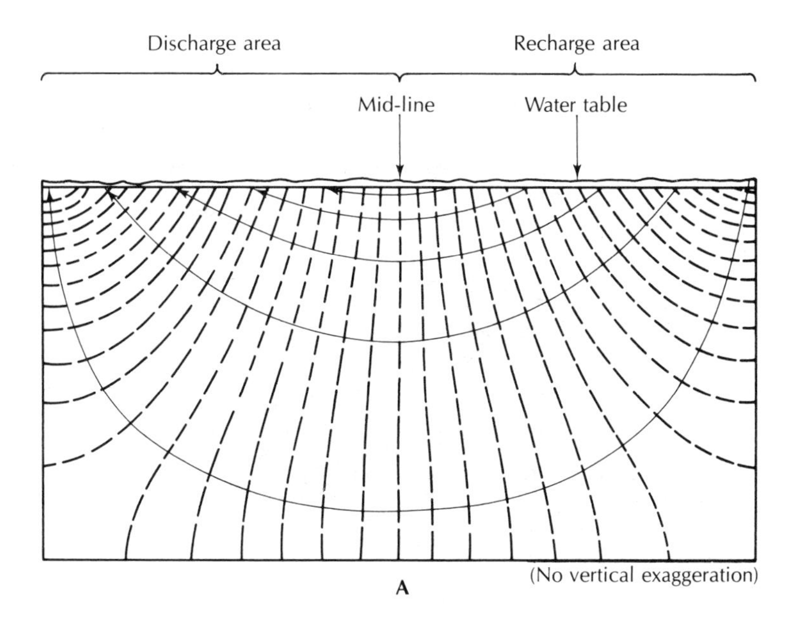


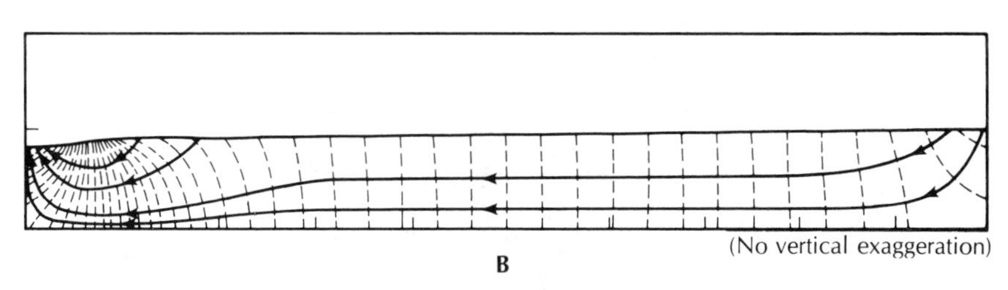


**FIGURE 7.3** **A.** Regional flow pattern in an area of sloping linear topography and water table. The flow pattern is symmetrical about the midline. **B.** Regional flow pattern in an area with a break in the regional topographic slope forming a major valley. The discharge area is controlled by the valley. Sources: Part A. J. A. Tóth, *Journal of Geophysical Research,* 67 (1962):4375–87. Part B. R. A. Freeze and P. A. Witherspoon, *Water Resources Research,* 3 (1967):623–34.

topography—hence, the water table—is linear, with a regional slope. However, if the surface topography has well-defined local relief, a series of **local ground-water flow systems** can form in humid regions (1, 4). This is due to the fact the topographic relief causes undulations in the water table. A local ground-water flow system has the recharge area at a topographic high spot and its discharge area at an adjacent topographic low. Figure 7.4A shows a flow net of a ground-water drainage basin with a series of local flow systems (2–4). The basin depth is one-twentieth of the basin length from the regional ground-water divide to the lowest part of the basin.

If the basin depth-to-width ratio increases, other flow systems may also develop. **Intermediate flow systems** have at least one local flow system between

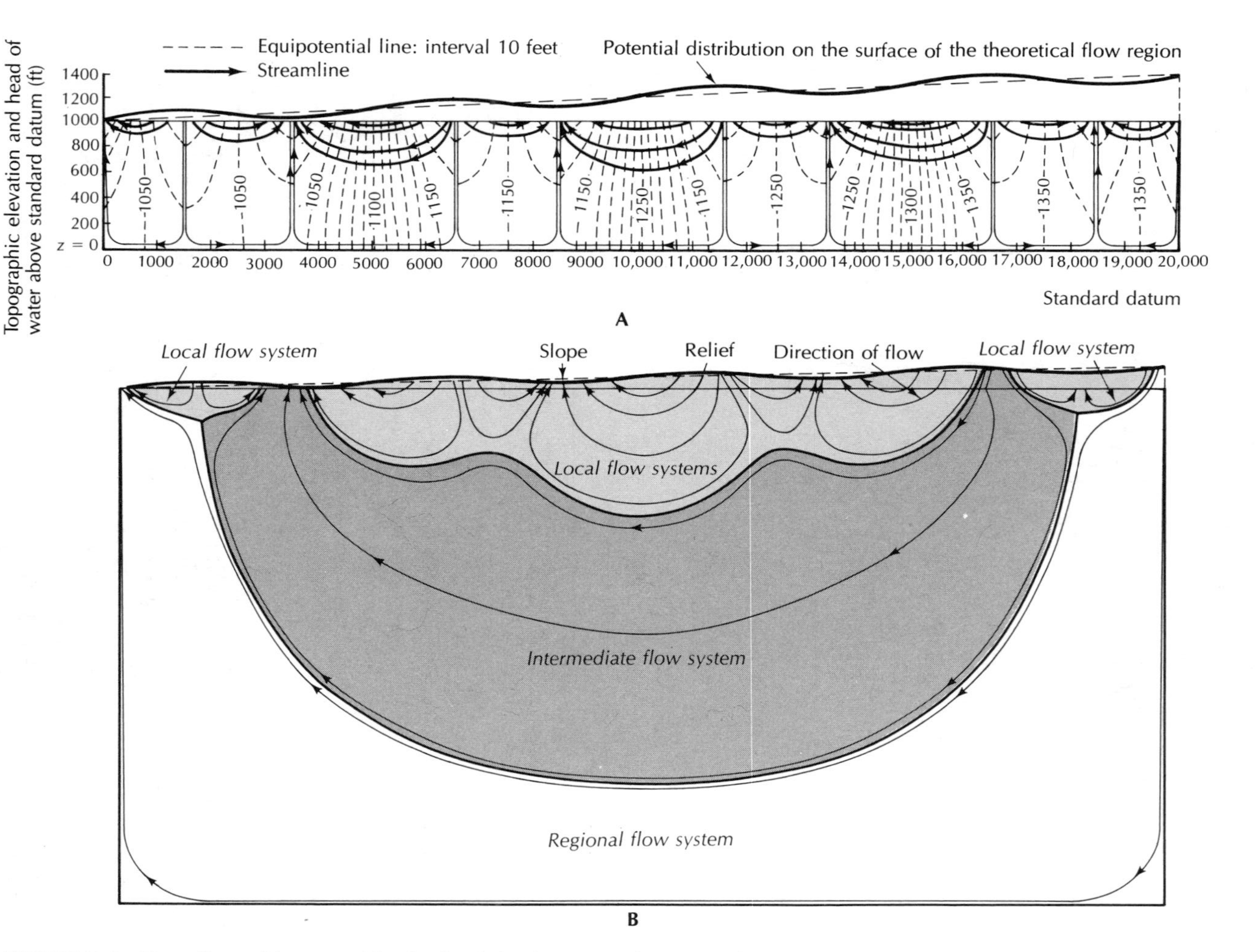


**FIGURE 7.4** The effect of increased basin depth is shown on these two figures. In Part A, the basin depth/length ratio is 1:20; in Part B, it is 1:2. The shallow basin has only local flow systems, while the deep basin has local, intermediate, and regional flow systems. The water-table configuration is the same for both basins. Source: J. A. Tóth, *Journal of Geophysical Research,* 68 (1963):4795–4811.

their recharge and discharge areas. **Regional flow systems** have the recharge area in the basin divide and the discharge area at the valley bottom (Figure 7.4B). Depending upon the drainage basin topography and the basin-shape geometry, flow systems may have regional; local; local and intermediate; or local, intermediate, and regional components.

In addition to the influence of the drainage basin depth/length ratio, it has been shown that the more pronounced the relief of the undulating water table, the deeper the local flow systems extend. If some basins, both local and regional flow systems may exist, while in other basins with a similar depth/length ratio but with a more pronounced water-table relief, only deep local flow systems develop. This is illustrated in Figure 7.5A, which has local and intermediate flow systems, and in Part B of the figure, where the more pronounced relief of the water table has resulted in the exclusive formation of local flow systems.

One of the features of complex flow systems is the presence of **stagnation points** in the flow field (2). At a stagnation point, the magnitudes of the vectors in the flow field are equal but opposite in direction and cancel each other. The value of the hydraulic potential is higher at the stagnation point than at any part of the surrounding region. Ground-water flow paths diverge around stagnation points, which are found at the juncture of local and regional flow systems. Figure 7.6 illustrates the potential field and flow paths at a stagnation point. Stagnation points can exist in materials that are completely isotropic and homogeneous.

It has been suggested that "dead cells" or stagnation points might be appropriate areas in which to inject waste fluids for permanent disposal (6). Diffusion would then be the only physical mechanism to disperse the fluid. However, if the waste fluid were not of the exact density and temperature as the native ground water, the original potential field might be disrupted, causing flow in the area of the stagnation point and resultant movement of the waste fluid. Likewise, ground-water pumpage could change the potential field, shifting or eliminating the locations of stagnation points.

If regional flow systems develop, the flow paths are long compared with those of local flow systems (2). In aquifers composed of soluble rock material, the degree of mineralization is a function of both the initial chemistry of the water and the length of time it is in contact with the aquifer (7). Referring back to Figure 7.4B, we see the boundaries of local, intermediate, and regional flow systems for a deep aquifer with an undulating water table. The surface area where recharge to the regional flow system takes place is quite small in relation to the volume of water stored in that region of the aquifer. The water moves slowly and circulates deeply within the aquifer, as the flow paths are long. At the point of discharge, the water from the regional flow system is likely to have relatively high mineralization and an elevated temperature due to the geothermal gradient. (The temperature of the earth increases with depth at a more or less constant rate of 1° C per 100 meters of depth.)

Local flow systems are shallower, with short flow paths. The size of the recharge area is much greater with respect to the volume of water in the aquifer.

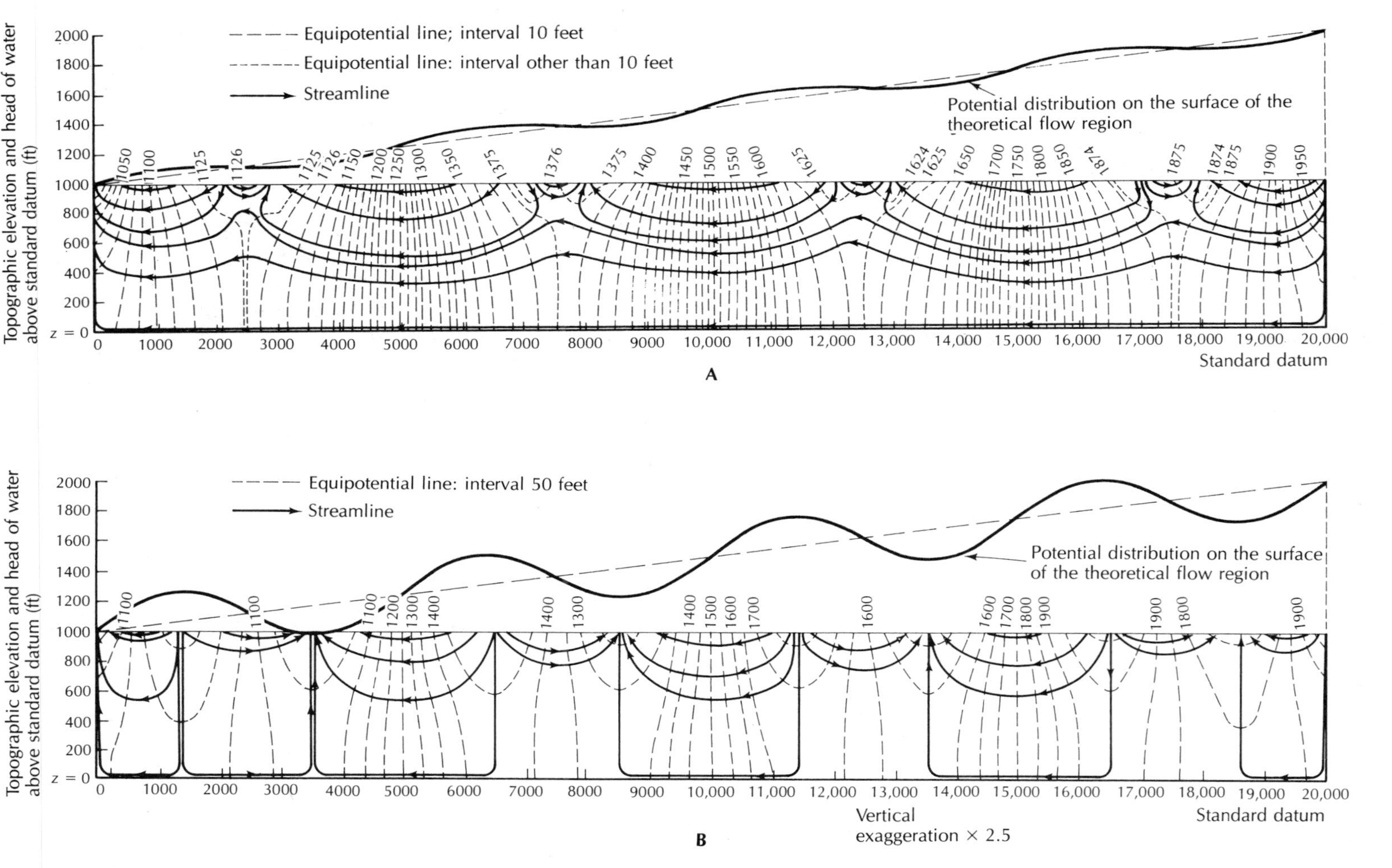


**FIGURE 7.5** The amplitude of the undulations of the water table controls the depth of local flow systems. For shallow basins, this can determine whether both local and regional flow systems will develop (Part A), or, with deeper undulations, only local flow systems will form (Part B). Source: J. A. Tóth, *Journal of Geophysical Research,* 68 (1963):4795–4811.

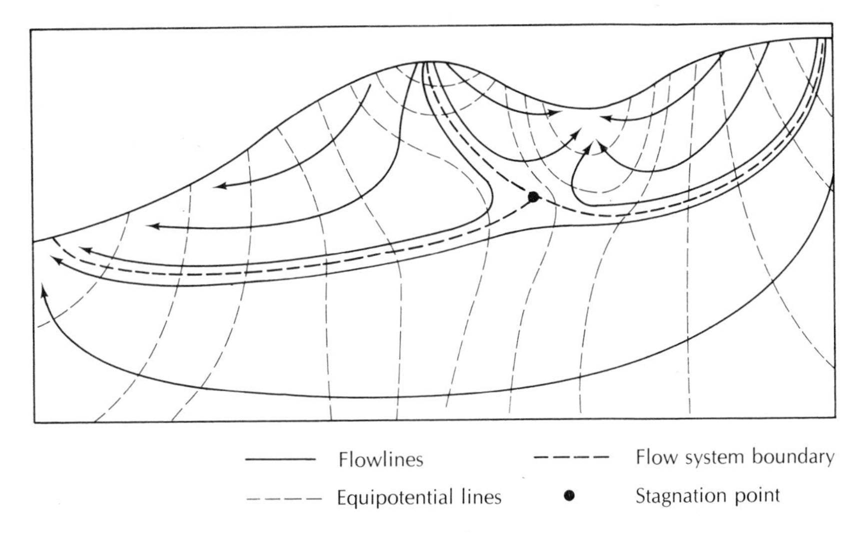


**FIGURE 7.6** The potential field and flowlines in the vicinity of a stagnation point, which will develop at the intersection of three flow systems.

Thus, water has a shorter contact time with the rocks and is potentially mineralized to a lesser degree than that of the regional system. The temperature of water discharging from the local flow systems is close to the mean annual air temperature. Local flow systems are areas of rapid circulation of ground water; therefore, ground water in these systems is much more active in the hydrologic cycle than ground water in regional flow systems (2). Spring discharge of local flow systems is closely related to recharge of precipitation and shows wide fluctuations (8). This is illustrated in Figure 7.7. Intermediate flow systems have properties falling between those of local and regional flow systems.

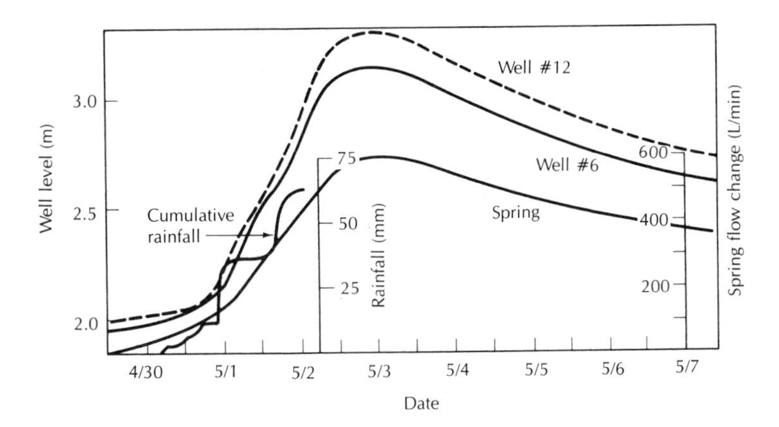


**FIGURE 7.7** The water table and spring flow of a local flow system will fluctuate with recharge from rainfall. Source: R. S. Sartz et al., *Water Resources Research,* 13 (1977):524–30.

If a flow system has extended areas with a flat water table, the potential is the same in all parts of the field. Neither local nor regional flow systems can develop, and the ground water is stagnant. Evapotranspiration is the only method of ground-water discharge. Ground water under such conditions is likely to be highly mineralized owing to a long contact time with the aquifer rocks and the concentration of dissolved salts by evaporation.

### 7.2.3 HETEROGENEOUS AQUIFERS

Piezometers may sometimes yield water levels that are apparently anomalous with respect to the expected regional flow pattern (1). A set of piezometers at various depths may show a water elevation equal to the water table for a shallow well, a lower water elevation for a piezometer of intermediate depth, and then a water elevation equal to the water-table elevation for a deep well. Geologic logs of these piezometers might show the shallow one to end in a fine, silty sand; the one of intermediate depth to end in coarse sand; and the deepest one to end in the fine, silty sand. Figure 7.8A shows a cross section of the potential distribution where a body of material of high hydraulic conductivity is surrounded by material with a lower conductivity. The high-conductivity zone acts as a conduit for flow, attracting water from much of the aquifer. The result is that the potential field bends away from the high-conductivity zone on either end. Flow will thus converge toward the high-conductivity zone on the upstream end and diverge away from it on the downstream side.

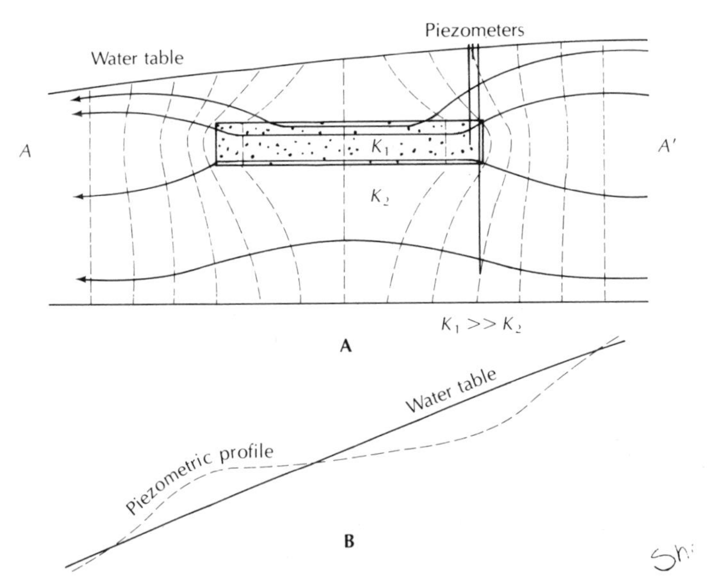


FIGURE 7.8 **A.** Equipotential field and flowlines in a region where a high-conductivity body is buried in a lower-conductivity aquifer. **B.** The water table and the potentiometric profile of a line of piezometers, each ending at the same elevation along line $A$–$A'$ of Part A.

A line of piezometers of equal depth, extending to line *A-A'*, would have a potentiometric profile that would differ from the water-table profile (Figure 7.8B). Upstream of the midpoint of the high-conductivity layer, the potentiometric profile would be lower than the water table. It would cross the water table at the midpoint and be higher than the water table below the midpoint. Such a profile would not occur in a homogeneous aquifer.

If a lens of low-permeability material is buried in an aquifer, it acts as a partial barrier to ground-water flow. The ground-water streamlines diverge around the lens. While a modicum of the flow is carried in the low-conductivity layer, the majority of the flow tends to be in the aquifer.

Layered aquifers are especially prevalent in sedimentary basins, with individual hydrostratigraphic* units having different hydraulic conductivities. If a lower formation has a substantially higher conductivity than the surface layer, it acts as the major conduit of flow (4). Figure 7.9A shows the potential distribution in a layered aquifer when the lower unit has a conductivity ten times that of the upper. Flow in the lower unit is horizontal, while the upper unit has vertical flow components in the recharge and discharge areas. As the difference in conductivity between the upper and lower layer increases, the components of vertical flow in the upper unit increase as more of the flow is carried in the lower unit.

If a high-conductivity layer overlies a unit of substantially lower conductivity, the potential field is very similar to that of an isotropic aquifer. Most of the flow is carried in the upper, more conductive, layer, as Figure 7.9B illustrates. The potential field of Figure 7.9B is quite similar to that of Figure 7.3B, which is homogeneous.

### 7.2.4 ANISOTROPIC AQUIFERS

There is considerable evidence that many aquifers are anisotropic. For deposits as uniform as glacial outwash, the horizontal permeability may be two to twenty times as great as the vertical (9). Figure 7.10 shows the potential distribution in an aquifer in which horizontal conductivity is ten times as great as the vertical. In anisotropic aquifers, the flowlines do not cross the equipotential lines at right angles; the correct angles can be obtained graphically (10). Figure 7.3B shows the same section in an isotropic medium. The vertical components of flow are more pronounced in the anisotropic aquifer. The greatest variation in the potential field occurs at the extreme ends of the ground-water basin.

---

*Hydrostratigraphic units are comprised of geologic units grouped together on the basis of similar hydraulic conductivity. Several geologic formations may be grouped into a single aquifer, for example. A single geologic formation may be divided into both aquifers and confining units.

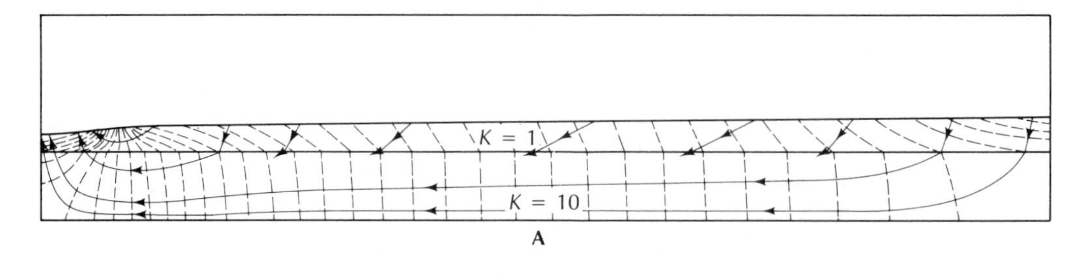


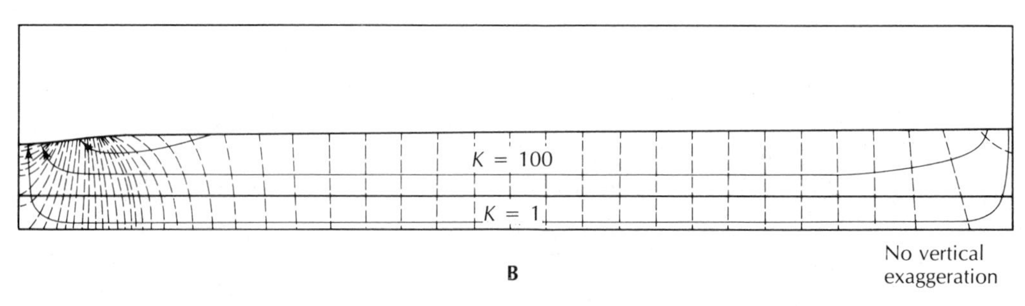


**FIGURE 7.9** Regional ground-water flow in layered aquifers. The greater proportion of the flow occurs in the layer with higher hydraulic conductivity. Source: R. A. Freeze and P. A. Witherspoon, *Water Resources Research,* 3 (1967):623–34.

## 7.3 CONFINED AQUIFERS

Aquifers that are overlain by a layer of substantially lower hydraulic conductivity are confined. The hydraulic gradient is generally greater in the confining bed than in the aquifer. Since the frictional resistance to flow is so much greater in the confining layer, most of the available energy of the potential field is dissipated there.

Confined aquifers may be either sloping or flat. If the aquifer crops out

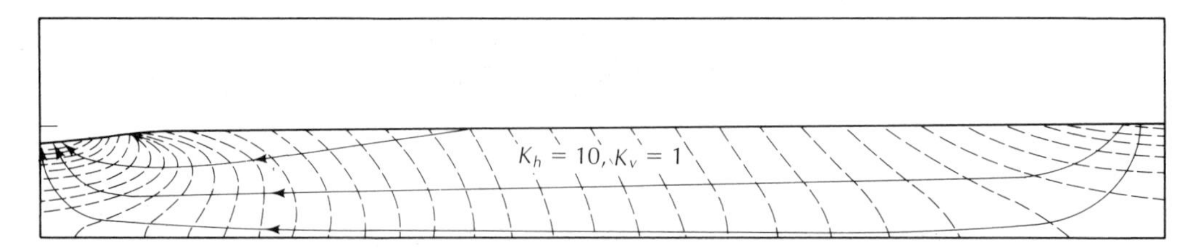


**FIGURE 7.10** Effect of anisotropy on regional ground-water flow. This figure is the same as Figure 7.3B, except that, here, the horizontal hydraulic conductivity is ten times the vertical. Source: R. A. Freeze and P. A. Witherspoon, *Water Resources Research,* 3 (1967):623–34.

near a topographic high, substantial recharge takes place in the outcrop area. In the sloping aquifer shown in Figure 7.11A, the confining layer retards flow. Wells drilled through it to the underlying aquifer would yield artesian flow. Streamlines refract as they cross the confining layer. Discharge of the regional flow system is concentrated in the valley bottom.

The flat-lying confined aquifer in Figure 7.11B does not have an outcrop area. Recharge to the aquifer occurs by downward flow through the confining layer. Almost all of the energy of the potential field is consumed as flow moves through the confining layer in recharge and discharge areas. Only one equipotential line crosses the aquifer. The volume of water flowing through the buried aquifer is less than it would be if the aquifer cropped out in the recharge area. This is due to the use of much of the available potential energy in forcing recharge through the low-conductivity layer. If a well were drilled in the discharge area, artesian conditions would occur.

Unless the confined aquifer is capped by a completely impermeable layer, there will be some discharge from the aquifer in the form of upward leakage in the area of upward hydraulic gradient. Many confined aquifers have heads above the land surface when the first wells are drilled (11). The amount of upward flow occurring through the confining beds is typically small, and the water does not circulate rapidly. Ground-water withdrawals in many confined regional aquifers have lowered the potentiometric head. This actually increases the rate of lateral ground-water flow in the aquifer, as the hydraulic gradient between the recharge area and the well-field area is increased.

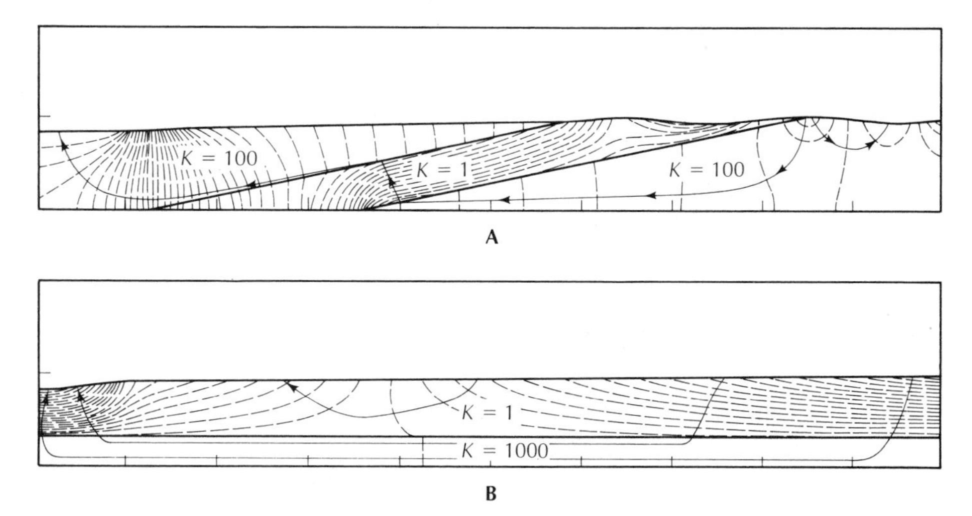


**FIGURE 7.11** Regional ground-water flow in confined aquifers: **A.** Aquifer confined by a sloping confining layer. **B.** Aquifer confined by a flat-lying confining layer. Source: R. A. Freeze and P. A. Witherspoon, *Water Resources Research,* 3 (1967):623–34.

## 7.4 TRANSIENT FLOW IN REGIONAL GROUND-WATER SYSTEMS

The systems we have considered have been in a state of **dynamic equilibrium.** The amount of water recharging the aquifer is balanced by an equal amount of natural discharge, and the potential field is more or less constant. If a well field is established in the ground-water basin, the withdrawal of well water increases the discharge from the system, disrupting the equilibrium. Thus, a new equilibrium must be established (12).

In the case of an unconfined aquifer, the water table around the well field will be drawn down. As the discharge exceeds the recharge, the difference comes from gravity drainage of ground water stored in the aquifer. The cone of depression around the well field will slowly expand until it affects the flow system enough to create a new equilibrium condition. This will occur when the area of the cone of depression is large enough to intercept sufficient aquifer recharge to supply the well discharge. This will reduce natural discharge somewhere else, and a new condition of dynamic equilbrium will be reached. Should the rate of withdrawal be so great that the cone of depression reaches the boundaries of the aquifer without intercepting sufficient recharge, the aquifer will not reach equilibrium and eventually could be drained.

In confined and leaky confined aquifers, pumping will reduce the heads near the wells. As a result, the potentiometric surface will decline. The cone of depression will expand rapidly owing to the small value of storativity of confined aquifers. Initially, the water being pumped comes from storage in the aquifer. The cone of depression in a leaky confined aquifer will stabilize when enough downward leakage is induced to balance pumpage. This, of course, will upset the natural equilibrium in the overlying aquifer that is furnishing the water.

In a confined aquifer, the cone of depression will grow until it reaches either the recharge area of the aquifer, or the discharge area, or both. The resulting change in the potential field will induce either increased recharge, or decreased natural discharge, or both. If this is sufficient to balance recharge and discharge, the aquifer will again be in dynamic equilibrium. If not, the water levels will continue to decline.

## 7.5 NONCYCLICAL GROUND WATER

There is a certain amount of water in the ground that is not encompassed by the hydrologic cycle. When sediments are deposited, water is present in the pores. The same may be true for undersea volcanic rocks. Later geologic events may bury the sediment or rock and its contained pore water. Water buried with the rock is termed **fossil water** (13). Interstitial water that was not buried with the rock but which has been out of contact with the atmosphere for an appreciable part of a geologic period is called **connate water** (13).

**Magmatic water** is associated with a magma. It may be in part **juvenile**

**water,** having never before circulated in the hydrologic cycle (14). However, most magmatic water comes from the recycling of connate or fossil water. Magmatic water can re-enter the hydrologic cycle through volcanic eruptions or thermal springs.

## 7.6 SPRINGS

Springs have played a role in the settlement pattern of many lands, where they have served as a local water supply. Mineralized and thermal springs have been thought to have therapeutic value. The importance of springs is evident from the many localities named for the springs found there (e.g., Tarpon Springs, Florida; Palm Springs, California; Hot Springs, Arkansas; Steamboat Springs, Colorado).

A spring may have a discharge that is fairly constant, or the discharge may vary. Springs can be permanent or ephemeral. The water may contain dissolved minerals of many different types or certain dissolved gases or petroleum. The temperature of the water may be close to the mean annual air temperature or be lower or higher—even boiling. Flow may range from a barely perceptible seepage to 1000 cubic feet (30 or more cubic meters) or more per second.

Topographic low spots provide the simplest mechanism for the formation of springs. **Depression springs** are formed when the water table reaches the surface (15). The change in topography creates a corresponding undulation in the water-table configuration. A local flow system is thus created, with a spring formed at the local discharge zone (Figure 7.12A).

Where permeable rock units overlie rocks of much lower permeability, a **contact spring** may result (15). A lithologic contact is often marked by a line of springs, which may be either in the main water table or in a perched water table. It is not necessary for the underlying layer to be impermeable, merely that the difference in hydraulic conductivity be great enough to preclude transmission of all of the water that is moving through the upper horizon (Figure 7.12B).

A classic occurrence of contact springs is found along the eastern side of Chuska Mountain, New Mexico. A sandstone cliff rises 197 to 492 feet (60 to 150 meters) above a terrace composed of shale, which also underlies the sandstone. More than thirty springs are found at the foot of the cliff at the contact of the sandstone and shale (16). One of the most spectacular series of springs in the world is in the Snake River Canyon below Shoshone Falls in Idaho. Along a 40-mile (64-kilometer) reach of the canyon, there are eleven springs with a discharge of more than 100 cubic feet (2.8 cubic meters) per second. The springs issue from permeable basalt flows, total spring flow is about 5000 cfs (140 $m^3$/sec) in this reach of the Snake River (17).

Faulting may also create a geologic control favoring spring formation. A faulted rock unit that is impermeable may be emplaced adjacent to an aquifer. This can form a regional boundary to ground-water movement and force water in the aquifer to discharge as a **fault spring** (Figure 7.12C).

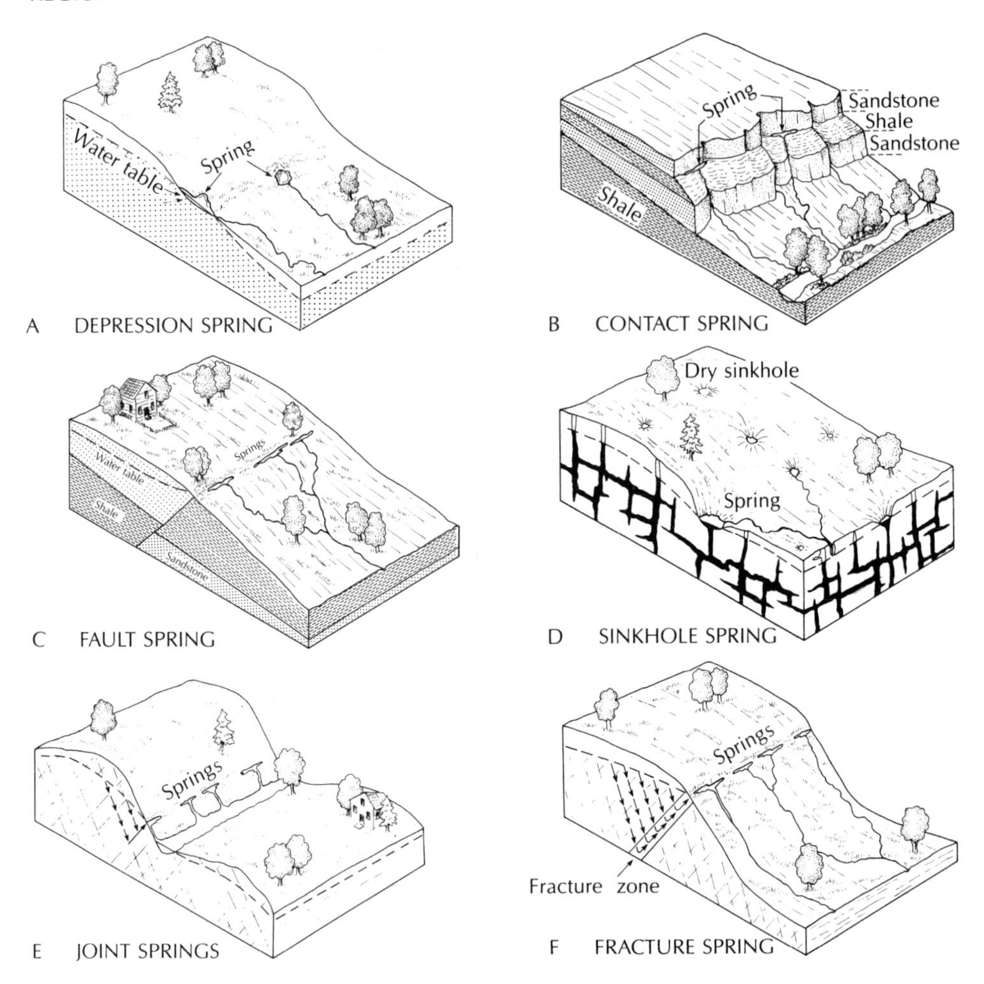


**FIGURE 7.12** Types of springs.

Some of the largest springs are found in areas of limestone bedrock. In such areas, the runoff may be carried in part or totally as subterranean flow. It may be either diffused flow in pores and fractures in the rock or channelized flow in caverns. Springs may be found where a cavern is connected to a shaft that rises to the surface. Many of the famous springs of Florida cover an area of several acres in which water rises to the surface through sinkholes (Figure 7.12D). The water in these **sinkhole springs** is under artesian pressure and comes from the principal artesian aquifer, or Floridan aquifer, which underlies Florida (18). This aquifer is in Tertiary-age limestones.

**Joint springs** or **fracture springs** may occur from the existence of jointed or permeable fault zones in low-permeability rock. Water movement through such rock is principally through fractures, and springs can form where these fractures intersect the land surface at low elevations (Figure 7.12E,F).

Springs in limestone terrane can be interconnected to topographic depressions caused by collapsed caverns (sinkholes) at higher elevations. Water level in the sinkholes may rise and fall owing to variations in runoff (19). Discharge of these springs, known as **karst springs,** may correspond with the elevation of water in the sinkholes.

## 7.7 GEOLOGY OF REGIONAL FLOW SYSTEMS

Several case studies illustrate different types of regional flow systems.

### CASE STUDY: REGIONAL FLOW SYSTEMS IN THE GREAT BASIN

The Basin and Range Province contains a number of topographically closed basins. These intermontane basins are characterized by an accumulation of relatively permeable clastic sediments. The mountains surrounding the basins are composed of bedrock, which also underlies the basins at depth. The hydraulic conductivity of the bedrock types is extremely variable (20).

Annual precipitation is greatest in the mountains and least in the valleys (21). Below an elevation of 6000 feet (1800 meters), annual precipitation is less than 8 inches (20 centimeters). This is almost all evaporated, with virtually no recharge of ground water. Above 9000 feet (2750 meters), there may be more than 20 inches (50 centimeters) of precipitation, with up to 5 inches (13 centimeters) of ground-water recharge. The areas of greatest precipitation and recharge are in topographic highs, which are good recharge zones.

Those mountain areas formed by crystalline rocks or low-permeability sedimentary rocks have near-surface permeability due to fracturing. Such mountain areas have many small springs and perennial streams, as the ground water is discharged by local flow systems (6). Mountains underlain by highly permeable carbonate rocks are generally dry. The ground water appears as the discharge of relatively large springs at the foot of the mountains or in the intermontane valleys. In the areas of carbonate aquifers, the water table is relatively flat, and may extend with a regional slope beneath topographic divides (6, 20, 21). A block diagram of single-valley hydrologic systems and regional flow systems is shown in Figure 7.13.

In the White River area of southeastern Nevada, a regional interbasin groundwater flow system has been identified (21). There are thirteen topographic basins: seven of them are closed; the other six were drained by the White River during the Pleistocene. The mountains are 8000 to 10,000 feet (2450 to 3050 meters) high, with the valley bottoms 2000 to 4000 feet (600 to 1220 meters) lower. The principal water-bearing units are Paleozoic limestone and dolomites, up to 30,000 feet (9150 meters) thick. There are some volcanic rocks (tuffs and welded tuffs) which can form locally perched aquifers. The valleys are filled with Tertiary-age clastic sedimentary rocks and evaporites.

Ground water is discharged by means of several large springs. The flow of Muddy River Springs, which is the largest, is highly uniform, suggesting a regional flow system as the source (22). A longitudinal profile of the area, which shows both the topography and the potentiometric surface, reveals that the regional hydraulic gradient is unaffected by crossing topographic divides (Figure 7.14).

The amount of ground-water recharge is much greater than discharge in the topographically higher basins. The water balance is reversed for the topographically lower basins where the large springs are located. However, the regional water budget is balanced when all thirteen of the basins are included (23).

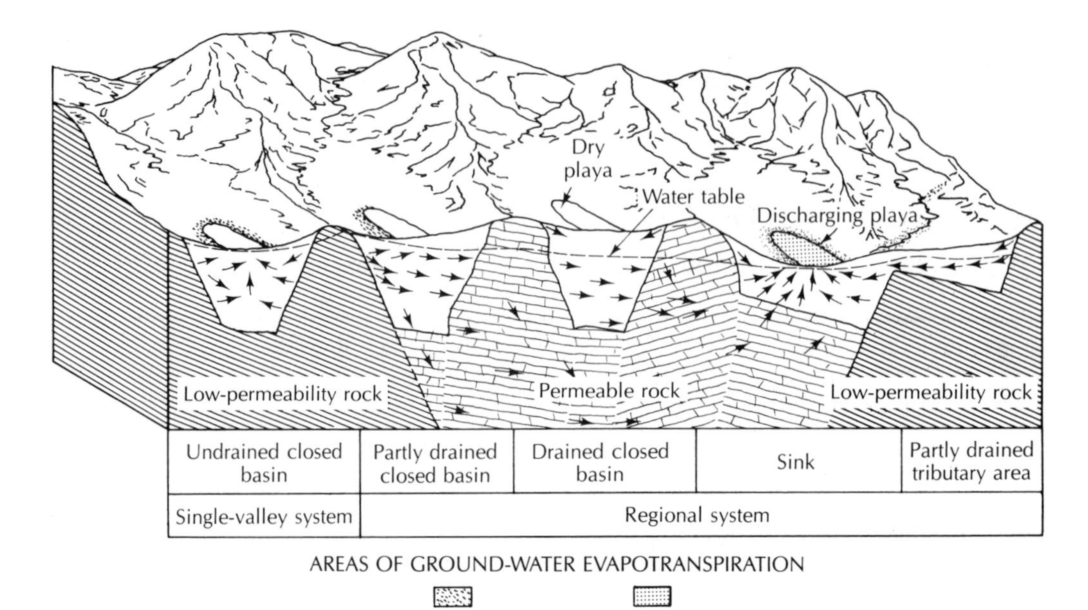


**FIGURE 7.13** Flow systems of the Great Basin Region. Source: T. E. Eakin et al., U.S. Geological Survey Professional Paper 813-G, 1976.

In the Great Basin area, local flow systems have small drainage areas and short flow paths within the same topographic basin. Springs have a wide fluctuation in discharge. The temperature of the spring water is about the same as mean annual air temperature, and the dissolved ion content is relatively low. Regional flow systems have long flow paths, which often cross basin divides. The drainage basin area is large. Springs are large, with fairly constant discharge, elevated temperatures, and higher concentrations of dissolved salts (6).

Another large ground-water flow system has been delineated in southern Nevada at the Nevada Test Site (24). The area has typical basin and range topography. Rocks range in age from Recent sediments in the valleys to Precambrian. A basement unit of Cambrian and Precambrian siltstones and quartzites is a regional confining layer. Above this layer is a major regional aquifer—the lower carbonate aquifer of Lower Paleozoic age. There are solution openings in the aquifer, but the hydraulic conductivity is due primarily to fractures. The saturated thickness ranges from a hundred to several thousand feet or more. There are several caves in the outcrop area. One of them, Devils Hole, is partially filled with water and extends vertically more than 300 feet (100 meters) below the water table. The lower carbonate aquifer feeds many large springs.

No other aquifer has a regional extent, although several are important locally. There is a regional confining layer that overlies the lower carbonate aquifer. This is the welded tuff confining layer of Tertiary age. The clay-rich beds of this layer restrict ground-water movement. The three regional units control ground-water flow. The welded tuff confining layer permits slow downward movement of water to the lower carbonate aquifer in recharge zones and upward movement in discharge zones. The lower carbonate aquifer underlies most of the valleys and ridges.

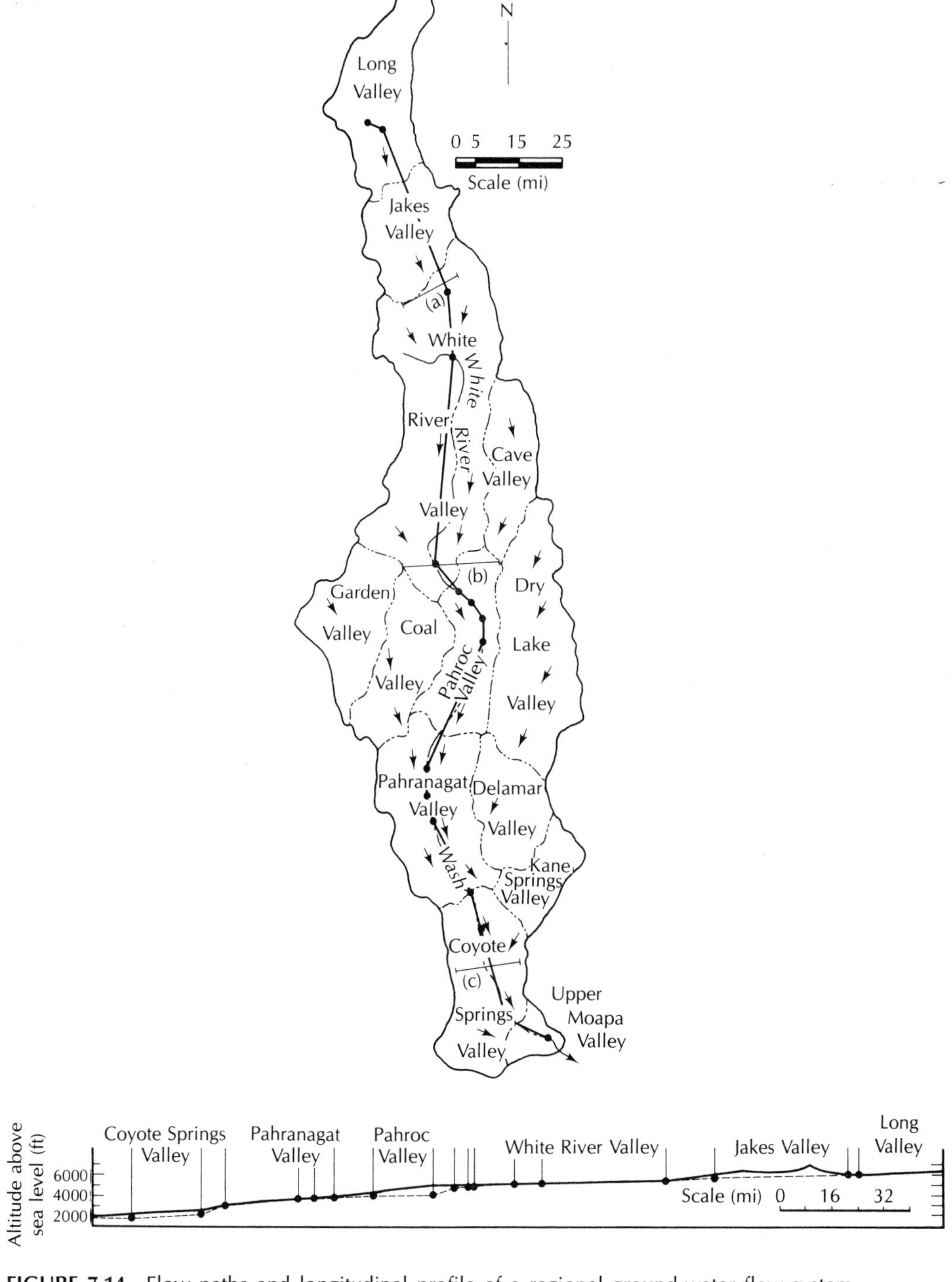


**FIGURE 7.14** Flow paths and longitudinal profile of a regional ground-water flow system in Nevada. Source: T. E. Eakin, *Water Resources Research*, 2 (1966):251–71.

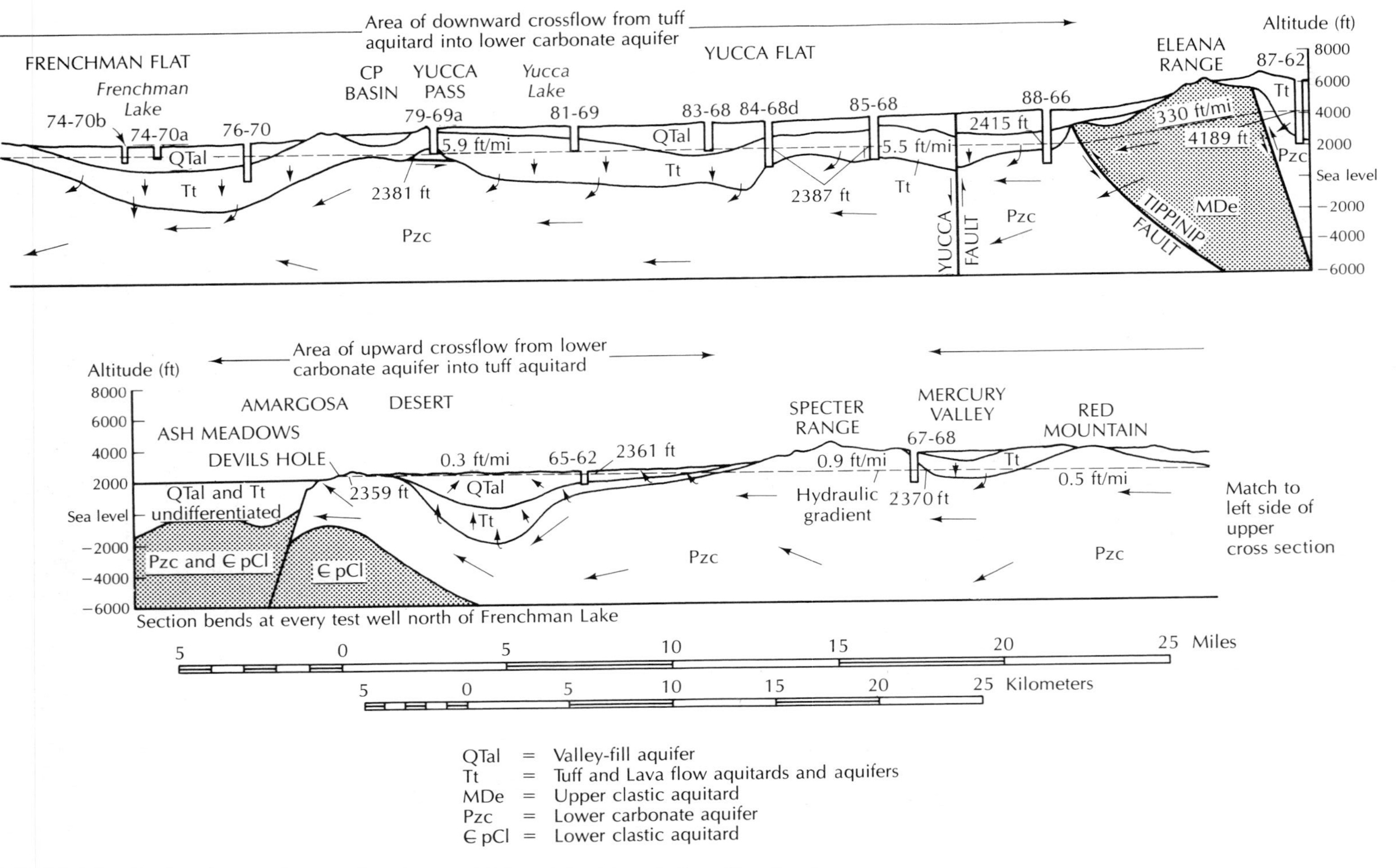


**FIGURE 7.15** Regional ground-water flow in the vicinity of the Nevada Test Site, Southern Nevada. Source: I. J. Winograd and W. Thordarson, U.S. Geological Survey Professional Paper 712-C, 1975.

In some of the upland valleys, such as Yucca Flats and Frenchman Flats, the water table lies from 700 to 2000 feet (210 to 610 meters) below the valley floor (Figure 7.15). Water drains from these valleys downward to the lower carbonate aquifer. Water flows laterally in this unit until it reaches the Ash Meadows area of the Amargosa Desert Basin. Running across Ash Meadows is a normal fault with a displacement of at least 500 feet (160 meters). This has formed a hydraulic barrier that impedes the ground-water flow. The result is a line of springs discharging at the outcrop of the lower carbonate aquifer. The average discharge of all the springs is 10,000 gallons per minute (630 liters per second).

The inferred ground-water drainage basin contains ten intermontane valleys and is on the order of 4500 square miles (11,650 square kilometers) in area. The Ash Meadows Basin may be hydrologically interconnected with other intermontane valleys from which ground-water flow may be obtained. There may also be underflow past the spring line at Ash Meadows. This spring line feeds into the central Amargosa Desert. The lack of wells and the extremely complex structure of the area make exact delineation of flow systems difficult.

## CASE STUDY: REGIONAL FLOW SYSTEMS IN THE COASTAL ZONE OF THE SOUTHEASTERN UNITED STATES

Ground-water flow regimes in humid regions characteristically have a much thinner unsaturated zone than those in arid regions. The saturated zone is close to the surface in recharge areas as well as in discharge areas. Because of the greater amounts of precipitation, the volume of recharge is much higher, and proportionally more water circulates through ground-water flow systems.

The coastal zone of the southeastern United States is underlain by extensive aquifers contained in Tertiary- and Quaternary-age limestones, forming some of the most productive aquifers in the United States (18, 25, 26). Sediments on the coastal plain dip seaward, and the units thicken as the coast is approached. Figure 7.16 shows the isopach map of the Cenozoic sediments and sedimentary rocks of the region. Starting at a featheredge near the interior border of the Atlantic Coastal Plain, they thicken to more than 5500 feet (1680 meters) in southwestern Florida. The structure is disrupted by the Ocala uplift north of Tampa, resulting in a thinning of the sediments.

There are a number of hydrostratigraphic units in the region. There is a surfical aquifer system, an upper confining unit, the upper Floridan aquifer, a middle confining unit, the lower Floridan aquifer, and a lower confining unit (26, 27).

Most of the region has permeable materials at or near the surface. These are predominantly sand, but also include gravel, sandy limestone, and limestone. In most places the sands are relatively thin and are termed the **surficial aquifer.** In southwest Florida the surficial material consists of highly permeable sands and limestones that are as much as 200 feet (60 meters) thick and occur in a wedge shape that thins to the northwest. It has been termed the **Biscayne aquifer.** In the western part of the panhandle of Florida the surficial material consists of a thick deposit of sand and gravel termed the **sand and gravel aquifer.** The surficial aquifer system is mostly unconfined and receives recharge directly from precipitation. It is underlain by either the upper confining unit or, if that unit is missing, by the upper Floridan aquifer. The surficial aquifer materials are Pleistocene to Recent in age. Figure 7.17 shows the surficial hydrostratigraphic units of the area.

The **upper confining layer** consists of low-permeability rocks that are primarily

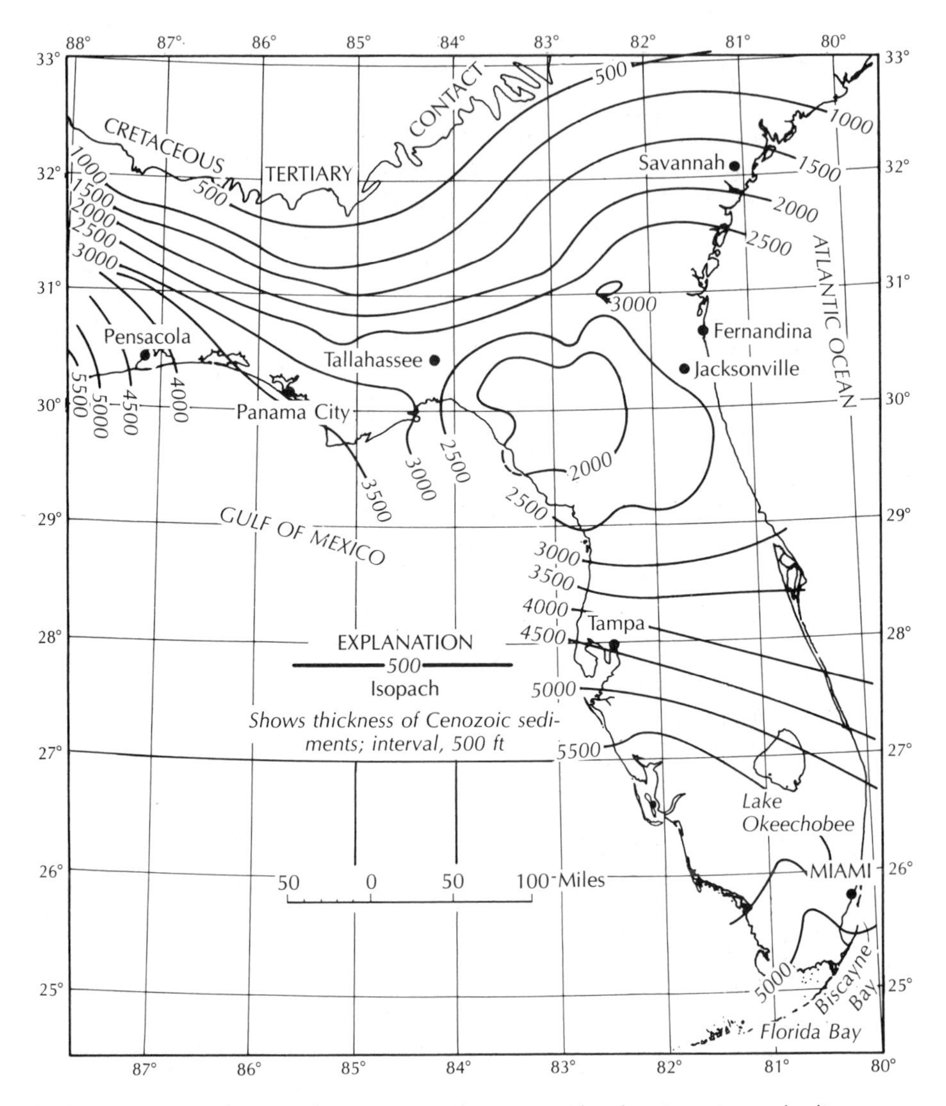


**FIGURE 7.16** Isopach map of Cenozoic sediments in Florida, Georgia, and adjacent parts of Alabama and South Carolina. Source: V. T. Stringfield, U.S. Geological Survey Professional Paper 517, 1966.

clastic. It basically consists of the Hawthorn formation of Miocene age, although in some areas the lower Miocene Tampa Limestone and the Oligocene Suwannee Limestone are included if the permeability of these limestones is comparatively low. The thickness and lithology of this unit are variable and hence the degree to which it retards vertical flow differs from place to place. In some areas where it is thin the Hawthorn has been breached by sinkholes so that direct pathways for water movement between the surficial deposits and the Floridan aquifer are present. In some places it is so thick, as much as 500 feet

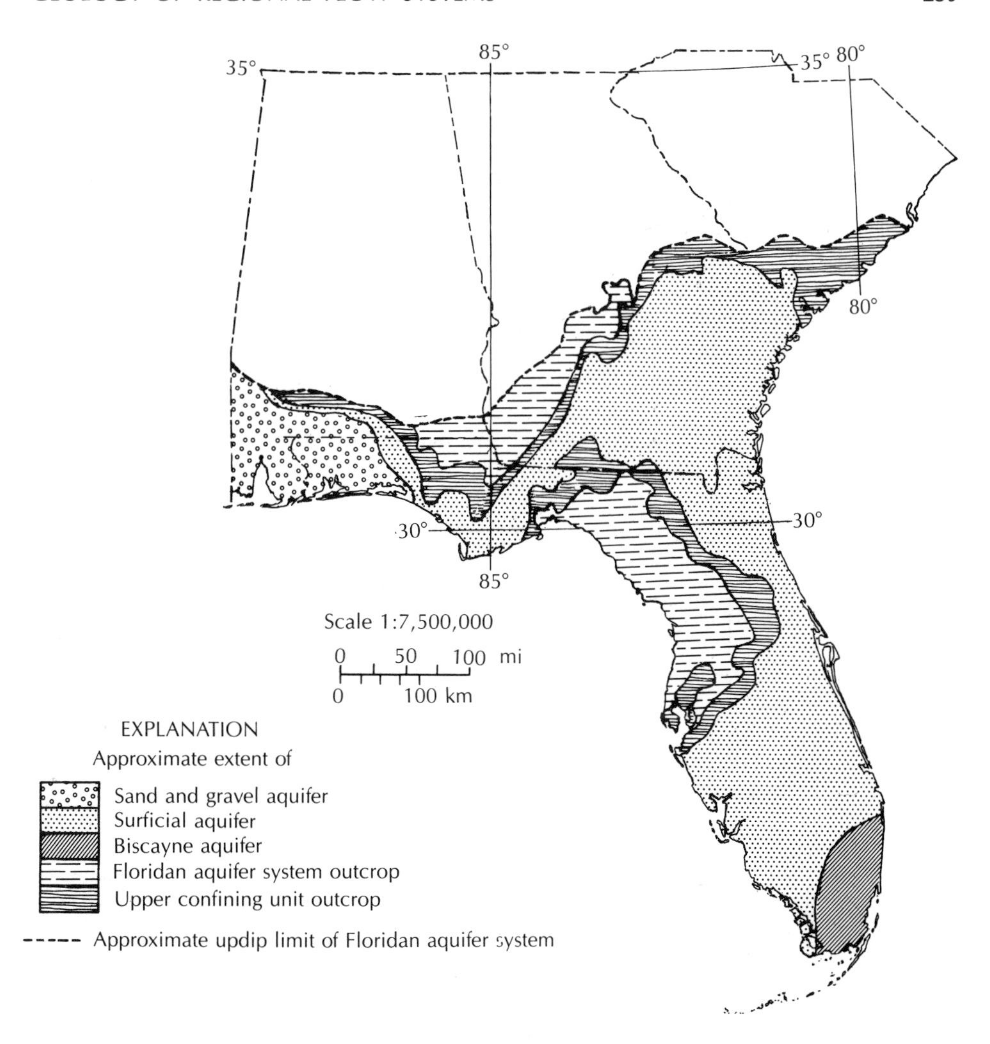


**FIGURE 7.17** Approximate extent of the surface aquifer, sand-and-gravel aquifer, Biscayne aquifer, and outcrop area of the upper confining unit in the southeastern United States. Source: J. A. Miller, "Hydrogeologic Framework of the Floridan Aquifer System in Florida and Parts of Georgia, Alabama and South Carolina," U.S. Geological Survey Professional Paper 1403B (1986):B41.

(150 meters), that it directs virtually all of the water in the surficial aquifer laterally toward the coastline. The Hawthorn Formation has been removed by erosion in the area of the Ocala uplift of Florida so that the underlying Floridan aquifer system forms the surface material (Figure 7.17).

The **Floridan aquifer system** is quite thick and consists of Eocene- to Miocene-age limestones of different permeability. The units contained in the Floridan aquifer system consist of the early Miocene Tampa Limestone, where it is permeable; the Oligocene Suwannee Limestone; and the Eocene Ocala Limestone, Avon Park Formation, and Oldsmar

Formation. The hydraulic conductivity of the rocks of the Floridan aquifer ranges from one to several orders of magnitude greater than that of the overlying confining layer. The Floridan aquifer system is unconfined in areas where the confining layer is missing and is semi-confined or confined elsewhere, depending upon the thickness and vertical hydraulic conductivity of the Hawthorn Formation. It is recharged in upland areas either directly, where the Hawthorn Formation is missing or is breached by sinkholes, or by leakage through the Hawthorn Formation. Water discharging from the Floridan aquifer forms some of the major springs of Florida. Water also discharges as subsea outflow in coastal areas. The Floridan aquifer is subdivided into an upper aquifer unit, a middle confining layer, and a lower aquifer unit (Figure 7.18). The **middle confining layer** is discontinuous and is represented by several different geologic units. In places the **upper Floridan aquifer** and the **lower Floridan aquifer** are joined into one unit. The upper aquifer unit is generally highly transmissive, although there is a complex relationship between the transmissivity of the limestone units and their postdepositional erosional history. In areas where the upper Floridan aquifer is unconfined, transmissivities of as much as $1 \times 10^6$ ft$^2$/day ($9 \times 10^4$ m$^2$/day) occur because of sinkholes and caves. Where the aquifer is confined by a thick zone of the Hawthorn formation, the transmissivity is much less, being as little as $5 \times 10^4$ ft$^2$/day ($5 \times 10^3$ m$^2$/day) in south Florida. The upper Floridan aquifer typically has good water quality and is extensively used as a water supply. The lower Floridan aquifer frequently contains saline water. It has some very highly transmissive areas. The **Boulder zone** of South Florida is a paleokarst zone developed in early Eocene rocks. It contains saline water and is extensively used for disposal of treated municipal sewage.

The Floridan aquifer is bounded on the bottom by a **lower confining unit.** This is a low-permeability unit that consists of the Cedar Keys Formation, which contains massive anhydrite beds as well as other units with a much lower hydraulic conductivity than the Floridan aquifer.

Precipitation in Florida averages about 55 inches (140 centimeters) annually with about 39 inches (100 centimeters) of evaporation (28). The 16-inch (40-centimeter) difference represents the recharge to aquifers and overland flow into streams. The vast majority of the available water, 14 inches (35 centimeters), flows into surface water, with only 1 to 2 inches (3 to 5 centimeters) recharging the Floridan aquifer. The major recharge areas are discernible as the areas of high elevation on the potentiometric map (Figure 7.19). The Floridan aquifer has two major hydraulic regions separated by a hydrologic divide (28). This is shown as a dashed line in Figure 7.19.

The northern portion of the Floridan aquifer is recharged in the outcrop area of central Georgia, where the elevations are 100 to 230 feet (30 to 70 meters) above sea level. Recharge to the aquifer also occurs in a region extending from north of Valdosta, Georgia, to east of Gainesville, Florida. This is an area of high topography with many sinkhole lakes. The sinkholes breach the Hawthorn Formation, a confining layer, so recharge occurs through the sinkhole lakes into the underlying Floridan aquifer. Water flows from the recharge areas, following the regional hydraulic gradient, and discharges along the coastlines. There are also lesser amounts of recharge from downward leakage through the Hawthorn Formation over much of Florida.

The southern portion of the Floridan aquifer is hydraulically separated from the north by the potentiometric divide. The Withlacoochee River and the St. Johns River act as ground-water drains to lower the potentiometric surface and isolate the southern aquifer. In the Polk City area, north of Winter Haven, the potentiometric surface is as much as 115 feet (35 meters) above sea level (29). This is another area in which sinkhole lakes as deep as 200 feet (60 meters) are present and can serve as recharge conduits.

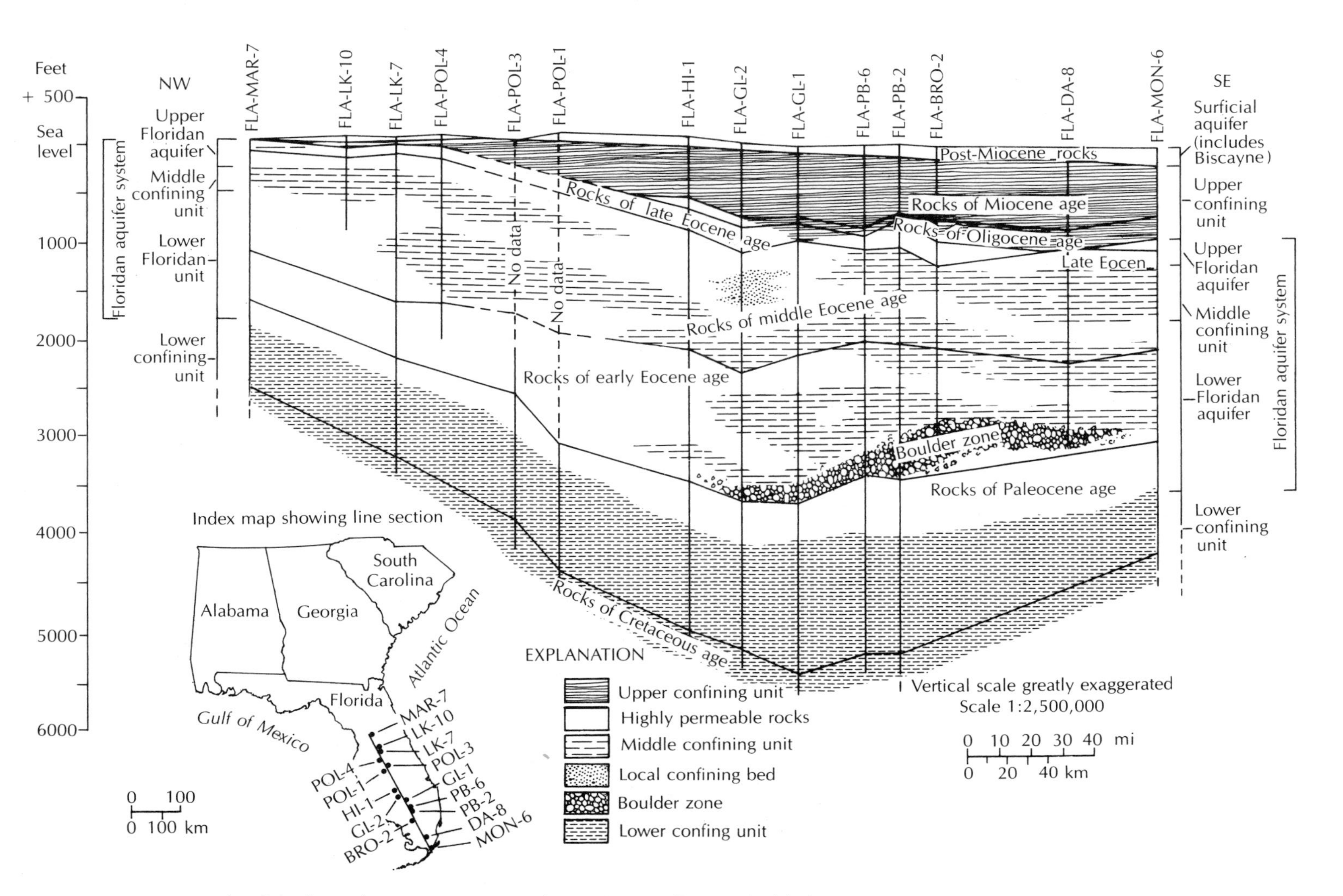


**FIGURE 7.18** Generalized hydrogeologic cross section from Monroe County to Marion County, Florida. Source: J. A. Miller, "Hydrogeologic Framework of the Floridan Aquifer System in Florida and Parts of Georgia, Alabama and South Carolina," U.S. Geological Survey Professional Paper 1403B (1986):B78.

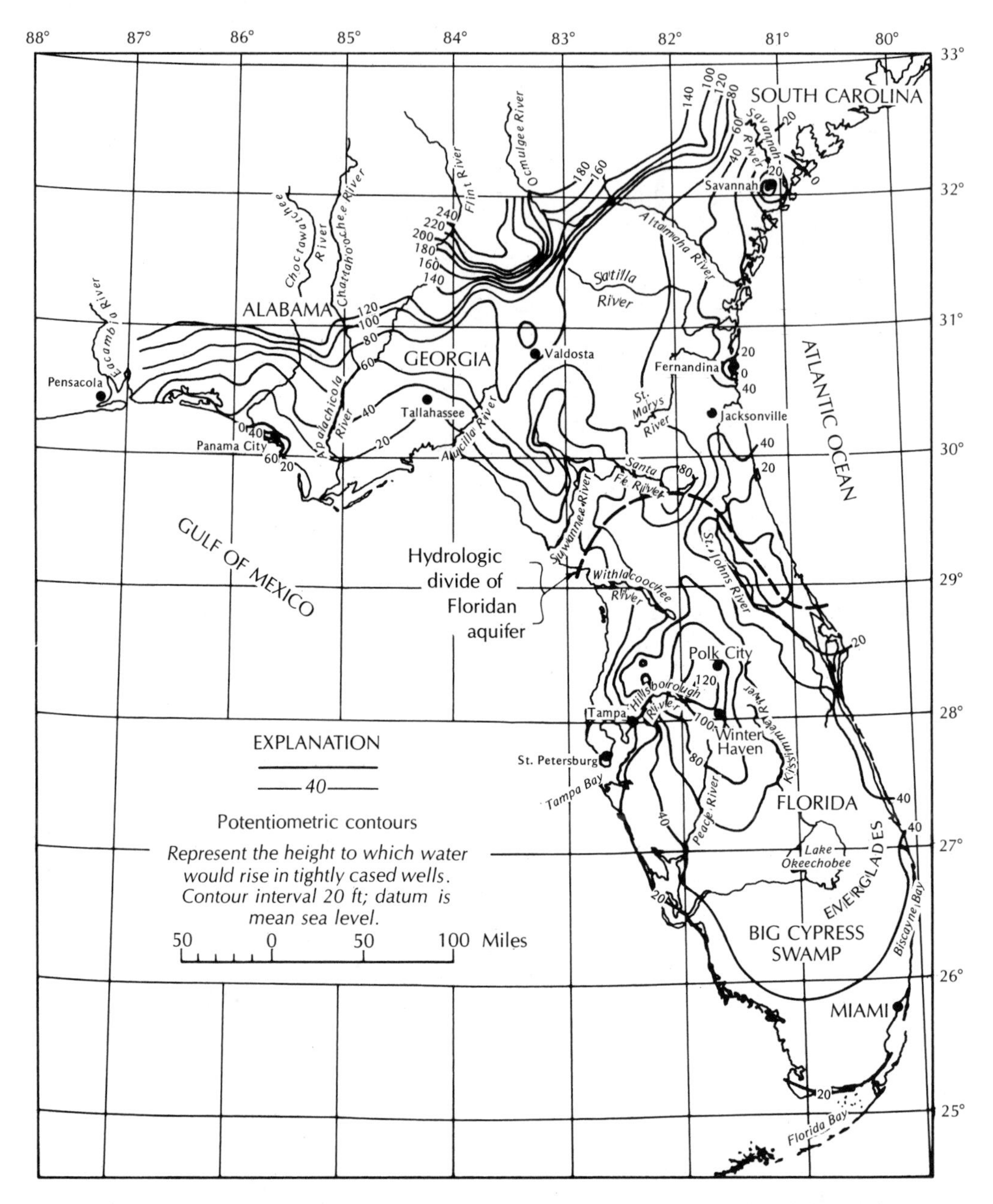


**FIGURE 7.19** Potentiometric surface of water in the principal artesian aquifer of the southeastern United States. Source: V. T. Stringfield, U.S. Geological Survey Professional Paper 517, 1966.

The general circulation of ground water in the Floridan aquifer is downward in the recharge areas and then laterally toward the coasts. To the west, discharge is by upward leakage through the Hawthorn Formation into the Gulf of Mexico or into coastal springs located in sinkholes. Flow eastward can reach the Atlantic Ocean floor, where some units of the Floridan aquifer crop out. There is also discharge into the rivers that drain the coastal area. Major rivers, such as the Suwannee and St. Johns in northern Florida, have a profound effect on the potentiometric surface, as they are regions of major discharge from the aquifer.

Zones of high-salinity ground water are found in the lower parts of the Floridan aquifer. These are naturally occurring, but heavy pumping of overlying fresh water is causing an upconing of the saline water. In a number of areas in the southeastern Coastal Plain, the principal artesian aquifer has been heavily pumped, with a subsequent lowering of the potentiometric surface. In an area east of Tampa, the potentiometric surface declined by as much as 60 feet (18 meters) from 1949 to 1969 (29). In the Savannah area, the cone of depression resulting from ground-water pumping has reversed the regional hydraulic gradient and has caused salt-water encroachment (30, 31). This is becoming common throughout much of the southeastern Coastal Plain.

The estimated predevelopment discharge of water from the Floridan aquifer system is shown in Figure 7.20. Water from the aquifer discharges into known springs as well as lakes and streams in areas where the aquifer is unconfined or loosely confined. As of 1980 the total discharge from some 300 known springs was about 13,000 $ft^3/sec$ (370 $m^3/sec$) with an additional 7000 $ft^3/sec$ (200 $m^3/sec$) estimated discharge into the lakes and streams. The remainder of the water from the Floridan aquifer, about 2500 $ft^3/sec$ (70 $m^3/sec$), diffused upward across the upper confining layer into surface aquifers.

Development of the Floridan aquifer system has resulted in the withdrawal of some 4000 $ft^3/sec$ (110 $m^3/sec$) of ground water by pumping. Overall, the discharge from the Floridan aquifer system is now estimated to be 24,100 $ft^3/sec$ (680 $m^3/sec$), which is 12 percent more than under predevelopment conditions. Figure 7.21 shows the estimated 1980 discharge from the Floridan aquifer. The creation of cones of depression in the potentiometric surface of the Floridan aquifer has enlarged the recharge area and resulted in increased recharge. Long-term regional water-level declines of more than 10 feet (3 meters) have occurred in the Savannah-Hilton Head Island area, Brunswick-Jacksonville-Fernandina beach area, Fort Walton beach area, and west-central Florida. Most of the discharge is still through the major springs, with some 18,000 $ft^3/sec$ (510 $m^3/sec$) of flow into known springs and surface-water bodies. There still remains a large amount of water in the Floridan aquifer that could be developed, especially in southeast Georgia and north-central Florida (27). However, future development could be hindered by salt-water encroachment, which is also a threat to those areas that have already developed deep cones of depression (31).

## CASE STUDY: REGIONAL FLOW SYSTEM OF THE HIGH PLAINS AQUIFER

The **High Plains aquifer** occurs in the states of Colorado, Kansas, Nebraska, New Mexico, Oklahoma, South Dakota, Texas, and Wyoming. It is a water-table aquifer system consisting primarily of near-surface sand and gravel deposits that were formed as alluvium deposited by streams draining from the Rocky Mountains, which lie to the west. The **Ogalalla Formation,** which consists of alluvium, is the principal hydrogeologic unit; but the aquifer system also includes the fractured and thus permeable zones of the Brule Formation, a

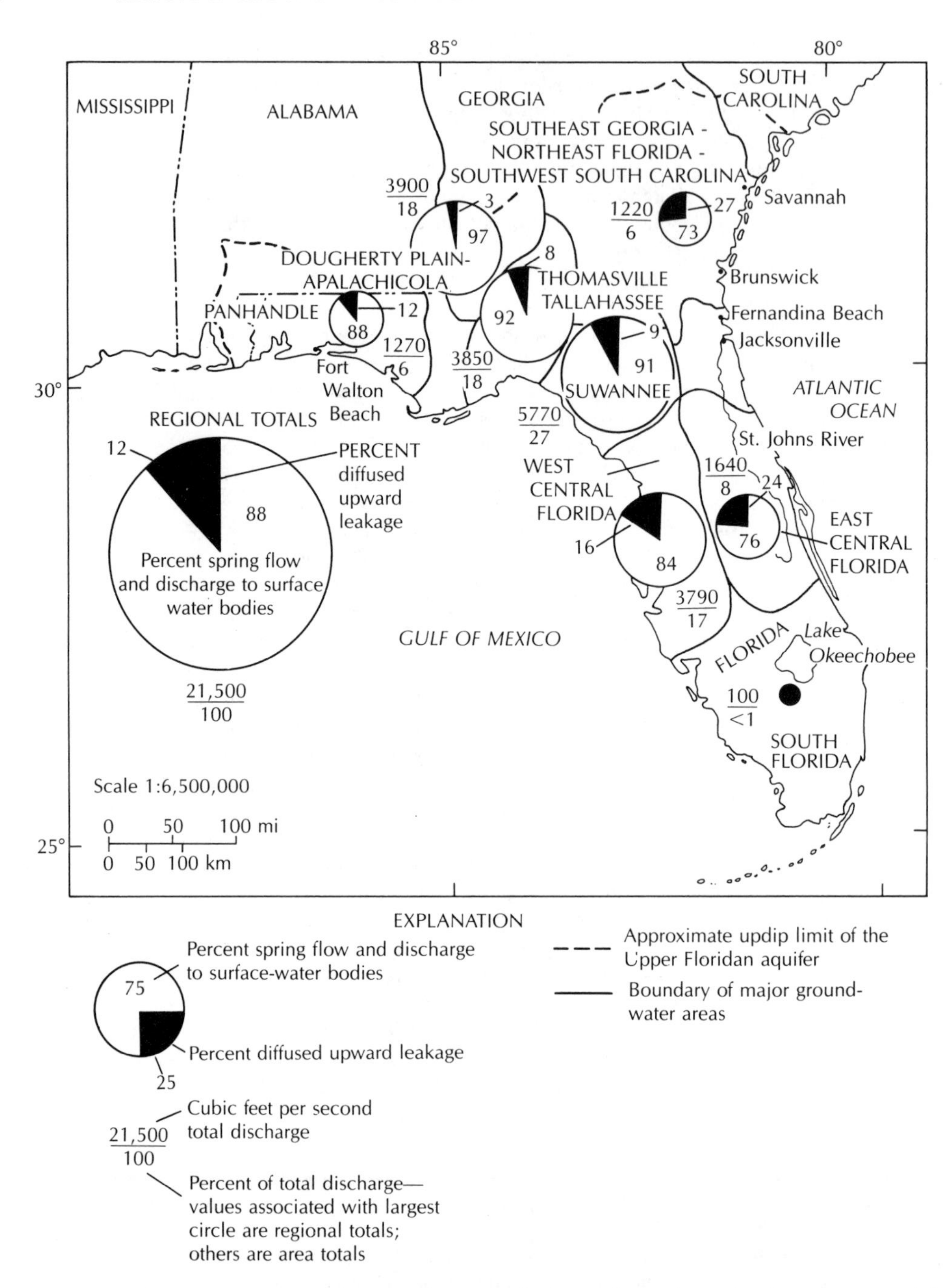


**FIGURE 7.20** Estimated predevelopment discharge for major ground-water areas of the upper Floridan aquifer. Source: P. W. Bush, and R. H. Johnson, "Floridan Regional Aquifer-System Study," in *Regional Aquifer-System Analysis Program of the U.S. Geological Survey, Summary of Projects, 1978–84,* U.S. Geological Survey Circular 1002 (1986):17–29.

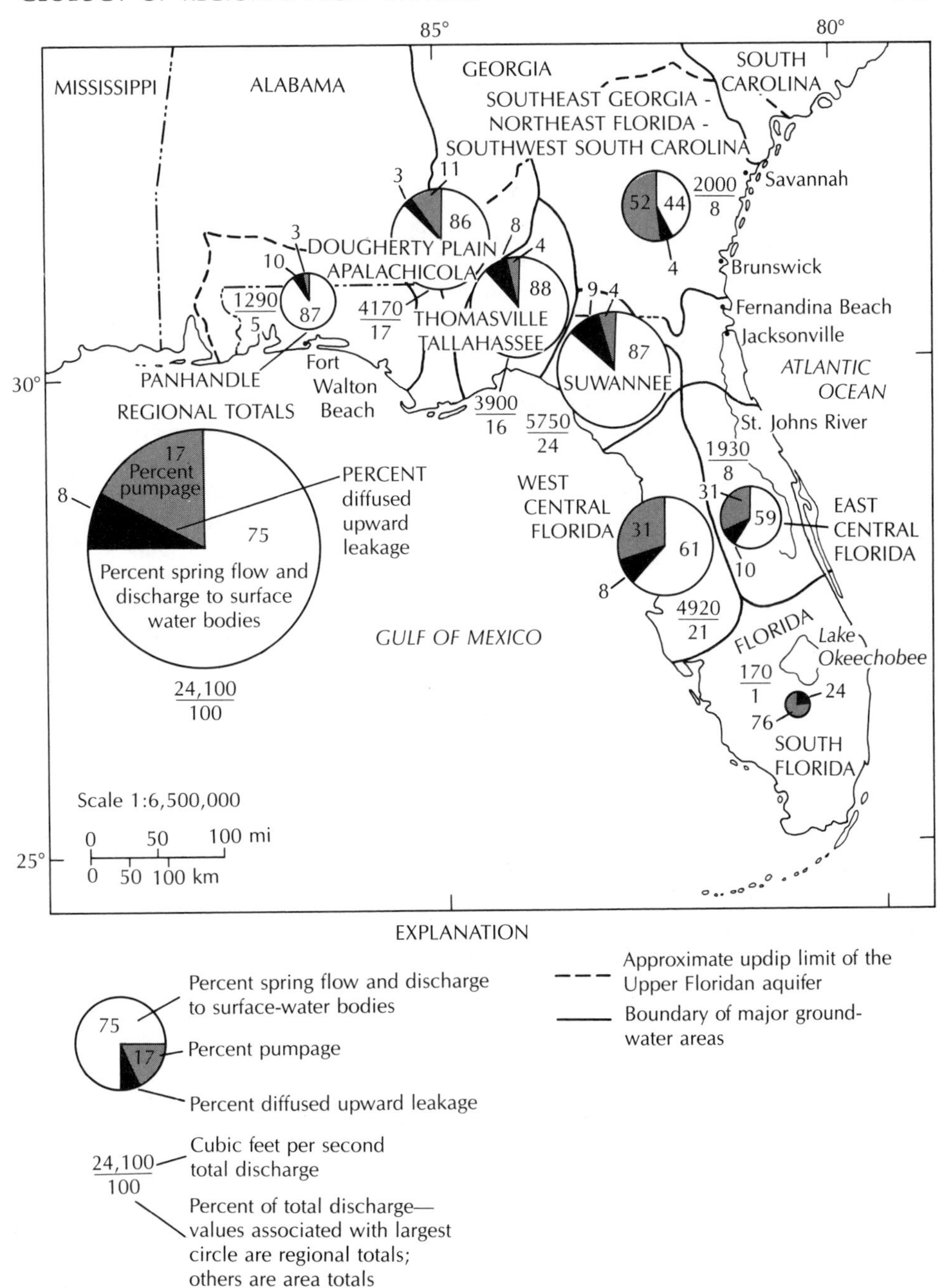


**FIGURE 7.21** Estimated 1980 discharge from major ground-water areas of the upper Floridan aquifer. Source: P. W. Bush and R. H. Johnson, "Floridan Regional Aquifer-System Study," in *Regional Aquifer-System Analysis Program of the U. S. Geological Survey, Summary of Projects, 1978–84,* U.S. Geological Survey Circular 1002 (1986):17–29.

massive siltstone, and the Arikaree Group, a sandstone. Dune sands of Quaternary age are also an important part of the High Plains aquifer, especially in Nebraska. Figure 7.22 is a geologic map showing the principal geologic units of the aquifer system (32, 33).

The High Plains aquifer is recharged by rainfall, which varies from 16 inches (41 centimeters) annually in the western part of the aquifer to 28 inches (71 centimeters) in the extreme eastern part. However, potential evapotranspiration greatly exceeds the available precipitation. Class A pan evaporation ranges from 60 inches (152 centimeters) in the northern part of the aquifer to 105 inches (267 centimeters) in the southern. As a result, net annual recharge to the aquifer system is as low as 0.024 inch (0.061 centimeter) per year in Texas to 6 inches (15 centimeters) per year in south-central Kansas. Recharge rates are greatest in areas of surficial sand dunes. Except in dune areas the long-term annual recharge is an inch (2.5 centimeters) per year or less (32).

The hydrogeologic characteristics of the High Plains aquifer materials vary widely because of the differing grain-size distribution of the sediments. The specific yield varies from less than 5 to more than 30 percent and averages 15 percent. Hydraulic conductivity varies from 25 to 300 ft/day (7.6 to 91 m/day) and averages 60 ft/day (18 m/day). There is a complex areal variability of the hydrogeologic characteristics because of the fluvial depositional history of the formations that comprise the aquifer (Figure 7.23).

The water table slopes from west to east (Figure 7.24). This is due to the regional topographic slope in this direction. Ground water is flowing from west to east at an average rate of about 1 ft/day (0.3 m/day). It discharges naturally into streams and springs as well as by evapotranspiration.

The saturated thickness of the aquifer ranges from zero at the western edge to about 1000 feet (305 meters) in west-central Nebraska and averages 200 feet (60 meters). Prior to development there were about 3.42 billion acre-feet ($4.22 \times 10^{12}$ cubic meters) of drainable water in storage in the aquifer system. Starting in the late nineteenth century, the aquifer was tapped by wells for irrigation. In 1978 there were some 170,000 wells pumping 23 million acre-feet ($2.84 \times 10^{10}$ $m^3$) of water annually (32). In some areas the amount of annual pumpage is from 2 to 100 times greater than the annual recharge. This has resulted in declines in the water table of more than 100 feet (30 meters) (Figure 7.25).

The volume of water in storage in the aquifer has decreased by about 166 million acre-feet ($2.05 \times 10^{11}$ cubic meters) per year, with most of the decline occurring in Kansas and Texas. Some 3.25 billion acre-feet ($4 \times 10^{12}$ cubic meters) of drainable water still remain in the aquifer system, but the costs of obtaining it increase as the depth to water increases. In some places, the natural quality of the water precludes its use for drinking water, although it is generally usable for irrigation. Careful management of this resource will help to extend the period of time that it will continue to supply water for irrigation in those areas of low recharge and high use.

## 7.8 GROUND-WATER–LAKE INTERACTIONS

One of the important aspects of lake hydrology is the interaction between lakes and ground water. This interaction plays a critical role in determining the water budget for a lake (34). Lakes may be classified hydrogeologically on the basis of domination of the annual hydrologic budget by surface water or ground water. Surface-water-dominated lakes typically have inflow and outflow streams, while seepage lakes are ground-water dominated (35).

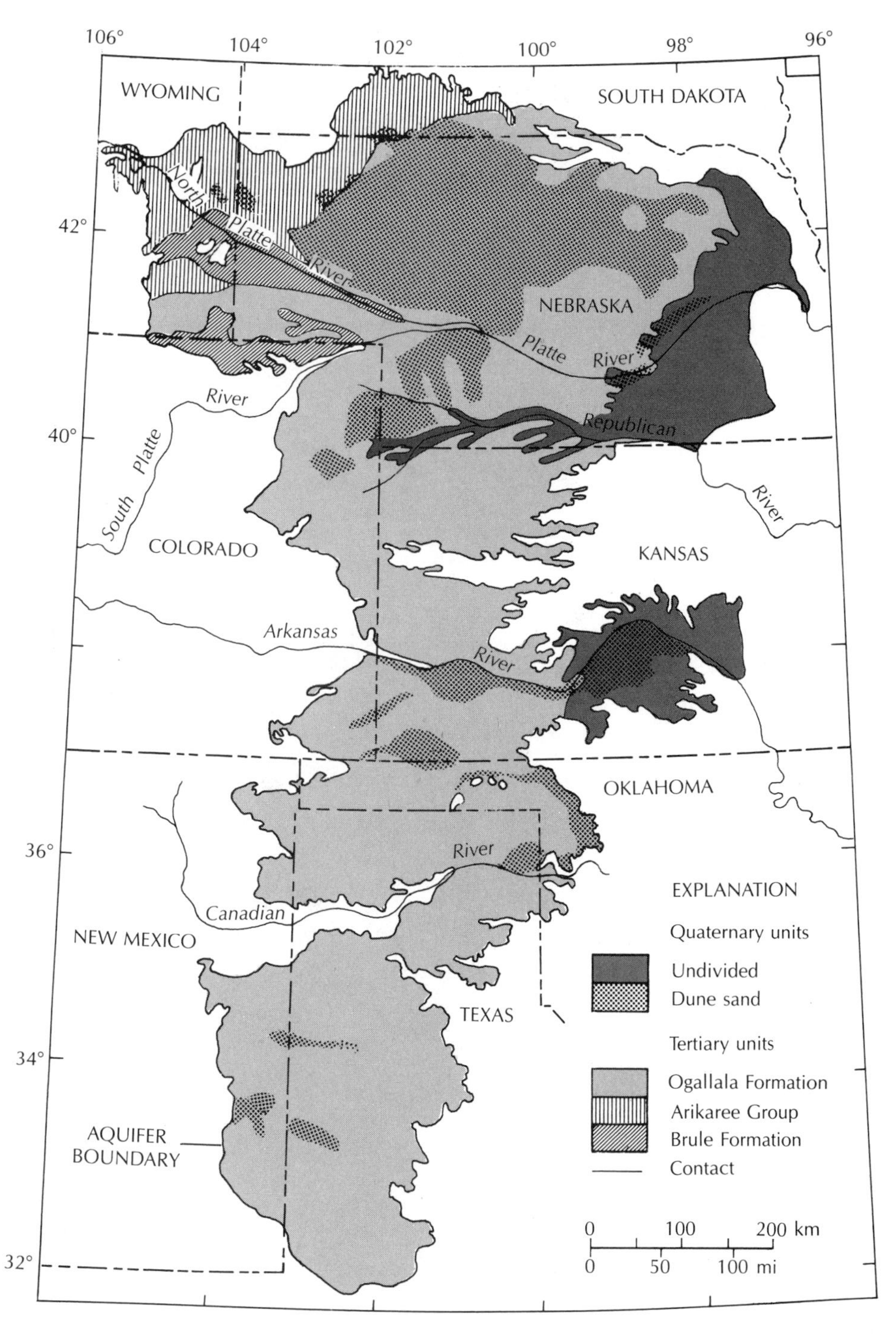


**FIGURE 7.22** Principal geologic units of the High Plains aquifer. Source: E. D. Gutentag, F. J. Heimes, N. C. Krothe, R. R. Luckey, and J. B. Weeks, U.S. Geological Survey Professional Paper 1400-B, 1984.

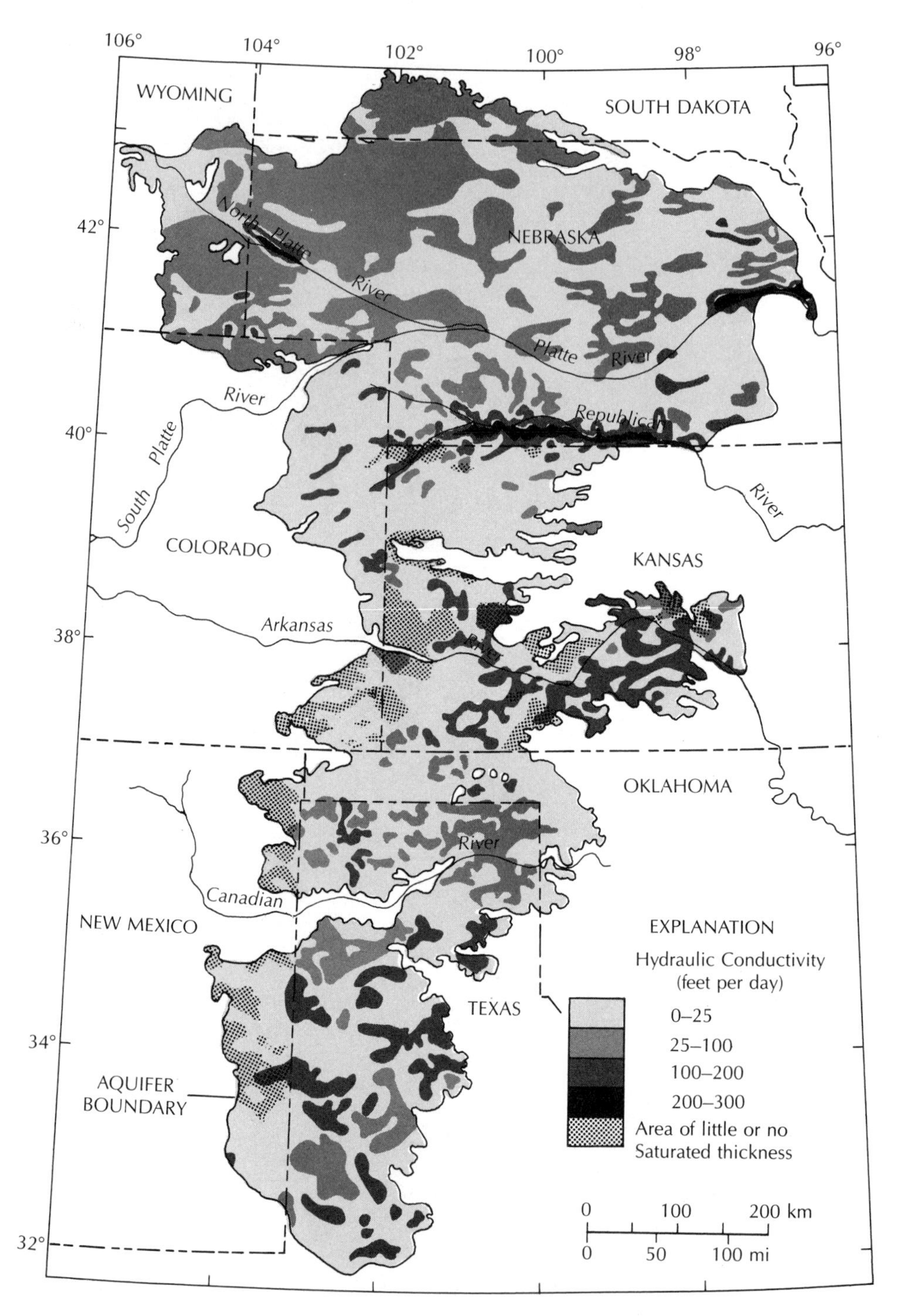


**FIGURE 7.23** Areal distribution of hydraulic conductivity in the High Plains aquifer. Source: E. D. Gutentag, F. J. Heimes, N. C. Krothe, R. R. Luckey, and J. B. Weeks, U.S. Geological Survey Professional Paper 1400-B, 1984.

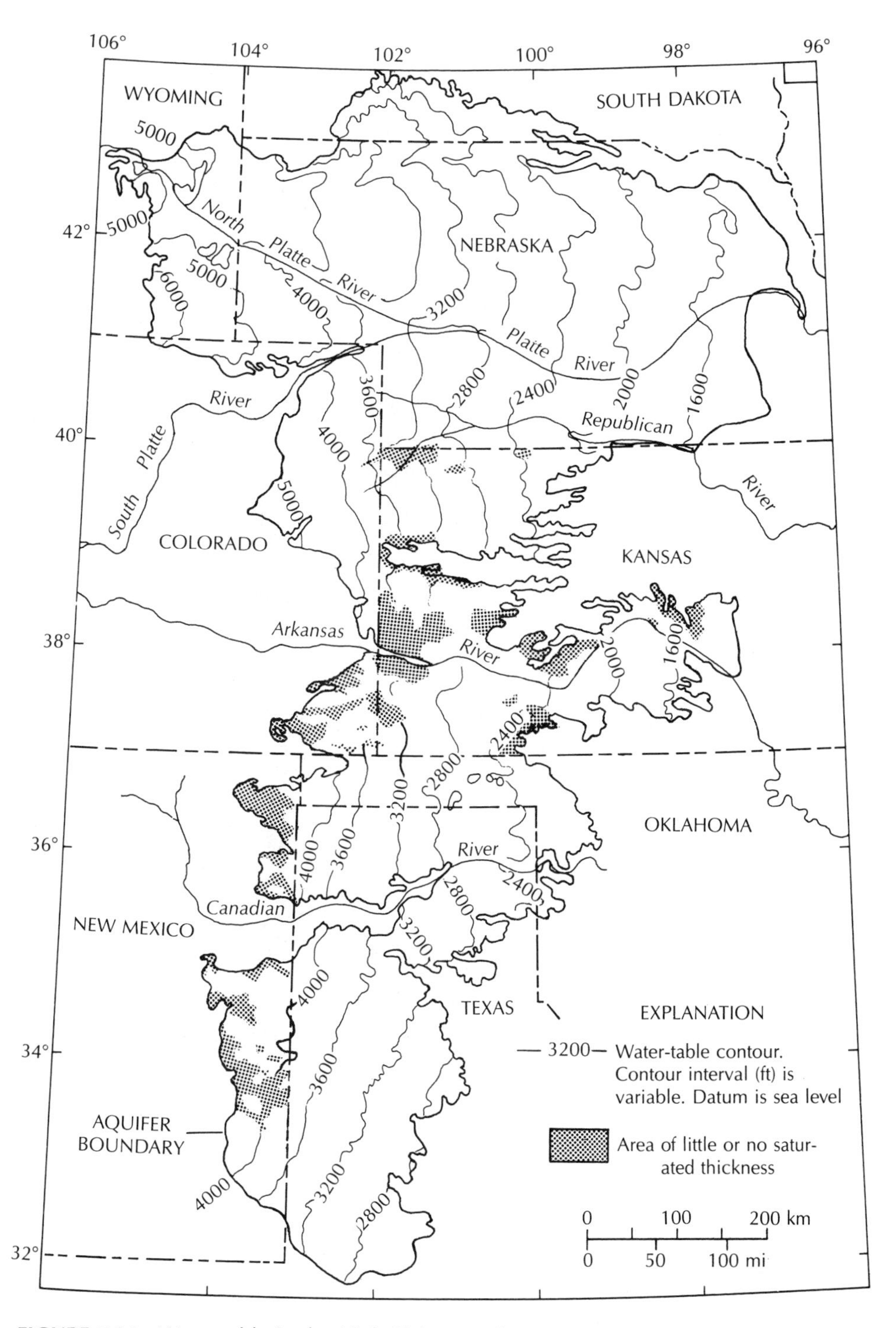


**FIGURE 7.24** Water table in the High Plains aquifer, 1980. Source: E. D. Gutentag, F. J. Heimes, N. C. Krothe, R. R. Luckey, and J. B. Weeks, U.S. Geological Survey Professional Paper 1400-B, 1984.

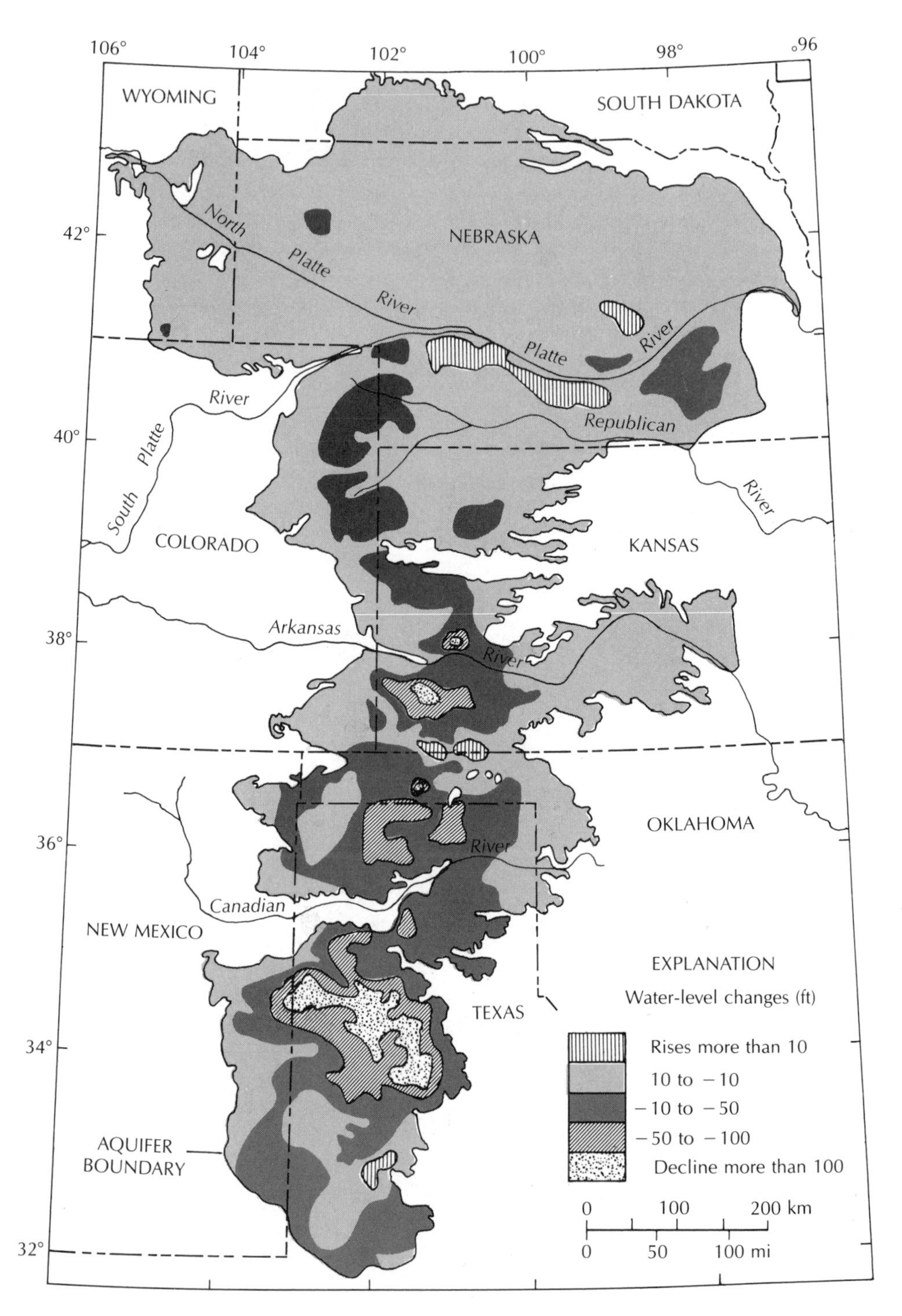


**FIGURE 7.25** Water-level changes in the High Plains aquifer, predevelopment to 1980. Source: E. D. Gutentag, F. J. Heimes, N. C. Krothe, R. R. Luckey, and J. B. Weeks, U.S. Geological Survey Professional Paper 1400-B, 1984.

In a field study of the ground-water regime of permanent lakes in hummocky moraine of western Canada, both local and intermediate flow systems were identified (36). The lakes were seepage-type (no surface outlet) and received inflow from either local or local plus regional flow systems. Figure 7.26 shows the water table and two lakes. Interlake areas are recharge areas, with the higher-elevation lake receiving water from local flow systems; the lower lake is fed by both local and regional flow systems.

In early spring, the water table was high, and most lakes were like those of Figure 7.26A. However, as the water table fell during the dry season, the ground-water divide disappeared between some lakes. These lakes became flow-through lakes, with ground water flowing in on one side and out on the other. The upper lake in Figure 7.26B illustrates this condition.

The flow conditions close to each lake were complicated by a growth of willow trees around the lake. Willows are **phreatophytes**—plants that use large amounts of ground water. During the growing season, the water table may be

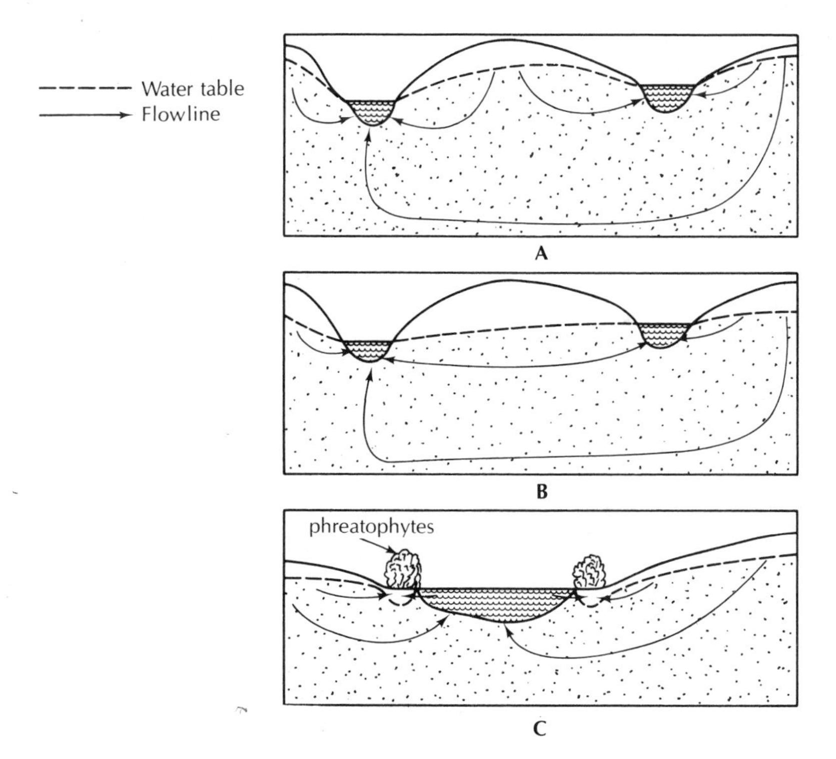


**FIGURE 7.26** Ground-water–lake interactions: **A.** High water table and interlake ground-water divide. **B.** Low water table and no interlake divide. **C.** Depressed water table due to fringe of phreatophytes. Source: Redrawn from P. Meyboom, *Journal of Hydrology,* 5 (1967):117–42. Used with permission of Elsevier Scientific Publishing Company, Amsterdam.

locally depressed below lake level beneath the phreatophyte fringe (Figure 7.26C). This can cause outflow at the margins of the lake, still leaving inflow beneath the bottom. Phreatophyte water usage is diurnal, with resulting fluctuations in the seepage balance of the lake (36).

A very interesting aspect of lake hydrology has been revealed through the numerical modeling of ground-water–lake systems. If the water table is higher than the lake level on all sides of a seepage lake, ground water will seep into the lake from all sides. Figure 7.27 shows a cross section through a seepage lake, with the water table higher than the lake surface. Potential distribution is based on a two-dimensional steady-state numerical model (37). There is a stagnation zone beneath the lake, indicating both local and regional ground-water flow. Upward seepage takes place throughout the lake bottom. As long as the stagnation zone is present, the lake will not lose water through the bottom. However, should an aquifer or high-conductivity layer underlie the lake, the stagnation zone could be eliminated (37). Without the stagnation zone, the lake will lose water through part or all of the bottom, even with the presence of a water-table mound downslope (Figure 7.28). Flow patterns of considerable complexity can form in multiple-lake systems, with some lakes having inflow through the bottom and some outseepage (Figure 7.29).

Three-dimensional numerical analysis of ground-water–lake systems has

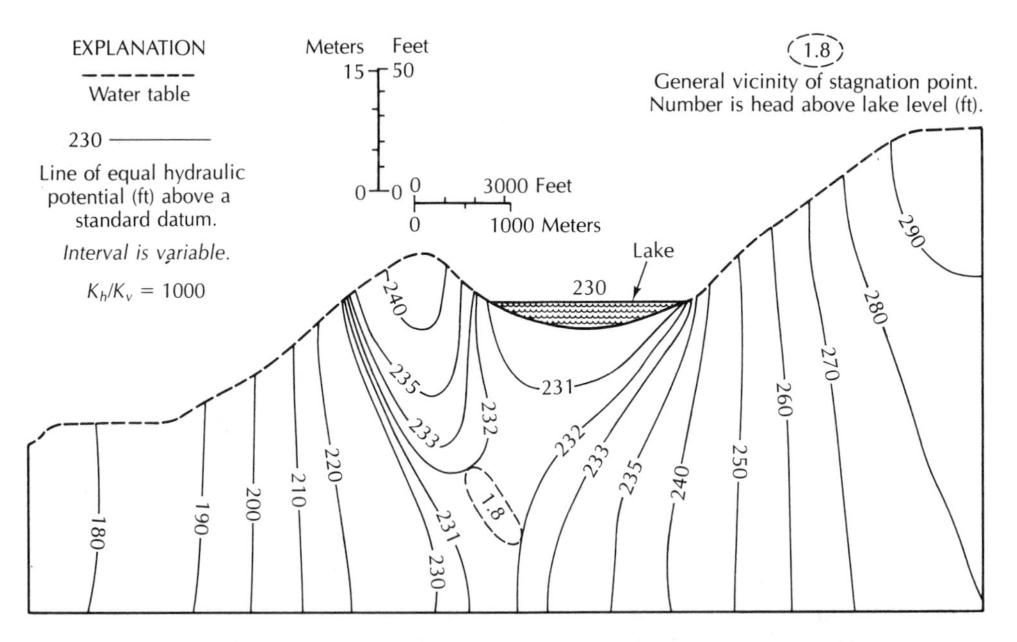


**FIGURE 7.27** Hydrogeologic cross section showing head distribution in a one-lake system with a homogeneous, anisotropic aquifer system. Results are based on a two-dimensional steady-state numerical-simulation model. Source: T. C. Winter, U.S. Geological Survey Professional Paper 1001, 1976.

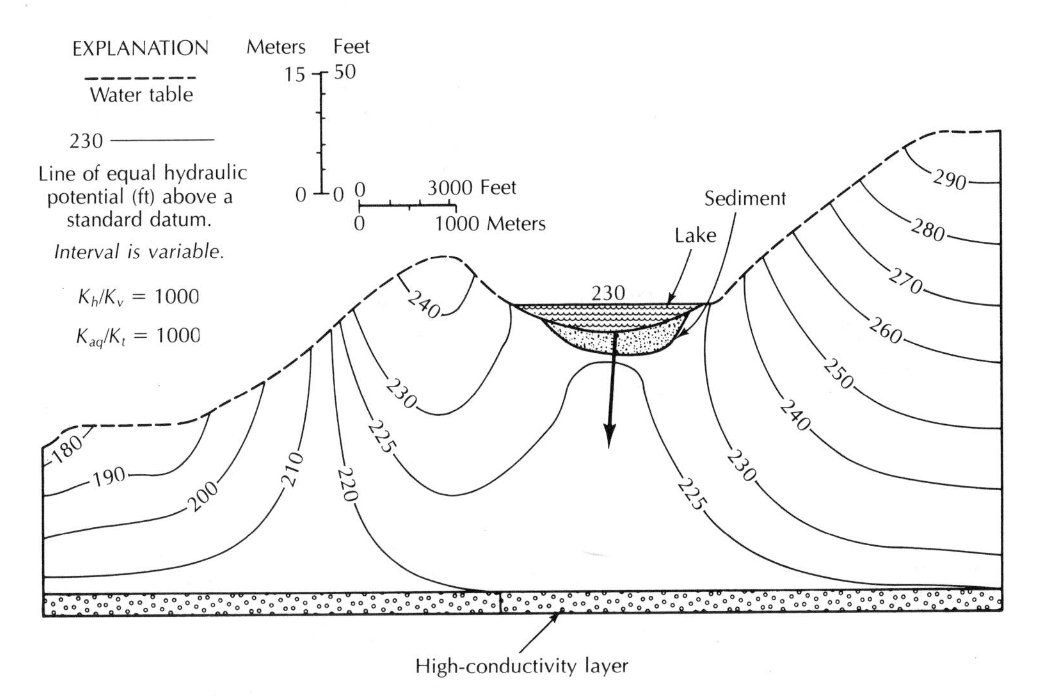


**FIGURE 7.28** Hydrogeologic cross section showing head distribution in a one-lake system with a layered aquifer system. The high-conductivity layer has a conductivity 1000 times as great as the low-conductivity layer. The lake loses water to the aquifer. Source: T. C. Winter, U.S. Geological Survey Professional Paper 1001, 1976.

shown that the stagnation zone, if present, is beneath the downgradient side of the lake and follows the lakeshore (38). The area of outseepage, if present, can be estimated from the three-dimensional model. The one-lake system in Figure 7.30 is shown in top view and two cross sections. Outseepage occurs through the center of the lake, while seepage into the lake takes place all around the edges.

Whether or not a stagnation zone will develop is largely determined by the following factors: (a) the height of adjacent water-table mounds relative to the lake level, (b) the position and conductivity of highly conductive layers in the ground-water system, (c) the ratio of horizontal to vertical hydraulic conductivity, (d) the regional slope of the water table, and (e) the lake depth (37).

As the downgradient ground-water mound height increases, the less is the likelihood that the lake will leak. If the ratio of horizontal to vertical hydraulic conductivity is low (less than 100), the model studies have indicated that stagnation points typically are present. As the ratio increases, so does the likelihood of the stagnation point disappearing and outseepage developing. Also, the presence of high-conductivity layers increases the possibility of outseepage, as does the presence of deeper lakes. Flow systems with a low regional water-table

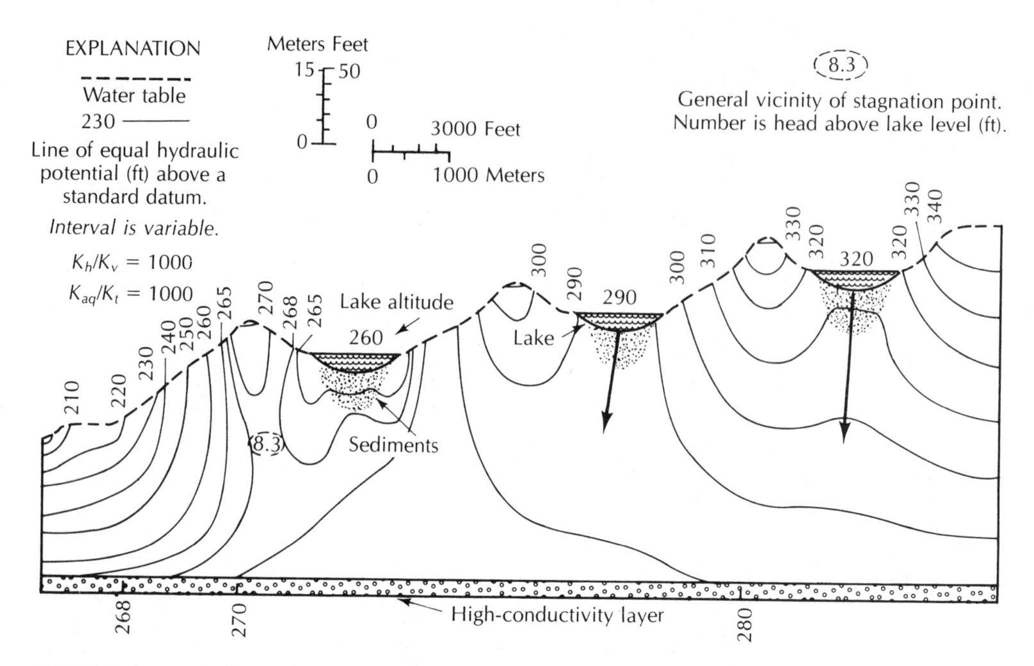


FIGURE 7.29 Hydrogeologic cross section through a three-lake system with a complex aquifer. Local and regional ground-water flow systems are present. Source: T. C. Winter, U.S. Geological Survey Professional Paper 1001, 1976.

gradient, as well as thin aquifers, are more likely to have only local flow systems—hence, lakes that do not leak if there is a downgradient water-table divide.

Numerical modeling studies have shown that, of all of the factors that affect ground-water flow systems near a lake, the water-table configuration is the most dynamic and hence the most important to monitor during field studies. The slope of the water table beyond the immediate drainage basin of a lake has a strong impact on lake-bed seepage. This is particularly true for the slope between the ground-water mound on a lake's downgradient side and a regional discharge area. The steeper the slope, the greater the likelihood that lake-bed seepage will occur (39).

Because of the complex nature of ground-water recharge, localized ground-water mounds may develop temporarily. These typically form in areas where the unsaturated zone is thin; for example, near lakes. They can create temporary flow reversals. Thus it is possible for ground-water mounds to temporarily develop on the downgradient side of a lake so that lakes that normally have outseepage could undergo transient metamorphosis to lakes with no outseepage (40).

In field studies of ground-water–lake interactions, there is no substitute for piezometers in defining the flow field. There must be several shallow piezom-

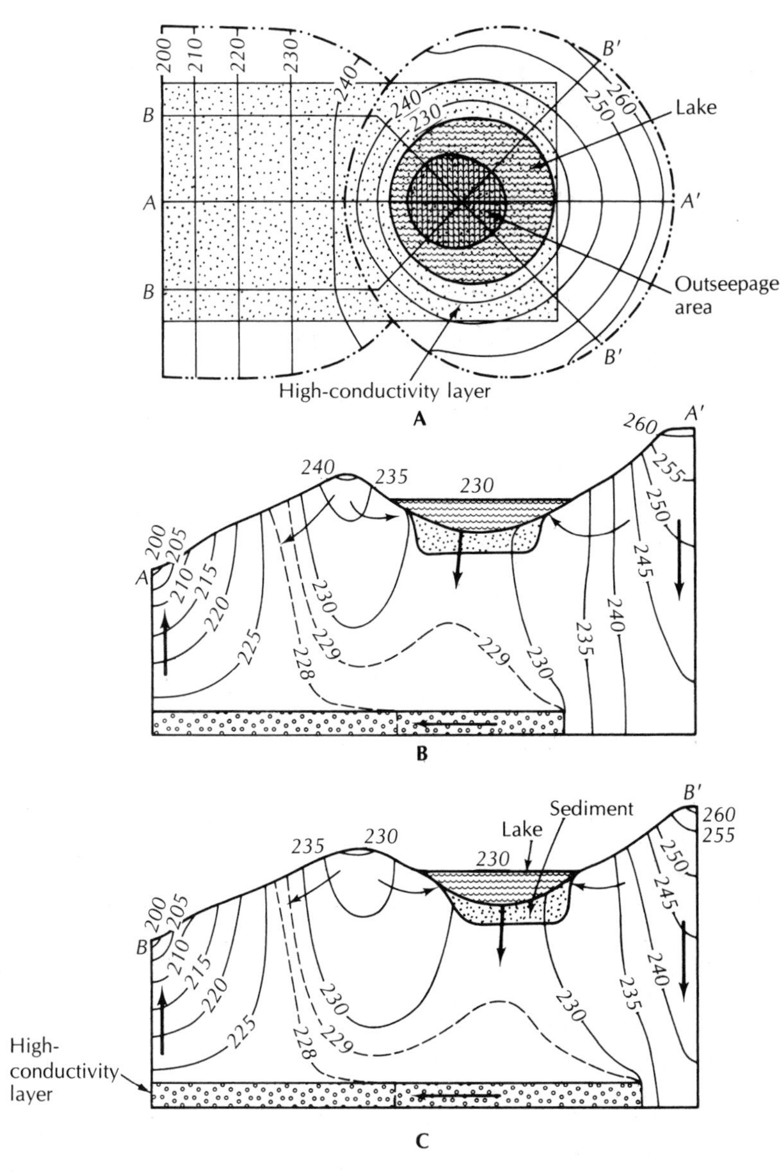


**FIGURE 7.30** Three-dimensional analysis showing the area of outseepage in the center of the lake. Seepage into the lake occurs over the rest of the bottom. Areal extent of the high-conductivity layer is indicated by stipples. Source: T. C. Winter, *Water Resources Research,* 14, no. 2 (1978):245–54.

eters all around the lake in order to define the water-table configuration (37, 38). It is necessary to determine whether a ground-water divide exists on the downgradient side of a lake or between two lakes. If a divide exists, then the maximum water-table elevation must be found. The existence of a stagnation point can be determined by placing a nest of closely spaced piezometers of different depths below the shoreline on the downslope side of the lake. If the head at all depths is greater than the elevation of the lake surface, then a stagnation point is indicated.

In order to calculate the seepage rate into a lake, the seepage per unit area can be measured at the sediment–lake interface (41). This is done with a device made of half an oil drum, with the open part pushed into the sediments. The rate of flux across the sediments is measured. Inflow generally occurs in the littoral zone of a lake (38, 41). Outseepage in lakes with no downgradient water-table divide can occur in both the downgradient littoral zone and the deeper part of the lake. If the lake has a downgradient water-table divide and there still is outseepage, it will be through the deeper part of the lake bottom. This area typically contains low-conductivity sediments, limiting the rate of water loss (37).

## REFERENCES

1. TÓTH, J. A. "A Theory of Ground-Water Motion in Small Drainage Basins in Central Alberta, Canada." *Journal of Geophysical Research,* 67, no. 11 (1962):4375–87.
2. TÓTH, J. A. "A Theoretical Analysis of Ground-Water Flow in Small Drainage Basins." *Journal of Geophysical Research,* 68, no. 16 (1963):4795–4811.
3. FREEZE, R. A., and P. A. WITHERSPOON. "Theoretical Analysis of Regional Groundwater Flow: 1. Analytical and Numerical Solutions to the Mathematical Model." *Water Resources Research,* 2, no. 4 (1966):641–56.
4. FREEZE, R. A., and P. A. WITHERSPOON. "Theoretical Analysis of Regional Groundwater Flow: 2. Effect of Water-Table Configuration and Subsurface Permeability Variation." *Water Resources Research,* 3, no. 2 (1967):623–34.
5. HUBBERT, M. K. "The Theory of Ground-Water Motion." *Journal of Geology,* 48, no. 8 (1940):785–944.
6. MAXEY, G. B. "Hydrogeology of Desert Basins." *Ground Water,* 6, no. 5 (1968):10–22.
7. BACK, W., and B. B. HANSHAW. "Comparison of the Chemical Hydrogeology of the Carbonate Peninsulas of Florida and Yucatan." *Journal of Hydrology,* 10 (1970):330–68.
8. SARTZ, R. S., et al. "Hydrology of Small Watersheds in Wisconsin's Driftless Area." *Water Resources Research,* 13, no. 3 (1977):524–30.
9. WEEKS, E. P. "Determining the Ratio of Horizontal to Vertical Permeability by Aquifer-Test Analyses." *Water Resources Research,* 5, no. 1 (1969):196–214.
10. LIAKOPOULOS, A. C. "Variation of the Permeability Tensor Ellipsoid in Homogeneous, Anisotropic Soils." *Water Resources Research,* 1, no. 1 (1965):135–41.

11. WEIDMAN, S., and A. R. SCHULTZ. *The Underground and Surface Water Supplies of Wisconsin.* Wisconsin Geological and Natural History Survey Bulletin 35, 1915, 664 pp.

12. THEIS, C. V. "The Significance and Nature of the Cone of Depression in Ground-Water Bodies." *Economic Geology,* 38 (1938):889–902.

13. WHITE, D. E. "Magmatic, Connate and Metamorphic Water." *Bulletin, Geological Society of America,* 68, no. 12 (1957):1659–82.

14. WHITE, D. E. "Thermal Waters of Volcanic Origin." *Bulletin, Geological Society of America,* 68, no. 12 (1957):1637–58.

15. BRYAN, K. "Classification of Springs." *Journal of Geology,* 27 (1919):522–61.

16. GREGORY, H. E. *The Navajo Country.* U.S. Geological Survey Water-Supply Paper 380, 1916, 219 pp.

17. MEINZER, O. E. *Large Springs of the United States.* U.S. Geological Survey Water-Supply Paper 557, 1927, 94 pp.

18. STRINGFIELD, V. T. *Artesian Water in Tertiary Limestone in the Southeastern States.* U.S. Geological Survey Professional Paper 517, 1966, 226 pp.

19. BROOK, G. A. "Surface and Groundwater Hydrogeology of a Highly Karsted Sub-Arctic Carbonate Terrain in Northern Canada." In *Karst Hydrogeology,* ed. J. S. Tolson and F. L. Doyle. Huntsville, Ala.: UAH Press, 1977, pp. 99–108.

20. MIFFLIN, M. D. *Delineation of Ground-Water Flow Systems in Nevada* (Technical Report Ser. H-W, Pub. No. 4). Reno: University of Nevada, Desert Research Institute, 1968, 54 pp.

21. EAKIN, T. E. "A Regional Interbasin Groundwater System in the White River Area, Southeastern Nevada." *Water Resources Research,* 2, no. 2 (1966): 251–71.

22. EAKIN T. E., and D. O. MOORE. *Uniformity of Discharge of Muddy River Springs.* U.S. Geological Survey Professional Paper 501-D, 1964, pp. 171–76.

23. EAKIN, T. E., et al. *Summary Appraisals of the Nation's Groundwater Resources—Great Basin Region.* U.S. Geological Survey Professional Paper 813-G, 1976, 37 pp.

24. WINOGRAD, I. J., and W. THORDARSON. *Hydrogeologic and Hydrochemical Framework, South Central Great Basin, with Special Reference to the Nevada Test Site.* U.S. Geological Survey Professional Paper 712-C, 1975, 126 pp.

25. STRINGFIELD, V. T., and H. E. LE GRAND. *Hydrology of Limestone Terranes in the Coastal Plain of the Southeastern United States.* Geological Society of American Special Paper 93, 1960, 46 pp.

26. MILLER, J. A. *Hydrogeological Framework of the Floridan Aquifer System in Florida and in Parts of Georgia, Alabama and South Carolina.* U.S. Geological Survey Professional Paper 1403-B, 1986, 91 pp.

27. BUSH, P. W., and R. H. JOHNSON. "Floridan Regional Aquifer—System Study." In *Regional Aquifer-System Analysis Program of the United States Geological Survey, Summary of Projects, 1978–84.* U.S. Geological Survey Circular 1002 (1986):17–29.

28. PARKER, G. G. "Water and Water Problems in the Southwest Florida Water Management District and Some Possible Solutions." *Water Resources Bulletin,* 11 (1975):1–20.

**29.** STEWART, J. W., et al. *Potentiometric Surface of Floridan Aquifer, Southwest Florida Water Management District.* U.S. Geological Survey Hydrologic Investigations Atlas HA-440, 1971.

**30.** COUNTS, H. B., and E. DONSKY. *Salt-Water Encroachment, Geology and Ground Water Resources of the Savannah Area, Georgia and South Carolina.* U.S. Geological Survey Water-Supply Paper 1611, 1963, 100 pp.

**31.** BUSH, P. W., J. A. MILLER, and M. L. MASLIA. "Floridan Regional Aquifer System, Phase II Study. In *Regional Aquifer-System Analysis Program of the U.S. Geological Survey, Summary of Projects, 1978–84.* U.S. Geological Survey Circular 1002 (1986):248–54.

**32.** GUTENTAG, E. D., F. J. HEIMES, N. C. KROTHE, R. R. LUCKEY, and J. B. WEEKS. *Geohydrology of the High Plains Aquifer in Parts of Colorado, Kansas, Nebraska, New Mexico, Oklahoma, South Dakota, Texas and Wyoming.* U.S. Geological Survey Professional Paper 1400-B, 1984 63 pp.

**33.** WEEKS, J. B. "High Plains Regional Aquifer Study." In *Regional Aquifer-System Analysis Program of the U.S. Geological Survey, Summary of Projects, 1978–84.* U.S. Geological Survey Circular 1002, 1986, 30–49.

**34.** WINTER, T. C. "Classification of the Hydrogeologic Settings of Lakes in the North Central United States." *Water Resources Research,* 13 (1977):753–67.

**35.** BORN, S. M., S. A. SMITH, and D. A. STEPHENSON. "The Hydrogeologic Regime of Glacial-Terrain Lakes, with Management and Planning Applications." *Journal of Hydrology,* 43, no. 1/4 (1979):7–44. (Special issue: *Contemporary Hydrogeology—The George Maxey Memorial Volume,* ed. William Back and D. A. Stephenson.)

**36.** MEYBOOM, P. "Mass-transfer Studies to Determine the Ground-Water Regime of Permanent Lakes in Hummocky Moraine of Western Canada." *Journal of Hydrology,* 5 (1967):117–42.

**37.** WINTER, T. C. *Numerical Simulation Analysis of the Interaction of Lakes and Groundwaters.* U.S. Geological Survey Professional Paper 1001, 1976, 45 pp.

**38.** WINTER, T. C. "Numerical Simulation of Steady-State Three-Dimensional Groundwater Flow near Lakes." *Water Resources Research,* 14 (1978):245–54.

**39.** WINTER, T. C. "Effects of Water-Table Configuration on Seepage through Lakebeds." *Limnology and Oceanography* 26 (1981):925–34.

**40.** WINTER, T. C. "The Interaction of Lakes with Variably Saturated Porous Media." *Water Resources Research* 19 (1983):1203–18.

**41.** LEE, D. R. "A Device for Measuring Seepage Flux in Lakes and Estuaries." *Limnology and Oceanography,* 22 (1976):140–47.

# EIGHT

# Geology of Ground-Water Occurrence

Let us not forget in this connection that every stream of water whenever it comes from a higher point and flows to a delivery tank through a short length of pipe, not only comes up to its measure but yields, moreover, a surplus; but whenever it comes from a low point, that is, under a less head, and is conducted a tolerable long distance, it will actually shrink in measure by the resistance of its own conduit.

*De aquis urbis Romae,* libri II, Sextus Julius Frontinus (ca. A.D. 35–104)

## 8.1 INTRODUCTION

Ground water inevitably occurs in geological formations. Knowledge of how these earth materials formed and the changes they have undergone is vital to the hydrogeologist. The earth is basically heterogeneous, and geological training is prerequisite to understanding the distribution of geologic materials of varying hydraulic conductivity and porosity.

A ground-water study of an area necessarily includes a review of previous geologic studies. In an area of limited geologic knowledge, a detailed geologic study must be made along with the ground-water study. The hydrogeologist is concerned with the distribution of earth materials as they affect the porosity and hydraulic conductivity of the earth. The results of a geologic study conducted by a ground-water geologist may well differ substantively from a study in the same area by a paleontologist or a petrologist. The methods employed by each would be very similar, however. Rock outcrops would be examined, certain properties noted, and the results mapped. Test borings and geophysical methods would be used to examine the subsurface. Earth materials would be grouped according to similar properties. The hydrogeologist most often begins with the proper formation names that have been assigned to an area. But these formations may then be grouped or split according to hydraulic characteristics rather than by lithology or fossil species present. For example, some bedding planes in a limestone may have been enlarged by solution, so as to transmit significant amounts of water. A hydrogeologic study might identify these on the basis of outcrop and subsurface data and map their locations. Other geological studies might not even note their occurrence.

A locality almost always has a variety of geologic materials resulting from different processes. For example, the Keweenaw Peninsula of Michigan has Precambrian bedrock famous for copper mineralization in the Portage Lakes Lava Series and Copper Harbor Conglomerate. Some water wells obtain a small supply of water from fractures in the dense bedrock, which has a very low primary permeability. One community obtains a supply from an old mine adit originally constructed to intercept seepage into the lower working levels of the mine. Sediments deposited as glaciofluvial deposits and as deltaic or long-shore deposits in Lake Superior (especially during the Pleistocene, when lake levels were higher) also serve as aquifers for community water supplies (1). A hydrogeological study of ground-water exploration in this area must examine a number of different geologic settings. A study focused on bedrock aquifers might locate wells yielding 1 to 15 gallons per minute (0.06 to 1 liter per second) from fractured bedrock. Sand deposited by long-shore currents along Lake Superior can yield up to 100 gallons per minute (6.3 liters per second) to wells. If present, this is a preferred alternative to bedrock wells.

The hydrogeologist should consider all alternatives before selecting one or more sites for detailed exploration. The preliminary survey usually consists of examination of air photos, topographic and geologic maps, and logs and reports of existing wells and then a "walking survey" of the area. This is followed by detailed study using methods such as test borings, fracture-trace analyses, and geophysical surveys. If the results of the detailed study are favorable, one or more test wells are usually installed and pumping tests made.

Hydrogeologic studies are also made to evaluate the suitability of a site for such uses as sanitary landfills, land treatment of wastewater, seepage ponds for wastewater disposal and spray irrigation of wastewater, or nuclear power plant siting. The basic methods of study and evaluation are similar. However, the target geologic terrane will differ according to the application. Landfills are usually placed in areas of low-permeability material in order to minimize the possibility of ground-water pollution. On the other hand, seepage ponds for artificial aquifer recharge would be located in as coarse a material as possible. While this chapter will focus on ground-water as a resource, the basic principles of occurrence are the same, regardless of the application.

## 8.2 UNCONSOLIDATED AQUIFERS

Materials ranging in texture from fine sand to coarse gravel are capable of being developed into a water-supply well. Material that is well sorted and free from silt and clay is best. The hydraulic conductivities of some deposits of unconsolidated sands and gravels are among the highest of any earth materials. The filterlike nature of many fine- to medium-grain sediments removes such particulate matter as bacteria and viruses, so that epidemiological water quality is usually good. However, dissolved contaminants such as nitrate, chloride, and tetrahydrofuran can travel through most host sediment and rock with little attenuation other than dilution. Unconsolidated materials are also often close to a source of recharge, such as a stream or lake. Shallow unconsolidated aquifers are in the region of rapid circulation of water—usually in local flow systems.

Unconsolidated deposits range in thickness up to tens of thousands of feet in sedimentary basins such as the Gulf of Mexico. Local well-construction regulations often call for about 30 feet (10 meters) of surface casing to prevent surface water or very shallow ground water from entering a well. An unconsolidated deposit must extend more than 30 feet deep to be useful for a water supply. Even for a domestic well, at least 3 to 6 feet (1 to 2 meters) of slotted casing, well screen, or a well point is needed below the minimum casing depth.

### 8.2.1 GLACIATED TERRANE

The midcontinental area of North America has been covered a number of times in the past by great sheets of glacier ice. The moving ice eroded and deposited as it waxed and waned across the land surface. As a result, deposits of glacial drift from less than a few feet to scores of feet thick mantle the bedrock. This covering of glacial drift is a potential source of ground water, although the hydrogeology of glaciated areas can be very complex (Figure 8.1).

Glacially related sediments have a wide range of hydraulic conductivity values. Materials carried by the glaciers ranged in size from clay to large boulders. The mode of deposition determined how well the sediment was size-sorted. Glacial till was deposited directly from the glacial ice without significant sorting by running water. As a result, it can contain particles of any size. If the ice held a wide range of particle sizes, so will the till it deposited. Generally the till will have a low hydraulic conductivity, especially if it is clay-rich. Mountain glaciers, in particular, have deposited some very coarse till materials.

Several studies have resulted in published and unpublished hydraulic conductivity values of glacial till deposits (2–7). Permeability of glacial tills depends upon the clay content, mode of deposition, and degree of weathering. **Ablation till** is deposited as the ice melts and is typically slightly water-sorted; this should make it more permeable than **basal till,** which is laid down beneath the ice. In one study of the Vandalia Till in Illinois, the hydraulic conductivity was determined by slug tests (Cooper et al. method) (6). The unaltered basal till had a mean hydraulic conductivity of $1.35 \times 10^{-7}$ centimeter per second; the basal till that was weathered and fractured had a mean hydraulic conductivity of $5.86 \times 10^{-6}$ centimeter per second; and the weathered ablation till had a mean hydraulic conductivity of $3.81 \times 10^{-5}$ centimeter per second. In New York State a homogeneous till was tested that contained 50 percent clay, 27 percent silt, 10 percent sand, and 13 percent gravel (5). The mean hydraulic conductivity as determined from slug tests (Hvorslev method) was $2 \times 10^{-8}$ centimeter per second and ranged from $1 \times 10^{-8}$ to $3 \times 10^{-8}$ centimeter per second. Falling-

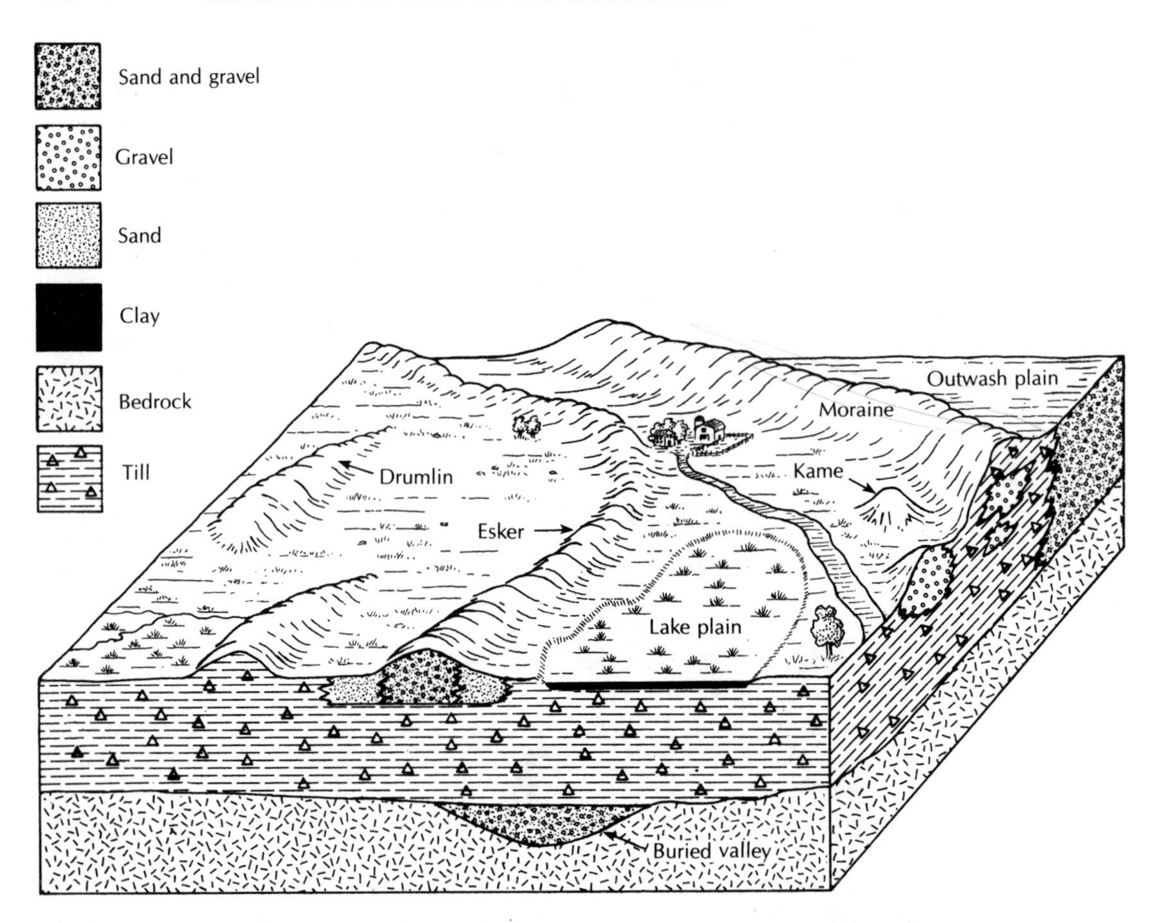


**FIGURE 8.1** Distribution of sediments in a glaciated terrane.

head permeameter tests on the same till had an average vertical hydraulic conductivity of $6.2 \times 10^{-8}$ centimeter per second and horizontal hydraulic conductivity of $3.8 \times 10^{-8}$ centimeter per second. A till that consisted of an upper, weathered zone and an unweathered lower zone was studied in Alberta (4). The unweathered till had a hydraulic conductivity of about $10^{-8}$ centimeter per second. The weathered till had both small-scale fractures with a hydraulic conductivity of $10^{-7}$ centimeter per second and large-scale fractures with a hydraulic conductivity of $10^{-5}$ centimeter per second. In Wisconsin a till that averaged 83 percent clay and silt had a mean hydraulic conductivity of $3 \times 10^{-8}$ centimeter per second on the basis of laboratory tests (7). However, slug tests in the field yielded a hydraulic conductivity of $1 \times 10^{-5}$ centimeter per second. This till is highly fractured and the matrix conductivity as revealed by the permeameter test is much less than the bulk conductivity, which is measured by the slug test. Other reports of till permeability indicate that it ranges from as little as $1.9 \times 10^{-9}$ to $2.3 \times 10^{-3}$ centimeter per second (2, 3). In most cases, if both laboratory and field tests are performed on the same till, the field test indicates one to three orders of magnitude more permeability. The reason for this is that the field tests measure the properties of a larger sample of the material, which may have fractures or sand and silt seams with higher conductivity values than the clay matrix.

Glacial lacustrine clays and silts have hydraulic conductivity values similar to those of tills. Lake Michigan bottom sediments were shown to have a hydraulic conductivity ranging from $2 \times 10^{-5}$ to $5 \times 10^{-8}$ centimeter per second (8). Fine-grained glacial lacustrine sediments in southern Indiana had lab permeability values ranging from $3.3 \times 10^{-6}$ to $2.7 \times 10^{-8}$ centimeter per second and averaged $5.9 \times 10^{-7}$ centimeter per second (9). A sandy lacustrine layer at the same site had a higher hydraulic conductivity of $3.3 \times 10^{-4}$ centimeter per second.

Glacial drift may also have been well sorted by running water—usually the meltwater from the glacier. Such sediments can be seen forming in the braided stream deposits of meltwater streams draining modern mountain glaciers. The coarse gravel is close to the terminus, sand is deposited farther downstream, and the silt and clay are carried away to be deposited far downstream in lakes or the ocean. Coarser, water-sorted glacial materials can have very high hydraulic conductivities. The general range for these sediments is 2.8 to 2835 ft/day ($10^{-3}$ to 1 centimeter per second). The materials with lower hydraulic conductivities are not as well sorted. Some silty outwash materials have hydraulic conductivities as low as $2.8 \times 10^{-2}$ ft/day ($10^{-5}$ centimeter per second). Well-sorted glacial deposits may be found associated with recognizable geomorphic features such as moulin kames, kame terraces, and eskers. However, these features are topographically high and, except for the base, may be unsaturated (Figure 8.1). If they are buried, then they can be excellent sources of water. Such buried gravel deposits can sometimes be located by electrical resistivity methods. Outwash plains and valley-train deposits are not as obvious geomorphic features, although they are often found near terminal moraines. Outwash deposits are usually well-sorted sand or gravel or mixtures of both. Some glacial outwash materials are tens of feet thick and make prolific aquifers. The city of Tacoma, Washington, obtains ground water from outwash deposits in the valley of the North Fork of the Green River. Six production wells were tested at rates of 7500 to 8600 gallons per minute (470 to 540 liters per second) with total drawdowns of 1.9 to 7.3 feet (0.58 to 2.23 meters). **Specific capacities*** as high as 4400 gallons per minute/ft (0.9 square meter per second) were obtained (10).

A key mark of glaciated terrane is the variability of hydraulic conductivity. A single test boring might reveal sequences of glacial deposits with a variation of hydraulic conductivity of 280 feet/day ($10^{-1}$ centimeter per second) for well-sorted glacial sands to 0.003 foot/day ($10^{-7}$ centimeter per second) for in-

*The specific capacity of a well is the yield divided by the drawdown of the water level from the nonpumping level. The units are thus flow/distance or $m^3/s/m$ ($m^2/s$) or gpm/ft. One gpm/ft = $2.07 \times 10^{-4}$ $m^2/s$.

terbedded tills or clays. Figure 8.2 shows glacial stratigraphic cross sections for three areas in the Mesabi Iron Range of Minnesota with a high potential for ground-water development (11).

During the Pleistocene, meltwaters from continental glaciers flowed across much of the North American landscape. This was true even in areas south of the limit of glaciation. These rivers carried large volumes of sediment. Well-sorted glaciofluvial sediments can provide excellent supplies of ground water.

The pre-Pleistocene landscape of the midcontinent was a bedrock erosional surface. Deeply incised rivers drained the land with a well-developed drainage network. The glacial sculpting of the North American landscape resulted in many changes of the preglacial drainage patterns. Many of the deep bedrock valleys were filled with sediment. Layers of till, lacustrine silts, and clays alternate with glaciofluvial deposits. Modern rivers follow the courses of some of the buried channels, whereas other former channels lie inconspicuously beneath the farmland of the Midwest. The courses of buried bedrock valleys can be determined by use of a number of remote-sensing and geophysical methods along with test drilling. In one study in Kansas (12) tonal patterns on springtime *LANDSAT* imagery and winter/summer anomalies in soil temperature were used along with seismic refraction, earth resistivity, and gravity measurements to define the channels of two bedrock valleys that had been filled with glacial drift. Test drilling was used to confirm the results of the remote techniques.

## CASE STUDY: HYDROGEOLOGY OF A BURIED VALLEY AQUIFER AT DAYTON, OHIO

There is a classic buried valley running beneath the city of Dayton, Ohio. One of the factors promoting the growth of Dayton has been the ready availability of a source of high-quality ground water (13). Permeable layers of glacial drift in the bedrock valley furnish water in large quantities to wells. In turn, these are recharged by infiltration of precipitation, as well as by water from the Miami River and its tributaries.

The bedrock in the area is the Richmond Shale of Ordovician age. During the Tertiary, an erosional surface developed, which was cleft by deeply incised rivers. During the late Tertiary, the main river draining the area was the Teays. This drainage system cut a number of valleys in the bedrock. Early Pleistocene glaciation dammed the rivers, so that lacustrine silts are found filling many parts of the Teays system. During the Kansan-Illinoian interglacial period, a radically different drainage system, the Deep Stage, prevailed in southwestern Ohio. Bedrock valleys were deeply entrenched during this time. The Deep Stage Valley passed through the present site of Dayton. Glacial processes of Illinoian and Wisconsinan age filled the Deep Stage Valley with layers of till and outwash. The character of the sediment is quite heterogeneous, as revealed by a test hole in the bedrock valley south of Dayton (Figure 8.3). The valley-train gravel aquifers alternate with confining till sheets. The aquifer layers beneath the till sheets are recharged where the till is missing because of either nondeposition or river-channel erosion. A cross section of the valley just south of Dayton shows the upper and lower aquifers separated by a discontinuous till sheet (Figure 8.4). The potentiometric surface in the lower aquifer lies below the bed of the Mi-

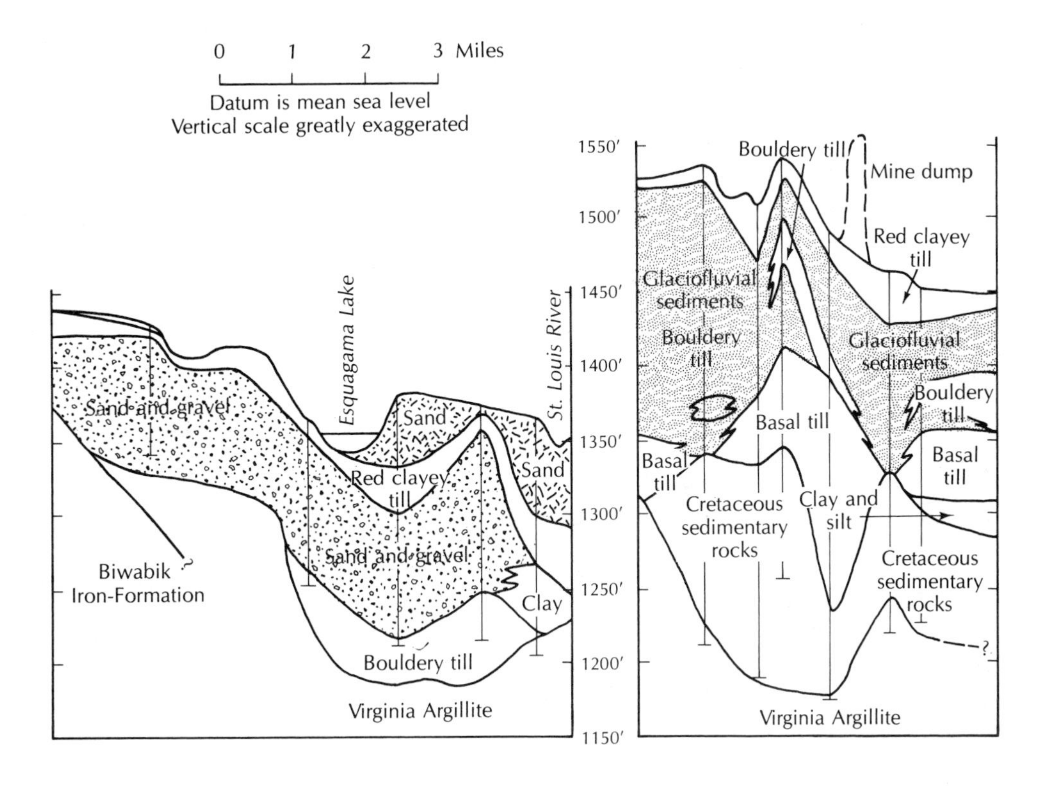


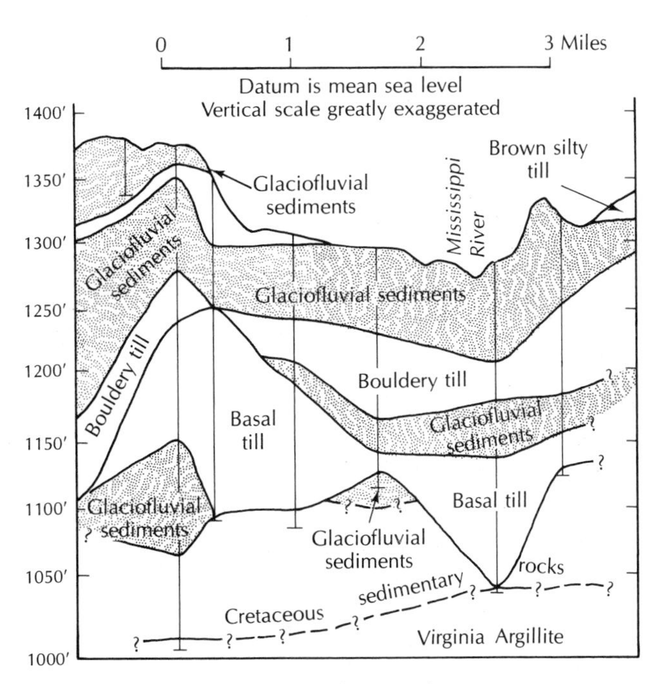


**FIGURE 8.2** Complex glacial stratigraphy in the Mesabi Iron Range, Minnesota. Sand and gravel and glaciofluvial sediments are potential aquifers. Source: T. C. Winter, U.S. Geological Survey Water-Supply Paper 2029-A, 1973.

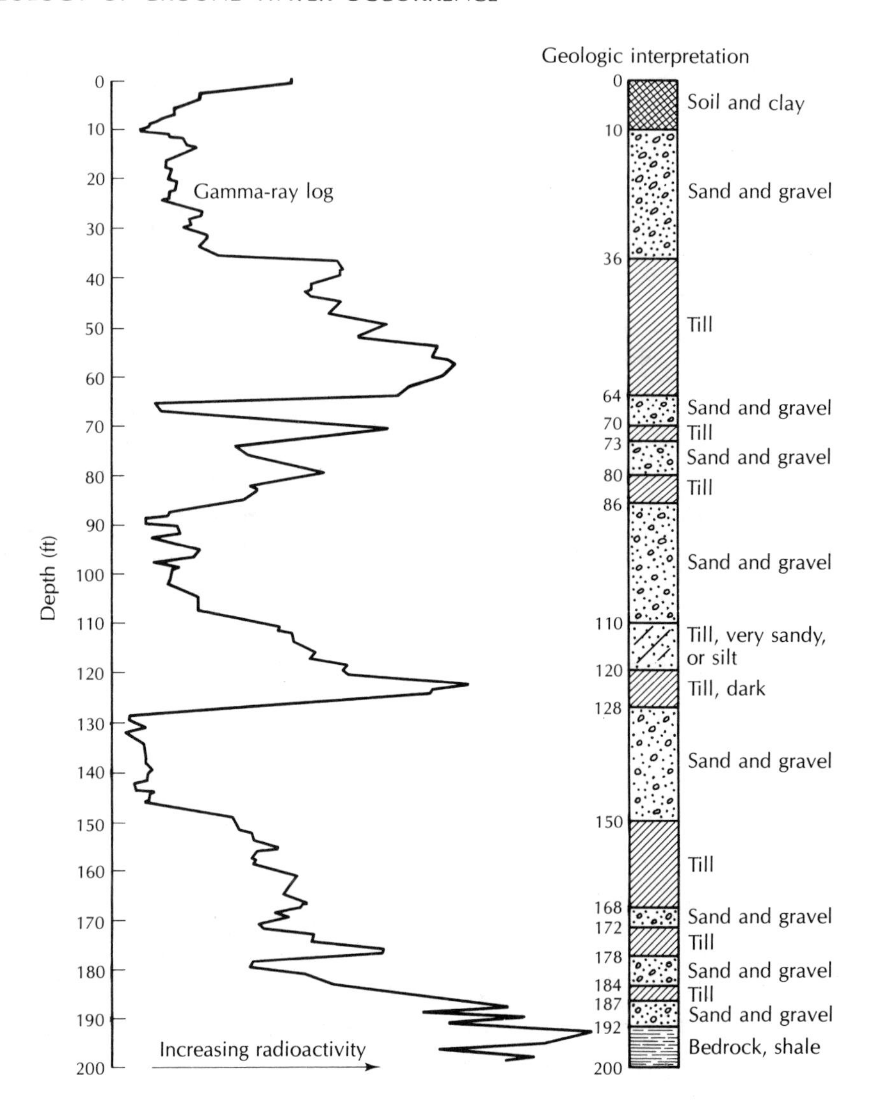


**FIGURE 8.3** Well log and gamma-ray log of uncased test hole in glacial deposits filling a buried bedrock valley south of Dayton, Ohio. Source: S. E. Norris and A. M. Spieker, U.S. Geological Survey Water-Supply Paper 1808, 1966.

ami River. The potentiometric surface rises sharply near the river when the discharge and stage of the Miami River rise. This indicates a good hydraulic connection between the river and the aquifer.

The Rohrers Island Well Field is located on an island in the Mad River, a tributary of the Miami River. The upper sand and gravel aquifer yields up to 90 million gallons per day (3940 liters per second). The upper aquifer, which is up to 65 feet (20 meters) thick, is artificially recharged with river water that floods onto about 20 acres (8 hectares) of infiltration lagoons during periods when the river turbidity is low. The lagoon bottoms are an-

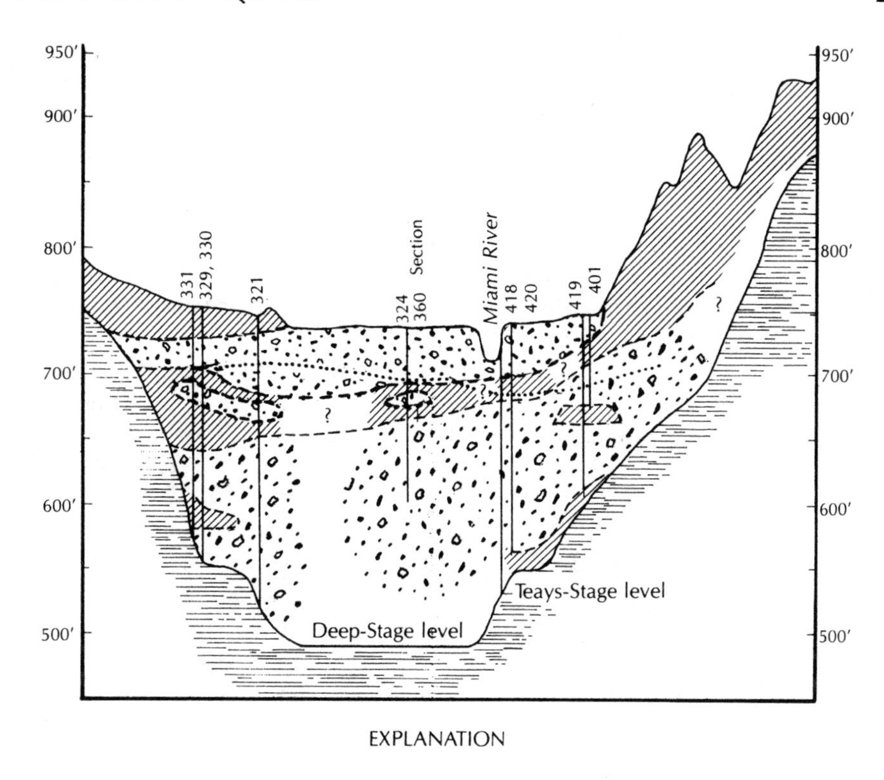


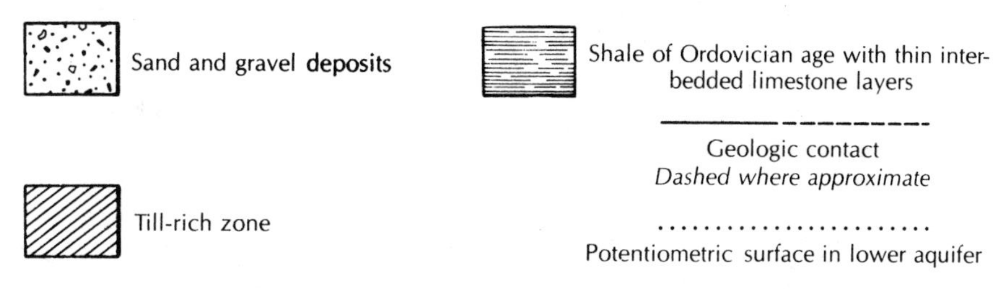


**FIGURE 8.4** Cross section of buried bedrock valley at Dayton, Ohio, showing upper (water-table) aquifer and lower (confined) aquifer. Source: Modified from S. E. Norris and A. M. Spieker, U.S. Geological Survey Water-Supply Paper 1808, 1966.

nually cleaned of silt and clay. This is a classic example of induced stream infiltration used as a water resource management technique. Virtually all of the water pumped from this aquifer comes indirectly from the river. The filtration through the sediments reduces turbidity and removes pathogens.

### 8.2.2 ALLUVIAL VALLEYS

Flowing rivers deposit sediment, generally termed **alluvium.** During periods of flooding, alluvium is deposited in the channel as well as in the floodplain. As the flood peak passes, flow velocities start to drop, the energy available to transport sediment decreases, and deposition begins. Coarse gravel is deposited in the stream channel, sand and fine gravel form natural levees along the banks, and silt and clay come to rest on the floodplain. Point bars are formed by deposition on the inside of a bend in the river. Point-bar formation is not limited to floods.

The alluvium may be reworked by a meandering stream, even during quiescent periods for the river. As the channel swings back and forth across the floodplain, point-bar deposits of sand and coarse gravel are left behind. If the stream is aggrading, the general land level subsiding, or both, the alluvial deposits will thicken with time (Figure 8.5).

Rivers draining glaciers, either modern or Pleistocene, have very heavy sediment loads. Braided streams and gravel bars in the river are typical and connote the aggrading nature of such rivers. Thick deposits of sand, gravel, or mixtures of both are formed. Downcutting by a river through previously deposited sediment can form terraces on the sides of the lower-stage floodplain. Many modern rivers, even some outside of glaciated regions, have terraces formed of Pleistocene-age gravel deposits.

Alluvial valleys can be excellent sources of water. There are zones of gravel in old channel and point-bar deposits with a very high conductivity. The task of the hydrogeologist is to find these gravel zones, since silt and clay floodplain deposits may also be present, obscuring their locations. A knowledge of fluvial processes is helpful in this regard. Surface geophysical surveys can often be used to locate sand and gravel deposits where they are surrounded by cohesive sediments. Test drilling is usually necessary to confirm the preliminary conclusions of the hydrogeologist and geophysicist.

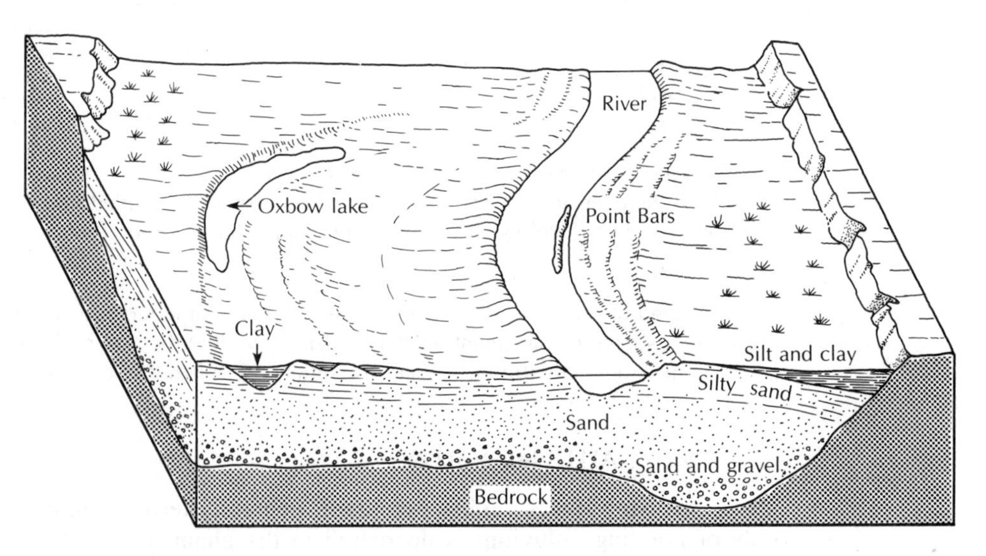


**FIGURE 8.5** Distribution of alluvium in a flowing river.

In evaluating the water-yielding potential of stream-terrace deposits, it should be kept in mind that erosion, as well as deposition, can result in terrace formation. If a terrace represents an erosional surface on bedrock, it is not underlain by potentially permeable alluvium.

### 8.2.3 ALLUVIUM IN TECTONIC VALLEYS

Many major valley systems are products of tectonic activity rather than of fluvial or glacial erosion. During mountain-building episodes, the uplift of mountain masses will result in intermontane basins being formed. Fault-block valleys can also be created by down-dropping of large crustal pieces along faults. Erosion of the mountains creates sediment that is carried into the valleys, forming talus slopes, alluvial fans, and alluvial and lacustrine deposits. These sediments can be very coarse, with high hydraulic conductivities—alluvial fan gravels and channel deposits, for example. Lacustrine clays, on the other hand, can be fine, with low conductivity. Gravel aquifers confined by lacustrine sediment are quite typical of such basins.

Intermontane valleys of the interior basins of southeastern California and the Great Basin area of Nevada and Utah are typically tectonic. The unconsolidated sediments in the basins may be part of either local or regional flow systems (Figure 8.6). In regions that are semiarid to arid, ground water is recharged by precipitation in the mountains. Bedrock beneath the mountains may receive direct recharge and then feed the valley-fill sediments. Streams originating in the mountains may also lose water to the alluvium when the flow goes across the valley bottoms (Figure 8.7).

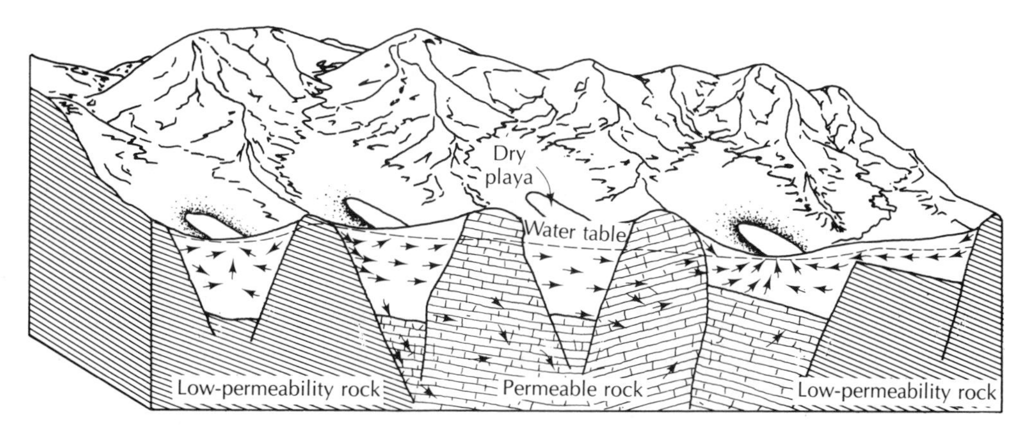


**FIGURE 8.6** Common ground-water flow systems in tectonic valleys filled with sediment. Basins bounded by impermeable rock may form local or single-valley flow systems. If the interbasin rock is permeable, regional flow systems may form. In closed basins, ground-water discharges into playas, from which it is discharged by evaporation and transpiration by phreatophytes. Source: Modified from T. E. Eakin, D. Price, and J. R. Harrill, U.S. Geological Survey Professional Paper 813-G, 1976.

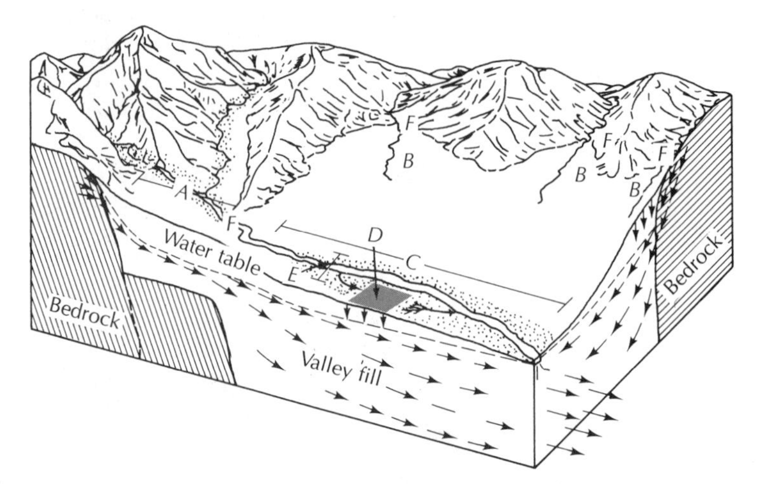


*A*, Gaining reach, net gain from ground-water inflow, although in localized areas stream may recharge wet meadows along floodplain. Hydraulic continuity is maintained between stream and ground-water reservoir. Pumping can affect streamflow by inducing stream recharge or by diverting ground-water inflow that would have contributed to streamflow.

*B*, Minor tributary streams, may be perennial in the mountains but become losing ephemeral streams on the alluvial fans. Pumping will not affect the flow of these streams because hydraulic continuity is not maintained between streams and the principal groundwater reservoir. These streams are the only ones present in arid basins.

*C*, Losing reach, net loss in flow due to surface-water diversions and seepage to ground water. Local sections may lose or gain depending on hydraulic gradient between stream and ground-water reservoir. Gradient may reverse during certain times of the year. Hydraulic continuity is maintained between stream and groundwater reservoir. Pumping can affect streamflow by inducing recharge or by diverting irrigation return flows.

*D*, Irrigated area, some return flow from irrigation water recharges ground water.

*E*, Floodplain, hydrologic regimen of this area dominated by the river. Water table fluctuates in response to changes in river stage and diversions. Area commonly covered by phreatophytes (shown by random dot patterns).

*F*, Approximate point of maximum stream flow.

**FIGURE 8.7** Ground-water–surface-water relationships in valley-fill aquifers located in arid and semiarid climates. Source: T. E. Eakin, D. Price, and J. R. Harrill, U.S. Geological Survey Professional Paper 813-G, 1976.

The water table is generally closer to the surface in the lower elevations of intermontane basins than it is at the edges next to the mountains, so wells in the former locations would have a smaller pumping lift—hence, lower energy use. The intermontane deposits typically slope downward from the mountain flanks. Beneath the upper parts of alluvial fans, the depth to ground water may be hundreds of feet. Surface streams flowing onto high alluvial fans may disappear as they lose water to the coarse sediment.

In the Great Basin region of the western United States, saturated valley-fill deposits over 1000 feet (300 meters) thick are common in the large valleys (14). Artesian conditions are often found in the lower elevations of these basins.

Pleistocene lacustrine clays near the surface overlying coarse alluvium create this situation.

In tectonic valleys, ground-water outflow can occur by transpiration, evaporation from surface water or saturated soils, spring discharge, and/or underflow into adjacent basins. Water pumped and used consumptively for irrigation is also an outflow. If the basin is topographically closed, there is no stream discharge. In such a case, evapotranspiration and underflow are the natural drainage methods.

Wells in tectonic valleys should be located where the aquifer material is coarse, the depth to water is not great, and a source of recharge water is available. These criteria are often met in well fields located near modern rivers. However, there may not be any surface streams near areas where wells are needed. Artesian aquifers overlain by thick lacustrine deposits may be too deep to find by surficial geophysical methods, such as electrical resistivity. Test drilling may be the only recourse in such cases, albeit an expensive one. Ground-water quality in tectonic valleys can have significant spatial variation. In general, ground water that is not actively circulating may be high in dissolved solids. Shallow ground water, with a high rate of direct evaporation, may also have high salinity. In the Great Basin area, individual wells have been developed with yields of up to 8600 gallons per minute (540 liters per second) with specific capacities of up to 3000 gallons per minute per foot (0.6 square meter per second). These are prodigious wells. The average yield is lower, but still substantial at 1000 gallons per minute (65 liters per second) (14).

## CASE STUDY: TECTONIC VALLEYS—SAN BERNARDINO AREA

The San Bernardino area is in the upper Santa Ana Valley of the coastal area of southern California (15). The valley is tectonic in origin and bounded on the north, east, and south by mountains of consolidated rocks. The alluvial valley is subdivided into separate ground-water basins by faults in the alluvial materials, which are barriers to ground-water movement. The Bunker Hill Basin has mountains on three sides and the San Jacinto Fault on the fourth side (Figure 8.8).

The alluvium in the basin consists of Pleistocene-age deposits overlain by Recent alluvium, river-channel deposits, and dune sands. The alluvium is from 690 to 1400 feet (210 to 425 meters) thick at the center of the basin. The older deposits consist of alluvial fans, terrace deposits, and stream channels. The deposits can be highly permeable, but facies of low permeability exist and form confining layers. Faulting cuts the older alluvium, and the faults are generally barriers to ground-water movement. They may impede ground-water flow by (a) offsetting of gravel beds against clay layers, (b) folding of impermeable beds upward along the fault, (c) cementation by carbonate formation, and (d) formation of clayey gouge.

Recent alluvium includes highly permeable river-channel deposits through which much of the ground-water recharge occurs. They underlie active and abandoned stream channels and are above the water table, for the most part. The Recent floodplain material is less permeable than the stream channels. It is 50 to 100 feet (15 to 30 meters) thick and is not known to be faulted.

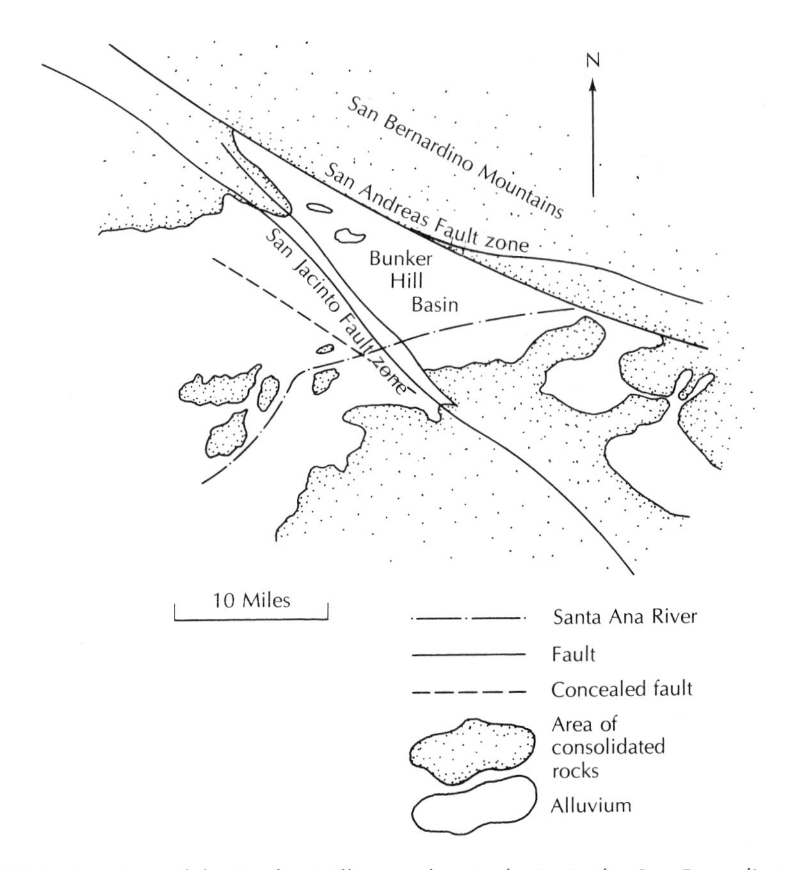


**FIGURE 8.8** Location of the Bunker Hill ground-water basin in the San Bernadino area of southern California. Source: Modified from L. C. Dutcher and A. A. Garrett, U.S. Geological Survey Water-Supply Paper 1419, 1963.

The lower parts of the valley floor receive 12 to 17 inches (30 to 43 centimeters) of precipitation per year. Owing to orographic effects, precipitation in the San Bernardino Mountains to the east is as much as 28 inches (71 centimeters). Runoff from the mountains flows into the Santa Ana River and its tributaries. In the upper reaches of the basin, the river is a losing stream. Ground-water recharge by floodwater takes place through the permeable river-channel deposits.

A cross-section of the Bunker Hill area is shown in Figure 8.9. Ground-water recharge takes place in the upper parts of the basin and flows toward lower elevations. Well hydrographs on the upstream side of the San Jacinto Fault show that the head in a shallow well is lower than in deeper wells, indicating upward flow. In the San Bernardino area, the younger alluvium is confining, and when wells were first drilled in the area they were flowing.

At one time, considerable ground-water outflow occurred from the Bunker Hill Basin in the area where the Santa Ana River crosses the San Jacinto Fault Zone (Colton Narrows). Although the fault zone does not offset the Recent alluvium, it forces deeper

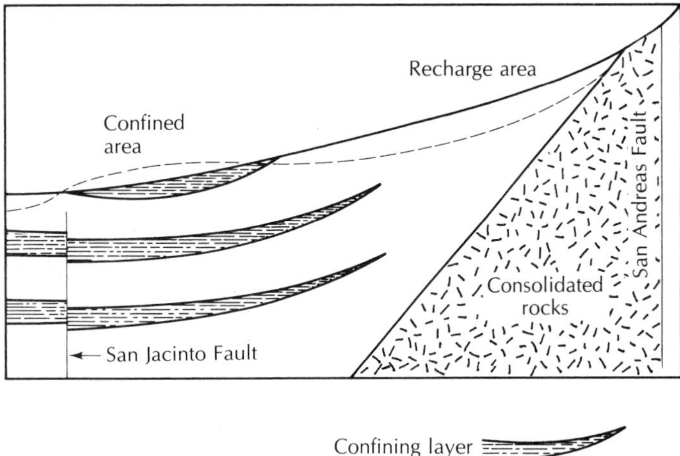


**FIGURE 8.9** Cross section through the unconsolidated deposits of the Bunker Hill groundwater basin. The section is along the course of the Santa Ana River, which provides recharge to the aquifers. Source: Modified from L. C. Dutcher and A. A. Garrett, U.S. Geological Survey Water-Supply Paper 1419, 1963.

water to move upward. Except for Recent river-channel deposits beneath the Santa Ana River, the surficial deposits are, for the most part, impermeable. At the turn of the century, large springs made this area marshy, and ponds were present. Heavy ground-water withdrawals for irrigation have lowered the water table and reduced the amount of outflow. Figure 8.10 shows the water-table contours where the Santa Ana River crosses the San Jacinto Fault as they existed in 1939. Ground-water contours northeast (upstream) of the fault indicate that ground water is moving toward the surface. Well hydrographs in the area also indicate upward ground-water movement. The water-table contours on either side of the fault indicate a very steep gradient across the fault zone.

By the 1970s the basin had undergone extensive ground-water overdrafts. The water level in the swampy area along the Santa Ana River dropped 50 to 150 feet (15 to 46 meters) below the land surface. This drained the wetland, and the land became dry. Extensive urban development occurred, with the site of the former swamp becoming downtown San Bernardino.

A computer model of the ground-water basin was developed (16). This model predicted that if ground-water pumpage were to be reduced or natural recharge were to increase, ground-water levels could rise. In the first edition of this book the author pointed out that this would be especially severe in the San Jacinto fault area, which is the point of natural ground-water discharge from the Bunker Hill Basin, and indicated that "Abandoned and forgotten wells could begin to flow again. Basements would be flooded and soils waterlogged." This prediction came true. By 1982 several years of above-average precipitation in the Bunker Hill Basin resulted in ground-water levels that were high enough to cause severe problems. Basements were flooded; abandoned wells began to flow (one beneath a building split the foundation); and springs at the land surface made a new water

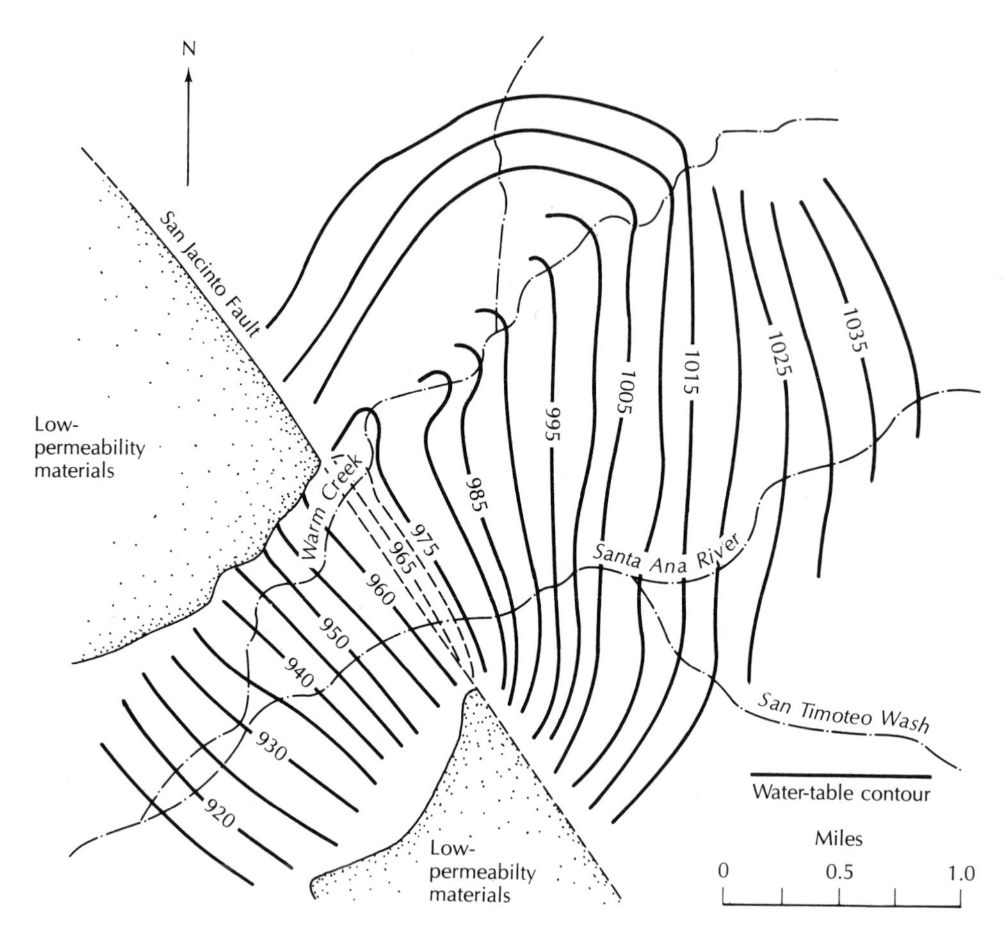


**FIGURE 8.10** Water-table contours in March 1939 of shallow deposits at the Colton Narrows area, where the Santa Ana River crosses the San Jacinto Fault. The contours for 970 and 965 are shown, but the exact positions were not known. Source: L. C. Dutcher and A. A. Garrett, U.S. Geological Survey Water-Supply Paper 1419, 1963.

hazard in the fifth fairway of the San Bernardino Golf Club (17). An even greater fear is that the high ground-water levels might result in the failure of building foundations owing to soil liquefaction during an earthquake (17). A decrease in recharge or an increase in pumpage is the only solution to reducing such high ground-water levels.

## 8.3 LITHIFIED SEDIMENTARY ROCKS

Clastic sedimentary rocks are typically composed of silicate, carbonate, or clay minerals. Chemically precipitated sedimentary rocks are primarily limestone, dolomite, salt, or gypsum. Coal and lignite can also be considered to be sedimentary deposits, with the original sediments being organic matter.

Sedimentary rock sequences were formed with younger beds laid down upon older ones. The original sediments may have been subaqueous or terrestrial. Rarely do sedimentary rocks occur as a single unit; there is typically a sequence of many beds. The original layered sequence may be undisturbed or it may be extensively folded and faulted.

## CASE STUDY: SANDSTONE AQUIFER OF NORTHEASTERN ILLINOIS–SOUTHEASTERN WISCONSIN

The sandstone or deep aquifer of northeastern Illinois–southeastern Wisconsin is comprised of formations of Cambrian and Ordovician age, primarily sandstones and dolomites. The formations strike north-south and dip to the east. The aquifer is confined by the Maquoketa Shale (Figure 8.11) and makes a classic confined artesian system in clastic rock. The aquifer extends from the area north of Milwaukee, Wisconsin, to south and west of the Chicago, Illinois, area (18–20).

The Maquoketa Shale has been eroded away in the western part of the area so that the upper formations of the sandstone aquifer can be recharged directly (Figure 8.12). The average direct recharge rate to the sandstone aquifer in Illinois is estimated to be 20,400 gallons per day per square mile (30 cubic meters per day per square kilometer) (20).

In the area where the aquifer is confined by the Maquoketa Shale the amount of

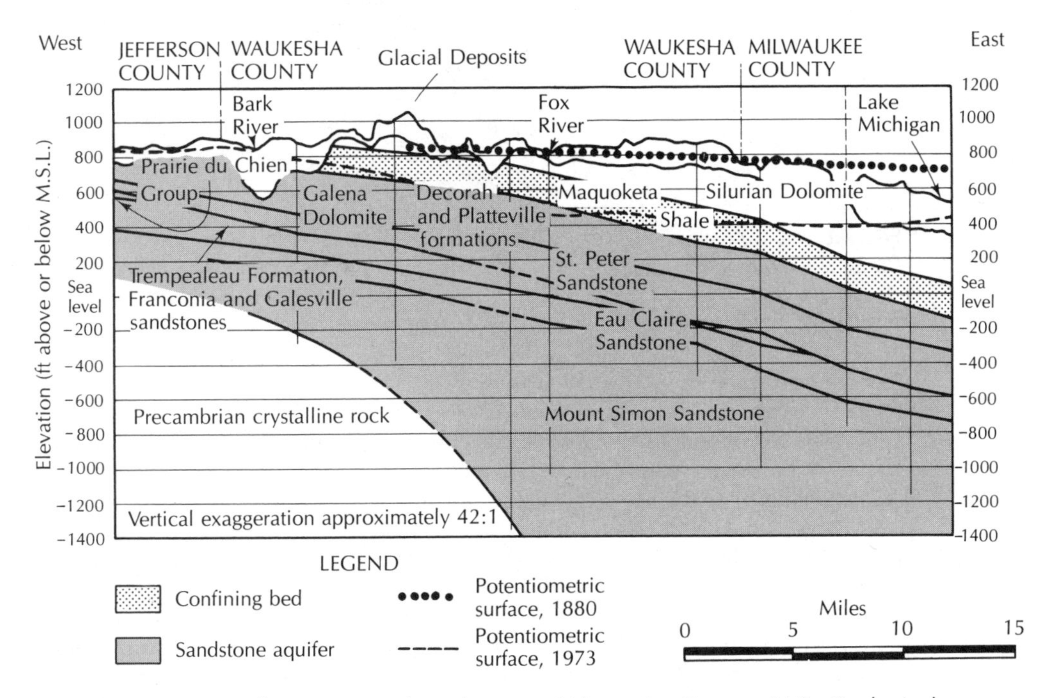


**FIGURE 8.11** Artesian flow system of southeastern Wisconsin. Source: U.S. Geological Survey.

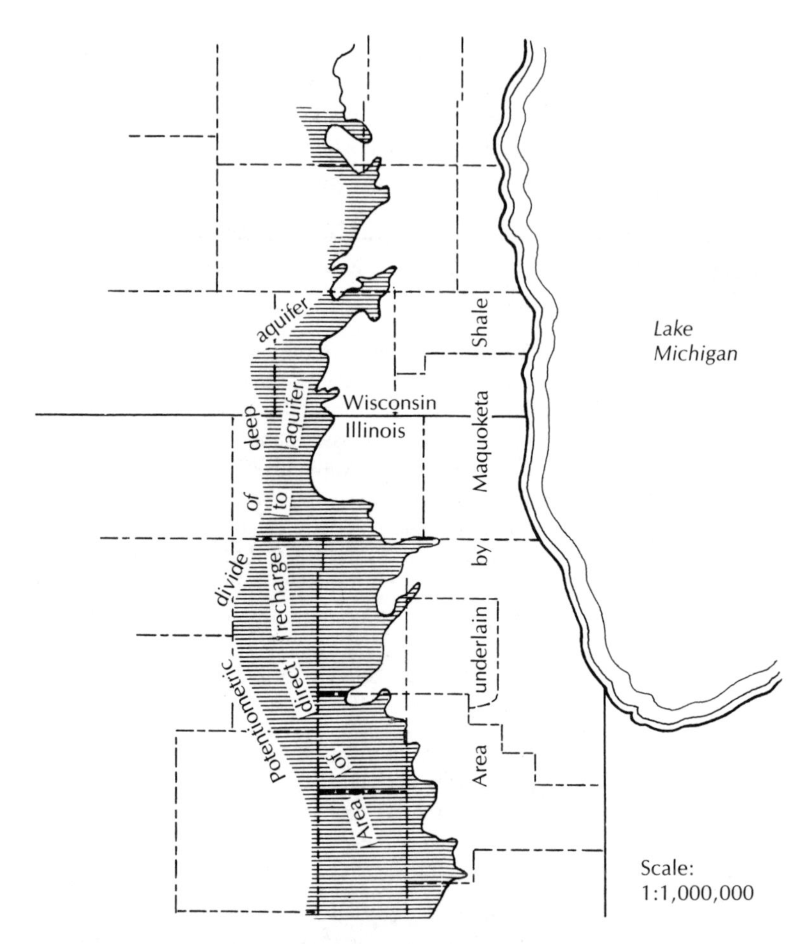


**FIGURE 8.12** Area where Maquoketa shale is thin or missing and direct recharge to the Cambrian-Ordovician aquifer can occur. Source: C. W. Fetter, Jr., *Ground Water* 19 (1981):201–13.

recharge across the shale is determined by the thickness and vertical hydraulic conductivity of the shale and the hydraulic gradient across it.

Before the sandstone aquifer was developed, there was a small amount of recharge in the western part of the region, with lateral flow to the east and upward flow from the deep aquifers to the overlying shallow aquifer in the vicinity of Lake Michigan. Figure 8.13A shows the potentiometric surface of this aquifer system prior to development. When wells were first drilled in eastern Wisconsin the artesian head in the sandstone aquifer was 130 to 200 feet (40 to 60 meters) above the level of Lake Michigan. As water was pumped from the deep sandstone aquifer, the potentiometric surface fell and the upward gradient in the eastern part of the aquifer disappeared. Eventually, substantial downward hydraulic gradients were developed, and the amount of downward recharge across the Maquoketa Shale increased. The maximum rate of recharge across the shale layer in Illinois is estimated to be about 3000 gallons per day per square mile (4.4 cubic meters per day per

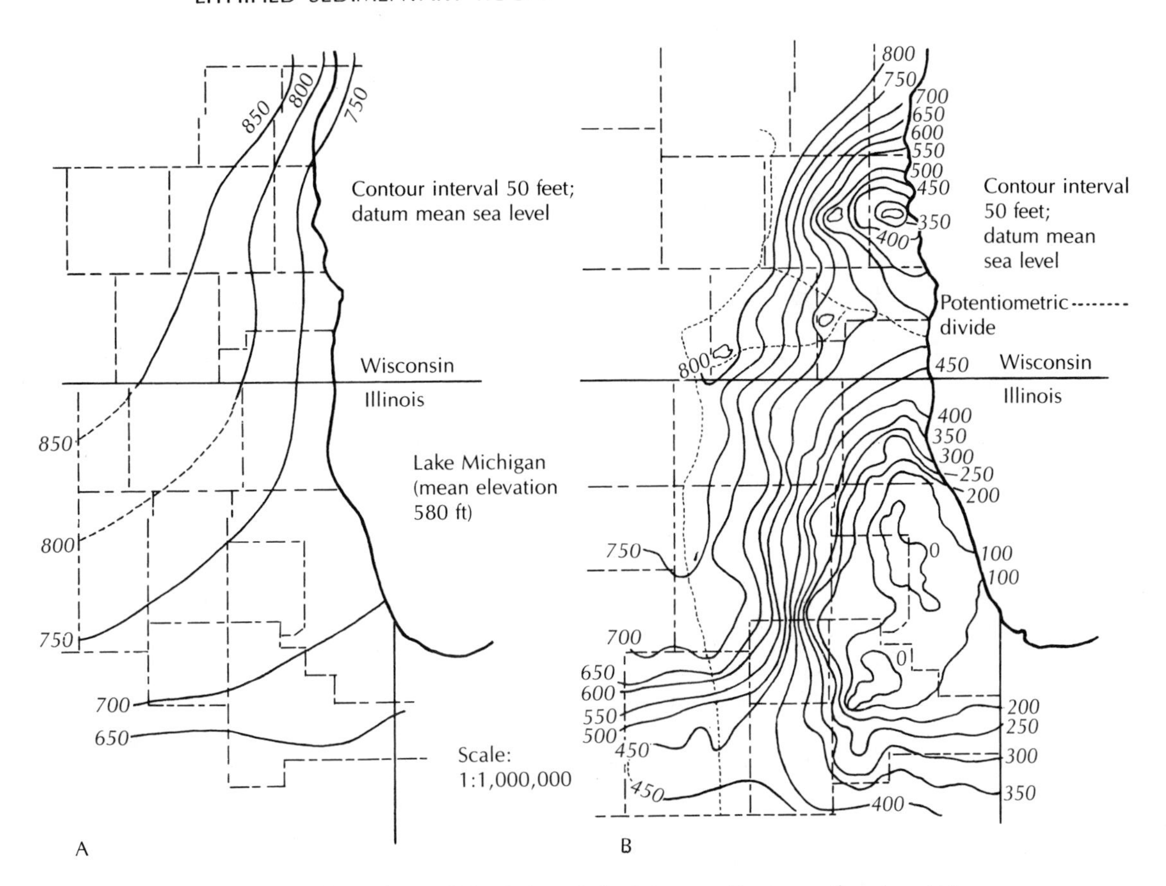


**FIGURE 8.13** Potentiometric surface of Cambrian-Ordovician aquifer in southeastern Wisconsin and northeastern Illinois. **A.** About 1865–1880. **B.** In 1973. Source: C. W. Fetter, Jr., *Ground Water* 19 (1981):201–13.

square kilometer) in Illinois (20) and 4200 gallons per day per square mile (6.1 cubic meters per day per square kilometer) in Wisconsin (19).

The practical sustained yield of the sandstone aquifer in Illinois has been estimated to be 65 million gallons (246,000 cubic meters) per day (20). This consists of 50 million gallons (189,240 cubic meters) a day of direct recharge, 12 million gallons (45,400 cubic meters) per day of downward leakage across the Maquoketa Shale and 3 million gallons (11,360 cubic meters) per day of upward leakage from deeper fresh-water aquifers (Mt. Simon aquifer). In Wisconsin the practical sustained aquifer yield has been estimated to be 34 million gallons (128,700 cubic meters) per day—25 million gallons (94,600 cubic meters) per day of direct recharge and 9 million gallons (34,100 cubic meters) per day of recharge across the shale layer (19).

Since 1958 the ground-water withdrawals from this aquifer system in Illinois have exceeded the practical sustained yield and in Wisconsin the pumpage first exceeded the practical sustained yield about 1980. As a result of the ground-water withdrawals, the piezometric pressure in the aquifer has declined and the potentiometric surface has declined. Figure 8.13B shows the potentiometric surface as it existed in 1973. Deep pumping cones had developed in the Milwaukee and Chicago areas. A ground-water divide formed in southern Wisconsin, with flow north of the divide going toward Milwaukee and flow south of the divide going toward Chicago.

The potentiometric surface in the Milwaukee area has fallen more than 400 feet (120 meters); in the Chicago region it has declined to below sea level, a distance of more than 900 feet (275 meters) (21). The potentiometric surface in parts of northeastern Illinois has fallen below the Maquoketa Shale so that the aquifer, which was formerly confined, is now unconfined. In some areas, parts of the Galena-Platteville unit have been dewatered along with the upper part of the Ancell unit. The great decline in the water levels indicates that much of the water that has been withdrawn has come from storage. Unless ground-water withdrawals are reduced to the amount of recharge in this circumstance, water levels will continue to decline until the aquifers can no longer transmit water from the recharge area to the well fields. Even if the pumpage were to be reduced to the practical sustained yield, ground-water levels would not recover to their former positions.

### 8.3.1 COMPLEX STRATIGRAPHY

The fact that a formation may change in lithology from one locality to another accounts for some of the difficulties associated with studies of sedimentary rock units. For example, a sandstone may grade into a shaly sandstone or siltstone, yet still have the same fossil fauna assemblage and the same stratigraphic nomenclature. The Eau Claire Formation of Cambrian age is a sandstone in east-central Wisconsin, but grades into a shale and siltstone in northern Illinois. In Wisconsin, it has high conductivity and is an aquifer; in Illinois it becomes a confining layer (22, 23).

Complex stratigraphy can be a very real hindrance to ground-water exploration. On the Hualapai Plateau of northwestern Arizona, the best potential aquifer is the stratigraphic sequence of sedimentary rocks in the Rampart Cave Member of the Muav Limestone Formation (24). Springs discharge from this member into the Grand Canyon along the flanks of the plateau. The Bright Angel Shale stratigraphically underlies the Rampart Cave Member of the Muav Limestone, causing the ground water to be perched (Figure 8.14). However, because these units were being deposited in a transgressing sea, the various beds are interfingering. As a result, at some localities the Bright Angel Shale can be both above and below the Rampart Cave Member (Figure 8.15). Drillers in the area have had unsuccessful wells because they stopped drilling when the Bright Angel Shale was first reached (25). Unfortunately, the target formation had not yet been penetrated. A well drilled at Location *A* of Figure 8.15 would not hit Bright Angel Shale until the Rampart Cave Member was penetrated. However, at location *B*, three shale members alternating with limestone would have to be penetrated before the Rampart Cave Member was reached. The area of these interfingering members extends for tens of miles. The hydrogeologist working in areas of such complex stratigraphy must be aware of the possible hydrogeologic consequences.

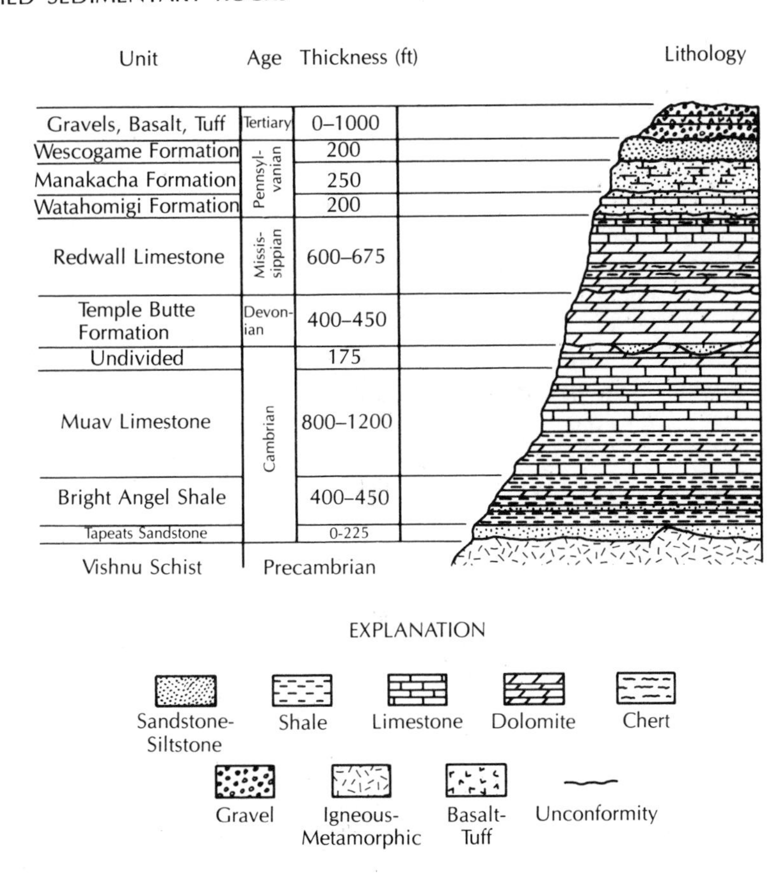


**FIGURE 8.14** Stratigraphy of the Grand Canyon area. Source: P. W. Huntoon, *Ground Water,* 15 (1977):426–33.

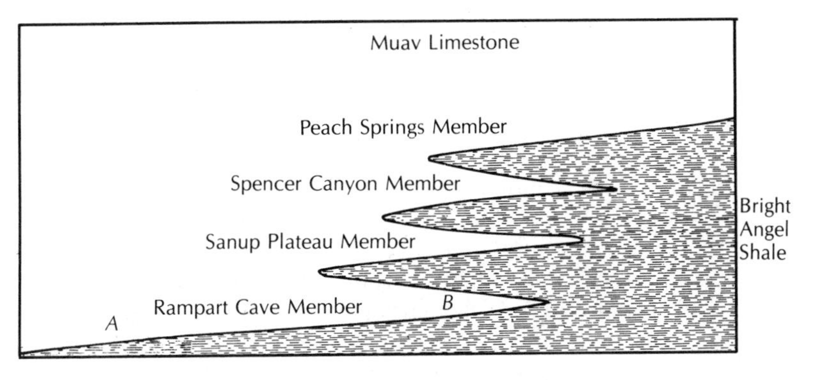


**FIGURE 8.15** Interfingering of sedimentary rock units of the Hualapai Plateau area. Source: Adapted from P. W. Huntoon, *Ground Water,* 15 (1977):426–33.

### 8.3.2 FOLDS AND FAULTS

Folding and faulting of sedimentary rocks can create very complex hydrogeologic systems, in which determination of the locations of recharge and discharge zones and flow systems is confounded. Not only must the hydrogeologist determine the hydraulic characteristics of rock units and measure ground-water levels in wells to determine flow systems, but detailed geology must also be evaluated. In most cases, the basic geologic structure will have already been determined; however, logs of test wells and borings must be reconciled with the pre-existing geologic knowledge.

Fault zones can act either as barriers to ground-water flow or as ground-water conduits, depending upon the nature of the material in the fault zone. If the fault zone consists of finely ground rock and clay (gouge), the material may have a very low hydraulic conductivity. Significant differences in ground-water levels can occur across such faults (see Figures 8.9 and 8.10). Impounding faults can occur in unconsolidated materials with clay present, as well as in sedimentary rocks where interbedded shales, which normally would not hinder lateral ground-water flow, can be smeared along the fault by drag folds. **Clastic dikes** are intrusions of sediment that are forced into rock fractures. If they are clay-rich, they can act as ground-water barriers in either sediments or in a lithified sedimentary rock. Clastic dikes are known to occur in alluvial sediments, glacial outwash, and lithified sedimentary rock.

Faults in consolidated rock units can act either as pathways for water movement or as flow barriers. If there has been little displacement along the fault, then the fault is more likely to develop fracture permeability because there is less opportunity for the formation of soft, ground-up rock, called **gouge,** to form between the moving surfaces (26). Fault gouge can have a matrix of rock breccia encased in clay and can have a wider range of permeability (27). Faults in poorly consolidated rocks in South Carolina had greater permeability than those in well-consolidated rocks (27).

If the fault zone has a high porosity and hydraulic conductivity, it can serve as a conduit for ground-water movement. Springs discharging into the Colorado River in Marble Canyon are controlled by a vertical fault zone, the Fence fault. The springs discharge where the faults intersect the river. The fault zones provide for vertical movement of recharging ground water from the land surface as well as lateral movement toward the river. The geochemistry of the spring water indicates that some of the water discharging on one side of the river originated in the ground-water basin on the opposite side of the river, indicating the fault zone was conducting some ground-water flow beneath the river even though it is a regional discharge zone (28).

Faults may contain ground water at great pressures at depths where tunnels or mines may be constructed. One of the dangers of hard-rock tunnelling is the possibility of breaching an unexpected fault zone. Damaging and dangerous flooding can occur if the fault contains ground water with a high hydraulic head. In Utah a well being drilled through an anticlinal structure that in an unfaulted state creates a regional confining layer encountered an exceptionally high-permeability zone created by normal faults and associated extensional joints that imprinted joint permeability on the brittle rocks. The well was being drilled to initiate solution mining of an abandoned underground potash mine. Unfortunately, ground water under high pressure in the fracture zone rushed into the well boring when the mine level was reached and flooded it with mineralized water (26).

Overthrust faulting can create conditions in which a rock, normally found as an impermeable basement unit, is overlying the sedimentary rock units, typically a ground-water source. In such a case, the hydrogeologist might recommend drilling through the "basement" rocks to attempt to obtain a ground-water supply from younger sedimentary units, provided there is an opportunity for recharge to occur and the water is not known to have a high mineral content (Figure 8.16).

The major artesian basins of Wyoming consist of sedimentary rock units bounded by major thrust faults. The fault-severed margins of the basins have good permeability, adequate recharge, and good-quality water in the sedimentary rock aquifers of the hanging wall. However, the foot-wall rocks receive little recharge, have poor-quality water, and have permeabilities often many orders of magnitude lower than those of the adjacent hanging wall segments of the same formation (29).

Folding can affect the hydrogeology of sedimentary rocks in several ways. The most obvious is the creation of confined aquifers at the centers of synclines. The nature of the fold will affect the availability of water. A tight, deeply plunging fold might carry the aquifer too deep beneath the surface to be economically developed. Deeply circulating ground water is also typically

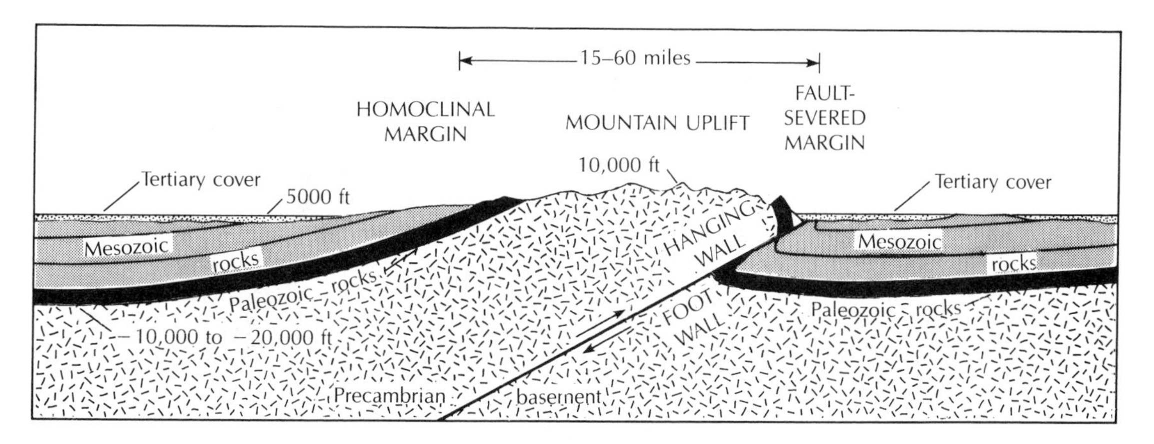


**FIGURE 8.16** Schematic cross section through a typical Wyoming mountain uplift showing the style of deformation that results in approximately equal percentages of fault severed and homoclinal basin perimeters. Source: P. W. Huntoon, *Ground Water,* 23 (1985):176–81.

warmed by the geothermal gradient and may be highly mineralized. A broad, gentle fold can create a relatively shallow confined aquifer that extends over a large area. This might be a good source of water if sufficient recharge can occur through the confining layer or if the aquifer can transmit enough water from areas where the confining layer is absent.

Another effect of folding is to create a series of outcrops of soluble rock, such as limestone, alternating with rock units that are not as permeable. Smaller streams flowing across the limestone might sink at the upper end, only to reappear at the lower outcrop. The type of trellis drainage that can develop on folded rocks is shown in Figure 8.17. Surface streams follow the strike of rock outcrops, usually along fault or fracture traces. In folded sedimentary rocks with solutional conduits in carbonate units, ground-water flow may be along the conduits that parallel the strike of the fold and not down the dip.

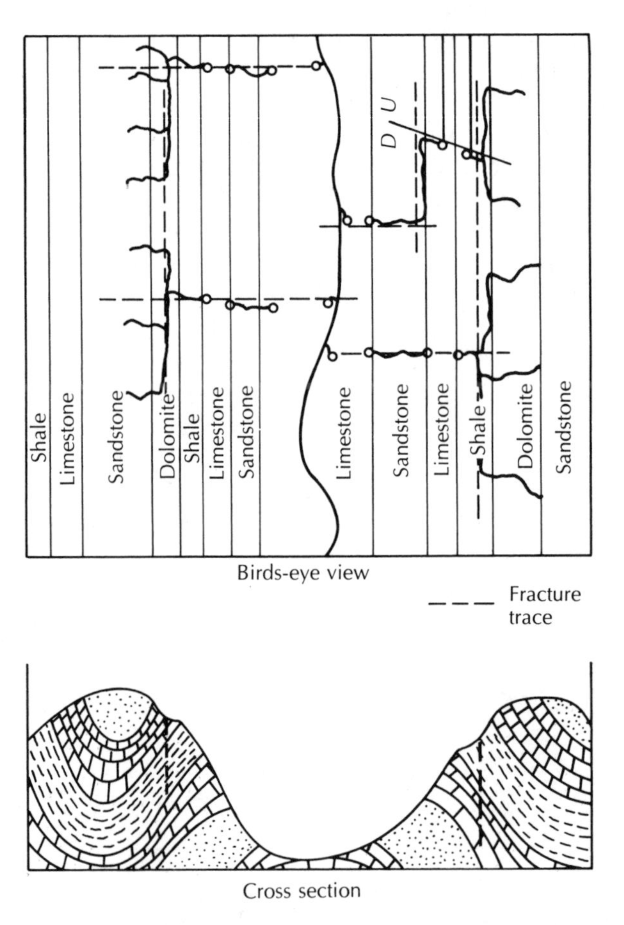


FIGURE 8.17 Drainage pattern developed in an area of longitudinally folded rock strata: **A.** Top view. **B.** Cross section.

In areas of homoclinal folds, the outcrop areas usually have bands of sedimentary rocks, with resistant rocks forming ridges and more easily erodable rocks forming valleys. The ridges may create ground-water divides, with aquifers outcropping in the valleys. The outcrop area of an aquifer will have local water-table flow systems with relatively large amounts of water circulating. These areas also serve as the recharge zones for the more distal parts of the aquifer, which are downdip in the basin and are confined. There is a limited amount of natural discharge from the confined portions of the aquifer; this is typically upward leakage into overlying beds with lower hydraulic head. Because of poor ground-water circulation, the confined portions of the aquifer may have low hydraulic conductivity and poor water quality (30).

Complex folded and faulted sedimentary rock units are a challenge to the hydrogeologist. Competency in geology is necessary in order to construct geologic cross sections based on well logs, drill-core samples, and outcrops. Regional flow systems can be controlled by the structural and stratigraphic relations of the confining beds and aquifer units. In addition, distribution of hydraulic potential must be determined, very often from limited data.

### 8.3.3 CLASTIC SEDIMENTARY ROCKS

Hydraulic conductivity of clastic sedimentary rocks, based on primary permeability, is a function of the grain size, shape, and sorting of the original sediment. The same factors that affect the permeability and porosity of loose sediments also are important in sedimentary rocks. **Cementation,** in which parts of the voids are filled with precipitated material such as silica, calcite, or iron oxide, can reduce the original porosity. Solution of the original material may occur during and after diagenesis, resulting in an increase in porosity.

Consolidated rocks also contain secondary porosity and permeability due to fracturing. Microfractures may add very little to the original hydraulic characteristics; however, major fracture zones may have localized hydraulic conductivities several orders of magnitude greater than that of the unfractured rock. Fracturing can occur through several geologic processes. Rock at depth is under great pressure owing to the weight of the overburden. As uplift and erosion bring the consolidated rock to the surface, it expands as the pressure is reduced. The expansion can cause fracturing of the rock, with the majority of the expansion fractures occurring within about 300 feet (100 meters) of the surface. Vertical fractures carry recharging precipitation downward and provide a very important function in bypassing low-permeability layers near the surface. Wells located in surface fracture zones are generally highly successful.

Fracturing may also be associated with tectonic activity. Rock deformed by faulting or folding may fracture when it is subjected to tension or compression. Such activity can take place at substantial depths; thus, secondary permeability is not strictly a near-surface phenomenon. However, greater pressure on the deeper fractures does not permit them to be as open (have as great a porosity) as shallow fractures.

The yield to wells is proportional to the transmissivity of the aquifer. This, in turn, is proportional to the aquifer thickness if the hydraulic conductivity is uniform throughout the aquifer. Sedimentary aquifers were deposited in sedimentary basins in which units gradually thicken. Variable thickness of a sedimentary aquifer may also be due to the deposition of the aquifer material over an eroded surface with high relief or a dissection of the top of the aquifer after deposition (Figure 8.18). Higher well yields will be obtained from thicker sections of the aquifer. The relationship between the specific capacity of wells and the uncased thickness of two sandstone aquifers in northern Illinois is shown in Figure 8.19.

Wells in sandstone aquifers should be located in such a manner as to penetrate the maximum saturated thickness of the aquifer. If one area of the aquifer is known to have a higher hydraulic conductivity than other areas, the combination of hydraulic conductivity and thickness should be considered in order to locate the well in the area of greatest aquifer transmissivity.

The yield of sandstone wells can sometimes be increased by the deto-

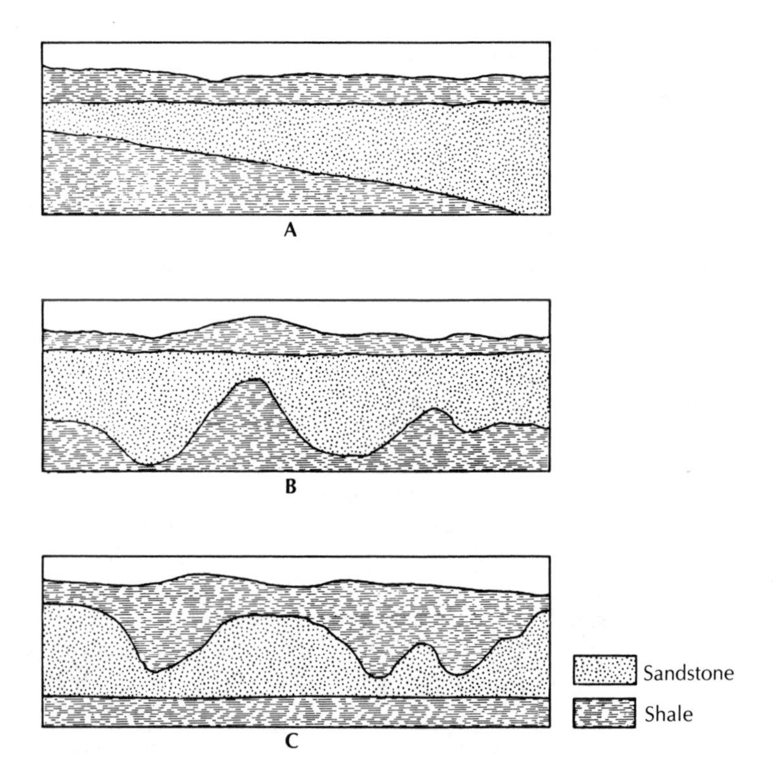


**FIGURE 8.18** Sedimentary conditions producing a sandstone aquifer of variable thickness: **A.** Sandstone deposited in a sedimentary basin. **B.** Sandstone deposited unconformably over an erosional surface. **C.** Surface of sandstone dissected by erosion prior to deposition of overlying beds.

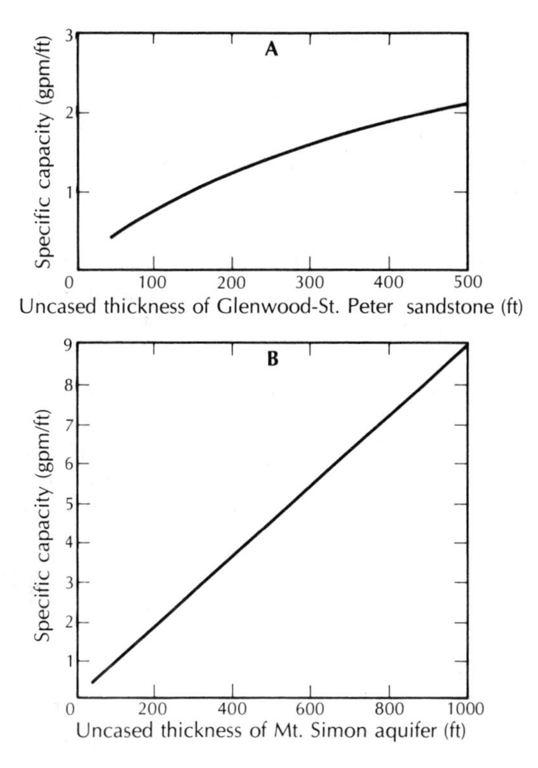


**FIGURE 8.19** Relation between the specific capacity of a well (gallons per minute of yield per foot of drawdown) and the uncased thickness of the sandstone aquifer: **A.** Glenwood-St. Peter Sandstone. **B.** Mt. Simon Sandstone. Both of northern Illinois. Source: W. C. Walton and S. Csallany, Illinois State Water Survey Report of Investigation 43, 1962.

nation of explosives in the uncased hole. The shots are generally located opposite the most permeable zones of the sandstone. The loosened rock and sand is bailed from the well prior to the installation of the pump. The shooting process has two effects on the borehole: it enlarges the diameter of the well in the permeable zones and also breaks off the surface of the sandstone, which may have been clogged by fine material during drilling. Fractures near the well may be opened all the way to the borehole by shooting. Old wells may be rehabilitated by shooting if the yield has decreased owing to mineral deposition on the well face. Shooting has increased the specific capacities of sandstone wells in northern Illinois by an average of 22 to 38 percent, depending upon the formation (31).

### 8.3.4 CARBONATE ROCKS

The primary porosity of limestone and dolomite is variable, If the rock is clastic, the primary porosity can be high. Chemically precipitated rocks can have a very low porosity and permeability if they are crystalline. Bedding planes can be zones of high primary porosity and permeability.

Limestone and (to a much lesser extent) dolomite are soluble in water that is mildly acidic. In general, if the water is unsaturated with respect to calcite or dolomite, it will dissolve the mineral until it reaches about 99+ percent saturation with respect to calcite (32, 33). The rate of solution is linear with respect to increasing solute concentration until somewhere between 65 and 90 percent saturation, at which value the rate decreases dramatically. Figure 8.20 shows the general nature of the solution rate as a function of degree of saturation.

Massive chemically precipitated limestones can have very low primary porosity and permeability. Secondary permeability in carbonate aquifers is due to the solutional enlargement of bedding planes, fractures, and faults (34). The rate of solution is a function of the amount of ground water moving through the system and the degree of saturation (with respect to the particular carbonate rock present) but it is nearly independent of the velocity of flow (33). The width of the initial fracture is one of the factors controlling how long the flow path is until the water reaches 99+ percent saturation and dissolution ceases (33).

Initially, more ground water flows through the larger fractures and bedding planes, which have a greater hydraulic conductivity. These become enlarged with respect to lesser fractures; hence, even more water flows through them. Solution mechanisms of carbonate rocks favor the development of larger openings at the expense of smaller ones. Carbonate aquifers can be highly anisotropic and nonhomogeneous if water moves only through fractures and bed-

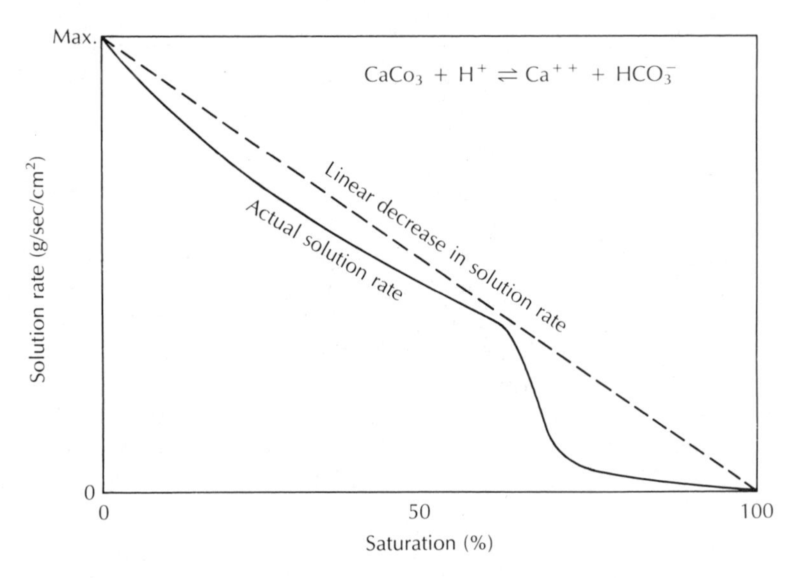


**FIGURE 8.20** Solution rate vs. degree of saturation. Instead of decreasing linearly, the solution rate drops sharply to a low level at 65–90 percent saturation. Source: A. N. Palmer, *Journal of Geological Education,* 32 (1984):247–53.

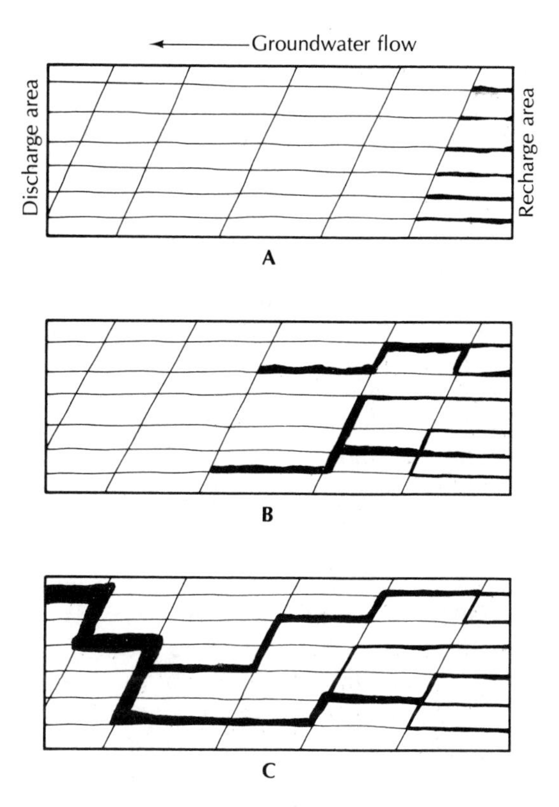


FIGURE 8.21 Growth of a carbonate aquifer drainage system starting in the recharge area and growing toward the discharge area. **A.** At first, most joints in the recharge area undergo solution enlargement. **B.** As the solution passages grow, they join and become fewer. **C.** Eventually, one outlet appears at the discharge zone.

ding planes that have been preferentially enlarged. Water entering the carbonate rock is typically unsaturated. As it flows through the aquifer, it approaches saturation, and dissolution slows and finally ceases. It has been shown experimentally that solution passages form from the recharge area to the discharge area and that, as they follow fracture patterns, many smaller solution openings join to form fewer but larger ones (35) (Figure 8.21). Eventually, many passages join to form one outlet. Greater ground-water movement—hence, solution—takes place along the intersection of two joints or a joint and a bedding plane. Ground water moving along a bedding plane tends to follow the strike of intersecting joints.

A second mode for the entry of unsaturated water into a carbonate aquifer occurs near valley bottoms. In **karst*** regions, flow in valleys with permanent

*Karst is a term applied to topography formed over limestone, dolomite, or gypsum; characterized by sinkholes, caverns, and lack of surface streams.

streams is usually discharged from carbonate aquifers recharged beneath highlands. Water tables in many karst areas are almost flat owing to the high hydraulic conductivity. Floodwaters from surface streams can enter the carbonate aquifers and reverse the normal flow. If the floodwaters are unsaturated with respect to the mineral in the aquifer, solution will occur (36).

**Swallow holes,** or shafts leading from surface streams, can carry surface water underground into caverns. Swallow holes can drain an entire stream or only a small portion of one.

Geochemical studies have shown that there are two types of ground water found in complex carbonate aquifer systems (37). The joints and bedding planes that are not enlarged by solution contain water that is saturated with respect to calcite (or dolomite). Because of the low hydraulic conductivity of these openings, this water mass moves slowly. Another mass of water, generally undersaturated, moves more rapidly through well-defined solution channels close to the water table. It is this second body that forms the passageways.

Cave systems can be formed above, at, or below the water table. They form when free-flowing water enlarges a fracture or bedding plane sufficiently for non-Darcian flow to occur. This can be above the water table if a surface stream enters the ground in the unsaturated zone **(vadose cave),** below the water table if the joint or bedding plane through which flow is occurring dips below the water table **(phreatic cave),** or at the water table itself **(water-table cave)** (34). The pattern of cave passages is controlled by the pattern and density of the joints and/or bedding planes in the carbonate rock (34). Figure 8.22 shows the influence of fissure density and orientation on cave formation. With widely spaced fissures, the cave can develop below the potentiometric surface because the fissure pattern is too coarse to allow the cave development to parallel the water table (Figures 8.22A and 8.22B). If the fissure density is great enough, cave development can occur along the water table (Figures 8.22C and 8.22D). Vertical shafts can form in the vadose zone by undersaturated infiltrating water trickling down the rock surface (38). Caves that are presently dry were formed below the water table when the regional water table was higher. The regional base level of a karst region is typically a large river. If the river is downcutting, the regional water table will be lowering. The result will be a series of dry caves at different elevations, each formed when the regional water table was at a different level.

Carbonate aquifers show a very wide range of hydrologic characteristics. There are, to be sure, a number of "underground rivers" where a surface stream disappears and flows through caves as open channel flow. At the other extreme, some carbonate aquifers behave almost like a homogeneous, isotropic porous medium. Most lie between these extremes.

Three conceptual models for carbonate aquifers have been proposed (36). **Diffuse-flow carbonate aquifers** have had little solutional activity directed toward opening large channels; these are to some extent homogeneous. **Free-flow carbonate aquifers** receive diffused recharge, but have well-developed solution channels along which most flow occurs. Ground-water flow in free-flow

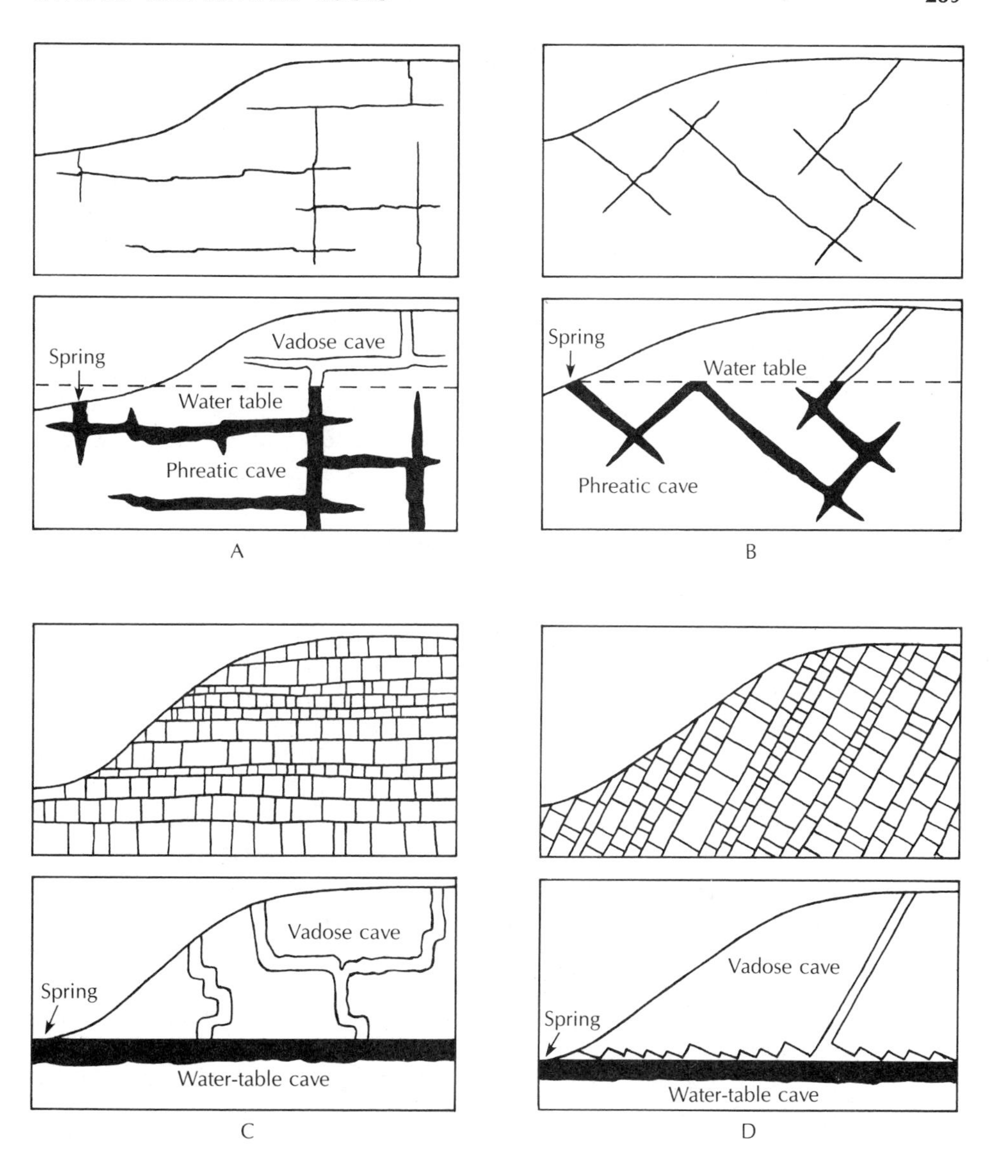


**FIGURE 8.22** Effects of fissure density and orientation on the development of caverns. Source: Modified from D. C. Ford and R. O. Ewers, *Canadian Journal of Earth Science*, 15 (1978).

aquifers is controlled by the orientation of the bedding planes and fractures that determine the locations of solutional conduits, but not by any confining beds. **Confined-flow carbonate aquifers** have solution openings in the carbonate units, but low-permeability noncarbonate beds exert control over the direction of ground-water movement.

Diffuse-flow aquifers are typically found in dolomitic rocks or shaly limestones, neither of which is easily soluble. Water movement is along joints and bedding planes that have been only modestly affected by solution. Moving ground water is not concentrated in certain zones in the aquifer and, if caves are present, they are small and not interconnected. Discharge is likely to be through a number of small springs and seeps. The Silurian-age dolomite aquifer of the Door Peninsula of Wisconsin is an example. Well tests have shown that the horizontal flow of water is along seven different bedding planes in the dolomite. Vertical recharge is through fractures. The bedding-plane zones can be identified by borehole geophysical means (caliper logs) and correlated across several miles (39). Because water movement takes place along broad bedding planes, the yield of wells is fairly constant from place to place. Wells in vertical fractures have a higher yield, as they possess both vertical and horizontal conductivity. The water table in diffuse-flow aquifers is well defined and can rise to a substantial elevation above the regional base level.

Free-flow carbonate aquifers have substantial development of solution passages. Not only are many joints and bedding planes enlarged, but some have formed large conduits. While all of the openings are saturated, the vast majority of flow occurs in the large channels; the flow behaves hydraulically as pipe flow. Velocities are similar to those of surface streams. Flow is turbulent, and the stream may carry a sediment load—as suspended material, bedload, and suspended bedload. Water quality is similar to that of surface water, and the regional discharge may occur through a few large springs. Because of the rapid drainage, the water table is nearly flat, having only a small elevation above the regional base level. The very low hydraulic gradient indicates that diffused flow through the unenlarged joints and fractures is exceedingly slow. Recharge to the subterranean drainage system is rapid, as water drains quickly through the vadose zone. The water level in the open pipe network may rise rapidly in a recharge event (40). The spring discharge will also increase in response to the amount of recharge, so that the spring hydrograph may resemble the flood peak of a surface stream. The water levels in the open-pipe network will also fall rapidly as the water drains. Caving expeditions have been known to end tragically when a "dry" cave passage became filled with surcharged water during a rapid recharge event.

The depth of major solution openings below base level is probably less than 200 feet (60 meters), unless artesian flow conditions are present (36). However, in areas where the regional base level was formerly at a lower level (for example, where a buried bedrock valley is present), cavern development may have taken place graded to that base level. In the coastal aquifers of the southeastern United States, the drilling fluid may suddenly drain from a well being drilled when it is 300 or more feet (100 meters) deep. Cavernous zones found at these depths are well below the present water table. They formed when mean sea level and the regional water table were lower during the Pleistocene. The development of a sinkhole in Hernando County, Florida, was initiated by drilling in the Suwannee Limestone. The drilling fluid was lost several times, and the drill-bit would drop through small caverns. At 200 feet (62 meters), the drill broke into a cavern and, within ten minutes, a large depression had formed, with the drill rig sinking into the ground and the drillers narrowly escaping the same fate. The present-day water level is close to the land surface (41).

Sinking surface streams may also feed the pipe-flow network of a karst region. Lost River of southern Indiana is a typical headwater surface stream flowing across a thick clay layer formed as a weathering residuum on the St. Louis and St. Genevieve limestones. Where the clay thins, a karst landscape is present, and Lost River sinks beneath an abandoned surface channel. It appears as a large spring some miles away. There is no surface drainage in the karst region other than some ephemeral streams flowing into sinkholes (42).

If the carbonate rock beneath the uplands between regional drainage systems is capped by a clastic rock, karst landforms will not form. Dry caves in the uplands capped with clastics are less likely to have collapsed than caves in areas that are not capped. Recharge to the phreatic zone occurs through vertical shafts located at the edge of the caprock outcrop (38). These shafts, which may be as large as 30 feet (10 meters) in diameter and more than 300 feet (100 meters) deep, extend only to the water table. Water flows from them through horizontally oriented drains.

The central Kentucky karst region, including the Mammoth Cave area, is a capped carbonate aquifer system in some areas (43, 44). A cross section through the Mammoth Cave Plateau is shown in Figure 8.23. The plateau is capped by the Big Clifty Sandstone Member of the Golconda Formation, with cavern development in the underlying limestone formations, including the Girkin, St. Genevieve, and St. Louis limestones. Contact springs are found at the margins of the top of the plateau, as there is a thin shale layer at the top of the Girkin Formation. Recharge to the main carbonate rock aquifer takes place through vertical shafts formed in the plateau where the shale layer is absent. Karst drainage in the Pennyroyal Plain also contributes to the regional water table. Drainage is to the Green River through large springs, such as the River Styx outlet and Echo River outlet. These streams may also be seen underground where they flow in cave passages. Wells in the area draw water from the Big Clifty Sandstone, which yields enough for domestic supplies. There are some perched water bodies in the limestone, but the amount of water in storage is limited. A more permanent supply is reached if the well penetrates the regional water table, below the level of the Green River. The water level in these deep wells can rise 25 feet (8 meters) in a few hours during heavy rains and then can fall almost as fast (43).

If a carbonate rock is confined by strata of low hydraulic conductivity, those strata may control the rate and direction of ground-water flow (36). Such confined systems may have ground water flowing to great depths, with solution openings that are much deeper than those found in free-flow aquifers. Flow is not localized, and a greater density of joint solution takes place. The cavern formation of the Pahasapa Limestone of the Black Hills is apparently of this type, as water flows through the equivalent Madison Limestone aquifer eastward

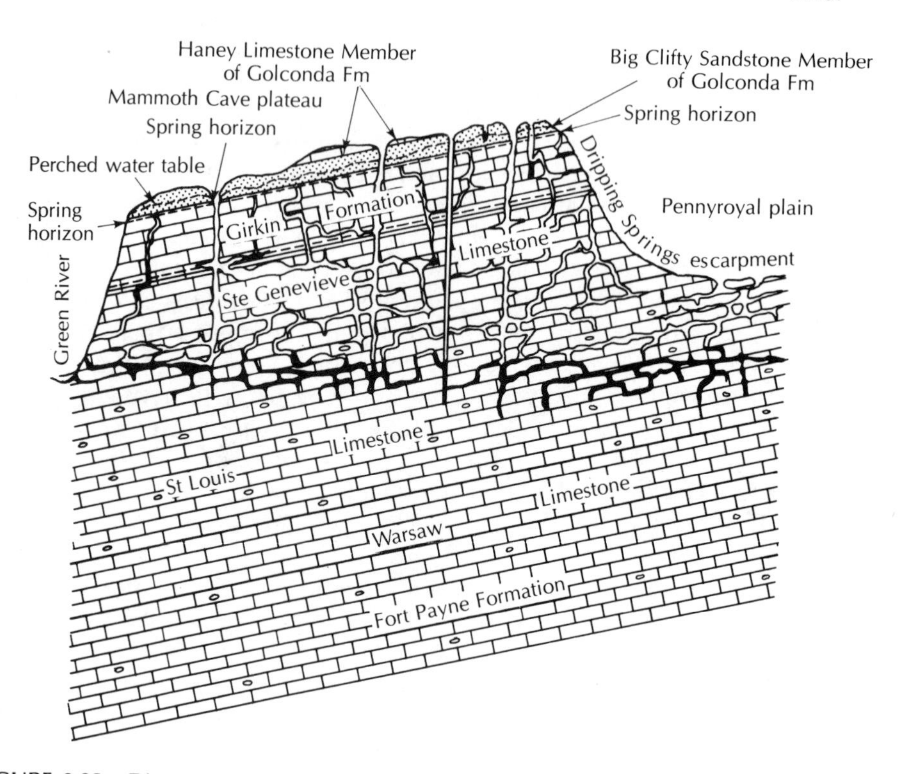


**FIGURE 8.23** Diagrammatic cross section through the Mammoth Cave Plateau. Ground-water flow in the carbonate aquifer is from south to north. Source: R. F. Brown, U.S. Geological Survey Water-Supply Paper 1837, 1966.

from a recharge area at the eastern side of the Black Hills. It has been suggested that because of the low gradient (0.00022) of the potentiometric surface, the Madison Limestone is highly permeable, owing to solution openings, for at least 130 miles (200 kilometers) east of the Black Hills (45).

In the preceding discussion of karst hydrology, it was assumed that the various types of carbonate rock aquifers were isolated; however, this may not actually be the case. Highly soluble carbonate rock may be adjacent to a shaly carbonate unit with only slight solubility. The slope and position of the water table is a reliable indicator of the relative hydraulic conductivity of different carbonate rock units. In general, the water table will have a steeper gradient in rocks of lower hydraulic conductivity (46). This may be due to either a change in lithology or the degree of solution enlargement of joints. Figure 8.24 illustrates some conditions that might result in a change in the water-table gradient. In Part A, the hilltop is capped by sandstone, with a low-permeability shale between the sandstone and the underlying limestone. Only a limited amount of recharge oc-

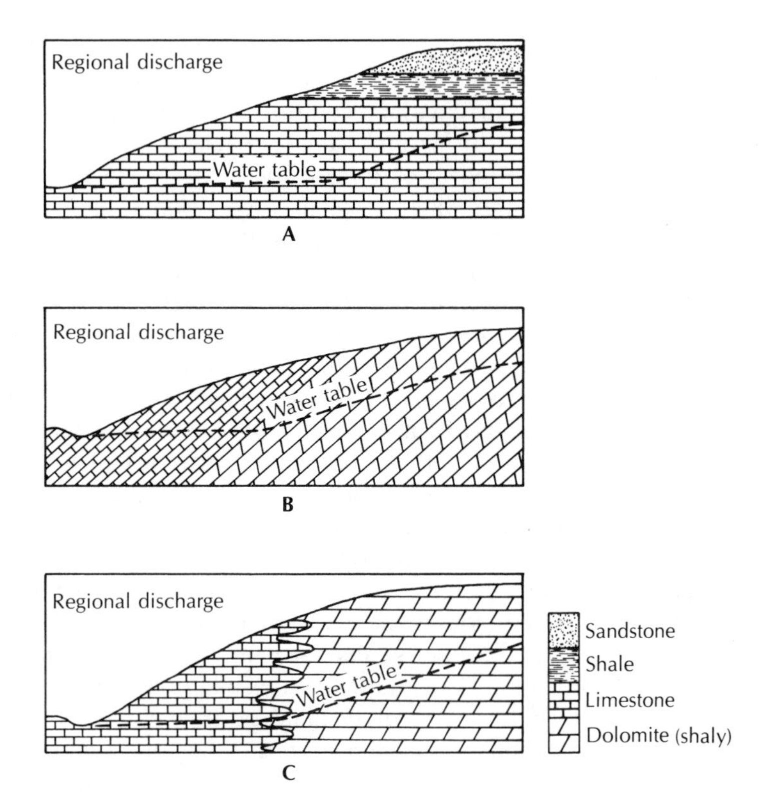


**FIGURE 8.24** Geologic conditions resulting in a difference in hydraulic conductivity: hence, a difference in the water-table gradient.

curs through the shale. A spring horizon exists at the sandstone-shale contact, with small streams flowing across the shale outcrop area only to sink into the limestone terrane. The much greater amount of water circulating through the limestone in the area where the shale is absent has created a highly permeable, cavernous unit. Beneath the caprock, the limestone is less dissolved, owing to lower ground-water recharge. Because of the lower hydraulic conductivity, the water-table gradient is steeper. This type of situation has been reported in the central Kentucky karst, with the hydraulic gradient beneath the caprock near the drainage divide being much steeper (0.01) than that near the Green River (0.0005) (47). If an area has two rock units, one of which is more soluble, the more soluble rock may develop large solution passages and, hence, have greater conductivity. Parts B and C of Figure 8.24 illustrate two situations in which the upland is underlain by shaly dolomite and the lowland by cavernous limestone. The difference in rock solubility creates a change in hydraulic gradient.

The general concept of a water table in free-flowing karstic regions may be different from the model water table found in sandstones or sand and gravel aquifers. Because of the extremely high conductivity of some limestones, the water table can occur far beneath the land surface in mountains. Water can "perch" in solution depressions above the main water table. The level of free-flowing streams in caves is controlled by the regional water table, and the streams can have losing and gaining reaches, just as surface streams. Finally, because the solution of carbonate rock can be isolated along such features as fracture zones, the water table may be discontinuous. Wells drilled between fractures may not have any water in them, whereas nearby wells of the same depth may be in a fracture and therefore measure a water table (Figure 8.25).

The selection of well locations in carbonate terrane is one of the great challenges for the hydrogeologist. As the porosity and permeability may be localized, it is necessary to find the zones of high hydraulic conductivity. One of the most productive approaches to the task is the use of **fracture traces** (48, 49). Fracture traces (up to 1 mile, or 1.5 kilometers) and lineaments (1 to 100 miles, or 1.5 to 150 kilometers) are found in all types of geologic terrane. As they represent the surface expression of nearly vertical zones of fracture concentrations, they are often areas with hydraulic conductivity 10 to 1000 times that of adjacent rock. The fracture zones are from 6 to 65 feet (2 to 20 meters) wide and have surface expressions such as swales and sags in the land surface; vegetation differences, due to variations in soil moisture and depth to the water table; alignment of vegetation type; straight stream and valley segments; and alignment of sinkholes in karst. The surface features can reveal fracture traces covered by up to 300 feet (100 meters) of residual or transported soils. Fracture traces are found over carbonate rocks, siltstones, sandstones, and crystalline rocks.

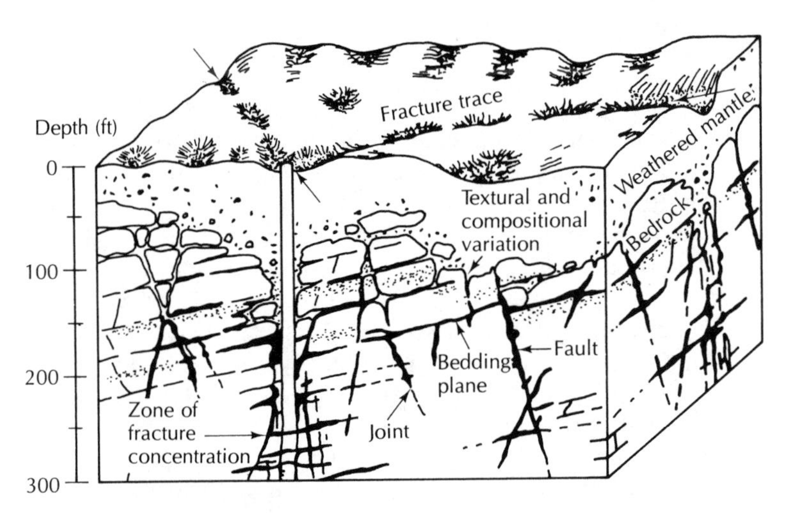


**FIGURE 8.25** Concentration of ground water along zones of fracture concentrations in carbonate rock. Wells that do not intercept an enlarged fracture or a bedding plane may be dry, thus indicating a discontinuous water table. Source: L. H. Lattman and R. R. Parizek, *Journal of Hydrology,* 2 (1964):73–91. Used with permission of Elsevier Scientific Publishing Company, Amsterdam.

Because of differential solution in carbonate rocks, if a fracture zone has somewhat higher conductivity than that of the unfractured rock, flowing ground water will eventually create a much larger conductivity difference. Wells located in a fracture trace, or especially at the intersection of two fracture traces, have a statistically significant greater yield than wells not located on a fracture trace (50). The same relationship is apparently true for wells located on lineaments, as opposed to those not on lineaments (49).

In areas where topography is influenced by structure, valleys may form along fracture traces. In central Pennsylvania 60 to 90 percent of a valley may be underlain by fracture traces (50). Under such conditions, valley bottoms are good places to prospect for ground water. In the Valley and Ridge Province of Pennsylvania, the valleys are structural, and wells in valley bottoms are statistically more productive. In Illinois, just the reverse is true. For the shallow dolomite aquifer, the yields of wells in bedrock uplands are greater than those in bedrock valleys (51).

In central Pennsylvania, the yield of wells drilled into anticlines is greater than the yield of wells drilled into synclines. However, the proximity to a fault trace and rock type are more significant than structure in determining yield (50). In other karst areas, synclines have been noted as major water producers (52). The relation between structure and well yield is not clear; thus, local experience must be used as a guide.

One of the integral parts of carbonate terrane hydrogeology is the **regolith**—the layer of soil and weathered rock above bedrock. The regolith can be composed of weathering residuum of insoluble minerals remaining after solution of the carbonate minerals. This is typically reddish in color, owing to iron oxides, and contains a high proportion of clay minerals. The regolith may also include transported materials, such as glacial drift. If recharge must first pass through a low-permeability regolith, the rapid response of a carbonate aquifer will be reduced or eliminated. As the regolith slowly releases water from storage, spring discharge from areas overlain by a thick regolith will be more constant than if the regolith were absent. The regolith may also be a local aquifer. The weathering residuum of the Highland Rim area of Tennessee contains localized zones of chert, which can yield water in small amounts (53, 54).

### 8.3.5 COAL AND LIGNITE

**Coal** contains bedding planes cut by fractures that are termed **cleat.** Cleat is similar to joint sets in other rock. It is formed as a response to local or regional folding of the coal. There are typically two trends of cleat—normal to the bedding planes and cutting each other at about a 90-degree angle (55). Coal is often an aquifer and yields water from the cleat and bedding. The quality of water from coal aquifers is variable and sometimes can be poor. Such coals are typically anisotropic, with the maximum hydraulic conductivity oriented along the face cleat, which develops perpendicular to the axis of folding.

There is not a great deal of information on the hydraulic characteristics of coals. The maximum hydraulic conductivity of the Felix No. 2 coal of the Wasatch Formation was determined to be 0.88 foot per day ($3.1 \times 10^{-4}$ centimeter per second), with a storativity of $1.2 \times 10^{-3}$ (55). The mean hydraulic conductivity of lignite from four sites in western North Dakota is 1.1 feet per day ($3.9 \times 10^{-4}$ centimeter per second) (56). The Sawyer-A coal aquifer (Fort Union Formation, Montana) exhibits horizontal anisotropy, with a maximum hydraulic conductivity of 3.3 feet per day ($1.2 \times 10^{-3}$ centimeter per second) and a minimum hydraulic conductivity of 0.85 foot per day ($3.0 \times 10^{-4}$ centimeter per second). The storativity was $3.4 \times 10^{-4}$. The Anderson coal aquifer of the same formation had a maximum horizontal hydraulic conductivity of 0.66 foot per day ($2.3 \times 10^{-4}$ centimeter per second) and a minimum value of 0.23 foot per day ($8.1 \times 10^{-5}$ centimeter per second) (57).

Thick coal beds, such as those of the Powder River Basin of Wyoming, may be important regional aquifers. Well yields of 10 to 100 gallons per minute (0.6 to 6 liters per second) are possible from these coals (58). Conflicts arise between water supply from coal aquifers and energy development. Strip mining takes place at the outcrop areas of the coal. If these are recharge areas for the coal aquifer, mine dewatering will adversely affect the potentiometric level in the coal aquifer. If the infiltration capacity of the area is reduced by mining, spoil disposal, or the like, this can also reduce the available recharge to the aquifer.

## 8.4 IGNEOUS AND METAMORPHIC ROCKS

### 8.4.1 INTRUSIVE IGNEOUS AND METAMORPHIC ROCK

Intrusive igneous and highly metamorphosed crystalline rocks generally have very little, if any, primary porosity. In order for ground water to occur, there must be openings developed through fracturing, faulting, or weathering. Fractures can be developed by tectonic movements, pressure relief due to erosion of overburden rock, loading and unloading of glaciation, shrinking during cooling of the rock mass, and the compression and tensional forces caused by regional tectonic stresses.

In general, the amount of fracturing in crystalline rocks decreases with depth (59). However, two deep test wells drilled in northern Illinois as exploratory holes for a possible pumped hydroelectric storage project have shown this is not always the case (60–63). These wells were drilled in an area of Paleozoic bedrock overlying crystalline bedrock comprised of biotite granite. Crystalline rocks were penetrated from a depth of 2179 to 5443 feet (664 to 1669 meters) in one hole and 2179 to 5273 feet (664 to 1607 meters) in the other. Even at these great depths, fractures were found in the crystalline rock. Porosity ranged from

1.42 to 2.15 percent and intrinsic permeability, determined from in situ packer tests,* from $10^{-4}$ darcy in a more highly fractured zone near the top of the crystalline basement to $10^{-8}$ darcy in an area with few fractures. Brines were found in the fractures of the rock at these depths. Brines have also been found in mines at depths in excess of 3000 feet (900 meters) in a number of areas of the Precambrian shield of North America (64). As crystalline rocks are a potential medium for the construction of mined repositories for high-level nuclear waste, the presence of fracture permeability and porosity at great depths is significant.

One study of crystalline rock wells in the eastern United States has shown that the yield, expressed in gallons per minute divided by the depth of the saturated zone penetrated by the well, decreases rapidly with depth (59) (Figure 8.26). However, jointed crystalline rock in the Piedmont of the eastern United States is known to be fractured to depths of 500 feet (150 meters) (65). In Wisconsin, fracture zones in crystalline rock are known to be present and can be delineated from air photos. One well drilled on a fracture trace to a depth of 353 feet (108 meters) was test pumped for twelve hours at 200 gallons per minute (13 liters per second) with a drawdown of 134 feet (41 meters); another well drilled to a depth of 400 feet (122 meters) had a yield of 80 gallons per minute (5 liters per second) (66).

Chemical weathering of crystalline rock can produce a weathering product called **saprolite.** This material has porosities of 40 to 50 percent and a specific yield of 15 to 30 percent. It acts as a reservoir, storing infiltrated water and releasing it to wells intersecting fractures in the underlying crystalline rock (67).

The probability of obtaining a high-yield well in crystalline rock areas can be maximized if drilling takes place in an area where fractures are localized. It has been observed that zones of high conductivity in crystalline rock areas underlie linear sags in the surface topography (68). Such sags are the surface features that overlie major zones of fracture concentration. These show as fracture traces and lineaments on aerial and satellite photographs (48). If, in drilling a water-supply well in a crystalline rock area, sufficient water is not encountered in the first 300 feet (100 meters) of drilling, in most situations other than where deep tectonic fracturing is suspected a new location should be sought rather than drilling any deeper. Because most fractures are vertical, or nearly so, an angled borehole will be more likely to intersect fractures and create a successful well. Well yields in some areas of crystalline rock are greater when the wells are located on valley bottoms (69). Many of the valley bottoms probably developed along fracture traces.

---

*A **packer test** is performed in an open borehole through rock. Inflatable seals, called packers, are used to isolate a particular segment of the open borehole. A pump test or a slug test is then performed on the isolated segment of the borehole.

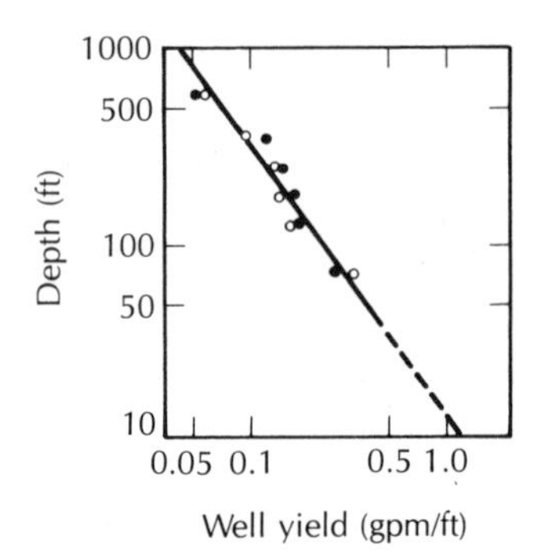


**FIGURE 8.26** Yields of wells in crystalline rock in the eastern United States. Open circles represent grouped mean yields of granite rock wells and black dots represent grouped mean yield for schist wells. Source: S. N. Davis and L. J. Turk, *Ground Water,* 2 (1964):6–11.

### 8.4.2 VOLCANIC ROCKS

Because volcanic rocks crystallize at the surface, they can retain porosity associated with lava-flow features and pyroclastic deposition. Hydraulic conductivity of volcanic rocks such as lava flows and cinder beds is typically quite high. However, ash beds, intrusive dikes, and sills may have a much lower hydraulic conductivity. Younger basalt flows tend to have greater conductivity than older ones. Flow features such as clinker zones and gas vesicles in **aa** (a type of lava) and lava tubes and gas vesicles in **pahoehoe** (another lava type), as well as vertical contraction joints and surface irregularities and stream gravels buried between successive flows, contribute to overall conductivity (70). Some of the most productive aquifers are located in basalt flows, as is described in the following case studies.

## CASE STUDY: VOLCANIC PLATEAUS—COLUMBIA RIVER BASALTS

The Columbia Plateau area of Washington, Oregon, and Idaho consists of a very thick sequence of Miocene-age basaltic lava flows that erupted from fissures. The lava was very fluid: individual flows are 150 to 500 feet (50 to 150 meters) thick, and some can be traced for 125 miles (200 kilometers). For the most part, the basalt flows are either flat-lying or gently tilted. In some places, the basalts have been folded by later tectonic activity. The basalt flows of the Columbia River Group of east-central Washington are as thick as 10,000 feet (3000 meters), although in most areas a thickness of 4600 feet (1400 meters) is more typical (71). The basalt flows dip at angles of 1 to 2 degrees from northwest to southeast. Water occurs in distinct zones, probably related to interflow boundaries.

A number of irrigation wells have been drilled in the uppermost 1000 feet (300 meters) of the basalt flows. Typical well yields of 1000 to 2000 gallons per minute (60 to 120 liters per second) are obtained from confined aquifers located at a depth of between 500 and 1000 feet (150 and 300 meters). Shallower aquifers are not as productive, but have a hydraulic head up to 100 feet (30 meters) greater than those of the deeper aquifers.

Uncased well bores are draining water from the shallower aquifers into the deeper ones (72). Recharge to the aquifer systems comes from precipitation and loss from ephemeral streams. Precipitation—hence, recharge—increases to the north and east. As these are also topographical high areas, regional ground-water flow is away from them, to the southwest. Age determinations based on carbon-14 dating of ground water of the basalts range from modern to as old as 32,000 years B.P. Age appears to increase with depth, but relationships are obscured by mixing of water of different ages coming from different aquifer zones (72).

The classic conceptual model of a series of confined aquifers corresponding to individual basalt flows might not be valid in all parts of the Columbia Plateau. Vertical joints form in basalt as it shrinks during cooling. These can provide substantial vertical conductivity, producing unconfined aquifer conditions in thick sequences of lava flows (73). In Horse Heaven Plateau, Washington, there is reported to be an unconfined aquifer consisting of the Saddle Mountains Basalt and the upper part of the underlying Wanapum Basalt. A deeper basalt, the Grande Ronde, is confined by a saprolite layer, which formed by weathering of a basalt flow (74).

## CASE STUDY: VOLCANIC DOMES—HAWAIIAN ISLANDS

In contrast to the very large area of similar geology of the Columbia Plateau Basalts, lava flows associated with some volcanic eruptions can have very heterogeneous aquifer systems. The Hawaiian Islands provide the classic example (70, 75–77).

Each of the Hawaiian Islands consists of shield volcanoes forming from one to five volcanic domes. Each dome is composed of thousands of individual basaltic lava flows coming from either craters or fissures. Lava flows cooling above sea level are thin bedded, highly fractured, or composed of vesicular and very permeable basalt. Those cooled under water are more massive and less permeable. However, owing to lowered sea levels during the Pleistocene and isostatic sinking of the islands, highly permeable, air-cooled basalt is found below present-day sea level. Interbedded with the lava flows are ash beds, which have a lower permeability. In the zones in which lava flows originated, igneous dikes with low porosity and permeability cut across the lava beds. The original dome structure may be partially eroded. Sediments have accumulated along some of the coastal areas. These coastal plain sediments consist of both terrestrial and marine deposits. The cross section of Figure 8.27 is through an idealized Hawaiian volcanic dome, with both the original and eroded states illustrated.

The Hawaiian Islands are typical examples of oceanic islands where fresh ground water is underlain by salty ground water. Because fresh water is less dense than salty water, the fresh ground water beneath an oceanic island can be thought of as "floating" as a thin lens in the salty ground water. The fresh ground water grades into salty ground water in a zone of mixing.

Ground water in the Hawaiian Islands is contained in the highly permeable basalt flows. It is recharged by rainfall, which can average as much as 20.8 feet (635 centimeters) per year on Oahu. In the interior of the islands, the basalt flows are isolated by cross-cutting igneous dikes, which are ground-water dams. The ground water is trapped at a high elevation behind these dams. High-level ground water is also found as perched water in lava beds overlying low-permeability ash beds. If the infiltration is not trapped on an ash bed or behind a dike dam, it moves downward to the basal ground-water body, which is a fresh-water lens in dynamic balance with salty ground water. The basal ground water may

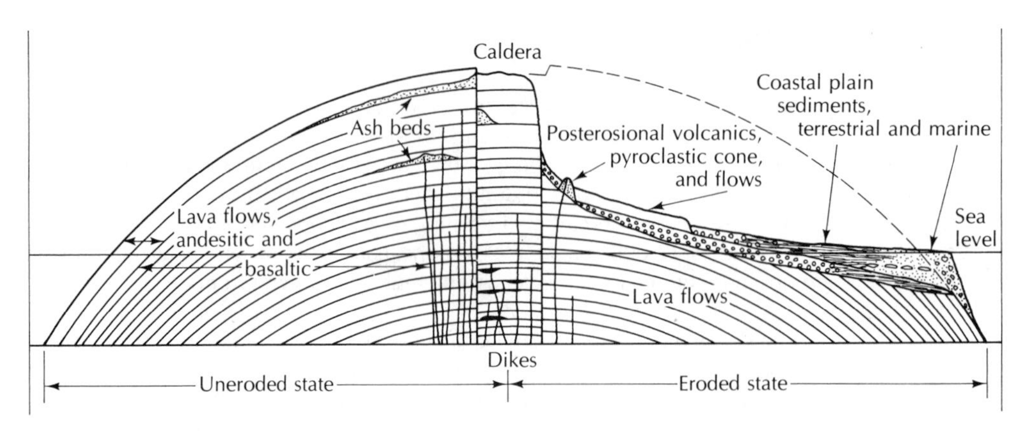


**FIGURE 8.27.** Geologic structure of an idealized Hawaiian volcanic dome. Source: D. C. Cox, *Hawaiian Planters Record,* 54 (1954). Used with permission.

be unconfined or, in areas of coastal plain sediments, may be confined by the low-permeability sediments locally termed **caprock.** A cross section of the occurrence of ground water is shown in Figure 8.28. A birds-eye view of the island of Oahu indicates areas of high-level water bodies impounded by dikes (Figure 8.29).

Springs issue from ash-bed perched aquifers and also from dike-dammed water bodies. Some of these are 300 feet (100 meters) or more above sea level. High-level water is developed by tunnels into the tops of ash beds or penetrating dike dams. Most ground water is developed from the basal ground-water body. Unconfined basal water is collected in horizontal skimming tunnels, called **Maui tunnels,** which are at sea level and slightly below. These skimming tunnels can develop water where the fresh-water lens is very thin and conventional wells would pump brackish or salt water. Maui tunnels are capable of producing up to $3.1 \times 10^4$ gallons per minute ($2 \times 10^3$ liters per second), although in most cases the yield is much less. Where the basal water is confined by coastal plain sediments, conventional wells are used for ground-water development.

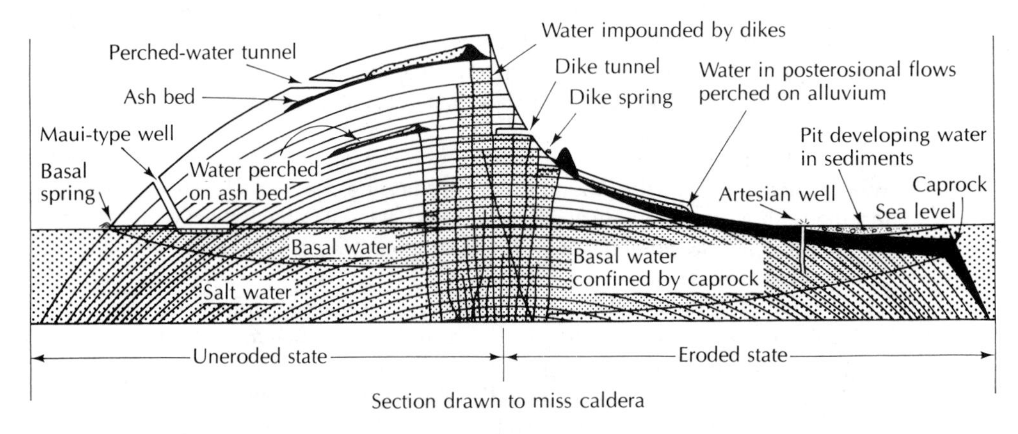


**FIGURE 8.28** Occurrence and development of ground water in an idealized Hawaiian volcanic dome. Source: D. C. Cox, *Hawaiian Planters Record* 54 (1954). Used with permission.

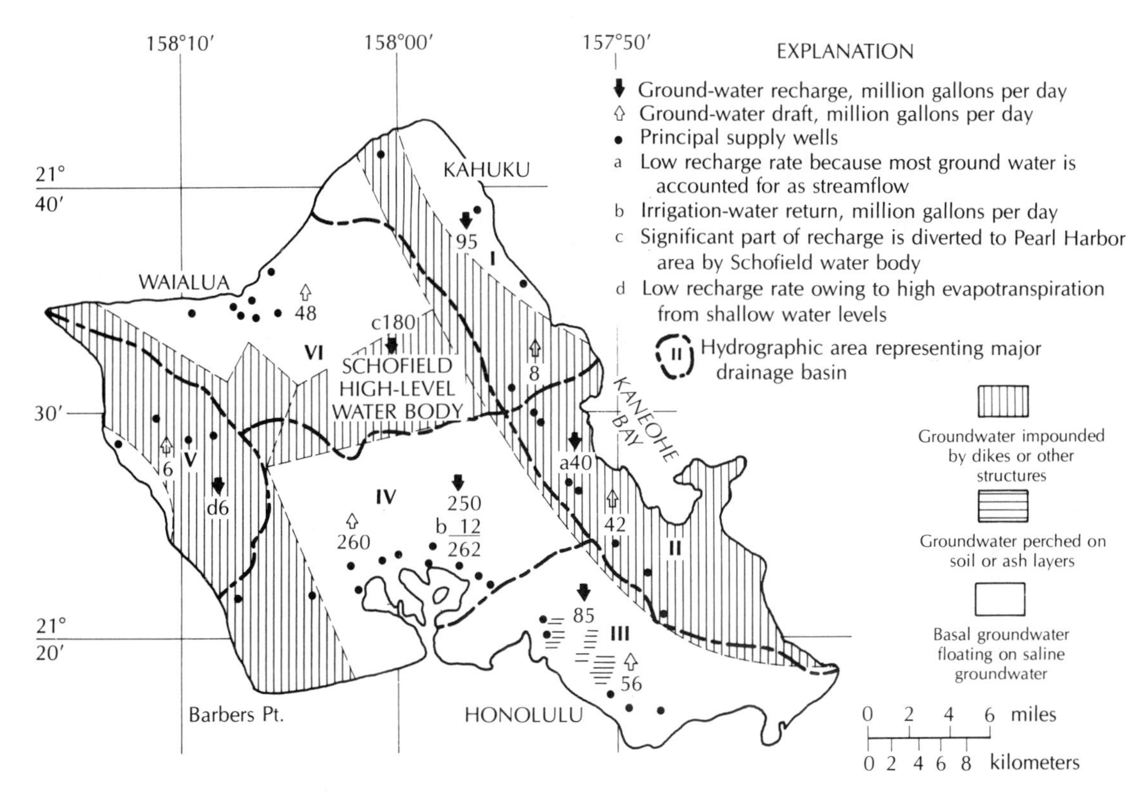


**FIGURE 8.29** Map of the island of Oahu showing the approximate outline of ground-water reservoirs, recharge, 1975 draft, and principal supply wells by hydrographic areas representing major drainage basins. Source: K. J. Takasaki. U.S. Geological Survey Professional Paper 813-M, 1978.

## 8.5 GROUND WATER IN PERMAFROST REGIONS

In polar latitudes and high mountains, the mean annual temperatures may be sufficiently low for the ground to be at a temperature below 0° C. If this temperature persists for two or more years, the condition is known as **permafrost** (78–80). During the summer, warm temperatures may cause the upper 3 to 6 feet (1 to 2 meters) of the soil or rock to thaw. This is called the **active layer,** but underneath it the ground may be frozen to depths to 1300 feet (400 meters).

The magnitude of the annual temperature fluctuation of the soil is greatest at the surface and diminishes with depth until, at some depth, there is zero annual temperature amplitude. The depth at which the maximum annual soil temperature is 0° C is the **permafrost table.** In some areas, the permafrost may occur in layers, with zones of unfrozen ground between them. This condition is usually the result of past climatic events, and the permafrost distribution is not congruent with the present climatic and thermal regime. The local depth of permafrost is a function of the geothermal gradient and the mean annual air temperature (81).

The insulating cover of glacier ice and large lakes may prevent the formation of permafrost. Permafrost is likewise not present beneath the ocean. At higher elevations and on north-facing slopes, where the mean temperature is lower, permafrost may be thicker (Figure 8.30A). Lakes and streams also affect the permafrost. Shallow lakes—6 feet (2 meters) deep or less—freeze to the bottom in winter and have little effect on the permafrost table, but the permafrost may be warmer; hence, not as thick. Deeper lakes are unfrozen at the bottom and have an insulating effect. Small deep lakes create a saucer-shaped depression in the permafrost table, which, in turn, creates an upward indentation

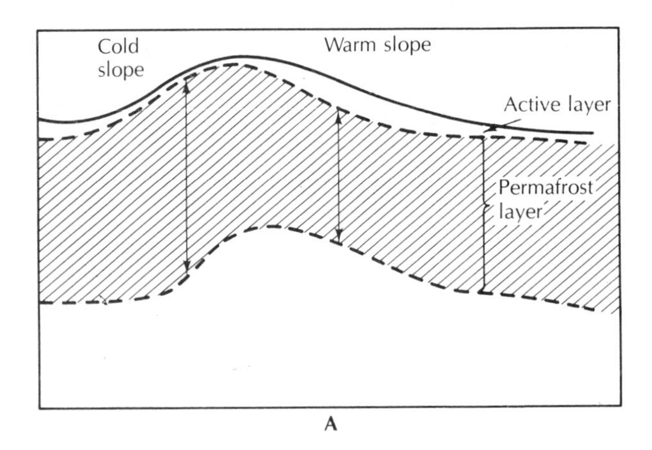


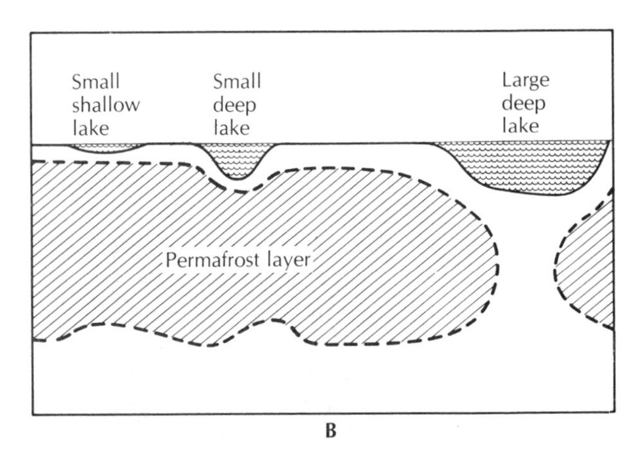


**FIGURE 8.30** **A.** Effect of topography on the permafrost layer with the thinnest layer found beneath a warm, south-facing slope. **B.** Effect of lakes on the permafrost layer.

in the bottom of the permafrost (82). Permafrost is absent beneath large deep lakes—even in the continuous permafrost zone (Figure 8.30B). The effects of large rivers on distribution of permafrost are similar to those of large deep lakes.

The permafrost table creates perched water in the active layer. This results in poorly drained soils and the typical muskeg and marsh vegetation of tundra regions. Deeper aquifers are recharged only in the absence of permafrost, as the permafrost layer acts as a confining layer.

Water below the permafrost layer is confined. The potentiometric surface may be in the permafrost layer or even above land surface (80). Subpermafrost water may discharge into large rivers and lakes, beneath which unfrozen conduits are open. Saline ground water may exist beneath permafrost, but it is typically not actively circulating (80). Discharge of ground water at the surface, especially in winter, may result in the development of sheets of surface ice or large conical hills called **pingos.** Pingos have ice cores and are formed by the upward arch of the ground surface due to hydraulic pressure of ground water confined by permafrost.

Alluvial river valleys are good sources of ground water in permafrost regions. The permafrost beneath and along the river may be thin, or absent in places, and alluvial gravel deposits are good aquifers. Large rivers will have more unfrozen ground—hence, more available water—than smaller tributaries and headwater streams.

In the far northern parts of Alaska, Canada, and the Soviet Union, permafrost is present nearly everywhere. It is in this continuous permafrost region that maximum permafrost depths are found. To the south, the permafrost layer is discontinuous. It may be up to 600 feet (180 meters) thick locally, but elsewhere there could be unfrozen ground. In the continuous permafrost regions of Alaska, alluvial valleys of large rivers may be the only water available. As the active layer freezes in winter, baseflow to even the largest rivers may be reduced to zero. Subpermafrost ground water is typically saline or brackish. In the discontinuous permafrost region, the permafrost is thinner, and the areas free of permafrost are much more extensive in alluvial river valleys.

Alluvial fans are found in Alaska at the margins of many of the mountain ranges. The fans are composed of glaciofluvial deposits, which tend to be coarse sand and gravel. Ground water can be obtained from these deposits, either below the permafrost or near rivers or lakes, where the permafrost may be thin or absent. In the more southerly alluvial fans, the water table may be below the permafrost layer. In this case, the permafrost has little impact on the hydrogeology other than preventing recharge to areas where it is present.

The distribution of permafrost in consolidated rocks is similar to that in unconsolidated deposits. If a rock unit has significant hydraulic conductivity in both the horizontal and vertical directions, ground-water hydrology should be similar to that of unconsolidated deposits. In highly anisotropic aquifers, even discontinuous permafrost bodies could act as ground-water dams, preventing horizontal flow and significantly reducing or eliminating vertical recharge. For example, ground water in fracture zones is replenished by downward recharge.

If the fracture zone were covered by a patch of permafrost, recharge would be difficult, even if the fracture zone extended below the permafrost layer.

### CASE STUDY: ALLUVIAL AQUIFERS—FAIRBANKS, ALASKA

In the area of Fairbanks, Alaska, the Chena and Tanaua rivers have alluvial deposits up to 800 feet (240 meters) thick. The alluvium consists of interbedded gravel, sand, and silt. The floodplains are interspersed with terraces from 3 to 25 feet (1 to 8 meters) in height. The distribution of permafrost in the area is irregular. Permafrost is absent or nearly so in the alluvium beneath the river channels. Thin permafrost may be found under islands. The youngest, low-terrace deposits are underlain by unfrozen alluvium, with some isolated permafrost bodies up to 80 feet (24 meters) thick. Higher terraces have more continuous permafrost which can be 200 feet (60 meters) deep. The older terraces have nearly continuous permafrost up to 280 feet (85 meters) deep (80). Permafrost thickness increases beneath progressively older terraces. Wells in the alluvial valley obtain water from unfrozen areas or, if there is permafrost, from either above or below it (83).

## 8.6 COASTAL PLAIN AQUIFERS

Coastal plains are found on all of the continents, with the exception of Antarctica. They are regional features, bounded on the continental side by highlands and seaward by a coastline. Coastal plains exist in areas of both stable basement rock as well as in those areas where the basement is sinking. The coastal plain may include large areas of former sea floor, and the geology of the coastal zone may be very similar to that of the adjacent continental shelf.

Sediment and sedimentary rocks of coastal plains were formed as either terrestrial or marine deposits. The terrestrial deposits tend to be landward and the marine deposits seaward, although fluctuating sea levels have resulted in alternating continental and marine strata—units deposited at a given time grade from continental to marine. Individual units have a variety of shapes, although the overall sequence usually thickens seaward. Coastal plains almost always contain Quaternary sediments; many also contain Tertiary- and Mesozoic-age deposits as well. The older rock units tend to crop out on the landward side, whereas younger, Quaternary and Recent, rock and sediments are found at the surface near the coast.

The coastal plain sediments may be unconsolidated or lithified, with no relationship to age. The Magothy aquifer of the northeastern United States is an unconsolidated sand of Cretaceous age, while the Biscayne aquifer of southeastern Florida consists of Pleistocene-age marine limestone. Aquifers typically are continental sands, gravels, and sandstones or marine sands or limestones. Confining beds consist of marine and continental silts and clays.

Typically, alternating and interfingering facies of different lithology are encountered. Some units are thick and extend for hundreds of miles; others are often not traceable for more than a few miles. One or more aquifers may be found at most locations. Baton Rouge, Louisiana, has ten aquifers in the first 3000 feet (900 meters) of sediments (84). Some of the world's most prolific aquifers are found in coastal plain deposits. A notable example is the Atlantic and Gulf coastal plain of North America (85–88).

The sediments of coastal plain areas were deposited either adjacent to or in shallow marine waters. Thus, pore water was originally saline. Fluctuating sea levels during the Pleistocene inundated many areas that are now land. As a result, saline waters occupied many contemporary fresh-water coastal plain aquifers in the not-too-distant geologic past. Relative sea level was up to 300 feet (90 meters) lower during the last Wisconsinan glaciation than present sea level, uncovering more of the continental shelf than is now exposed. During this period, saline water was flushed from inland aquifers to considerable depths.

Because of the seaward-sloping nature of coastal plain strata, deep aquifers are recharged in inland areas. Fresh water flows downgradient and then discharges by several mechanisms to the coastal waters. The amount of aquifer flushing that has occurred is a function of the volume of aquifer recharge and the amount of fresh water that can escape downgradient via the available mechanisms. Saline water has been flushed to a depth of 5900 feet (1800 meters) in Karnes County, Texas (89), and 3495 feet (1070 meters) below sea level in St. Tammany Parish, Louisiana (90), although this is unusual. Fresh water is found at depths of 1000 feet (300 meters) or more below sea level in many areas of the coastal plain aquifer (89). Fresh water has been found to occur in deep, confined coastal plain sediments many miles at sea. In a deep test well on Nantucket Island, located 40 miles (64 kilometers) off the New England coast, fresh water was found in confined aquifers at depths of 730 to 820 feet (223 to 250 meters) and 900 to 930 feet (274 to 283 meters). These are Crectaceous-age sands and are confined by clays (91). They represent aquifers that were recharged during the Pleistocene, when sea level was lower (91, 92).

Fresh water can discharge from a coastal aquifer via several natural mechanisms: (a) evapotranspiration, (b) direct seepage into springs, streams, tidal water, and the ocean floor, (c) mixing with saline ground water in a zone of diffusion, (d) flow across a semipermeable layer under the influence of a hydraulic gradient, and (e) flow across a semipermeable layer due to osmotic pressure caused by a salinity gradient. Examples of the various discharge mechanisms in a coastal area are shown in Figure 8.31. Mechanisms (a) and (b) are very efficient in discharging water from unconfined aquifers. Fresh-water springs in the sea bottom occur in unconfined aquifers or confined aquifers where the confining layer is breached. Deep confined aquifers utilize only the last three methods—(c), (d), and (e)—which are not as efficient. Assuming that all aquifers have equal transmissivities, the deeper the aquifer, the less fresh water it can discharge because of the reduction in the number and efficiency of discharge methods. Confined coastal aquifers can contain fresh water, even though overlying aquifers are salty.

On Long Island, New York, there is an unconfined aquifer and two deep confined aquifers, all recharged by precipitation at the center of the island. Most

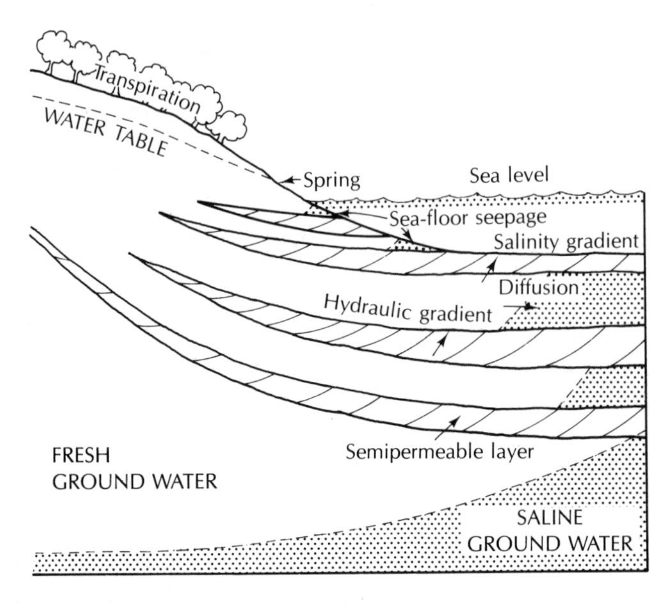


**FIGURE 8.31** Typical fresh-water–salt-water relationship in a layered coastal aquifer.

of the fresh ground water discharges through the unconfined aquifer, with much lower volumes flowing in the confined aquifers. Wells drilled in a barrier island 28,600 feet (8700 meters) from the coastline show fresh water the full depth of both confined aquifers, with 5 to 10 feet (1.5 to 3 meters) of artesian head in the upper confined aquifer and 20 feet (6 meters) in the lower. This is a distance of at least 11 to 12 miles (18 to 19 kilometers) from the closest part of the recharge area for the confined aquifers (93).

The most characteristic type of water-quality degradation occurring in coastal plain aquifers is saline-water intrusion. Sources of saline water are found as connate water below inland fresh-water aquifers, as subsurface sea water below island aquifers, on the seaward edges of coastal aquifers, and as surface tidal waters in natural estuaries and artificial canals.

The shape and position of the boundary between saline ground water and fresh ground water is a function of the volume of fresh water discharging from the aquifer (excluding intrinsic aquifer characteristics). Any action that changes the volume of fresh-water discharge results in a consequent change in the salt-water–fresh-water boundary. It should be noted that minor fluctuations in the boundary position occur with tidal actions and seasonal and annual changes in the amount of fresh-water discharge. For this reason, the boundary is in a state of quasi-equilibrium. Natural changes in the equilibrium position can result from long-term changes in climatic patterns or the position of relative sea level.

Human action that results in saline ground water entering a fresh-water aquifer is termed **saline-water encroachment.** It occurs as a result of a diversion of fresh water that previously had discharged from a coastal aquifer. Salt-water encroachment can be either **active** or **passive** (94).

Passive saline-water encroachment occurs when some fresh water has been diverted from the aquifer—yet the hydraulic gradient in the aquifer is still sloping toward the salt-water–fresh-water boundary. In this case, the boundary will slowly shift landward until it reaches an equilibrium position based on the new discharge conditions. Passive saline-water encroachment is taking place today in many coastal plain aquifers where ground-water resources are being developed. It acts slowly and, in some areas, it may take hundreds of years for the boundary to move a significant distance (Figure 8.32).

The consequences of active saline-water encroachment are considerably more severe, as the natural hydraulic gradient has been reversed and fresh water is actually moving away from the salt-water–fresh-water boundary (Figure 8.33). This occurrence is due mainly to the concentrated withdrawals of ground water, creating a deep cone of depression. The boundary zone moves much more rap-

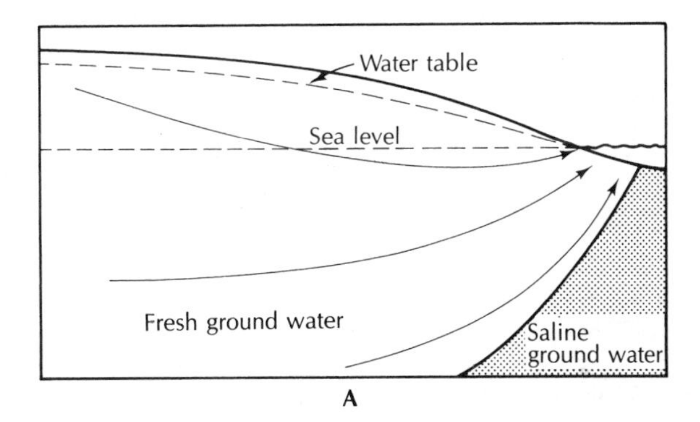


A

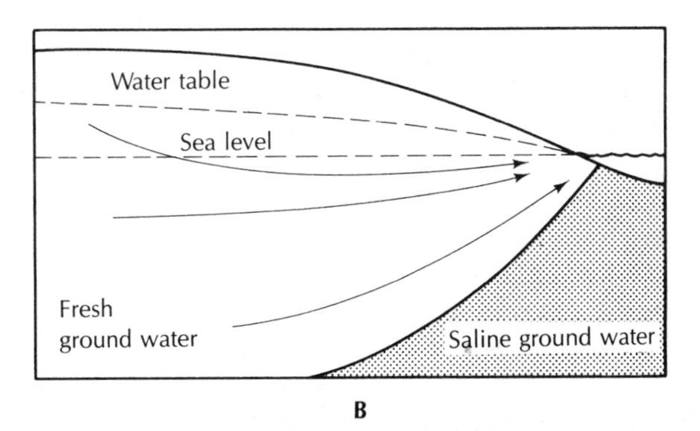


B

FIGURE 8.32 A. Unconfined coastal aquifer under natural ground-water discharge conditions. B. Passive saline-water encroachment due to a general lowering of the water table. Flow in the fresh-water zone is still seaward.

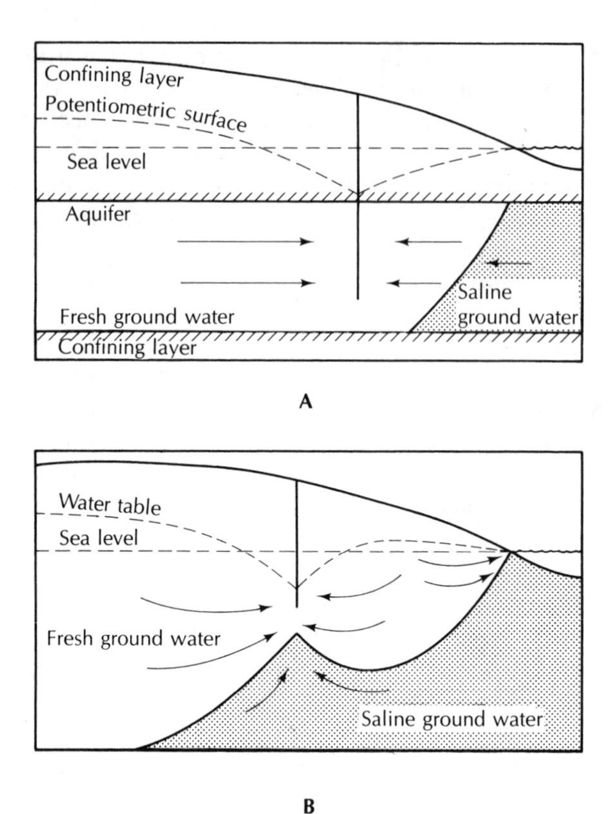


**FIGURE 8.33** **A.** Active saline-water encroachment in a confined aquifer with the potentiometric surface below sea level. **B.** Active saline-water encroachment in an unconfined aquifer with the water table drawn below sea level.

idly than it does during passive saline-water encroachment. Furthermore, it will not stop until it has reached the low point of the hydraulic gradient: the center of pumping. It is this type of rapid encroachment that destroyed the aquifers beneath Brooklyn, New York, in the 1930s, when the water table was lowered 30 to 50 feet (9 to 15 meters) below sea level.

Both types of saline-water encroachment can occur in areas of inland connate saline waters as well as in coastal areas. In Baton Rouge, Louisiana, heavy industrial pumpage has lowered the potentiometric surface in the "600-foot sand" aquifer. This has caused connate saline ground water in the aquifer to move northward toward the well field (84).

Saline-water encroachment is also a serious problem in the Miami area of Florida. The unconfined Biscayne aquifer is the sole source of ground water for this area, as the Floridan aquifer is salty here. The encroachment began with the 1907 installation of inland drainage canals to lower the water table. This diverted fresh water that had been flowing through the Biscayne aquifer to Biscayne Bay. Passive saline-water encroachment resulted as the salt-water–fresh-water boundary sought a new equilibrium position. Because of the high transmissivity of the aquifer, saline water occupied many areas of the Biscayne aquifer beneath Miami in only a few years (95). Sea water also moved up the canals toward the aquifer during high tides. Dams have now been built across the canals in order to keep salt water from traveling up them. Water impounded beyind the dams is fresh, adding to the available recharge to the Biscayne aquifer. This fresh water had formerly been lost to the sea during flooding. New canals must be constructed with salinity-control dams.

## 8.7 GROUND WATER IN DESERT AREAS

Desert areas receive 10 inches (25 centimeters) of precipitation or less each year. Cold deserts, such as Antarctica, can have great accumulations of water. In warm deserts, however, the potential evapotranspiration may be many times the annual precipitation. Under such conditions, there is often virtually no local ground-water recharge. Yet, some warm deserts can have large volumes of fresh water stored beneath them.

Both bedrock and unconsolidated aquifers are known in desert areas. As the dry climate promotes mechanical weathering of rock, the unconsolidated materials are often rather coarse and permeable. Alluvial fans and talus slopes at the bases of mountains and escarpments can be productive. In South Yemen on the Arabian peninsula, ground water is obtained from shallow wells in alluvial materials (96). Recharge occurs during flooding that accompanies the infrequent precipitation.

Vast bedrock aquifers are known to occur in the sedimentary basins of Egypt, Jordan, and Saudi Arabia (97). The Sahara Desert is underlain by the Nubian aquifer, a sandstone up to 3000 feet (900 meters) thick. Several younger aquifers overlie it. The ground water in the Nubian Sandstone is confined, and there are large initial heads. High-capacity wells now tap this aquifer. Several major aquifer systems, primarily sandstone, also underlie the Arabian peninsula. Carbonate aquifers may not have high-solution permeability owing to the lack of actively circulating ground water. Those that are permeable are a result of primary permeability or solution permeability developed during moister periods of the past.

Recharge to aquifers can occur through runoff from adjacent mountains, which receive relatively more water. However, under much of North Africa and Arabia, the ground water is old, exceeding 35,000 radiocarbon years B.P. (98). It has been shown by model studies that the water in these aquifers was probably recharged during wet climatic periods during the Pleistocene. Under such conditions, ground water is mined from the aquifers, as there is no modern recharge.

Arid zones may have significant water-quality problems. The slow circulation of ground water results in mineralization. In addition, evaporation of ground water from discharge areas results in salt deposition, with consequent high salinity in the soil and shallow ground water.

## 8.8 GROUND-WATER REGIONS OF THE UNITED STATES

One useful generalization in the study of hydrogeology is the concept of ground-water regions. These are geographical areas of similar occurrence of ground water. If an area is subdivided into several smaller regions, useful comparisons can be made between areas of well-known hydrogeology and areas that are geologically similar but have not been as well studied.

The ground-water regions of the United States have been classified by several different authorities. In 1923, O. E. Meinzer, who could be considered to be the father of modern hydrogeology in the United States, proposed a classification system based on 21 different ground-water provinces (99). Thomas proposed a system based on 10 ground-water regions in 1952 (100) and McGuinness revised Thomas' system in 1963 (101). In 1984 Heath published yet another revision of Thomas' basic system (102). Heath based his classification system on five features of ground-water systems:

(1) The components of the system and their arrangement,
(2) the nature of the water-bearing openings of the dominant aquifer or aquifers with respect to whether they are of primary or secondary origin,
(3) the mineral composition of the rock matrix of the dominant aquifers with respect to whether it is soluble or insoluble,
(4) the water storage and transmission characteristics of the dominant aquifer or aquifers, and
(5) the nature and location of recharge and discharge areas.

Eleven of the ground-water regions are based on physiography and are not necessarily contiguous. The locations of these areas are shown on Figure 8.34. Area 12 is comprised of alluvial valleys throughout the United States and is shown on Figure 8.35. Area 13 is Hawaii, area 14 is Alaska, and area 15 is Puerto Rico and the Virgin Islands.

### 1. WESTERN MOUNTAIN RANGES

*This region consists of mountains with thin soils over fractured rocks and alternating narrow alluvial valleys, some of which have been glaciated.*

The mountains of the western United States are the headwater areas of many rivers, including the Columbia, Snake, San Joaquin, Sacramento, Missouri, Platte, Colorado, Arkansas, and Rio Grande. The mountains tend to be well watered, receiving higher levels of precipitation than the surrounding lowlands. The rocks in the western mountains are predominantly crystalline, although there are some sedimentary ranges. Intermontane valleys may contain aquifers of alluvial deposits of glacial outwash. Most consolidated rock aquifers are of local rather than regional extent. Mountain ranges included in this region are the northern Coast Ranges, the Sierra Nevada and Cascade Range, parts of the physiographic Basin and Range Province, and the Rocky Mountains.

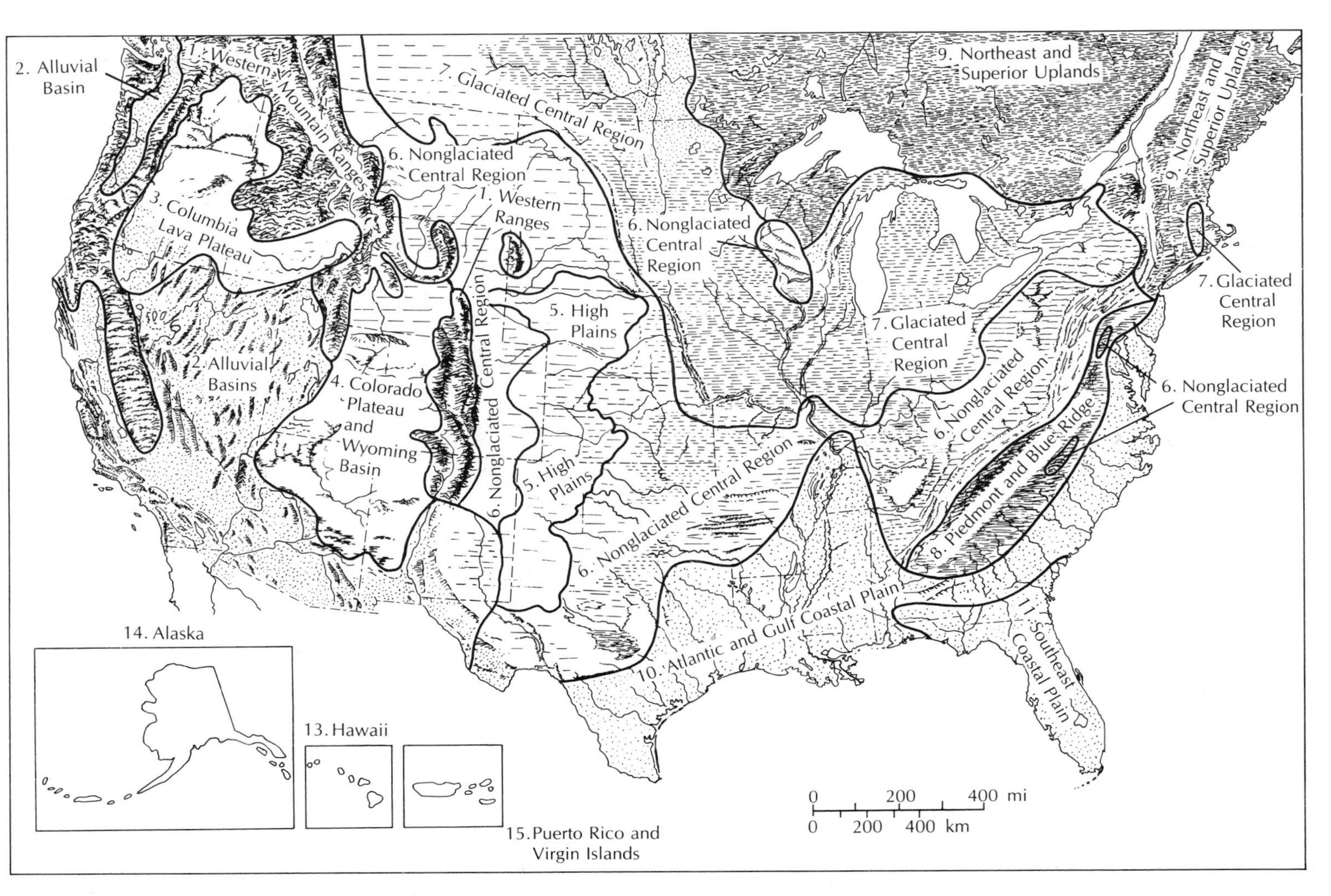


**FIGURE 8.34** Ground-water regions of the United States. Source: R. C. Heath, U.S. Geological Survey Water-Supply Paper 2242, 1984.

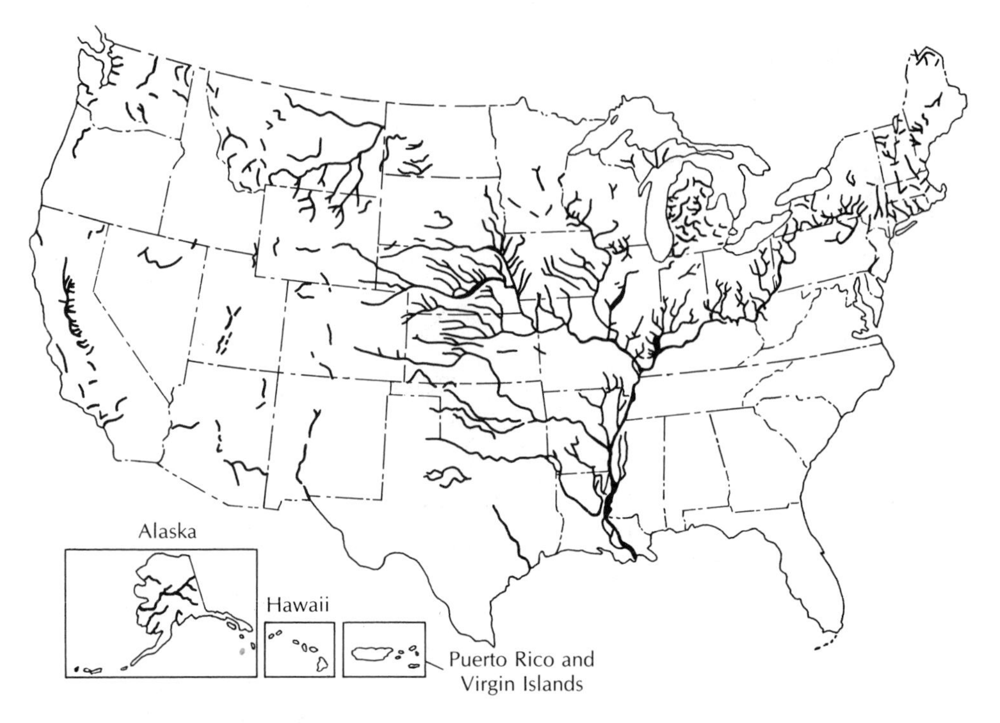


**FIGURE 8.35** Alluvial Valleys ground-water region. Source: R. C. Heath, U.S. Geological Survey Water-Supply Paper 2242, 1984.

## 2. ALLUVIAL BASINS

*This region consists of thick alluvial deposits in basins and valleys that are bordered by uplifted mountain ranges.*

Lying between the ranges of the western mountains are tectonic troughs that have been partially filled with erosional products from the adjacent uplands and mountains. The sediment filling the valleys has been sorted into various size fractions by running water. Sand and gravel layers are found interspersed between silt and clay deposits. The valley-fill material is porous; hence, there are many permeable zones—especially, close to the mountains, where coarser debris was deposited in coalescing alluvial fans.

Runoff coming from the mountains can soak into the coarse deposits; however, rainfall over the basins themselves may be rather low. The deep sedimentary basins hold vast volumes of water in storage. Many of the aquifers can yield large amounts of water to individual wells. Demands for water are high, and the amount of ground-water recharge is relatively low. As a result, the water in storage is being withdrawn and ground-water levels are falling in many of the basins. Land subsidence due to the lowered ground-water levels in some of the basins is also a problem.

## 3. COLUMBIA LAVA PLATEAU

*This region consists of thin soils overlying a thick sequence of basalt flows with interbedded sediments.*

The Columbia Lava Plateau is formed of intrusive volcanic rock with interbedded alluvium and lake sediments. The plateau is somewhat arid, but some of the higher areas and surrounding mountains are more humid.

The lava flows are nearly horizontal, with some layers and interflow surfaces highly permeable. Shrinkage cracks create vertical permeability. The amount of water in storage is high; it is replenished by local precipitation, runoff from nearby mountains, and excess irrigation water applied to the surface.

The plateau is deeply dissected by the Columbia and Snake rivers and their tributaries. There are many large springs flowing from the permeable basalt layers into incised rivers; e.g., the Thousand Springs area of the Snake River. Ground water is obtained from wells in the basalt, many of them 1000 feet (300 meters) deep or more. Glacial outwash gravel deposits along many of the rivers also are productive aquifers.

## 4. COLORADO PLATEAU AND WYOMING BASIN

*This region consists of thin soils over consolidated sedimentary rocks.*

The Colorado Plateau and Wyoming Basin are areas of extensive sedimentary rocks. The strata are flat-lying for the most part, but some areas have been tilted, folded, or faulted. At different levels, there are a number of plateaus that are deeply dissected by the Colorado River system or the Missouri River system. The sediments are chiefly shales and low-permeability sandstones.

The rate of ground-water recharge is very low, as the warm and arid climate causes most of the precipitation to be quickly evaporated. Some of the higher plateaus, such as the southern edge of the Colorado, do receive somewhat more precipitation. The most productive consolidated rock aquifer of the region, the Coconino Sandstone (Permian), crops out here. Other aquifers include the Dakota Sandstone (Cretaceous), the Entrada and Junction Creek sandstones (Jurassic), and the Wingate Sandstone (Triassic). Small amounts of water can be obtained from sandstone and limestone aquifers, which can receive local recharge, as well as from some alluvial deposits, such as the Uinta Basin at the edge of the Uinta Range. In general, the Colorado Plateau-Wyoming Basin is a water-poor region.

## 5. HIGH PLAINS

*This region consists of thick alluvial deposits over sedimentary rocks.*

The High Plains ground-water region corresponds with the High Plains part of the physiographic Great Plains Province. This area is underlain by sedimentary rocks that dip gently to the east and are upturned at their contact with the Rocky Mountains and other places at which dome mountains, such as the Black Hills, have pushed through them. The entire surface is mantled with an alluvial apron representing material eroded from the Rocky Mountains. This alluvial cover, known as the Ogallala Formation, slopes to the east, but has been dissected in some areas.

The area of Nebraska between the Platte and Niobrara rivers is mantled with deposits of windblown sand—the Sand Hills. Both the Sand Hills and the Ogallala Formation (Pliocene) are highly porous and permeable. Rivers, which have cut all the way into bedrock, isolate these unconsolidated aquifers from recharge by streams draining from the Rocky Mountains. Recharge is by means of local precipitation, and the Sand Hills receives enough to be relatively well supplied compared with the rest of the High Plains. The Ogallala Formation is the major aquifer through most of the High Plains. It had a large volume of water in storage, but with heavy ground-water development for irrigation, water levels are rapidly falling—particularly in Texas and New Mexico.

Consolidated rock aquifers in Nebraska include units of the Arikaree Group (Miocene), Chadron Formation (Oligocene) and Dakota Group (Cretaceous). The Dakota Group is overlain by up to 3000 feet (1000 meters) of impermeable Pierre Shale and is recharged in the Black Hills outcrop area. Water is also obtained from Quaternary-age river alluvium along modern rivers, which can furnish a local source for recharge.

## 6. NONGLACIATED CENTRAL REGION

*This region consists of thin regolith over sedimentary rocks.*

The Nonglaciated Central Region encompasses a large area of the central United States extending from the Rocky Mountains to the Appalachian Plateau and Valley and Ridge physiographic province and includes several small outlying sections such as the driftless area of Wisconsin and the Triassic sedimentary basins of the East. The bedrock underlying the region consists of level or gently tilted and folded sedimentary rocks ranging in age from Paleozoic to middle Tertiary. The land surface encompasses plains and plateaus and is gently rolling to sharply dissected. Climate ranges from semiarid to humid, but most of the area is subhumid to humid.

Consolidated rock aquifers include the Edwards Limestone (Cretaceous) of southwestern and central Texas, from which a flowing well in San Antonio yielded 24 million gallons per day (1050 liters per second) when first drilled. Sands of the Trinity Group (early Cretaceous) and Woodbine Sand (late Cretaceous) are productive aquifers in north-central Texas and southern Oklahoma. These units include primarily clay, sand, and slightly consolidated sandstone. The Dakota Sandstone (late Cretaceous), Minnelusa Formation (Pennsylvanian and Permian), and Pahasapa Limestone (early Mississippian) are bedrock aquifers in South Dakota. In North Dakota, the Dakota Sandstone and Lakota Formation (early Cretaceous) are important. In eastern Montana, the Hell Creek Formation (late Cretaceous), Fox Hills Sandstone (late Cretaceous), and Fort Union Formation (Paleocene) are widespread. Other aquifers in Montana that are more restricted are the Judith River Formation (late Cretaceous) and the

Kootenai, Dakota, and Lakota formations (Cretaceous). Fresh water may be available at depth from the Madison Limestone (Mississippian). The Roswell Basin of New Mexico includes the San Andres Limestone (Permian), which is confined by the Artesia Group. The initially high artesian heads in the Roswell Basin have long since declined owing to the many flowing wells, as has the head in the Dakota Sandstone elsewhere. Limestone aquifers also occur in Arkansas, Missouri, Kentucky, and Tennessee. These units tend to be cavernous and many yield large springs, but generally they are not very productive because of the sporadic occurrence of water and the low storage capacity.

Carbonate aquifers, such as the Knox Dolomite of the Valley and Ridge physiographic province, provide ground water in many parts of the eastern sectors of this region.

## 7. GLACIATED CENTRAL REGION

*This region consists of glacial deposits over sedimentary rocks.*

The Glaciated Central Region extends from the Catskill Mountains in New York to the northern Great Plains in Montana and includes the Triassic Valley of central Connecticut and Massachusetts. Only the eastern part of the region is underlain by uplands. The bedrock aquifers of the area are contiguous with those in the adjacent nonglaciated areas.

The principal feature of the Glaciated Central Region is the yield of ground water from glacial drift. Outwash plains and buried bedrock valleys are the principal sources. Most of the drift is fine grained, but there are numerous sand and gravel deposits. The region ranges from semiarid in North Dakota to humid in the Great Lakes states.

Major bedrock aquifers include limestones and sandstones of Paleozoic age. Some of these are widespread and include the following: Mt. Simon Sandstone, Galesville Sandstone, Ironton Sandstone, Franconia Sandstone, Jordan Sandstone, and Trempealeau Formation (all Cambrian); Prairie du Chien Formation and St. Peter Sandstone (both Ordovician). These are the bedrock aquifers of eastern Iowa, southern Minnesota, southern Wisconsin, and northern Illinois. The bedrock aquifers are collectively known locally as the deep aquifer, sandstone aquifer, or Cambrian-Ordovician aquifer. Elsewhere in this region the Paleozoic bedrock aquifers are not regional in extent, and saline water in them tends to be a problem. In areas underlain by Precambrian crystalline rocks, the yield to wells is generally small. Alluvium, especially glaciofluvial deposits, is a major source of ground water.

## 8. PIEDMONT–BLUE RIDGE REGION

*This region consists of a thick mantle of weathering residuum over fractured crystalline and metamorphic rock.*

The Appalachian Mountains are a part of this region. Bedrock in the area includes metamorphic and crystalline rocks of the Piedmont and Blue

Ridge. Precipitation ranges from 30 inches (76 centimeters) up to 75 inches (190 centimeters).

Weathered zones in the metamorphic rocks of the Piedmont yield small to moderate amounts of water almost anywhere, with larger-yield wells possible on fracture traces. More-productive wells are usually in valleys rather than on hilltops. Some sandstone units that lie in belts parallel to the structural trend are also moderately good aquifers. There is little ground water in the Blue Ridge; for the most part, the bedrock is impermeable.

### 9. NORTHEAST AND SUPERIOR UPLANDS

*This region consists of glacial deposits over fractured crystalline rock.*

There are two separate areas in this region. One is in northeast New York State and northern New England and the other in northern Wisconsin and Minnesota. The region extends into Canada as the Canadian shield. Glacial deposits of varying thickness mantle crystalline bedrock. Ground water is obtained from fractures in the bedrock or from permeable glacial deposits. Glacial outwash in the form of sand and gravel deposits forms the best aquifer systems in this region. The area is humid, with 30 to 45 inches (76 to 114 centimeters) of precipitation.

### 10. ATLANTIC AND GULF COASTAL PLAIN

*This region consists of complex sequences of interbedded sand, silt, clay, and limestone.*

The Atlantic and Gulf coastal plain starts at Cape Cod and includes Long Island and southern New Jersey, most of each of the Atlantic coastal states, part of Alabama and Mississippi, parts of Missouri and Arkansas, all of Louisiana, and southeastern Texas. At the interior edge, the deposits thin to a featheredge and thicken toward the coast. They consist of unconsolidated to consolidated continental and marine sediments. Almost all of the coastal plain has a humid climate, with ample water available to recharge the aquifers.

Cape Cod and Long Island are covered with glacial deposits, including outwash, which is a good aquifer. Pleistocene sediments are also the most productive aquifer in Delaware. Elsewhere, the coastal plain deposits contain many excellent aquifers, including sands, sandstones, dolomites, and limestones. Saline water is present at the coasts and at depth in many of the areas. Otherwise, water quality is generally good.

Major aquifers of the Atlantic coastal plain include the Magothy Formation and Lloyd Sand Member of the Raritan Formation (both Cretaceous) of Long Island and northern New Jersey, Cheswold and Fredonia aquifers (Miocene) in Delaware, and Castle Hayne Limestone of North Carolina. A number of good aquifers exist in Texas and include the Wilcox Group and Carrizo Sand (Eocene); Catahoula Sandstone, Oakville Sandstone, and sand units of Lagarto

Clay (Miocene); and Goliad Sand, Willis Sand, and sand units of the Lissie Formation (Pliocene to Pleistocene). Other states have equally productive aquifers, either unnamed or of local extent.

### 11. SOUTHEAST COASTAL PLAIN

*This region consists of thick layers of sand and clay over semiconsolidated and consolidated carbonate rock.*

This area includes the Florida Peninsula and parts of coastal South Carolina, Georgia, and Alabama. The surface is underlain by unconsolidated deposits of clay, sand, gravel, and shell beds. Deeper layers consist of alternating layers of semiconsolidated and consolidated limestones and dolomites. The Floridan aquifer is located in the carbonate units and is one of the most prolific aquifers in the world. It is confined by the overlying Hawthorn Formation. Water may also be obtained from the surficial sand and gravel deposits. This region is described in greater detail as a case study in Chapter 7.

### 12. ALLUVIAL VALLEYS

*This region consists of thick sand and gravel deposits beneath floodplains and stream terrace deposits.*

This region is not one of geographic continuity, but rather one of similar geologic origin. Many river systems have deposited thick sequences of sand and gravel, which are highly porous and permeable. Much of the sand and gravel is glacial outwash deposited by streams that carried water away from the melting ice front during the Pleistocene. These sediments were deposited far beyond the extent of the glaciers in some instances. They represent long, narrow aquifer systems and, as in most cases, are the foundations for modern river systems, with an ample supply of water to recharge that removed by pumping. The Miami River of Dayton, Ohio, is an example.

### 13. HAWAIIAN ISLANDS

*This region consists of lava flows interbedded with ash deposits and segmented by dikes.*

The occurrence of ground water in the Hawaiian Islands has already been discussed as a case study in this chapter. Ground water occurs in several different types of aquifers, with some boundary conditions that are unique to the Hawaiian Islands. Different types of ground-water development schemes, such as Maui tunnels and inclined wells, are necessary to capture the ground water by skimming, thereby preventing saline-water encroachment.

### 14. ALASKA

*This region consists of glacial and alluvial deposits, occupied in part by permafrost and overlying bedrock of various types.*

Alaska is placed in a single ground-water region for reasons of convenience. There are many different terranes in which ground water occurs in Alaska. The state is thinly populated and not a great deal of ground-water exploration has taken place. The occurrence of ground water is to a large extent controlled by the permafrost. In areas of continuous permafrost, ground-water supplies are limited to small, discontinuous thawed areas beneath large lakes and rivers. In the area of discontinuous permafrost, the water in the sand and gravel deposits is frequently not frozen and is available for development. Most inhabited areas of Alaska have glacial outwash or river alluvium that can serve as an aquifer.

### 15. PUERTO RICO

*This region consists of alluvium and limestones overlying and bordering fractured igneous rocks.*

This region contains the Islands of Puerto Rico and the U.S. Virgin Islands. These islands are generally hilly and underlain by both limestones and volcanic and intrusive igneous rocks. These islands receive high amounts of rainfall; ground-water recharge averages almost 6 feet (2 meters) annually in Puerto Rico. Alluvium, which occurs in stream valleys and along the coast areas, is an effective aquifer where sand and gravel are present. The limestone areas are also aquifers; however, the volcanic rocks are metamorphosed and contain water mainly in fractures, as do other dense crystalline rocks.

## REFERENCES


1. DOONAN, C. J., G. E. HENDRICKSON, and J. R. BYERLAY. *Ground Water and Geology of Keweenaw Peninsula, Michigan*. Michigan Department of Natural Resources, Geological Survey Division, Water Investigation 10, 1970, 40 pp.
2. NORRIS, S. E. *Permeability of Glacial Till.* U.S. Geological Survey Professional Paper 450-E, 1963, pp. 150-51.
3. SHARP, J. M., JR. "Hydrogeologic Characteristics of Shallow Glacial Drift Aquifers in Dissected Till Plains (North-Central Missouri)." *Ground Water,* 22 (1984):683–89.
4. HENDRY, M. J. "Hydraulic Conductivity of a Glacial Till in Alberta." *Ground Water,* 20 (1982):162–69.
5. PRUDIC, D. E. "Hydraulic Conductivity of a Fine-Grained Till, Cattaraugus County, New York." *Ground Water,* 20 (1982):194–204.
6. HERZOG, B. L., and W. J. MORSE. "A Comparison of Laboratory and Field Determined Values of Hydraulic Conductivity at a Hazardous Waste Disposal Site." In *Proceedings, Seventh Annual Madison Waste Conference—Municipal and Industrial Waste*. University of Wisconsin, 1984, pp. 30–52.
7. GORDON, M. E., and P. M. HUEBNER. "An Evaluation of the Performance of Zone of Saturation Landfills in Wisconsin." In *Proceedings, Sixth Annual Madison*

*Waste Conference—Municipal and Industrial Waste*. University of Wisconsin, 1983, pp. 23–53.

8. BRADBURY, K. R., and R. W. TAYLOR. "Determination of the Hydrogeologic Properties of Lakebeds Using Offshore Geophysical Surveys." *Ground Water,* 22 (1984):690–95.
9. FETTER, C. W., JR. "Final Hydrogeologic Report, Seymour Recycling Corporation Hazardous Waste Site, Seymour, Indiana." Report to United States Environmental Protection Agency, 1985, 171 pp. and appendices.
10. CARR, J. R. "Tacoma's Well Field Might Be World's Most Productive." *Johnson Drillers Journal* (September and October 1976):1–4.
11. WINTER, T. C. *Hydrogeology of Glacial Drift, Mesabi Iron Range, Northeastern Minnesota*. U.S. Geological Survey Water-Supply Paper 2029-A, 1973, 23 pp.
12. DENNE, J. E., and others. "Remote Sensing and Geophysical Investigations of Glacial Buried Valleys in Northeastern Kansas." *Ground Water,* 22 (1984):56–65.
13. NORRIS, S. E., and A. M. SPIEKER. *Ground-Water Resources of the Dayton Area, Ohio*. U.S. Geological Survey Water-Supply Paper 1808, 1966, 167 pp.
14. EAKIN, T. E., D. PRICE, and J. R. HARRILL. *Summary Appraisals of the Nation's Ground Water Resources—Great Basin Region*. U.S. Geological Survey Professional Paper 813-G, 1976, 37 pp.
15. DUTCHER, L. C., and A. A. GARRETT. *Geological and Hydrologic Features of the San Bernardino Area, California*. U.S. Geological Survey Water-Supply Paper 1419, 1963, 114 pp.
16. HARDT, W. F., and C. B. HUTCHINSON. "Model Aids Planners in Predicting Rising Ground-Water Levels in San Bernardino, California." *Ground Water,* 16 (1978):424–31.
17. HOFFMAN, J. "Flooded San Bernardino Fears Even a Minor Quake." *Los Angeles* (California) *Herald Examiner,* April 17, 1983.
18. YOUNG, H. *Digital Computer Model of the Sandstone Aquifer in Southeastern Wisconsin*. Southeastern Wisconsin Regional Planning Commission, Technical Report 16, 1976, 42 pp.
19. FETTER, C. W., JR. "Interstate Conflict over Ground Water: Wisconsin-Illinois." *Ground Water,* 19 (1981):201–13.
20. VISOCKY, A. P., M. G. SHERRILL, and K. CARTWRIGHT. *Geology, Hydrology and Water Quality of the Cambrian and Ordovician Systems in Northern Illinois*. Cooperative Ground Water Report 10, Illinois State Geological Survey and Illinois State Water Survey, 1985, 136 pp.
21. YOUNG, H. L., and others. "Northern Midwest Regional Aquifer System Study." In *Regional Aquifer-System Analysis Program of the U. S. Geological Survey*. U.S. Geological Survey Circular 1002, 1986, pp. 72–87.
22. WALTON, W. C. *Groundwater Recharge and Runoff in Illinois*. Illinois State Water Survey Report of Investigation 48, 1965, 55 pp.
23. SCHICHT, R. J., J. R. ADAMS, and J. B. STALL. *Water Resources Availability, Quality and Cost in Northeastern Illinois*. Illnois State Water Survey Report of Investigation 83, 1976, 90 pp.
24. TWENTER, F. R. *Geology and Promising Areas for Ground-Water Development*

*in the Hualapai Indian Reservation, Arizona*. U.S. Geological Survey Water-Supply Paper 1576-A, 1962, 38 pp.

**25.** HUNTOON, P. W. "Cambrian Stratigraphic Nomenclature and Ground-Water Prospecting Failures in the Hualapai Plateau, Arizona." *Ground Water,* 15 (1977):426–33.

**26.** HUNTOON, P. W. "Incredible Tale of Texasgulf Well 7 and Fracture Permeability, Paradox Basin, Utah." *Ground Water,* 24 (1986):643–53.

**27.** SNIPES, D. S., and others. "Ground-Water Problems in the Mesozoic Pax Mountain Fault Zone." *Ground Water,* 24 (1986):375–81.

**28.** HUNTOON, P. W. "Fault Controlled Ground-Water Circulation under the Colorado River, Marble Canyon, Arizona." *Ground Water,* 19 (1981):20–27.

**29.** HUNTOON, P. W. "Fault Severed Aquifers along the Perimeters of Wyoming Artesian Basins." *Ground Water,* 23 (1985):176–81.

**30.** LE GRAND, H. E., and W. A. PETTYJOHN, "Regional Hydrogeological Concepts of Homoclinal Flanks." *Ground Water,* 19 (1981):303–10.

**31.** WALTON, W. C, and S. CSALLANY. *Yields of Deep Sandstone Wells in Northern Illinois*. Illinois State Water Survey Report of Investigation 43, 1962, 47 pp.

**32.** PLUMMER, N. L., T. M. L. WIGLEY, and D. L. PARKHURST. "The Kinetics of Calcite Dissolution in $CO_2$-water systems at 5° to 60° C and 0.00 to 1.0 Atm. $CO_2$." *American Journal of Science,* 278 (1978):176–216.

**33.** PALMER, A. N. "Recent Trends on Karst Geomorphology." *Journal of Geological Education,* 32 (1984):247–53.

**34.** FORD, D. C., and R. O. EWERS. "The Development of Limestone Cave Systems in the Dimensions of Length and Depth." *Canadian Journal of Earth Science,* 15 (1978):1783–98.

**35.** EWERS, R. O., et al. "The Origin of Distributary and Tributary Flow within Karst Aquifers." Geological Society of America, *Abstracts with Programs,* 10, no.7 (1978):398–99.

**36.** WHITE, W. B. "Conceptual Models for Carbonate Aquifers." *Ground Water,* 7, no. 3 (1969):15–22.

**37.** SHUSTER, E. T., and W. B. WHITE. "Seasonal Fluctuations in the Chemistry of Limestone Springs: A Possible Means for Characterizing Carbonate Aquifers." *Journal of Hydrology,* 14 (1971):93–128.

**38.** BRUCKER, R. W., J. W. HESS, and W. B. WHITE. "Role of Vertical Shafts in the Movement of Groundwater in Carbonate Aquifers." *Ground Water,* 10, no. 6 (1972):5–13.

**39.** SHERRILL, M. G. *Geology and Ground Water in Door County, Wisconsin, with Emphasis on Contamination Potential in the Silurian Dolomite*. U.S. Geological Survey Water-Supply Paper 2047, 1978, 38 pp.

**40.** FOX, I. A., and K. R. RUSHTON. "Rapid Recharge in a Limestone Aquifer." *Ground Water,* 14, no. 1 (1976):21–27.

**41.** BOATWRIGHT, B. A., and D. W. AILMAN. "The Occurrence and Development of Guest Sink, Hernando County, Florida." *Ground Water,* 13, no. 4 (1975):372–75.

**42.** RUHE, R. V. "Summary of Geohydrologic Relationships in the Lost River Watershed, Indiana, Applied to Water Use and Environment." In *Hydrologic Problems in Karst Regions,* ed. R. R. Dilamarter and S. Csallany. Bowling Green, Ky: Western Kentucky University, 1977, pp. 64–78.

**43.** BROWN, R. F. *Hydrology of the Cavernous Limestones of the Mammoth Cave Area, Kentucky*. U.S. Geological Survey Water-Supply Paper 1837, 1966, 64 pp.

**44.** WHITE, W. B., et al. "The Central Kentucky Karst." *Geographical Review,* 60 (1970):88–115.

**45.** SWENSON, F. A. "New Theory of Recharge to the Artesian Basin of the Dakotas." *Bulletin, Geological Society of America,* 79 (1968):163–82.

**46.** LE GRAND, H. E., and V. T. STRINGFIELD. "Water levels in Carbonate Rock Terranes." *Ground Water,* 9, no. 3 (1971):4–10.

**47.** CUSHMAN, R. V. *Recent Developments in Hydrogeologic Investigation in the Karst Area of Central Kentucky*. International Association of Hydrogeologists Memoirs, Congress of Istanbul, 1967, pp. 236–48. Cited in Reference 46.

**48.** LATTMAN, L. H., and R. R. PARIZEK. "Relationship between Fracture Traces and the Occurrence of Groundwater in Carbonate Rocks." *Journal of Hydrology,* 2 (1964):73–91.

**49.** PARIZEK, R. R. "On the Nature and Significance of Fracture Traces and Lineaments in Carbonate and Other Terranes." In *Karst Hydrology and Water Resources,* ed. V. Yevjevich. Fort Collins: Colo.: Water Resources Publications, 1976, pp. 47–108.

**50.** SIDDIQUI, S. H., and R. R. PARIZEK. "Hydrogeologic Factors Influencing Well Yields in Folded and Faulted Carbonate Rocks in Central Pennsylvania." *Water Resources Research,* 7 (1971):1295–1312.

**51.** CSALLANY, S., and W. C. WALTON. *Yields of Shallow Dolomite Wells in Northern Illinois*. Illinois State Water Survey Report of Investigation 46, 1963, 43 pp.

**52.** LAMOUREAUX, P. E., and W. J. POWELL. *Stratigraphic and Structural Guides to the Development of Water Wells and Well Fields in Limestone Terrane*. International Association of Scientific Hydrology, Pub. 52, 1960, pp. 363–75.

**53.** MARCHER, M. V., R. H. BINGHAM, and R. E. LOUNSBURY. *Ground Water Geology of the Dickson, Lawrenceburg and Waverly Areas of the Western Highland Rim, Tennessee.* U.S. Geological Survey Professional Paper 1764, 1964, 50 pp.

**54.** ZURAWSKI, A. *Summary Appraisals of the Nation's Groundwater Resources—Tennessee Region*. U.S. Geological Survey Professional Paper 813-L, 1978, 35 pp.

**55.** STONE, R., and D. F. SNOEBERGER. "Cleat Orientation and Areal Hydraulic Anisotropy of a Wyoming Coal Aquifer." *Ground Water,* 15 (1977):434–38.

**56.** REHM, B. W., G. H. GROENEWOLD, and S. R. MORAN. "The Hydraulic Conductivity of Lignite and Associated Geological Materials and Strip Mine Spoils, Western North Dakota." Geological Society of America, *Abstracts with Programs,* 10, no. 7 (1978):477.

**57.** STONER, J. B. "Horizontal Anisotropy Determined by Pumping in Two River Basin Coal Aquifers, Montana." *Ground Water,* 19 (1981):34–40.

**58.** KEEFER, W. R., and R. F. HADLEY. *Land and Natural Resource Information* *and Some Potential Environmental Effects of Surface Mining of Coal in the Gillette Area, Wyoming*. U.S. Geological Survey Circular 743, 1976, 27 pp.

**59.** DAVIS, S. N., and L. J. TURK. "Optimum Depth of Wells in Crystalline Rocks." *Ground Water,* 2, no. 2 (1964):6–11.

**60.** COATES, M. S., C. B. HAIMSON, W. J. HINZE, and W. R. VAN SCHMUSS. "Introduction to the Illinois Deep Hole Project." *Journal of Geophysical Research,* 88, B9 (1983).

**61.** HAIMSON, B. C., and T. W. DOE. "State of Stress, Permeability and Fractures in the Precambrian Granite of Northern Illinois." *Journal of Geophysical Research,* 88, B9 (1983):7355–71.

**62.** DANIELS, J. J., G. R. OLHOEFT, and J. H. SCOTT. "Interpretation of Core and Well Log Physical Property Data from Drill Hole UPH-3, Stephenson County, Illinois." *Journal of Geophysical Research,* 88, B9 (1983):7346–54.

**63.** COUTRE, R. A., M. G. STEITZ, and M. J. STEINDLER. "Sampling of Brine in Cores of Precambrian Granite from Northern Illinois." *Journal of Geophysical Research,* 88, B9 (1983):7331–34.

**64.** BRACE, W. "Permeability of Crystalline and Argillaceous Rocks." *International Journal of Rock Mechanics and Geomechanical Abstracts,* 17 (1980):241–51.

**65.** STEWART, J. W. *Relation of Permeability and Jointing in Crystalline Metamorphic Rocks Near Jonesboro, Georgia*. U.S. Geological Survey Professional Paper 450-D, 1962, pp. 168–70.

**66.** SOCHA, B. J. *Fracture Trace Analysis for Water-Well Site Locations in Precambrian Igneous and Metamorphic Rock in Central Wisconsin*. Wisconsin Geological and Natural History Survey Misc. Paper 83-5, 1983, 37 pp.

**67.** WELBY, C. W. "Ground-Water Yields and Inventory for Land-Use Planning in Crystalline Rock Areas of Wake County, North Carolina." *Water Resources Bulletin,* 20 (1984):875–82.

**68.** LE GRAND, H. E. "Perspective on Problems in Hydrogeology." *Bulletin, Geological Society of America,* 73 (1962):1147–52.

**69.** LE GRAND, H. E. *Geology and Groundwater in the Statesville Area, North Carolina*. North Carolina Department of Conservation and Development, Division of Mineral Resources Bulletin 68, 1954, 68 pp.

**70.** PETERSON, F. L. "Water Development on Tropic Volcanic Islands—Type Example: Hawaii." *Ground Water,* 10, no. 5 (1972):18–23.

**71.** LUZIER, J. E., and J. A. SKRVIAN. *Digital Simulation and Projection of Water Level Declines in Basalt Aquifers of the Odessa-Lind Area, East Central Washington*. U.S. Geological Survey Water-Supply Paper 2036, 1975, 48 pp.

**72.** NEWCOMB, R. C. *Quality of the Ground Water in Basalt of the Columbia River Group, Washington, Oregon and Idaho*. U.S. Geological Survey Water-Supply Paper 1999-N, 1972, 71 pp.

**73.** FOXWORTHY, B. L. "Pacific Northwest Region." In *Ground Water Resources of the United States,* ed. David K. Todd, Berkeley, Calif.: Premier Press, 1983, pp. 590–629.

**74.** BROWN, J. C. "Definition of the Basalt Groundwater Flow System, Horse Heaven Plateau, Washington." Geological Society of America, *Abstracts with Programs,* 11, no. 7 (1979):395.

75. COX, D. C. "Water Development for Hawaiian Sugar Cane Irrigation." *Hawaiian Planters Record,* 54 (1954):175–97.

76. VISHER, F. N., and J. F. MINK. Ground Water Resources, Southern Oahu, Hawaii. U.S. Geological Survey Water-Supply Paper 1778, 1964, 133 pp.

77. TAKASAKI, K. J. *Summary Appraisals of the Nation's Ground Water Resources—Hawaii Region.* U.S. Geological Survey Professional Paper 813-M, 1978, 29 pp.

78. CEDERSTROM, D. J., P. M. JOHNSON, and S. SUBITZKY. *Occurrence and Development of Ground Water in Permafrost Regions.* U.S. Geological Survey Circular 275, 1953, 30 pp.

79. BRANDON, L. V. "Evidences of Ground-Water Flow in Permafrost Regions." *Proceedings of the 1963 International Conference on Permafrost,* Lafayette, Indiana, 1965, pp. 176–77.

80. WILLIAMS, J. R. *Ground Water in the Permafrost Regions of Alaska.* U.S. Geological Survey Professional Paper 696, 1970, 83 pp.

81. TERZAGKI, K., "Permafrost." *Boston Society of Civil Engineers Journal,* 39 (1950):1–50.

82. LACHENBRUCH, A. H., M. C. BREWER, G. W. GREENE, and B. V. MARSHALL. "Temperatures in Permafrost." In *Temperature and Its Measurement and Control in Science and Industry,* vol. 3. American Institute of Physics. New York: Reinhold Publishing Corporation, 1962, pp. 791–803.

83. CEDERSTROM, D. J. *Ground-Water Resources of the Fairbanks Area, Alaska.* U.S. Geological Survey Water-Supply Paper 1590, 1963, 84 pp.

84. KAZMANN, R. G. *The Present and Future Ground Water Supply of the Baton Rouge Area.* Louisiana Water Resources Research Institute Bulletin 5, 1970, 44 pp. and appendices.

85. MILLER, J. A. *Hydrogeologic Framework of the Floridan Aquifer System in Florida and in Parts of Georgia Alabama, and South Carolina.* U.S. Geological Professional Paper 1403-B, 1986, 91 pp.

86. GRUBB, H. F. "Gulf Coastal Plain Regional Aquifer-System Study." In *Regional Aquifer-System Analysis Program of the United States Geological Survey,* U.S. Geological Survey Circular 1002, 1986, 152–61.

87. WAIT, R. L., and others, "Southeastern Coastal Plain Regional Aquifer-System Study." In *Regional Aquifer-System Analysis Program of the United States Geological Survey,* U.S. Geological Survey Circular 1002, 1986, pp. 205–22.

88. MEISLER, H. and others. "Northern Atlantic Coastal Plain Regional Aquifer-System Study." In *Regional Aquifer-System Analysis Program of the United States Geological Survey.* U.S. Geological Survey Circular 1002, 1986, pp. 168–94.

89. MC GUINNESS, C. L. *The Role of Groundwater in the National Water Situation.* U.S. Geological Survey Water-Supply Paper 1800, 1963, 1121 pp.

90. ROLLO, J. R. *Ground Water in Louisiana.* Louisiana Geological Survey and Louisiana Department of Public Works Water Resources Bulletin 1, 1960, 84 pp.

91. KOHOUT, F. A., et al. "Fresh Groundwater Stored in Aquifers under the Continental Shelf: Implications from a Deep Test Well, Nantucket Island, Massachusetts. *Water Resources Bulletin,* 13, (1977):373–86.

92. COLLINS, M. A. "Discussion." *Water Resources Bulletin,* 14 (1978):484–85.

**93.** PLUHOWSKI, E. J., and I. H. KANTROWITZ. *Hydrology of the Babylon-Islip Area, Suffolk County, Long Island, New York.* U.S. Geological Survey Water-Supply Paper 1768, 1964, 119 pp.

**94.** FETTER, C. W., JR. "Water Resources Management in Coastal Plain Aquifers." *Proceedings of the International Water Resources Association, First World Congress on Water Resources,* 1973, pp. 322–31.

**95.** PARKER, G. G., R. H. BROWN, D. G. BOGART, and S. K. LOVE. "Salt Water Encroachment." In *Water Resources of Southeastern Florida.* U.S. Geological Survey Water-Supply Paper 1255, 1955, pp. 571–711.

**96.** CEDERSTROM, D. J. "Ground Water in the Aden Sector of Southern Arabia." *Ground Water,* 9, no. 2 (1971):29–34.

**97.** HARSHBARGER, J. W. "Ground-Water Development in Desert Areas," *Ground Water,* 6, no. 5 (1968):2–4.

**98.** LLOYD, J. W., and M. H. FARAG. "Fossil Groundwater-Gradients in Arid Regional Sedimentary Basins." *Ground Water,* 16 (1978):388–93.

**99.** MEINZER, O. E. *The Occurrence of Groundwater in the United States, with a Discussion of Principles.* U.S. Geological Survey Water-Supply Paper 489, 1923, 321 pp.

**100.** THOMAS, H. E. "Ground Water Regions of the United States—Their Storage Facilities." U.S. 83rd Congress, House Interior and Insular Affairs Committee, *The Physical and Economic Foundation of Natural Resources,* vol. 3, 1952, 78 pp.

**101.** MC GUINESS, C. L. *The Role of Ground Water in the National Water Situation.* U.S. Geological Survey Water-Supply Paper 1800, 1963, 1121 pp.

**102.** HEATH, R. C. *Ground Water Regions of the United States.* U.S. Geological Survey Water-Supply Paper 2242, 1984, 78 pp.

# NINE

# Water Chemistry

They [clouds] are often wafted about and borne by the winds from one region to another, where by their density they become so heavy that they fall in thick rain; and if the heat of the sun is added to the power of the element of fire, the clouds are drawn up higher still and find a greater degree of cold, in which they form ice and fall in storms of hail. Now the same heat which holds up so great a weight of water as is seen to rain from the clouds, draws them from below upwards, from the foot of the mountains, and leads and holds them within the summits of the mountains and these, finding some fissure, issue continously and cause rivers.

Leonardo da Vinci (1452–1519),
in *The Literary Works of Leonardo da Vinci*, J. P. Richter, ed., 1939

## 9.1 INTRODUCTION

For most of its uses, the chemical properties of water are as important as the physical properties and available quantity. In the next chapter, we will consider water quality, which involves the type and amount of substances dissolved in the water. In this chapter, our focus will be the chemical reactions that occur between water and the solids and gases it contacts.

Natural waters are never pure; they always contain at least small amounts of dissolved gases and solids. The composition of the aqueous solution is a function of a multiplicity of factors; for example, the initial composition of the water, the partial pressure of the gas phase, the type of mineral matter the water contacts, and the pH and oxidation potential of the solution. Water containing a biotic assemblage has an even more complex chemistry owing to the life processes of the biota.

The detailed study of water chemistry is far beyond the scope of this chapter. We will concentrate on the aspect of solubility of gases and liquids in dilute aqueous solutions. Further, we will assume that all reactions take place at a temperature of 25° C and a pressure of 1 atmosphere. Small deviations from this assumption (a few atmospheres pressure and $\pm 10°$ to $15°$ C) will not lead to significant error (1). The systems considered will be presumed to be abiotic.

It is very difficult to collect a sample of ground water that is actually representative of the chemistry of the water as it exists in the ground. The process of drawing the sample up from the aquifer by means of a well and pump can change the pressure of the water. The sample may also be exposed to atmospheric oxygen during the sampling process. As a result, the *Eh*, pH, and equilibrium conditions of the water can change. Ground-water monitoring and methods of collecting representative ground-water samples are discussed in Chapter 10.

## 9.2 UNITS OF MEASUREMENT

Chemical analysis of an aqueous solution yields the amount of solute in a specified amount of water. There are several ways in which this can be reported.

Weight per weight units are dimensionless ratios of the weight of the solute divided by the weight of the solvent; for example, **parts per million** (ppm) and **parts per billion** (ppb). These units are no longer commonly used.

Weight per volume units are the more commonly used units today. They are expressed in terms of weight of solute per volume of water. Common units are **milligrams per liter** (mg/L) and **micrograms per liter** (μg/L). As a liter of pure water contains one million milligrams at 3.89° C, the temperature where it is most dense, it is commonly assumed that 1 ppm is equal to 1 mg/L. The density and hence weight of a liter of water will change with temperature and dissolved mineral matter. However, as a practical matter, the density corrections are necessary only if the dissolved solids of the water are in excess of 7000 mg/L (1).

**Equivalent weight** units are very handy when the chemical behavior of the solute is being considered. The equivalent, or combining, weight of a dissolved ionic species is the formula weight divided by the electrical charge. By dividing a concentration in milligrams per liter by the equivalent weight of the ion, the result is a concentration expressed in **milliequivalents per liter** (meq/L). Dissolved species such as silica, which is not ionic, cannot be expressed in meq/L.

In chemical thermodynamics, units of **molality** are useful. One **mole** of a compound is its formula weight in grams. A solution of one mole per 1000 grams of solvent is a one-**molal** solution; a solution of one mole of solute per liter of solvent is a one-**molar** solution. For dilute solutions, up to about 0.01 molal, the concentration expressed in either molality or molarity is equal.

For dilute solutions, it is not necessary to make density corrections; the following conversion factors may be used (2):

$$\text{Molality} = \frac{\text{milligrams per liter} \times 10^{-3}}{\text{formula weight in grams}} \qquad \textbf{(9-1)}$$

$$\text{Molality} = \frac{\text{milliequivalents per liter} \times 10^{-3}}{\text{valence of ion}} \quad \textbf{(9-2)}$$

**EXAMPLE PROBLEM**

**Part A:** What is the weight of NaCl in a 0.01-molal solution?

Atomic weight of sodium = 22.991 g

Atomic weight of chlorine = 35.457 g

One mole of NaCl = 58.448 g

0.01 mole = 0.01 × 58.448 = 0.58448 g

0.01-molal solution = 0.58448 g NaCl in 1000 g $H_2O$

**Part B:** What is the concentration of NaCl in a 0.01-molal solution at 25° C?

At 25° C, the density of water is 0.99707 gram per milliliter. The volume of 1000 grams is 1000 g/0.99707 g/mL, or 1002.94 mL. The concentration is 0.58448 g in 1.00294 L or 582.8 mg/L.

## 9.3 TYPES OF CHEMICAL REACTIONS IN WATER

Chemical reactions in an aqueous solution are either **reversible** or **irreversible.** Those that are reversible can reach equilibrium with their hydrochemical environment and are amenable to study by kinetic and thermodynamic methods.

The simplest aqueous reaction is the dissociation of an inorganic salt. If the salt is present in excess, it will tend to form a saturated solution:

$$NaCl \rightleftharpoons Na^+ + Cl^- \quad \textbf{(9-3)}$$

Natural systems always tend toward equilibrium; thus, if the solution is undersaturated, more salt will dissolve. If it is supersaturated, salt will crystallize, although for kinetic reasons the solution may remain supersaturated. Notice that the water molecule does not actively participate in this reaction.

Water molecules can actively bond with either a gas or solid in a reversible reaction:

$$CaCO_3 + H_2O \rightleftharpoons Ca^{++} + HCO_3^- + OH^- \quad \textbf{(9-4)}$$

$$CO_2 + H_2O \rightleftharpoons HCO_3^- + H^+ \quad \textbf{(9-5)}$$

In this type of reaction, the water molecule breaks into $H^+$ and $OH^-$ radicals when combining with the species in solution.

Reversible oxidation-reduction reactions may also involve the transfer of electrons from one ion to another. When this happens, the species undergo a valence change:

$$4Fe^{++} + 3O_2 + 8e^- \rightleftharpoons 2Fe_2O_3 \quad \textbf{(9-6)}$$

In this example, the ferrous iron is oxidized to ferric iron by the transfer of an electron from the iron to the oxygen.

A gas or solid may also dissolve in an aqueous solution without dissociation:

$$O_{2\ (gas)} \rightleftharpoons O_{2\ (aqueous)} \tag{9-7}$$

## 9.4 LAW OF MASS ACTION

If a reversible reaction can go in either of two directions, which way will it go? The answer to this basic question is found in the **law of mass action,** which suggests that the reaction will strive to reach equilibrium. In an aqueous mixture, both reactions are occurring simultaneously:

$$A + B \rightharpoonup C + E \tag{9-8}$$

and

$$C + D \rightharpoonup A + B \tag{9-8}$$

At chemical equilibrium, the two rates are equal; thus, if the mixture is not at chemical equilibrium, it will proceed in the direction that produces equilibrium. A chemical reaction may be expressed as

$$c\mathrm{C} + d\mathrm{D} \rightleftharpoons x\mathrm{X} + y\mathrm{Y} \tag{9-9}$$

where capital letters represent chemical constituents and lowercase letters represent coefficients. The **equilibrium concentration** of each chemical formula is $[X]$, and the **equilibrium constant,** $K$, for the given reaction is

$$K = \frac{[X]^x[Y]^y}{[\mathrm{C}]^c[\mathrm{D}]^d} \tag{9-10}$$

where $[X]$ represents the molal concentration of the $X$-ion. An equilibrium constant is valid only for a specific chemical reaction. It is either experimentally determined or calculated from thermodynamic properties. In equilibrium studies, the value of the concentration of a pure liquid or solid is defined as 1.

If AgCl is dissolved in water, it will eventually saturate the water and no more will dissolve. The reaction is

$$AgCl \rightleftharpoons Ag^+ + Cl^- \tag{9-11}$$

The equilibrium reaction is given by

$$K_{sp} = \frac{[Ag^+][Cl^-]}{[AgCl]} \tag{9-12}$$

The equilibrium constant for a slightly soluble salt is termed the **solubility product,** $K_{sp}$. The experimentally determined value of $K_{sp}$ for the reaction is $10^{-9.8}$. Since [AgCl] is defined as 1,

$$K_{sp} = [Ag^+][Cl^-] = 10^{-9.8}$$

---

**EXAMPLE PROBLEM** What is the solubility of $Ag^+$ at equilibrium?

The two ions have equal solubility:

$$[Ag^+] = [Cl^-] = \text{Solubility}$$

The product of $[Ag^+][Cl^-]$ is the square of the solubility of either ion; thus, the solubility of either ion is the square root of the equilibrium constant:

$$\text{Solubility} = \sqrt{K_{sp}} = \sqrt{10^{-9.8}} = 1.26 \times 10^{-5} \text{ mole}$$

The solubility of $Ag^+$ is $1.26 \times 10^{-5}$ mole per liter.

---

The situation is more complex if it involves a salt, such as $PbCl_2$. The reaction is

$$PbCl_2 \rightleftharpoons Pb^{++} + 2Cl^-$$

and the solubility product is given by

$$K_{sp} = \frac{[Pb^{++}][Cl^-]^2}{[PbCl_2]} \quad \textbf{(9-13)}$$

One mole of $PbCl_2$ yields one mole of $Pb^{++}$ and two moles of $Cl^-$. To solve the equation for the solubility, $X$, of $PbCl_2$, use the expression

$$K_{sp} = [X][2X]^2 \quad \textbf{(9-14)}$$

The value of $K_{sp}$ is $10^{-4.8}$ and, $X$, the solubility of $PbCl_2$, is found from

$$K_{sp} = 4X^3$$
$$X = \sqrt[3]{K_{sp}/4} = \sqrt[3]{0.25 \times 10^{-4.8}}$$
$$= 0.0158 \text{ mole}$$

---

**EXAMPLE PROBLEM** One thousand grams of water will dissolve $1.0 \times 10^{-4}$ mole of $PbSO_4$. Calculate $K_{sp}$.

$$[Pb^{++}] = [SO_4^{=}] = 1 \times 10^{-4} \text{ mole}$$
$$K_{sp} = [Pb^{++}][SO_4^{=}] = 10^{-8}$$

---

## 9.5 COMMON ION EFFECT

If the solvent contains another source for an ion also present in a salt, the **common ion effect** will reduce the solubility of the salt. This applies to any salt in equilibrium with its saturated solution. If we dissolve AgCl in two solutions, one pure water and one containing 0.1 mole of NaCl, less of the AgCl will dissolve in the solution of NaCl. The solubility of NaCl is many orders of magnitude greater than that of the AgCl and is not affected. The total amount of the common ion in solution controls the amount of the less soluble salt that can dissolve. For example, consider the solution of AgCl in the 0.1-molal solution of NaCl. There will be $X$ moles of AgCl and 0.1 mole of $Cl^-$ from the NaCl. Thus, there are $X$ moles of $Ag^+$ and $X + 0.1$ moles of $Cl^-$:

$$K_{sp} = [Ag^+][Cl^-] = [X][X + 0.1] = 10^{-9.8}$$

and

$$[0.1X] + [X^2] = 10^{-9.8}$$

Since $[X]$ is small, $[X^2]$ is very small and can be ignored; hence,

$$[X] = 10^{-8.8} = 1.58 \times 10^{-9}$$

The solubility of AgCl in pure water is $1.26 \times 10^{-5}$ mole, while in a 0.1-molal solution of NaCl, it is only $1.58 \times 10^{-9}$ mole. In general, ground and surface waters contain ions from many sources, so that the common ion effect must be considered.

## 9.6 CHEMICAL ACTIVITIES

In very dilute aqueous solutions, the molal concentrations can be used to determine equilibrium and solubility. For the general case, chemical activities must be computed from the concentration before the law of mass action can be applied. This is due to the fact that electrostatic forces cause the behavior of the solutes to be nonideal.

The **chemical activity** of an ion is equal to the molal concentrations times a factor known as the **activity coefficient:**

$$\alpha = \gamma m \qquad \textbf{(9-15)}$$

where

$\alpha$ is the chemical activity
$m$ is the molal concentration
$\gamma$ is the activity coefficient

In order to compute the activity coefficient of an individual ion, the **ionic strength** of the solution must be determined. For a mixture of electrolytes in solution, the ionic strength is given by

$$I = \frac{1}{2} \sum m_i z_i^2 \quad \textbf{(9-16)}$$

where

$I$ is the ionic strength
$m_i$ is the molality of $i$th ion
$z_i$ is the charge of $i$th ion

The ionic strength of 0.2-molal solution of $CaCl_2$ is

$$\begin{aligned} I &= \tfrac{1}{2}(m_{Ca^{++}} \times 2^2 + m_{Cl^-} \times 1^2) \\ &= \tfrac{1}{2}(0.2 \times 2^2 + 0.4 \times 1^2) = 0.6 \end{aligned}$$

**EXAMPLE PROBLEM** Compute the ionic strength of ground water from a Cambrian-age sandstone in Neenah, Wisconsin.

Chemical Analysis (mg/L)
(major ions only)

| $Ca^{++}$ | $Mg^{++}$ | $HCO_3^-$ | $SO_4^=$ |
|---|---|---|---|
| 234 | 39 | 290 | 498 |

The concentrations must be converted to molality by Equation 9-1:

Chemical Analysis (molalities)

| $Ca^{++}$ | $Mg^{++}$ | $HCO_3^-$ | $SO_4^=$ |
|---|---|---|---|
| 0.00584 | 0.0016 | 0.00475 | 0.00518 |

The ionic strength is then computed using Equation 9-16:

$$\begin{aligned} I &= \tfrac{1}{2}(0.00584 \times 2^2 + 0.0016 \times 2^2 + 0.00475 \times 1^2 + 0.00518 \times 2^2) \\ &= 0.0276 \end{aligned}$$

Once the ionic strength of a solution of electrolytes is known, the activity coefficient of the individual ion can be determined from the **Debye-Hückel equation:**

$$-\log \gamma_i = \frac{A z_i^2 \sqrt{I}}{1 + a_i B \sqrt{I}} \quad \textbf{(9-17)}$$

where

$\gamma_i$ is the activity coefficient of ionic species $i$

$z_i$ is the charge of ionic species $i$

$I$ is the ionic strength of the solution

$A$ is a constant equal to 0.5085 at 25° C

$B$ is a constant equal to 0.3281 at 25° C

$a_i$ is the effective diameter of the ion (Table 9.1)

The Debye-Hückel equation is valid for solutions with an ionic strength of 0.1 or less (approximately 5000 milligrams per liter).

Whereas Equation 9–10 is valid only for very dilute solutions where $\gamma \simeq 1$, the law of mass action, expressed in terms of activity coefficients, is valid for solutions with any ionic strength:

$$K = \frac{(\alpha_X)^x(\alpha_Y)^y}{(\alpha_C)^c(\alpha_D)^d} \tag{9-18}$$

where $cC + dD \rightleftharpoons xX + yY$ and $\alpha_x$ is the activity of the $X$-ion.

**EXAMPLE PROBLEM** Determine $\gamma_i$ and $\alpha$ for $Ca^{++}$ in a solution where the molal concentration of $Ca^{++}$ is 0.00584 and $I = 0.0276$ at 25° C. The value of $a_i$ for $Ca^{++}$ is 6.

The activity coefficient can be determined using Equation 9-17:

$$-\log \gamma_i = \frac{Az_i^2\sqrt{I}}{1 + a_iB\sqrt{I}}$$

$$\log \gamma_i = -\frac{0.5085(2)^2\sqrt{0.0276}}{1 + (6)(0.3281)\sqrt{0.0276}}$$

**TABLE 9.1** Values of the parameter $a_i$ in the Debye-Hückel equation

| $a_i$ | Ion |
|---|---|
| 11 | $Th^{+4}$, $Sn^{+4}$ |
| 9 | $Al^{+3}$, $Fe^{+3}$, $Cr^{+3}$, $H^+$ |
| 8 | $Mg^{+2}$, $Be^{+2}$ |
| 6 | $Ca^{+2}$, $Cu^{+2}$, $Zn^{+2}$, $Sn^{+2}$, $Mn^{+2}$, $Fe^{+2}$, $Ni^{+2}$, $Co^{+2}$, $Li^+$ |
| 5 | $Fe(CN)_6^{-4}$, $Sr^{+2}$, $Ba^{+2}$, $Cd^{+2}$, $Hg^{+2}$, $S^{-2}$, $Pb^{+2}$, $CO_3^{-2}$, $SO_3^{-2}$, $MoO_4^{-2}$ |
| 4 | $PO_4^{-3}$, $Fe(CN)_6^{-3}$, $Hg_2^{+2}$, $SO_4^{-2}$, $SeO_4^{-2}$, $CrO_4^{-3}$, $HPO_4^{-2}$, $Na^+$, $HCO_3^-$, $H_2PO_4^-$ |
| 3 | $OH^-$, $F^-$, $CNS^-$, $CNO^-$, $HS^-$, $ClO_4^-$, $K^+$, $Cl^-$, $Br^-$, $I^-$, $CN^-$, $NO_2^-$, $NO_3^-$, $Rb^+$, $Cs^+$, $NH_4^+$, $Ag^+$ |

Source: J. Kielland, "Individual Activity Coefficients of Ions in Aqueous Solutions," *American Chemical Society Journal*, 59 (1937):1676–78.

$$= -\frac{(0.5085)(4)(0.166)}{1 + (6)(0.3281)(0.166)}$$
$$= -0.255$$
$$\gamma_i = 0.556$$

The activity of calcium is then found from Equation 9-15:

$$\alpha = \gamma m$$
$$= (0.556)(0.00584)$$
$$= 0.00325$$

---

The **ion activity product** ($K_{iap}$), which is the product of the measured activities, can be calculated for any aqueous solution in order to test for saturation. The value of $K_{iap}$ for a mineral equilibrium reaction in a natural water may be compared with the value of $K_{sp}$, the solubility product of the mineral. If the value of $K_{iap}$ is equal to or greater than $K_{sp}$, the natural water is saturated or supersaturated with respect to the mineral. If $K_{iap}$ is less than $K_{sp}$, the solution is undersaturated with respect to the mineral, and the mineral may be actively dissolving. For the case where the mineral C is being dissolved according to the reaction $cC \rightleftharpoons xX + yY$, $K_{iap}$ is given by

$$K_{iap} = (\alpha_X)^x(\alpha_Y)^y \qquad \textbf{(9-19)}$$

Solubility products for a number of compounds are given in Appendix 11.

## 9.7 IONIZATION CONSTANT OF WATER AND WEAK ACIDS

Water undergoes a dissociation into two ionic species:

$$H_2O \rightleftharpoons H^+ + OH^-$$

In reality, a hydrogen ion ($H^+$) cannot exist; it must be in the form $H_3O^+$, the hydronium ion, formed by the interaction of water with the hydrogen ion. For convenience, however, we will represent it as $H^+$. The equilibrium constant for water is

$$K_{eq} = \frac{\alpha_{H^+}\alpha_{OH^-}}{\alpha_{H_2O}} \qquad \textbf{(9-20)}$$

For water that is neutral, there are exactly the same concentrations of $H^+$ and $OH^-$ radicals, $10^{-7}$. The negative logarithm of the concentration of $H^+$ ions in an aqueous solution is called the pH of the solution. For all aqueous solutions, either acidic or basic, the product $\alpha_{H^+}\alpha_{OH^-}$ is always $10^{-14}$ (at about 25° C). Since a neutral solution has equal amounts of $H^+$ and $OH^-$ radicals, the pH is 7. If there are more $H^+$ ions than $OH^-$ ions, the solution is acidic and the pH is less than 7.0. Basic solutions have more $OH^-$ than $H^+$ ions and a pH between 7 and 14.

---

**EXAMPLE PROBLEM** What is the $[H^+]$ and $[OH^-]$ of an aqueous solution of pH 3.2?

Since pH is the negative logarithm of $[H^+]$, the value of $[H^+]$ is $10^{-3.2}$. Since the product $[H^+][OH^-] = 10^{-14}$,

$$[OH^-] = 10^{-14}/[H^+] = 10^{-14}/10^{-3.2} = 10^{-10.8}$$

---

It is apparent that by measuring the pH of an aqueous solution, we can obtain the numerical value of both $[H^+]$ and $[OH^-]$. If the solution is nonideal, the pH meter measures the activity of $H^+$, since $\alpha_{H^+}\alpha_{OH^-} = 10^{-14}$.

An acid is a substance that can add $H^+$ (more properly, $H_3O^+$) ions to aqueous solutions. Strong acids will completely dissociate in water to release $H^+$ ions. Since the $[H^+][OH^-]$ product is constant at a given temperature, the concentration of $OH^-$ ions decreases. A 1-molal solution of HCl will have a pH of 0 and a $[H^+]$ of 1. A 0.01-molal solution will have a pH of 2, and a $[H^+]$ of $10^{-2}$. On the other hand, a 0.01-molal solution of $H_2CO_3$ will have a higher pH, as it is a weak acid. In dilute aqueous solution, the $H_2CO_3$ is only slightly broken down into ions. Weak acids with more than one $H^+$ per molecule ionize in steps; for example,

$$H_2CO_3 \rightleftharpoons H^+ + HCO_3^- \quad \textbf{(9-21a)}$$

$$HCO_3^- \rightleftharpoons H^+ + CO_3^= \quad \textbf{(9-21b)}$$

The equilibrium constants at 25° C are

$$\frac{[H^+][HCO_3^-]}{[H_2CO_3]} = K_1 = 10^{-6.4} \text{ (first ionization constant)} \quad \textbf{(9-22a)}$$

and

$$\frac{[H^+][CO_3^=]}{[HCO_3^-]} = K_2 = 10^{-10.3} \text{ (second ionization constant)} \quad \textbf{(9-22b)}$$

The value of $K_{eq}$ for water varies significantly with temperature. Table 9.2 lists the equilibrium constants for the dissociation of water at temperatures between 0° and 60° C. At 0° C, a neutral solution has a pH of 7.5; at 60° C, neutrality occurs at pH 6.6.

**TABLE 9.2** Equilibrium constants for dissociation of water

| Temperature (°C) | $K_{eq}$ | pH of a Neutral Solution |
|---|---|---|
| 0 | $0.1139 \times 10^{-14}$ | 7.47 |
| 5 | $0.1846 \times 10^{-14}$ | 7.37 |
| 10 | $0.2920 \times 10^{-14}$ | 7.27 |
| 15 | $0.4505 \times 10^{-14}$ | 7.17 |
| 20 | $0.6809 \times 10^{-14}$ | 7.08 |
| 25 | $1.008 \times 10^{-14}$ | 7.00 |
| 30 | $1.469 \times 10^{-14}$ | 6.92 |
| 35 | $2.089 \times 10^{-14}$ | 6.84 |
| 40 | $2.919 \times 10^{-14}$ | 6.77 |
| 45 | $4.018 \times 10^{-14}$ | 6.70 |
| 50 | $5.474 \times 10^{-14}$ | 6.63 |
| 55 | $7.297 \times 10^{-14}$ | 6.57 |
| 60 | $9.614 \times 10^{-14}$ | 6.51 |

**EXAMPLE PROBLEM**

What is the pH of a 0.01-molal solution of $H_2CO_3$ at 25° C?

There are five ionic species present: $H^+$, $OH^-$, $H_2CO_3$, $HCO_3^-$, and $CO_3^=$. The total of the carbonate species, $H_2CO_3$, $HCO_3^-$, and $CO_3^=$, is 0.01 mole. The 0.01-molal solution is obtained by dissolving 0.01 mole of $CO_2$ in one liter of water. For most geologic applications, some assumptions can be made to simplify the problem. From the values of $K_1$ and $K_2$, we see that $K_2$ is $10^4$ smaller than $K_1$, so almost all of the $H^+$ ions come from $H_2CO_3 \rightleftharpoons H^+ + HCO_3^-$. There will also be a very small value of $CO_3^=$, since $K_2$ is so small. Likewise, there will be relatively few $OH^-$ ions since it is an acid solution.

From the dissociation reactions, we must balance the electrical charges:

$$[H^+] = [OH^-] + [HCO_3^-] + 2[CO_3^=]$$

Since $OH^-$ and $CO_3^=$ are relatively small,

$$[H^+] \simeq [HCO_3^-]$$

From the equilibrium equation,

$$\frac{[H^+][HCO_3^-]}{[H_2CO_3]} = K_1 = \frac{[H^+]^2}{[H_2CO_3]} = 10^{-6.4}$$

Since the solution has 0.01 mole $CO_2$, total,

$$[HCO_3^-] + [H_2CO_3] + [CO_3^=] = 0.01 \text{ mole/1000 g}$$

With a small value for $CO_3^{=}$, and $HCO_3^{-}$ equal to $H^+$,

$$[H_2CO_3] + [H^+] \simeq 0.01$$

Since this is a weak acid, $[H^+]$ is very small compared with $[H_2CO_3]$, so that $[H_2CO_3] \simeq 0.01$. Putting these two results together,

$$\frac{[H^+]^2}{[H_2CO_3]} = 10^{-6.4} \text{ and } [H_2CO_3] = 0.01$$

$$[H^+]^2 = 0.01 \times 10^{-6.4} = 0.01 \times 3.98 \times 10^{-7} = 3.98 \times 10^{-9}$$

$$[H^+] = 6.31 \times 10^{-5} = 10^{-4.2}$$

Thus, pH = 4.2. Concentrations of other ions would be

$$[HCO_3^-] = [H^+] = 10^{-4.2}$$

$$[OH^-] = \frac{10^{-14}}{[H^+]} = 10^{-9.8}$$

$$[CO_3^{=}] = \frac{10^{-10.3}[HCO_3^-]}{[H^+]} = 10^{-10.3}$$

These values are accurate to ±1 percent (3).

While this type of problem can promote understanding of weak acids, in a real-world situation there may be many other ionic species present, increasing the ionic strength. This would necessitate the use of chemical activities and might also introduce the common ion effect.

## 9.8 CARBONATE EQUILIBRIUM

In hydrogeologic studies, the equilibrium of calcium carbonate in contact with natural water, either surface or ground water, is one of the most important geochemical reactions. Neutral water exposed to $CO_2$ in the atmosphere will dissolve $CO_2$ equal to the partial pressure. The $CO_2$ will react with $H_2O$ to form $H_2CO_3$, a weak acid, and the resulting solution will have a pH of about 5.7. Soil $CO_2$ from organic decomposition is another source of even more importance in ground-water studies. As calcite and dolomite are soluble in acid solution, even rainwater will dissolve carbonate rocks. Likewise, a change in pH can result in a precipitation of $CaCO_3$ from a solution that was at equilibrium prior to the pH shift. The equilibrium constants for the various reactions in carbonate equilibrium are given in Table 9.3.

**TABLE 9.3** Carbonate equilibrium constants at 25° C (4)

| |
|---|
| 1.* $CaCO_3 \rightleftharpoons Ca^{++} + CO_3^{=}$ |
| $K_{CaCO_3} = \dfrac{\alpha_{Ca^{++}}\,\alpha_{CO_3^{=}}}{\alpha_{CaCO_3}} = 10^{-8.35}$ |
| 2. $H_2CO_3 \rightleftharpoons H^+ + HCO_3^-$ |
| $K_{H_2CO_3} = \dfrac{\alpha_{H^+}\,\alpha_{HCO_3^-}}{\alpha_{H_2CO_3}} = 10^{-6.4}$ |
| 3. $HCO_3^- \rightleftharpoons H^+ + CO_3^{=}$ |
| $K_{HCO_3^-} = \dfrac{\alpha_{H^+}\,\alpha_{CO_3^{=}}}{\alpha_{HCO_3^-}} = 10^{-10.3}$ |
| 4.† $H_2O + CO_2 \rightleftharpoons H_2CO_3$ |
| $K_{CO_2} = \dfrac{\alpha_{H_2CO_3}}{P_{CO_2}} = 10^{-1.5}$ |

*The solubility product for calcite is $10^{-8.35}$. For aragonite, it is $10^{-8.22}$.

†$P_{CO_2}$ is the partial pressure of carbon dioxide, which for most hydrogeologic conditions is equal to the gas activity.

### 9.8.1 CARBONATE EQUILIBRIUM IN WATER WITH FIXED PARTIAL PRESSURE OF $CO_2$

Water in streams and lakes is in contact with the atmosphere, in which $CO_2$ is present. The gas is dissolved in the water, adding to the carbonate content of the water. The system is described by

$$H_2O + CO_2 \rightleftharpoons H_2CO_3 \quad \textbf{(9-23a)}$$

$$H_2CO_3 \rightleftharpoons H^+ + HCO_3^- \quad \textbf{(9-23b)}$$

$$HCO_3^- \rightleftharpoons H^+ + CO_3^{=} \quad \textbf{(9-23c)}$$

$$CaCO_3 \rightleftharpoons Ca^{++} + CO_3^{=} \quad \textbf{(9-23d)}$$

$$H_2O \rightleftharpoons H^+ + OH^- \quad \textbf{(9-23e)}$$

The system is electrically neutral, so that

$$2m_{Ca^{++}} + m_{H^+} = 2m_{CO_3^{=}} + m_{HCO_3^-} + m_{OH^-} \quad \textbf{(9-23f)}$$

The partial pressure of $CO_2$ in the atmosphere is $10^{-3.5}$. The value of the activity of $H_2CO_3$ can be computed from the activity product of $CO_2$:

$$K_{CO_2} = \frac{\alpha_{H_2CO_3}}{P_{CO_2}} = 10^{-1.5}$$

Therefore,

$$\alpha_{H_2CO_3} = P_{CO_2} \times 10^{-1.5} = 10^{-3.5} \times 10^{-1.5} = 10^{-5.0}$$

The activity values for the remaining ionic species can be determined in relationship to $\alpha_{H^+}$. This enables computation of the pH of a solution of calcite that is in equilibrium with atmospheric $CO_2$. The $H_2CO_3$ will dissociate into $H^+$ and $HCO_3^-$:

$$K_{H_2CO_3} = \frac{\alpha_{H^+}\alpha_{HCO_3^-}}{\alpha_{H_2CO_3}} = 10^{-6.4}$$

As we know, $\alpha_{H_2CO_3} = 10^{-5.0}$:

$$\alpha_{HCO_3^-} = (10^{-6.4} \times 10^{-5.0})/\alpha_{H^+} = 10^{-11.4}/\alpha_{H^+}$$

The $HCO_3^-$ will further dissociate into $H^+$ and $CO_3^=$:

$$K_{HCO_3^-} = \frac{\alpha_{H^+}\alpha_{CO_3^=}}{\alpha_{HCO_3^-}} = 10^{-10.3}$$

This can be rearranged to give $\alpha_{CO_3^=}$:

$$\alpha_{CO_3^=} = \frac{(\alpha_{HCO_3^-})(10^{-10.3})}{\alpha_{H^+}}$$

$$= \frac{10^{-11.4}}{\alpha_{H^+}} \times \frac{10^{-10.3}}{\alpha_{H^+}} = \frac{10^{-21.7}}{(\alpha_{H^+})^2}$$

The equilibrium constant for calcite is $10^{-8.3}$. For the solid phase of a substance at saturation, $\alpha = 1$; therefore $\alpha_{CaCO_3} = 1$:

$$K_{CaCO_3} = \frac{\alpha_{Ca^{++}}\alpha_{CO_3^=}}{\alpha_{CaCO_3}} = 10^{-8.3}$$

The activity of $Ca^{++}$ can be found as

$$\alpha_{Ca^{++}} = \frac{10^{-8.3}}{\alpha_{CO_3^=}} = \frac{10^{-8.3}}{10^{-21.7}/(\alpha_{H^+})^2} = 10^{13.4}(\alpha_{H^+})^2$$

By definition, $\alpha_{OH^-} = 10^{-14}/\alpha_{H^+}$.

For very dilute solutions $\gamma_i \simeq 1$; therefore $m_i = \alpha_i$, and the equation for electrical neutrality can be expressed as

$$2\alpha_{Ca^{++}} + \alpha_{H^+} = 2\alpha_{CO_3^=} + \alpha_{HCO_3^-} + \alpha_{OH^-}$$

Expressions for $\alpha_{Ca^{++}}$, $\alpha_{CO_3^=}$, $\alpha_{HCO_3^-}$, and $\alpha_{OH^-}$ have been determined with respect to $\alpha_{H^+}$. The equation for electrical neutrality can be determined to be

$$2[10^{13.4}(\alpha_{H^+})^2] + \alpha_{H^+} = 2[10^{-21.7}/(\alpha_{H^+})^2] + 10^{-11.4}/\alpha_{H^+} + 10^{-14}/\alpha_{H^+}$$

Solution of the above yields $\alpha_{H^+} = 10^{-8.4}$.

Thus, the pH of a solution open to atmospheric pH and in equilibrium with calcite is 8.4. This is lower than the pH of a solution of calcite with no external source of $CO_2$, which is about 9.9. This suggests that field measurements of pH should always be made, especially if the water is from a source with no external $CO_2$. Exposure of such a sample to the atmosphere for more than a few minutes would result in a lowering of the pH. Ground-water samples should always be tested for pH in the field as soon as the sample is collected. As a practical matter it is quite difficult to collect representative ground-water samples that don't undergo reactions during the process of collection.

### 9.8.2 CARBONATE EQUILIBRIUM WITH EXTERNAL pH CONTROL

In most ground-water and many surface-water bodies, there are ionic species other than $H_2CO_3$ that influence or control the pH. The hydrogeologist often has a set of chemical analyses and a measured pH for the total solution. If pH, total calcium, total carbonate, and ionic strength are known, the ion activity product, $K_{iap}$, can be calculated and compared with the solubility product, $K_{sp}$, to determine whether or not the water is in equilibrium with calcite. If $K_{iap}/K_{sp}$ is greater than 1, the solution is supersaturated; if less than 1, it is undersaturated; if equal to 1, the solution is in equilibrium.

**EXAMPLE PROBLEM** Determine whether the sample of ground water represented by the following analysis is saturated with respect to calcite. The field pH is 7.15. The total dissolved solids (TDS) is 371 milligrams per liter.

| | $Ca^{++}$ | $Mg^{++}$ | $Na^{+}$ | $K^{+}$ | $HCO_3^{-}$ | $SO_4^{=}$ | $Cl^{-}$ | $NO_3$ |
|---|---|---|---|---|---|---|---|---|
| Concentration (mg/L) | 82 | 9 | 25 | 7.6 | 252 | 17 | 40 | 38 |
| Molality × $10^3$ | 2.046 | 0.37 | 1.087 | 0.194 | 4.13 | 0.177 | 1.128 | 0.613 |

1. Calculate the ionic strength.

$$I = \tfrac{1}{2}(0.002046 \times 2^2 + 0.00037 \times 2^2 + 0.001087 + 0.000194 + 0.00413 + 0.000177 \times 2^2 + 0.001128 + 0.000613)$$
$$= 0.0088$$

2. Compute $\gamma_i$ for $Ca^{++}$, $HCO_3^-$, and $CO_3^=$. For water at 25° C, the Debye-Hückel equation is

$$\log \gamma_i = -\frac{0.5085 z_i^2 \sqrt{I}}{1 + 0.3281 a_i \sqrt{I}}$$

Values of $a_i$ (from Table 9.1) are

$$Ca^{++} = 6 \qquad HCO_3^{-} = 4 \qquad CO_3^{=} = 5$$

$$\log \gamma_{CA^{++}} = -\frac{0.5085(2)^2\sqrt{0.0088}}{1 + 0.3281(6)\sqrt{0.0088}}$$

$$= -0.161$$

$$\gamma_{Ca^{++}} = 0.690$$

$$\alpha_{Ca^{++}} = m_{Ca^{++}}\gamma_{Ca^{++}} = 0.00246 \times 0.690 = 0.0017 = 10^{-2.77}$$

$$\log \gamma_{HCO_3^-} = -\frac{0.5085(1)\sqrt{0.0088}}{1 + 0.3281(4)\sqrt{0.0088}}$$

$$= -0.0425$$

$$\gamma_{HCO_3^-} = 0.907$$

$$\alpha_{HCO_3^-} = m_{HCO_3^-}\gamma_{HCO_3^-} = 0.00413 \times 0.907 = 0.003746 = 10^{-2.43}$$

$$\log \gamma_{CO_3^=} = -\frac{0.5085(2)^2\sqrt{0.0088}}{1 + 0.3281(5)\sqrt{0.0088}}$$

$$= -0.165$$

$$\gamma_{CO_3^=} = 0.683$$

The molality of $CO_3^=$ is below detectable limits, but activity can be computed since

$$\frac{\alpha_{H^+}\alpha_{CO_3^=}}{\alpha_{HCO_3^-}} = 10^{-10.3}$$

From the pH, $\alpha_{H^+} = 10^{-7.15}$

$$\alpha_{CO_3^=} = \frac{10^{-10.3} \times 10^{-2.43}}{10^{-7.15}} = 10^{-5.58}$$

3. The calculated ion activity product is

$$K_{iap} = \alpha_{Ca^{++}}\alpha_{CO_3^=}$$
$$= 10^{-2.77} \times 10^{-5.58} = 10^{-8.35}$$

4. The value of $K_{sp}$ for calcite is $10^{-8.35}$. By comparing the ratio $K_{iap}/K_{sp}$, we determine whether or not the solution is saturated:

$$K_{iap}/K_{sp} = 10^{-8.35}/10^{-8.35} = 10^0 = 1.00$$

The water is saturated with respect to calcite.

## 9.9 FREE ENERGY

One of the functions in chemical thermodynamics is termed **free energy** (also known as **Gibbs free energy**). It is a measure of the driving energy of a reaction.

At standard conditions, the standard Gibbs free energy of a reaction $\Delta G_r^0$ is the difference between the sum of the free energy of the products and the sum of the free energy of the reactants:

$$\Delta G_r^0 = \sum \Delta_r^0 \text{ products} - \sum \Delta G_r^0 \text{ reactants} \quad \textbf{(9-24)}$$

It is related to the equilibrium constant by the formula

$$\Delta G_r^0 = -RT \ln K_{sp} \quad \textbf{(9-25)}$$

where

$R$ is the gas constant, which is a conversion factor equal to 0.00199 kcal/(mole · °K)

$T$ is the temperature in degrees Kelvin

$\Delta G_r^0$ is in kcal/mole

This is a useful relationship. Since the values of $\Delta G_r^0$ have been measured for many reactions, the value of $K_{sp}$ can be computed if $\Delta G_r^0$ is known. At 1 atmosphere pressure, 25° C, and in base 10 logs (1),

$$\log K_{sp} = -\frac{\Delta G_r^0}{1.364} \quad \textbf{(9-26)}$$

## 9.10 OXIDATION POTENTIAL

For chemical reactions in which electrons are transfered from one ion to another (**oxidation-reduction reactions**, or redox reactions), the oxidation potential of an aqueous solution is called the *Eh*. A transfer of electrons is an electrical current; therefore a redox equation has an electrical potential. At 25° C and 1 atmosphere pressure, the standard potential, $E^0$ (in volts), has been measured for many reactions. The sign of the potential is positive if the reaction is oxidizing and negative if it is reducing. The absolute value of $E^0$ is a measure of the oxidizing or reducing tendency.

The oxidation potential of a reaction is given by the **Nernst equation:**

$$Eh = E^0 + \frac{RT}{nF} \ln K_{sp} \quad \textbf{(9-27)}$$

where

$R$ is the gas constant, 0.00199 kcal/(mole · °K)

$T$ is the temperature (degrees Kelvin)

$F$ is the Faraday constant, 23.1 kcal/volt

$n$ is the number of electrons

Oxidation potential is measured with a specific ion electrode meter. A positive value indicates that the solution is oxidizing; a negative value indicates that it is chemically reducing.

If the pH and *Eh* of an aqueous solution are known, the stability of minerals in contact with the water may be determined. This stability relationship is best represented on an *Eh*-pH diagram. Water, itself, is stable only in a certain part of the *Eh*-pH field. Figure 9.1 shows the framework of aqueous *Eh*-pH fields. Water in nature at near-surface environments is usually between pH 4 and pH 9, although values that are more acid or more basic can occur.

The *Eh*-pH diagram can be used to show the fields of stability for both solid and dissolved ionic species. It has been used very effectively for iron. The *Eh*-pH diagram depends upon the concentrations of all ionic species present. For the simple ions and hydroxides of iron, the fields depend upon the molality of the iron in solution. Figure 9.2 shows the stability-field diagram for a $10^{-7}$-molal solution of iron. The iron may be either in the $Fe^{+++}$ or $Fe^{++}$ valence state,

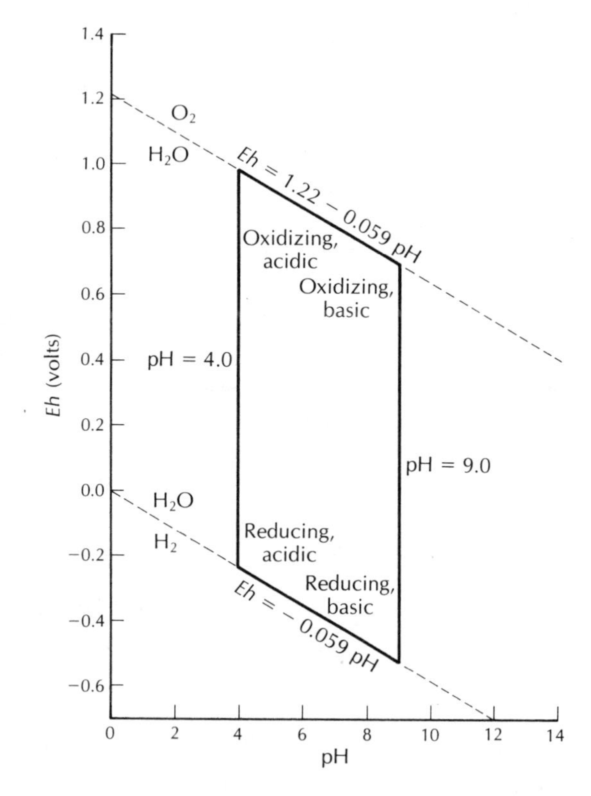


**FIGURE 9.1** The *Eh*-pH field where water is a stable component. The usual limits of *Eh* and pH for near-surface environments are also indicated by solid lines. Source: K. Krauskopf, *Introduction to Geochemistry* (New York: McGraw-Hill Book Company, 1967). Used with permission.

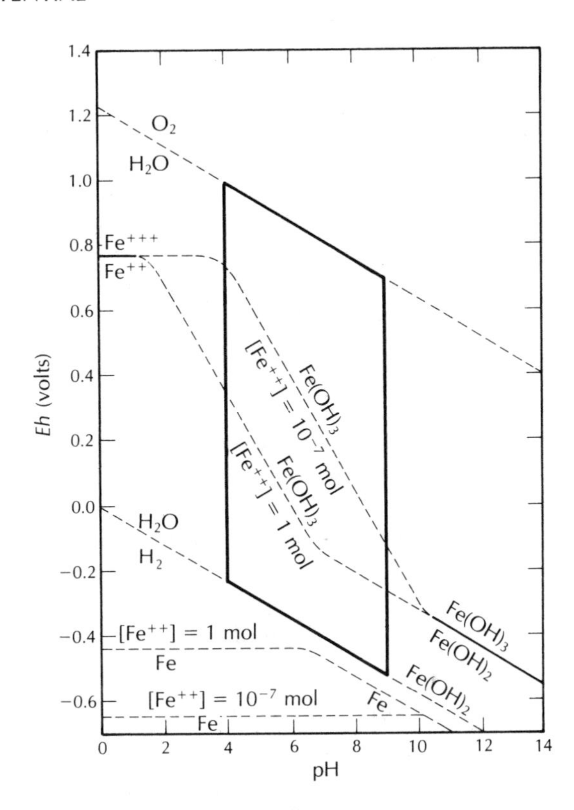


**FIGURE 9.2** Stability-field diagram for a $10^{-7}$-molal solution of iron. Source: K. Krauskopf, *Introduction to Geochemistry* (New York: McGraw-Hill Book Company, 1967). Used with permission.

depending upon its position in the stability field. As the iron concentration increases, the line separating the ferrous and ferric state shifts to the left. This is demonstrated by a dashed line in Figure 9.2, which represents a 1-molal iron concentration. Ferrous iron can exist as $Fe^{++}$, $Fe(OH)^{+}$, or $Fe(OH)_2$, depending upon the *Eh* and pH of the solution; ferric iron can be in the forms $Fe^{+++}$, $Fe(OH)_2^{+}$, and $Fe(OH)_3$. The pH at which these species change is also a function of the total amount of iron present. Procedures are available to compute an *Eh*-pH field for iron of any molality (5).

Of practical concern is the great difficulty in measuring the *Eh* of ground water under field conditions. Even for spring water, the measured *Eh* has been shown to be too great for the amount of ferrous iron in the sample (5). With very careful work, oxygen can be excluded from the sampling procedure and field *Eh* measured (6).

High *Eh* is generally the direct result of dissolved oxygen in the water. For deep ground-water systems, the *Eh* is usually sufficiently low that, for a pH of less than about 8, iron is present as the soluble $Fe^{++}$ ion. Near a recharge zone, the ground water may have sufficient dissolved oxygen to elevate the *Eh*. As the water travels through the aquifer, the oxygen is chemically reduced by contact with reducing species, and the *Eh* is lowered. The oxygen can react with the small amount of ferrous iron to form ferric hydroxide, $Fe(OH)_3$. Interestingly, the ferric hydroxide thus formed may be colloidal, and can move through the aquifer with the ground water. In the *Eh*-pH range where $Fe^{++}$ exists, large amounts of dissolved iron can be present.

Natural waters contain many ionic species. Again, using iron as an example, an *Eh*-pH diagram can be used to show the stable iron minerals in a mixed aqueous solution with iron, sulfur, and carbonate present. This is done in Figure 9.3; the given concentrations are iron 56 μg/L ($10^{-6.00}$ molar), 96 mg/L dissolved sulfur as $SO_4^{2-}$, and 61 mg/L dissolved carbon dioxide as $HCO_3^-$. Shaded areas indicate *Eh*-pH domains where solid species would be thermodynamically stable. The stable ionic species are also indicated on the figure. For a thorough discussion of *Eh*-pH diagrams, references 3–5, 7, and 8 are suggested.

Because of the difficulty of measuring *in situ Eh* in ground water, there is not a great deal of information on the *Eh*-pH range of natural ground waters. Some data do suggest that a range of *Eh* from − 0.2 to +0.7 volt can occur (6, 8). In one study, the measured *Eh* ranged from −0.04 to +0.7 volt in ground water found in a single county in Maryland (6). The ground water was in a coastal plain aquifer with a regional flow pattern toward the sea. The highest oxidation potentials were in shallow ground water of recharge areas. *Eh* was found to decrease with increasing length of flow from the aquifer recharge area. As might be expected, an inverse relationship was found between the oxidation potential and the amount of iron in solution. In the same study, field pH was found to range from 3.20 to 7.79, although, in general, higher and lower values are possible. For example, water draining from mineral deposits or mines can have a pH as low as 2 (8).

Surface waters are usually oxidizing, although low *Eh* can occur in the anaerobic depths of some lakes. The pH of surface waters typically is in the range of 4 to 10 (8).

## 9.11 SURFACE PHENOMENA

The ionic species present in aqueous solution can react with the surfaces of solids and the rock and solid particles. The surface area of a soil or sediment increases with decreasing grain size, so that the clays are the most reactive. There are two separate processes: **ion exchange,** in which an ion in the mineral lattice is replaced by one of the ions in the aqueous solution; and **adsorption,** in which the solid surface attracts and retains a layer of ions from the solution. These processes may be important in determining the chemical composition of natural ground waters. In addition, potential contaminants from such sources as septic tanks, sanitary landfills, and wastewater may be attenuated, at least in part, by passing through soil (9).

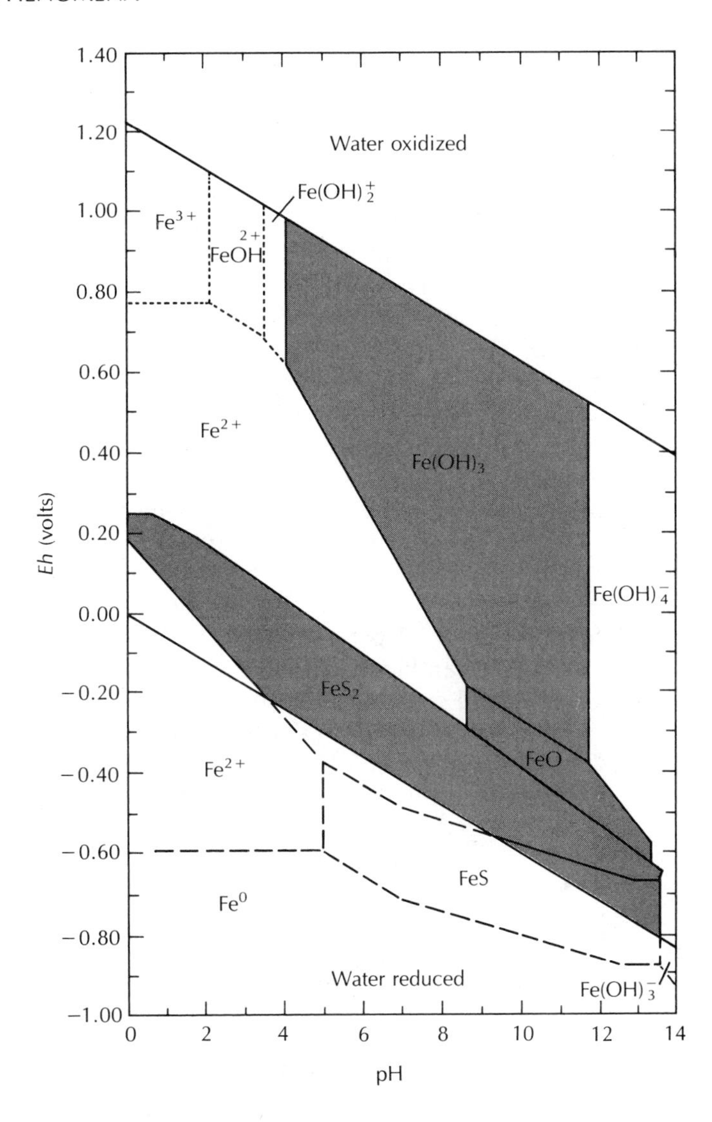


**FIGURE 9.3** Stability fields based on *Eh* and pH for solid and dissolved forms of iron in an aqueous solution of 56 μg/L iron, 96 mg/L of sulfur as $SO_4^{-2}$, 61 mg/L carbon dioxide as $HCO_3^-$ at 25° C, and one atmosphere pressure. Source: J. D. Hem, U.S. Geological Survey Water-Supply Paper 2254, 1985.

### 9.11.1 ADSORPTION

The surfaces of solids, especially clays, have an electrical charge due to isomorphous replacement, broken bonds, or lattice defects (10). The electrical charge is imbalanced, and may be satisfied by adsorbing a charged ion. The adsorption may be relatively weak, essentially a physical process caused by van der Waals forces. It may be stronger if chemical bonding occurs between the surface and the ion. Clays tend to be strong adsorbers, since they have both a high surface area per unit volume and significant electrical charges at the surface.

Most clay minerals have an excess of imbalanced negative charges in the crystal lattice. Adsorptive processes in soils thus favor the adsorption of cations. Divalent cations are usually more strongly adsorbed than monovalent ions. Some positively charged sites exist, but they are not as abundant as negative sites. In addition, some common negatively charged ions, such as $HCO_3^-$, $SO_4^=$, and $NO_3^-$, are too large to be effectively adsorbed. Chloride ions are larger than the common cations (1).

The adsorptive capacity of specific soils or sediments is usually determined experimentally. It is a function of mineralogy, particle size, ambient temperature, soil moisture, tension, pH, *Eh*, and activity of the ion (11–15). Equal weights of air-dried soil are shaken in solutions of varying activities of a particular ion. The amount of the ion adsorbed will be proportional to the activity. Analysis of the aqueous solution in equilibrium with the soil yields an equilibrium concentration, $C$. Knowing the initial concentration of the solute, the weight of ion removed per unit dry weight of soil, $C^*$, can be determined.

The concentration of solute remaining in solution, $C$, is a function of the amount adsorbed onto the solid surface, $C^*$. A graphical plot of $C$ as a function of $C^*$ is known as an **adsorption isotherm.** When plotted on log-log paper the data for many trace level solutes in contact with geologic media plot on a straight line, with the resulting curve described by the equation

$$\log C^* = b \log C + \log K_d \tag{9-28}$$

where

$b$ = the slope of the line

$K_d$ = the intercept of the line with the axis

Both $K_d$ and $b$ are coefficients that are a function of the solute, soil type, and equilibrium conditions in the solute/soil system. Equation 9-28 can also be expressed as

$$C^* = K_d C^b \tag{9-29}$$

Equation 9-29 is known as the **Freundlich isotherm.** It is one of many isotherms known to describe adsorption phenomena (16).

If the value of $b$ is 1.0, the isotherm is linear, and the data will plot on a straight line on arithmetic paper; Equation 9-29 then becomes

$$C^* = K_d C \tag{9-30}$$

In this case the coefficient $K_d$ is known as the **distribution coefficient** and is the slope of the isotherm. This will be a very useful relationship in our study of mass transport of solutes in flowing ground water.

A second isotherm may be determined by plotting $C/C^*$ versus $C$ on arithmetic graph paper. If this falls on a straight line, it is the **Langmuir adsorption isotherm** (17) and is given by

$$\frac{C}{C^*} = \frac{1}{\beta_1 \beta_2} + \frac{C}{\beta_2} \quad \textbf{(9-31)}$$

where

$C$ is the equilibrium concentration of the ion in contact with the soil (mg/L)

$C^*$ is the amount of the ion adsorbed per unit weight of soil (mg/g)

$\beta_1$ is an adsorption constant related to the binding energy

$\beta_2$ is the adsorption maximum for the soil

A plot of $C/C^*$ as a function of $C$ is made on rectilinear scales. The data points will fall on a straight line. Some experiments yield two straight-line segments—one at lower concentrations of the ion and one at higher concentrations with a lower slope (18). In some cases, the soil may have adsorbed some of the ion under natural conditions prior to the laboratory test. If this is the case, a correction must be made (19).

The maximum ion adsorption, $\beta_2$, is the reciprocal of the slope of the straight line. The binding energy constant, $\beta_1$, is the slope of the line divided by the intercept. The experimental procedure usually tests for only a single ion at a time. Natural waters are more complex, and field reactions may differ from those determined by laboratory study. The Langmuir adsorption isotherm can be used for both anions and cations.

---

**EXAMPLE PROBLEM**

A calcareous glacial outwash sediment was present at the proposed site for an artificial recharge basin for wastewater (9). The phosphorus-removal capacity of the soil was determined by laboratory adsorption studies. Equal weights of soil were shaken in various concentrations of disodium phosphate. The soil already had 0.016 milligram of phosphorus per gram adsorbed prior to the test, and this was added to the value of $C^*$ adsorbed during the test to determine total $C^*$ in equilibrium with the solution. The equilibrium concentration of the solute is $C$.

A plot of $C/C^*$ as a function of $C$ is given in Figure 9.4. There are two straight-line segments, indicating that one type of adsorption is taking place at low activities of phosphorus and another type of adsorption, with a higher bonding energy, is occurring at greater concentrations. The slope of the straight line

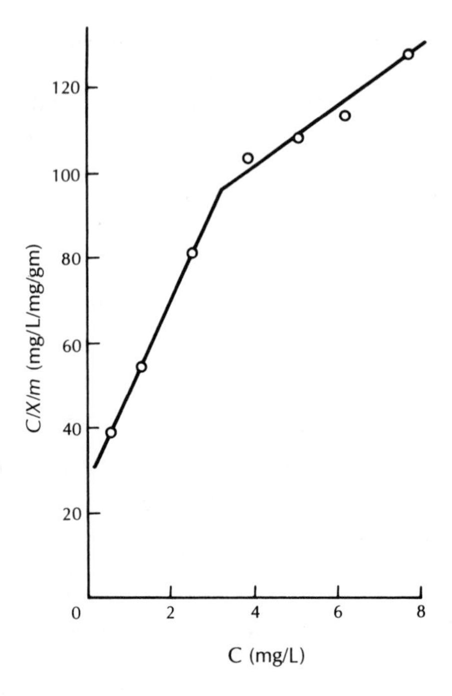


**FIGURE 9.4** Langmuir adsoption isotherm for phosphorus adsorbed on calcareous glacial outwash. Source: C. W. Fetter, Jr. *Ground Water,* 15 (1977):365–71.

at lower concentrations is $\frac{20\ (\text{mg/L})/(\text{mg/g})}{1\ \text{mg/L}}$, so the reciprocal is 0.05 milligram of phosphorus per gram of soil, which is the adsorption maximum ($\beta_2$). For higher concentrations, the adsorption maximum is 0.16 milligram of phosphorus per gram of soil.

### 9.11.2 ION EXCHANGE

Under certain conditions, the ions attracted to a solid surface may be exchanged for other ions in aqueous solution. This process is known as ion exchange. Both cation exchange and anion exchange can occur, but in some natural soils cation exchange is the dominant process. The presence of exchange sites is a function of the same general conditions affecting adsorption sites. The ion-exchange process can be conceptualized as the preferential adsorption of selective ions with concomitant loss of other ions. Ion-exchange sites are found primarily on clays and soil organic materials (20), although all soils and sediments have some ion-exchange capacity.

The ion-exchange reactions of different soils must be studied individually in the laboratory. Results are reported in terms of milliequivalents per 100 grams of soil. In one study of exchange capacities for stream sediments (21), the results shown in Table 9.4 were reported.

A general ordering of cation exchangeability for common ions in ground water is

$$Na^+ > K^+ > Mg^{++} > Ca^{++}$$

The divalent ions are more strongly bonded and tend to replace monovalent ions. However, it is a reversible reaction and, at high activities, the monovalent ions can replace divalent ions. This is the concept behind the home water softener. The divalent $Ca^{++}$ and $Mg^{++}$ ions replace the monovalent $Na^+$ ions on the exchange media. The exchange medium is regenerated when a brine solution with very high $Na^+$ activity is forced through the softener. The $Na^+$ replaces the $Ca^{++}$ and $Mg^{++}$ at the exchange sites. Ion-exchange capacities of organic colloids and clays can also remove heavy metal cations and thus provide some protection to ground-water supplies (22); but enough cases of ground-water pollution from heavy metals have been documented to demonstrate that such protection is limited to areas with clay in the soils.

One particularly well-studied ion-exchange reaction is the replacement of calcium in the soil with sodium. If water used for irrigation is high in sodium and low in calcium, the cation-exchange complex may become saturated with sodium. This can destroy the soil structure owing to dispersion of the clay particles. A simple method of evaluating the danger of high-sodium water is the **sodium-adsorption ratio, SAR** (23):

$$SAR = \frac{(Na^+)}{\left[\dfrac{(Ca^{++}) + (Mg^{++})}{2}\right]^{0.5}} \tag{9-32}$$

A low SAR (2 to 10) indicates little danger from sodium; medium hazards are between 7 and 18, high hazards between 11 and 26, and very high hazards above that. The lower the ionic strength of the solution, the greater the

**TABLE 9.4** Ion-exchange values for stream sediments (21)

| Size Fraction (μm) | Ion Exchange (meq/100 grams soil) |
|---|---|
| 4 | 14–65 |
| 4–61 | 4–30 |
| 61–1000 | 0.3–13 |

sodium hazard for a given SAR. Anions present in the water can affect calcium replacement (24, 25).

If the ion-exchange process is controlled by a reversible equilibrium process, the following equation applies:

$$b[\overline{\mathrm{A}}] + a[\mathrm{B}] \rightleftharpoons a[\overline{\mathrm{B}}] + \mathrm{b}[\mathrm{A}] \quad \textbf{(9-33)}$$

where

A and B are chemically exchanging species, A with a valence of $a$ and B with a valence of $b$

[A] is the concentration of solute A in terms of mass per unit volume of liquid

$[\overline{\mathrm{A}}]$ is the amount of solute A adsorbed by ion exchange on a unit mass of sediment or soil

When the exchanged ions are in equilibrium, the concentration of products and reactants at equilibrium is described by the ion-exchange selectivity coefficient, $K_s$:

$$K_s = \frac{[\overline{\mathrm{B}}]^a\,[\mathrm{A}]^b}{[\overline{\mathrm{A}}]^b\,[\mathrm{B}]^a} \quad \textbf{(9-34)}$$

The **cation-exchange capacity** (CEC) is defined as $[\overline{\mathrm{A}}] + [\overline{\mathrm{B}}]$ and the total solute concentration, $C_0$, is equal to [A] + [B]. When the concentration of one of the exchanging ions is very low, the adsorbed phase of the other (dominant) ion is approximately equal to CEC, and the total solute concentration is almost entirely that of the dominant ion. Equation 9-34 can be rewritten under these conditions, if A is the major species, as

$$K_s = \frac{[\overline{\mathrm{B}}]^a\,C_0^{\mathrm{b}}}{CEC^b\,[\mathrm{B}]^a} \quad \textbf{(9-35)}$$

The **ion-exchange distribution coefficient,** $K_d$, is the ratio of the adsorbed species concentration to the concentration of the solute:

$$K_d = \frac{[\overline{\mathrm{B}}]}{[\mathrm{B}]} \quad \textbf{(9-36)}$$

A standard laboratory test is available to determine the cation exchange capacity of soils. A 100-gram sample of dry soil is mixed with a solution of ammonium acetate to saturate the exchange sites with $NH_4^+$ ions. The pH of the pore water is adjusted to a value of 7.0. The soil is leached with a strong NaCl solution to replace the $NH_4^+$ on the exchange sites with $Na^+$ ions. The sodium content of the leaching solution is then determined and the CEC computed as the difference between the sodium in the original solution and the sodium in the leaching solution at equilibrium. It is reported in milliequivalents per 100 grams of soil. The CEC is frequently used as an indication of the potential of a soil to attenuate pollutants with exchangeable ions.

## 9.12 ISOTOPE HYDROLOGY

The use of isotopes in hydrology has not found widespread use in the United States, although their use is more widespread in Europe. Isotopes can be separated into environmental isotopes, which are found in the ground water, and isotopes that are introduced into the ground as a part of a ground-water study. The latter are most often used to trace ground-water flow direction or determine velocity.

Environmental isotopes can be either radioactive or stable. They can be used to determine the locations of ground-water recharge areas, circulation patterns in aquifers, sources of dissolved solids in ground water, and the age of ground water—the length of time it has been out of contact with the atmosphere (26–28).

### 9.12.1 TRITIUM

Tritium, $^3H$, is an unstable isotope of hydrogen with a half-life of 12.4 years. Tritium in the atmosphere is typically in the form of the molecule $H^3HO$ and enters the ground water as recharging precipitation. Prior to 1953, rainwater had less than 10 tritium units (TU). Starting in 1953, the manufacturing and testing of nuclear weapons has increased the amount of tritium in the atmosphere, with a resulting increase in tritium in the ground water. As a result, $^3H$ can be used in a qualitative manner to date ground water in the sense that ground water with less than 2 to 4 TU is dated prior to 1953 and if the amount is significantly greater than 10 to 20 TU it has been in contact with the atmosphere since 1953. Because of the great temporal and spatial variations in $^3H$ injected into the atmosphere since 1953, it cannot be used with more precision. Tritium has been used to trace the seepage of contaminated ground water from low-level nuclear-waste disposal areas (29).

### 9.12.2 RADIOCARBON DATING OF GROUND WATER

Radiocarbon dating methods can be applied to obtain the age of ground water. Carbon exists in several naturally occurring isotopes, $^{12}C$, $^{13}C$, and $^{14}C$. Carbon 14 is formed in the atmosphere by the bombardment of $^{14}N$ by cosmic radiation (30). The $^{14}C$ forms $CO_2$, so that the atmospheric $CO_2$ has a constant radioactivity due to modern $^{14}C$. If the $CO_2$ is incorporated into a form in which it is isolated from modern $^{14}C$, age determinations can be made from the $^{14}C$ radioactivity as a percent of the original. The half-life of $^{14}C$ is 5570 years, so that if one-fourth of the original activity is present, two half-lives, or 11,140 years, have elapsed. When precipitation soaks into the ground, it is saturated with respect to $CO_2$, with a known $^{14}C$ activity. Once the water has entered the soil, additional carbon may come from soil $CO_2$ and the solution of carbonate minerals. The modern carbon is diluted by the inactive carbon from carbonate minerals. The raw dates obtained must be adjusted for this dilution.

If $A$ is the measured $^{14}C$ radioactivity, and $A_0$ is the activity at the time the sample was isolated, then the following equation may be used:

$$A = QA_0 2^{-t/T} \qquad \textbf{(9-37)}$$

where

$t$ is the age

$T$ is the half-life of $^{14}C$

$Q$ is an adjustment factor to account for dilution by **dead carbon*** (31)

The equation requires an estimation of initial value, $A_0$, and the adjustment factor, $Q$.

The value of $A_0$ will depend on the carbonate equilibria established under an open system in which the ground water was exposed to an infinite reservoir of $CO_2$. This occurs in nature in the soil zone and in shallow ground water. When the ground-water system becomes closed with respect to $CO_2$, then any added carbon would be only from carbonate rocks; i.e., dead carbon. The value of $Q$ is generally in the range of 0.5 to 0.9. Carbon 14 dates of ground water thus tend to be somewhat less than raw dates as a result of the dilution by dead carbon from carbonate minerals. The determination of $A_0$ and $Q$ is somewhat complex, and several different methods are available (31, 32). Radiocarbon dates of ground water of up to 50,000 to 80,000 years may be obtained, although the accuracy under the best of conditions is on the order of $\pm 20$ percent (26).

### 9.12.3 STABLE ISOTOPES OF OXYGEN AND HYDROGEN

There are two stable isotopes of hydrogen, $^1H$ and $^2H$ (deuterium), as well as three stable isotopes of oxygen $^{16}O$, $^{17}O$ and $^{18}O$. There are nine different combinations of the above which make stable water molecules with atomic masses ranging from 18 to 22. The most abundant water molecule $^1H_2{}^{16}O$, which is the lightest, has a much higher vapor pressure than the heaviest form, $^2H_2{}^{18}O$. During phase changes of water between liquid and gas the heavier water molecules tend to concentrate in the liquid phase, which fractionates the hydrogen and oxygen isotopes. Water that evaporates from the ocean is isotopically lighter

*Dead carbon is carbon from a source old enough for any $^{14}C$ to have decayed below measurable limits.

than the water remaining behind, and precipitation is isotopically heavier; that is, it contains more $^2H$ and $^{18}O$ than the vapor left behind in the atmosphere.

The use of mass spectrometry can determine the ratio of isotopes in a water sample. The most important isotope ratios are $^{18}O/^{16}O$ and $^2H/^1H$. These isotopic ratios from an environmental water sample can be compared with the isotopic ratio of standard mean ocean water (SMOW). The comparison is made by means of the parameter δ, which is defined as

$$\delta^{18}O\ (‰) = \left[\frac{(^{18}O/^{16}O)_{sample}}{(^{18}O/^{16}O)_{SMOW}} - 1\right]10^3 \quad \textbf{(9-38)}$$

$$\delta^{2}H\ (‰) = \left[\frac{(^{2}H/^{1}H)_{sample}}{(^{2}H/^{1}H)_{SMOW}} - 1\right]10^3 \quad \textbf{(9-39)}$$

When $\delta^2H$ is plotted as a function of $\delta^{18}O$ for water found in continental precipitation, an experimental linear relationship was found which can be described by the equation (33)

$$\delta^2H = 8\delta^{18}O + 10 \quad \textbf{(9-40)}$$

This is known as the **meteoric water line.** Continental precipitation samples will tend to group close to this line. Precipitation falling in areas with lower temperatures or at higher latitudes will tend to have lower $\delta^2H$ and $\delta^{18}O$ values. Naturally, oceanic water will fall below the meteoric water line as it is isotopically enriched. Deviations from the meteoric water line can be interpreted as being caused by precipitation that occurred during a warmer or colder climate than at present or by geochemical changes that occurred when the water was underground (34). Geothermal water tends to be isotopically enriched with respect to $\delta^{18}O$ owing to equilibration of the oxygen in the ground water with respect to oxygen in the rocks (33). Figure 9.5 shows the meteoric water line and the results of stable isotope analyses for a number of spring water samples from the Meade thrust area of southeastern Idaho.

### 9.12.4 OTHER ISOTOPES

A number of other isotopes have been used or have been proposed to be used to date ground water. Chlorine 36 is one of them (35). It has been proposed as a method to date ground water that is older than water that can be dated with carbon 14 as it has a half-life of $3.01 \times 10^5$ years. A ratio of $^{36}Cl$/total Cl is determined; the higher the ratio, the younger the sample. Oceanic water is old enough that little, if any, $^{36}Cl$ is present. Young water near the coastline, which is likely to contain chloride produced as salt spray from the ocean, will appear to be much older owing to the large amount of "dead" chloride. Other isotopes with possible uses in age dating are $^{85}Kr$, $^{81}Kr$, $^{39}Ar$, and

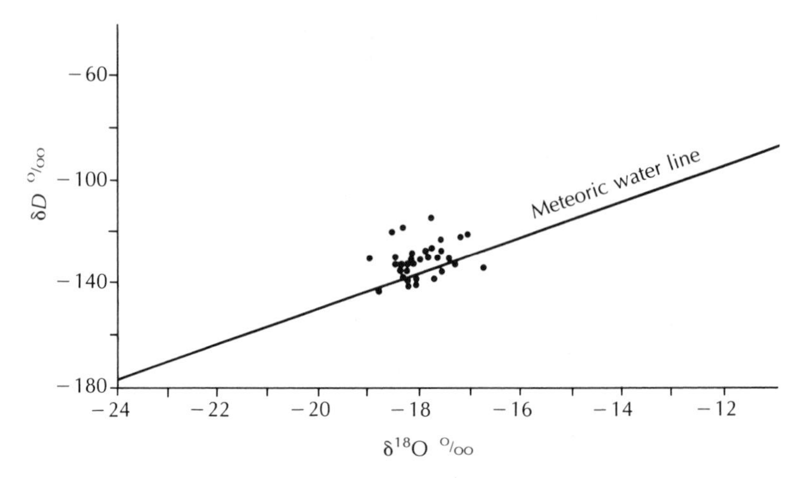


**FIGURE 9.5** $\delta D$ ($\delta^2H$) and $\delta^{18}O$ values from spring and well waters in the Meade thrust area, southeastern Idaho. Source: A. L. Mayo, Ground-Water Flow Patterns in the Meade Thrust Allochthon, Idaho-Wyoming Thrust Belt, Southeastern Idaho. Ph.D. thesis, University of Idaho, 1982.

$^{32}Si$ (26). Nitrogen 15 has been used in identifying the source of nitrate pollution in ground water (36), and sulfur 34 has potential to be used to identify sources of sulfate in ground water (37).

## 9.13 PRESENTATION OF RESULTS OF CHEMICAL ANALYSES

Tables of data are the most common form in which the results of an analysis of water chemistry are reported. The data can be expressed in milligrams per liter (mg/L), milliequivalents per liter (meq/L), or millimoles per liter. For many purposes, the data may be also displayed in graphical form. There have been a great number of graphical forms proposed, including bar graphs, vectors, pie diagrams, and nomographs, to name but a few (1). One of the more widely used is the **trilinear diagram.** Inasmuch as this method also forms the basis for a common classification scheme for natural waters, it will be described in detail.

The major ionic species in most natural waters are $Na^+$, $K^+$, $Ca^{++}$, $Mg^{++}$, $Cl^-$, $CO_3^=$, $HCO_3^-$, and $SO_4^=$. A trilinear diagram can show the percentage composition of three ions. By grouping $Ca^{++}$ and $Mg^{++}$ together, the major cations can be displayed on one trilinear diagram. Likewise, if $CO_3^=$ and $HCO_3^-$ are grouped, there are also three groups of the major anions. Figure 9.6 shows the form of a trilinear diagram that is commonly used in water-chemistry studies (38). Analyses are plotted on the basis of the percent of each cation (or anion).

Each apex of a triangle represents a 100 percent concentration of one of

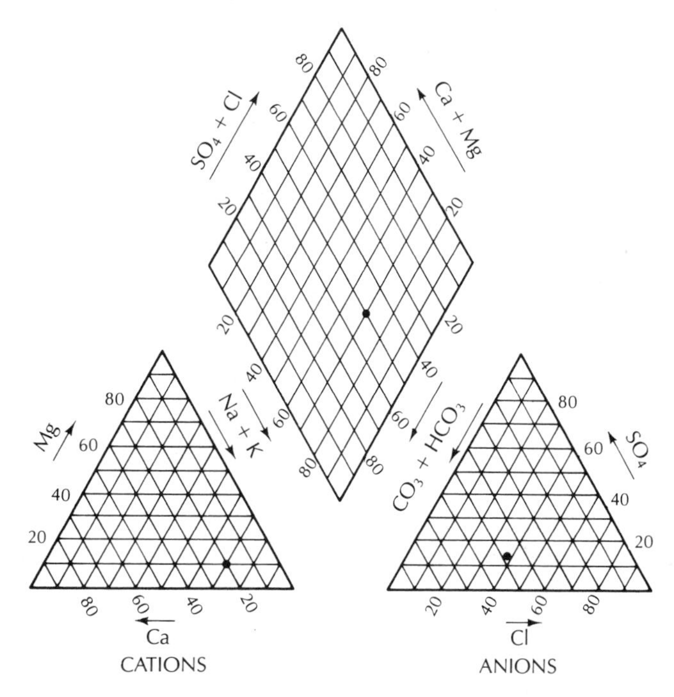


FIGURE 9.6 Trilinear diagram of the type used to display the results of water-chemistry studies. Results of the Example Problem from Section 9.13 are plotted on the diagram.

the three constituents. If a sample has two constituent groups present, then the point representing the percentage of each would be plotted on the line between the apexes for those two groups. If all three constituent groups are present, the analyses would fall in the interior of the field. The diamond-shaped field between the two triangles is used to represent the composition of water with respect to both cations and anions. The cation point is projected parallel to the sulfate axis. The intersection of the two points is then plotted.

As water flows through an aquifer it assumes a diagnostic chemical composition as a result of interaction with the lithologic framework The term **hydrochemical facies** is used to describe the bodies of ground water, in an aquifer, that differ in their chemical composition. The facies are a function of the lithology, solution kinetics, and flow patterns of the aquifer (39, 40). Hydrochemical facies can be classified on the basis of the dominant ions in the facies by means of the trilinear diagram (Figure 9.7).

**EXAMPLE PROBLEM**

Plot the results of the following analysis on a trilinear diagram:

| | $Ca^{++}$ | $Mg^{++}$ | $Na^{+}$ | $K^{+}$ | $HCO_3^{-}$ | $CO_3^{=}$ | $SO_4^{=}$ | $Cl^{-}$ |
|---|---|---|---|---|---|---|---|---|
| mg/L | 23 | 4.7 | 35 | 4.7 | 171 | 0 | 1.0 | 9.5 |
| meq/L | 1.15 | 0.39 | 1.52 | 0.12 | 2.80 | 0 | 0.02 | 0.27 |

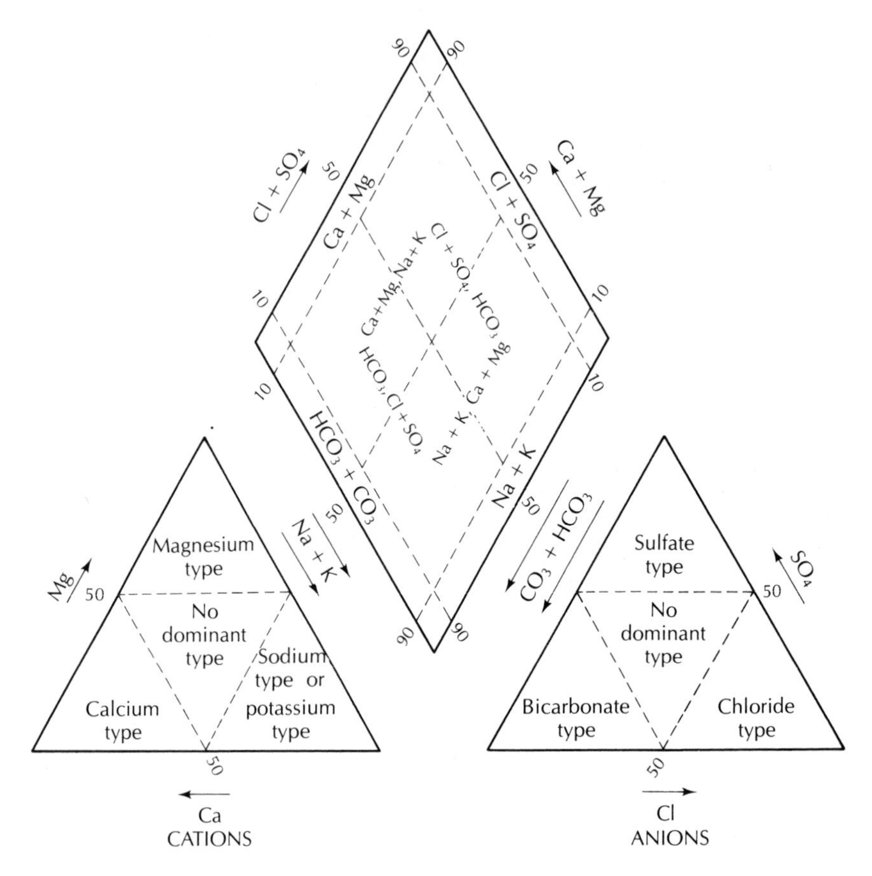


**FIGURE 9.7** Hydrogeochemical classification system for natural waters using the trilinear diagram.

The first step is to find the percent of each cation and anion group as a percentage of the total:

| Cations | meq/L | % of Total | Anions | meq/L | % of Total |
|---|---|---|---|---|---|
| $Ca^{++}$ | 1.15 | 36 | $Cl^-$ | 0.27 | 9 |
| $Mg^{++}$ | 0.39 | 12 | $SO_4^=$ | 0.02 | 1 |
| $Na^+ + K^+$ | 1.64 | 72 | $CO_3^= + HCO_3^-$ | 2.80 | 90 |
| Total | 3.18 | | Total | 3.09 | |

Note: Due to analytical error and unreported minor constituents, the total equivalents of anions and cations do not exactly match. Theoretically, the total equivalent weight of the anions should be exactly that of the cations, as equivalent weights are based on the amount of the ion that would combine with $O_2$.

Referring back to Figure 9.6, the points for both the cations and anions are plotted on the appropriate triangle diagrams. The positions of the points are projected parallel to the magnesium and sulfate axes, respectively, until they intersect in the center field.

---

A second type of graphical presentation of chemical analyses is the **Stiff pattern** (41). A polygonal shape is created from four parallel horizontal axes extending on either side of a vertical zero axis. Cations are plotted in milliequivalents per liter on the left of the zero axis, one to each horizontal axis, and anions are plotted on the right. Figure 9.8 shows several Stiff patterns. The use of the lower horizontal bar with iron and carbonate is optional as in many waters

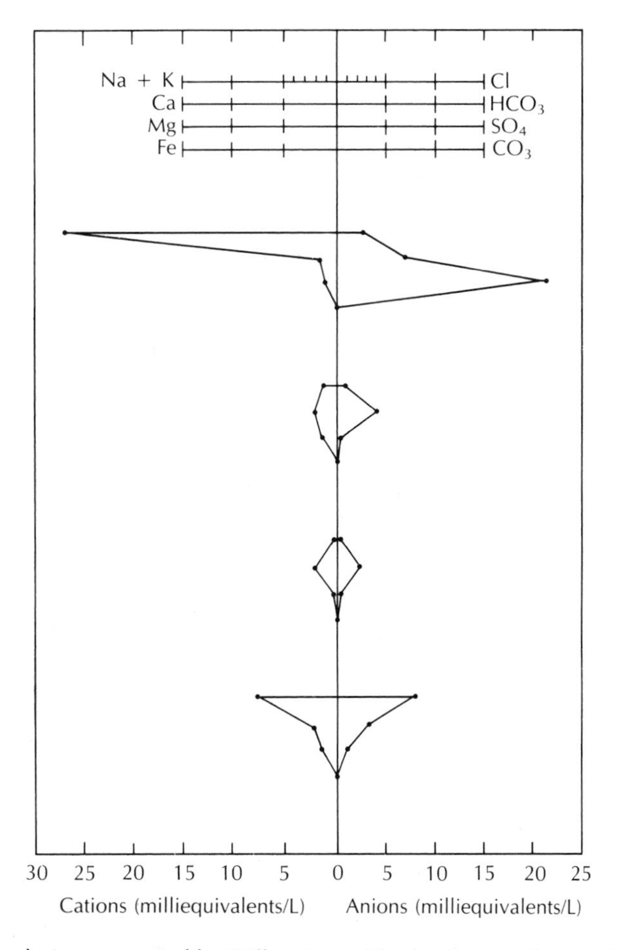


**FIGURE 9.8** Analysis represented by Stiff patterns. The horizontal distance from the vertical axis is based on the number of milliequivalents per liter of each anion or cation. Use of the lower bar for iron and carbonate is optional. Source: J. D. Hem, U.S. Geological Survey Water-Supply Paper 2254, 1985.

they are close to zero. Stiff patterns are useful in making a rapid visual comparison between water from different sources. The larger the area of the polygonal shape, the greater the concentrations of the various ions. Figure 9.9 shows the use of Stiff patterns in an area where mineralized water exists in a portion of an aquifer system. The isocon lines represent lines of equal total dissolved solids and the Stiff patterns are centered over the location of a particular well. In these patterns the lower horizontal line for iron and carbonate was not used.

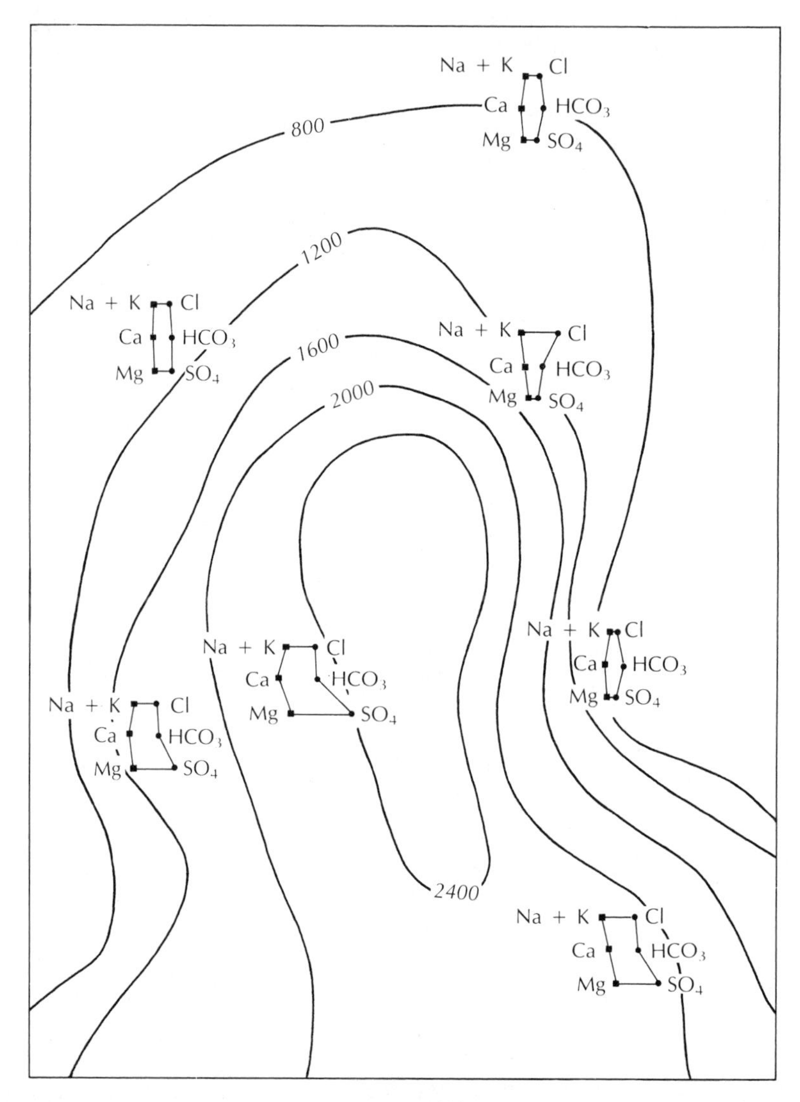


**FIGURE 9.9** Use of Stiff patterns to show the varying ground-water chemistry in an area containing mineralized water from a waste-disposal operation. Solid lines represent isocons of total dissolved solids. Stiff patterns are centered over locations of wells.

## CASE STUDY: CHEMICAL GEOHYDROLOGY OF THE FLORIDAN AQUIFER SYSTEM

The regional geohydrology of the Floridan aquifer was discussed in Section 7.7. The chemical geohydrology of this aquifer system is also well known. We will consider the chemical changes that take place as the water flows from the central recharge area at Polk City to the south. Data from five wells form the basis for the hydrochemical cross sections (42–44). Figure 9.10 indicates the locations of the wells on the potentiometric map of central Florida, with Polk City located at the southern edge of the recharge area. The five wells all tap the Floridan aquifer and lie approximately along the same flow path. Chemical analyses of water from the wells are given in Table 9.5 (p. 361). As water travels down the flow path, it increases in total dissolved solids, from 138 to 726 milligrams per liter. All ions except bicarbonate show a progressive increase along the flow path (Figure 9.11). Computation of ion-activity products from the analyses shows that both dolomite and calcite saturation increase along the flow path. For the most part, the $K_{iap}/K_{sp}$ of the water is greater than 1, indicating supersaturation. Hydrochemical cross sections along the flow path are shown in Figure 9.11. A trilinear plot (Figure 9.12) of the well analyses indicates that the

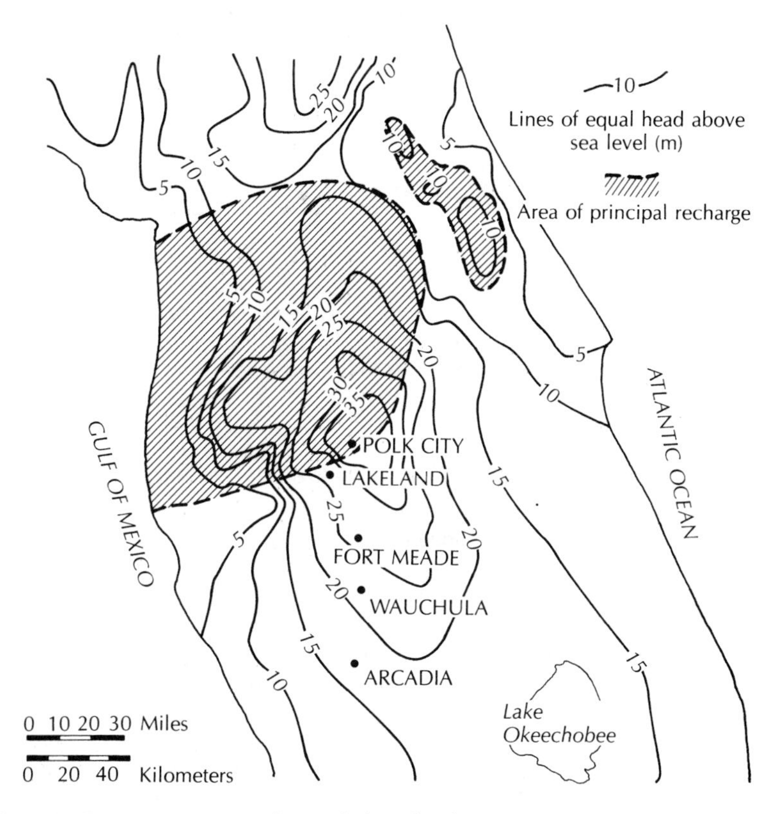


**FIGURE 9.10** Potentiometric surface of the Floridan aquifer in central Florida. Source: Adapted from L. N. Plummer, *Water Resources Research*, 13 (1977):801–12.

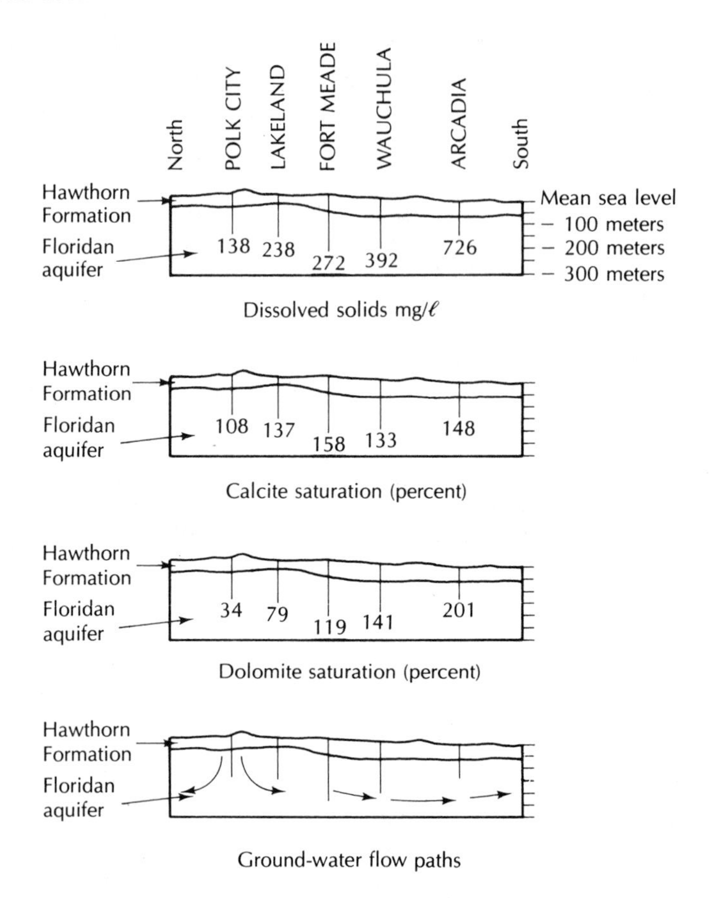


**FIGURE 9.11** Hydrogeochemical cross sections from Polk City through Arcadia through the Floridan aquifer. Sources: Data from Back, Cherry, and Hanshaw, 1966; Back and Hanshaw, 1970; and Plummer, 1970 (References 42, 43, and 44, respectively).

chemical composition of the water is shifting along the flow path. This is due to an increase in the $Mg^{++}/Ca^{++}$ and the $SO_4^{=}/HCO_3^{-}$ ratios with increasing distance from the recharge area. The change in these ratios is due to the solution of gypsum ($CaSO_4 \cdot 2H_2O$) and dolomite ($CaMg(CO_3)_2$) along the flow paths. It is important to note that these reactions involve only the solution of minerals in fresh water. If this water mixes with sea water, water chemistry would rapidly change and be dominated by sodium and chloride, which both are minor constituents in fresh water.

Calculation of the age of ground water from the Floridan aquifer on the basis of $^{14}C$ activity is complex owing to the solution of dead carbon from carbonate. Complicating the dating is the fact that the flow path from Polk City to Ft. Meade is partially open to soil $CO_2$, and then it is closed to $CO_2$ from Ft. Meade to Wauchula. However, from Wauchula to Arcadia, it is again open to $CO_2$, the source being the oxidation of lignite from sulfate reduction (44). Several authors have dealt with the $^{14}C$ dating of this water (30, 43, 44) and have cited dates ranging from 20,600 to 24,100 to 36,000 years B.P. Based on the

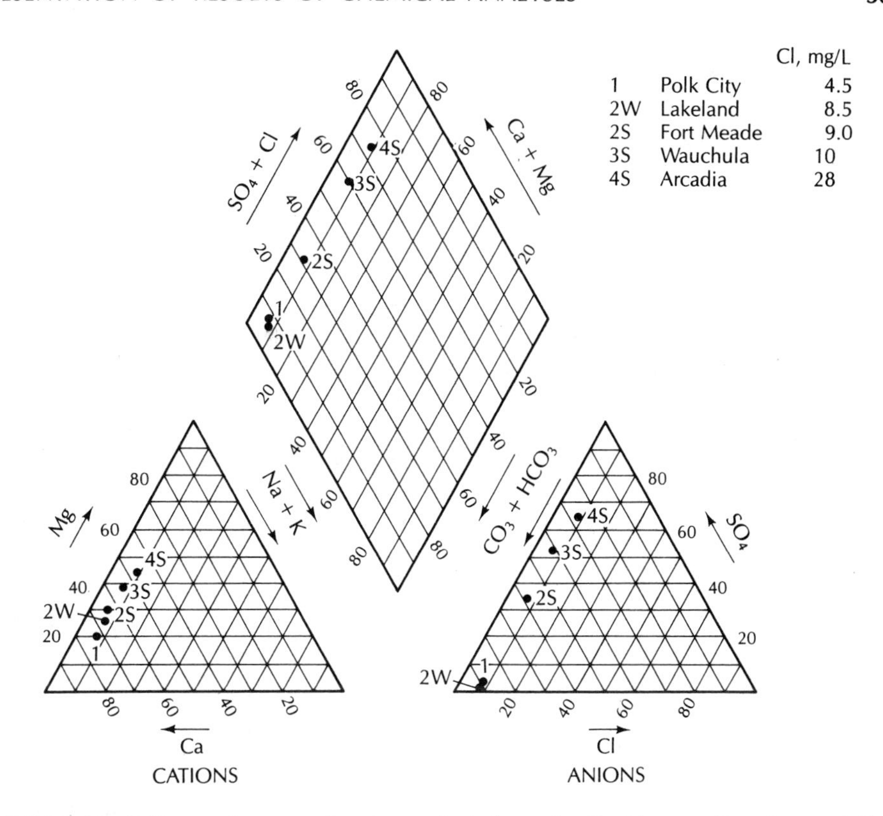


**FIGURE 9.12** Trilinear diagram of water analysis from the Floridan aquifer of central Florida. Locations are shown on Figure 9.10. Source: W. Back and B. B. Hanshaw, *Journal of Hydrology*, 10 (1970):330–68 (Amsterdam: North-Holland Publishing Company).

isotopic dating and the reaction coefficients, apparent rates of solution for several mineral species have been calculated (44).

This type of study illustrates what can be accomplished in chemical geohydrology. It can be a companion to the more commonly performed hydrodynamic study of flow paths and rates. The flow velocity as determined from hydrodynamic considerations can be compared with radiocarbon estimates.

**TABLE 9.5** Chemical analysis of water from Floridan aquifer in central Florida

| Well | Location | Temp % C | Field pH | Milligrams per liter | | | | | | | | |
|---|---|---|---|---|---|---|---|---|---|---|---|---|
| | | | | $SiO_2$ | $Ca^{++}$ | $Mg^{++}$ | $Na^+$ | $K^+$ | $HCO_3^-$ | $SO_4^=$ | $Cl^-$ | TDS |
| 1 | Polk City | 23.8 | 8.0 | 12 | 34 | 5.6 | 3.2 | 0.5 | 124 | 2.4 | 4.5 | 138 |
| 2W | Lakeland | 26.3 | 7.62 | 18 | 54 | 14 | 6.9 | 1.0 | 253 | 3.6 | 8.5 | 238 |
| 2S | Ft. Meade | 26.6 | 7.75 | 16 | 58 | 17 | 6.1 | 0.7 | 163 | 71 | 9.0 | 272 |
| 3S | Wauchula | 25.4 | 7.69 | 18 | 66 | 29 | 8.3 | 2.0 | 168 | 155 | 10 | 392 |
| 4S | Arcadia | 26.3 | 7.44 | 31 | 106 | 60 | 21 | 3.7 | 206 | 344 | 28 | 726 |

Source: Data from W. Back and B.B. Hanshaw, *Journal of Hydrology*, 10 (1970):330–68.

## REFERENCES


1. HEM, J.D. *Study and Interpretation of the Chemical Characteristics of Natural Water,* 3rd ed. U.S. Geological Survey Water-Supply Paper 2254, 1985, 263 pp.
2. BACK, W., and B. B. HANSHAW. "Chemical Geohydrology." In *Advances in Hydroscience,* vol. 2, ed. V. T. Chow. New York: Academic Press, 1965, pp. 49–109.
3. KRAUSKOPF, K. B. *Introduction to Geochemistry.* New York: McGraw-Hill, 1967, 721 pp.
4. GARRELS, R. M., and C. L. CHRIST. *Solutions, Minerals and Equilibria.* New York: Harper & Row, 1965, 450 pp.
5. HEM, J. D., and W. H. CROPPER. *Survey of the Ferrous-Ferric Chemical Equilibria and Redox Potential.* U.S. Geological Survey Water-Supply Paper 1459-A, 1959, 31 pp.
6. BACK, W., and I. BARNES. *Relation of Electrochemical Potentials and Iron Content to Groundwater Flow Patterns.* U.S. Geological Survey Professional Paper 498-C, 1965, 16 pp.
7. HEM, J. D. *Restraints in Dissolved Ferrous Iron Imposed by Bicarbonate Redox Potentials, and pH.* U.S. Geological Survey Water-Supply Paper 1459-B, 1960, 55 pp.
8. BASS BECKING, L. G. M., I. R. KAPLAN, and D. MOORE. "Limits of the Natural Environment in Terms of pH and Oxidation-Reduction Potential." *Journal of Geology,* 68 (1960): 243–84.
9. FETTER, C. W., JR. "Attenuation of Wastewater Elutriated through Glacial Outwash." *Ground Water,* 15 (1977):365–71.
10. WAYMAN, C. H. "Adsorption on Clay Mineral Surfaces." In *Principles and Applications of Water Chemistry,* ed. S. D. Faust and J. V. Hunter. New York: John Wiley & Sons, 1967, pp. 127–67.
11. BALLARD, R., and J. G. A. FISKELL. "Phosphorus Retention in Coastal Plain Soils: I. Relation to Soil Properties." *Soil Science Society of America, Proceedings,* 38 (1974):250–55.
12. BARROW, N. J., and T. C. SHAW. "The Slow Reactions between Soil and Anions: 2. Effect of Time and Temperature on the Decrease in Phosphate Concentration in the Soil Solution." *Soil Science,* 119 (1975):167–77.
13. BARROW, N. J., and T. C. SHAW. "The Slow Reactions between Soil and Anions: 3. The Effects of Time and Temperature on the Decrease in Isotopically Exchangeable Phosphate." *Soil Science,* 119 (1975):190–97.
14. MATTINGLY, G. E. G. "Labile Phosphate in Soils." *Soil Science,* 119 (1975):369–75.
15. VIJAYACHANDRAN, P. K., and R. D. HARTER. "Evaluation of Phosphorus Adsorption by a Cross Section of Soil Types." *Soil Science,* 119 (1975):119–26.
16. CHERRY, J. A., R. W. GILLHAM, and J. F. BARKER. "Contaminants in Groundwater: Chemical Processes." In *Groundwater Contamination.* Washington, D.C.: National Academy Press, 1984, pp. 46–63.
17. OLSEN, S. R., and F. S. WATANABE. "A Method to Determine a Phosphorus

Adsorption Maximum of Soils as Measured by the Langmuir Isotherm." *Soil Science Society of America, Proceedings,* 21 (1957):144–49.

**18.** SYERS, J. K., M. G. BROWMAN, G. W. SMILLIE, and R. B. COREY. "Phosphate Sorption by Soils Evaluated by the Langmuir Adsorption Equation." *Soil Science Society of America, Proceedings,* 37 (1973):358–63.

**19.** FITTER, A. H., and C. D. SUTTON. "The Use of the Freudlich Isotherm for Soil Phosphate Sorption Data." *Journal of Soil Science,* 26 (1975):241–46.

**20.** MITCHELL, J. "The Origin, Nature and Importance of Soil Organic Constituents Having Base Exchange Properties." *Journal of American Society of Agronomy,* 24 (1932):256–75.

**21.** KENNEDY, V. C. *Mineralogy and Cation Exchange Capacity of Sediments from Selected Streams.* U.S. Geological Survey Professional Paper 433-D, 1965, 28 pp.

**22.** WENTINK, G. R., and J. E. ETZEL. "Removal of Metal Ions by Soil." *Journal of the Water Pollution Control Federation,* 44 (1972):1561–74.

**23.** RICHARDS, L. A., ed. *Diagnosis and Improvement of Saline and Alkali Soil.* U.S. Department of Agriculture Agricultural Handbook 60, 1954.

**24.** PRATT, P. F., and F. L. BLAIR. "Sodium Hazard of Bicarbonate Irrigation Waters." *Soil Science Society of America, Proceedings,* 33 (1969):880–83.

**25.** BOWER, C. A., G. OGATA, and J. M. TUCKER. "Sodium Hazard of Irrigation Waters as Influenced by Leaching Fraction and by Precipitation on Solution of Calcium Carbonate." *Soil Science,* 106 (1968):29–34.

**26.** DAVIS, S. N., and H. W. BENTLEY. "Dating Groundwater, a Short Review." In *Nuclear and Chemical Dating Techniques: Interpreting the Environmental Record,* ed. Lloyd A. Curie. American Chemical Society Symposium Series No. 176, 1982, pp. 187–222.

**27.** FONTES, J. CH. "Environmental Isotopes in Groundwater Hydrology." In *Handbook of Environmental Isotope Hydrology, Volume 1, The Terrestrial Environment,* ed. P. Fritz and J. Ch. Fontes. Amsterdam: Elsevier Scientific Publishers, 1980, pp. 25–140.

**28.** MULLER, A. B., and A. L. MAYO. "Ground-Water Circulation in the Meade Thrust Allochthon Evaluated by Radiocarbon Techniques." *Radiocarbon,* 25 (1983):357–72.

**29.** FOSTER, J. B. "Lessons Learned in a Hydrogeological Case at Sheffield, Illinois." *Proceedings, Symposium on Low-Level Waste Disposal, Site Characterization and Monitoring.* Oak Ridge National Laboratory, NUREG/CP-0028, CONF-820674, vol. 2, 1982, pp. 237–44.

**30.** DE VRIES, H. "Measurement and Use of Natural Radiocarbon." In *Researches in Geochemistry,* vol. 1, ed. P. H. Abelson. New York: John Wiley & Sons, 1959, pp. 169–89.

**31.** WIGLEY, T. M. L. "Carbon 14 Dating of Ground-water from Closed and Open Systems." *Water Resources Research,* 11 (1975):324–28.

**32.** PLINES, P., D. LANGMUIR, and R. S. HARMON. "Stable Carbon Isotope Ratios and the Existence of a Gas Phase in the Evolution of Carbonate Ground Waters." *Geochimica et Cosmochimica Acta,* 38 (1974):1147–64.

**33.** MAYO, A. L., A. B. MULLER, and D. R. RALSTON. "Hydrogeology of the Meade Thrust Allochthon, Southeastern Idaho, U.S.A., and Its Relevance to Stratigraphic and Structural Groundwater Flow Control." *Journal of Hydrology,* 76 (1985):27–61.

34. CRAIG, H. "Isotopic Variations in Meteoric Water." *Science,* 133 (1961):1702–3.
35. BENTLEY, H. W., and S. N. DAVIS. "Feasibility of $^{36}Cl$ Dating of Very Old Ground Water. *EOS, American Geophysical Union Transactions,* 61 (1980):230.
36. KREITLER, C. W., S. E. RAGONE, and B. G. KATZ. "Nitrogen Isotope Ratios of Groundwater Nitrate, Long Island, New York." *Ground Water,* 16 (1978):404–9.
37. DAVIS, S. N., and others. "Ground-Water Tracers—a Short Review" *Ground Water,* 18 (1980):14–23.
38. PIPER, A. M. "A Graphic Procedure in the Geochemical Interpretation of Water Analyses." *Transactions, American Geophysical Union,* 25 (1944):914–23.
39. BACK, W. "Origin of Hydrochemical Facies in Groundwater in the Atlantic Coastal Plain." *Proceedings, International Geological Congress* (Copenhagen), I (1960):87–95.
40. BACK, W. *Hydrochemical Facies and Groundwater Flow Patterns in Northern Part of Atlantic Coastal Plain.* U.S. Geological Survey Professional Paper 498-A, 1966, 42 pp.
41. STIFF, H. A., JR. "The Interpretation of Chemical Water Analysis by Means of Patterns." *Journal of Petroleum Technology,* 3 (1951):15–17.
42. BACK, W., R. N. CHERRY, and B. B. HANSHAW. "Chemical Equilibrium between Water and Minerals of a Carbonate Aquifer." *National Speleological Society Bulletin,* 28 (1966):119–26.
43. BACK, W., and B. B. HANSHAW. "Comparison of Chemical Hydrogeology of the Carbonate Peninsulas of Florida and Yucatan." *Journal of Hydrology,* 10 (1970):330–68.
44. PLUMMER, L. N. "Defining Reactions and Mass Transfer in Part of Floridan Aquifer." *Water Resources Research,* 13 (1977):801–12.

## PROBLEMS

1. How much KCl is in a 0.1-molal solution?
2. The solubility product for CuCl is $10^{-5.9}$. What is the solubility of $Cu^{+}$ at equilibrium?
3. The solubility product of fluorite, $CaF_2$, is $10^{-10.5}$.
   a. What is the solubility of $Ca^{++}$ at equilibrium?
   b. If $CaF_2$ is dissolved in a solution of 0.001-m $F^{-}$, what is the solubility of $CaF_2$?
4. Given the following ground-water analysis:

| | | | |
|---|---|---|---|
| $Ca^{++}$ | 143 mg/L | $SO_4^{=}$ | 254 mg/L |
| $Mg^{++}$ | 35 mg/L | $HCO_3^{-}$ | 317 mg/L |
| $Na^{+}$ | 14 mg/L | $Cl^{-}$ | 4 mg/L |
| pH | 8.0 | | |

   with total dissolved solids = 652 mg/L,
   a. Convert all analyses to molal concentrations.
   b. Compute the ionic strength.

**c.** Compute the activity coefficient for each ion.
**d.** Find the activity of each ion.
**e.** Find the $K_{iap}$ of anhydrite ($CaSO_4$).
**f.** Compare $K_{iap}$ for anhydrite with the $K_{sp}$ of $10^{-4.5}$
**g.** Find the $K_{iap}$ of calcite ($CaCO_3$).
**h.** Compare the $K_{iap}$ for calcite with the $K_{sp}$ of $10^{-8.35.}$
**i.** Convert the analyses to milliequivalents per liter.
**j.** Do a cation/anion balance using the meq/L values.

**5.** What is the $[H^+]$ and $[OH^-]$ of an aqueous solution af pH 8.7?

**6.** What is the pH of a 0.0027-molal solution of HCl
**a.** at 0° C?
**b.** at 25° C?

**7.** What is the pH of a 0.0027-molal solution of $H_2CO_3$ at 25° C?

**8.** Given the following analysis of ground water,

| | | | |
|---|---|---|---|
| $Ca^{++}$ | 76.8 mg/L | $SO_4^{=}$ | 37 mg/L |
| $Na^+$ | 11.72 mg/L | $CL^-$ | 13.25 mg/L |
| $K^+$ | 1.67 mg/L | $HCO_3^-$ | 305.4 mg/L |
| $Mg^{++}$ | 20.54 mg/L | $NO_3^-$ | 2.03 mg/L |
| $Fe^{++}$ | 6.38 mg/L | | |

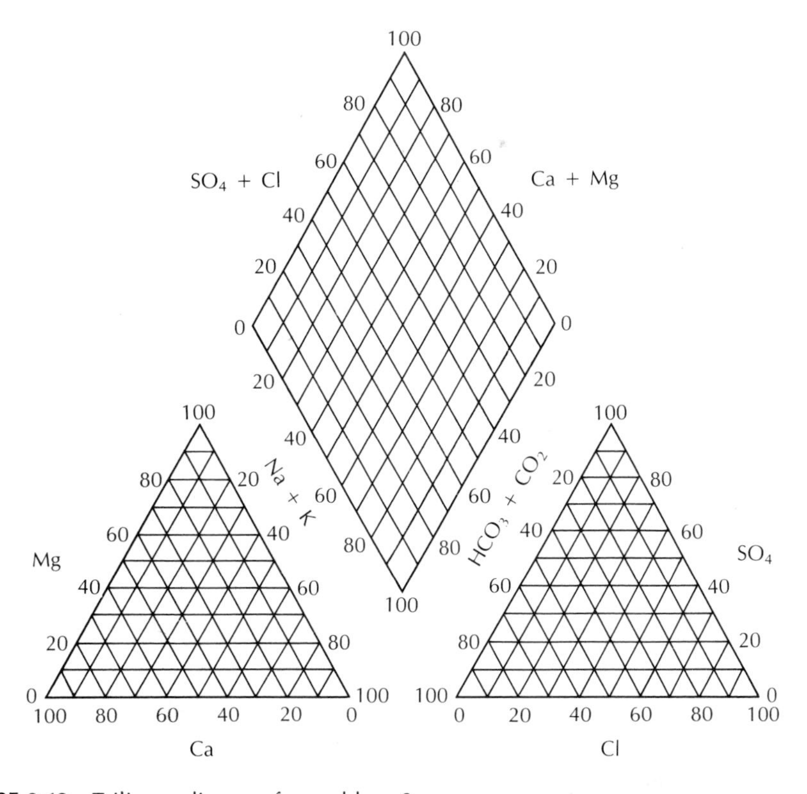


FIGURE 9.13 Trilinear diagram for problem 8c.

**a.** Convert all values to meq/L.
**b.** Do an anion/cation balance.
**c.** Plot the position on a trilinear diagram (Figure 9.13, p. 365).
**d.** Make a Stiff pattern of the analysis.

# TEN

# Water Quality and Ground-Water Contamination

And these waters, falling on these mountains through the ground and cracks, always descend and do not stop until they find some region blocked by stones or rock very close-set and condensed. And then they rest on such a bottom and having found some channel or other opening, they flow out as fountains or brooks or rivers according to the size of the openings and receptacles; and since such a spring cannot throw itself (against nature) on the mountains, it descends into the valleys. And even though the beginnings of such springs coming from the mountains are not very large, they receive aid from all sides, to enlarge and augment them; and particularly from the lands and mountains to the right and left of these springs.

*Discours admirables,* Bernard Palissy (ca. 1510–1590)

## 10.1 INTRODUCTION

The quality of water that we ingest as well as the quality of water in our lakes, streams, rivers, and oceans is a critical parameter in determining the overall quality of our lives. Water quality is determined by the solutes and gases dissolved in the water, as well as the matter suspended in and floating on the water. Water quality is a consequence of the natural physical and chemical state of the water as well as any alterations that may have occurred as a consequence of human activity. The usefulness of water for a particular purpose is determined by the water quality. If human activity alters the natural water quality so that it is no longer fit for a use for which it had previously been suited, the water is said to be **polluted** or **contaminated.** It should be noted that in many areas water quality has been altered by human activity, but the water is still usable.

One basic measure of water quality is the **total dissolved solids (TDS),** which is the total amount of solids, in milligrams per liter, that remain when a water sample is evaporated to dryness. Table 10.1 gives a classification scheme for water based on the total dissolved solids.

Water naturally contains a number of different dissolved inorganic constituents. The major anions are calcium, magnesium, sodium, and potassium; the

**Table 10.1** Classification of water based on total dissolved solids

| Class | TDS (mg/L) |
|---|---|
| Fresh | 0–1,000 |
| Brackish | 1,000–10,000 |
| Saline | 10,000–100,000 |
| Brine | >100,000 |

major cations are chloride, sulfate, carbonate, and bicarbonate. Although not in ionic form, silica can also be a major constituent. These **major constituents** constitute the bulk of the mineral matter contributing to total dissolved solids. In addition there may be **minor constituents** present, including iron, manganese, fluoride, nitrate, strontium, and boron. **Trace elements** such as arsenic, lead, cadmium, and chromium may be present in amounts of only a few micrograms per liter, but they are very important from a water-quality standpoint.

Dissolved gases are present in both surface and ground water. The major gases of concern are oxygen and carbon dioxide. Nitrogen, which is more or less inert, is also present. Minor gases of concern include hydrogen sulfide and methane. Hydrogen sulfide is toxic and imparts a bad odor, but is not present in water that contains dissolved oxygen.

Surface water may be adversely impacted by human activity. If organic matter, such as untreated human or animal waste, is placed into the surface-water body, **dissolved oxygen** levels diminish as microorganisms grow, using the organic matter as an energy source and consuming oxygen in the process. The total dissolved solids may increase owing to the disposal of wastewater, urban runoff, and increased erosion due to land-use changes in the drainage basin.

The concentration of dissolved solids in Lake Michigan increased by some 20 mg/L from 1895 to 1965 (1). This has been due to the discharge of waste products into the lake as well as changes in the land uses of the basin that have altered the quality of water draining from the land. Air pollution has resulted in an increase in the dissolved solids of precipitation into the lake. A large-volume lake such as Lake Michigan can accept some increased amount of common dissolved salts and unreactive sediment without significant water-quality degradation. However, when a lake is lacking in a mineral critical to plant growth, the addition of only a small amount of the **limiting nutrient** can overly stimulate plant growth and result in a dramatic increase in the rooted vegetation and floating algae. This process is known as **eutrophication** (2). In Lake Michigan there is a greater concentration of phosphorus in the water near Milwaukee, Wisconsin, than there is at midlake. There is also a larger concentration of diatoms, a type of algae, near shore, where the phosphorus content is high (3). The source of the increased phosphorus is agricultural and urban runoff, as well as sewage effluent carried into the lake by the Milwaukee River.

The natural quality of ground water varies substantially from place to place. It can range from total dissolved solids contents of 100 mg/L or less for some fresh ground water to more than 100,000 mg/L for some brines found in deep aquifers. The U.S. Environmental Protection Agency has developed a three-part classification system, taking this variability into account, for the ground waters of the United States (4):

Class I: *Special Ground Waters* are those that are highly vulnerable to contamination because of the hydrological characteristics of the areas under which they occur and that are also either an irreplaceable source of drinking water or ecologically vital in that they provide the baseflow for a particularly sensitive ecological system.

Class II: *Current and Potential Sources of Drinking Water and Waters Having Other Beneficial Uses* are all other ground waters except Class III.

Class III: *Ground Waters Not Considered Potential Sources of Drinking Water and of Limited Beneficial Use* because the salinity is greater than 10,000 mg/L or the ground water is otherwise contaminated beyond levels that can be removed using methods reasonably employed in public water-supply treatment.

The U.S. EPA uses this classification scheme in promulgating rules and regulations at the federal level. The highest degree of protection in given to Class I ground water.

Pollution of surface water frequently results in a situation where the contamination can be seen or smelt. However, contamination of ground water most often results in a situation that cannot be detected by human senses. Ground-water contamination can be due to bacteriological or toxic agents or simply to an increase in common chemical constituents to a concentration whereby the usefulness of the water is impaired. Figure 10.1 shows the extent of four plumes of contaminated water that have developed in the San Gabriel ground-water basin of southern California.

In the past, water contamination was primarily due to microbiological agents. Although many advances in public health have been made, incidences of waterborne disease still occur in the United States and appear to be increasing. Of 672 cases of waterborne disease in the United States from 1946 to 1980, 52 percent were due to unknown causes, 22 percent were due to bacteria, 12 percent were viral in nature, 7 percent were due to parasites, 4 percent were caused by inorganic chemicals, and 3 percent were caused by organic chemicals (5). Use of untreated, contaminated ground water was responsible for 35 percent of the disease outbreaks in public water-supply systems during this period, while only 8 percent were due to untreated, contaminated surface water (5). In addition, many of the remaining outbreaks were caused by the failure of systems designed to treat contaminated ground water.

Our understanding of the toxicology of carcinogenic compounds has increased along with the analytical capacity to detect low concentrations of or-

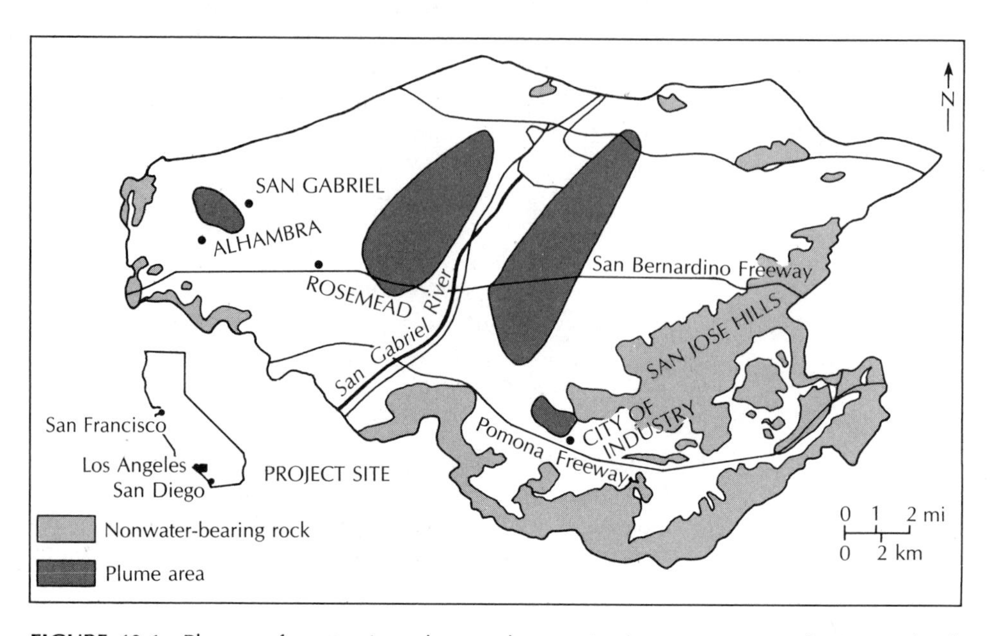


**FIGURE 10.1** Plumes of contaminated ground water in the 170-square-mile San Gabriel ground-water basin, southern California. Contaminants, which include trichloroethylene, perchloroethylene, carbon tetrachloride, and other suspected carcinogens, are found in more than 88 municipal supply wells to depths as great as 1000 feet (300 meters) Multiple sources of contamination are responsible for the common industrial solvents and degreasers found in the ground water. Source: J. J. Kosowatz, and M. J. Sponseller, *Engineering News Record,* 217, no. 21 (November 21, 1986):28–29.

ganic compounds in aqueous samples. Recent regulations have greatly increased the amount of ground-water monitoring required. As a result, numerous instances of ground-water contamination have been revealed. Legal cases involving ground-water contamination have resulted in corporations paying millions of dollars to clean up contaminated ground water as well as paying for damages to families suffering from illness and death alleged to have been caused by organic chemicals in well water ingested by the plaintiffs (6).

The chemical and microbiological agents that are adversely impacting the quality of ground water are coming from a variety of sources, including land application of agricultural chemicals; animal wastes; septic-tank disposal systems for sewage; sewage-treatment lagoons; land application of organic wastes; municipal landfills; toxic- and hazardous-waste landfills; leaking underground storage tanks; faulty underground injection wells, pits, ponds, and lagoons used for storage; treatment and disposal of various liquid compounds; and chemical and petroleum product spills. In this chapter we will consider water-quality standards that have been developed in the United States, methods of monitoring ground-water quality, mass transport of contaminants in flowing ground water, sources of ground-water contamination, and methods of aquifer restoration.

## 10.2 WATER-QUALITY STANDARDS

**Water-quality standards** are regulations that set specific limitations on the quality of water that may be applied to a specific use. **Water-quality criteria** are values of dissolved substances in water and their toxicological and ecological meaning. These data can be used to set water-quality standards (7).

In Public Law 92-500, Section 302, the U.S. Congress directed each state to establish water-quality standards for surface-water bodies. These water-quality standards specify maximum concentrations of substances in surface water for the purpose of protecting aquatic life, users of surface water, and consumers of aquatic life. States are of course free to establish ground-water standards and other types of water-quality standards.

The U.S. Environmental Protection Agency (EPA) has been directed by Congress to establish **drinking water standards** under provisions of Public Law 93-523, the Safe Drinking Water Act, and its amendments. The goal of the Safe Drinking Water Act is to determine **maximum contaminant level goals (MCLGs)** and **maximum contaminant levels (MCLs)** for materials that may be found in drinking water. There are three criteria for selection of contaminants for regulation under this act: (1) the analytical ability to detect a contaminant in drinking water, (2) the potential health risk, and (3) the occurrence or potential for occurrence in drinking water. Maximum contaminant-level goals* are nonenforceable health goals set at a level to prevent known or anticipated adverse effects with an adequate margin of safety. Maximum contaminant levels are enforceable standards that are to be set as close to the MCLGs as is feasible on the basis of the water-treatment technologies and cost. Compounds that are carcinogenic will have their MCLGs set at 0, whereas the MCLGs for chronically toxic compounds are based upon an acceptable daily intake, which takes into account total exposure from air, food, and drinking water. Risk from carcinogenic compounds is expressed in terms of additional cancer risk over a lifetime of exposure at a given level. A cancer risk of $10^{-6}$ means that there would be one additional cancer death out of a population of 1,000,000 people. In 1975 the EPA issued National Primary Drinking Water Standards, which established MCLs for several inorganic and organic substances as well as radioactivity and bacteriological standards. Primary MCLs have been promulgated for substances with a health risk and **secondary maximum contaminant levels (SMCLs)** for substances such as iron, chloride, sulfate, and TDS, which can affect the aesthetic quality of water by imparting taste and odor and staining fixtures. Table 10.2 lists the national interim drinking water standards.

---

*Prior to May 1986, MCLGs were known as **recommended maximum contaminant levels (RMCLs)**.

**Table 10.2** National interim drinking water standards

**Maximum contaminant levels for inorganic chemicals**

| Contaminant | Level, milligrams per liter (micrograms per liter in parentheses) |
|---|---|
| Arsenic | 0.05 (50 μg/L) |
| Barium | 1. (1000 μg/L) |
| Cadmium | 0.010 (10 μg/L) |
| Chromium | 0.05 (50 μg/L) |
| Fluoride | 2.2 |
| Lead | 0.05 (50 μg/L) |
| Mercury | 0.002 (2 μg/L) |
| Nitrate (as N) | 10. |
| Selenium | 0.01 (10 μg/L) |
| Silver | 0.05 (50 μg/L) |

**Secondary inorganic chemical and physical standard**

(1) Waters containing inorganic chemicals in quantities above the limits contained in this section are not hazardous to health but may be objectionable to an appreciable number of persons.

| Standard | Milligrams per liter (micrograms per liter in parenthesis) except as noted |
|---|---|
| Chloride | 250 |
| Color | 15 units |
| Copper | 1.0 (1,000 μg/L) |
| Corrosivity | Noncorrosive |
| Foaming agents | |
| MBAS (methylene-blue active substances) | 0.5 |
| Hydrogen sulfide | not detectable |
| Iron | 0.3 |
| Manganese | 0.05 (50 μg/L) |
| Odor | 3 (Threshold No.) |
| Sulfate | 250 |
| Total residue | 500 |
| Zinc | 5 (5,000 μg/L) |

**Maximum contaminant levels for organic chemicals**

| Contaminant | Level, milligrams per liter |
|---|---|
| (1) Chlorinated hydrocarbons: | |
| Endrin (1,2,3,4,10, 10-hexachloro- 6,7-expoxy-1,4, 4a,5,6,7,8,8a-octahydro-1,4-endo, endo-5,8- dimethano naphthalene). | 0.0002 |

**Table 10.2** Continued

| | |
|---|---|
| Lindane (1,2,3,4,5,6-hexachloro-cyclohexane, gamma isomer). | 0.004 |
| Methoxychlor [1,1,1-trichloro- 2, 2 - bis (*p*-methoxyphenyl) ethane]. | 0.1 |
| Toxaphene $C_{10}H_{10}Cl_8$-Technical chlorinated camphene, 67–69 percent chlorine). | 0.005 |
| (2) Chlorophenoxys: | |
| 2,4 - D (2,4-dichlorophenoxyacetic acid). | 0.1 |
| 2,4,5 - TP Silvex (2,4,5-trichlorophenoxypropionic acid). | 0.01 |
| (3) Total trihalomethanes [the sum of the concentrations of bromodichloromethane, dibromochloromethane, tribromomethane (bromoform), and trichloromethane (chloroform)]. | 0.10 |

**Maximum contaminant levels for radium-226, radium-228, and gross alpha particle radioactivity**

(1) Combined radium-226 and radium-228—5 pCi/L.

(2) Gross alpha particle activity (including radium-226 but excluding radon and uranium)—15 pCi/L.

**Maximum contaminant levels for beta particle and photon radioactivity from manmade radionuclides in community water systems**

(1) The average annual concentration of beta particle and photon radioactivity from manmade radionuclides in drinking water shall not produce an annual dose equivalent to the total body or any internal organ greater than 4 millirem/year.

(2) Except for the radionuclides listed below, the concentration of manmade radionuclides causing 4 mrem total body or organ dose equivalents shall be calculated on the basis of a 2-liter per day drinking water intake. If 2 or more radionuclides are present, the sum of their annual dose equivalent to the total body or to any organ shall not exceed 4 millirem/year.

Average annual concentrations assumed to produce a total body or organ dose of 4 mrem/yr

| Radionuclide | Critical Organ | pCi per liter |
|---|---|---|
| Tritium | Total body | 20,000 |
| Strontium-90 | Bone marrow | 8 |

Source: *Federal Register*, Feb. 1978, No. 266.

The process of establishing final maximum contaminant levels under the Safe Drinking Water Act will be a continuous process. The U.S. Congress will specify lists of compounds for which the EPA will issue proposed MCLGs, MCLs, and SMCLs. After public comments are received and analyzed, the EPA will issue final MCLGs, MCLs, and SMCLs. Periodically, Congress will assign additional compounds for which MCLGs and MCLs will be established. What was a fairly short list of Interim Drinking Water Standards will become a long list of Final Drinking Water Standards. Table 10.3 lists the contaminants that will be regulated under the Safe Drinking Water Act as of the date given on the table. Table 10.4 lists the status of the MCLGs and MCLs for various contami-

**TABLE 10.3** Contaminants regulated under Safe Drinking Water Act, 1986 amendments

**Volatile Organic Chemicals**
Trichloroethylene*
Tetrachloroethylene
Carbon tetrachloride*
1, 1, 1-Trichloroethane*
1, 2-Dichloroethane*
Vinyl chloride*
Methylene chloride
Benzene*
Chlorobenzene*
Dichlorobenzene(s)*
Trichlorobenzene(s)*
1, 1-Dichloroethylene*
trans-1, 2-Dichloroethylene*
cis-1, 2-Dichloroethylene*

**Microbiology and Turbidity**
Total coliforms*
Turbidity*
*Giardia lamblia**
Viruses*
Standard plate count
*Legionella*

**Inorganics**
Arsenic*
Barium*
Cadmium*
Chromium*
Lead*
Mercury*
Nitrate*
Selenium*
Silver
Fluoride*
Aluminum
Antimony
Molybdenum
Asbestos*
Sulfate
Copper*
Vanadium
Sodium
Nickel
Zinc
Thallium
Beryllium
Cyanide

**Organics**
Endrin
Lindane*
Methoxychlor*
Toxaphene*
2, 4-D*
2, 4, 5-TP*
Aldicarb*
Chlordane*
Dalapon
Diquat
Endothall
Glyphosphate
Carbofuran*
Alachlor*
Epichlorohydrin*
Toluene*
Adipates
2, 3, 7, 8-TCDD (Dioxin)
1, 1, 2-Trichloroethane
Vydate
Simazine
Polynuclear aromatic hydrocarbons (PAHs)
Polychlorinated biphenyls (PCBs)
Atrazine
Phthalates
Acrylamide*
Dibromochloropropane (DBCP)*
1, 2-Dichloropropane*
Pentachlorophenol*
Pichloram
Dinoseb
Ethylene dibromide*
Dibromomethane
Xylene*
Hexachlorocyclopentadiene

**Radionuclides**
Radium 226 and 228
Beta particle and photon radioactivity
Uranium
Gross alpha particle activity
Radon

*Included in USEPA proposed and final rules published in *Federal Register*, Nov. 13, 1985.

nants as of the date given on the table. The reader should be cognizant of the fact that these lists will be updated by the EPA from time to time and that the EPA should be consulted about any specific MCLG or MCL. Maximum contaminant levels will be enforced for all public water-supply systems by the various states. States are free to set MCLs that are more strict than the federal standards, but not less stringent.

Drinking-water standards are especially important for evaluating ground-water quality because many consumers utilize untreated ground water that is pumped directly from a well. Public water-supply systems that rely upon ground water are required to perform a complete analysis of the water for the drinking-water standards prior to the time a well is put into service and periodically thereafter. Private wells are often tested for bacteria and nitrate only when they are first drilled and then never tested again. It is important to maintain high quality in ground water in order to protect private well owners. State water-quality standards for ground water are sometimes based on the drinking-water standards (see Chapter 11).

## 10.3 COLLECTION OF WATER SAMPLES

The practicing hydrogeologist rarely will perform the chemical analysis of water samples; this is typically done in a specialized analytical laboratory. However, the hydrogeologist will usually be involved with the collection of water samples in the field. This section will focus on methods of collecting representative water samples for chemical analysis.

The program to sample both ground and surface water must be carefully planned. There are four basic steps involved:

1. Determination of the purpose of the sampling program. Is the objective to define the basic water chemistry, to determine if the water meets drinking-water standards, or to determine if there is contamination present? Will surface water, water in the vadose zone, or ground water be tested?

2. Deciding how many sampling points will be tested. Will all possible points be tested, or will only selected sampling points be involved? Will new sampling points, such as ground-water monitoring wells, be needed?

3. Determining which chemical constituents will be analyzed and the **quantification limits** that the lab will employ. Analytical instruments have a lower limit to the range in which the results can be quantified and below that a range where a compound can be detected, but not quantified. Results can be expressed as detected, but not quantified, or as not detected. In some cases, the detection limit will be more sensitive for some instrumental methods than others for the same compound. The quantification limits selected should be based on the purpose of the sampling program.

4. Development of a **quality assurance/quality control (QA/QC)** program. There are many aspects of QA/QC, the purpose of which is to assure that the analytical results reported by the laboratory accurately express the actual con-

**TABLE 10.4** USEPA Drinking-water standards and health goals

| Chemical | MCLG (μg/L) | MCL (μg/L) | SMCL (μg/L) |
|---|---|---|---|
| **Volatile Organic Chemicals** | | | |
| Trichloroethylene | 0# | 5† | |
| Carbon tetrachloride | 0# | 5† | |
| Vinyl chloride | 0# | 2† | |
| 1,2-Dichloroethane | 0# | 5† | |
| Benzene | 0# | 5† | |
| 1,1-Dichloroethylene | 7# | 7† | |
| 1,1,1-Trichloroethane | 200# | 200† | |
| p-Dichlorobenzene | 75† | 75† | |
| **Synthetic Organic Chemicals** | | | |
| Acrylamide | 0* | | |
| Alachlor | 0* | | |
| Aldicarb (including aldicarb sulfoxide & aldicarb sulfone) | 9* | | |
| Carbofuran | 36* | | |
| Chlordane | 0* | | |
| cis-1,2-Dichloroethylene | 0* | | |
| Dibromochloropropane (DBCP) | 0* | | |
| 1,2-Dichloropropane | 6* | | |
| o-Dichlorobenzene | 620* | | |
| 2,4-Dichlorophenoxyacetic acid | 70* | | |
| Epichlorohydrin | 0* | | |
| Ethyl benzene | 680* | | |
| Ethylene dibromide (EDB) | 0* | | |
| Heptachlor | 0* | | |
| Lindane | 0.2* | | |
| Methoxychlor | 340* | | |
| Monochlorobenzene | 60* | | |

centrations of the solutes in the water as it existed in the field (8–10). This is not a trivial problem. It is beyond the scope of this book to discuss laboratory methods of QA/QC. The hydrogeologist has two basic methods of checking on the **accuracy** and **precision** of the laboratory. Accuracy—the ability of the laboratory to report what is in the sample—can be measured by the use of **spiked samples,** where a set of samples with a known concentration of a solute is submitted to the lab. Precision—the ability of the laboratory to reproduce results—is determined by submitting **duplicate samples** from the same source. The duplicate sample should be thoroughly mixed before being split for shipment to the lab. A field

**TABLE 10.4** Continued

| Chemical | MCLG (μg/L) | MCL (μg/L) | SMCL (μg/L) |
|---|---|---|---|
| Pentachlorophenol | 220* | | |
| Polychlorobiphenyls (PCBs) | 0* | | |
| Styrene | 140* | | |
| Toluene | 2000* | | |
| Toxaphene | 0* | | |
| trans-1,2-Dichloroethylene | 70* | | |
| 2-(2,4,5-Trichlorophenoxy)-propionic acid (2,4,5-TP) | 52* | | |
| Xylene | 440* | | |
| **Inorganic Chemicals** | | | |
| Arsenic | 50* | | |
| Asbestos | 7.1 x $10^6$ long fibers per liter* | | |
| Barium | 1500* | | |
| Cadmium | 5* | | |
| Chromium | 120* | | |
| Copper | 1300* | | |
| Fluoride | 4000# | 4000* | 2000* |
| Lead | 20* | | |
| Mercury | 3* | | |
| Nitrate | 10000* | | |
| Nitrite | 1000* | | |
| Selenium | 45* | | |
| **Microbiological Parameters** | | | |
| *Giardia* | 0 organisms* | | |
| Total coliforms | 0 organisms* | | |
| Turbidity | 0.1 turbidity units* | | |
| Viruses | 0 organisms* | | |

#Final value. Published in *Federal Register*, Nov. 13, 1985.
*Proposed value. Published in *Federal Register*, Nov. 13, 1985.
†Final value. Published in *Federal Register*, Jul. 8, 1987.

sampling program should have duplicates submitted to the lab for 10 percent of the samples, and they should be submitted as **blind duplicates** so the lab doesn't know which samples are duplicates. **Field blanks** are used to assess the field sampling program. Highly purified water (HPLC grade) is taken into the field in a sealed container, run through the field sampling devices, placed into sample containers and shipped to the laboratory for analysis. If trace amounts of solutes are reported in both the samples and the field blanks, then they can be assumed to be a result of the sampling or lab methodology and are not actually present in the water in the field.

Sampling protocols have been developed by the U.S. EPA (11). They specify the type of sample that is needed (grab or composite), the type of container that is to be used for the sample, the method by which the sample container is cleaned and prepared, whether or not the sample is filtered, the type of preservative that is to be added to the sample in the field, and the maximum of time the sample can be held prior to analysis in the laboratory.

It is good field practice to thoroughly clean the sampling device prior to use. The method of cleaning should be such that no residue remains. For example, equipment rinsed in acetone should first be rinsed thoroughly with distilled water after the acetone has been used and then autoclaved to volatilize any acetone that may not have been rinsed away. Frequently the analytical methods employed will detect acetone in concentrations of a few parts per billion. The sampling devices and bottles should be rinsed with a sample of the water being sampled if they are not thoroughly dry. This will prevent the mixing of rinse water with the final sample.

## 10.4 GROUND-WATER MONITORING

### 10.4.1 PLANNING A GROUND-WATER MONITORING PROGRAM

The science of ground-water sampling has advanced greatly in recent years, not only in our understanding of the techniques to be used, but in the development of materials and equipment used in the sampling process (11–17). The first step in designing a ground-water monitoring program is to determine the purpose. There are at least four major reasons to monitor ground water: to determine the water quality and chemistry of a region, to determine the water quality and chemistry of a specific water-supply well or well field, to determine the extent of ground-water contamination from a known source, and to monitor a potential source of contamination to determine if the ground water becomes contaminated.

If the purpose of a study is to evaluate the existing water quality and chemistry of a region, then it is likely that only existing wells and springs would be sampled. In this event, it is necessary to know the construction details of the well and pump. One needs to know what aquifer the well is drawing water from in order to interpret the chemical analyses. Wells that tap more than one aquifer should not be used in these studies, which are usually designed to map the distribution of ions in a specific aquifer. A good geographical distribution of wells is needed. In studies of chemical hydrogeology, wells located in both recharge and discharge areas should be sampled. One must be careful that the water sample collected has not been altered by the well system. A common occurrence with existing private wells is that all of the water in the system passes through a water softener, which changes the water chemistry. Some older-model submersible pumps have capacitors that, if they leak, can introduce polychlorinated bi-phenols (PCBs) into the well water.

If the water quality of a potential well site is being investigated, nearby wells may be sampled to establish regional water quality. In most cases it is desirable to construct a test well before a permanent well is installed. As the test boring is being advanced, temporary test wells and screens can be installed at progressively deeper depths to sample multiple aquifers or potential water-yielding zones in the same aquifer. At each screen zone the drill column and bit are removed from the hole and a temporary casing and well screen are installed. The temporary well is pumped to develop it until the water is clear. Replicate water samples are then collected and analyzed. The hydraulic and water-quality results of the various tested zones are compared and the best one is selected for the permanent well construction. It has been shown that water samples from such temporary test wells are very similar to those from high-capacity municipal wells later constructed in the same aquifer (18). If there is only one potential aquifer zone, a small-capacity test well may be installed on a permanent basis. This well can later be used as an observation well for a pumping test on the permanent well and may be useful as a standby well for emergency use.

### 10.4.2 INSTALLING GROUND-WATER MONITORING WELLS

Most of the **ground-water monitoring wells** installed today are for the purpose of determining ground-water quality at localities such as waste-storage or -disposal facilities, underground storage tanks, mines, and areas of known or suspected ground-water contamination. Ground-water monitoring wells are installed for the specific purpose of determining the quality of the ground water in a specific aquifer and at a particular location.

The design of a typical ground-water monitoring well is given in Figure 10.2. There are a number of specific steps that need to be followed in order to properly install a ground-water monitoring well. One important consideration is to take care during the installation of the well that contaminating materials are not introduced into the ground (19).

1. A **boring** is advanced into the ground by means of a drilling rig. The drilling method needs to be selected on the basis of the local geology, size and depth of the monitoring well to be installed, and available expertise. Borings in unconsolidated materials are frequently installed by means of **hollow-stem augers,** which are rotated into the ground to bring up soil and create the boring (Figure 10.3). A plug at the bottom of the hollow stem keeps soil from going up into the interior of the stem.

2. Samples of the geologic materials are then collected so that a **geologic log** can be constructed. A **split-spoon sampler,** a hollow tube comprised of two halves, or a **Shelby tube,** a one-piece hollow tube, may be driven into the soil ahead of the **bit,** or cutting edge, at the bottom of the hollow-stem auger by

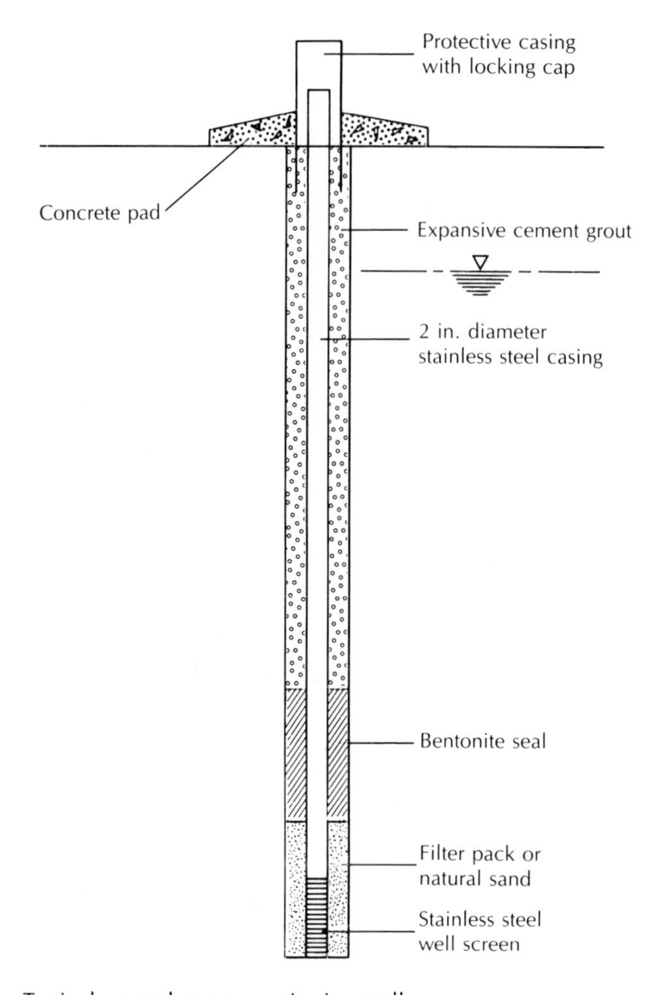


**FIGURE 10.2** Typical ground-water monitoring well.

temporarily removing the end plug. The sampling tube is retrieved and the soil sample is ejected and described by the geologist. It is also available for further testing, such as grain-size analysis.

3. Once the auger is at the desired final depth of the bottom of the monitoring well, the plug at the end is removed. A **knockout plug** made of some noncontaminating material such as stainless steel may also be used if soil samples are not being collected as the auger is being advanced.

4. The well consists of a **casing** and a **screen.** The casing is a piece of solid-wall pipe and the screen is a piece of pipe with holes, slots, gauze, or a continuous wire wrapped around it. The purpose of the screen is to allow water to enter the well but to keep out the soil. A commercially manufactured screen should be used rather than a pipe with slots cut into it. The screen is attached

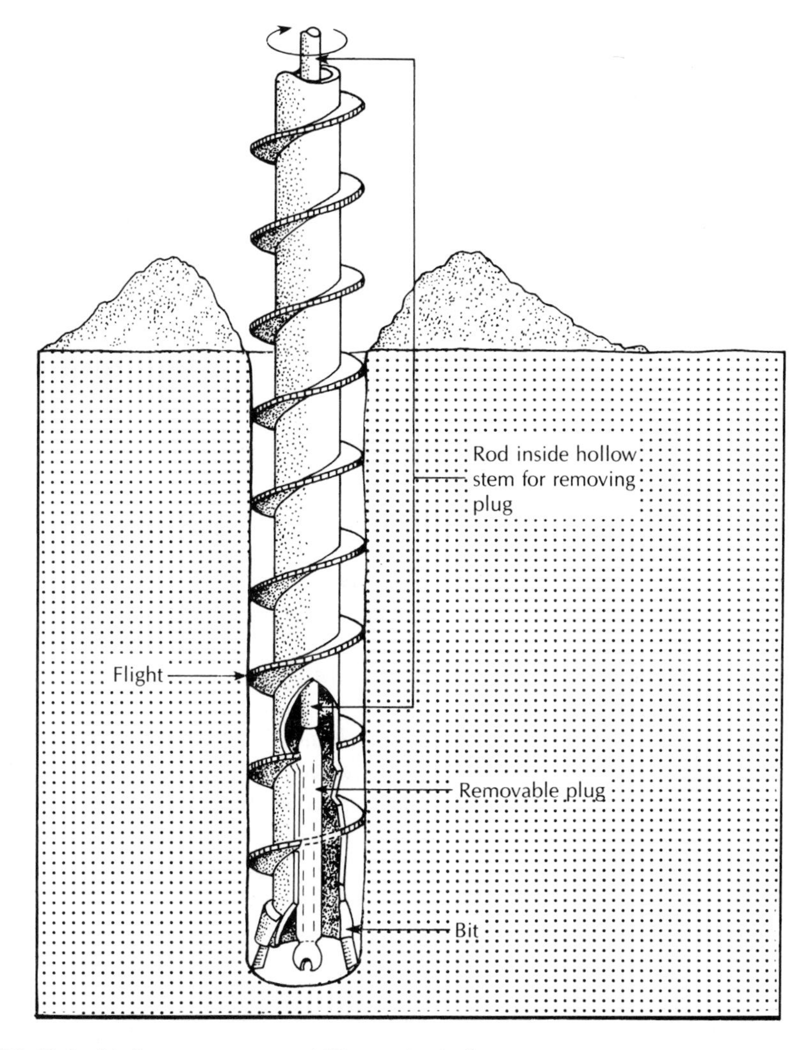


**FIGURE 10.3** Hollow-stem auger drilling. The hollow-stem, continuous-flight auger bores into soft soils, carrying the cuttings upward along the flights. When the desired depth is reached, the plug is removed from the bit and withdrawn from inside the hollow stem. A well point on a casing (1¼-in. or 2-in.) can then be inserted to the bottom of the hollow stem and the auger pulled out, leaving the small-diameter monitoring well in place. Source: M. L. Scalf and others, *Manual of Ground Water Sampling Procedures,* National Water Well Association, 1981.

to the end of the casing by means of threaded joints. Casings and screens made from several different materials are commercially available. The hydrogeologist must select a material on the basis of cost, durability, and potential reactivity with the water in the aquifer. Teflon is the most costly, least durable, but most inert material. Stainless steel is the most durable material; it has moderate cost and is also essentially inert. Rigid polyvinyl chloride (PVC) pipe, because of low cost, is frequently used. Only PVC pipe with threaded joints is acceptable; PVC pipe joined with solvent-welded joints should never be used because solvent can add organic contaminants to the water. Because of its structural weakness and high cost, for most uses Teflon is inferior to stainless steel as well-casing material. Stainless steel may react unfavorably with acidic or saline ground water; rigid PVC casing with threaded joints would be superior under such conditions.

In addition to monitoring wells, there are **multilevel sampling devices,** which can be installed in a single borehole to sample ground water. When monitoring wells are used to sample ground water from different depths, each monitoring well must be installed in a separate borehole. Drilling costs can be reduced substantially by the use of multilevel sampling devices. Figure 10.4 shows the design of a multilevel device used in sandy soil with a shallow water table and primarily horizontal flow (20). The device consists of a rigid PVC pipe inside of which are multiple tubes, each of which ends at a sampling port at a different depth. Such a device can be used to collect ground-water samples at elevations in the aquifer as close as one to two feet. Only a small sample of water is withdrawn, so that each sample represents water from a very small portion of the aquifer. As a result, a very detailed picture of the vertical distribution of ground-water contamination can be developed. The multilevel device is installed by augering a boring to the desired depth with hollow-stem augers. The PVC pipe is lowered through the hollow stem and, as the augers are withdrawn, the sand heaves into the annular space around the pipe and seals it off. One disadvantage of the device is that water levels cannot usually be measured.

A second type of multilevel device can be used in a borehole in bedrock or cohesive material such as dense glacial till. Figure 10.5 shows the design of such a device (21), which is similar to the one shown in Figure 10.4. However, inflatable packers are located above and below each sampling port. In this case the device is lowered into an open borehole and each zone to be sampled is isolated by inflating the packers above and below the sampling port. The device is intended to be permanently installed so that multiple samples over time can be taken. If needed, it could be removed from a well after a study is completed and reused.

Once a monitoring well or other device is installed, it is necessary to develop the well. The purpose of **well development** is to remove any fine material that may be blocking the well screen or port. A secondary purpose may be to remove any water from the aquifer that was introduced during well construction and which may not be representative of the local ground-water quality. Well development is usually accomplished by surging the well, that is, making the water in the well flow into and out of the well screen for a period of time. The well is then usually pumped for a while to remove the loose sediment from the well casing and screen. In order to maintain the integrity of the well for water-quality sampling, it is usually not good practice to add any water to the well during well development.

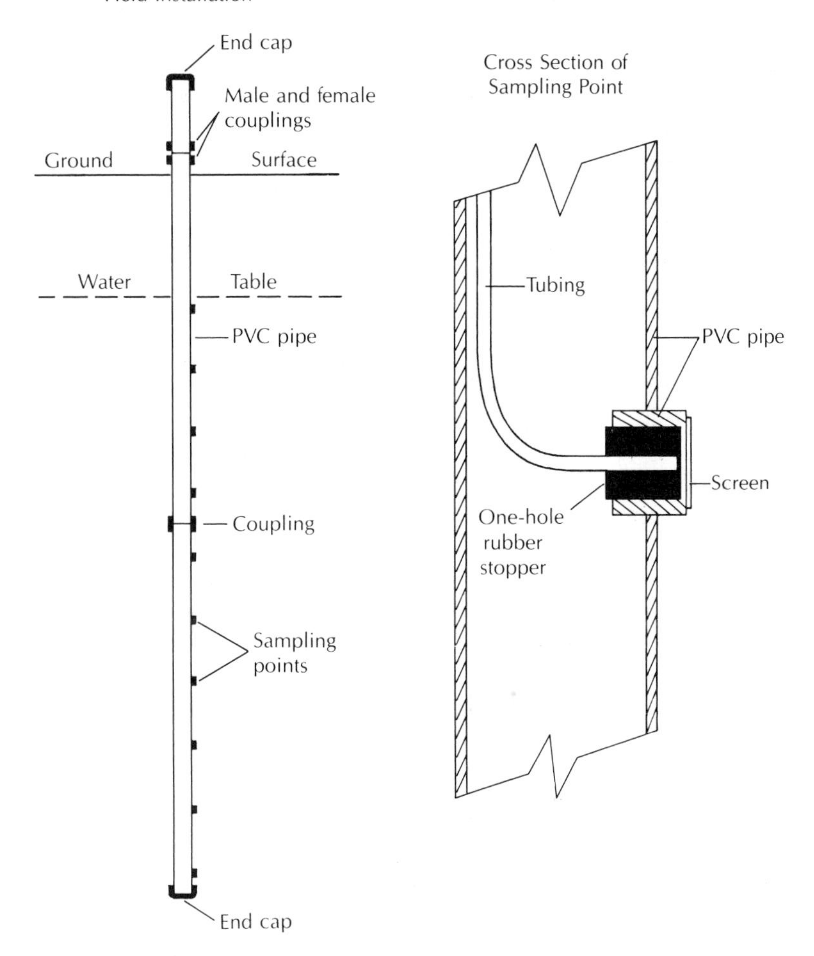


**FIGURE 10.4** Multilevel ground-water sampling device for use in sandy soil. Source: J. F. Pickens and others. *Ground Water Monitoring Review,* 1, no. 1 (1981):48–51.

### 10.4.3 WITHDRAWING WATER SAMPLES FROM MONITORING WELLS

Once the monitoring well is installed and developed, a method or removing the water from the well must be selected. There are a number of different ways to pump water from a monitoring well (22–24). There are a large number of different devices commercially available for this task. The basic considerations in selecting a pumping device will be (1) does it collect a representative sample, (2) can it be easily cleaned and decontaminated if it is to be used in more than one well, (3) will it work in the application that is at hand, (4) can it

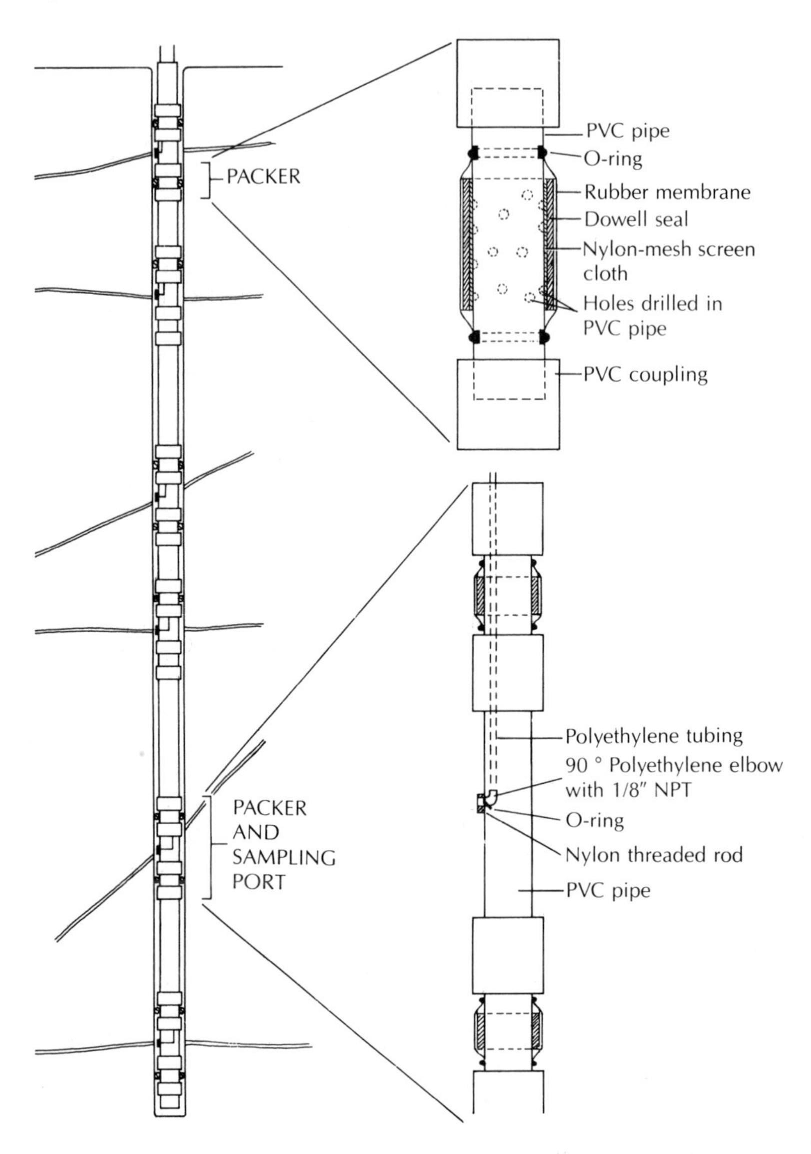


**FIGURE 10.5** Multilevel ground-water sampling device for use in fractured rock borehole. Source: J. A. Cherry and P. E. Johnson, *Ground Water Monitoring Review,* 2, no. 3 (1982):41–44.

lift the water from the water level in the well to the surface, (5) can it pump the well at a rate sufficient to purge it prior to sampling, (6) will the method of pumping or the materials from which the pump is made change the water chemistry of the sample, (7) how easy is it to use the device, (8) how reliable is the device, and (9) how much does it cost to buy, maintain, and operate?

In selecting a sampling device it is important to pick one that will not alter the chemistry of the sample as it is brought to the surface. This could occur if the materials from which the pump is constructed would either leach compounds into the sample or absorb compounds from the sample. One of the first considerations will be the selection of the type of material used in the construction of the pumping device. Teflon and stainless steel are inert materials that could be used for the rigid parts of the pump; Teflon and polypropylene are inert materials that could be used for the flexible parts of the pump. Polyvinyl chloride may be acceptable for some uses, but is not as inert as the other materials.

Changes in the pressure of the sample while it is being transported to the surface can cause loss of dissolved gases. This can result in a change in pH due to change in carbon dioxide, a change in *Eh* due to a change in oxygen, and a change in the dissolved volatile organic compounds due to a drop in pressure. The best sampling devices will not put the sample in contact with air or noninert gasses as it is brought to the surface and will maintain the sample under positive pressure. These considerations limit the usefulness of several sampling devices, such as suction pumps, peristaltic pumps, and air-lift pumps.

**Bailers** are tubes that can be made of any material and that have a check valve on the bottom. They are inexpensive and simple to operate as they can be lowered into the well on a wire or cord and yield a representative water sample if used carefully. **Bladder pumps** are positive-displacement devices that use a pulse of gas to push the sample to the surface. The gas does not come into contact with the sample and positive pressure is maintained at all times. The bladder pump yields a very representative water sample, provided the correct materials are used in its construction. The bladder pump is superior in performance to the bailer, but costs more than an order of magnitude more.

One concern in sampling a monitoring well is to be assured that the sample does not contain any water that was standing in the well casing. One way to assure this is to purge the well before pumping. Depending upon the design of the well and the type of pump, from one to five times the volume of water standing in the well casing and the screen should be removed prior to withdrawing the water sample. During the purging process it is good practice to monitor the pH and conductivity of the water until a stable condition is reached. It is also good practice to purge the well until any turbidity has cleared, although in some wells this may not be possible. If the well is slow to recharge, it is probably not a good idea to pump the well dry during the purging process. For such a well, one to two well volumes should be sufficient to purge any stagnant water.

Deep open-borehole wells in solid rock or mine shafts can present an opportunity to the hydrogeologist to obtain water-quality data at depths to which it might be uneconomical to construct dedicated monitoring wells. If the well is not pumped for some time and the borehole is not acting as a conduit for water flowing from one aquifer to another, the water quality at various levels in the borehole can reflect the ground-water quality at the same level. **Borehole geochemical probes** are water-quality monitoring devices that can be lowered into the well on a cable to measure such parameters as pH, *Eh,* temperature, and specific conductance. By lowering the probe and taking readings at discrete intervals, geochemical well logs can be constructed. In addition, the **Kemmerer sampler,** a sample collection device developed for drawing water samples from depths in lakes and oceans, can be used in deep wells. The Kemmerer sampler is lowered into the borehole in an open position until the desired depth is reached. At that point a weight is sent down on the cable, triggering a spring-loaded device on the sampler to close it. The closed Kemmerer sampler is then raised to the surface by the cable and the water sample is withdrawn. The Kemmerer sampler can be used to collect water samples at different depths so that a geochemical log can be developed.

A modification of the medical **syringe** can be used to draw a water sample from a specific depth in a monitoring well. The syringe is lowered into the well on the end of a length of tubing. A weight may be needed to sink the syringe and tubing. At the desired depth, a vacuum is applied to the tubing and the pressure of the water in the well will force the syringe plunger up, forcing a water sample into the syringe. The syringe can then be raised up by the tubing. The advantages of the syringe are that the sample does not undergo a pressure change or exposure to the atmosphere. The syringe can even be used as the sample storage container. However, the sample volume is small, and the syringe cannot be used to purge the well. If the syringe is used for sample storage, the chance of cross-contamination is very small if the tubing and other appurtenances are carefully cleaned between wells (24).

## 10.5 VADOSE-ZONE MONITORING

If a contaminant is detected in the ground water, it is an indication that a problem is at hand. Solutions to the problem may be difficult and expensive. It would be desirable to be able to determine if ground-water contamination is likely to occur at some time in the future so that corrective action could be taken early to prevent contamination from reaching the water table. For example, when a landfill or a lagoon is constructed, a ground-water monitoring system might be installed to detect leaks from the facilities. Storage facilities for toxic and hazardous materials can also be monitored for leaks. However, if a leak were to be detected prior to the time that the contaminant reaches the water table, corrective action, such as draining a lagoon and repairing the liner, might be possible before the ground water becomes contaminated. Leaks can potentially be detected by monitoring the water in the vadose (unsaturated) zone (25–29).

Contaminants in the unsaturated zone can move in both the liquid and vapor phases. Vapors can move in any direction depending upon their relative density and air-pressure gradients. Liquids can generally move downward only. Vadose-zone monitoring systems for liquids must be constructed beneath the facility to be monitored as that is the direction the liquid would be seeping. Therefore, such devices are usually installed prior to the time that the facility is constructed. It is difficult to install such devices after construction; for example, in order to put a collection device beneath an active landfill, either a vertical boring must be made through the landfill or an angular boring must be made from the side. Because of the fact that vapors can migrate laterally, gas-monitoring wells can be placed to the side of active facilities. Hence it is easier to retrofit a site with gas-monitoring devices than liquid-monitoring devices in the vadose zone.

Vapors can be detected by **gas-monitoring wells,** which are simply wells that terminate in the unsaturated zone (Figure 10.6). Gas-monitoring wells can be used to detect the movement of methane from municipal landfills. Methane has been implicated in explosions near landfills. Gas-monitoring wells can also detect the vapors of volatile organic chemicals in the vadose zone. These compounds move rapidly in ground water and many are known or suspected carcinogens. Gas-monitoring wells can be located beneath a landfill or lagoon if they

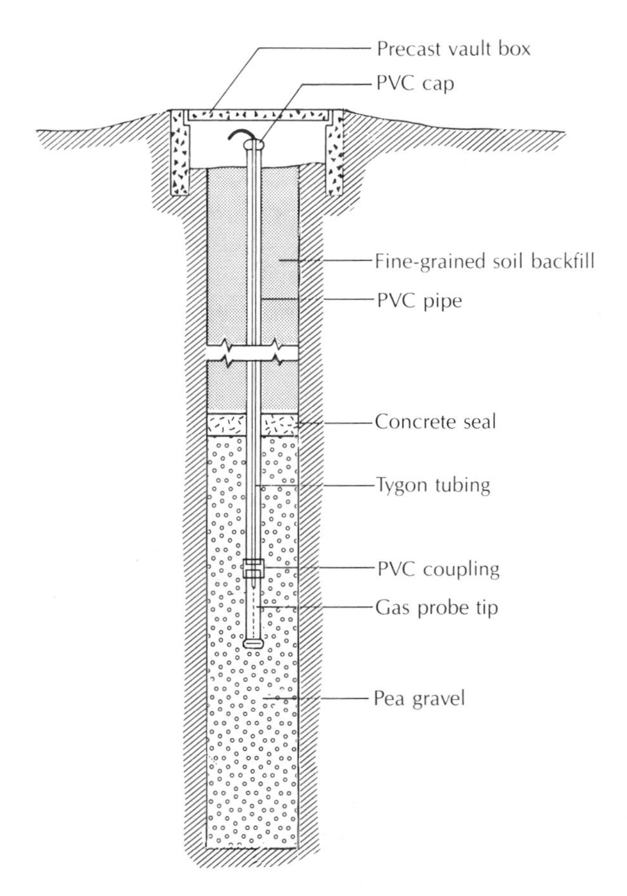


**FIGURE 10.6** Gas monitoring well in vadose zone. Source: L. S. Wilson, *Ground Water Monitoring Review,* 3, no. 1 (1983):155–66.

are installed before construction. After construction, it is easier to place them to the side of an active facility.

The vadose zone can be monitored by indirect means. For example, a lagoon that holds aqueous liquids could be monitored for changes in soil moisture. Changes in soil moisture could mean a leak. If the lagoon holds a brine, changes in the electrical conductivity of resistivity blocks buried in the soil beneath the landfill could mean a leak. Temperature changes in the vadose zone could also indicate a leak.

The vadose zone can also be sampled by direct sampling of the soil. Soil samples can be removed by shovel, by a sampling tube being pushed into the ground, or by augering a borehole and then pushing a sampling tube ahead of the auger. Soil samples can be tested for moisture content, the moisture can be extracted and tested for chemical composition, or the soil sample can be leached with a standard leaching solution and the leachate tested. If the soil sample is placed in a vial for volatile organic compound testing within a few seconds of being exposed to the atmosphere, most of the volatile organics in the soil will be retained and testing of the volatile organics in the headspace in the vial will give a good indication of the organics in the soil.

There are a number of devices that can be placed in the vadose zone to collect a sample of the liquid in the pores. These fall into two categories: (1) **suction lysimeters,** which are devices that apply tension to a porous ceramic cup to induce the pore water which is under negative pressure to enter the porous cup, whereby it can be transported to the surface, and (2) **collection lysimeters,** which have a membrane liner buried in the vadose zone so that liquid collects on the liner until the soil above it becomes saturated, forcing the liquid to flow to a wet well, where it can be sampled.

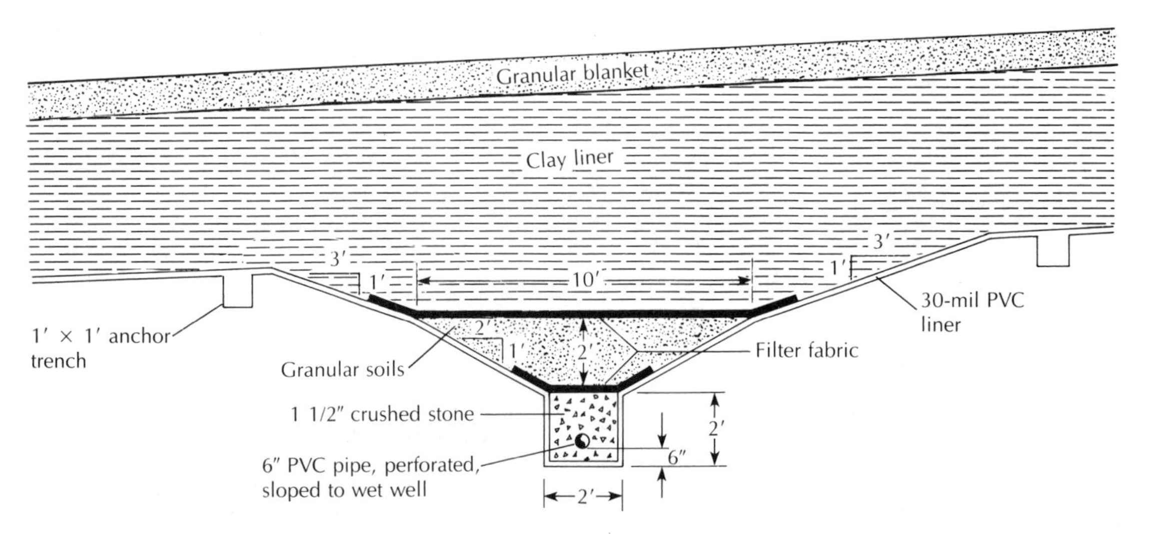


**FIGURE 10.7** Cross-section of leachate-collection lysimeter beneath clay liner of a landfill.

The suction lysimeter has been used for many years in agricultural applications. It has been used in landfills to monitor for leachate with mixed success. The vacuum that can be applied is limited to 0.5 to 0.8 bar. Consequently, in some applications, the lysimeter fails to collect liquids, probably because it becomes clogged. It also is somewhat limited in that it samples liquids from only a very small portion of the aquifer. If it doesn't happen to be located under a leak, that leak may go undetected.

Collection lysimeters are less prone to failure and have the added advantage that they can be made any size; they can, therefore, be placed under spots where failure of a system component is more likely, such as a seam. Figure 10.7 shows a collection lysimeter beneath a portion of a landfill. One landfill in Wisconsin was located over an aquifer where it appeared that ground-water monitoring would be very difficult to do reliably. The landfill was designed with a collection lysimeter beneath the entire landfill to serve as the primary means of detecting leaks in the clay liner of the landfill. Obviously, such a system must be installed as the facility is being constructed.

## 10.6 MASS TRANSPORT OF SOLUTES

### 10.6.1 INTRODUCTION

In studying ground-water contamination it is helpful to understand the basic theory behind the movement of solutes contained in ground water. In the study of water chemistry, the processes by which substances can become dissolved in water are examined. However, the processes by which these substances move through porous media are complex. They can be expressed mathematically, although in some instances we do not fully understand how to obtain the field data necessary to apply the theoretical equations.

There are two basic processes operating to transport solutes. **Diffusion** is the process by which both ionic and molecular species dissolved in water move from areas of higher concentration (i.e., chemical activity) to areas of lower concentration. **Advection** is the process by which moving ground water carries with it dissolved solutes. We will see how, as solutes are carried through porous media, the process of **dispersion** acts to dilute the solute and lower its concentration. Finally, there are chemical and physical processes that cause **retardation** of solute movement so that it may not move as fast as the advection rate would indicate.

### 10.6.2 DIFFUSION

The diffusion of a solute through water is described by Fick's laws. Fick's first law describes the flux of a solute under steady-state conditions:

$$F = -D\, dC/dx \qquad \textbf{(10-1)}$$

where

$F$ = mass flux of solute per unit area per unit time
$D$ = diffusion coefficient (area/time)
$C$ = solute concentration (mass/volume)
$dC/dx$ = concentration gradient (mass/volume/distance)

The negative sign indicates that the movement is from greater to lesser concentrations. Values for $D$ are well known for electrolytes in water. For the major cations and anions in water, $D$ ranges from $1 \times 10^{-9}$ to $2 \times 10^{-9}$ $m^2/s$.

For systems where the concentrations may be changing with time, Fick's second law may be applied:

$$dC/dt = D\, d^2C/dx^2 \qquad \textbf{(10-2)}$$

where

$dC/dt$ = change in concentration with time

Both Fick's first and second law as expressed above are for one-dimensional situations. For three-dimensional analysis, more general forms would be needed.

In porous media, diffusion cannot proceed as fast as it can in water because the ions must follow longer pathways as they travel around mineral grains. In addition, the diffusion can take place only through pore openings because mineral grains block many of the possible pathways. To take this into account, an effective diffusion coefficient must be used. This is termed $D^*$.

The value of $D^*$ can be determined from the relationship

$$D^* = wD \qquad \textbf{(10-3)}$$

where $w$ is an empirical coefficient that is determined by laboratory experiments. For species that are not adsorbed onto the mineral surface it has been determined that $w$ ranges from 0.5 to 0.01 (30). Berner (31) gave a nonempirical relationship between $D^*$ and $D$ that indicated that $D^*$ was equal to $D$ times the porosity divided by the square of the **tortuosity** of the flow path of the diffused species. Tortuosity is the actual length of the flow path, which is sinuous in form, divided by the straight-line distance between the ends of the flow path. Unfortunately, tortuosity cannot be determined in the field, and one is left with the experimental approach.

The process of diffusion is complicated by the fact that ions must maintain electrical neutrality as they diffuse. If we have a solution of NaCl, the $Na^+$ cannot diffuse faster than the $Cl^-$, unless there is some other negative ion in the region into which the $Na^+$ is diffusing.

It should also be mentioned at this point that if the solute is adsorbed onto the mineral surfaces of the porous medium, the net rate of diffusion will obviously be less than for a nonadsorbed species. This topic will be addressed more fully in the section on retardation.

It is possible for solutes to move through a porous medium by diffusion, even though the ground water is not flowing. Thus, even if the hydraulic gradient is zero, a solute could still move. In rock and soil with very low permeability, the water may be moving very slowly. Under these conditions, diffusion might cause a solute to travel faster than the ground water is flowing. Under such conditions, diffusion is more important than advection.

### 10.6.3 ADVECTION

The rate of flowing ground water can be determined from Darcy's law as

$$v_x = \frac{K}{n_e}\frac{dh}{dl} \tag{10-4}$$

where

$v_x$ = average linear velocity
$K$ = hydraulic conductivity
$n_e$ = effective porosity
$\frac{dh}{dl}$ = hydraulic gradient

Contaminants that are advecting are traveling at the same rate as the average linear velocity of the ground water.

### 10.6.4 MECHANICAL DISPERSION

As a contaminated fluid flows through a porous medium, it will mix with noncontaminated water. The result will be a dilution of the contaminant by a process known as dispersion. The mixing that occurs along the streamline of fluid flow is called longitudinal dispersion. Dispersion that occurs normal to the pathway of fluid flow is lateral dispersion.

There are three basic causes of longitudinal dispersion. (1) As fluid moves through pores, it will move faster through the center of the pore than along the edges. (2) Some of the fluid will travel in longer pathways than other fluid. (3) Fluid that travels through larger pores will travel faster than fluid moving in smaller pores. This is illustrated by Figure 10.8.

Lateral dispersion is caused by the fact that, as a fluid containing a contaminant flows through a porous medium, the flow paths can split and branch out to the side. This will occur even in the laminar flow conditions that are prevalent in ground-water flow (Figure 10.9).

The mechanical dispersion due to the above factors is equal to the prod-

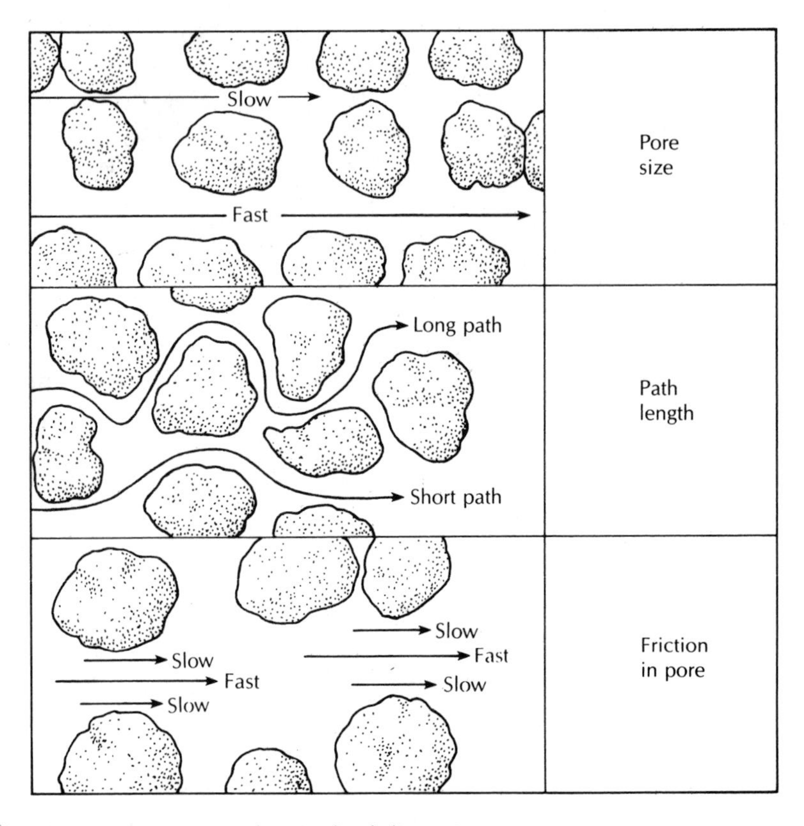


**FIGURE 10.8** Factors causing longitudinal dispersion.

uct of the average linear velocity and a factor called the dynamic dispersivity ($a_L$).

$$\text{Mechanical dispersion} = a_L v_x \tag{10-5}$$

### 10.6.5 HYDRODYNAMIC DISPERSION

The processes of molecular diffusion and mechanical dispersivity cannot be separated in flowing ground water. Instead, a factor termed the coefficient of hydrodynamic dispersion, $D_L$, is introduced. It takes into account both the mechanical mixing and diffusion. For one-dimensional flow it is represented by the following equation:

$$D_L = a_L v_x + D^* \tag{10-6}$$

where

$D_L$ = the longitudinal coefficient of dispersion
$a_L$ = the dispersivity

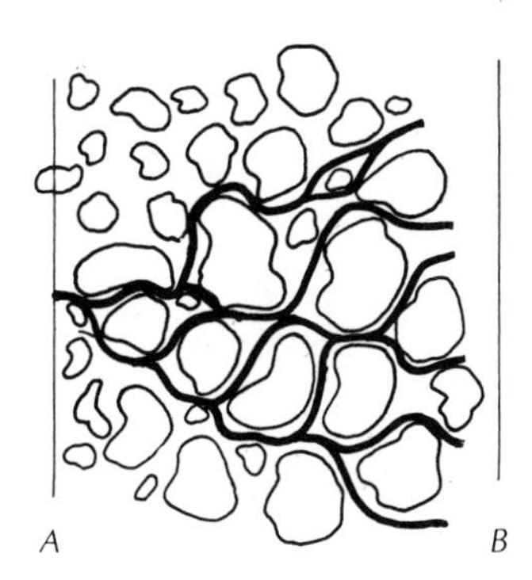


**FIGURE 10.9** Flow paths in a porous medium which cause lateral hydrodynamic dispersion.

$v_x$ = the average linear ground-water velocity
$D^*$ = the molecular diffusion

The process of longitudinal dispersion can be illustrated by the following simple experiment. A tube is filled with sand and set up so that distilled water is flowing through it at a constant rate. We then change the influent to a one percent saline solution and begin to monitor the effluent for chloride. The effluent has zero chloride initially, as distilled water is still flushing from the tube. Eventually we will begin to detect chloride in the effluent water. It arrives initially at a very low concentration. The "breakthrough" is not at the one percent concentration. This small amount gradually increases until the one percent saline concentration is reached. The first chloride ions to arrive traveled through the shortest flow paths. Diffusion in advance of the advecting water may have even caused some of the chloride to reach the outlet prior to the water that was advecting it. The initial chloride was being diluted by the distilled water that was arriving at the same time. The amount of distilled water available for dilution continually decreased until the one percent saline solution filled all the pores and the effluent water was at the influent concentration. Figure 10.10 illustrates this process.

The one-dimensional equation for hydrodynamic dispersion (32, 33) is given by

$$D_L \frac{\delta^2 C}{\delta x^2} - v_x \frac{\delta C}{\delta x} = \frac{\delta C}{\delta t} \qquad \textbf{(10-7)}$$

where

$D_L$ is the longitudinal dispersion coefficient
$C$ is the solute concentration
$v_x$ is the average ground-water velocity in the $x$-direction
$t$ is the time since start of solute invasion

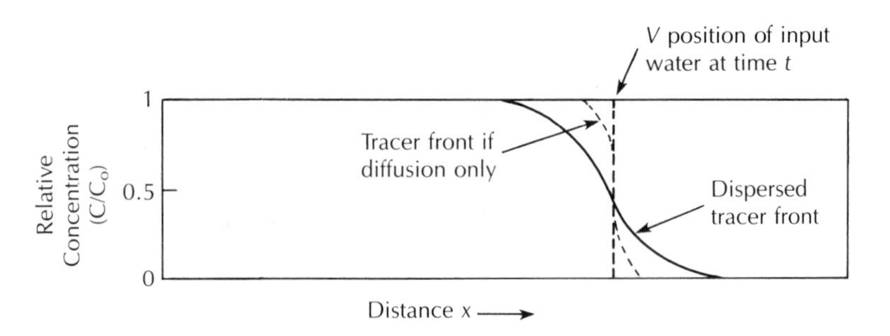


**FIGURE 10.10** Influence of dispersion and diffusion on "breakthrough" of a solute.

The concentration, $C$, at some distance, $L$, from the source at concentration, $C_0$, at time, $t$, is given by the following expression (34), where erfc is the complementary error function:

$$C = \frac{C_0}{2}\left[\operatorname{erfc}\left(\frac{L - v_x t}{2\sqrt{D_l t}}\right) + \exp\left(\frac{v_x L}{D_L}\right)\operatorname{erfc}\left(\frac{L + v_x t}{2\sqrt{D_l t}}\right)\right] \qquad \textbf{(10-8)}$$

It should be noted that the diffusion and mechanical dispersion caused the solute front to advance faster than would be predicted by the average linear velocity.

The coefficient of hydrodynamic dispersivity, $a_L$, is a function that is related to scale, i.e., the greater the area over which it is measured, the larger the value seems to be. Laboratory values are much smaller than field values. Laboratory values seem to be measured only in terms of centimeters. While little work has been done with field values, those that have been measured are in the range of meters. This is due to macroscopic dispersion caused by larger inhomogeneities in the rock strata (35). The limited data on field values for dispersivity indicate that the range for longitudinal dispersivity of alluvial sediments is from 39 to 200 feet (12 to 61 meters), while lateral dispersivity for the same rock type is from 13 to 98 feet (4 to 30 meters). One investigation of glacial deposits showed a longitudinal value of 69 feet (21 meters) and a lateral value of 13 feet (4 meters). Larger values were reported for fractured basalt, which has more heterogeneities (35). As a practical matter, the coefficient of longitudinal dispersivity, $a_L$, can be estimated to be one-tenth of the length of the flow path, in the same units (36). In extremely homogeneous materials, it appears that longitudinal dispersivity may become constant at values of 3 feet (1 meter) or less at distances of about 164 feet (50 meters) from the source.

Because of hydrodynamic dispersion, the concentration of a solute will decrease with distance from the source. The solute will spread in the direction of ground-water movement more than it will in the direction perpendicular to the flow. This is becuse the longitudinal dispersivity is greater than the lateral dispersivity. A continuous source will yield a plume, whereas a spill will yield a

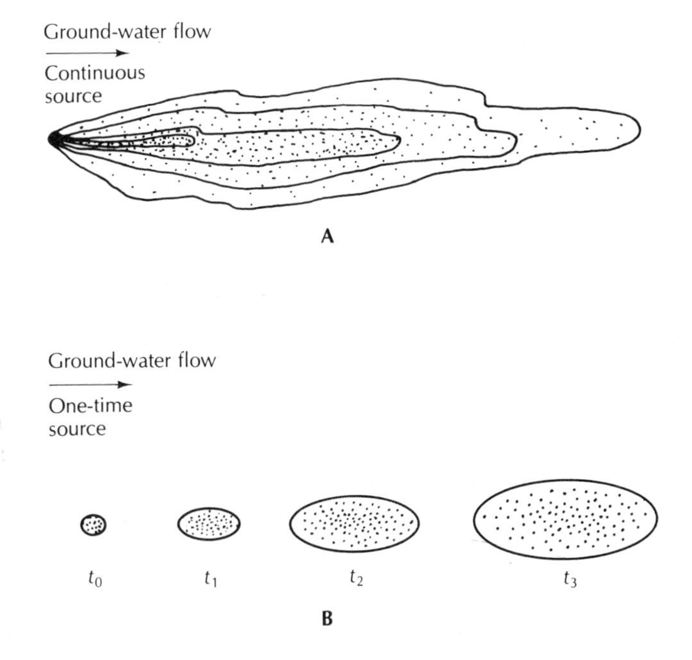


**FIGURE 10.11** **A.** The development of a contamination plume from a continuous point source. **B.** The travel of a contaminant slug from a one-time point source. Density of dots indicates solute concentration.

slug that grows with time as it moves down the ground-water flow path. This is illustrated by Figure 10.11.

Heterogeneities in the aquifer can cause the pattern of the solute movement to vary from what one might expect in homogeneous beds. Because flowing ground water always follows the most permeable pathways, those pathways will also have the most contaminant.

**EXAMPLE PROBLEM** A landfill is leaking leachate with a chloride concentration of 725 mg/L which enters an aquifer with the following properties:

$$\text{hydraulic conductivity} = 3 \times 10^{-3} \text{ cm/sec } (3 \times 10^{-5} \text{ m/sec})$$
$$dh/dl = 0.002$$
$$\text{effective porosity} = 0.23$$
$$D^* = 1 \times 10^{-9} \text{ m}^2\text{/sec (estimated)}$$

Compute the concentration of chloride in one year at a distance 15 meters from the point where the leachate entered the ground water.

**1.** Determine average linear velocity

$$v_x = \frac{K(dh/dl)}{n_e} = (3 \times 10^{-5}\ \text{m/sec} \times 0.002)/(0.23)$$

$$= 2.61 \times 10^{-7}\ \text{m/sec}$$

**2.** Determine the longitudinal dispersion coefficient

$$D_L = a_L \times v_x + D^*$$

**a.** Estimate the value of $a_L$ as 0.1 of the flow length, $L$, which is 15 meters; therefore, $a_L = 1.5$ meters.

**b.** $D_L = (1.5\ \text{m} \times 2.61 \times 10^{-7}\ \text{m/sec}) + 1 \times 10^{-9}\ \text{m}^2/\text{sec}$

$$= 3.91 \times 10^{-7}\ \text{m}^2/\text{sec}$$

**3.** Restate the one-year time of travel in seconds

$$t = 1\ \text{year} \times 60\ \text{sec/min} \times 1440\ \text{min/day} \times 365\ \text{day/year}$$

$$= 3.15 \times 10^7\ \text{sec}$$

**4.** Substitute values into Equation 10-7

$$C_0 = 725\ \text{mg/L}$$

$$L = 15\ \text{m}$$

$$t = 3.15 \times 10^7\ \text{sec}$$

$$D_L = 3.91 \times 10^{-7}\ \text{m/sec}$$

$$v_x = 2.61 \times 10^{-7}\ \text{m/sec}$$

$$C = \frac{725}{2}\ \text{erfc}\left(\frac{15\ \text{m} - (2.61 \times 10^{-7}\ \text{m/sec} \times 3.15 \times 10^7\ \text{sec})}{2 \times [3.91 \times 10^{-7}\ \text{m}^2/\text{sec} \times 3.15 \times 10^7\ \text{sec}]^5}\right) + \left[\exp\left(\frac{2.61 \times 10^{-7}\ \text{m/sec} \times 15\ \text{m}}{3.91 \times 10^{-7}\ \text{m}^2/\text{sec}}\right) \times \text{erfc}\left(\frac{(15\ \text{m} + (2.61 \times 10^{-7}\ \text{m/sec} \times 3.15 \times 10^7\ \text{sec})}{2 \times [3.91 \times 10^{-7}\ \text{m/sec} \times 3.15 \times 10^7\ \text{sec}]^{.5}}\right)\right]$$

$$C = 362.5\ \text{erfc}\left(\frac{15\ \text{m} - 8.22\ \text{m}}{7.02\ \text{m}}\right) + \exp(10.01) \times \text{erfc}\left(\frac{15\ \text{m} + 8.22\ \text{m}}{7.02\ \text{m}}\right)$$

$$C = 362.5\ \text{erfc}(0.966) + \exp(10.01) \times \text{erfc}(3.31)$$

The complementary error function can be determined from Appendix 13. Since the complementary error function of numbers greater than 3 is infinitesimally small, we may ignore the second term of the equation.

$$C = 362.5 \times 0.172$$
$$= 62 \text{ mg/L}$$

---

### 10.6.6 VALIDITY OF DISPERSION EQUATION

Recent work indicates that the given form of the dispersion equation may not be valid for field applications near the source of the contaminant (37). Thus for short time periods, or close distances, there is no good way to represent contaminant movement at the present. It appears that the equation becomes valid at tens to hundreds of feet from the source of contamination. In general, it will take long times for flow to reach these distances. Because of the impact of nonhomogeneities on the dispersion process, very detailed field studies will be necessary in performing the research necessary to develop methods of dealing with advection-dispersion close to the source.

### 10.6.7 RETARDATION

We can consider solutes in two broad classes: conservative and reactive. Conservative solutes do not react with the soil and/or native ground water or undergo biological or radioactive decay. The chloride ion is a good example of a conservative solute.

Reactive substances can undergo chemical, biological, or radioactive change that will tend to reduce the concentration of the solute. Chemical reactions include adsorption-desorption, cation exchange, precipitation-dissolution, and oxidation-reduction. Biological reactions may be either aerobic or anaerobic. Recent research (38, 39) has given strong evidence that many toxic organic chemicals can undergo microbial decay to more simple compounds. Unfortunately, the end product may be carcinogenic. Chlorinated solvents such as tetrachloroethane can become progressively dehalogenated, first to trichloroethane, then to dichloroethane, and finally to vinyl chloride, which is carcinogenic. Chloroethane is the end-product of the dehalogenation of 1,1,1-trichloroethane.

For most hazardous wastes, the adsorption-desorption and cation exchange reactions between the solutes and the solid matrix are very important. Many of the heavy metals are readily adsorbed onto solid surfaces or trapped by clays through ion exchange. The magnitude of adsorption-desorption or cation exchange between a specific solute and a given soil can be determined experimentally.

Laboratory studies of adsorption involve the use of a specific soil and given solutes. In one approach, batches of solutions of the ionic substances are prepared in differing concentrations. For each batch a preweighed mass of dry soil is mixed with measured portions of the solutions, each with a different initial concentration. In a second approach, different amounts of dry adsorbent are mixed with portions of the same solution, each with the same concentration. After equilibrium between the soil and the solution has been determined, the equilibrium concentration of the solute in each batch is determined. The amount of solute adsorbed by the soil will be in direct proportion to the chemical activity of the solute. The amount adsorbed onto the soil can be calculated as the difference in mass of the solute in solution before the test and in solution at equilibrium. As a result of the test for each batch, one knows the mass adsorbed per unit weight of soil, $C^*$, as a function of the equilibrium concentration of solute remaining in solution, $C$. This process is more fully described in Section 9.11, where Langmuir and Freudlich isotherms are introduced.

The adsorption characteristics are typically plotted on graph paper by showing the mass of solute adsorbed per unit mass of soil as a function of the concentration of the solute. Figure 10.12 shows an adsorption isotherm for lead. While either the Langmuir or Freundlich isotherm equation could be used to describe this relationship, the Freundlich equation is of special utility as it can easily be applied to retardation studies.

When an adsorption relationship can be plotted as a straight line on log-log paper, it is described by the Freundlich isotherm as

$$\log C^* = b \log C + \log K_f \tag{10-9}$$

or

$$C^* = K_f C^b \tag{10-10}$$

where

$C^*$ = mass of solute sorbed per bulk unit dry mass of soil
$C$ = solute concentration
$K_f, b$ = coefficients

The slope of the curve on the log-log paper is represented by $b$. In a plot of $C^*$ versus $C$, where the slope is a straight line, the relationship is linear, and $b$ has a value of one. Under these conditions, the derivative of $C^*$ with respect to $C$ yields the relationship

$$dC^*/dC = K_d \tag{10-11}$$

where $K_d$ is known as a **distribution coefficient.**

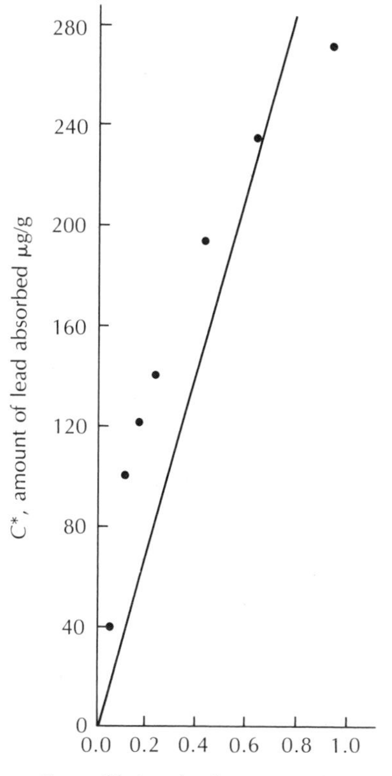


**FIGURE 10.12** Lead adsorption by Cecil clay loam at pH 4.5, and at 25°C described by a linear Freundlich equation through the origin. Source: W. R. Roy, I. G. Krapac, S. F. J. Chou, and R. A. Griffin, *Batch-Type Adsorption Procedures for Estimating Soil Attenuation of Chemicals*. Technical Resource Document, EPA/530-SW-87-006, 1987.

The $K_d$ value can be used to compute the retardation of the solute front as it passes through the soil by the following equation:

$$\text{Retardation factor} = 1 + (P_b/\theta)\,(K_d) \quad \textbf{(10-12)}$$

where

$P_b$ = dry bulk mass density of the soil
$\theta$ = volumetric moisture content of the soil
$K_d$ = distribution coefficient for the solute with the soil

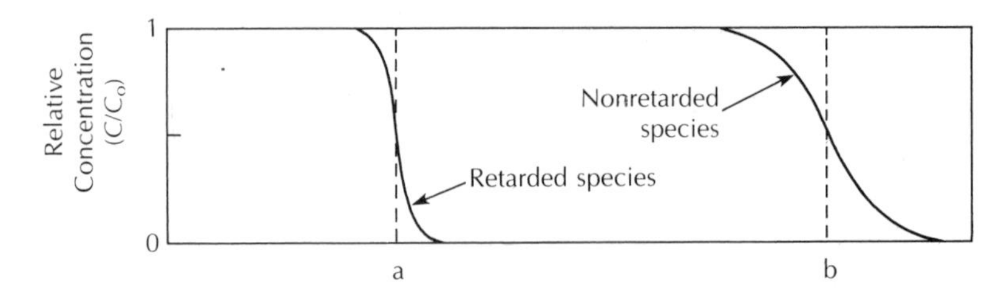


**FIGURE 10.13** Influence of retardation on movement of a solute front.

If a solute is reactive, it will travel at a slower rate than the ground water owing to adsorption. The rate of solute movement can be determined by the retardation equation

$$v_c = v_x/[1 + (P_b/\theta)(K_d)] \tag{10-13}$$

where

$v_x$ = average linear velocity

$v_c$ = velocity of the solute front where the solute concentration is one-half of the original value ($C/C_0 = 0.5$)

Figure 10.13 shows the impact of retardation on the movement of a solute front compared with a nonretarded species. It can be seen that the effect of retardation is to cause the solute front to advance more slowly.

**EXAMPLE PROBLEM** The following table shows the adsorption of varying amounts of lead by the Cecil clay. The aliquots of solution were each 200 mL and the amount of adsorbent in each case was 10.18 grams.

| Initial Conc. (mg/L) | C Equilibrium Conc. (mg/L) | C* Amount Adsorbed (μg/g) |
|---|---|---|
| 2.07 | 0.05 | 40 |
| 5.11 | 0.11 | 98 |
| 6.22 | 0.16 | 119 |
| 7.28 | 0.22 | 139 |
| 10.2 | 0.41 | 192 |
| 12.4 | 0.65 | 231 |
| 14.6 | 0.94 | 268 |

The value of $C^*$, the amount adsorbed, is found by taking the initial concentration, subtracting from that the equilibrium concentration, multiplying by the volume of the solution, and dividing the entire product by the weight of the adsorbent. In the first line of the above table, the computation is

$$C^* = \frac{(2.07 \text{ mg/L} - 0.05 \text{ mg/L}) \times 0.200 \text{ L}}{10.18 \text{ g}} = 0.040 \text{ mg/g} = 40 \ \mu\text{g/g}$$

Figure 10.14 shows the plot of the equilibrium concentration, $C$, versus the amount adsorbed, $C^*$, for the above data. The data can be reasonably approximated by a straight line drawn through the origin.

A statistical test called a linear regression can be used to find the straight-line equation that best fits the data set. Many scientific calculators have a built-in linear regression function. It can also be run on a personal computer with LOTUS 1-2-3. Otherwise, the method may be found in a statistics text.

The straight-line equation, $y = mx + b$, where $m$ is the slope of the line and $b$ is the intercept with the $y$ axis, simplifies to $y = mx$ when the intercept passes through the origin. In Figure 10.12 the data were forced through the ori-

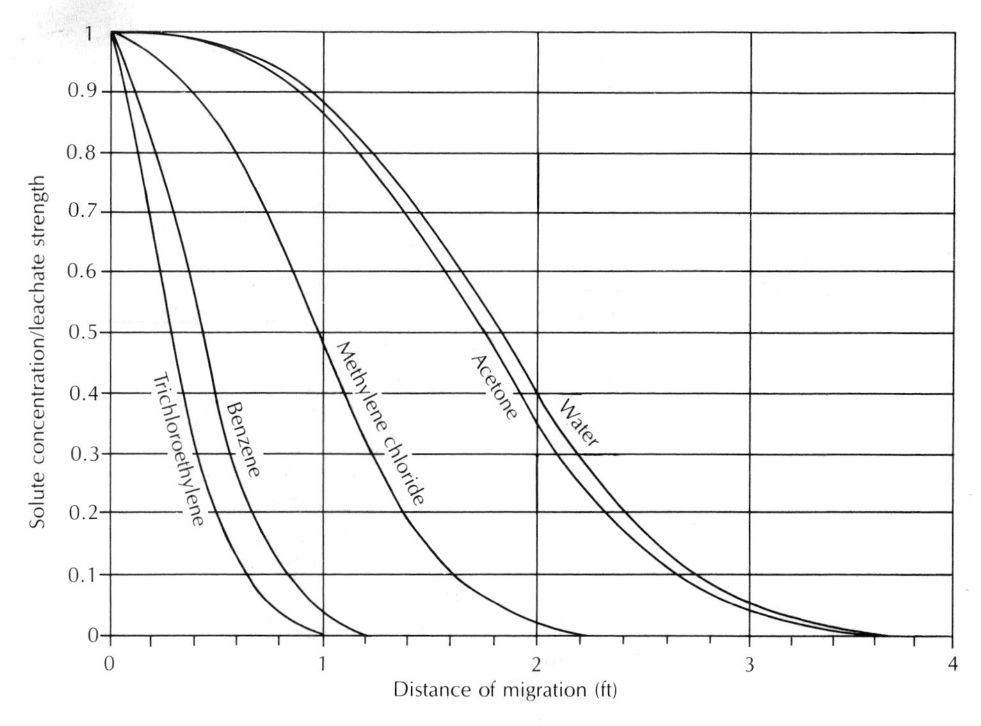


**FIGURE 10.14** Vertical migration, in feet per 100 years, of various synthetic organic compounds through a soil with a hydraulic conductivity of $1.6 \times 10^{-8}$ cm/sec, hydraulic gradient of 0.222, bulk density of 2.00 gm/cm$^3$, particle density of 2.65, effective porosity of 0.22, and soil organic carbon content of 0.5 percent.

gin; i.e., $b$ was set to zero. The equation for the straight line in Figure 10.12 is $C^* = 232\ C$ since the slope of the line, $dC^*/dC$, is 232. The value of $K_d$, the retardation coefficient, is thus 232.
Data from W. R., Roy, I. G. Krapac, S. F. J. Chou, and R. A. Griffin, "Batch-type soil adsorption procedures for estimating soil attenuation of chemicals," U.S. Environmental Protection Agency, Technical Resource Document EPA/530-SW-87-006, 1987.

---

Potential solutes in ground water have a wide range of soil-solute specific distribution coefficients. Because of the large surface areas and numerous ion-exchange sites, clays will have the largest $K_d$ values for specific inorganic solutes. Cations are more often strongly adsorbed than anions, and divalent cations will be adsorbed more readily than those of monovalent species. Substances such as chloride may be only very weakly adsorbed or even completely unattenuated by passage through a clay. Sodium is weakly attenuated; potassium, ammonia, magnesium, silicon, and iron are moderately attenuated; and lead, cadmium, mercury, and zinc can be strongly held.

Synthetic organic chemicals in solution can be adsorbed by the organic carbon in the soil. In determining the movement and retardation of organic chemicals when one does not have site-specific batch adsorption data, one must first consider the solubility of the organic compound in water. The relative tendency for an organic compound to remain dissolved in water rather than to be adsorbed onto soil organic carbon is related to the octanol-water partition coefficient of that chemical, which is the tendency for the chemical to be dissolved into either water or *n*-octanol when shaken in a solution of the two. The soil–water partition coefficient, $K_{oc}$, can be estimated from either the water solubility or the octanol-water partition coefficient. Table 10.5 gives the solubility and soil–water partition coefficient for a large number of organic chemicals that are potential ground-water contaminants. The $K_d$ for an organic compound in a specific soil can easily be estimated as it is merely the $K_{oc}$ for that compound times the percent organic carbon in the soil. If the soil is a pure silica sand there will be very limited retardation.

As a result of the processes of advection, dispersion, diffusion, and retardation, there is a pattern of solute distribution as one moves away from the source of the contamination. If the contaminant source contains multiple solutes, then each will have a retardation typified by the $K_d$ for that solute and there will be a number of solute fronts. The resulting plume can be very complex. Figure 10.14 shows the relative rates of movement for water and several organic compounds of differing mobility classes. The movement is through a low-permeability glacial till with about 0.5 percent soil organic carbon. The curves were obtained from the one-dimensional solute transport equation and were solved on a personal computer using an algorithm written for the LOTUS 1-2-3 program.

**TABLE 10.5** Solubility, $K_{oc}$, and mobility class for common organic pollutants

| Compound | Solubility (ppm) | $K_{oc}$ | Mobility Class |
|---|---|---|---|
| 1,4-dioxane | miscible | 1 | very high |
| 4-hydroxy-4-methyl-2-pentanone | miscible | 1 | very high |
| acetone | miscible | 1 | very high |
| tetrahydrofuran | miscible | 1 | very high |
| N,N′-dimethylformamide | | 1 | very high |
| N,N′-dimethylacetamide | | 2 | very high |
| 2-methyl-2-butanol | 140000. | 6 | very high |
| 2-butanol | 125000. | 6 | very high |
| ethyl ether | 84300. | 8 | very high |
| cyclohexanol | 56700. | 10 | very high |
| 3-methylbutanoic acid | 42000. | 12 | very high |
| benzyl alcohol | 40000. | 12 | very high |
| aniline | 34000. | 13 | very high |
| 2-hexanone (butylmethylketone) | 35000. | 14 | very high |
| 2-hydroxy-triethylamine | | 15 | very high |
| 2-methylphenol (*o*-cresol) | 31000. | 15 | very high |
| 2-methyl-2-propanol | | 16 | very high |
| 4-methylphenol (*p*-cresol) | 24000. | 17 | very high |
| pentanoic acid | 24000. | 17 | very high |
| cyclohexanone | 23000. | 18 | very high |
| 4-methyl-2-pentanone | 19000. | 20 | very high |
| 2,4-dimethyl phenol | 17000. | 21 | very high |
| 4-methyl-2-pentanol | 17000. | 21 | very high |
| methylene chloride | 13200. | 25 | very high |
| isophorone | 12000. | 26 | very high |
| phenol | 82000. | 27 | very high |
| 2-chlorophenol | 11087. | 27 | very high |
| hexanoic acid | 11000. | 28 | very high |
| chloroform | 7840. | 34 | very high |
| 1,2-dichloroethane | 8450. | 36 | very high |
| 1,2-trans-dichloroethene | 6300. | 39 | very high |
| chloroethane | 5700. | 42 | very high |
| 5-methyl-2-hexanone | 5400. | 43 | very high |
| chloromethane | 5380. | 43 | very high |
| 1,1-dichloroethane | 5100. | 45 | very high |
| 1,1,2-trichloroethane | 4420. | 49 | very high |
| 1,2-dichloropropane | 3570. | 51 | high |
| benzoic acid | 2900. | 64 | high |
| octanoic acid | 2500. | 70 | high |
| heptanoic acid | 2410. | 71 | high |
| 1,1,2,2-tetrachloroethane | 3230. | 88 | high |
| benzene | 1780. | 97 | high |

**TABLE 10.5** Continued

| Compound | Solubility (ppm) | $K_{oc}$ | Mobility Class |
|---|---|---|---|
| diethyl phthalate | 1000. | 123 | high |
| 2-nonanol | 1000. | 123 | high |
| bromodichloromethane | 900. | 131 | high |
| 3-methylbenzoic acid | 850. | 136 | high |
| trichloroethene | 1100. | 152 | moderate |
| 1,1,1-trichloroethane | 700. | 155 | moderate |
| di-*n*-butyl phthalate | 400. | 217 | moderate |
| 1,1-dichloroethene | 400. | 217 | moderate |
| carbon tetrachloride | 800. | 232 | moderate |
| 2-butanone (methylethylketone) | 353. | 235 | moderate |
| 4-methylbenzoic acid | 340. | 240 | moderate |
| toluene | 500. | 242 | moderate |
| tetrachloroethylene | 200. | 303 | moderate |
| chlorobenzene | 448. | 318 | moderate |
| 1,2-dichlorobenzene | 148. | 343 | moderate |
| *o*-xylene | 170. | 363 | moderate |
| 1,2,2-trifluoro-1,1,2-trichloroethane | | 372 | moderate |
| styrene | 162. | 380 | moderate |
| 1,3-dichlorobenzene | 118. | 463 | moderate |
| fluorotrichloromethane | 110. | 476 | moderate |
| 4,6-dinitro-2-methylphenol | | 477 | moderate |
| *p*-xylene | 156. | 552 | low |
| *m*-xylene | 146. | 588 | low |
| 1,4-dichlorobenzene | 79. | 594 | low |
| ethyl benzene | 150. | 622 | low |
| pentachlorophenol | 14. | 900 | low |
| *N*-nitrosodiphenylamine | 35.1 | 982 | low |
| 3,5-dimethylphenol | | 1038 | low |
| BHC-delta | 31.5 | 1052 | low |
| 2,6-dimethylphenol | | 1060 | low |
| 1,2,4-trichlorobenzene | 30. | 1080 | low |
| naphthalene | 31.7 | 1300 | low |
| 4-ethylphenol | | 1986 | low |
| dibenzofuran | 10. | 2140 | slight |
| hexachloroethane | 8. | 2450 | slight |
| acenaphthene | 7.4 | 2580 | slight |
| tri-*N*-propylamine | | 2610 | slight |
| BHC-alpha | 8.5 | 2627 | slight |
| BHC-beta | 2.7 | 3619 | slight |
| hexachlorobenzene | 0.035 | 3910 | slight |
| hexachlorobutadiene | 3.2 | 4330 | slight |

**TABLE 10.5** Continued

| Compound | Solubility (ppm) | $K_{oc}$ | Mobility Class |
|---|---|---|---|
| di-*n*-octyl phthalate | 3. | 4510 | slight |
| butyl benzyl phthalate | 2.9 | 4606 | slight |
| fluorene | 1.98 | 5835 | slight |
| 2-methylnaphthalene | 25.4 | 8500 | slight |
| bis(2-ethylhexyl)phthalate | 0.6 | 12200 | slight |
| toxaphene | 0.4 | 15700 | slight |
| heptachlor epoxide | 0.35 | 17087 | slight |
| endosulfan II | 0.28 | 19623 | slight |
| fluoranthene | 0.275 | 19800 | slight |
| 1,2-diphenylhydrazene (as azobenzene) | 0.252 | 20947 | immobile |
| endosulfan sulfate | 0.22 | 22788 | immobile |
| phenanthrene | 1.29 | 23000 | immobile |
| dieldrin | 0.188 | 25120 | immobile |
| anthracene | 0.073 | 26000 | immobile |
| BHC-gamma | 0.15 | 28900 | immobile |
| decanoic acid | | 39610 | immobile |
| chlordane | 0.056 | 53200 | immobile |
| pyrene | 0.135 | 63400 | immobile |
| PCB-1254 | 0.042 | 63914 | immobile |
| heptachlor | 0.03 | 78400 | immobile |
| endrin | 0.024 | 90000 | immobile |
| benzo(a)anthracene | 0.014 | 125719 | immobile |
| aldrin | 0.013 | 132000 | immobile |
| 4,4′-DDE | 0.01 | 155000 | immobile |
| 4,4′-DDT | 0.0017 | 238000 | immobile |
| 4,4′-DDD | 0.005 | 238000 | immobile |
| benzo(a)pyrene | 0.0038 | 282185 | immobile |
| PCB-1260 | 0.0027 | 349462 | immobile |
| chrysene | 0.022 | 420108 | immobile |
| benzo(*b*)fluoranthene | | 1148497 | immobile |
| benzo(*k*)fluoranthene | | 2020971 | immobile |

Source: R. A. Griffin, 1985, personal communication, and W. R. Roy, and R. A. Griffin, "Mobility of organic solvents in water-saturated soil materials," *Environmental Geology and Water Sciences,* 7 (1985):241–47.

**EXAMPLE PROBLEM** Compute the relative velocity of the solute front of a solute-soil system with a distribution coefficient of 10 mL/g, a $P_b$ value of 1.75 g/cm$^3$ and $\theta$ of 0.20:

$$\begin{aligned} v_c &= v_x/[1 + (P_b/\theta)(K_d)] \\ &= v_x/[1 + (1.75/0.20)(10)] \\ &= v_x/88.5 = 0.011\ v_x \end{aligned}$$

# 10.7 GROUND-WATER CONTAMINATION

## 10.7.1 INTRODUCTION

The specter of ground-water contamination looms over industrialized, suburban, and rural areas. The sources of ground-water contamination are many and the contaminants numerous. Common industrial solvents such as trichloroethylene, 1,1,1-trichloroethane, tetrachloroethane, benzene, and carbon tetrachloride have been found in widespread areas, with all indications being there are multiple sources (40). Suburban areas have ground water with high levels of nitrate due to the use of lawn fertilizers as well as septic tank discharges (41). Agricultural areas have not only high levels of fertilizers found in ground water (42), but specialized synthetic organic agricultural chemicals as well (43). Landfills in urban and rural areas are known sources of contamination (44, 45). Underground storage tanks holding petroleum products (46) and synthetic organic chemicals (47) have leaked and caused ground-water contamination.

It has been estimated that ground water has been contaminated in only one to two percent of the aquifers in the United States (48). However, these aquifers may well be in urban areas where the water is most needed. Contaminated ground water will in most cases not travel more than a few thousand feet from the source and in many cases not more than a few hundred feet. If there is a single source, then the contamination may be localized. If there are multiple sources, or if the contamination is a result of widespread land-use practices, then the contamination may cover a large area. Table 10.6 is a partial listing of contaminants reported to have been found in ground water.

Ground-water contamination is not an irreversible process. There are natural conditions that act to remove contaminants. Attenuation mechanisms include dilution, dispersion, mechanical filtration, volatilization, biological activity, ion exchange and adsorption on soil particle surfaces, chemical reactions, and radioactive decay. Even synthetic organic compounds can undergo biological decay (38), although the decay products may also be toxic. In recent years a number of techniques have been developed for restoring the quality of ground water that has been contaminated. This is discussed in Section 10.8.

## 10.7.2 SEPTIC TANKS AND CESSPOOLS

The disposal of domestic wastewater is accomplished in many areas through the use of septic tanks and drain tile fields. An estimated 800 billion gallons of water per year are discharged to the subsurface in the United States via septic tanks (49). Anaerobic decomposition of wastes takes place in the septic tank. The liquid waste is carried to a drain tile field, where it seeps through the vadose zone to the water table. An analysis of the typical septic tank effluent is given in Table 10.7.

Septic tank effluent contains bacteria and viruses. It is a major factor in the incidences of waterborne disease from private wells in the United States (50)

**TABLE 10.6** Chemicals and organisms known to have caused ground-water contamination (various sources)

| Metals | Nonmetals | Organics | Extractable Organic Compounds | Volatile Organic Compounds | Organisms |
|---|---|---|---|---|---|
| aluminum | acids | aldrin | tri-*n*-propylamine | benzene | *Giardia lamblia* |
| arsenic | ammonia | BOD | 3- and/or 4-methyl phenol | 1,2-dichloroethane | *Salmonella* sp. |
| barium | boron | chlordane | 4-methyl benzoic acid | 1,1,1-trichloroethane | *Shigella* sp. |
| cadmium | chloride | DDT | 1,4-dioxane | 1,1-dichloroethane | typhoid |
| chromium | cyanide | detergents | 4-methyl-2-pentanol | 1,1,2-trichloroethane | *Yersinin enterocolitica* |
| copper | fluoride | ethyl acrylate | *n,n*-dimethylformamide | chloroethane | viral hepatitis |
| iron | nitrate | gasoline | 2-hexanone | 1,1-dichloroethene | |
| lead | phosphate | hydroquinone | 4-methyl-2-pentanone | trans-1,2-dichloroethene | |
| lithium | radium | lindane | 1-methyl-2-pyrrolidinone | ethyl benzene | |
| manganese | selenium | paramethyl animophenol | 2-hexanol | methylene chloride | |
| mercury | sulfate | PBB | 3,5-dimethyl phenol and/or 4-ethyl phenol | tetrachloroethane | |
| molybdenum | various radioactive isotopes | PCB | benzoic acid | toluene | |
| nickel | | DCPD (dicyclopentadiene) | hexanoic acid | trichloroethene | |
| silver | | DIMP (diisopropyl-methyl-phosphonate) | cyclohexanol | vinyl chloride | |
| uranium | | DBCP (dibromochloropropane) | 2-ethyl hexanoic acid | tetrahydrofuran | |
| zinc | | | octanoic acid | acetone | |
| | | | pentanoic acid | 2-methyl-2-propanol | |
| | | | bis(2-ethylhexyl) phthalate | 2-butanone | |
| | | | di-*n*-butyl phthalate | 2-butanol | |
| | | | 2,4-dimethyl phenol | 2-propanol | |
| | | | isophorone | | |
| | | | phenol | | |
| | | | 1,2-dichlorobenzene | | |

**TABLE 10.7.** Effluent quality from six septic tanks*

| Site | Avg. Flow (gpd) | BOD (mg/L) | COD (mg/L) (unfiltered) | COD (mg/L) (filtered) | TSS (mg/L) | Fecal Coliforms (no./mL) | Fecal Strep (no./mL) | Total N (mg/L) | Ammonia N (mg/L) | Nitrate-Nitrogen (mg/L) | Total P (mg/L) | Ortho P (mg/L) |
|---|---|---|---|---|---|---|---|---|---|---|---|---|
| A | 75 | 131 | 325 | 249 | 69 | 2907 | 2.7 | 50.5 | 34.1 | 0.68 | 12.3 | 10.8 |
| B | 125 | 176 | 361 | 323 | 44 | 4127 | 39.7 | 57.8 | 42.5 | 0.46 | 14.1 | 13.6 |
| C | 245 | 272 | 542 | 386 | 68 | 27,931 | 1387 | 76.3 | 45.6 | 0.60 | 31.4 | 14.0 |
| D | 315 | 127 | 291 | 217 | 52 | 11,113 | 184 | 40.2 | 33.2 | 0.35 | 11.0 | 10.1 |
| E | 860† | 120 | 294 | 245 | 51 | 2310 | 20.7 | 31.6 | 20.1 | 0.16 | 11.1 | 10.5 |
| F | 150 | 122 | 337 | 281 | 48 | 3246 | 25.3 | 56.7 | 38.3 | 0.83 | 11.6 | 10.5 |

Source: R. J. Otis, W. C. Boyle, and D. K. Sauer, Small-Scale Waste Management Program, University of Wisconsin-Madison, 1973.
*All values are means.
†Includes 340-gpd sewer flow and 520 gpd from foundation drain.

and presumably elsewhere as well. The most important factor that influences the development of ground-water contamination from septic tanks is the density of septic tank systems in the area (51). Documented cases of widespread ground-water contamination from septic tank systems have been in areas where the lot sizes range from less than one-quarter of an acre to three acres (51).

Several cases of infectious disease outbreaks due to septic tanks have been reported. In Polk County, Arkansas, in 1971, an outbreak of viral hepatitis was traced to a well that was contaminated by seepage from a septic tank located 95 feet away (52). In 1972, typhoid in Yakima, Washington, was attributed to well water from driven well points. Waste water from the septic tank serving the home of a typhoid carrier was discharged into the ground 210 feet away from the contaminated well. A Norwalk-like virus was responsible for 400 cases of gastroenteritis at a resort camp in Colorado. A septic tank was located 50 feet above the spring supplying drinking water to the camp (53).

Septic tanks are most likely to contribute to ground-water contamination in areas where (1) there is a high density of homes with septic tanks, (2) the soil layer over permeable bedrock is thin, (3) the soil is extremely permeable, such as gravel, or (4) the water table is within a couple of feet of the land surface. Areas with high population density should not be served with septic tanks, and areas with thin soils, extremely permeable soils, and high water tables should be avoided.

### 10.7.3 LANDFILLS

Burial in a landfill is the most common means of disposing of municipal refuse, ashes, garbage, leaves, demolition debris, and sludges from municipal and industrial wastewater treatment facilities. Radioactive, toxic, and hazardous wastes have also been subjected to land burial as a means of disposal. Precipitation that infiltrates the waste can mix with liquids already present in the waste and leach compounds from the solid waste. The result is a liquid known as **leachate.** Leachate can move downward from the landfill into the water table and cause ground-water contamination. If the waste is buried below the water table, moving ground water can leach compounds from the waste and become contaminated. Table 10.8 gives a chemical analysis of the leachate from a municipal waste landfill in Du Page, Illinois. Table 10.9 gives the overall range and range of median values for municipal solid-waste leachate in Wisconsin. Landfill leachates can contain very high concentrations of both inorganic and organic compounds.

**TABLE 10.8.** Chemical analysis of landfill leachate at Du Page, Illinois

| | | |
|---|---|---|
| Na | 748. | mg/L |
| K | 501. | mg/L |
| Ca | 46.8 | mg/L |
| Mg | 233. | mg/L |
| Cu | <0.1 | mg/L |
| Zn | 18.8 | mg/L |
| Pb | 4.46 | mg/L |
| Cd | 1.95 | mg/L |
| Ni | 0.3 | mg/L |
| Hg | 0.0008 | mg/L |
| Cr | <0.10 | mg/L |
| Fe | 4.2 | mg/L |
| Mn | <0.1 | mg/L |
| Al | <0.1 | mg/L |
| $NH_4$ | 862. | mg/L |
| As | 0.11 | mg/L |
| B | 29.9 | mg/L |
| Si | 14.9 | mg/L |
| Cl | 3484. | mg/L |
| $SO_4$ | <0.01 | mg/L |
| $PO_4$ | <0.1 | mg/L |
| COD | 1340. | mg/L |
| Organic acids | 333. | mg/L |
| Carbonyls as acetophenone | 57.6 | mg/L |
| Carbohydrates as dextrose | 12. | mg/L |
| pH | 6.9 | |
| *Eh* | +7 mv | |
| Conductivity | 10.20 mmhos/cm | |

Source: K. Cartwright, R. A. Griffin, and R. H. Gilkeson, *Ground Water,* 15 (1977):294–305.

**TABLE 10.9** Overall summary from the analysis of municipal solid-waste leachates in Wisconsin

| Parameter | Overall Range* | Typical Range (range of site medians)* | Number of Analyses |
|---|---|---|---|
| TDS | 584–50430 | 2180–25873 | 172 |
| Specific conductance | 480–72500 | 2840–15485 | 1167 |
| Total susp. solids | 2–140900 | 28–2835 | 2700 |
| BOD | ND–195000 | 101–29200 | 2905 |
| COD | 6.6–97900 | 1120–50450 | 467 |
| TOC | ND–30500 | 427–5890 | 52 |
| pH | 5–8.9 | 5.4–7.2 | 1900 |
| Total alkalinity ($CaCO_3$) | ND–15050 | 960–6845 | 328 |
| Hardness ($CaCO_3$) | 52–225000 | 1050–9380 | 404 |
| Chloride | 2–11375 | 180–2651 | 303 |
| Calcium | 200–2500 | 200–2100 | 9 |
| Sodium | 12–6010 | 12–1630 | 192 |
| Total Kjeldahl nitrogen | 2–3320 | 47–1470 | 156 |
| Iron | ND–1500 | 2.1–1400 | 416 |
| Potassium | ND–2800 | ND–1375 | 19 |
| Magnesium | 120–780 | 120–780 | 9 |
| Ammonia-nitrogen | ND–1200 | 26–557 | 263 |
| Sulfate | ND–1850 | 8.4–500 | 154 |
| Aluminum | ND–85 | ND–85 | 9 |
| Zinc | ND–731 | ND–54 | 158 |
| Manganese | ND–31.1 | 0.03–25.9 | 67 |
| Total phosphorus | ND–234 | 0.3–117 | 454 |
| Boron | 0.87–13 | 1.19–12.3 | 15 |
| Barium | ND–12.5 | ND–5 | 73 |
| Nickel | ND–7.5 | ND–1.65 | 133 |
| Nitrate-nitrogen | ND–250 | ND–1.4 | 88 |
| Lead | ND–14.2 | ND–1.11 | 142 |
| Chromium | ND–5.6 | ND–1.0 | 138 |
| Antimony | ND–3.19 | ND–0.56 | 76 |
| Copper | ND–4.06 | ND–0.32 | 138 |
| Thallium | ND–0.78 | ND–0.31 | 70 |
| Cyanide | ND–6 | ND–0.25 | 86 |
| Arsenic | ND–70.2 | ND–0.225 | 112 |
| Molybdenum | 0.01–1.43 | 0.034–0.193 | 7 |
| Tin | ND–0.16 | 0.16 | 3 |
| Nitrite-nitrogen | ND–1.46 | ND–0.11 | 20 |
| Selenium | ND–1.85 | ND–0.09 | 121 |
| Cadmium | ND–0.4 | ND–0.07 | 158 |
| Silver | ND–1.96 | ND–0.024 | 106 |
| Beryllium | ND–0.36 | ND–0.008 | 76 |
| Mercury | ND–0.01 | ND–0.001 | 111 |

*All concentrations in mg/L except pH (std. units) and specific conductance (μmhos/cm). ND incidates no data.

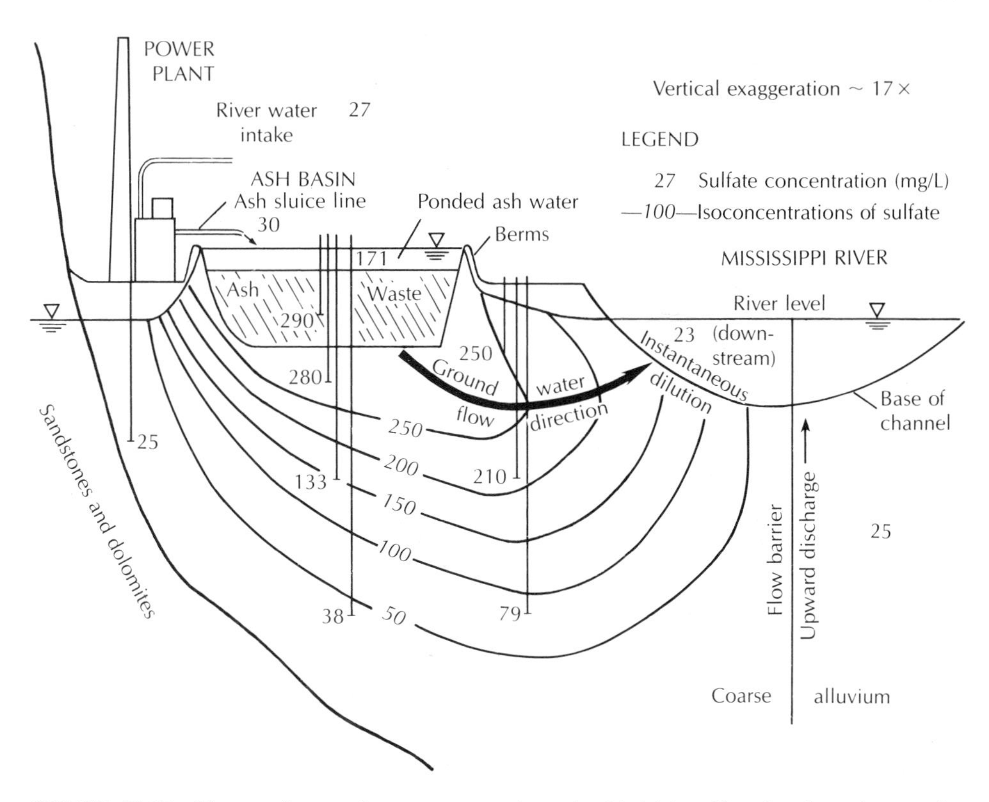


**FIGURE 10.15** Plume of ground water contaminated with high-sulfate leaching from a fly ash landfill located below the water table. Source: Daniel R. Viste, Proceedings, First Annual Conference of Applied Research and Practice on Municipal and Industrial Waste (Madison: University of Wisconsin, 1978): 327–40.

When leachate from a landfill mixes with ground water, it forms a plume that spreads in the direction of the flowing ground water. As one goes away from the source, the concentration decreases owing to hydrodynamic dispersion and retardation. Figure 10.15 shows in cross section a plume of ground water with high sulfate from a fly ash* disposal pond located with the waste pile below the water table. In this particular case, the plume of high-sulfate water is moving directly toward the Mississippi River, where it is discharging into the river. The location of a waste-disposal area adjacent to a ground-water discharge zone limits the amount of ground water that can become contaminated, although surface-water contamination could occur.

The volume of leachate that is produced is a function of the amount of water percolating through the refuse. Land disposal of solid waste in humid

*Fly ash is the ash that is produced from the burning of coal at a power plant. Large amounts are produced from power generation and this presents a major waste-disposal problem.

areas is more likely to produce large volumes of leachate than land disposal in arid zones. The vadose zone in arid regions may receive little or no charge. Under such conditions, solid-waste disposal is not likely to result in ground-water contamination.

Current state-of-the-art hydrogeologic and engineering practice for determining the locations of municipal waste landfills calls for careful geologic analysis of several alternative sites in order to select the site least likely to result in ground-water contamination. Landfills may be designed to minimize the formation of leachate as well as to minimize the amount of leachate that escapes from the landfill. Leachate may also be collected and treated.

It is desirable in most cases to construct landfills above the water table. Considerable attenuation of leachate may occur as it passes through the unsaturated zone. A **natural-attenuation** landfill is one that relies totally on natural processes to attenuate any leachate formed. Such landfills should be well above the water table to promote maximum attentuation in the vadose zone. Soils with the greatest potential attentuation are clays because they have the most ion exchange and adsorption sites. Unfortunately, in humid areas, the water table in clay soils tends to be close to the land surface; this means that much of the landfill should be above grade. Leachate generation may be reduced by capping the landfill with two to three feet of compacted clay soil or a synthetic membrane. If large amounts of leachate are generated in a natural-attenuation landfill located in a low-permeability clay, there is a tendency for leachate to come to the land surface and form leachate springs. Figure 10.16 shows the design of a natural-attenuation landfill.

A **lined** landfill is one designed to capture part of all of the leachate generated. Landfill liners are typically constructed of 3 to 10 feet (0.9 to 3 meters) of compacted clay soils. The permeability of the liner should be no greater than $3 \times 10^{-4}$ ft/day ($1 \times 10^{-7}$ cm/sec). Alternatively, a synthetic membrane such as HDPE (high-density polyethylene) could be used as the liner. Because leachate will collect on the liner, a **leachate-collection system** is also needed. The leachate-collection system consists of a blanket of sand or gravel, with perforated drainage lines, lying on the liner. The base of the liner is sloped toward the drain tiles. Leachate drains through the leachate-collection system to a holding tank or sewer and is ultimately removed and treated. Clay-lined systems can be designed to collect about 70 to 90 percent of the leachate produced (54). The remainder of the leachate will seep through the liner. A double liner and secondary leachate-collection system installed beneath the primary liner can be constructed to capture the leakage through the primary liner. A membrane or clay liner could be used for the secondary liner. Figure 10.17 shows a double-lined system with leachate collection.

In areas with a high water table and low-permeability soils ($3 \times 10^{-3}$ ft/day, or $10^{-6}$ cm/sec), a **zone-of-saturation** landfill could be constructed (Figure 10.18). An excavation below the water table is made. In clay soils this can easily be done because ground water seeps into the excavation at a very slow rate and evaporates; it does not accumulate. A recompacted clay liner with recompacted

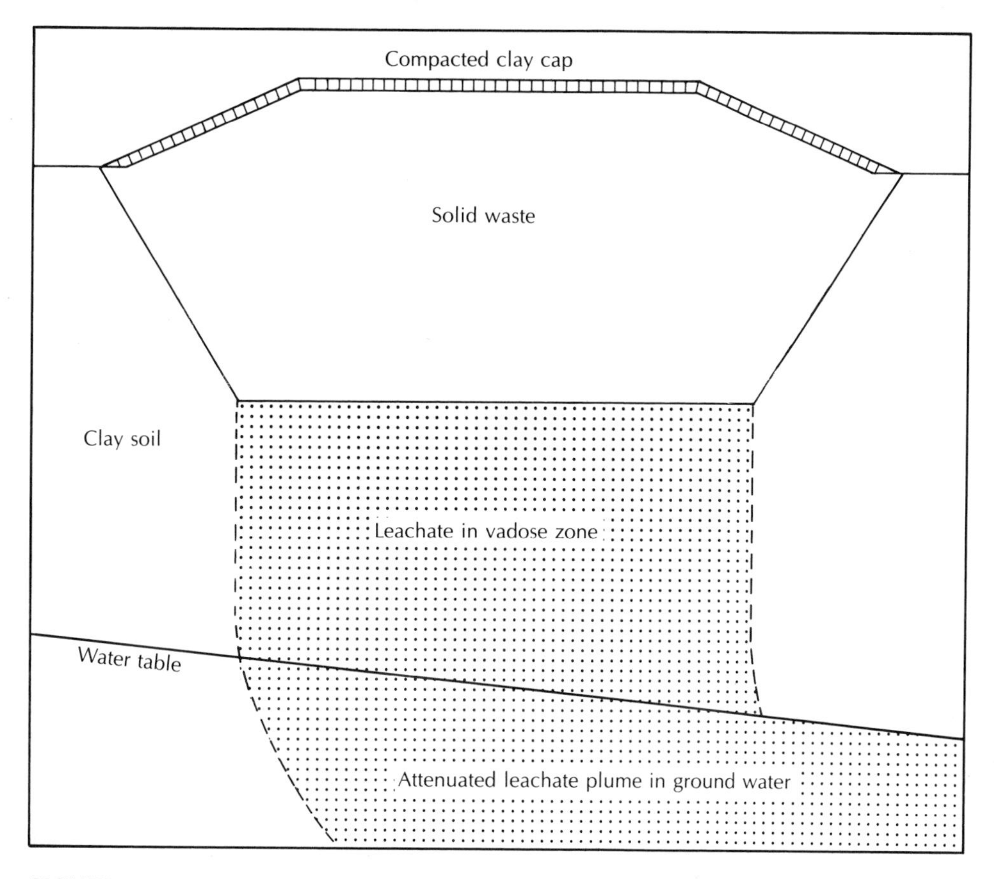


**FIGURE 10.16** Natural-attenuation landfill. Clay soils attenuate leachate to varying degrees depending upon soil type and leachate. Attenuated leachate passes through the unsaturated zone to the water table, where a plume is formed.

clay sidewalls is installed to further reduce the amount of seepage into the excavation. A leachate-collection system is installed. This collects not only the leachate that forms, but any ground water that seeps in as well. The zone-of-saturation landfill is more efficient than the lined landfill because no leachate escapes. However, it will be necessary to collect the leachate long after the landfill is closed because the waste will be below the water table.

As an alternative to the zone-of-saturation landfill, a **hydraulic-gradient-control** landfill can be constructed in areas of a high water table (Figure 10.19). Underdrains are placed beneath the liner to lower the water table so that it is below the liner. A leachate-collection system is installed above the liner. The water discharging from the underdrains must be monitored to determine if it is being impacted by the portion of the leachate that drains through the liner. If it is, then it must be treated before being discharged. As with all of the systems, a

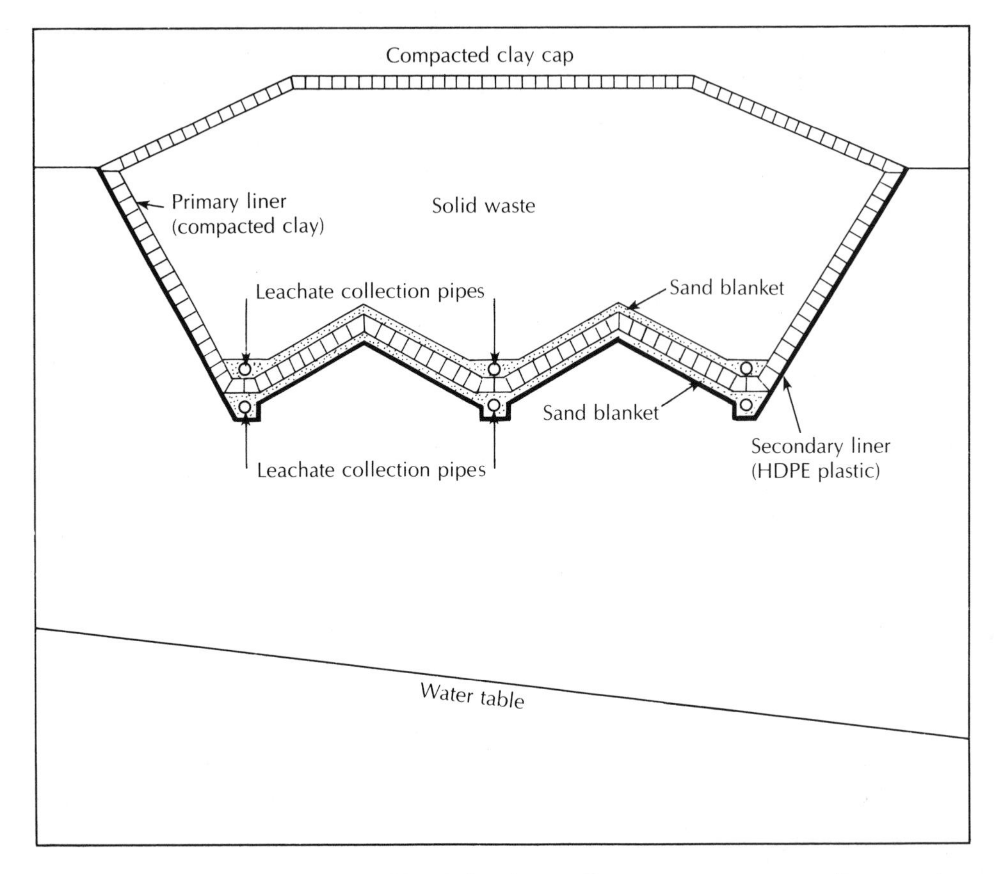


**FIGURE 10.17** Double-lined landfill with leachate collection system. Primary liner consists of five feet of compacted clay soil with hydraulic conductivity of no more than $1 \times 10^{-7}$ cm/sec. Secondary liner is flexible membrane such as 40 mil HDPE plastic. Leachate collection system consists of one-foot-thick sand layers with perforated pipes, which drain to a leachate collection tank.

well-designed, well-constructed cap is essential in reducing the amount of leachate formed, thereby reducing the amount of leachate that must be handled.

### 10.7.4 CHEMICAL SPILLS AND LEAKING UNDERGROUND TANKS

Ground-water contamination due to a variety of inorganic and organic compounds has occurred as a result of spills and leaks of toxic and hazardous chemicals. These discharges may be the result of a sudden action, such as a tank car accident, or may be the result of slow leakage. Typically more than one chemical may be released. As was seen in Section 10.6, different chemicals will

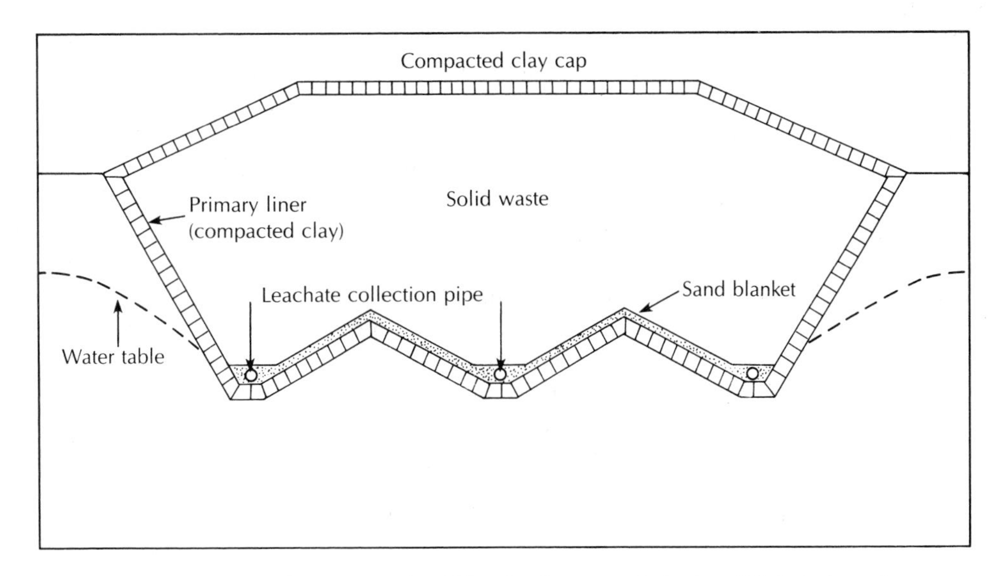


**FIGURE 10.18** Zone-of-saturation landfill. The bottom of the landfill is below the water table. Leachate-collection system collects both leachate and ground water that seeps into the landfill.

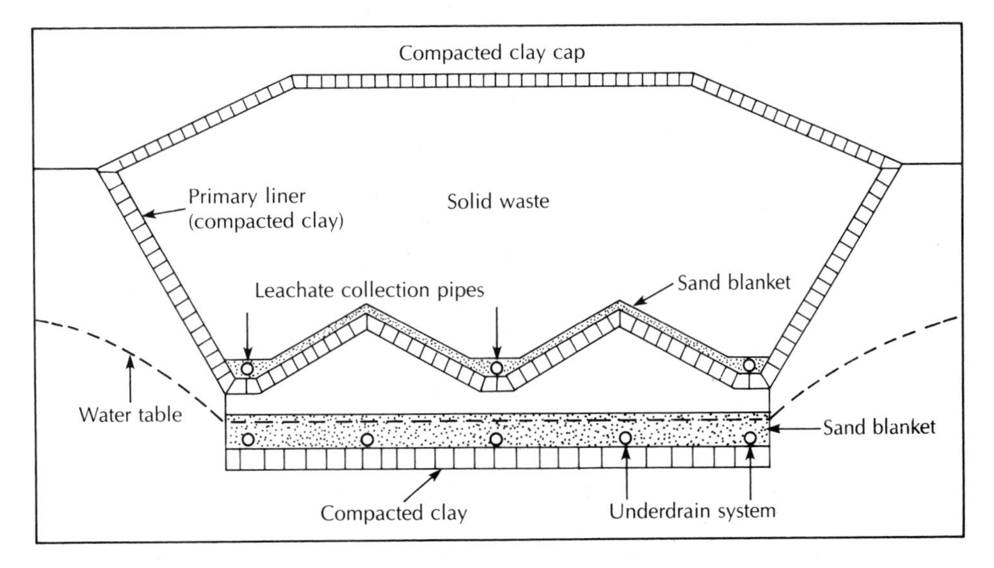


**FIGURE 10.19** Lined landfill with water-table control system. Underdrains below the liner keep water table from saturating the liner and waste. Leachate that seeps through the liner is collected by the underdrain system.

travel through the ground at different rates owing to retardation effects. As a result, complex plumes of contaminated water may result.

If the contaminant dissolves in the water, it will flow along with the ground water. However, if a liquid discharged into ground has a specific gravity less than that of water, it can float on the water table. This is what happens when a petroleum product leaks into the ground. Figure 10.20 shows a gasoline plume moving along the surface of the water table. Note that the gasoline will spread upgradient a short way and that there will be some soluble components of the gasoline, such as benzene, dissolved in the water below the water table.

Dense liquids may sink to the bottom of the aquifer. In general, the chlorinated hydrocarbons are heavier than water. They have various solubilities in water. Table 10.10 gives the density and solubility of some organic compounds. For those that are denser than water, the pure product will sink to the bottom of the aquifer. Some of the product will go into the solution so that there will also be a plume of ground water with dissolved product. Figure 10.21 shows this.

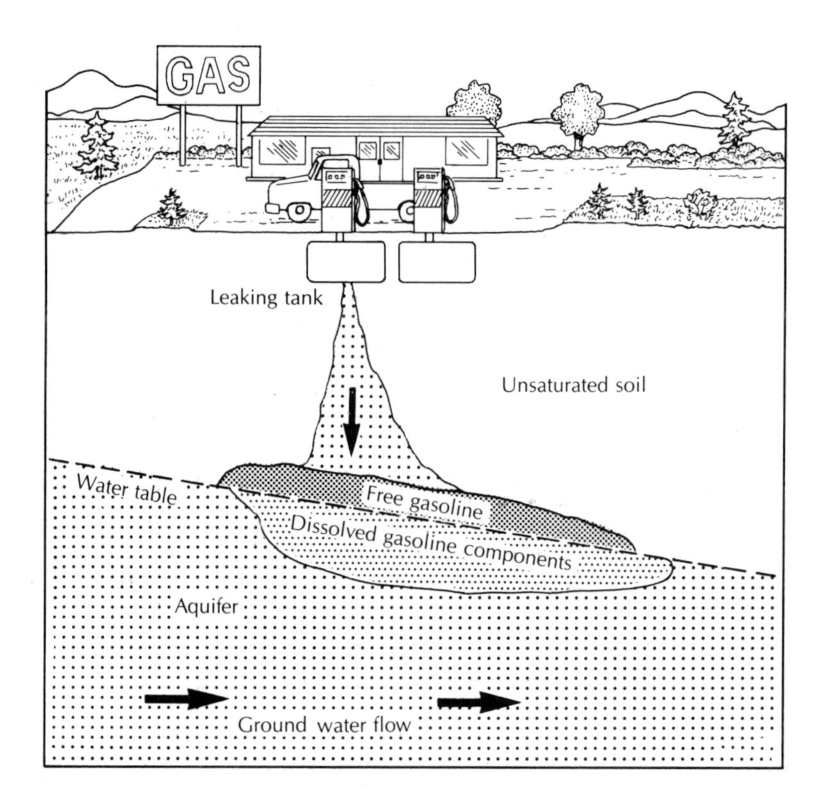


**FIGURE 10.20** Organic liquids, such as gasoline, which are only slightly soluble in water and are less dense than water, tend to float on the water table when a spill occurs.

**TABLE 10.10** Density and solubility in water of organic compounds

| Compound | Specific Gravity[a] | Solubility[b] Milligrams compound/liter water (@ °C Temperature) |
|---|---|---|
| Acetone | 0.79 | Infinite |
| Benzene | 0.88 | 1780 (20) |
| Carbon tetrachloride | 1.59 | 800 (20), 1160 (25) |
| Chloroform | 1.48 | 8000 (20), 9300 (25) |
| Methylene chloride | 1.33 | 20,000 (20), 16,700 (25) |
| Chlorobenzene | 1.11 | 500 (20), 488 (30) |
| Ethyl benzene | 0.87 | 140 (15), 152 (20) |
| Hexachlorobenzene | 1.60 | 0.11 (24) |
| Ethylene chloride | 1.24 | 9200 (0), 8690 (20) |
| 1, 1, 1-trichloroethane | 1.34 | 4400 (20) |
| 1, 1, 2-trichloroethane | 1.44 | 4500 (20) |
| Trichloroethylene | 1.46 | 1100 (25) |
| Tetrachloroethylene | 1.62 | 150 (25) |
| Phenol | 1.07 | 82,000 (15) |
| 2-Chlorophenol | 1.26 | 28,500 (20) |
| Pentachlorophenol | 1.98 | 5 (0), 14 (20) |
| Toluene | 0.87 | 470 (16), 515 (20) |
| Methyl ethyl ketone | 0.81 | 353 (10) |
| Naphthalene | 1.03 | 32 (25) |
| Vinyl chloride | 0.91 | 1.1 (25) |

[a]Source: R. Weast, "Handbook of Chemistry and Physics," 60th ed., CRC Press, Inc., 1979, 1980.
[b]From Verschueren, Karel. "Handbook of Environmental Data on Organic Chemicals." New York, Van Nostrand Reinhold, 1983. Numbers in parentheses = temperatures.

There are a large number of sites in the United States where ground water has been contaminated with organic compounds. In some cases, contaminated water has migrated from a spill area to a water-supply well. In a shallow aquifer in South Brunswick Township, New Jersey, 1,1,1-trichloroethane and tetrachloroethane migrated more than 3000 feet from a factory site to a water-supply well. At times the 1,1,1-trichloroethane was present in amounts in excess of 1000 μg/L and the tetrachloroethane in amounts ranging from 100 to 300 μg/L. The affected well was taken out of service (55). Figure 10.22 shows the plume of 1,1,1-trichloroethane. There are actually three plumes, each from a different source, although only one was affecting the supply well.

### 10.7.5 MINING

Extraction and processing of metallic ore and coal has been the source of both surface- and ground-water contamination. Ground water moving through mineralized rock zones may contain excessive amounts of heavy metals (56). Mining and milling expose overburden and waste rock to oxidation. Oxidation

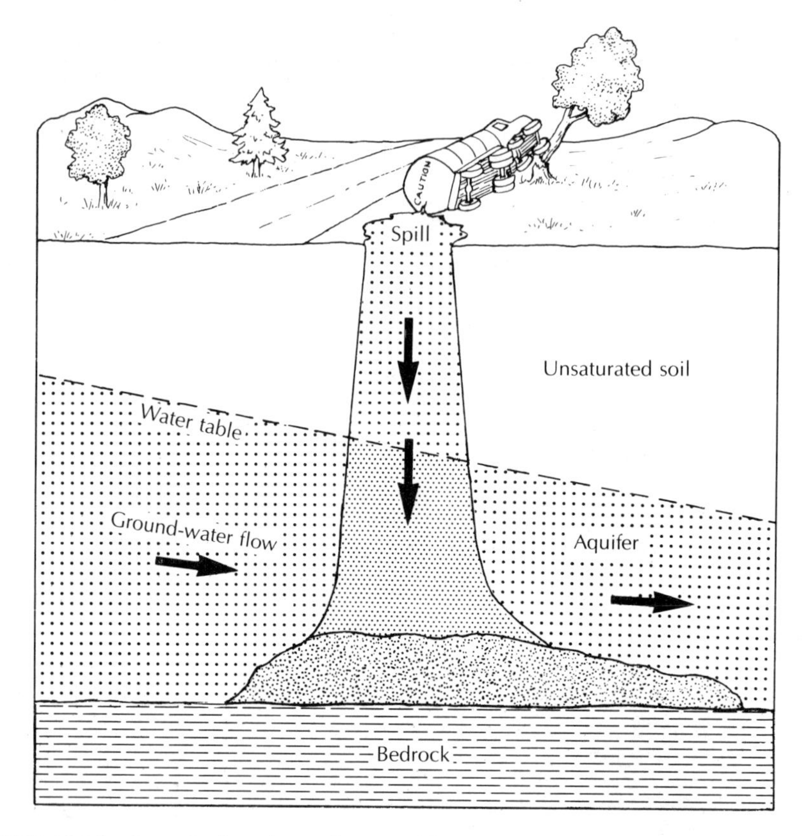


**FIGURE 10.21** Organic liquids, such as trichloroethylene, which are only slightly soluble in water and are more dense than water, may sink to the bottom of an aquifer when a spill occurs.

of pyrite, a common mineral, can produce sulfuric acid. In the Appalachian region of the eastern United States, 6000 tons of sulfuric acid are produced daily in this manner (57). This results in highly acidic water draining from spoil piles and tailings deposits; hence, the shallow ground water and surface water of the region tend to have a low pH. The low-pH water draining through spoil piles and tailings can also leach heavy metals (58, 59), as well as soluble calcium, magnesium, sodium, and sulfate (60). Uranium and thorium mining and milling operations can release radioactive isotopes to the atmosphere, surface water, and ground water (61).

## CASE STUDY: CONTAMINATION FROM URANIUM TAILINGS PONDS

In one study of the ground water beneath two unlined uranium tailings ponds in Utah, a comparison was made of baseline natural uranium activity before the tailings ponds were put into service and the natural uranium activity after 11 years of operation (62). The tail-

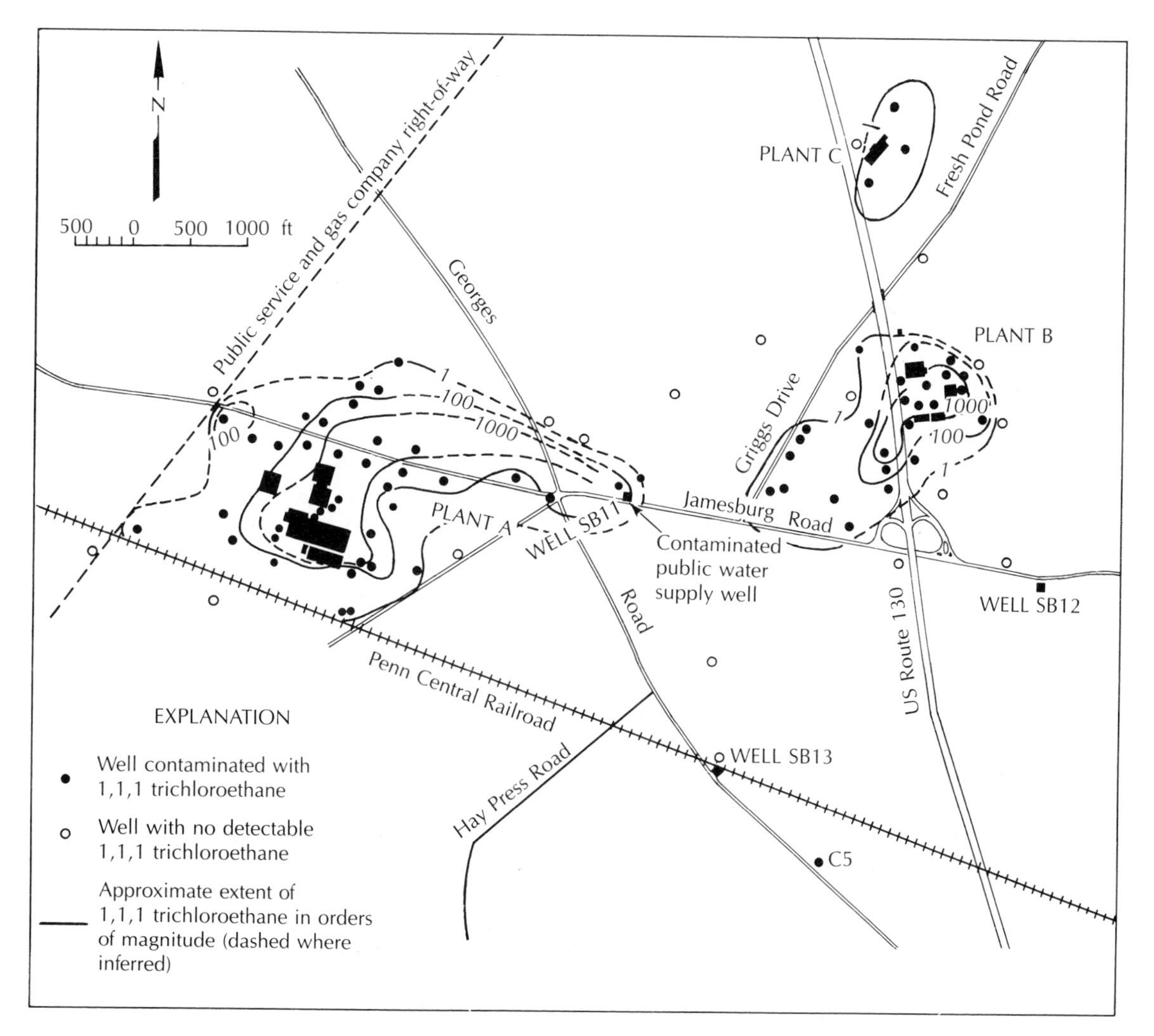


**FIGURE 10.22** Plumes of 1,1,1-trichloroethane in shallow ground water in South Brunswick Township, New Jersey. The plume to the southwest has traveled to the east and has reached a public water-supply well. Source: P. H. Roux and W. F. Althoff, *Ground Water,* 18 (1980):464–72.

ings ponds, each of which covers about 40 acres, were constructed by placing earthen dams across a small valley. The surface geology is a fine-grained sandstone interfingered with siltstones and claystones. Ground-water flow is fracture-controlled. Figure 10.23A shows the locations of the tailings ponds and the fracture patterns affecting the ground water. Figure 10.23B shows the potentiometric surface. It was found that the baseline water quality showed somewhat elevated levels of natural uranium activity due to natural leaching of uranium-bearing rocks by water moving along a fracture system. Figure 10.23C

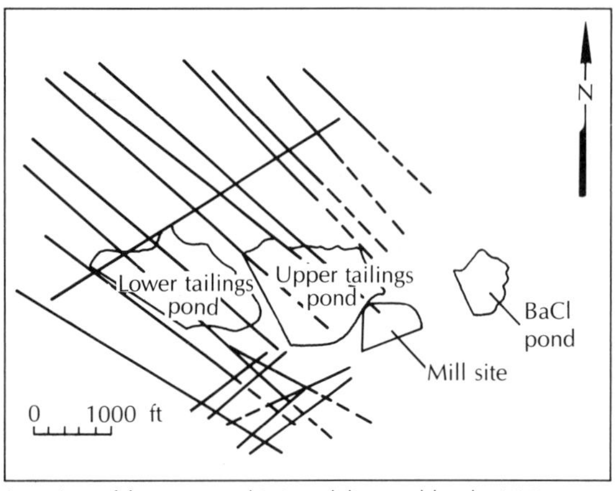


Location of fractures and joints delineated by the VLF survey

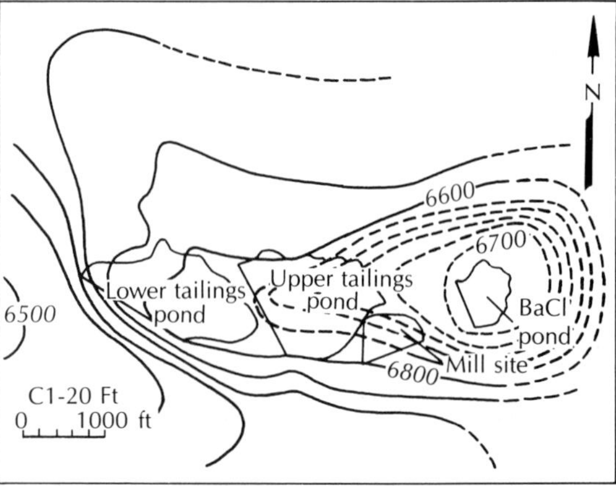


Elevation of the potentiometric surface in February 1984

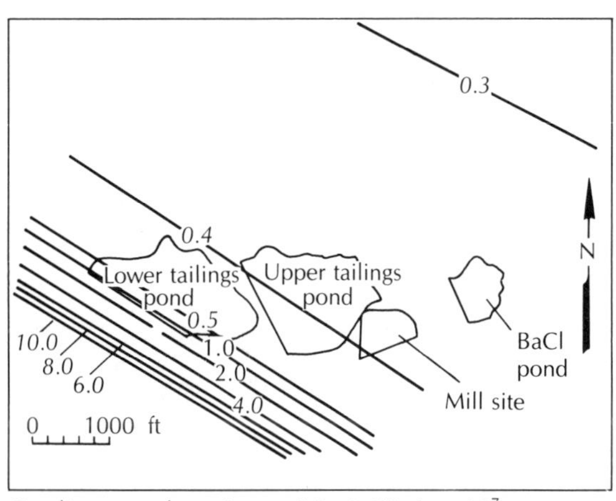


Baseline natural uranium activity (μCi/mL × $10^7$)

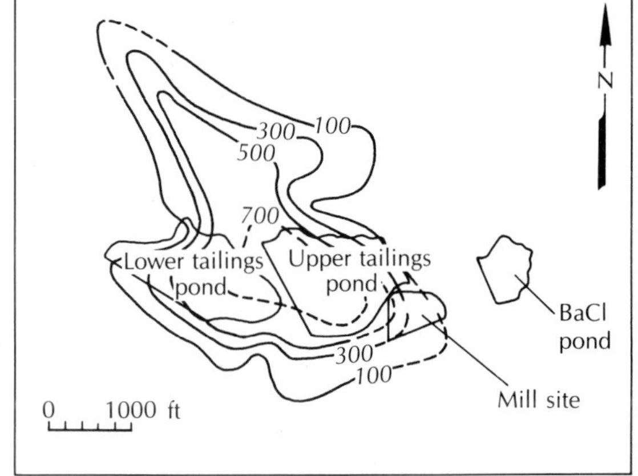


Elevated activity of natural uranium in the summer of 1983 (μCi/mL × $10^7$)

**FIGURE 10.23** Ground-water contamination by uranium beneath active uranium mill tailings ponds. Source: R. B. White and R. B. Gainer, *Ground Water Monitoring Review,* 5, no. 2 (1985):75–82.

shows the baseline natural uranium activity. It follows a linear pattern along the strike of a major fracture southwest of the lower tailings pond. After 11 years of operation, the natural uranium activity had increased significantly above the baseline measurements. Figure 10.23D shows the elevated activity, corrected for the baseline variation. The plume is spreading to the northwest, following the strike of the fractures. This illustrates one of the facets of ground-water contamination in fractured rock aquifers that are anisotropic. That is, the plume will tend to follow the fractures and not necessarily move normal to the potentiometric surface. If the aquifer were isotropic, from the potentiometric surface map one would expect that the plume would spread primarily to the southwest.

### 10.7.6 OTHER SOURCES OF GROUND-WATER CONTAMINATION

In a 1977 report to Congress, the U.S. Environmental Protection Agency listed a number of waste-disposal practices and their potential impact on ground water (49). Waste-disposal practices mentioned in that report and not already covered in this section include liquid industrial waste disposal in lagoons and injection wells, oil-field brine disposal in lagoons and wells, land-spreading of sewage and industrial sludges, leakage from municipal wastewater sewers and lagoons, and land disposal of animal waste from feed lots. Other major causes of ground-water contamination listed in the report to Congress were spills and leaks; mine drainage; salt-water intrusion; poorly constructed or abandoned water, oil, and gas wells; infiltration of contaminated surface water; agricultural activities; highway deicing salts; and atmospheric contaminants.

## 10.8 GROUND-WATER RESTORATION

### 10.8.1 INTRODUCTION

Once ground water has been contaminated, it may take many years after the source of contamination has been eliminated for natural processes to remove the contaminants from an aquifer. During the 1980s, methods of restoring ground-water quality by implementing various types of remedial measures were developed. However, most of these methods are time-consuming and extremely expensive. There are two broad categories of remedial measures: one must either remove or isolate the source and/or pump and treat the ground water (63–65).

### 10.8.2 SOURCE-CONTROL MEASURES

One extreme method of source control is to excavate and remove the source. This was done at a toxic chemical waste landfill in Wilsonville, Illinois, when the court ordered a landfill operator to exhume and remove a large number of drums of liquid waste that had been buried in a licensed landfill (66). A less drastic approach is to isolate the waste in place. If the waste is entirely above the water table, this is much more easily done than if the waste extends below the water table.

Changing the surface drainage so that runoff from upland areas does not cross the land surface above the waste will reduce the amount of surface water that infiltrates into the waste and produces leachate. The construction of a low-permeability cap above the waste can also be very effective in reducing the amount of infiltration through the waste. Caps can be constructed of compacted clay, synthetic membranes, concrete, asphalt, and other types of materials. The most effective caps have several layers and include coarse granular layers be-

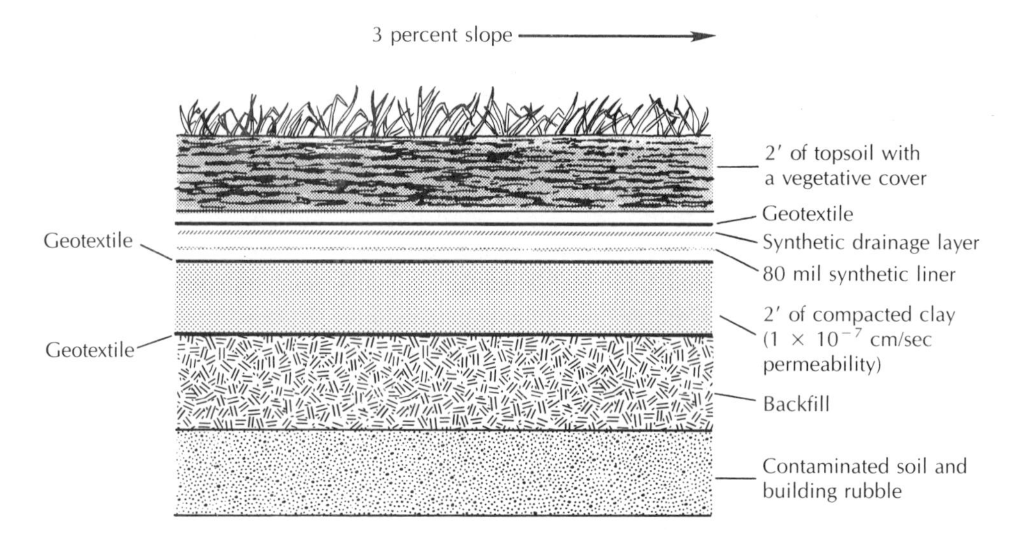


**FIGURE 10.24** Design of a low-permeability multimedia cap to cover waste. Fill material is used above waste to create a 3 percent slope if the waste material or land surface over the waste material is not sloped.

tween fine-grained layers to act as drains to divert infiltration away from the waste (67). Figure 10.24 shows a multimedia cap design.

If the waste extends below the water table, then it is necessary to keep the ground water from flowing through it. This can be accomplished by installing a low-permeability vertical barrier around the waste body (68–72). Vertical barriers can be constructed by digging a trench and backfilling it with a slurry-type mixture of water, soil, and bentonitic clay. This is called a **slurry wall.** Slurry wall construction is limited to the depths that the trench can be constructed. A **grout curtain** can be installed by injecting any of a number of compounds into boreholes around the site. The materials fill the pore spaces in the rock or soil and harden. Grout curtains can be installed to great depths. Interlocking metal **sheet piling** can also be driven into soil to form a cutoff wall.

Figure 10.25A shows a waste material buried beneath the water table. Figure 10.25B indicates the installation of a low-permeability slurry wall to lower the water table and divert flowing ground water from the waste source. In this case, one wall installed upgradient of the waste is sufficient. In other cases, the waste could be surrounded with walls. Grout can even be injected through the waste to form a bottom seal. Used in conjunction with walls around the waste, the waste can be totally encapsulated. Figure 10.25C shows this.

Hydraulic gradient control measures can be used to lower the water table where it is in contact with the waste (73–76). Figure 10.25D shows a pumping well installed upgradient of buried waste material. The water table is lowered so that it no longer is in contact with the waste.

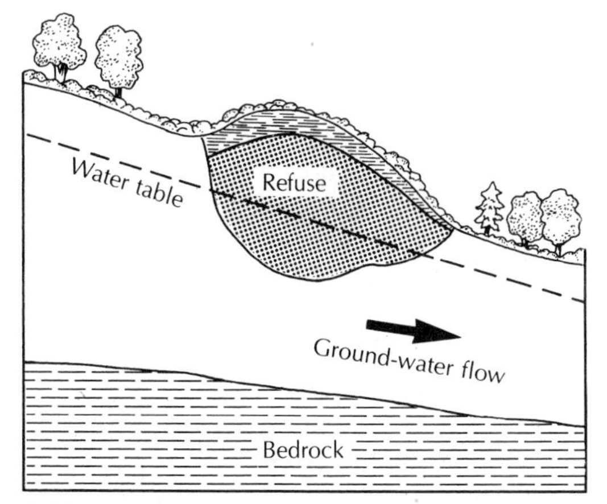


A No control measure

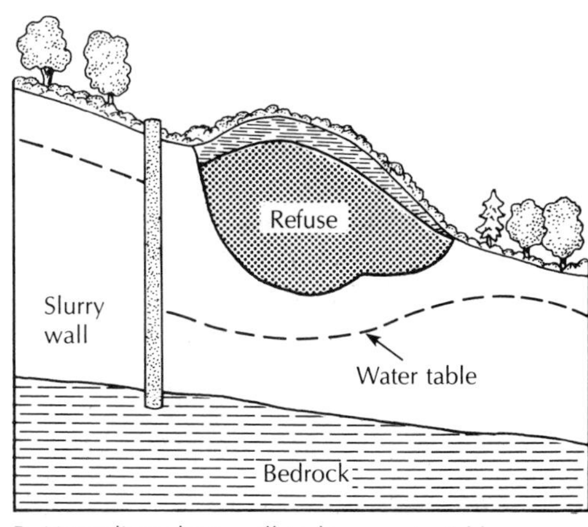


B Upgradient slurry wall to lower water table

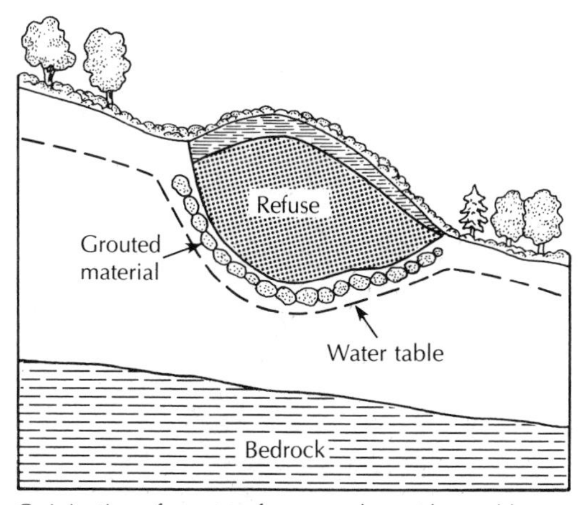


C Injection of grout to form a seal on sides and bottom

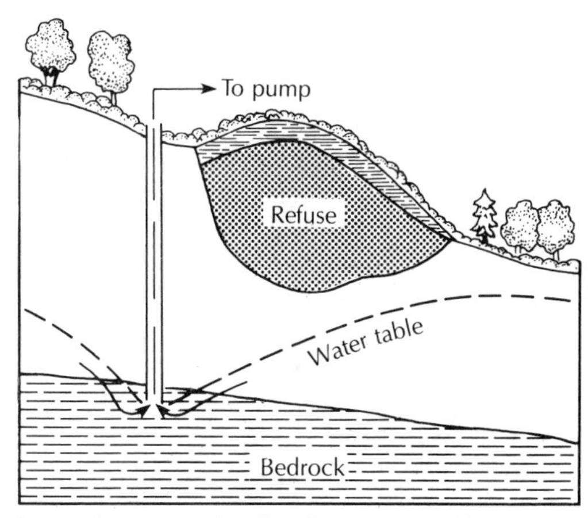


D Gradient control well to lower water table

**FIGURE 10.25** Isolation methods used to lower water table so that buried waste material is not in contact with ground water. **A.** No control measure. **B.** Upgradient slurry wall to lower water table. **C.** Injection of grout to form a seal on sides and bottom. **D.** Gradient control well to lower water table. Source: Modified from U.S. Environmental Protection Agency.

### 10.8.3 PLUME TREATMENT

Once the contamination source is isolated, the task remains to restore the quality of the ground water that has become contaminated. One option is to take no action and let the contaminants be flushed from the aquifer by natural recharge. This process could be made more efficient in some hydrogeologic settings by artificially increasing the amount of water that enters the aquifer and hence accelerating the natural process. This approach is usually not desirable if the aquifer is used as a source of drinking water. Additionally, if the plume will eventually discharge to a surface-water body, it might cause contamination of the surface water. The time for natural restoration might be tens to hundreds of years. Even if the aquifer is not presently a drinking-water source, in the future it could be put to such a use. If the waste source were no longer present, future generations could unsuspectingly drill wells into the contaminated aquifer.

Plume treatment can occur *in situ* or by extraction of the water via wells (77–81). *In situ* treatment can be chemical or biological. Figure 10.26A shows a method of injecting nutrients and oxygen into a plume. Some compounds, such as hydrocarbons, can be biologically treated by natural soil bacteria with the proper mix of nutrients. If a chemical treatment scheme were proposed, the upgradient injection well of Figure 10.26A could be used to inject the proper chemicals. For example, an oxidizing agent could be added to the ground water. Figure 10.26B shows a permeable treatment bed installed where the water table is shallow. An acidic leachate could be neutralized by installing a permeable treatment bed with limestone gravel in it. Materials with ion-exchange properties could possibly be used to remove heavy metals.

Where the plume contains water contaminated with chlorinated solvents, many of which are believed to be carcinogenic at the low parts per billion range, the best option in most cases is to use **extraction wells** to remove the contaminated water. Figure 10.27 shows plume removal by means of shallow gradient-control wells. The spacing and pumping rate for the extraction wells is typically determined by computer modeling. The extraction wells are designed to capture the plume while at the same time removing as little of the uncontaminated water as possible. Extraction wells can also be planned as **plume-stabilization wells.** In this case they are located somewhere within the plume and sized to reverse the hydraulic gradient beyond the edge of the plume. They then prevent further movement of the plume. Locating an extraction well outside of the plume will tend to expand the plume boundaries.

Contaminated water that has been removed from the ground is usually treated before being discharged. Naturally, the treatment will depend upon the types and concentrations of the contaminants. Synthetic organic compounds are usually removed from contaminated water by one or more of the following methods: volatilization to the atmosphere in an air-stripping column, adsorption on activated carbon, and biological treatment.

## 10.9 CASE HISTORY—GROUND-WATER CONTAMINATION AT A SUPERFUND SITE

### 10.9.1 BACKGROUND

From 1970 until 1979 a solvent-recovery and -recycling plant, Seymour Recycling Corporation, was operated in Seymour, Indiana. The company went bankrupt in 1979 and the owners abandoned 98 large tanks and approximately

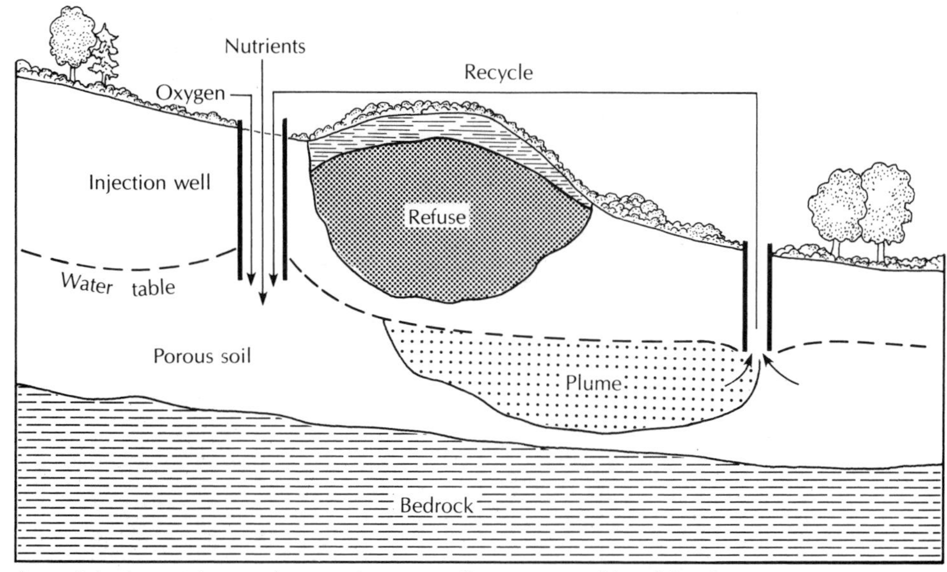


A. Injecting nutrients and oxygen into an aquifer to promote bioreclaimation.

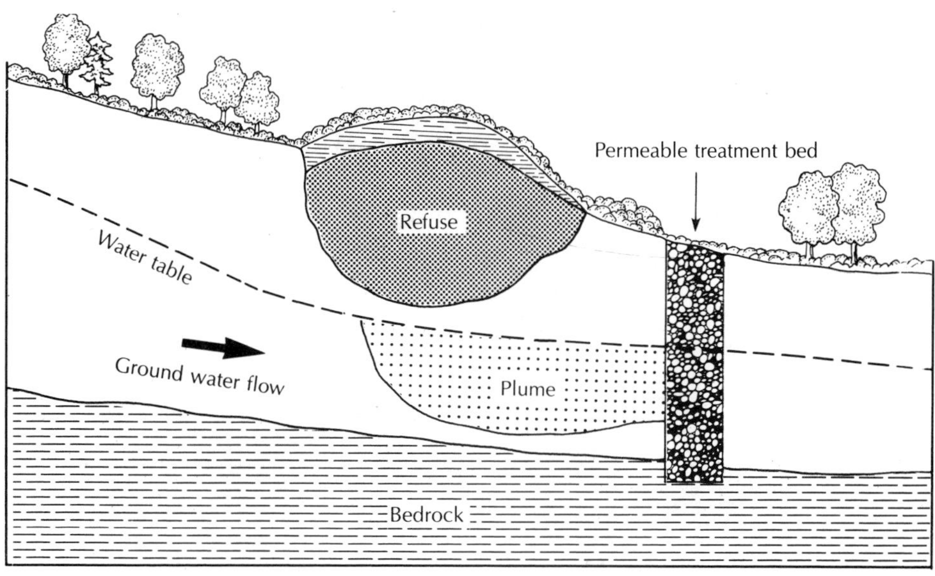


B. Installation of a permeable treatment bed in path of plume to provide contact-type treatment such as ion exchange or neutralization.

**FIGURE 10.26** *In situ* treatment methods. **A.** Injecting nutrients into an aquifer to promote bioreclamation. **B.** Installation of a permeable treatment bed in path of plume to provide contact-type treatment such as ion exchange or neutralization. Source: Modified from U.S. Environmental Protection Agency.

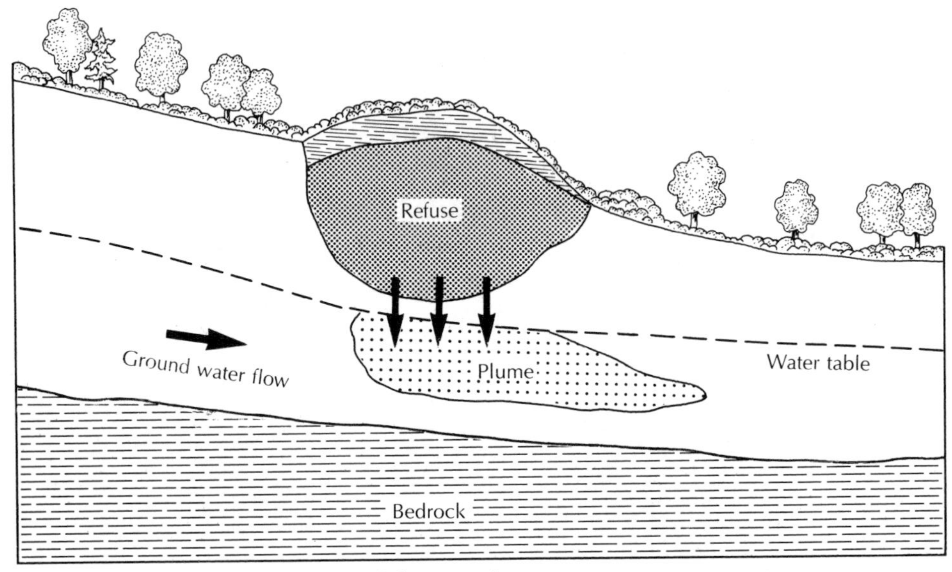


Before pumping

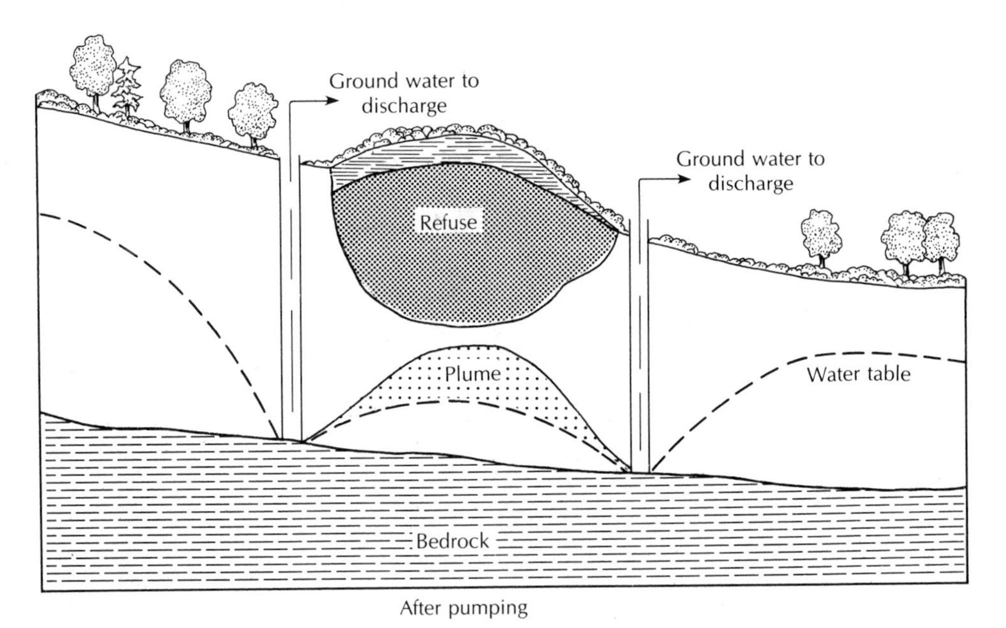


After pumping

**FIGURE 10.27** Use of extraction wells to remove contaminated ground water. Source: U.S. Environmental Protection Agency.

50,000 drums, all filled with organic chemicals. Many of the drums were in a deteriorated condition, and between 1979 and 1983, when they were all removed, an unknown amount of liquid synthetic organic compounds leaked into the soil. Recharging precipitation washed a substantial amount of the organic compounds into the ground-water reservoir.

### 10.9.2 GEOLOGY

The Seymour Recycling Corporation hazardous-waste site is located on a nearly level plain at the edge of a glacial sluiceway now occupied by the East Fork of the White River at Seymour, Indiana. Figure 10.28 shows the location of the site. There are about 80 feet of unconsolidated glaciofluvial sediments overlying shale bedrock at the site. The shale is reported to have very low permeability and is not an aquifer. No bedrock wells are located in the area. A deposit of sand and gravel overlies the bedrock or a basal clay. The sand and gravel is called the deep aquifer. It is overlain by a silty clay that acts as a confining layer. There is a shallow aquifer consisting of sandy glacial outwash, which is relatively thin, 8 feet (2.4 meters), in the south part of the study area; it thickens to as much as 40 feet (12 meters) at the northern part of the area. The shallow aquifer is covered by a layer of dune sand and loess, which is 6 to 10 feet (2 to 3 meters) thick and acts as a semiconfining layer for the shallow aquifer. Figure 10.29 is a northwest-southeast geologic cross section.

### 10.9.3 HYDROLOGY

Ground water in the shallow aquifer is moving to the north-northwest at a rate estimated to be between 81 and 362 feet per year. Figure 10.30 shows the potentiometric surface of the shallow aquifer. There is a shallow stream, the ''east-west'' creek, lying 1000 feet north of the site. During periods of high ground-water levels, the shallow aquifer discharges into this stream. During periods of diminished precipitation, ground-water levels fall and the creek dries up. During such periods, the ground water in the shallow aquifer flows beneath the ''east-west'' creek and toward the East Fork of the White River and an intervening intermittent tributary, Von Fange Ditch.

The water table in the shallow aquifer is higher than the potentiometric surface in the deep aquifer. Figure 10.31 is the potentiometric map of the deep aquifer. As a result there is a downward hydraulic gradient across the confining layer. Sediments in the confining layer have a wide variety of grain sizes and a wide range of vertical hydraulic conductivities. The average linear ground-water velocity from the shallow to the deep aquifer varies from less than 0.1 foot per year to more than 450 feet per year at different locations. In addition, the confining layer ranges in thickness from as little as 7 to as many as 50 feet.

The deep aquifer flows to the south, which is the opposite direction of the shallow aquifer. The potentiometric surface between the site and the ''east-west'' creek is very flat and at times there might be a flow reversal so that during those times water in the deep aquifer would be flowing away from the site to

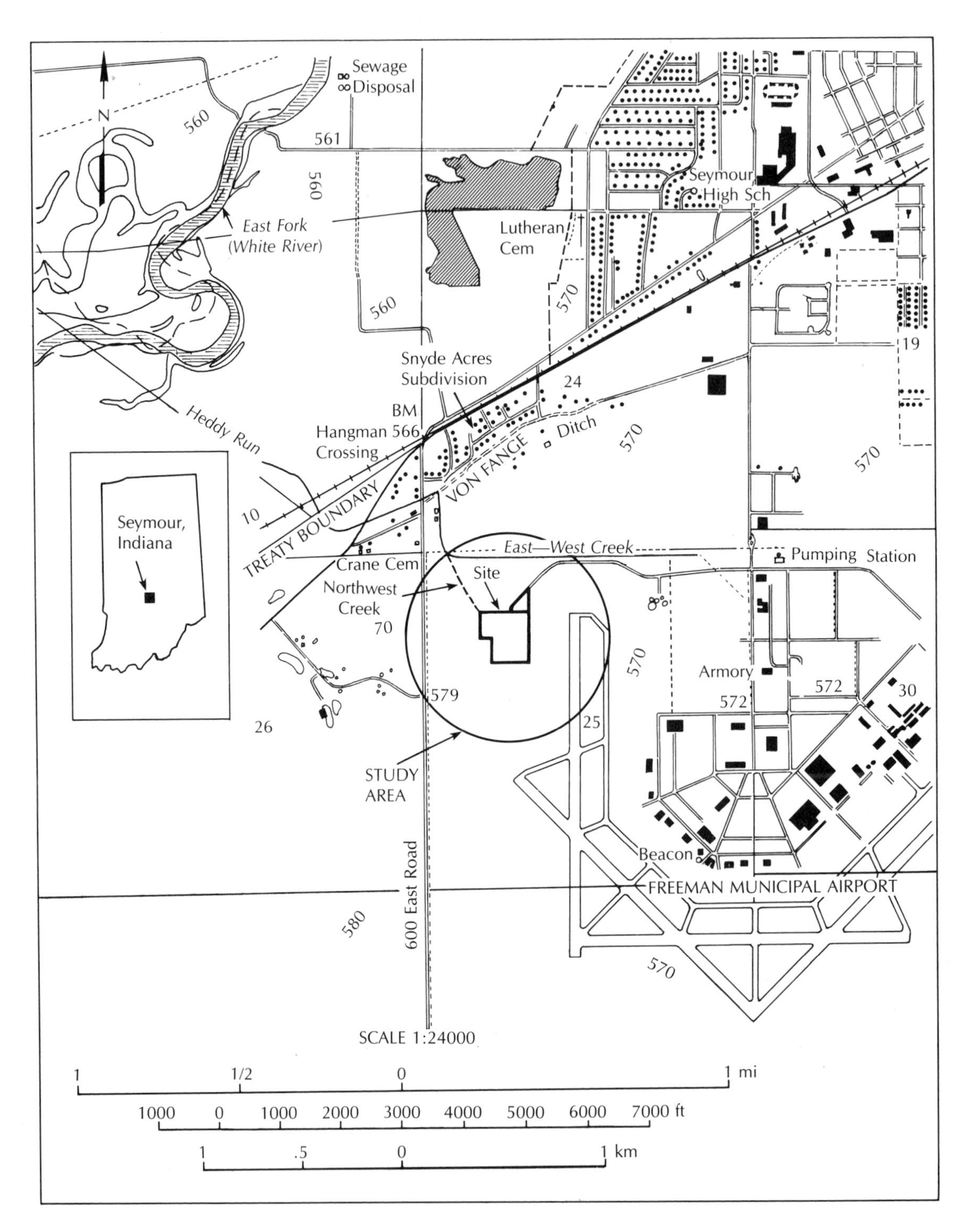


**FIGURE 10.28** Location of the Seymour Recycling Corporation Superfund site. Source: CH2M-Hill, *Feasibility Study Report, Seymour Recycling Corporation Hazardous Waste Site*, U.S. Environmental Protection Agency, 1986.

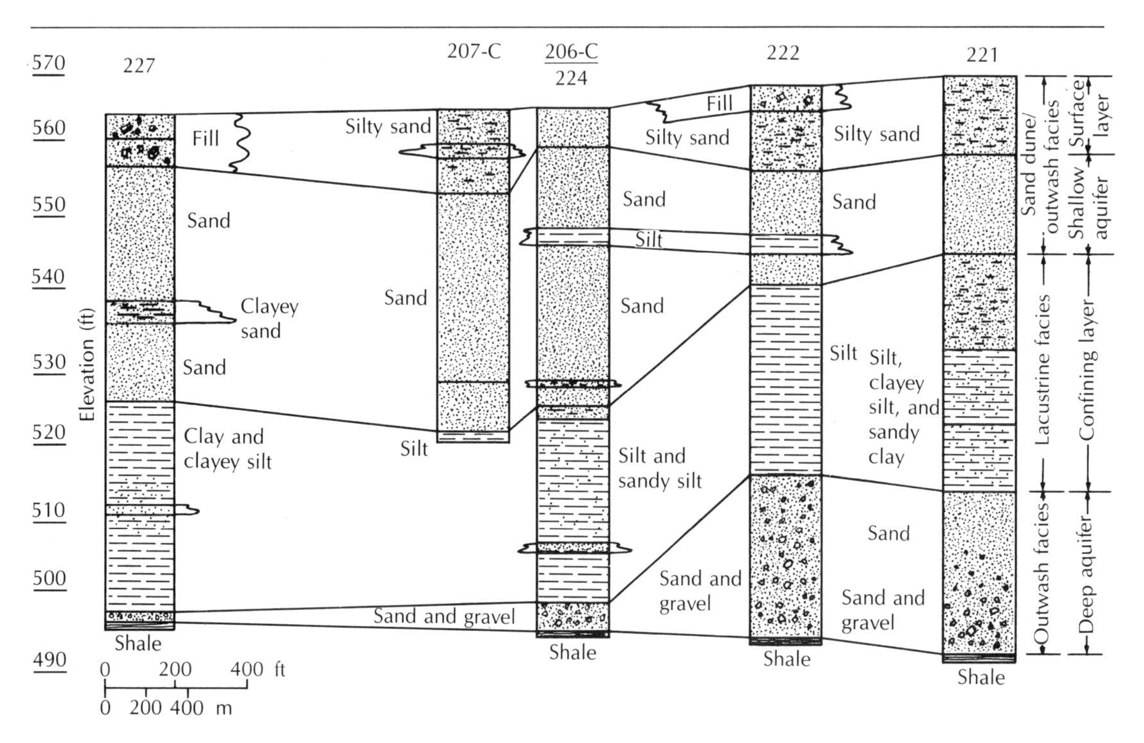


**FIGURE 10.29** Geologic cross section through the unconsolidated sediments beneath the Seymour Recycling Corporation Superfund site showing both the shallow and the deep aquifers. Source: C. W. Fetter, *Final Hydrogeologic Report, Seymour Recycling Corporation Hazardous Waste Site*. U.S. Environmental Protection Agency, 1985.

both the north and the south. The flow rate in the deep aquifer is estimated to fall in the range of 147 to 796 feet per year.

### 10.9.4 GROUND-WATER CONTAMINATION

Ground water in the shallow aquifer beneath the site is highly contaminated with organic compounds, including 14 different volatile priority pollutants and 6 extractable priority pollutants. Concentrations of hazardous organic substances of up to 370,000 μg/L have been reported from the on-site wells. Figure 10.32 is a cross section showing the vertical distribution of the plume. As there are a large number of compounds, and the plume is not homogeneous, it is mapped as total organic compounds as detected by GC/MS* analysis. Figure

---

*Organic compounds in water may be analyzed by a method called gas chromotography (GC) which may be combined for certain purposes with mass spectroscopy (MS), in which case the method is abbreviated GC/MS. The sample is treated in different ways to determine the concentration of volatile organic compounds and organic compounds, which can be extracted from the water sample by use of an organic solvent under acidic, neutral, and basic pH conditions. Volatiles and extractables are reported separately.

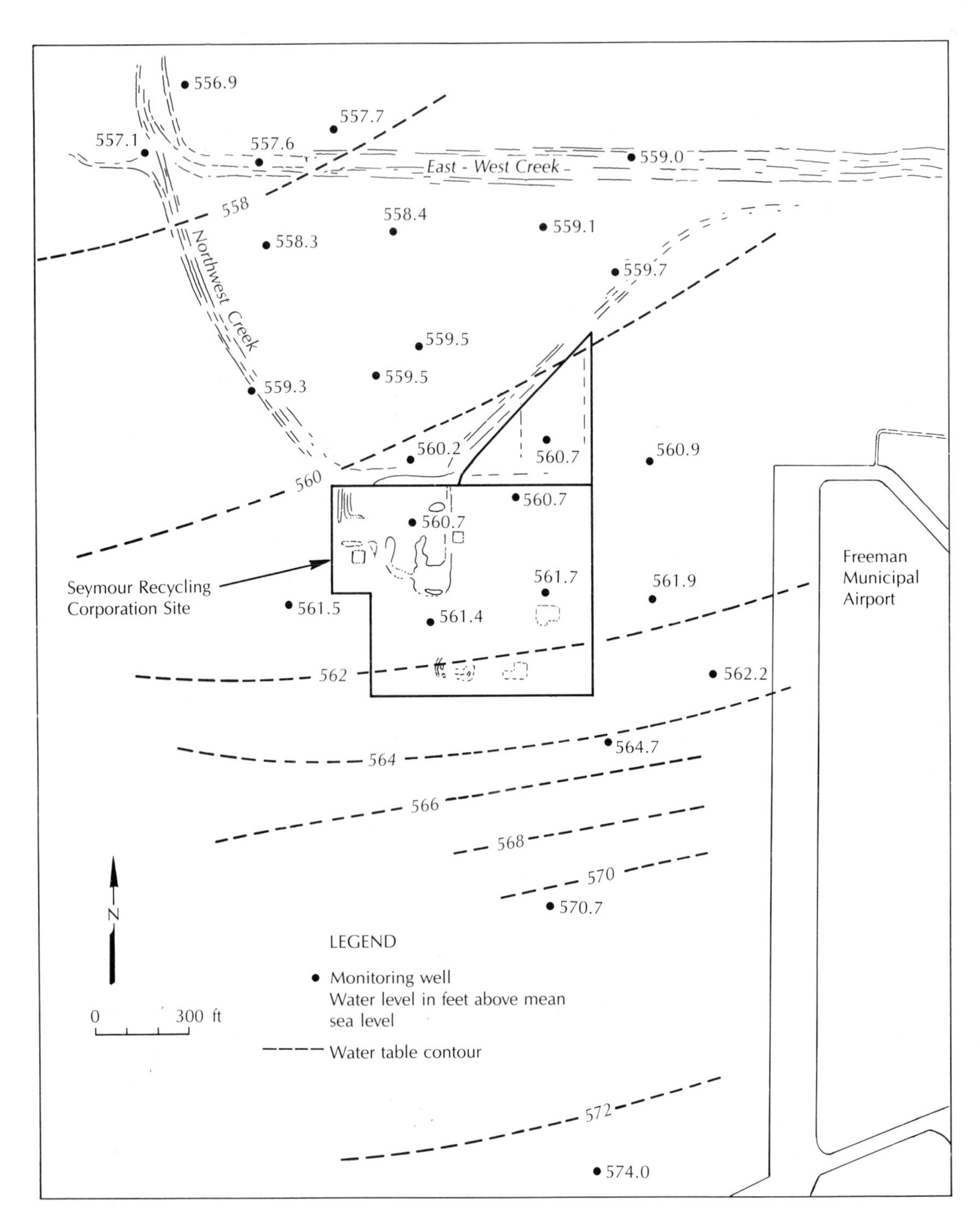


**FIGURE 10.30** Potentiometric surface of the shallow aquifer at the Seymour Recycling Corporation Superfund site. Source: C. W. Fetter, *Final Hydrogeologic Report, Seymour Recycling Corporation Hazardous Waste Site,* U.S. Environmental Protection Agency, 1985.

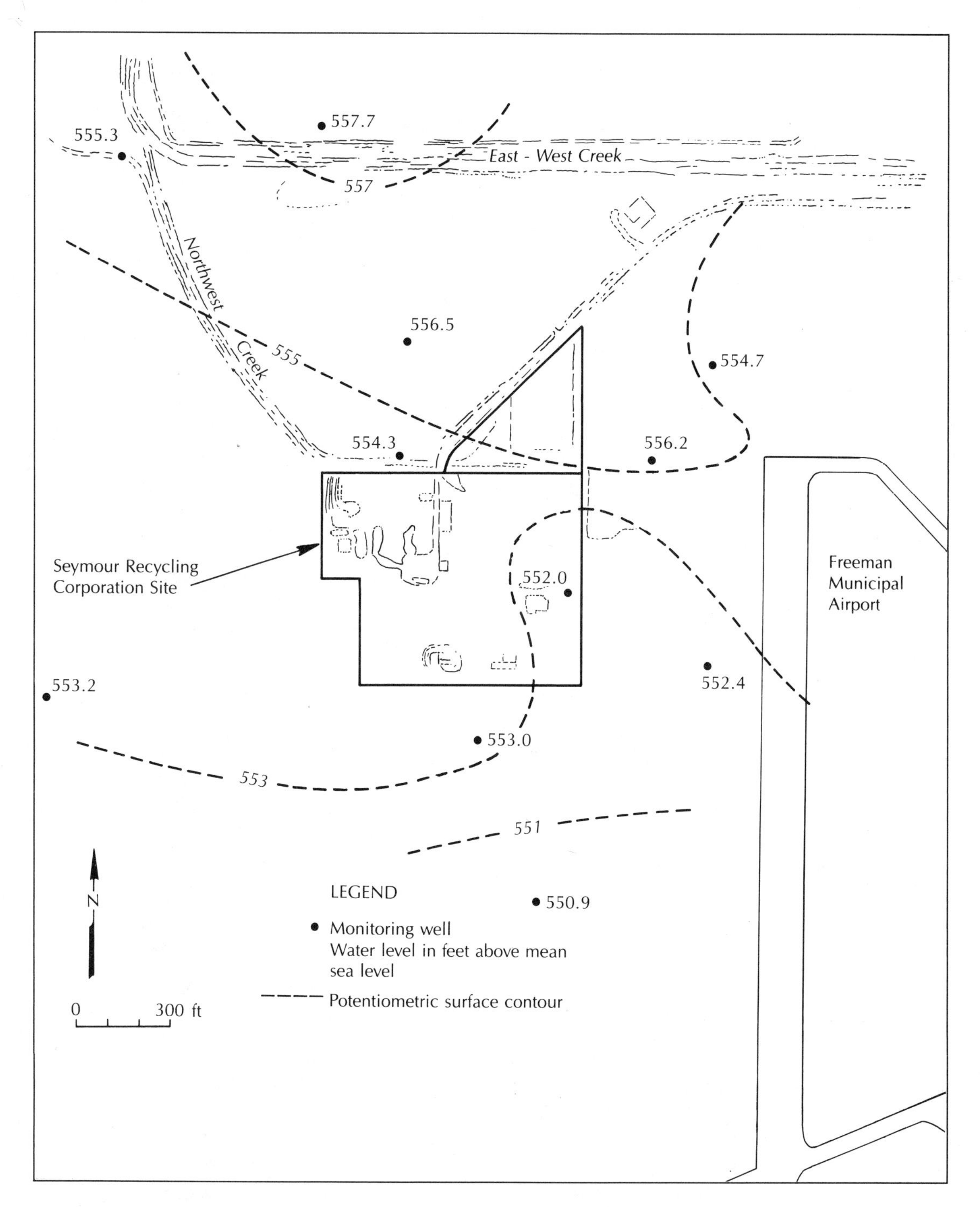


**FIGURE 10.31** Potentiometric surface of the deep aquifer at the Seymour Recycling Corporation Superfund site. Source: C. W. Fetter, *Final Hydrogeologic Report, Seymour Recycling Corporation Hazardous Waste Site*. U.S. Environmental Protection Agency, 1985.

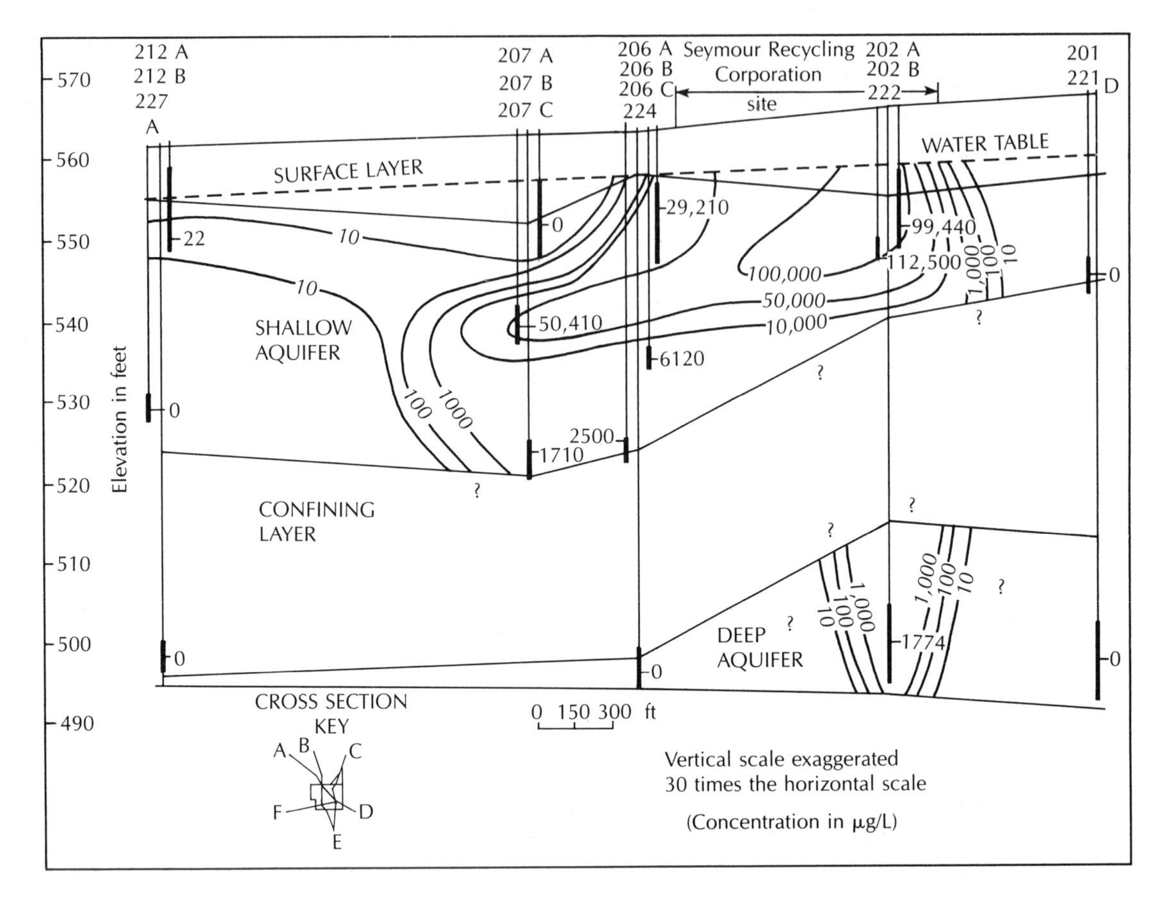


**FIGURE 10.32** Northwest-southeast cross section showing the plume of ground-water contaminated with synthetic organic compounds at the Seymour Recycling Corporation Superfund site. Concentrations are expressed in micrograms per liter of total GC/MS organics. Source: C. W. Fetter, *Final Hydrogeologic Report, Seymour Recycling Corporation Hazardous Waste Site*. U.S. Environmental Protection Agency, 1985.

10.33 is a cross section showing the plume of a single compound, chloroethane. It may be seen to be similar in configuration to the plume of total GC/MS organic compounds. Hazardous compounds with the greatest concentrations include trans-1,2-dichloroethene, 1,2-dichloroethane, vinyl chloride, chloroethane, 2-butanone, 1,1,1-trichloroethane, and 1,1-dichloroethane. Transformation of the chlorinated ethanes and ethenes to lower-molecular-weight compounds may be occurring. For example, the concentration of chloroethane, which is the degradation product of the chlorinated ethanes, appears to be increasing with time.

There is a plume of contaminated ground water in the shallow aquifer moving to the north-northwest. As of December 1984, the plume had reached a distance of about 400 feet from the site, with concentrations in excess of 50,000

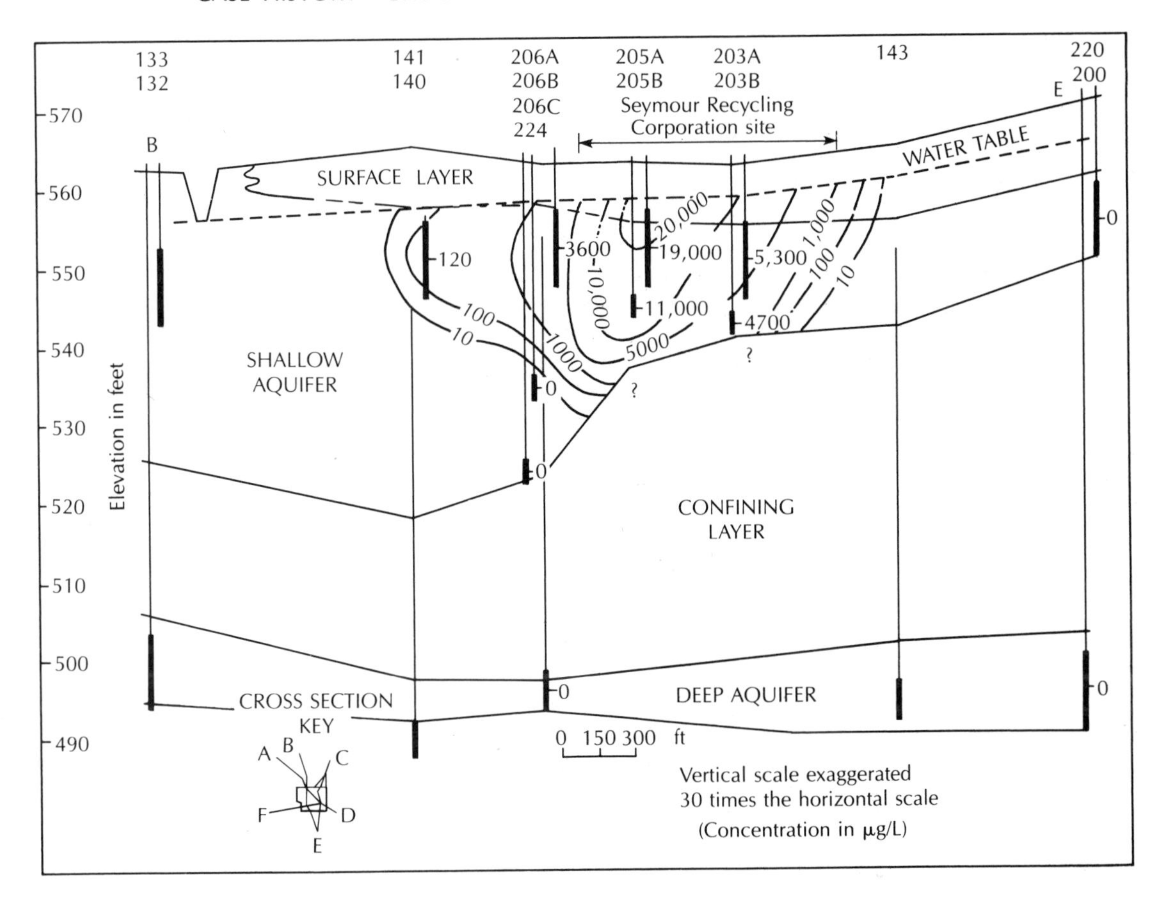


**FIGURE 10.33** North-south cross section showing the distribution of chloroethane in ground water at the Seymour Recycling Corporation Superfund site, Concentrations expressed in micrograms per liter. Source: C. W. Fetter, *Final Hydrogeologic Report, Seymour Recycling Corporation Hazardous Waste Site*. U.S. Environmental Protection Agency, 1985.

μg/L of total GC/MS organic compounds. Lesser concentrations are found as far as 1100 feet from the site. Figure 10.34 indicates the areal extent of the plume in December 1984.

Contamination of the deep aquifer by organic compounds has occurred. Well 222, which is located on the site, had organic compounds present all three times that it was sampled. The total concentration averaged about 950 μg/L. Well 224, located just north of the site, had organic compounds present the first time that it was sampled, but not the two succeeding times. It is suspected that the development of this well, which occurred three weeks prior to the first round of ground-water sampling, may have drawn contamination into the well. Although several other deep-aquifer wells had low levels of organic contamination reported from time to time, it is not clear if they represent contamination of the deep aquifer by the site.

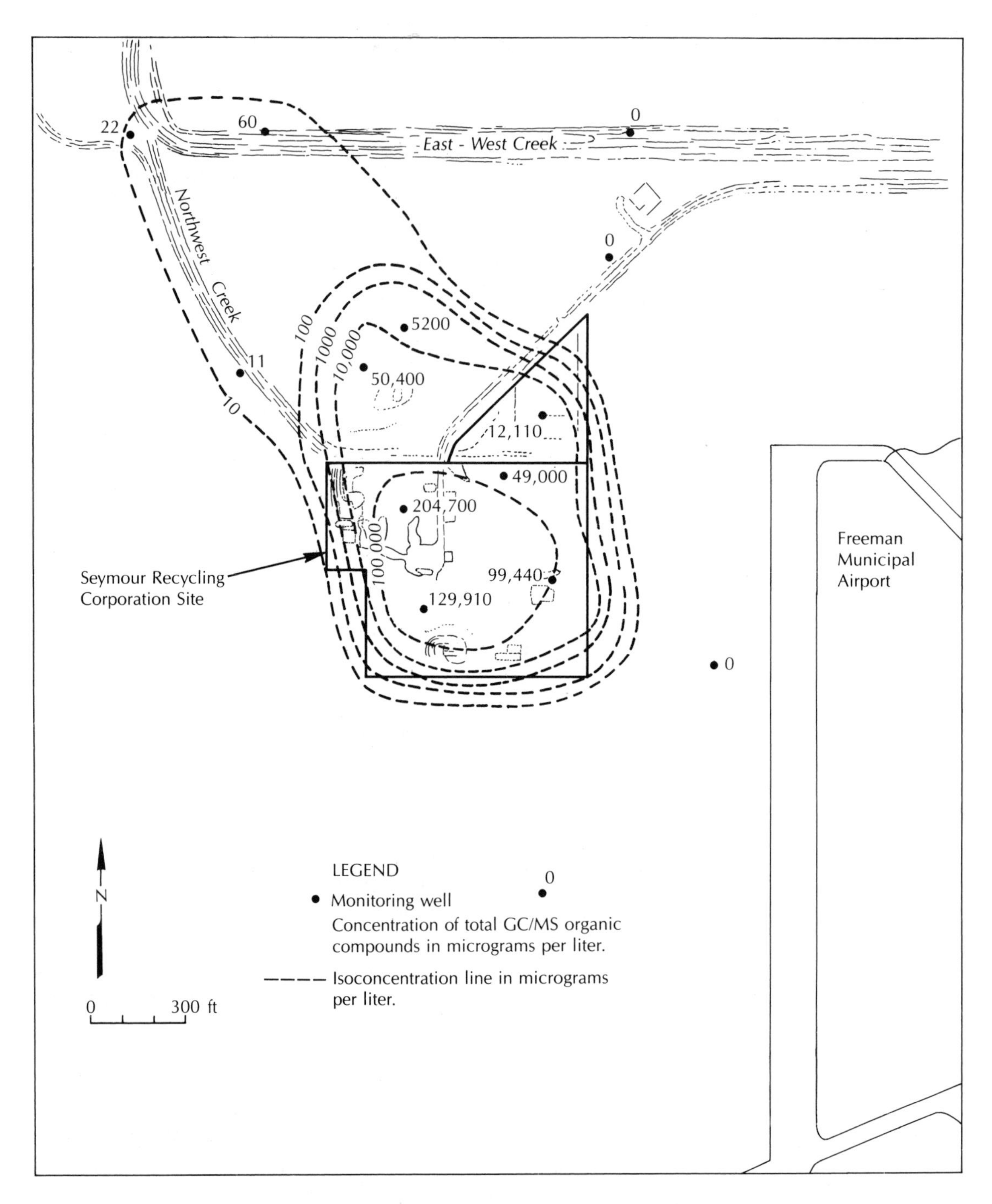


**FIGURE 10.34** Plan view of the plume of contaminated ground water at the Seymour Recycling Corporation Superfund site. Concentrations expressed in micrograms per liter of total GC/MS organic compounds. Source: C. W. Fetter, *Final Hydrogeologic Report, Seymour Recycling Corporation Hazardous Waste Site,* U.S. Environmental Protection Agency, 1985.

### 10.9.5 REMEDIAL ACTION PLAN

Private water-supply wells were located to the north of the site in the shallow aquifer. As a part of the remedial effort at the site, the U.S. Environmental Protection Agency has extended public water supply to the residents of this area. While this alleviated an obvious danger, as the shallow aquifer discharges into surface waters and there is no institutional control over new shallow wells being drilled, restoration of the shallow aquifer was necessary. There are municipal wells in the deep aquifer, one only 1250 feet northeast of the site. Although there was no indication that, under the present rate of pumping from the deep aquifer, contaminated ground water was flowing toward the well field, it was deemed prudent to continue monitoring the lower aquifer as well.

In order to prevent the spread of the contamination plume in the shallow aquifer, the preliminary remedial action plan developed by the EPA calls for a shallow well to be installed about 500 feet northwest of the site. When pumped at about 80 gallons per minute, the resulting cone of depression will capture the downgradient edge of the plume of contaminated water. This will stabilize the plume until the final remedial action can be implemented.

The contaminated ground water will be removed from the aquifers by a series of four shallow wells and one deep well. Figure 10.35 shows the locations of the withdrawal wells. Computer modeling has estimated that a total of 165 gallons per minute will be pumped and that it will be about 30 years until the cancer risk level for the most carcinogenic compound will be reduced to the generally accepted level of $10^{-6}$. In order to best flush the heavily contaminated ground water beneath the site, a strong hydraulic gradient will be established by placing an injection well in the shallow aquifer in the center of the site. This will operate in concert with the withdrawal wells at the edges of the site.

The long period of pumpage is required because dilution and dispersion, as well as the low rate of ground-water movement, mean that many pore volumes of water must move through the area in order to reduce the contaminants to the planned level. In addition, a reservoir of organic compounds still exists in the soil above the water table. In order to reduce the strength of the source, a vapor-extraction system for the shallow soil zone and upper part of the shallow aquifer is planned. The ground-water withdrawal wells around the site will be used to lower the water table beneath the site by about 12 feet. Twenty-foot-deep vapor-extraction wells will be installed roughly 50 feet on center throughout the areas of soil contamination. These are two-inch diameter wells that are screened in the unsaturated zone, which will be 20 feet thick when the site has been dewatered. A vacuum will be applied and the soil gas, which contains volatile organic compounds, will be drawn out. The volatile organic compounds, which are by far the most mobile in soil and ground water, will thus be removed. It is anticipated that soil vapor extraction will last for one to three years. The ground-water injection well in the center of the site will not be installed until the end of the vapor-extraction phase.

All ground water pumped from the area will be treated to remove the

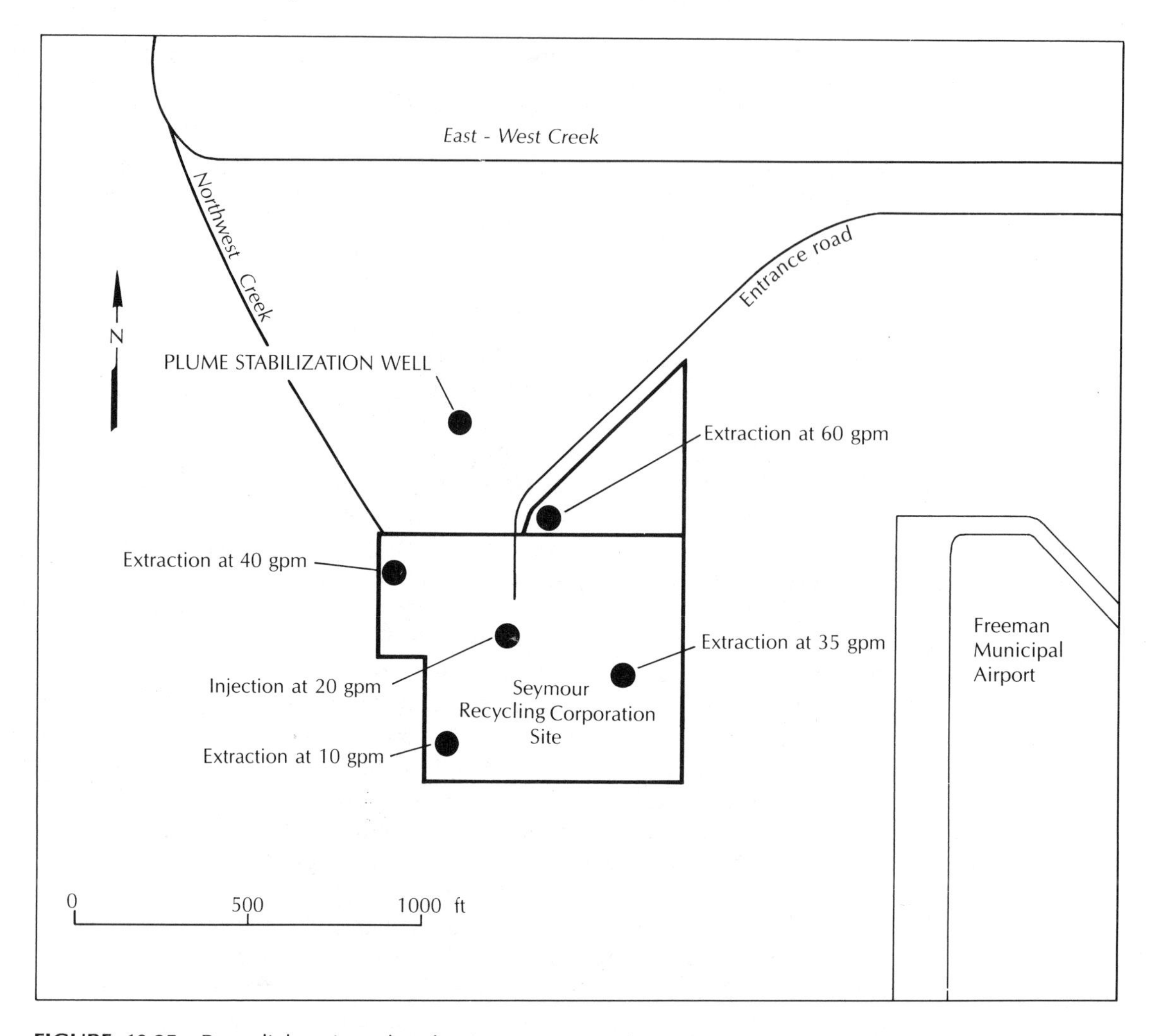


**FIGURE 10.35** Remedial action plan for Seymour Recycling Corporation superfund site showing the locations of the plume stabilization well, ground-water extraction wells for the shallow and deep aquifers, and shallow aquifer injection well. Source: CH2M-Hill, *Feasibility Study Report, Seymour Recycling Corporation Hazardous Waste Site.* U.S. Environmental Protection Agency, 1986.

organic compounds. Treatment will consist of spraying the water through an air-stripping column to volatilize part of the organic matter. The water will then be put through a sand filter to remove precipitated iron, which is naturally present in the aquifer. It will then be subjected to filtration through a granular activated-carbon bed, which will absorb most of the remaining organic matter. Final treatment will occur biologically as the water is sent to the municipal sewage treatment plant.

As a final measure, a low-permeability cap (Figure 10.24) will be placed over the site. This will seal off the residual contamination in the soil and greatly reduce the amount of precipitation that infiltrates through the unsaturated zone to the water table.

## REFERENCES


1. BEETON, A. M. "Eutrophication of the St. Lawrence Great Lakes." *Limnology and Oceanography,* 10 (1965):240–54.
2. BEETON, A. M. "Indices of Great Lakes Eutrophication." Great Lakes Research Division Publication No. 15, University of Michigan, 1966, 8 pp.
3. HOLLARD, R. E., and A. M. BEETON. "Significance to Eutrophication of Spatial Differences on Nutrients and Diatoms in Lake Michigan." *Limnology and Oceanography,* 17 (1972):88–96.
4. U.S. ENVIRONMENTAL PROTECTION AGENCY. *A Ground Water Protection Strategy for the Environmental Protection Agency.* Washington, D.C., August 1984.
5. LIPPY, E. C., and S. C. WALTRIP. "Waterborne Disease outbreaks—1946–1980: a Thirty-Five Year Perspective." *Journal, American Water Works Association,* 76 (1984):60–67.
6. ANON. "Millions to Plaintiffs in Groundwater Contamination Suits." *Mainstream,* American Water Works Association, 30, no. 11 (November 1986):12.
7. U.S. ENVIRONMENTAL PROTECTION AGENCY. *Quality Criteria For Water.* Washington, D.C.: U.S. Government Printing Office, 1976, 256 pp.
8. KEITH, L. H., and others. "Principles of Environmental Analysis." *Analytical Chemistry,* 55 (1983):2210–18.
9. KEITH, S. J., and others. "Dealing with the Problem of Obtaining Accurate Ground-Water Quality Analytical Results." In *Proceedings of the Third National Symposium on Aquifer Restoration and Ground-Water Monitoring.* Worthington, Ohio: National Water Well Association, 1983, pp. 272–82.
10. KIRCHMER, C. J. "Quality Control in Water Analyses." *Environmental Science and Technology,* 17, no. 4 (1983):178A–181A.
11. FORD, P. A., P. J. TURINA, and D. E. SEELY. *Characterization of Hazardous Waste Sites—A Methods Manual, Volume II, Available Sampling Methods.* U.S. Environmental Protection Agency, EPA-600/4-83-040, 1983, 215 pp.
12. DUNLAP, W. J., and others. *Sampling for Organic Chemicals and Microorganisms in the Subsurface,* U.S. Environmental Protection Agency, EPA-600/2-77-176, 1977, 27 pp.
13. FENN, D., and others. *Procedures Manual for Ground Water Monitoring at Solid Waste Disposal Facilities.* U.S. Environmental Protection Agency, SW-611, 1977, 269 pp.
14. GIBB, J. P., R. M. SCHULLER, and R. A. GRIFFIN. *Procedures for the Collection of Representative Water Quality Data from Monitoring Wells.* Illinois State Geological Survey and Illinois State Water Survey Cooperative Ground Water Report 7, 1981, 61 pp.

15. SCALF, M. R., and others. *Manual of Ground Water Quality Sampling Procedures*. Worthington, Ohio: National Water Well Association, 1981.

16. BARCELONA, M. J., J. P. GIBB, and R. A. MILLER. *A Guide to the Selection of Materials for Monitoring Well Construction and Ground Water Sampling*. Illinois State Water Survey Contract Report 327, EPA-600/52-84-024, 1983, 78 pp.

17. BARCELONA, M. J., and others. *Practical Guide for Ground-Water Monitoring*. Illinois State Water Survey Contract Report 374, 1985, 94 pp.

18. FETTER, C. W., JR. "Use of Test Wells as Water Quality Predictors." *Journal of American Water Works Association,* 76 (1975):516–18.

19. FETTER, C. W., JR. "Potential Sources of Contamination in Ground Water Monitoring." *Ground Water Monitoring Review,* 2 (1983):60–64.

20. PICKENS, J. F., and others. "A Multilevel Device for Ground Water Sampling." *Ground Water Monitoring Review,* 1, no. 1, (1981):48–51.

21. CHERRY, J. A., and P. E. JOHNSON. "A Multilevel Device for Monitoring in Fractured Rock." *Ground Water Monitoring Review,* 2, no. 3 (1982):41–44.

22. BARCELONA, M. J., and others. "A Laboratory Evaluation of Ground Water Sampling Mechanisms." *Ground Water Monitoring Review,* 4, no. 2 (1984):32–41.

23. SLAWSON, G. C., JR., K. E. KELLY, and L. G. EVERETT. "Evaluation of Ground-Water Pumping and Bailing Methods—Application in the Oil Shale Industry." *Ground Water Monitoring Review,* 2, no. 3 (1982):27–31.

24. NIELSEN, D. M., and G. L. YEATES. "A Comparison of Sampling Mechanisms Available for Small-Diameter Ground Water Monitoring Wells." *Ground Water Monitoring Review,* 5, no. 2 (1985):83–99.

25. EVERETT, L. G. "Monitoring in the Vadose Zone." *Ground Water Monitoring Review,* 1, no. 2 (1981):44–51.

26. JOHNSON, T. M., K. CARTWRIGHT, and R. M. SCHULLER. "Monitoring of Leachate Migration in the Unsaturated Zone in the Vicinity of Sanitary Landfills." *Ground Water Monitoring Review,* 1, no. 3 (1981):55–63.

27. WILSON, L. G. Monitoring in the Vadose Zone: Part III." *Ground Water Monitoring Review,* 3, no. 1 (1983):155–66.

28. EVERETT, L. G., and others. "Constraints and Categories of Vadose Zone Monitoring Devices." *Ground Water Monitoring Review,* 4, no. 1 (1984):26–32.

29. ROBBINS, G. A., and M. M. GEMMELL. "Factors Requiring Resolution in Installing Vadose Zone Monitoring Systems." *Ground Water Monitoring Review,* 5 no. 3 (1985):75–80.

30. FREEZE, R. A., and J. A. CHERRY. *Groundwater*. Englewood Cliffs, N.J.: Prentice Hall, Inc., 1979, 604 pp.

31. BERNER, R. A. *Principles of Chemical Sedimentology*. New York: McGraw-Hill, 1971, 240 pp.

32. BERUCH, J. C., and R. A. STREET. "Two-dimensional Dispersion." *Journal, Sanitary Engineering Division, American Society of Civil Engineers,* 93, SA6 (1967):17–39.

33. HOOPES, J. A., and D. R. F. HARLEMAN. "Wastewater Recharge and Dispersion in Porous Media." *Journal, Hydraulics Division, American Society of Civil Engineers,* 93, HY5 (1967):51–71.

34. OGATA, A. *Theory of Dispersion in a Granular Medium*. U.S. Geological Survey Professional Paper 411-I, 1970.

35. ANDERSON, M. P. "Using Models to Simulate the Movement of Contaminants Through Groundwater Flow Systems." *Critical Reviews in Environmental Control*. 9, no. 2 (1979):97–156.

36. GRIFFIN, R. A. Illinois State Geological Survey, personal communication, 1985.

37. ANDERSON, M. P. "Movement of Contaminants in Groundwater: Groundwater Transport—Advection and Dispersion." In *Groundwater Contamination*. Washington, D.C.: National Academy Press, 1984, pp. 37–45.

38. CLINE, P. V., and D. R. VISTE. "Migration and Degradation Patterns of Volatile Organic Compounds." Proceedings, Seventh Annual Madison Waste Conference, University of Wisconsin, Madison, Wisconsin, 1984, pp. 14–29.

39. PARSONS, F., P. R. WOOD, and J. DeMARCO. "Transformations of Tetrachloroethane and Trichloroethane in Microorganisms and Ground Water." *Journal, American Water Works Association,* 76, no. 2 (1984):56–59.

40. FUSILLO, T. V., J. J. HOCHREITER, JR., and D. G. LORD. "Distribution of Volatile Organic Compounds in a New Jersey Coastal Plain Aquifer System." *Ground Water,* 23 (1985):354–60.

41. FLIPSE, W. J., JR., and others. "Sources of Nitrate in Ground Water in a Sewered Housing Development, Central Long Island, New York." *Ground Water,* 22 (1984):418–26.

42. PIONKE, H. B., and J. B. URBAN. "Effect of Agricultural Land Use on Ground Water Quality in a Small Pennsylvania Watershed." *Ground Water,* 23 (1985):68–80.

43. ROTHSCHILD, E. R., R. J. MANSER, and M. P. ANDERSON. "Investigation of Aldicarb in Ground Water in Selected Areas of the Central Sand Plain of Wisconsin." *Ground Water,* 20 (1982):437–45.

44. NOSS, R. R., and E. T. JOHNSON. "Field Monitoring of the Adams, Massachusetts, Landfill Leachate Plume." *Proceedings, Fourth National Symposium and Exposition on Aquifer Restoration and Ground Water Monitoring*. National Water Well Association, 1984, pp. 356–62.

45. MC LEOD, R. S. "Evaluation of 'Superfund' Sites for Control of Leachate and Contaminant Migration." *Proceedings, The 5th National Conference on Management of Uncontrolled Hazardous Waste Sites,* Hazardous Materials Control Research Institute, 1984, pp. 114–21.

46. KRAMER, W. H. "Ground Water Pollution from Gasoline." *Ground Water Monitoring Review,* 2, no. 2 (1982):18–22.

47. OLIVEIRA, D. P., and N. SITAR. "Ground Water Contamination from Underground Solvent Storage Tanks, Santa Clara, California." *Proceedings, Fifth National Symposium and Exposition on Aquifer Restoration and Ground Water Monitoring*. National Water Well Association, 1985, pp. 691–708.

48. LEHR, J. H. "A Problem Yes—A Disaster No." *Ground Water,* 19 (1981):2–3.

49. U.S. ENVIRONMENTAL PROTECTION AGENCY. *The Report to Congress—Waste Disposal Practices and Their Effects on Ground Water*. Washington, D.C.: U.S. Environmental Protection Agency, 1977, 512 pp.

**50.** CRAUN, G. F. "Outbreaks of Waterborne Disease in the United States: 1971-78." *Journal, American Water Works Association,* 73 (1981):360–69.

**51.** YATES, M. V. "Septic Tank Density and Ground-Water Contamination." *Ground Water,* 23 (1985):586–91.

**52.** CRAUN, G. F. "Waterborne-Disease Status Report Emphasizing Outbreaks in Ground-Water Systems." *Ground Water,* 17 (1979):183–91.

**53.** CRAUN, G. F. "Health Aspects of Ground Water Pollution." In *Groundwater Pollution Microbiology,* ed. G. Britton and C. P. Gerba. New York: John Wiley & Sons, 1984.

**54.** KMET, P., K. J. QUINN, and C. SLAVIK. "Analysis of Design Parameters Affecting the Collection Efficiency of Clay Lined Landfills." *Proceedings, Fourth Annual Madison Waste Conference.* Madison, Wis.: University of Wisconsin, 1981.

**55.** ROUX, P. H., and W. F. ALTHOFF. "Investigation of Organic Contamination of Ground Water in South Brunswick Township, New Jersey." *Ground Water,* 18 (1980):464–71.

**56.** KLUSMAN, R. W., and K. W. EDWARDS. "Toxic Metals in Ground Water of the Front Range, Colorado." *Ground Water,* 15 (1977):160–69.

**57.** AHMAD, M. U. "Coal Mining and Its Effect on Water Quality." *American Water Resources Proceedings,* no. 18, 1974, pp. 138–48.

**58.** NORBECK, P. N., L. L. MINK, and R. E. WILLIAMS. "Ground Water Leaching of Jig Tailing Deposits in the Coeur d'Alene District of Northern Idaho." *American Water Resources Association Proceedings,* no. 18, 1974, pp. 149–57.

**59.** RALSTON, D. R., and A. G. MORILLA. "Ground Water Movement through an Abandoned Tailings Pile." *American Water Resources Association Proceedings,* no. 18, 1974, pp. 174–83.

**60.** McWHORTER, D. B., R. K. SKOGERBOE, and G. V. SKOGERBOE. "Potential of Mine and Mill Spoils for Water Quality Degradation." *American Water Resources Association Proceedings,* no. 18, 1974, pp. 123–37.

**61.** KAUFMAN, R. F., G. G. EADIE, and C. R. RUSSELL. "Effects of Uranium Mining and Milling on Ground Water in the Grants Uranium Belt, New Mexico." *Ground Water,* 14 (1976):296–308.

**62.** WHITE, R. B., and R. B. GAINER. "Control of Ground Water Contamination at an Active Uranium Mine." *Ground Water Monitoring Review,* 5, no. 2 (1985):75–81.

**63.** JRB ASSOCIATES, INC. *Handbook, Remedial Actions at Waste Disposal Sites.* U.S. Environmental Protection Agency, EPA-625/6-82-006, 1982, 497 pp.

**64.** CANTER, L. W., and R. C. KNOX. *Ground Water Pollution Control.* Chelsea, Mich.: Lewis Publishers, Inc., 1985, 526 pp.

**65.** CANTER, L. W. "Overview of Aquifer Restoration." *Proceedings, Second National Symposium on Aquifer Restoration and Ground Water Monitoring.* National Water Well Association, 1982, pp. vii–x.

**66.** JOHNSON, T. J., and others. "Hydrologic Investigations of Failure Mechanisms and Migration of Organic Chemicals at Wilsonville, Illinois." *Proceedings, Third National Symposium on Aquifer Restoration and Ground Water Monitoring.* National Water Well Association, 1983, pp. 413–20.

67. HERZOG, B. L., K. CARTWRIGHT, T. M. JOHNSON, and H. J. H. HARRIS. "A Study of Trench Covers to Minimize Infiltration at Waste Disposal Sites." Illinois State Geological Survey Contract Report No. 1981-5, Nuclear Regulatory Commission, NUREG/CR-2478, 1982, 245 pp.

68. AYRES, J. E., D. C. LAGER, and M. J. BARVENIK. "The First EPA Superfund Cutoff Wall: Design and Specifications." *Proceedings, Third National Symposium on Aquifer Restoration and Ground Water Monitoring*. National Water Well Association, 1983, pp. 13–22.

69. BRUNSING, T. P., and J. CLEARY. "Isolation of Contaminated Ground Water by Slurry-Induced Ground Displacement." *Proceedings, Third National Symposium on Aquifer Restoration and Ground Water Monitoring*. National Water Well Association, 1983, pp. 28–36.

70. FITZWATER, P. L., C. L. BRASSOW, and C. W. FETTER, JR. "Assessment of Ground-Water Contamination and Remedial Action for a Hazardous Waste Facility in the Gulf Coast." *Proceedings, Third National Symposium on Aquifer Restoration and Ground Water Monitoring*. National Water Well Association, 1983, pp. 135–41.

71. DRUBACK, G. W., and S. V. ARLOTTA. "Subsurface Pollution Containment Using a Composite System Vertical Cut-Off Barrier." *Proceedings, Fifth National Symposium on Aquifer Restoration and Ground Water Monitoring*. National Water Well Association, 1985, pp. 400–11.

72. LYNCH, E. R., and others. "Design and Evaluation of In-Place Structures Utilizing Ground Water Cutoff Walls. *Proceedings, Fourth National Symposium on Aquifer Restoration and Ground Water Monitoring*. National Water Well Association, 1984, pp. 1–7.

73. KEELY, J. F. "Optimizing Pumping Strategies for Contaminant Studies and Remedial Actions." *Proceedings, Fourth National Symposium on Aquifer Restoration and Ground Water Monitoring*. National Water Well Association, 1984, pp. 33–42.

74. SCHAFER, J. M. "Determining Optimum Pumping Rates for Creation of Hydraulic Barriers to Ground Water Pollutant Migration." *Proceedings, Fourth National Symposium on Aquifer Restoration and Ground Water Monitoring*. National Water Well Association, 1984, pp. 50–63.

75. CAMPBELL, P., R. C. BOST, and R. W. JACOBSEN. "Subsurface Organic Recovery and Contaminant Migration Simulation." *Proceedings, Fourth National Symposium on Aquifer Restoration and Ground Water Monitoring*. National Water Well Association, 1984, pp. 82–91.

76. POULOS, S. J., and A. C. LAWS. "Gradient Control for Containment of Pollutants." *Proceedings, Fifth National Symposium on Aquifer Restoration and Ground Water Monitoring*. National Water Well Association, 1985, pp. 390–99.

77. LENZO, F. C. "Air-Stripping for VOCs in Water: Pilot, Design, Construction." *Proceedings, Fourth National Symposium on Aquifer Restoration and Ground Water Monitoring*. National Water Well Association, 1984, pp. 100–10.

78. FLATHMAN, P. W., J. R. QUINCE, and L. S. BOTTOMLEY. "Biological Treatment of Ethylene Glycol-Contaminated Ground Water at the Naval Air Engineering Center, Lakehurst, New Jersey." *Proceedings, Fourth National Symposium on Aquifer Restoration and Ground Water Monitoring*. National Water Well Association, 1984, pp. 111–19.

79. BRENOEL, M., and R. A. BROWN. "Remediation of a Leaking Underground Storage Tank with Enhanced Bioreclamation." *Proceedings, Fifth National Symposium on Aquifer Restoration and Ground Water Monitoring*. National Water Well Association, 1985, pp. 527–37.

80. YANIGA, P. M., C. MATSON, and D. J. DEMKO. "Restoration of Water Quality in a Multiaquifer System Via In-Situ Biodegradation of the Organic Comtaminants." *Proceedings, Fifth National Symposium on Aquifer Restoration and Ground Water Monitoring*. National Water Well Association, 1985, pp. 510–26.

81. FLATHMAN, P. W., and others. "In-Situ Physical/Biological Treatment of Methylene Chloride (Dichloromenthane) Contaminated Ground Water." *Proceedings, Fifth National Symposium on Aquifer Restoration and Ground Water Monitoring*. National Water Well Association, 1985, pp. 571–97.

82. FETTER, C. W., JR. Final Hydrogeologic Report, Seymour Recycling Corporation Hazardous Waste Site. Report to U.S. Environmental Protection Agency, 1985.

83. CH2M-HILL, INC. Remedial Investigation, Seymour Recyling Corporation. Report to U.S. Environmental Protection Agency, 1985.

84. CH2M-HILL, INC. Feasibility Study Report, Seymour Recycling Corporation Hazardous Waste Site. Report to U.S. Environmental Protection Agency, 1986.

# ELEVEN

# Ground-Water Development and Management

I approach now an account of the experiments that I carried out at Dijon together with Engineer Charles Ritter, to determine the laws of flow of water through sand. . . . Each experiment consisted of establishing a specified pressure in the upper chamber of the column by adjustment of the inflow tap; then, when it was established by means of two observations that the flow had become essentially uniform, the outflow from the filter during a certain time was noted, and the mean outflow per minute was calculated from it.

*Les fontaines publiques de la ville de Dijon,* Henry Darcy, 1856

## 11.1 INTRODUCTION

The area of water resources development and management is so broad that we will make no attempt to cover all aspects in this book. Instead we will focus on ground water, since it is this aspect of the general topic to which hydrogeology most closely relates. However, as ground water is not isolated from surface water, a study of ground-water development necessarily encompasses many aspects of surface-water flow.

Surface water occurs in readily discernible drainage basins. The boundaries are topographic and may be easily delineated on a topographic map. The water conveniently flows in the direction in which the land surface is sloping. Moreover, surface water does not cross topographic divides (except, perhaps, during floods) and the locations of the drainage divides are fixed. Ground water, on the other hand, occurs in aquifers that are hidden from view. The boundaries of an aquifer are physical: it can crop out, abut an impermeable rock unit, grade into a lower-permeability deposit, or thin and disappear. At a given location, the land surface may be underlain by several aquifers. Each aquifer may have different chemical makeup and different hydraulic potential; each may be recharged in a different location and flow in a different direction. Moreover, ground-water divides do not necessarily coincide with surface-water divides. Clearly, the development and management of ground water is more complicated than that of surface water, simply on the basis of the mode of occurrence.

## 11.2 DYNAMIC EQUILIBRIUM IN NATURAL AQUIFERS

Under natural conditions, an aquifer is usually in a state of **dynamic equilibrium** (1). A volume of water recharges the aquifer and an equal volume is discharged. The potentiometric surface is steady and the amount of water in storage in the aquifer is a constant. The aquifer transmits the water from the recharge areas to the discharge zones. The maximum amount of water any section of the aquifer can transmit is a function of the transmissivity and the maximum gradient of the potentiometric surface. If the water table is close to the surface of an unconfined aquifer, the aquifer is full and is transmitting the maximum amount of water. If, however, the water table is far below the surface, the aquifer is not transmitting water at full capacity.

The amount of water that recharges an unconfined aquifer is determined by three factors: (a) the amount of precipitation that is not lost by evapotranspiration and runoff and is thus available for recharge; (b) the vertical hydraulic conductivity of surficial deposits and other strata in the recharge area of the aquifer, which determines the volume of recharged water capable of moving downward to the aquifer; and (c) the transmissivity of the aquifer and potentiometric gradient, which determine how much water can move away from the recharge area. Should an aquifer be transmitting the maximum volume of water, it is more than likely that some potential recharge is being rejected in the recharge area. This is often the case in humid areas. Should the water table be low, indicating that the aquifer is not flowing at full capacity, there is probably either a lack of potential recharge or low vertical hydraulic conductivity in the recharge area, retarding downward movement. Aquifers in arid regions typically have deep water tables in the recharge areas, indicating a deficiency in the amount of potential recharge.

Recharge to confined aquifers can occur in places in which the confining layer is absent. Under such conditions, the three factors affecting unconfined aquifer recharge are controlling. If there is a hydraulic gradient across a leaky confining layer in a direction that promotes flow into the aquifer, then recharge can occur across the confining layer. In this case, the vertical hydraulic conductivity of the confining layer, the thickness of the confining layer, and the head difference across it control the amount of recharge. Recharge to a confined aquifer may come from both downflow from a higher aquifer or upflow from a lower aquifer.

When a well begins to pump water from an aquifer, the water is withdrawn from storage around the well and from vertical leakage (1). As the cone of depression grows, an increasingly larger portion of the aquifer will be contributing water from storage. The amount of water discharging naturally from the aquifer will remain at the predevelopment rate until the pumping cone reaches the recharge or discharge area. When the pumping cone reaches a discharge area, the potentiometric gradient toward the discharge area is lowered and the amount of natural discharge proportionally reduced. If the pumping cone reaches the recharge area of an aquifer, it may induce additional recharge of water that was previously rejected. It is even possible for a section of the aquifer to change from a discharge area to a recharge area. For example, drawdown near a river may eliminate ground-water discharge to the river and induce infiltration from the river into the aquifer, reversing the prior direction of flow.

In any event, the pumping cone will continue to grow until it has sufficiently reduced natural discharge or increased recharge to balance the volume of water removed by pumping. With this occurrence, a new condition of dynamic equilibrium is reached (1). However, it is important to note that in order for this to happen a pumping cone must form. This means the potentiometric surface in parts of the aquifer must be lowered. The resultant cone of depression is usually a complex surface resulting from the withdrawals of hundreds of wells. Should the sum of the remaining natural discharge and the pumping withdrawals exceed the available recharge, the cone of depression will not stabilize and water levels will continue to fall.

The rate at which the cone of depression spreads is a function of the storativity of a confined aquifer or the specific yield of a water-table aquifer. As storativity values are 100 to 1000 times smaller than specific yields, the cone of depression will spread 100 to 1000 times faster in an artesian aquifer than in a water-table aquifer. It can take many years for the cone of depression to influence recharge or discharge areas sufficiently for an aquifer to regain dynamic equilibrium.

## CASE STUDY: DEEP SANDSTONE AQUIFER OF NORTHEASTERN ILLINOIS

The deep sandstone aquifer beneath northeastern Illinois is confined by the Maquoketa Shale, a leaky layer. It has an outcrop area that receives direct recharge west of the limit of the shale. Under predevelopment conditions, the potentiometric surface stood at or above land level and had a very gentle slope toward Lake Michigan of about $3 \times 10^{-4}$ foot per foot. Flow through the aquifer from the recharge area was about 0.9 million gallons (3400 cubic meters) per day (mgd) (2). Natural discharge was by slow upward leakage through the shale into Silurian limestone beneath Lake Michigan. Most of the ground-water recharge in the recharge area of the regional aquifer was circulating in local flow systems. It could not move as a part of regional flow owing to the low carrying capacity of the aquifer caused by the gentle slope of the potentiometric surface. As ground-water development began, the well discharge soon exceeded the natural discharge and the pumping cone rapidly grew. As the hydraulic gradient steepened, increasingly greater amounts of water were drawn into the regional flow system. By 1958, an estimated 19 million gallons (72,000 cubic meters) per day were moving into the pumping cone (2). At that time, pumpage from the aquifer was 43 million gallons (163,000 cubic meters) per day. The difference between pumpage and recharge was coming from storage as the cone of depression grew. By 1975, ground-water pumpage was 150 million gallons (568,000 cubic meters) per day, and the hydraulic gradient was $4 \times 10^{-3}$ foot per foot, with water levels falling several feet per year.

If pumpage from the deep sandstone aquifer in the Chicago region had been limited to 46 million gallons (174,000 cubic meters) per day, the cone of depression would have eventually stabilized. The period to reach a stable configuration would have been about 50 years from the time pumpage first reached 46 million gallons (174,000 cubic meters) per day (3). However, so long as the rate of deep-well pumpage exceeds 46 million gallons per day, the cone of depression will continue to grow (4).

## 11.3 GROUND-WATER BUDGETS

Some knowledge of the amount of natural recharge to an aquifer is mandatory in a ground-water development program. There are several methods available to make such an estimate. In Section 3.7, a method was described for estimating annual ground-water recharge from baseflow-recession curves. This method is useful for areas in which the ground-watershed and the river basin coincide.

A second method is to determine the flow in the aquifer across a vertical plane at the boundary of the recharge area and the discharge area (5). If there is sufficient knowledge of the potential field of the aquifer, the area(s) of recharge and discharge can be determined. The rate of steady flow from recharge areas to discharge areas is determined using either Darcy's law for a confined aquifer or the Dupuit equation for an unconfined aquifer. The flow from the recharge areas is equal to the rate of recharge for aquifers in dynamic equilibrium.

A **water budget** for the recharge area of an aquifer is a very useful means of determining ground-water recharge. An advantage of the water-budget method is that the aquifer does not have to be in dynamic equilibrium in order to use it. Many of the parameters used for a hydrologic budget are measured directly: precipitation, streamflow, transported water, and reservoir evaporation. Ground-water inflow, outflow, and change in storage are computed from the hydraulic aquifer characteristics and measured potentiometric data. The amount of evapotranspiration could be measured in lysimeters, but it is more typically computed using an appropriate formula such as the Thornthwaite method (Section 2.3). Basinwide evapotranspiration may also be determined by a water-budget analysis as follows:

$$\begin{aligned}\text{Evapotranspiration} = {} & (\text{precipitation} + \text{surface-water inflow} \\ & + \text{imported water} + \text{ground-water inflow}) \\ & - (\text{surface-water outflow} + \text{ground-water} \\ & \quad \text{outflow} + \text{reservoir evaporation} \\ & \quad + \text{exported water}) \\ & \pm \text{changes in surface-water storage} \\ & \pm \text{changes in ground-water storage}\end{aligned} \qquad \textbf{(11-1)}$$

The natural recharge to an undeveloped aquifer may be determined by a water-budget analysis of the recharge area:

$$\begin{aligned}\text{Ground-water recharge} = {} & (\text{precipitation} + \text{surface-water inflow} \\ & + \text{imported water} + \text{ground-water inflow}) \\ & - (\text{evapotranspiration} + \text{reservoir} \\ & \text{evaporation} + \text{surface-water outflow} \\ & + \text{exported water} + \text{ground-water outflow}) \\ & \pm \text{changes in surface-water storage}\end{aligned} \qquad \textbf{(11-2a)}$$

Equation 11-2 can account for ground-water recharge not only from precipitation, but also from losing streams, irrigation water, unlined canals, and so forth. Its usefulness may be limited, however, if evapotranspiration cannot be determined. Detailed knowledge of all the factors is necessary if the computation of recharge is to be accurate.

If the land surface of the recharge area is developed for agriculture, industry, urban growth, and so forth, additional computations of recharge may be necessary. Water used for many purposes may be recharged to the ground-water reservoir. Excess irrigation water is one example. On Long Island, New York, all ground water pumped for cooling or air conditioning must be returned to the aquifer from which it was removed. Water used for domestic purposes is often recharged by septic tank drain fields or cesspools. The increasing emphases on the use of land systems* for municipal wastewater treatment means that treated sewage effluent that formerly flowed into rivers may now be recharging aquifer systems. The amount of recharge from such sources can be determined by a supplemental water-budget analysis:

$$\begin{aligned}&\text{Additional} \\ &\text{ground-water recharge} = {} (\text{industrial use} + \text{municipal use} \\ & \qquad + \text{domestic use} + \text{irrigation use}) \\ & \qquad - (\text{cooling-water evaporation} \\ & \qquad + \text{irrigation-water evapotranspiration} \\ & \qquad + \text{water exported in products} \\ & \qquad + \text{sewage discharge into surface waters})\end{aligned} \qquad \textbf{(11-2b)}$$

The determination of the additional ground-water recharge is often more complicated than an analysis of the basic amount of recharge. Water-use records of dozens or hundreds of individual water users and sewage dischargers must be collected. These records may range from excellent to nonexistent. Many of the factors must be estimated. For example, the owners of private wells for home

*Land systems of treatment include, among other techniques, spraying wastewater as irrigation water, recharging it through seepage basins, and allowing it to flow across the land surface. Biological, chemical, and physical processes act to remove pollutants and purify the water.

use will almost never know how much water they pump. An accurate accounting of this type involves long and tedious inventory analysis.

Should an attempt be made to balance the additions to the water supply of an area with the depletions, the result should be accurate to within the accuracy of measurement or estimation of the various parameters. Each parameter will have an accuracy of estimation dependent upon how precisely the measurement can be made. Measurement of streamflow might be accurate to ±5 percent for a good measurement. If total streamflow is 30 cubic feet per second, then there is a variability of 30 cubic feet per second × 0.05, or ±1.5 cubic feet per second. The overall accuracy of the estimate can be determined by taking a weighted average of the individual variability.

**EXAMPLE PROBLEM** A water-budget analysis of a watershed indicates an estimated total outflow of 90 cubic feet per second. The accuracy of estimation of each of the individual components is known. What is the accuracy of the estimate of the total discharge?

| Factor | Estimated Flow ($ft^3/sec$) | Accuracy of Estimate (%) | Variability of Factor ($ft^3/sec$) |
|---|---|---|---|
| Evapotranspiration | 60 | ±25 | ±15 |
| Surface outflow | 20 | ± 5 | ± 1 |
| Ground-water outflow | 5 | ±10 | ± 0.5 |
| Exported water in canal | 5 | ± 2 | ± 0.1 |
| TOTAL | 90 | | ±16.6 |

The sum of the variability of each factor is ±16.6 cubic feet per second; therefore, the accuracy of the total flow is 16.6/90 or ±18 percent.

The amount of water available for use from an aquifer is not the natural recharge. It is the increase in recharge or leakage from adjacent strata induced by development, along with the reduction in discharge. As water levels fall to accommodate the development, there will also be some water available from storage. A water-budget analysis is helpful in determining the amount of natural recharge and discharge. Further hydrologic analysis is then necessary to evaluate the effects that pumping will have on these figures.

## 11.4 MANAGEMENT POTENTIAL OF AQUIFERS

Aquifers can play many roles in the overall development of the water resources of an area. Some of the functions have been recognized for many years; others have been recognized only recently. The most obvious use of an aquifer is to supply water to wells—the **supply function.** One of the more vexing problems in ground-water management has been to determine how much water an aquifer can supply. This problem will be discussed in Section 11.5.

Aquifers also transmit water from one location to another. This has been called a **pipeline function** (6). Many communities are dependent upon ground-water sources that are recharged elsewhere. In this case, the aquifer acts as an aqueduct. However, when the user community does not have zoning and land-use control, as, for example, when the aquifer is transmitting water some distance, there may be difficulty in protecting the recharge area of the aquifer from overdevelopment or contamination* (7). Aquifers are not as efficient in carrying large volumes of water as are surface canals. However, surface canals can lose large amounts of water by evaporation; furthermore, they require capital for construction.

Ground water can be mined in the same manner as minerals. Whenever ground water is withdrawn at a rate greater than the rate of replenishment, mining is occurring. In some aquifers, the rate of replenishment is so low that it is almost nonexistent. For example, the average annual recharge to the Ogallala aquifer of the southern High Plains of New Mexico and Texas is 1.5 inches (3.8 centimeters) per year (8). Based on the average rate of withdrawal for the 1951–1960 period, the ground-water mine will be exhausted in 100 years. That pumping rate is twenty-eight times the natural recharge. Property owners of land in the High Plains are permitted an income-tax depletion allowance to compensate for the loss of property value due to a falling water table (9).

The unsaturated zone overlying an aquifer can act as a waste-treatment system. This has been called the **"filter-plant" function** of aquifers (7). However, the unsaturated zone can do much more than act as a physical filter to remove bacteria and viruses. It is also effective in removing phosphorus and heavy metals (10). Passage of water through the saturated zone can also improve the quality. Degradation of native water quality can occur if the treatment potential of soil systems is exceeded.

Ground water can also have an **energy-source function.** The ground-water heat pump is a viable alternative to conventional heat pumps in some localities (11). As the thermal energy of the aquifer is removed by a heat pump in colder climates, it is another type of mining. In Wisconsin, the thermal impact is very small (12). In states farther south, where heating and air conditioning demands are more equal, the net impact on ground-water temperatures is even less.

Ground-water reservoirs sometimes also have a **storage function.** This is not true for an undeveloped aquifer in dynamic equilibrium if the recharge zone is rejecting potential recharge. However, if the ground-water reservoir has unused storage capacity, it can effectively store water from wet periods for use

---

*The 1986 Safe Drinking Water Act in the United States enables communities to establish ground-water protection zones around well fields.

during time of drought (13–16). Aquifers with available storage capacity may be either those that are not filled by the natural recharge to them or those in which the potentiometric surface has been lowered by pumping. Water put into storage could be from natural sources, especially extreme precipitation or flood events (13). Increasingly, treated wastewater effluent is being used to replenish aquifers (17).

When ground water stored in aquifers can be used to replace surface-water reservoirs, a number of benefits accrue (18). The expense of surface-water reservoirs is circumvented, as there are not capital costs involved. There are no evaporative losses from ground-water storage, nor are there any infiltration losses. Surface-water reservoirs sometimes create great ecological disruption with the destruction of riverine and floodplain environments. Productive farmland may also be lost beneath surface reservoirs. Such disruption may be avoided with ground-water storage; furthermore, there is no worry over dam safety. On the other side of the coin, a person cannot waterski in a ground-water reservoir!

Aquifers are also used for the storage of natural gas. The aquifer must be confined, and a structural or stratigraphic feature, such as an anticline, is required to hold the gas in place. Wells are drilled through the structure and the gas is pumped into the aquifer under pressure. It displaces water and forms a bubble in the aquifer. Fresh water could also be stored in salt-water aquifers, as the former is less dense and would float as a bubble in the saline water.

## 11.5 PARADOX OF SAFE YIELD

It is a natural inclination of scientists to compare and classify phenomena in quantitative terms. Thus, it is to be expected that hydrogeologists have attempted to define the amount of water that could be developed from a ground-water reservoir. The term **safe yield** was apparently used in this regard as early as 1915 (19). At that time, safe yield was regarded as the amount of water that could be pumped "regularly and permanently without dangerous depletion of the storage reserve." Later, other factors that need to be considered were added, such as economics of ground-water development (20), protection of the quality of the existing store of ground water (21), and protection of existing legal rights and potential environmental degradation (22). Synonyms for safe yield appear in the literature, including "potential sustained yield" (23), "permissive sustained yield" (24), and "maximum basin yield" (25). A composite definition, based on the ideas of many authors, could be expressed as follows: Safe yield is the amount of naturally occurring ground water that can be withdrawn from an aquifer on a sustained basis, economically and legally, without impairing the native ground-water quality or creating an undesirable effect such as environmental damage.

The concept of ground-water withdrawals causing environmental damage warrants more than a mention. Many surface-water systems are dependent upon natural ground-water discharge. It has been shown by model studies that ground-water development may reduce streamflow and, as a consequence, lower lake levels and dry wetlands (26). As these may be environmentally sensitive areas, the danger of environmental harm is real. Likewise, ground-water withdrawals have been linked to subsidence of the land surface (27). This has resulted in land-surface cracking and damage to structures, highways, pipelines, dams, and tunnels. The gradients of irrigation canals have been changed—even reversed—and low areas have become flooded by sea water. In a broader sense, environmental impacts include ecological, economical, social, cultural, and political values (28).

Many authorities are uneasy with the concept of safe yield. For some, the term is too vague (29). Obviously, the amount of ground water that can be produced will vary under varied patterns of pumping and development. In addition, the question of what would constitute an undesirable result to be avoided is open to debate (30). The abandonment of the term safe yield has been proposed on the grounds that it does not take into account the interrelationship of ground water and surface water and may preclude the development of the storage functions of an aquifer (6).

However, in spite of the reservations of many hydrogeologists with regard to the concept of safe yield and its implications, the basic concept must be applied whenever the use of an aquifer is planned and managed. Ground-water management programs obviously imply that water must be pumped from the ground (31). If there is no evaluation of the hydrologic and environmental impacts of various withdrawal programs, it is possible that uncontrolled withdrawals will exceed prudent levels.

A single value for the safe yield of an aquifer cannot be provided in the same sense as a quantity such as mean annual precipitation. Safe-yield values are based on a number of constraints; such values must be determined by a team of professionals, in the same manner that an environmental impact statement is prepared. Economists, engineers, engineering geologists, plant and wildlife ecologists, and lawyers might all participate with the hydrogeologist in preparing a safe-yield determination for an aquifer or a ground-water basin. The safe-yield evaluation should include a statement of the legal and economic constraints that were considered, as well as the limiting values of environmental damage that were considered. Indeed, such a study should provide a series of safe-yield values and the different factors that applied to each determination. This is obviously not a simple matter. Computer models of ground-water flow systems are ideal tools for estimating the series of values. All of the hydraulic factors can be evaluated. R. A. Freeze has shown how a computer model can compute a "maximum basin yield" (25).

The safe yield of an aquifer system is only one facet of a ground-water management program. Artificial augmentation of precipitation or recharge could increase the amount of water that can be withdrawn on a sustained basis. The use of ground-water reservoirs for cyclic storage means that in drought years it is necessary, and desirable, to pump water on a temporary basis far in excess of the safe yield. Under these conditions, the ground-water supply would replace surface-water supplies that might be critically low or be used to irrigate crops normally watered by rainfall. In wet years, the ground-water reservoir would be replenished by above-average recharge and pumping at rates below the safe yield.

The underlying principle of ground-water development is that by withdrawing water from an aquifer, some of the natural discharge may be made available for use (31).

## 11.6 WATER LAW

### 11.6.1 LEGAL CONCEPTS

The development and management of water resources must take place within a framework of legal obligations, rights, and constraints. Naturally these factors differ from country to country; even in the United States, each individual state has its own body of water law. Because of this, it is not possible to fully discuss all water laws that might be applicable to a particular situation. However, we will look at some general principles.

There are two basic aspects of the legal framework: common law and legislative law (32). **Common laws** are the traditional legal precepts laid down by court decisions. They are based primarily on precedent, but can be overturned by later courts if it is felt that societal needs have changed. In the United States, common law derives its legitimacy from the U.S. Constitution and the constitutions of the various states. The final arbitrator of common law is either the highest court in a state or the U.S. Supreme Court, depending upon whether an action is brought in state court or federal court.

**Legislative law** has two arms: **statutory law** and **administrative law.** Congress or state legislatures may pass laws that regulate water, thus creating statutory law. In addition, legislative bodies may enable their administrative bodies to write rules and regulations that have the power of law, thus creating administrative law.

Water law has traditionally been concerned with the quantity of water available. In this regard, a complex body of common law has risen. In recent years there has been a trend for legislation to be written to allocate water, especially in areas where it is scarce, rather than to rely upon common law. Society is acting to remedy what the majority see as inadequacies and inequities in the common law.

The concept of protecting the quality of the water is relatively recent. Most common law has arisen with respect to water quantity. Water-quality issues that have been addressed under common law have generally been ones of some specific episode of pollution in which one or more of several common law theories (e.g., trespass, negligence, private nuisance, public nuisance, or strict liability) have been applied. Legislative law that addresses the issue of water quality has been passed at the federal, state, and local levels.

### 11.6.2 LAWS REGULATING QUANTITY OF SURFACE WATER

A **water right,** as defined by law, is not legal title to the water, but the legal right to use it in a manner dictated by state law.

In the eastern United States, the **riparian doctrine** of ownership of surface-water rights is recognized (33). This concept holds that the riparian landowner—the property owner adjacent to a surface-water body—has the first right to withdraw and use the water. This right is often controlled by the state in the sense that application to the state for a permit to withdraw is necessary. Ownership of the water right is held with ownership of the land, and all riparian owners have equal rights to the water. Water withdrawals are limited to "reasonable" use in comparison with other riparians. As all riparians have equal rights to use reasonable amounts of water—even new users—it is fortunate that there is a large amount of surface water available in the eastern United States. As eastern riparian owners generally return most of the water to the stream, albeit sometimes more polluted, there have been few crises due to lack of water except during droughts. Water problems in these areas typically involve quality rather than quantity of water.

Water in rivers was used in large quantities for mining during the 1869 California gold rush. Mining law recognized that the first to stake a claim to a property was the owner of the mineral therein. It was only natural that the same legal concept be applied to the water needed to extract, process, and transport the ore. Thus arose the **prior appropriation doctrine** (34), which is followed in 17 western states of the United States. This was first applied in *Irwin v. Phillips* (35). The right to use water is separate from other property rights. The water-right holder does not necessarily need to be riparian, and riparian owners may have no water rights. The first person to divert water from a surface course has the primary water right, and it passes to successive owners. Junior users have lower rights and in time of drought may not receive any water, even though senior users always get their full share. However, some states have established a priority of use, with domestic use receiving top priority, irrigation receiving second priority, and industrial and commercial receiving lowest priority. Most states have some limitation on the transfer of appropriate rights, sometimes going so far as to link the water right to a specific use and piece of land (33). Some states provide for the forfeiture of a water right if it is not exercised for a given time period.

Courts have recently placed limitations on absolute ownership of surface water rights. In 1983 the California Supreme Court ruled in *National Audubon Society v. Superior Court of Alpine County* (36) that the Los Angeles Department of Water and Power must limit its diversions from Mono Lake, even though the diversions were legal when initiated. The **public trust doctrine** was invoked. This basically held that the private right to use water was limited by the need to preserve the environmental aspects of a unique scenic, recreational, and scientific area, which benefited all (37). In 1984 the California State Water Control Board held that the Imperial Irrigation District of southeastern California could not permit Colorado River water to drain into the Salton Sea as return flow from irrigation (38). As this water was then not available for use by other parties, the Water Control Board ruled that the Irrigation District was wasting water, which is illegal under California law. (The Salton Sea is in a closed basin and has been formed by drainage of Colorado River water that had been diverted to California. Hence, there is no public trust to preserve the Salton Sea because it is not a natural feature.)

The **Winters Doctrine** also limits the doctrine of prior appropriation. In 1908 the U.S. Supreme Court held that when Congress created reservations for the Indian Nations, the water rights to develop those reservations and make the land productive were reserved (39). This included water on nonreservation land that had been opened to settlement. Indian water rights are senior to those granted by state law in most water basins (40, 41). Most of the Indian water rights have never been fully utilized and it is estimated that in total they may amount to a quantity of water three times that of the annual flow of the Colorado River (34). Resolution of water rights under the Winters Doctrine will be an issue for many years to come (42). In 1985 the state of Montana reached agreement on Indian water rights with the Assiniboine and Sioux tribes on the Fort Peck Reservation (43). The Fort Peck-Montana Compact is one of the very few Indian water rights issues to have been settled. Although the Winters Doctrine was originally applied to surface water, the Supreme Court has applied the doctrine to ground water as well (44). The Tohono O'odham (Papago Nation) of Arizona, along with the Indians of the Fort Peck Reservation of Montana, are authorized to lease Indian water for off-reservation uses (45, 46). This may be a way for non-Indian water users to maintain a source of water without infringing upon Indian rights under Winters.

Inasmuch as most major rivers cross state boundaries, it is not surprising the surface-water law has a strong federal component. The U.S. Congress, for example, apportioned the water from the Colorado River among various western states. The appropriation of surface water in the Colorado River Basin has been made on the basis of a concatenation of events that included the Colorado River Compact (1922), the Boulder Canyon Project Act (1928), a treaty with Mexico (1944), the Upper Colorado River Compact (1948), the Colorado River Storage Project Act (1956), a Supreme Court decision, *Arizona v. California* (1963), and the Colorado River Basin Project Act (1968) (47).

In addition, Congress has also appropriated most of the money to develop the surface-water resources of the western United States. The Reclamation Act of 1902 is the linchpin of federal funding for western water projects. As a result, about 15 percent of all water in the west is supplied by the U.S. Bureau of Reclamation, all at a subsidized price (34). The Central Arizona Project is a massive canal system built mostly with federal funds to transfer the Colorado River water appropriated to Arizona to the users in Phoenix and Tucson (46).

Another example of the federal role in surface water is the case of *Wisconsin et al.* v. *Illinois et al.* (48). Illinois began to divert surface water from the Lake Michigan Basin at Chicago in the early part of this century by reversing the flow of the Chicago River to divert sewage flows from the water system intakes in Lake Michigan. The diversion resulted in a lowering of lake levels; this impacted upon shipping and power production at Niagara Falls. The riparian states on the Great Lakes sued Illinois in federal court in a common law action starting in 1925. The U.S. Supreme Court ruled in 1967 that Illinois could not divert more than 3200 cubic feet per second from the lake for all purposes.

### 11.6.3 LAWS REGULATING QUANTITY OF GROUND WATER

There are several different types of state law governing ground-water use, with a number of variations of the basic concepts. The right of absolute ownership of the water under a property holder's land is known as the **English Rule.** In 1843, an English court held that a landowner could pump ground water at any rate, even if an adjoining property owner were harmed (49). The idea of absolute ownership was transported across the Atlantic and planted in the United States (50). In 1903, a Wisconsin court ruled in *Huber* v. *Merkel* that a property owner could pump ground water, even with malicious intent to harm a neighbor, the water being put to no good use (51). However, this ruling was overturned in 1974 with the opinion of Wisconsin Justice Landry that "the Huber case and its misconceived progeny can no longer be relied upon as conferring to the owner of land an absolute right to use with impunity all of the water that can be pumped from the subsoil underneath" (52). Although Wisconsin courts now hold the English Rule in disfavor, it is still the law of the land in Texas and in many other states (53).

Even before the 1903 *Huber* v. *Merkel* decision—as early as 1862—an American state court had ruled that a landowner had the right to use only a reasonable amount of ground water. The rights of adjacent property owners were also recognized (54). This is known as the **American Rule,** and, as evidenced by the Wisconsin case, a state can change from the English Rule to the American Rule by judicial action. As case law moves with glacially slow progress, it is unusual for a state to shift from one rule to the other by court action.

Some western states have ground-water law dictating that waters within the state boundaries belong to the public. Individuals can establish water rights to ground water by use; the first to draw from an aquifer has established a senior right. New Mexico is a state with this doctrine for both ground and surface water (53). However, the state engineer of New Mexico can regulate the withdrawal of ground water from "declared" underground water basins.

One area of conflict arises when surface water is fed by discharge of ground water. Development of the ground-water reservoir typically depletes the baseflow of the stream. Holders of surface-water rights may find that the allocated surface water is diminishing in volume as ground-water withdrawals increase. On the South Platte River Basin of Colorado, the stream depletion due to ground-water withdrawals is about equal to the consumptive use of the ground-water pumped from the basin for irrigation (55). In some western states, the surface-water right can be usurped from the holder by ground-water pumpage. In the case of Colorado, this conflict is being resolved by integrating the claims of both surface-water and ground-water users with regard to the same basin (56). In some ground-water basins of New Mexico, would-be ground-water developers must acquire sufficient surface-water rights to compensate for the reduced flow from the stream (53).

In California, a ground-water system based on correlative rights is followed. The right to use ground water belongs to the owner of the overlying land, so long as the use is reasonable. Water in excess of that used by the overlying owners can be allocated by appropriation. Thus, there are two systems of ownership of water rights. The determination of available water is based on the average recharge to the aquifer, termed the safe yield (although it is not safe yield as defined in Section 11.5). If withdrawals are less than the average annual recharge, the excess can be apportioned. If an **overdraft** (defined as pumping in excess of the average annual recharge) exists for at least five years, the ground water is apportioned among all users in amounts proportional to their individual pumping rates, with the total pumping set at the average annual recharge. This is the **doctrine of mutual prescription,** which was first adjudicated in the case of the Raymond Basin (57).

The rights of states to regulate ground water was limited in 1982 by the federal courts in *Sporhase v. Nebraska* (58, 59). Sporhase and Moss owned land in both Nebraska and Colorado. A well located in Nebraska pumped water from the underlying Ogallala aquifer, which was used to irrigate both the Nebraska and the Colorado land. Nebraska had a regulation that one who wished to pump ground water and export it to another state was required to first apply for and obtain a permit. Sporhase and Moss did not apply for the permit and were sued in Nebraska court. The Nebraska court ruled against them, but was overruled by the Supreme Court. Justice Stevens wrote:

> Although water is indeed essential for human survival, studies indicate that over 80% of our water supplies is used for agricultural purposes. The agricultural markets supplied by irrigated farms are worldwide. They provide the archetypal example of commerce among the several States for which the framers of our Constitution intended to authorize federal regulation. The multi-state character of the Ogallala aquifer—underlying tracts of land in Colorado and Nebraska, as well as parts of Texas, New Mexico, Oklahoma and Kansas—confirms the view that there is a significant federal interest in conservation as well as in fair allocation of the diminishing resource.

The high court ruled that the states can regulate water within their boundaries, but do not own it and cannot prohibit export except in times of severe water shortages within the state. This ruling was applied in 1983 in *City of El Paso v. Reynolds*. El Paso, Texas, wished to withdraw water from two large and scarcely used aquifers in New Mexico. The court ruled that New Mexico did not need the aquifer for human use and therefore New Mexico could not prohibit the transfer of water to El Paso (60).

If one wishes to artificially recharge and store water in an aquifer, can someone else pump out that water? Can a public agency use the aquifer under a property holder's land for storage of imported water? These are questions central to the use of aquifers as storage reservoirs, and they must be answered if cyclic storage is to become an important procedure in water management. Two California cases dealing with aquifer storage have apparently settled these questions. They also form a basis for resolution of these problems in other states (61).

The first California case involved the rights of the landowner to exploit an aquifer. Alameda County has a water district formed to protect the aquifer, especially from salt-water intrusion. As a conservation practice, local and imported water is injected into the aquifer to maintain the seaward hydraulic gradient. A sand and gravel pit was operating and, by 1969, the sand-mining operation extended 80 feet into the aquifer. The injected water was flooding the sand pit, requiring the mining company to pump large amounts of water from it. The mining company sued the water district to enjoin it from recharging the aquifer. The mining company lost the case, as the court held that public agencies can store water in an aquifer, up to the historic ground-water level, even if it decreases the usefulness of land to the property holders (62).

In the second case, a public entity injected water for underground storage with the intention of withdrawing it at a later time. This right to recapture injected waters was recognized by a court ruling in 1943 (63) and then reaffirmed in 1975 (64). In both instances, the aquifers in question were in the Los Angeles area, where several cities were competing for use of ground water that comes from natural sources and artificial recharge.

## CASE STUDY: ARIZONA'S GROUND-WATER CODE

In 1980 the state of Arizona passed a comprehensive new ground-water law designed to reduce the severe overdraft of ground water as well as to allocate the limited ground-water resources of the state (65). A new state agency, the Department of Water Resources, was established to administer the code. Geographical areas known as Active Management Areas (AMAs) were established for those areas with severe overdraft problems. Within the designated AMAs, existing and future use of ground water is regulated. Ground-water users at the time the AMA was formed could claim a grandfather right. Persons can apply for a ground-water withdrawal permit for a new or expanded use for almost any purpose but irrigation. They must show that there is sufficient ground water available for the permit and that no water is available from other sources, such as the grandfather ground-water rights or the Central Arizona Project. Management plans must also be established for each AMA. For the three urban AMAs, the management plan has a goal of reducing ground-water withdrawals to the safe yield by 2025. In the agricultural AMA the goal is to preserve irrigated agriculture as long as possible while still preserving future water supplies for nonagricultural uses. Water conservation will be necessary to meet the goals of the management plans. The difference in urban-area water usage between Phoenix (267 gallons, or 1011 liters, per capita per day) and Tucson (160 gallons, or 606 liters, per capita per day) illustrates the possible scope of savings via conservation. However, the biggest potential savings is in agriculture; it accounts for 89 percent of the total water usage in the state. The real strength of the code is the provision that prohibits the establishment of new urban areas unless there is a 100-year supply of water available. This prevents new development of water-short areas and protects home buyers.

### 11.6.4 LAWS REGULATING THE QUALITY OF WATER

Common law can be applied to a situation where damages to an individual have occurred because of contamination of ground or surface water. In such a case, a suit would be filed and one or more theories of common law advanced along with expert testimony about the technical facts surrounding the alleged contamination.

A number of laws have been written with the intent of preventing the contamination of ground and surface water. There are a number of relevant federal laws, which apply in all states, as well as specific state laws. As a general principle, state laws can be more strict, but cannot be less strict, than federal laws if they address the same topics.

#### National Environmental Policy Act of 1969 (P.L. 91-190)

Title I of the act establishes a national environmental policy and environmental goals. Environmental-impact statements are required for major federal projects. The Supreme Court has ruled that the policies and goals of Title I are not enforceable standards and the federal agencies are required only to consider them when making decisions. An environmental-impact statement is required to disclose the environmental impact of a proposed action, unavoidable adverse consequences, alternatives to the proposed action, the relationship of short-term uses of the environment to long-term productivity, and irretrievable commitments of resources.

#### Federal Water Pollution Control Act of 1972 (P.L. 92-500)
#### Clean Water Act Amendments of 1977 (P.L. 95-217)

The objective of this act and its amendments is "to restore and maintain the chemical, physical and biological integrity of the nation's water." National goals to achieve a degree of water quality that would make the waters of the United States "fishable and swimmable" by 1983 and to eliminate the discharge of pollutants into the water of the United States by 1985 were established. The act sets water-quality standards, establishes minimum national effluent standards, requires pollution discharge permits, and has a construction grant program for public sewage-treatment plants. As a result of this program, surface-water quality has increased dramatically in many areas of the United States. The goal of eliminating discharges into surface waters has placed emphasis on land treatment and disposal of wastewater and sludges. In some cases this can pose a threat of ground-water contamination.

### Safe Drinking Water Act of 1974 (P.L. 93-523) and Amendments

This law was passed to set standards for safe drinking water, protect "sole source" aquifers, and protect drinking-water aquifers from contamination resulting from underground injection of waste.

The Safe Drinking Water Act requires the U.S. Environmental Protection Agency (EPA) to set drinking-water standards to protect public health and welfare. The 1986 amendments to the act substantially enlarge the list of substances to be regulated. Maximum contaminant levels and maximum contaminant-level goals are to be established for toxic and carcinogenic compounds as well as secondary maximum contaminant levels for substances with no health risk but which do create aesthetic concerns.

The "Gonzales Amendment" authorized EPA to designate aquifers that are especially valuable because they are the only source of drinking water in an area. No federal financial assistance may be given to a project that might contaminate one of these "sole source" aquifers so as to create a significant hazard to public health. In the 1986 amendments, Congress directed the states to develop plans to protect the surface area around public water-supply wells from potential contamination from such sources as hazardous wastes, pesticides, and leaking underground storage tanks. These wellhead protection plans will need a very strong input from hydrogeologists if they are to be successful.

The underground injection of hazardous wastes and other materials is also regulated under the Safe Drinking Water Act. Current drinking-water aquifers as well as all other aquifers with total dissolved solids less than 10,000 mg/L are protected by regulating injection of liquid waste into deep boreholes. Class I wells dispose of hazardous waste into isolated strata below current or potential drinking-water aquifers. Class II wells are for the recirculation of oil-field brines. Class III wells are used in solution-mining of ores. Wastes are not to be injected into or above potable water aquifers.

### Resource Conservation and Recovery Act of 1976 (RCRA) (P.L. 94-580)

The Resource Conservation and Recovery Act was the first major federal effort to deal with hazardous solid waste. It is a management system designed to regulate hazardous waste from the time it is created and continuing through its final disposition. The keystone of the system is a permit and manifest system used to keep track of hazardous waste. Waste generators and transporters, as well as managers of hazardous waste storage, treatment, and disposal facilities are all regulated. Facilities may not be placed in the recharge zones of sole-source aquifers. New facilities are required to have ground-water monitoring systems, and old facilities must be retrofitted for ground-water monitoring. Hazardous-waste landfills and lagoons are required to monitor leachate and to have double-liner systems. Land disposal of liquid hazardous waste is prohibited, and land disposal of certain other hazardous wastes is to be phased out.

### Comprehensive Environmental Response, Compensation and Liability Act of 1980 (CERCLA) (P.L. 96-510)

The Comprehensive Environmental Response, Compensation and Liability Act of 1980 is commonly referred to as "Superfund." This act is targeted at the cleanup of releases of hazardous substances in the air, on land, or in the water. Parties who release contaminants from hazardous wastes are required to clean them up. If the responsible parties do not do so, the EPA can do the work and bill the cost to the responsible parties. Potentially responsible parties include the generators, transporters, and disposers of the waste. A provision known as "joint and several liability" permits the government to recover the full costs of the remedial action from any of the responsible parties, even if they were responsible for only part of the waste. The EPA is empowered to respond immediately to the release of a hazardous substance and then recover costs at some time in the future from the responsible parties—if any can be found and if they have any money!

CERCLA requires the EPA to establish a National Priorities List of sites to be targeted for remedial action. From time to time new sites are added to the National Priority List. Many of them are abandoned sites where the responsible owners are bankrupt or assetless. Other sites may be owned by Fortune 500 companies.

### Superfund Amendments and Reauthorization Act of 1986 (SARA)

Superfund was reauthorized in 1986 with additional funds added to the program and additional mandates from Congress relating to the restoration of contaminated sites. Section 121 of SARA places emphasis on clean-up remedies that will reduce the volume, toxicity, and mobility of hazardous substances and contaminants to the maximum extent practicable.

### Surface Mining Control and Reclamation Act (SMCRA) (P.L. 95-87)

The Surface Mining Control and Reclamation Act was passed with the intent of preventing imminent danger to the health and safety of the public and significant, imminent environmental harm that might be caused by both underground- and surface-mining operations. One aspect of the law is that it requires that a hydrogeological study be performed prior to the covering or burial of acid-forming or toxic waste materials from mining or when any mine is to be filled with waste material. These waste materials are typically overburden spoils or mill tailings. The mine operator must prove that there will be only minimal disturbance of the hydrologic regime around the mine. If a mining activity contaminates, diminishes, or interrupts the ground-water or surface-water supply of an adjacent land owner, SMCRA requires the mine operator to replace the water supply.

#### Uranium Mill Tailings Radiation and Control Act of 1978 (UMTRCA) (P.L. 95-604 as amended by P.L. 95-106 and P.L. 97-415)

The Uranium Mill Tailings Radiation Control Act regulates the storage and disposal of mill tailings at both active and inactive uranium mill operations. The act provides that uranium mill tailings should be stabilized, controlled, and disposed of in an environmentally sound and safe manner. Title I of the act addresses remedial actions that must be taken at unsafe abandoned sites. Title II provides for the regulation of the handling and disposal of waste materials at active sites. Correction of adverse impacts on ground or surface water from uranium mill tailings is a feature of UMTRCA.

#### Toxic Substances Control Act (TOSCA) (P.L. 94-469 amended by P.L. Law 97-129)

The purpose of this act is to protect human health and the environment by requiring testing and use restrictions for chemicals that may present an "unreasonable" risk to health and the environment. This is an umbrella act that regulates toxic compounds during research and development, manufacturing, processing, and distribution. It has been interpreted to permit controls on the use, disposal, or storage of toxic compounds where ground-water contamination from such compounds has occurred or may be expected to occur. Federal agencies and state governments come under the jurisdiction of this act.

#### Federal Insecticide, Fungicide and Rodenticide Act (FIFRA) (P.L. 92-516, amended by P.L. 94-140, P.L. 95-396, P.L. 96-539, and P.L. 98-201)

The primary purpose of FIFRA is to regulate the manufacture, use, and disposal of pesticides. It requires pesticides to be labeled and restricts their use where appropriate. Use restrictions can be applied in areas where ground-water contamination by pesticides has occurred or could be expected to occur.

## CASE STUDY: WISCONSIN'S GROUND-WATER PROTECTION LAW

State of Wisconsin Act 410 was enacted to protect the quality of the state's ground water. There are five areas that the act addresses: establishment of numerical standards for ground-water quality, a laboratory certification program, an expanded ground-water monitoring program, establishment of an environmental repair fund for remedial work at contaminated sites, and a compensation fund to pay part of the cost of drilling new wells for property owners whose wells have become contaminated with substances other than bacteria and nitrate. The most far-reaching provision is the numerical standards program. This is set out in administrative rules (NR 140). Ground-water quality standards are set out for substances that have been detected in or have a reasonable probability of entering ground water in Wisconsin. An enforcement standard is set for each substance on the basis of federal standards indicating a concentration above which there is a threat to public health or the environment. Enforcement standards are applicable to ground water that has been impacted by some regulated activity, such as a landfill or a hazardous-waste storage area. Enforcement standards apply at any point beyond the property boundary of a regulated facility, at any point of ground-water use, and at any point beyond what is known as a "design management zone" for a regulated activity. A design management zone is a three-dimensional portion of the earth beneath a regulated facility within which natural attenuation may be used to reduce the concentration of a contaminant. If the site is small, the property boundary is coincident with the design management zone boundary. If the site is larger, the design management zone is a designated horizontal distance from the edge of the facility. For example, for land treatment of wastewater, the design management zone is 250 feet (76 meters); for wastewater treatment lagoons it is 100 feet (30 meters) and for hazardous waste disposal facilities it is 0 feet. The law also specifies preventative action limits (PALs), which are set at some percentage of the enforcement standard. For substances of public health concerns, PALs are set at 20 percent of the enforcement standard, except for carcinogenic and mutagenic compounds, where the PAL is 10 percent of the enforcement standard. For public welfare–related substances the PAL is set at 50 percent of the enforcement standard. When a PAL is exceeded anywhere within the site, including the design management zone, a wide range of responses is available. These range from no action to requiring additional monitoring to revising operating procedures at the managed facility to remedial action to restore ground-water quality. If an enforcement standard is violated, some action must be taken. These include requiring changes in operation of a facility, requiring redesign of a facility, requiring alternative methods of waste treatment and disposal, requiring closure of a facility, and requiring restoration of ground-water quality. Leeway has been given to the Wisconsin Department of Natural Resources in determining the appropriate response and the timeframe for it.

This water-quality protection law, in conjunction with other rules under which Wisconsin regulates waste treatment and disposal activities, creates a strong framework to protect the water quality of the state. A survey of water quality in both community and private wells in Wisconsin was conducted prior to the implementation of the ground-water protection law. It found that 65 of 1174 community wells tested and 82 of 617 private wells had detectable levels of volatile organic compounds. However, only 5 community wells and 14 private wells had volatile organic compounds above health advisory levels. In one area of Wisconsin, the central sand plains, about 900 wells were sampled for aldicarb, a pesticide. A substantial number, 201, had detectable aldicarb levels and 70 contained aldicarb in concentrations in excess of Wisconsin's recommended health advisory level of 10 μg/L (66). It would appear that, in general, Wisconsin has high-quality ground water, although some areas, such as the central sand plains, have substantial problems.

## 11.7 ARTIFICIAL RECHARGE

Resource management sometimes is interpreted as limiting the development and use of a resource in order to conserve it. Ground-water management has a somewhat broader scope, in that artificial recharge can be used to expand the amount of available water. **Water spreading** has been used in the western United States for decades to capture additional runoff. Much of the recharge of alluvial basin-fill aquifers comes from stream-bed infiltration during the wet season. In order to increase the amount of infiltration, stream water is diverted onto empty land below the mouths of the canyons carrying the streams (21).

Surface spreading is feasible given the following circumstances: the upper soil layers are permeable, the water table is not close to the surface, the land is relatively flat, and the aquifer to be recharged has a transmissivity great enough to carry the water away from the spreading area. It is extensively practiced simply by placing low dams across the ephemeral streams draining from the mountains. The dams act to flood the land along the stream channel.

Surface spreading tends to raise the water table over a rather extensive area. The same is true of other diffused types of artificial recharge. Irrigation is also a form of artificial recharge. Because of problems of salt accumulation in the soil, the amount of applied irrigation water is in excess of the needs of plants for evapotranspiration. The unevaporated water percolates through the unsaturated zone and recharges the water table. When spray irrigation is used as a means of wastewater treatment, the amount of water applied is far in excess of the plant requirement (67). The result may be a substantial increase in the elevation of the water table (68). Similar diffused sources of artificially recharged water include septic tank drain fields. Wastewater from these sources percolates through the unsaturated zone and recharges the water table (69, 70).

**Recharge basins** are frequently used to recharge unconfined aquifers, especially where land costs are high. Basins are advantageous in that a substantial hydraulic head can be maintained in order to increase the infiltration rate. They are inexpensive to construct and operate. On Long Island, New York, basins are used to recharge storm-water runoff from roadways and parking lots (71, 72). They are also used in Peoria, Illinois (73, 74), and Fresno, California (75, 76), to recharge aquifers with river water. The infiltration capacity through coarse river gravels at Peoria is very high, with rates of 20 to 100 feet (6.1 to 30.5 meters) per day being typical annual averages.

Recharge basins concentrate a large volume of infiltrating water in a small area. As a result, a ground-water mound forms beneath the basin. As the recharge starts, the mound begins to grow; when the recharge ceases, the mound decays as the water spreads through the aquifer. The growth and decay can be described mathematically (77, 78). Digital-computer models have been used to evaluate the impact of recharge basins on the water table (79, 80). It is important to maintain the unsaturated zone beneath the recharge basins in order to help maintain high infiltration rates, while still providing water-quality improvements (10, 17).

The infiltration capacity of recharge basins is initially high and then declines as recharge progresses. This is due to surface clogging by fine sediments (81) and biological growths in the uppermost few inches of the soil (82). It has been found that the operation of recharge basins with alternating flooding and drying-out periods will maintain the best infiltration rates (17). Fine surface sediments may occasionally need to be removed mechanically (83).

If the intent of a management program is to recharge a confined aquifer, then **recharge wells** must be used. The design of a well for artificial recharge is similar to that of a supply well. The principal difference is that water flows out of the recharge well and into the surrounding aquifer under either a gravity head or a head maintained by an injection pump.

Injection wells for artificial recharge are prone to clogging. A large amount of water is being pushed through a small volume of aquifer near the well face. Clogging can be due to a number of factors; for example,

1. air entrainment caused by aeration of water falling into the well,
2. filtration of suspended sediment and organic matter,
3. development of bacterial growths in the aquifer,
4. formation of precipitates due to geochemical reactions between the recharge water and native ground water,
5. swelling of clay colloids in the aquifer,
6. dispersal of clay particles due to ion exchange between recharging water and aquifer materials,
7. precipitation of iron from native ground water due to recharge of water with a pH and *Eh* in the range of ferric iron,
8. growth of iron bacteria, or
9. mechanical compaction of aquifer materials due to high injection pressures.

Frequent maintenance of injection wells may be necessary. This typically consists of pumping water out of the wells to redevelop the flow capacities. Chlorination may also be used to reduce biological growths (17). Wisconsin is one state that does not permit the use of injection wells for any purpose. The reason for this prohibition is to prevent well disposal of wastes that may result in ground-water contamination.

## 11.8 PROTECTION OF WATER QUALITY IN AQUIFERS

One of the most important aspects of ground-water management is the protection of the water quality in an aquifer. There are a number of artificial sources of potential ground-water contamination. Pollution from such sources as septic tanks, sanitary landfills, land-treatment systems for municipal wastewater, waste injection wells, toxic chemical disposal sites, cemeteries, mine tailings, acid mine drainage, water softener regeneration salt, highway deicing salt, oil-field brines, agricultural chemicals and fertilizers, and accidental oil, gasoline, and chemical spills have been well documented in scientific literature. An incident such as the burial in Michigan of slaughtered cattle that had been given feed contaminated by PBB, a chemical fire retardant, points out that the earth is typically thought of as the most convenient repository for material that society does not know how to handle.

Ground-water contamination can occur also when water of poor quality is drawn into a well field that originally has been developed in high-quality water. The best example of this is salt-water intrusion in coastal areas. Heavy pumping of coastal aquifers can cause a landward migration of the interface separating fresh and salty ground water.

One critical aspect of preventing ground-water pollution is the identification of the recharge areas of aquifers. In such areas, protection of the aquifer is vital. Recharge areas should be zoned as water-quality conservation areas, with close control of potential sources of contamination. The hydrogeology of sites for such facilities as sanitary landfills and land-treatment systems should be intensively studied to ensure that the condition of local soils and the rate of direction of local ground-water flow preclude possible aquifer contamination. Hazardous- and toxic-waste storage and disposal should be barred from aquifer recharge areas. Sanitary sewers for the collection of domestic wastewater are preferable to septic tanks in aquifer recharge areas. Under most conditions, land areas underlain by extensive confining layers or situated in the discharge areas of unconfined aquifers are preferable for uses that might contaminate the ground water. Agricultural practices in recharge areas should also be regulated. Overapplication of chemicals and fertilizers should be discouraged and controlled collection and disposal of animal waste encouraged.

One of the most difficult aspects of aquifer protection is the control of abandoned wells. Many states require that abandoned wells be backfilled. This regulation can be enforced for a newly drilled water, oil, or gas well that proves to be unproductive. However, in the case of wells that fall into disuse and casual abandonment, the owners might not be aware of the danger they can pose to ground-water quality. Likewise, improper well construction can cause shallow ground water or surface water to migrate downward into the aquifer. Abandoned wells may also be used for the disposal of liquid or solid waste, an obvious threat to water quality. Education is more efficient than regulation in preventing contamination of this type.

Salt-water encroachment can be prevented by regulating the spacing and withdrawal rates of wells. The objective is to avoid establishing a hydraulic gradient that slopes from the zone of contaminated water toward the well field. If the aquifer contains fresh water overlying salty water, pumping rates must be low to avoid drawing up salty water from below, a phenomenon known as **up-**

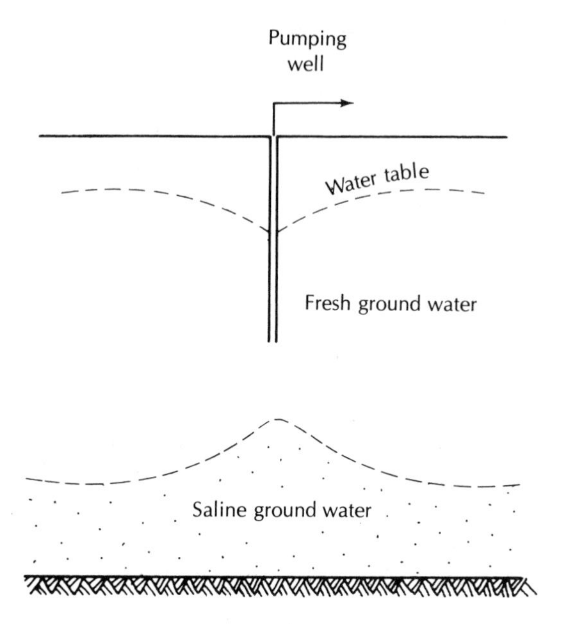


FIGURE 11.1 Upconing of saline water caused by a well pumping from an overlying fresh-water zone.

**coning** (Figure 11.1). Saline-water aquifers underlie up to two-thirds of the land area of the conterminous United States, so the problems of water encroachment and upconing are not limited to coastal areas (84).

Control of sea-water intrusion in coastal areas has been practiced in a number of localities. In coastal areas of southern California, where there is a confined aquifer containing salt water and fresh water, a pressure-ridge system has been used (85). A line of injection wells parallel to the coast injects water

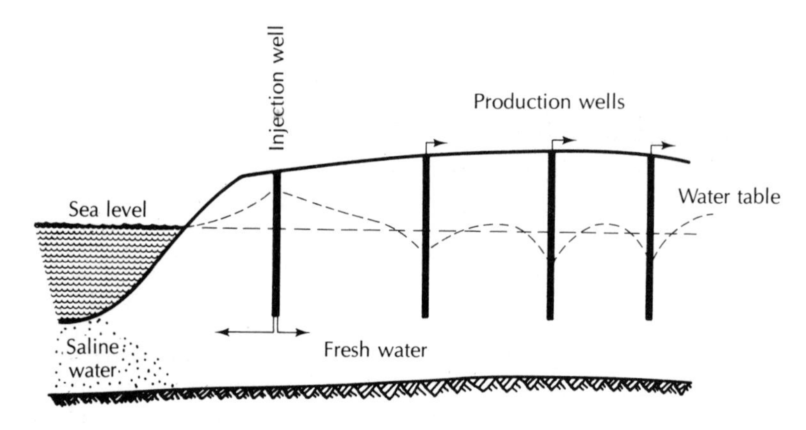


FIGURE 11.2 Use of injection wells to form a pressure ridge to prevent salt-water intrusion in an unconfined coastal aquifer.

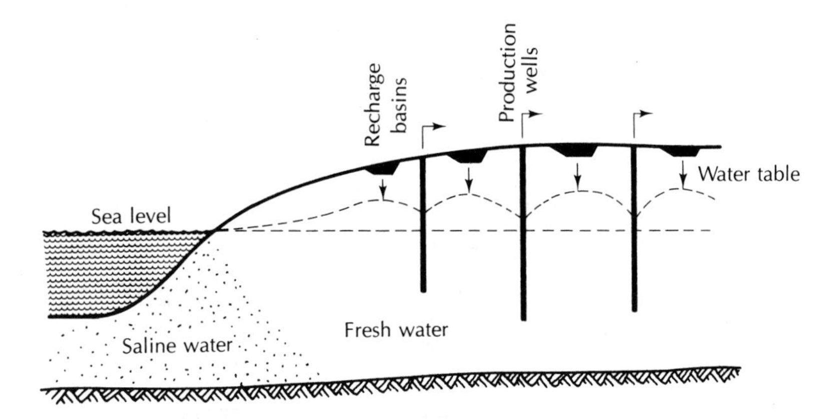


**FIGURE 11.3** Use of artificial recharge in the area of production wells in an unconfined coastal aquifer. The artificially recharged water maintains the water table above sea level to prevent salt-water intrusion.

into the aquifer. The result is a ridge in the potentiometric surface. Figure 11.2 shows this for an unconfined aquifer system. Water levels behind the barrier can be drawn down below sea level with no fear of salt-water encroachment. Similar barriers could be used in unconfined aquifers using artificial recharge from wells, pits, or trenches. Artificial recharge in the area of pumping-well fields could also be used to maintain the elevations of the potentiometric surface above sea level (Figure 11.3). A row of pumping wells could be installed parallel to the coast in either a confined or an unconfined aquifer. They could create a trough in the potentiometric surface lower than either sea level or the well-field areas behind the trough. The trough wells would pump salty water, which would not be suitable for most uses. However, wells behind the trough would pump fresh water (Figure 11.4).

## 11.9 GROUND-WATER MINING AND CYCLIC STORAGE

From the previous sections it is apparent that in order to develop the ground-water reservoir a cone of depression must be created. This means that water is removed from storage or is, in a sense, mined. Likewise, in order to utilize many aquifers for cyclic storage, the water levels must be drawn down in order to create storage space. Again, some water may have to be mined in order to prepare the aquifer to perform an essential service (86). Ground-water management programs with an objective to maintain natural equilibrium water levels may not be as efficient as other management plans.

The most dramatic need for limited ground-water mining is to provide supplemental water for use when precipitation and surface supplies are limited by drought. Normal rates of industrial and agricultural productivity can be main-

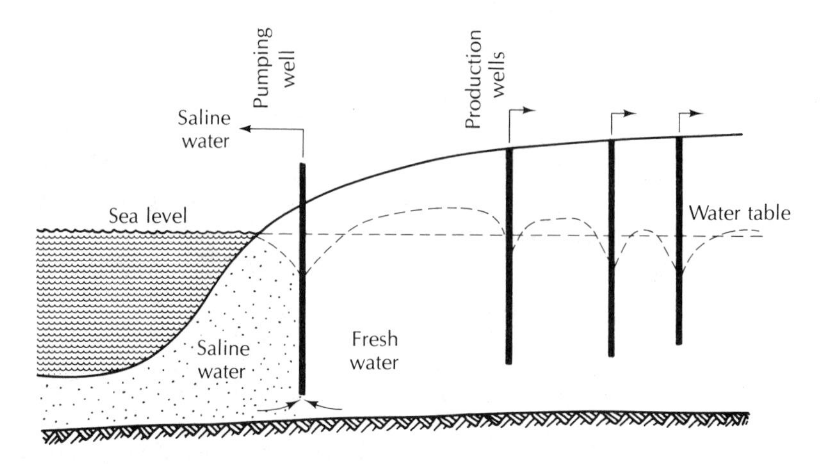


FIGURE 11.4 Use of pumping wells at the coastline to form a trench in the water table; the trench acts as a barrier to further salt-water encroachment.

tained by the use of supplemental ground water (15). Water managers may find it difficult to convince themselves that the best time to deplete the ground-water reservoir artificially is when the amount of natural recharge is lowest. Ground-water levels will be falling at a seemingly alarming rate. However, the drawdown will create space for storage when nature provides an especially wet year (13). Water managers must be ready to capture this excess by artificial recharge to supplement the natural infiltration. It has been suggested that major recharge events that can result in a rapid rise in ground-water levels can be expected every 10 to 15 years (13). If water levels have not been depleted by "overpumping" during drought, much of the potential recharge may be lost owing to lack of aquifer storage space. This concept of **cyclic storage** is not applicable to aquifers to which the amount of possible recharge is limited by their capacity to transport and store ground water rather than by the amount of water available to them.

In some arid lands, considerable ground water is known to exist. Furthermore, there may be rather large hydraulic gradients, which could mean that this ground water is moving through the aquifers. This interpretation is at variance with the extremely low rates of recharge known to occur at present. Model studies have suggested that the gradients are "fossil" (87); i.e., decayed remnants of higher ground-water gradients of 10,000 years ago, when pluvial periods during the late Pleistocene provided large amounts of recharge. Management programs, however, cannot be based on these extremely long cycles between recharge events. For practical purposes, such aquifers must be considered to be unreplenished, and all ground-water pumpage to be simply mining when recharge events are millennia apart.

The mining of ground water is occurring in many localities. The Ogallala aquifer of the southern High Plains of Texas and New Mexico is an example of a mined aquifer in arid regions, and the deep sandstone aquifer of northeastern Illinois is a humid-region example. Mining the deep sandstone aquifer appears to be an integral part of the areawide development of the ground-water and surface-water resources of northeastern Illinois (88, 89). Water levels will be drawn down by overdrafts until a "critical level" is reached. Ground-water pumping will then be curtailed to the extent that no further drawdown will occur.

Many areas of the world are underlain by aquifers containing saline ground water. Although desalinization is technically feasible, it is not economic except in some very limited applications. In water-short areas of the western United States, there is increasing competition between energy development and agricultural interests for the available water (90). Some of this water is used for power-plant cooling, with subsequent evaporational losses. It has been suggested that saline ground-water resources could be mined to provide the necessary cooling water (14). In this approach, it would be necessary to ensure that the available saline ground-water supply is sufficient for the design life of the power plant. Furthermore, the problem of the disposal of brine or salt would have to be dealt with. Saline ground water could also serve as a source of emergency cooling water for nuclear reactors.

## 11.10 CONJUNCTIVE USE OF GROUND AND SURFACE WATER

It is obvious that in stream-aquifer systems it is counterproductive to consider surface-water management and ground-water management as separate actions. When water can flow from the stream to the aquifer, or vice versa, it is tantamount to having two separate policies or plans for the same water. No one would open a joint checking account with a total stranger. However, the legal separation of ground water and surface water, or the administration of ground-water and surface-water management by separate agencies, could have a similar result: a rapid depletion of the total resources due to overuse.

As is the usual case, the need for management of ground water and surface water as elements of an interrelated system is most critical when demand exceeds supply. The greatest application of such management in the United States has come in the arid west. Overpumping of aquifers near a river could deplete the surface resources, while development of a losing river could reduce the normal recharge to an aquifer system. On a regional basis, the incremental drawdown caused by any one pumper is small. It is the cumulative effect of many pumpers that can cause a depletion of the total resource. The individual user has no economic incentive to reduce pumping (91). The situation is exacerbated when ground-water pumping results in a reduction in the flow of surface waters. Ground-water users may feel social pressure if a neighbor's well is adversely affected; however, this pressure is not experienced when reduction in availability of surface water affects an unknown user many miles away. Individual litigation among tens or hundreds of water users along a stream-aquifer system would present a legal nightmare. Some type of institutional control offers the most workable management approach.

One management plan might be a total ban on ground-water withdrawals during such times that the flow of the river falls below a specified value. This, however, would not prevent the streamflow from falling even more, as the response time of an aquifer is quite long (92). In order to fully protect the rights of senior surface-water users, ground-water pumping might be prohibited altogether. While this would serve one purpose, it would also eliminate the use of the ground-water reservoir for storage of excess water. In addition, ground-water development usually provides water supplies in excess of streamflow alone.

The most extreme case occurs when a river fed by mountain runoff flows across an aquifer system that receives no recharge from precipitation. Under such conditions, the river is the only source of water for both ground-water and surface-water users. Even in this extreme case, the river and aquifer system can be managed conjunctively to maximum benefit. If the water is used for irrigation, there is a demand only part of the year. The remainder of the year, the streamflow passes downstream. The maximum benefit would accrue if some of the water flowing in the nongrowing season could be stored for use when it is needed. This can be accomplished by taking advantage of the delayed response of the aquifer system (92). Wells located away from the river and pumped during the irrigation season would draw water from storage at that time, and then, when the cone of depression reaches the river, infiltration would be induced mostly during the nongrowing season (Figure 11.5). Wells located closer to the river would be pumped only near the end of the irrigation season.

While the benefits of conjunctive ground-water and surface-water use are obvious, the implementation of management is not easy. The legal framework for each state is different. In New Mexico, ground-water users in designated ground-water basins fall under the jurisdiction of the New Mexico state engineer. New appropriations of ground water are made only on the condition that the ground-water user acquire sufficient surface-water flow (53, 93). In the state of Colorado, the Water Right Determination and Administration Act of 1969 is enforced by the state engineer. In part, this act reads ". . . it shall be the policy of this state to integrate the appropriation, use and administration of underground water tributary to a stream with the use of surface water in such a way as to maximize the beneficial use of all the waters of this state." However, the Colorado state engineer does not have unlimited license to maximize available water, as the act also states that "no reduction of any lawful diversion because of the operation of the priority system shall be permitted unless such reduction would increase the amount of water available to and required by water rights having senior priorities." A management plan that would increase the total amount of water but would result in junior users receiving more water at the expense of senior users is apparently not what the Colorado legislature intended.

It has been pointed out that, in most cases, a number of different management plans might be possible (94). For example, an applicant for ground water might be allocated a large amount of water, but it would be available only

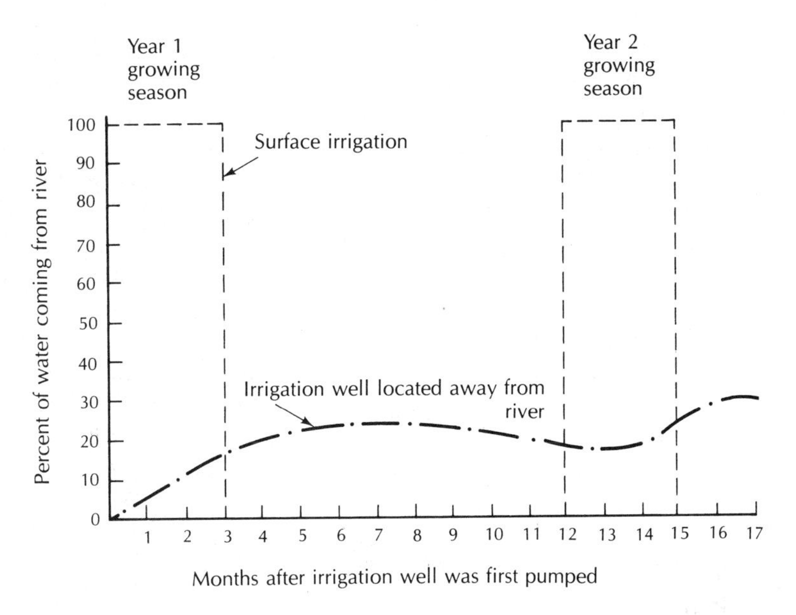


**FIGURE 11.5** Comparison of withdrawal of irrigation water from a surface stream with induced infiltration to an irrigation well. The irrigation well delivers the same amount of water during the irrigation season as the surface withdrawal. Source: R. A. Young and J. D. Bredehoeft, *Water Resources Research*, 8, no. 3 (1972):533–56.

during the nongrowing season. A total volume not nearly as great might be allocated if it were to be pumped during the irrigation season. Both plans could protect the senior water rights equally.

Conjunctive ground-water and surface-water management lends itself to the use of computer models (92–97). The hydrologic interaction between the stream-aquifer system can be described by partial differential equations, subsequently solved by numerical methods (Chapter 13). Techniques of operations theory can be used to maximize the amount of water available while minimizing the cost of development. The models can be established so that the holders of senior water rights are protected and water is allocated to the most beneficial use.

## 11.11 TRENDS IN WATER RESOURCES MANAGEMENT

Water is increasingly held to be a public rather than a private resource. In states where the right to use water is held to be a property right, it is protected by the U.S. Constitution. However, even in England, where the English Rule of absolute ownership of ground water originated, the Water Resources Act of 1963 changed concepts of ownership of water. Withdrawal of surface water and ground water in England and Wales is subject to approval of a permit application (98).

In arid states of the United States West, competition for water between public and private users is increasing—especially as population shifts from rural to urban areas (99). Municipalities are able to acquire water rights in the same manner as other entities: they may find an unallocated source or purchase a water right from its owner. However, they are also able to garner water rights by means not available to individuals or corporations: the exercise of the power of eminent domain. Through legal action, a municipal entity may acquire private property, including water rights (100). The use must generally be for a public purpose, and the owner must be compensated for the loss of property. As with any case of public condemnation of private property, the question of what constitutes "adequate compensation" is full of thorns. In Colorado, the compensation is set as "fair market value." This can be established on the basis of prior voluntary sales of water rights. Compensation can also include the loss of value to land that is separated from a water right necessary to put it to a given use (100).

Traditionally, water rights have been considered in connection with off-stream purposes such as irrigation or municipal use. However, there are a number of in-stream uses that are important for public purposes. This includes the maintenance of water quality and fish and wildlife habitat. In Montana, there are increasing demands for water for energy development. In order to protect the ecological integrity of the Yellowstone River, the Montana Fish and Game Commission and the Montana State Water Quality Bureau have formally requested the appropriation of large amounts of water for in-stream flows (101, 102). This is a further example of the supersession of private needs by public in the allocation of a scarce resource.

Both the states and the federal government are moving to protect the quality of ground water. Federal laws such as the Resource Conservation and Recovery Act and the Comprehensive Environmental Response, Compensation and Liability Act have provisions for regulating activities that could degrade ground-water quality. CERCLA also has provisions to restore the quality of degraded aquifers. The Safe Drinking Water Act amendments of 1986 enable states to establish programs to regulate activities in the recharge areas of aquifers in order to protect ground-water quality. It is likely that more state laws, such as Wisconsin's ground-water protection law, will be enacted to protect the quality of ground water.

## REFERENCES


1. THEIS, C. V. "The Significance and Nature of the Cone of Depression in Ground-water Bodies." *Economic Geology,* 38 (1938):889–902.
2. SUTER, M., et al. *Preliminary Report of Ground-water Resources of the Chicago Region, Illinois.* Cooperative Ground-water Report 1, Illinois State Water Survey and Illinois State Geological Survey, 1959, 89 pp.
3. FETTER, C. W., JR., and H. YOUNG. Unpublished results of a computer model

study of the Deep Sandstone Regional Aquifer, 1978; and Harley Young, *Digital-Computer Model of the Sandstone Aquifer in Southeastern Wisconsin.* Southeast Wisconsin Regional Planning Commission Technical Report 16, 1976, 42 pp.

4. SASMAN, R. T., et al. *Water Level Decline and Pumpage in Deep Wells in the Chicago Region, 1971–75.* Illinois State Water Survey Circular 125, 1977, 35 pp.
5. WALTON, W. C. *Leaky Artesian Aquifer Conditions in Illinois,* Illinois State Water Survey Report of Investigation 39, 1960, 27 pp.
6. KAZMANN, R. G. "'Safe Yield' in Ground-water Development, Reality or Illusion?" *Proceedings of the American Society of Civil Engineers,* 82, IR3 (1956):1103:1–1103:12.
7. HORDON, R. M. "Water Supply as a Limiting Factor in Developing Communities; Endogenous vs. Exogenous Sources." *Water Resources Bulletin,* 13 (1977):933–39.
8. BRUTSAERT, W., G. W. GROSS, and R. M. MC GEHEE. "C. E. Jacob's Study on the Prospective and Hypothetical Future of the Mining of the Ground Water Deposited under the Southern High Plains of Texas and New Mexico." *Ground Water,* 13 (1975):492–505.
9. SELLERS, J. H. "Tax Implications of Ground-water Depletion." *Ground Water,* 11, no. 4 (1973):27–35.
10. FETTER, C. W., JR. "Attenuation of Waste-water Elutriated through Glacial Outwash." *Ground Water,* 15 (1977):365–71.
11. GASS, T. E., and J. LEHR. "Ground-water Energy and the Ground-water Heat Pump." *Water Well Journal,* 31, no. 4 (1977):42–47.
12. ANDREWS, C. B. "The Impact of the Use of Heat Pumps on Ground-water Temperatures." *Ground Water,* 16 (1978):437–43.
13. AMBROGGI, R. P. "Underground Reservoirs to Control the Water Cycle." *Scientific American,* 236, no. 5 (1977):21–27.
14. GREYDANUS, H. W. "Managment Aspects of Cyclic Storage of Water in Aquifer Systems." *Water Resources Bulletin,* 14 (1978):477–83.
15. LEHR, J. H. "Ground Water: Nature's Investment Banking System." *Ground Water,* 16 (1978):143–43.
16. THOMAS, H. E. "Cyclic Storage, Where Are You Now?" *Ground Water,* 16 (1978): 12–17.
17. FETTER, C. W., JR., and R. G. HOLZMACHER. "Ground Water Recharge with Treated Waste Water." *Journal of Water Pollution Control Federation,* 46 (1974):260–70.
18. HELWIG, O. J. "Regional Ground-water Management." *Ground Water,* 16 (1978): 318–21.
19. LEE, C. H. "The Determination of Safe Yield of Underground Reservoirs of the Closed Basin Type." *Transactions, American Society of Civil Engineers,* 78 (1915):148–51.
20. MEINZER, O. E. *Outline of Ground-water Hydrology, with Definitions.* U.S. Geological Survey Water-Supply Paper 494, 1923.
21. CONKLING, H. "Utilization of Ground-water Storage in Stream System Development." *Transactions, American Society of Civil Engineers,* 3 (1946):275–305.
22. BANKS, H. O. "Utilization of Underground Storage Reservoirs." *Transactions, American Society of Civil Engineers,* 118 (1953):220–34.

23. FETTER, C. W., JR. "The Concept of Safe Groundwater Yield in Coastal Aquifers." *Water Resources Bulletin,* 8 (1972):1173–76.
24. American Society of Civil Engineers. *Groundwater Basin Management.* Manual of Engineering Practices 40, 1961, 160 pp.
25. FREEZE, R. A. "Three-Dimensional, Transient, Saturated-Unsaturated Flow in a Groundwater Basin." *Water Resources Research,* 7 (1971):347–66.
26. COLLINS, M. A. "Ground-Surface Water Interactions in the Long Island Aquifer System. *Water Resources Bulletin,* 6 (1972):1253–58.
27. BOUWER, H. "Land Subsidence and Cracking Due to Ground-water Depletion." *Ground Water,* 15 (1977):358–64.
28. FETTER, C. W., JR. "Hydrogeology of the South Fork of Long Island, New York: Reply." *Bulletin, Geological Society of America,* 88 (1977):896.
29. THOMAS, H. E. *Conservation of Ground Water.* New York: McGraw-Hill Book Company, 1951, 327 pp.
30. ANDERSON, M. P., and C. A. BERKEBILE. "Hydrogeology of the South Fork of Long Island, New York: Discussion." *Bulletin, Geological Society of America,* 88 (1977):895.
31. PETERS, H. J., "Ground-water Management." *Water Resources Bulletin,* 8 (1972):188–97.
32. TANK, R. W. *Legal Aspects of Geology.* New York: Plenum Press, 1983, 583 pp.
33. HIRSHLEIFER, J., J. C. DE HAVEN, and J. W. MILLIMAN. *Water Supply.* Chicago: University of Chicago Press, 1960, 378 pp.
34. WILKINSON, C. F. "Western Water Law in Transition." *Journal, American Water Works Association,* 78, 10 (1986):34–47.
35. Irwin v. Phillips, 5 Cal. 140 (1855).
36. National Audubon Society v. Superior Court of Alpine County. 33 Cal.3d 419, 658 P.2d 709, 189 Cal. Rptr. 346 (1983): Cert. denied, 104 S.Ct. 413 (1983).
37. GOLDFARB, W. "Mono Lake and the Public Trust Doctrine." *Bulletin, American Water Resources Association,* 20 (1984):292–93.
38. CALIFORNIA WATER RESOURCES CONTROL BOARD. Alleged Waste and Unreasonable Use of Water by Imperial Irrigation District, Decision 1600, 1984.
39. Winters v. United States, 207 US 564 (1908).
40. BERRY, M. P. "The Importance of Perceptions in the Determination of Indian Water Rights." *Water Resources Bulletin,* 10 (1974):137–43.
41. COLLINS, R. B. "Indian Reservation Water Rights. *Journal, American Water Works Association,* 78, no. 10 (1986):48–54.
42. FOSTER, K. E. "The Winters Doctrine: Historical Perspective and Future Applications of Reserved Water Rights in Arizona." *Ground Water,* 16 (1978):186–91.
43. State of Montana/Assiniboine and Sioux Tribes of Fort Peck Indian Reservation Compact, ratified in S.B. 467, 49th Leg., 1985 Montana Laws.
44. Cappaert v. United States, 426 US 128 (1976).
45. P. L. 97-293, Title III, 96 Stat. 1274 (1982), San Xavier Papago Reservation of Arizona.

46. WEATHERFORD, G. D., and S. J. SCHUPE. "Reallocating Water in the West." *Journal, American Water Works Association,* 78, no. 10 (1986):63–71.

47. JACOBY, G. C., JR., G. D. WEATHERFORD, and J. W. WEGNER. "Law, Hydrology and Surface Water Supply in the Upper Colorado River Basin." *Water Resources Bulletin,* 12 (1976):973–84.

48. Wisconsin et al. v. Illinois et al., 388 US 426 (1967).

49. Acton v. Blundell, 12 M&W. 324, 354 (1843).

50. Frazier v. Brown, 12 Ohio St. 294 (1861).

51. Huber v. Merkel, 117 Wis. 355, 94 N.W. 354 (1903).

52. Wisconsin v. Michaels Pipeline Constructors, Inc., 63 Wis.2d 278 (1974).

53. THOMAS, H. E. "Water-Management Problems Related to Ground-water Rights in the Southwest." *Water Resources Bulletin,* 8 (1972):110–17.

54. Basset v. Salisbury Manufacturing Co., Inc., 43 N.H. 569, 82 Am. Dec. 179 (1862).

55. DANIELSON, J. A., and A. R. QAZI. "Stream Depletion by Wells in the South Platte Basin—Colorado." *Water Resources Bulletin,* 8 (1972):359–66.

56. PEAK, W. "Institutionalized Inefficiency: The Unfortunate Structure of Colorado's Water Resources Management System." *Water Resources Bulletin,* 13 (1977):551–62.

57. Pasadena v. Alhambra, 33 Calif. (2d) 908, 207 Pac. (2d) 17 (1949).

58. Sporhase v. Nebraska, 102 S.Ct. 3456 (1982).

59. BARNETT, P. M., "Mixing Water and the Commerce Clause: The Problems of Practice, Precedent and Policy in Sporhase v. Nebraska." *Natural Resources Journal,* 24 (1984):161–94.

60. El Paso v. Reynolds, Civ. No. 80-730 HB.

61. GLEASON, V. E. "The Legalization of Ground Water Storage." *Water Resources Bulletin,* 14 (1978):532–41.

62. Niles Sand and Gravel Co., Inc., v. Alameda County Water District, 37 Calif. App. 3d 924 (1974): Cert. denied 419 US 869.

63. City of Los Angeles v. Glendale, 23 C2d 68 (1943).

64. City of Los Angeles v. City of San Fernando, 14 Calif. 3d 199 (1975).

65. FERRIS, K. "Arizona's Groundwater Code: Strength in Compromise." *Journal, American Water Works Association,* 78, no. 10 (1980):79–84.

66. KRILL, R. M., and W. C. SONZOGNI. "Chemical Monitoring of Wisconsin's Ground Water." *Journal, American Water Works Association,* 78, no. 9 (1986):70–75.

67. KARDOS, L. T. "Waste-water Renovation by the Land—A Living Filter." In *Agriculture and the Quality of Our Environment,* ed. N.C. Brady. American Association for the Advancement of Science Publication 85, 1967, pp. 241–50.

68. PARIZEK, R. R., and E. A. MEYERS. "Recharge of Groundwater from Renovated Sewage Effluent by Spray Irrigation." *American Water Resources Association, Proceedings of the Fourth Conference,* 1967, pp. 426–43.

69. DUDLEY, J. G., and D. A. STEPHENSON. *Nutrient Enrichment of Ground-water from Septic Tank Disposal Systems.* Inland Lake Renewal and Shoreland Man-

agement Demonstration Project Report, University of Wisconsin—Madison, 1973, 131 pp.

70. BEATTY, M. T., and J. BOUMA. "Application of Soil Surveys to Selection of Sites for On-Site Disposal of Liquid Household Wastes." *Geoderma,* 10 (1973):113–22.

71. SEABURN, G. E. *Preliminary Results of Hydrologic Studies at Two Recharge Basins on Long Island, New York.* U.S. Geological Survey Professional Paper 627-C, 1970, pp. 1–17.

72. SEABURN, G. E. *Preliminary Analysis of Rate of Movement of Storm Runoff through the Zone of Aeration beneath a Recharge Basin on Long Island, New York.* U.S. Geological Survey Professional Paper 700-B, 1970, pp. 196–98.

73. SMITH, H. F. *Artificial Recharge and Its Potential in Illinois.* International Association of Scientific Hydrology Publication No. 72 (Haifa), 1967, pp. 136–42.

74. HARMESON, R. H., R. L. THOMAS, and R. L. EVANS. "Coarse Media Filtration for Artificial Recharge." *Journal American Water Works Association,* 60 (1968):1396–1403.

75. NIGHTINGALE, H. I., and W. C. BIANCHI. "Ground-water Recharge for Urban Use: Leaky Acres Project." *Ground Water,* 11 no. 6 (1973):36–43.

76. SALO, J. E., D. HARRISON, and E. M. ARCHIBALD, "Removing Contaminants by Ground Water Recharge Basins." *Journal American Water Works Association,* 78, no. 9 (1986):76–81.

77. HANTUSH, M. S. "Growth and Decay of Ground-water Mounds in Response to Uniform Percolation." *Water Resources Research,* 3 (1967):227–34.

78. SINGH, R. "Prediction of Mound Geometry under Recharge Basins." *Water Resources Research,* 12 (1976):775–80.

79. BIANCHI, W. C., and HASKELL, E. E., JR. "Field Observations Compared with Dupuit-Forcheimer Theory for Mound Heights under a Recharge Basin." *Water Resources Research,* 4 (1968):1049–57.

80. HUNT, B. W. "Vertical Recharge of Unconfined Aquifer." *Journal of Hydraulics Division, American Society of Civil Engineers,* 97 (1971):1017–30.

81. BEHNKE, J. J. "Clogging in Surface Spreading Operations for Artificial Groundwater Recharge." *Water Resources Research,* 5 (1969):870–76.

82. MORAVCOUA, V., L. MASINOVA, and V. BERNATOVA. "Biological and Bacteriological Evaluation of Pilot Plant Artificial Recharge Experiments." *Water Research,* 2 (1968):265–76.

83. SNIEGOCKI, R. T., and R. F. BROWN. "Clogging in Recharge Wells, Causes and Cures." *Proceedings of the Artificial Ground-water Recharge Conference, The Water Resources Association* (England), 1970, pp. 337–57.

84. BRUINGTON, A. E. "Saltwater Intrusion into Aquifers." *Water Resources Bulletin,* 8 (1972):150–60.

85. BRUNINGTON, A. E., and F. D. SEARES. "Operating a Sea Water Barrier Project." *Journal, Irrigation and Drainage Division, American Society of Civil Engineers,* 91 (1965):117–40.

86. RALSTON, D. R. "Administration of Ground-water as Both a Renewable and Non-renewable Resource." *Water Resources Bulletin,* 9 (1973):908–17.

**87.** LLOYD, J. W., and M. H. FARAG. "Fossil Ground-water Gradients in Arid Regional Sedimentary Basins." *Ground Water,* 16 (1978):388–93.

**88.** SCHICHT, R. J., J. R. ADAMS, and J. B. STALL. *Water Resources Availability, Quality, and Cost in Northeastern Illinois.* Illinois State Water Survey Report of Investigation 93, 1976, 90 pp.

**89.** SCHICHT, R. J., and J. R. ADAMS. *Effects of Proposed 1980 and 1985 Lake Water Allocations in the Deep Sandstone Aquifer in Northeastern Illinois.* Illinois State Water Contract Report for Illinois Division of Water Resources, 1977.

**90.** PLOTKIN, S. E., H. GOLD, and I. L. WHITE. "Water and Energy in the Western Coal Lands." *Water Resources Bulletin,* 15 (1979):94–107.

**91.** BREDEHOEFT, J. D., and R. A. YOUNG. "The Temporal Allocation of Groundwater—A Simulation Approach." *Water Resources Research,* 6 (1970):3–21.

**92.** YOUNG, R. A., and J. D. BREDEHOEFT. "Digital Computer Simulation for Solving Management Problems of Conjunctive Ground-water and Surface-water Systems." *Water Resources Research,* 8 (1972):533–56.

**93.** BRUTSAERT, W. F., and T. G. GEBHARD, JR. "Conjunctive Availability of Surface and Ground-water in the Albuquerque Area, New Mexico: A Modeling Approach." *Ground Water,* 13 (1975):345–53.

**94.** MOREL-SEYTOUX, H. J. "A Simple Case of Conjunctive Surface–Ground-water Management." *Ground Water,* 13 (1975):506–15.

**95.** MADDOCK, T., III. "The Operation of a Stream-Aquifer System under Stochastic Demands." *Water Resources Research,* 10 (1974):1–10.

**96.** HAIMES, Y. Y., and Y. C. DREIZIN. "Management of Ground-water and Surface-water via Decomposition." *Water Resources Research,* 13 (1977):69–77.

**97.** FLORES, E. Z., A. L. GUTJAHR, and L. W. GELHAR. "A Stochastic Model of the Operation of a Stream-Aquifer System." *Water Resources Research,* 14 (1978):30–38.

**98.** TRELEASE, F. J. "New Water Laws for Old and New Countries." In *Contemporary Developments in Water Law,* ed. C. W. Johnson and S. H. Lewis. University of Texas Center for Research in Water Resources, Symposium No. 4, 1970, pp. 40–54.

**99.** ANDERSEN, R. L., and N. I. WENGERT. "Developing Competition for Water in the Urbanizing Areas of Colorado." *Water Resources Bulletin,* 13 (1977):769–73.

**100.** RADOSEVICH, G. E., and M. B. SABEY. "Water Rights, Eminent Domain, and the Public Trust." *Water Resources Bulletin,* 13 (1977):747–57.

**101.** THOMAS, J. L., and D. KLARICH. "Montana's Experience in Reserving Yellowstone River Water for Instream Beneficial Uses—Legal Framework." *Water Resources Bulletin,* 15 (1979):60–74.

**102.** THORSON, J. E. "Public Rights at the Headwaters." *Journal American Water Works Association,* 78, no. 10 (1986):72–79.

# TWELVE

# Field Methods

The rocks that form the crust of the earth are in few places, if anywhere, solid throughout. They contain numerous open spaces, called voids or interstices, and these spaces are the receptacles that hold the water that is found below the surface of the land and is recovered in part through springs and wells. There are many kinds of rocks, and they differ greatly in the number, size, shape and arrangement of their interstices and hence in their properties as containers of water. The occurrence of water in the rocks of any region is therefore determined by the character, distribution, and structure of the rocks it contains—that is by the geology of the region.

*The Occurrence of Ground Water in the United States,* Oscar Edward Meinzer, 1923

## 12.1 INTRODUCTION

The day is past when the only activity of the hydrogeologist was to locate and design a water well. Today, hydrogeologists are involved in many phases of resource management, including environmental impact analysis, as integral members of a multidisciplinary team. Hydrogeological studies are necessary and generally required by regulatory agencies for site studies prior to construction of such projects as sanitary landfills, land-treatment systems for wastewater, surface mines, power plants, artificial-recharge lagoons, nuclear-waste repositories, dams, and reservoirs.

In this chapter, we will introduce several techniques that can be applied both to the exploration for ground-water supplies and to various aspects of environmental hydrogeology. This includes the use of aerial photographs and other remote sensing data, as well as both surface and borehole geophysical methods. The methods of site evaluation will be examined.

## 12.2 FRACTURE-TRACE ANALYSIS

One technique that has been gaining acceptance among hydrogeologists is the use of **fracture-trace analysis.** As we discussed in Chapter 8, ground water is known to be concentrated in fracture zones found in many different rock types. The fracture traces are located by study of linear features on aerial or satellite photographs. On air photos, natural linear features consist of tonal variation in soils, alignment of vegetative patterns, straight stream segments or valleys, aligned surface depressions, gaps in ridges, or other features showing a linear orientation (1). Some linear features may be visible on the ground; for instance, surface sags or straight stream segments. Others, such as variation in soil tone, or alignment or height of vegetation of a certain type, may not be noticeable except on aerial photographs (1). Many of these natural linear features consist of interrupted segments, which may be of different types. For instance, a straight stream segment in a floodplain may align with a row of trees in a nearby woods. Natural linear features from 1000 feet (300 meters) to around 4300 feet (1500 meters) in length are fracture traces. Those greater than 1000 feet are termed **lineaments** (1). Some lineaments are up to 90 miles (150 kilometers) long (2).

Fracture traces are surface expressions of joints, zones of joint concentration, or faults (3). It is generally believed that the joint sets tend to be nearly perpendicular (2). They are known to extend to a depth of 3300 feet (1000 meters) at one Arizona location (3), although this is probably far deeper than is typical. Mapped fracture traces have been traced to cliffs where the fracture zone can be seen in cross section (2, 3). Under these observed conditions, the fracture zones dip at approximately 87 to 89 degrees. Figure 12.1 shows an exposure of a fracture in a cliff in central Pennsylvania.

These fracture zones are less resistant to erosion than rock, which is less fractured. Hence, valley and stream segments tend to run along them. They may be zones of ground-water drainage, so that soils over them have a deeper water table or are not as moist as soils in surrounding areas. The soil color or vegetation may appear to be different from that of surrounding soils. If they are zones of concentrated ground-water discharge, there may be a line of springs or seeps. Fracture traces in carbonate rocks are typically areas of solution. Aligned sinkholes or surface sags are typical surface expressions.

Fracture traces may be related to regional tectonic activity. They tend to be oriented at a constant angle to the regional structural trend; however, the orientation appears to be independent of local folds (3). Lineaments are known to cut across rocks of many ages and cross folds and faults (2). They have been observed to be parallel to the major joint sets in flat-lying or gently dipping strata, but this is not the case if the strata are steeply dipping. If surface areas are separated by major faults, the individual fault blocks may have fracture traces of different orientation (2). The majority of fracture traces in an area appear to be grouped into two subparallel sets that are approximately perpendicular. Streams developed in rocks where fracture control is evident have been

**FIGURE 12.1** Cross section of a zone of fracture concentration revealed by a fracture trace near Spring Creek in Centre County, Pennsylvania. Photo courtesy of R. R. Parizek.

described as having a "stair-step" pattern (4). In the area of central Pennsylvania shown in Figure 12.2, the valley development follows fracture traces. Most fractures are generally N-S or E-W, with lesser numbers running NW-SE and SW-NE.

Statistical studies of wells in carbonate terrane have shown that those located on fracture traces, either intentionally or accidentally, have a greater yield than those not on fracture traces (5). Figure 12.3 illustrates that the productivity of fracture-trace wells is significantly above that of other wells not on fracture traces. The use of fracture-trace analysis has resulted in a very successful location of well yields in areas where random location has yielded very erratic results (6). The greatest yields come from wells located at the intersection of two fracture traces. Caliper logs of wells on fracture traces in carbonate-rock terrane showed many more cavernous openings and enlarged bedding planes than logs of those wells drilled in interfracture areas (Figure 12.4).

Many hydrogeologists are successfully using fracture-trace analysis to locate high-yield wells. The technique has been applied to carbonate-rock terrane (6) but is also applicable to most other rock types (2). It is reportedly usable even if the bedrock is mantled by up to 170 feet (50 meters) of glacial drift (7). Fracture-trace analysis is also widely used in selecting sites for sanitary landfills. Naturally, the landfill locations are most suitable if they fall in interfracture areas. Other uses include analysis of foundation and dam sites, evaluation of potential water problems in mines and tunnels, and control of mine drainage (2).

Fracture-trace analysis is also very useful in determining the locations

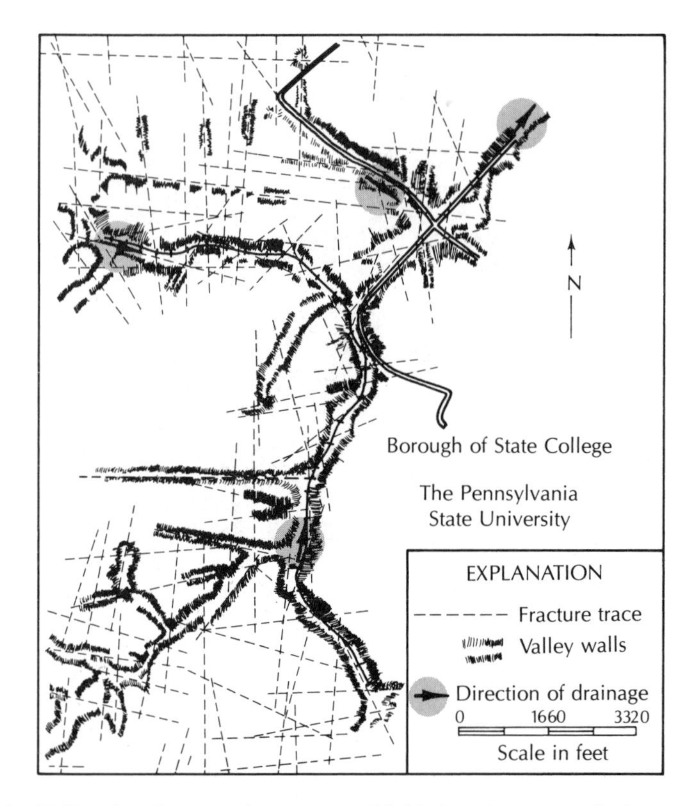


**FIGURE 12.2** Valley development in an area of folded carbonate rocks in central Pennsylvania. The valleys tend to follow fracture traces. Source: R. R. Parizek, *Hydrogeologic Framework of Folded and Faulted Carbonates—Influence of Structure,* Mineral Conservation Series Circular 82, College of Earth and Mineral Sciences, Pennsylvania State University, 1971, pp. 28–65.

of ground-water monitoring wells. Because ground-water flow preferentially follows the most permeable pathway, monitoring wells should be located on fracture traces. For example, if a hazardous-waste storage lagoon is located in an area of fractured bedrock, at least one of the downgradient monitoring wells required under the Resource Conservation and Recovery Act should be located on a fracture trace.

In the identification of fracture traces on aerial photographs, a low-magnification stereoscope is generally used (1). Possible fracture traces are indicated by drawing directly on the photograph. One problem in identification is the confusion of linear features of human origin (fences, cowpaths, roads, power lines, plow and harvest patterns, etc.) with natural linear features. These is also a tendency to map fracture traces at oblique angles to regular grid systems on the photographs. As section lines almost always appear on air photos, especially in cultivated areas, there is a tendency to preferentially map NW-SE and NE-SW

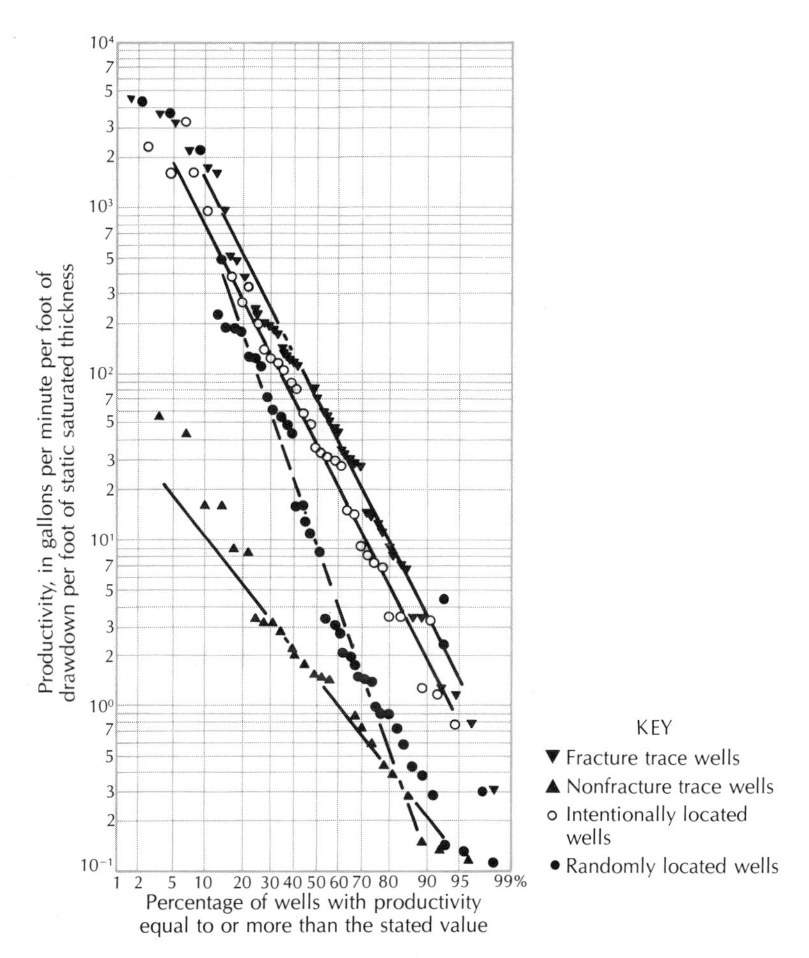


**FIGURE 12.3** Production-frequency graph for water wells grouped according to whether or not they fall on a fracture trace. Source: S. H. Siddiqui and R. R. Parizek, *Water Resources Research,* 7 (1971):1295–1312.

features as fracture traces. Following the stereoscopic mapping, the photos should be checked without use of the stereoscope to see if any other features are noticed. A typical air-photo scale for fracture-trace analysis is 1:20,000. Linear features that show up in more than one expression, and those crossing roads or fields, are more likely to represent fracture traces. In Figure 12.5, subtle fracture traces are indicated in an area of farmland in central Pennsylvania.

Following the mapping of linear features on air photos, it is necessary to make a field check. Some mapped features will usually turn out to be due to human activity. The more inexperienced the geologist, the more likely it is that this will occur. If a suspected fracture trace has a surface expression, it will be easier to locate in the field. Those without obvious surface expression must be

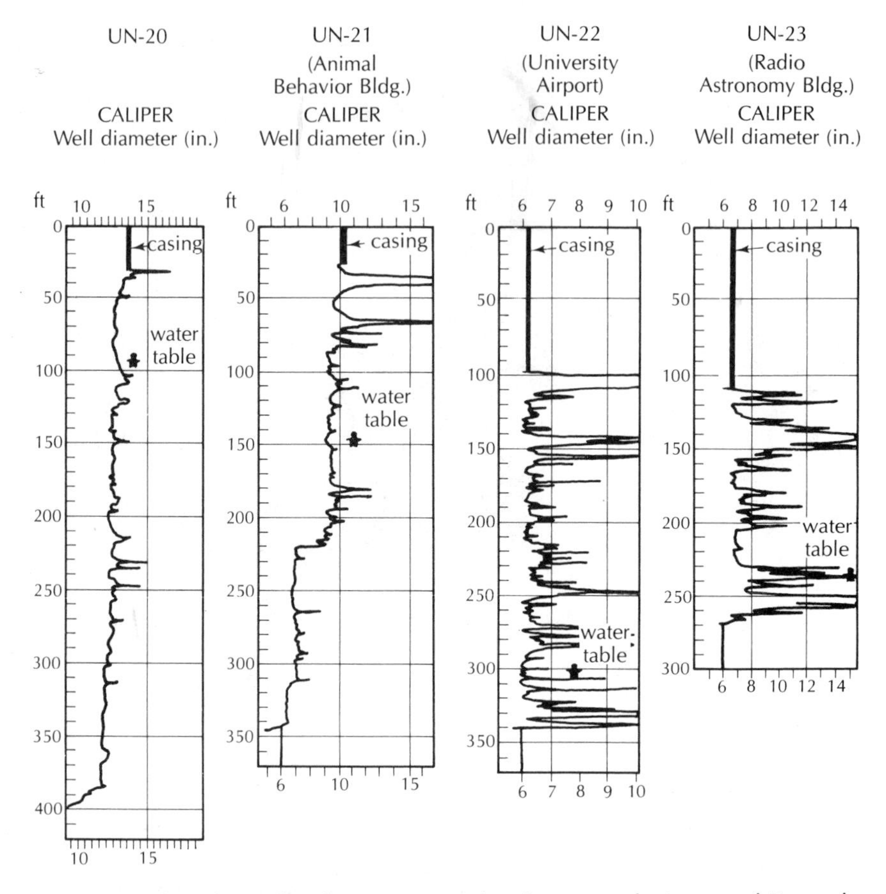


**FIGURE 12.4** Caliper logs of wells in an area of carbonate rocks in central Pennsylvania. Wells UN-20 and UN-21 were drilled in interfracture areas; Wells UN-22 and UN-23 were located on fracture traces. Source: L. H. Lattman and R. R. Parizek, *Journal of Hydrology* (Elsevier Scientific Publishing Company) 2 (1964):73–91. Used with permission.

located by virtue of their spatial relation to individual trees or buildings that are visible on the photographs and can be identified on the ground. In urbanizing areas, it may be possible to use older photographs taken before extensive development to map fracture traces. This makes the field location of the fracture traces even more difficult.

## 12.3 SURFICIAL METHODS OF GEOPHYSICAL INVESTIGATIONS

Geophysical surveys have been used by the mining and petroleum industries for many decades. Ground-water geologists soon discovered the usefulness of these

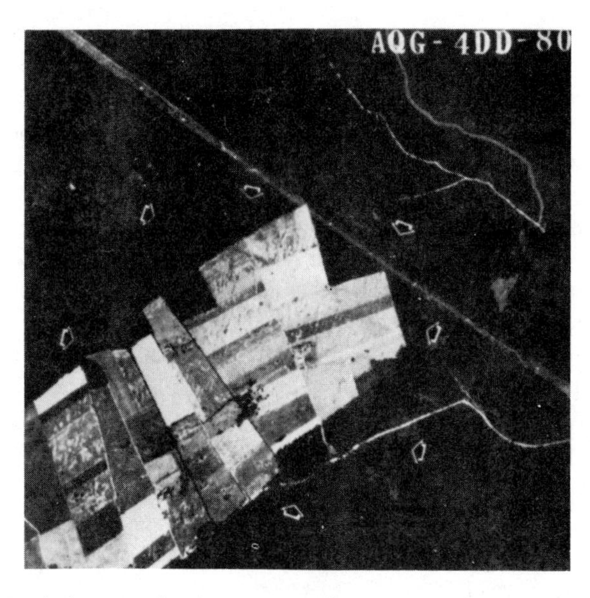


**FIGURE 12.5** Aerial photograph of an area with 16 to 260 feet (5 to 80 meters) of transported sediments overlying folded and faulted dolomite, limestone, and sandy dolomite, Centre County, Pennsylvania. The line of each fracture trace is indicated by arrows at both ends. Photo courtesy of R. R. Parizek.

surveys in exploring the shallow subsurface (within a few hundred feet), where ground-water supplies are usually found (8–10). A number of different techniques are used, the most common of which are direct-current resistivity, seismic refraction, and gravity and magnetics methods. Seismic reflection is less widely used, although it is the preferred method in petroleum exploration.

Geophysical methods may be used to determine indirectly the extent and nature of the geologic materials beneath the surface. The thickness of unconsolidated surficial materials, the depth to the water table, the location of subsurface faults, and the depth of the basement rocks can all be determined. In some instances, the location, thickness, and extent of subsurface bodies, such as gravel deposits or clay layers, can also be evaluated. The correlation of geophysical data with well logs or test-boring data is generally more reliable than either type of information used by itself. As with all hydrogeological investigations, a careful definition of the problem and determination of the best type of information to solve the problem should be made before geophysical work is done. The geophysical survey should then be planned to yield the greatest amount of useful data for the budgeted cost.

### 12.3.1 DIRECT-CURRENT ELECTRICAL RESISTIVITY

Of the several electrical geophysical methods, **direct-current electrical resistivity** has found the greatest application to hydrogeology (11). A commutated direct current or a current of very low frequency (less than 1 cycle per second) is generated in the field or provided by storage batteries. It is introduced into the ground by means of two metal electrodes. If the soil is dry, water may be needed around the electrodes to establish a good connection. The voltage in the ground is measured between two other metal electrodes, also driven into the ground. By knowing that current flowing through the ground and the potential differences or voltage between two electrodes, it is possible to compute the resistivity of the earth materials between the electrodes. The resistivity of earth materials varies widely, from $10^{-6}$ ohm-meter for graphite to $10^{12}$ ohm-meters for quartzite. Dry materials have a higher resistivity than similar wet materials, as moisture increases the ability to conduct electricity. Gravel has a higher resistivity than silt or clay under similar moisture conditions, as the electrically charged surfaces of the fine particles are better conductors.

Electrical resistivity, $R$, is equal to the expression

$$R = \frac{A}{L}\frac{\Delta V}{I} \tag{12-1}$$

where

$A$ is the cross-sectional area of current flow
$L$ is the length of the flow path
$\Delta V$ is the voltage drop
$I$ is the electrical current

Electrical resistivity is measured in units of ohm-meters or ohm-feet. The four electrodes used can be designated as follows:

$A$ is the positive-current electrode
$B$ is the negative-current electrode
$\left.\begin{matrix} M \\ N \end{matrix}\right\}$ are the potential electrodes

If $\overline{XY}$ indicates the distance between Electrode $X$ and Electrode $Y$, Equation 121 can be expressed as (11)

$$\overline{R} = \left( \frac{2\pi}{\dfrac{1}{\overline{AM}} - \dfrac{1}{\overline{BM}} - \dfrac{1}{\overline{AN}} + \dfrac{1}{\overline{BN}}} \right) \frac{\Delta V}{I} \tag{12-2}$$

As earth materials are almost never homogeneous and electrically isotropic, the resistivity found by Equation 12-2 is an apparent resistivity, $\overline{R}$.

There are several electrode configurations in common usage. The **Wenner array** consists of the four electrodes spaced equal distances apart in a straight line: $\overline{AM} = \overline{MN} = \overline{NB} = a$. A current electrode is on each end (Figure

12.6A). In using the Wenner array, the apparent resistivity, $\overline{R}$, may be found from the expression

$$\overline{R} = 2\pi a \frac{\Delta V}{I} \tag{12-3}$$

which is solved from Equation 12-2.

A second configuration is the **Schlumberger array.** It is a linear array, with potential electrodes placed close together (Figure 12.6B). Typically, $\overline{AB}$ is set equal to or greater than five times the value of $\overline{MN}$. The apparent resistivity is given by

$$\overline{R} = \pi \frac{(\overline{AB}/2)^2 - (\overline{MN}/2)^2}{\overline{MN}} \frac{\Delta V}{I} \tag{12-4}$$

The **dipole-dipole** array is particularly convenient for making electrical **soundings,** which measure the changes in electrical properties with depth. The dipole-dipole configuration has a pair of current electrodes separated from a pair of potential electrodes. The same spacing, $a$, is used between the current electrodes as between the potential electrodes, and the distance between the electrode pairs, $na$, which is a multiple, $n$, of $a$, is much greater than the electrode spacing (Figure 12.6C). The apparent resistivity for the dipole-dipole array is given by

$$\overline{R} = n(n + 1)(n + 2)\, a \frac{\Delta V}{I} \tag{12-5}$$

Geophysical instruments are available to measure the value of $\Delta V$ for a known $I$. The appropriate formula for the electrode array is used to compute the apparent resistivity.

Resistivity surveys are made in two fashions. An **electrical sounding** will reveal the variations of apparent resistivity with depth. **Horizontal profiling** is used to determine lateral variations in resistivity. When the electrode spacing is expanded in making an electrical sounding, the distance between the potential electrodes and the current electrodes increases. This means that the current will travel progressively deeper through the ground and will measure apparent resistivity to greater depths. Either the Wenner or the Schlumberger array may be used; however, the latter is more convenient for electrical sounding. This is because, for each incremental measurement, only the outer current electrodes must be moved every time. The inner electrodes are spread only occasionally. In the Wenner array, all four electrodes must be moved for each incremental measurement. The sounding is begun with the electrodes close together. After each reading, the electrodes are repositioned with $a$, or $\overline{AB}/2$, increased and a new measurement made. The apparent resistivity is plotted on logarithmic paper as a function of electrode spacing. For a number of reasons (11), the Schlum-

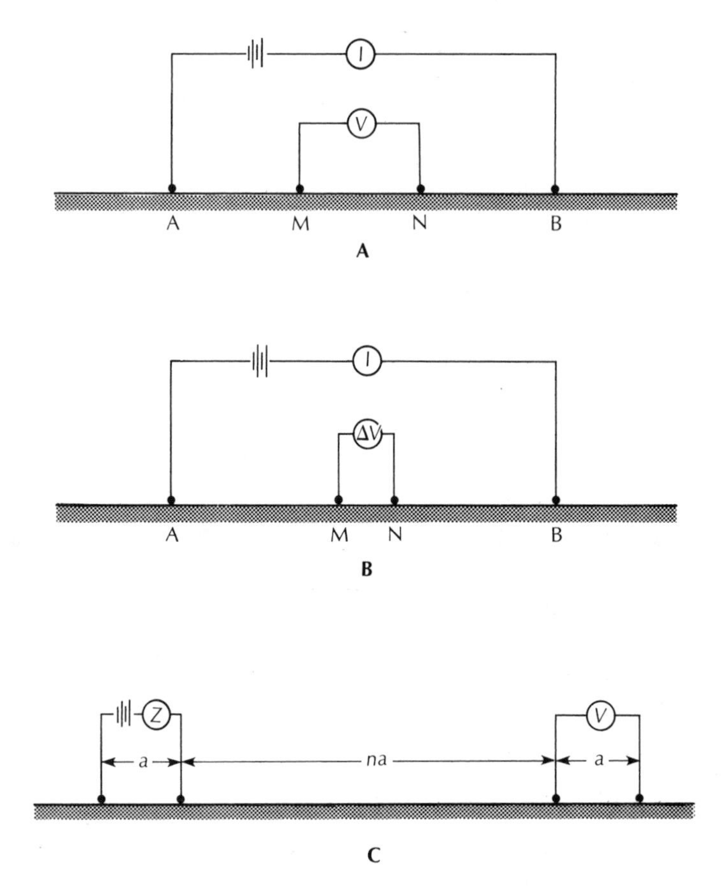


FIGURE 12.6 **A.** Wenner electrode array. **B.** Schlumberger electrode array. **C.** Dipole-dipole electrode array.

berger array is superior to the Wenner array for electrical soundings. There is, however, a set of theoretical type curves of Wenner apparent resistivity for two-, three-, and four-layer earth models (12). This could be helpful in interpreting the results of a Wenner-array electrical sounding.

For a homogeneous earth, there is a definite relationship between the electrode spacing and the percent of the current that penetrates to a given depth (13). For a nonhomogeneous and layered earth, the exact relationship cannot be easily determined. It is safe to assume that the greater the electrode spacing, the deeper the stratum influencing the apparent-resistivity curve. There are a number of possible earth models that could produce a given curve. In Figure 12.7, there are three possible theoretical interpretations expressed as resistivity and a test-boring log. The rise in apparent resistivity indicates a shallow zone of high resistivity. The test boring shows this to be a layer of silty gravel and boulders from 5 to 23 feet. It should be noted that the apparent-resistivity curve peaks at

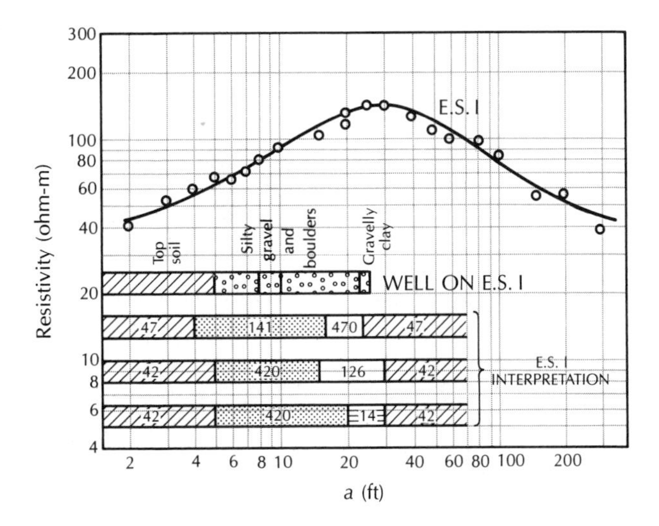


**FIGURE 12.7** Wenner electrical-sounding curve of apparent resistivity as a function of electrode spacing. Three possible interpretations and a test boring are also given. Source: A. A. R. Zohdy, *Ground Water*, 3, no. 3 (1965):41–48.

30 feet. An interpretation that the layer of maximum resistivity lay at 30 feet would have been wrong.

In horizontal profiling, the electrode spacing is kept at a constant value. The electrodes are moved in a grid pattern over the land surface. The apparent resistivity of each point on the grid is marked on a map and isoresistivity contours are drawn. Figure 12.8 shows an apparent-resistivity map based on a large number of horizontal resistivity measurements. An area of buried stream-channel gravels is delineated where the apparent resistivity exceeds 80 ohm-meters (13). A geologic cross section based on test borings and electrical resistivity soundings is shown in Figure 12.9. The location of the cross section is indicated in Figure 12.8 as line *AB*.

Geoelectrical methods are useful in ground-water studies for such purposes as defining buried stream channels and areas of saline versus fresh ground water. Saline water has a much lower resistivity, as it is a better electrical conductor. Interpretation for such cases is relatively simple and may be done qualitatively. Layers of very low resistivity, such as clay, can also be discerned on sounding curves. It is often impossible to pick out the water table on an electrical sounding (11), although it is frequently attempted.

Electrical resistivity methods have been applied to many ground-water situations where the resistance of the fluid in the ground varies. For example, this occurs where there is a plume of saline ground water. Such a plume may be the result of salt-water intrusion, saline water seeping from a brine pit, or leachate from a landfill. The dissolved solids in the ground water can conduct electricity more readily and thus will have a lower apparent resistivity. A map of the

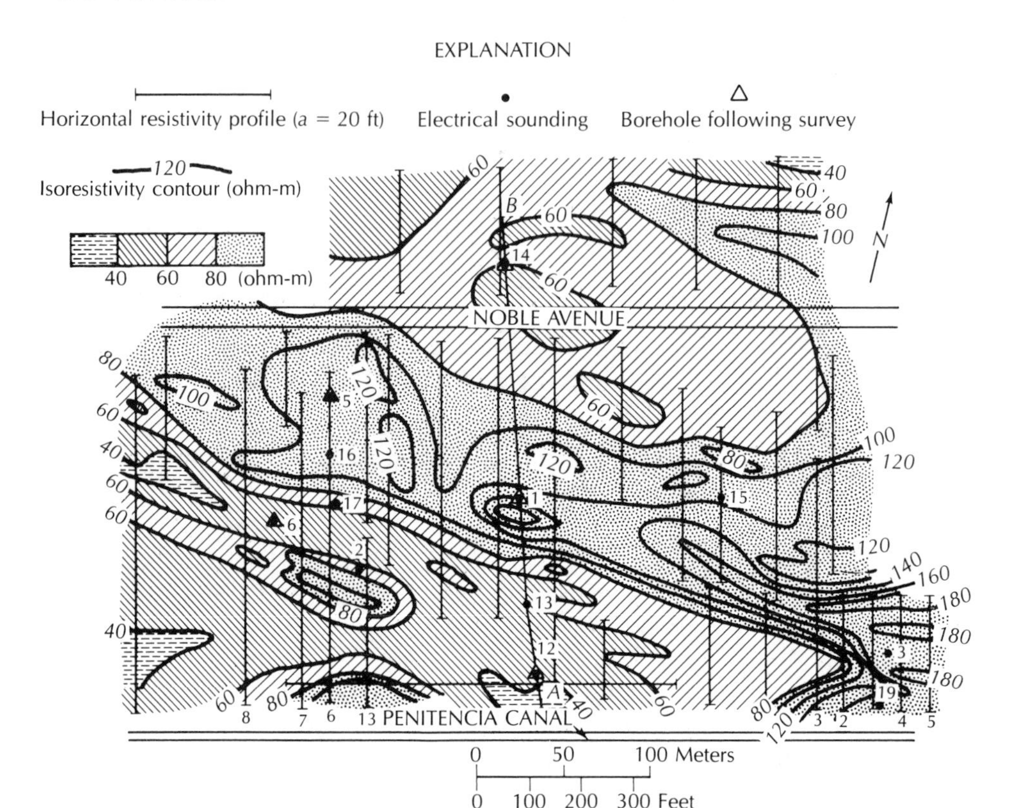


**FIGURE 12.8** Apparent-resistivity map of Penitencia, California. Locations of resistivity profiles, soundings, and boreholes are shown. The location of Borehole and Electrical Sounding 1 is in the center of the high-resistivity zone. Source: A. A. R. Zohdy, *Ground Water,* 3, no. 3 (1965):41–48.

apparent resistivity created by horizontal profiling can often show areas of contaminated ground water (14–18). Figure 12.10 shows the apparent resistivities around oil field brine ponds in Illinois. The data were obtained by horizontal profiling using a Wenner electrode array of 20 feet (19). The areas of low apparent resistivity indicate where the ground water contains higher levels of dissolved solids. These areas occur around both an active holding pond and abandoned ponds. There is a ground-water mound beneath the active brine pond, indicating that seepage from it is the cause of part of the ground-water contamination. However, the low apparent resistivity in the vicinity of the abandoned ponds indicates that there is also residual ground-water contamination.

### 12.3.2 ELECTROMAGNETIC CONDUCTIVITY

Electricity traveling through sedimentary units is transmitted more readily by the fluids in the pore spaces and by the fluid–grain interfaces than through

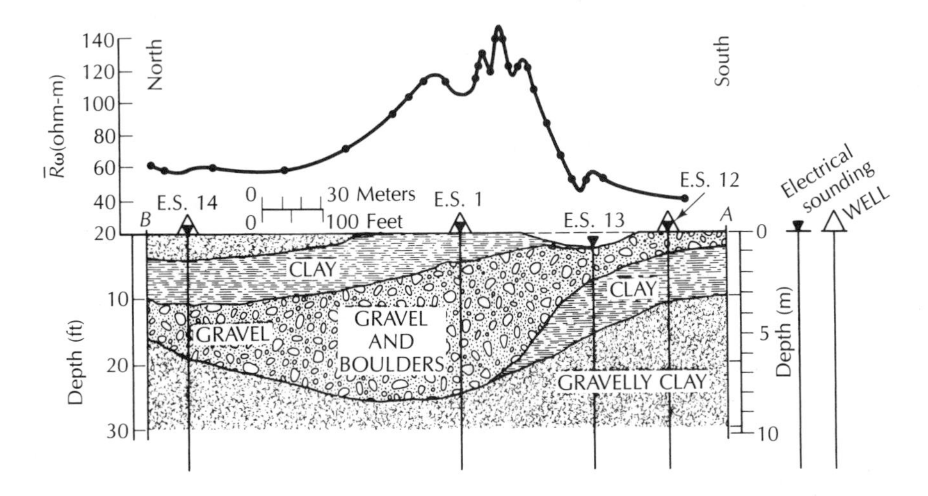


**FIGURE 12.9** Geologic cross section based on test borings and electrical soundings. The position of the cross section is shown as line *AB* on Figure 12.8. Source: A. A. R. Zohdy, *Ground Water,* 3, no. 3 (1965):41–48.

the mineral grains themselves. As a result, the pore surface area and the pore fluid conductivity are important factors in determining overall bulk conductivity of earth units (20). Electromagnetic conductivity is the inverse of electrical resistivity. Field studies have shown similar results when both methods have been used at the same site (21).

Electromagnetic methods use an electromagnetic field generated by a transmitter coil through which an alternating current is passed. This generates a magnetic field around the transmitter coil. When the transmitter coil is held near the earth, the magnetic field induces an electrical field in the earth. The electrical field will travel through the ground at different strengths depending upon the ground conductivity. The field strength is measured in a passive receiver coil. Changes in the phase, amplitude, and orientation of the primary field can be measured either with time or distance by using the receiver. These changes are related to the electrical properties of the earth.

There are several different electromagnetic methods available. They all have the advantage of being rapid, as none of them require the insertion of electrodes into the ground. Electromagnetic methods are not inherently more accurate than direct resistivity methods, but are likely to be more cost-effective because field work can often be completed more rapidly. They can be used to detect changes in earth conductivity related to contaminant plumes, buried metallic waste such as drums, or salt-water interfaces (20–23).

Geonics, Ltd., of Mississisauga, Ontario, Canada, is the manufacturer of a line of electromagnetic equipment. This is not an endorsement of Geonics, Ltd. products; the brand name is given because their products are generally referenced by model number in the hydrogeological literature.

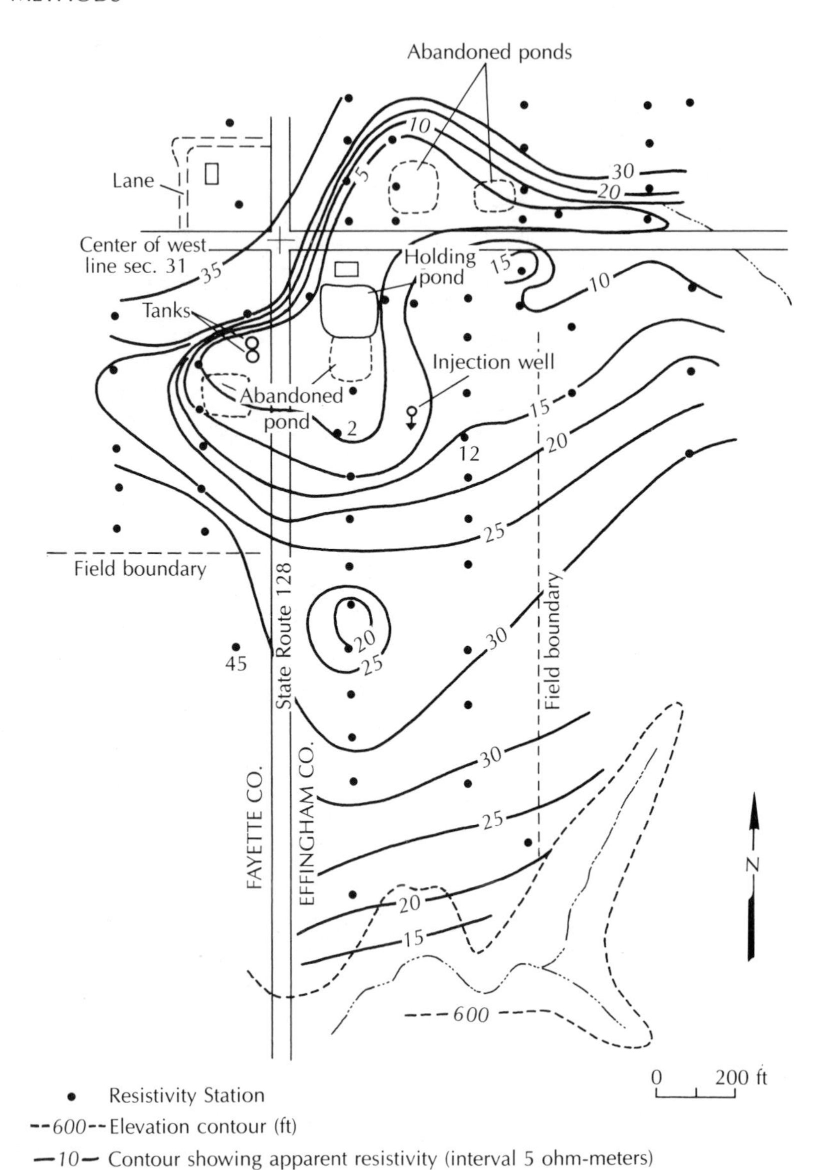


**FIGURE 12.10** Apparent resistivity from horizontal profiling with a 20-foot spacing in the vicinity of oil-field brine holding ponds in Illinois. Source: Modified from P. C. Reed, K. Cartwright, and D. Osby, Illinois State Geological Survey Environmental Geology Note 95, 1981, p. 17.

The Geonics EM-31 has the transmitter and receiver coils in the same unit. The coils are mounted on a long pole so that they have a fixed separation distance of 12 feet (3.66 meters). The unit can be used by a single operator, who can walk along a line and note the meter readings at stations every 10 feet (3 meters) or so. The output from the instruments is apparent conductivity in millimhos per meter. It can also be read continuously, allowing small-scale heterogeneities to be determined with greater precision than practical with electrical resistivity methods, where the electrodes must be moved for each separate reading. As the distance between the transmitter and the receiver cannot be varied, the depth of penetration of the electrical field is constant and relatively shallow, about 20 feet (6 meters).

The Geonics EM-34-3 has separate units for the transmitter and the receiver coils. Two operators are needed, one for each coil. The coils can be held either horizontally or vertically. They are separated by a distance, $L$. With the coils oriented horizontally, the effective depth of penetration is about 0.75 $L$. If the coils are oriented vertically, the effective depth of penetration is about 1.5 $L$ and the readings are less influenced by the near-surface layers. The EM-34-3 can be operated at three different intercoil spacings: 32.8, 65.6, and 131.1 feet (10, 20, and 40 meters). The EM-34-3 unit can be used to study earth conductivity to much greater depths than the EM-31 as the spacing between transmitter and receiver can be much greater than the fixed 12 feet of the EM-31. Figure 12.11 illustrates the use of both the EM-31 and EM-34-3 in terrain conductivity mapping at a sanitary landfill site. The area was mapped by both electrical resistivity and electromagnetic conductivity using the Wenner array with a 20-meter electrode spacing (Figure 12.11A), the EM-31 (Figure 12.11B), the EM-34-3 with a 15-meter vertical coplaner coil spacing (Figure 12.11C), and the EM-34-3 with a 31-meter vertical coplaner spacing (Figure 12.11D). The earth conductivity was converted to resistivity, in ohm-meters, by inverting and then dividing by 1000. The term **inductive resistivity** refers to conductivity converted to resistivity in this manner. Note that the terrain conductivity surveys took about one-sixth of the person days that were required for the resistivity survey.

Terrain conductivity surveys can be distorted by conductors such as buried pipelines and steel tanks. They can also be adversely impacted by high-tension power lines and electrical storms.

### 12.3.3 SEISMIC METHODS

Seismic methods using artificially created seismic waves traveling through the ground are quite commonly employed in hydrogeology. These methods are useful in determining depth to bedrock, slope of the bedrock, depth to water table and, in some cases, the general lithology. Applied seismology has been highly developed in the petroleum industry, where the **seismic reflection method** is used almost exclusively. Structural and formational boundaries can be indicated to great depth.

Hydrogeological studies often involve finding the thickness of unconsol-

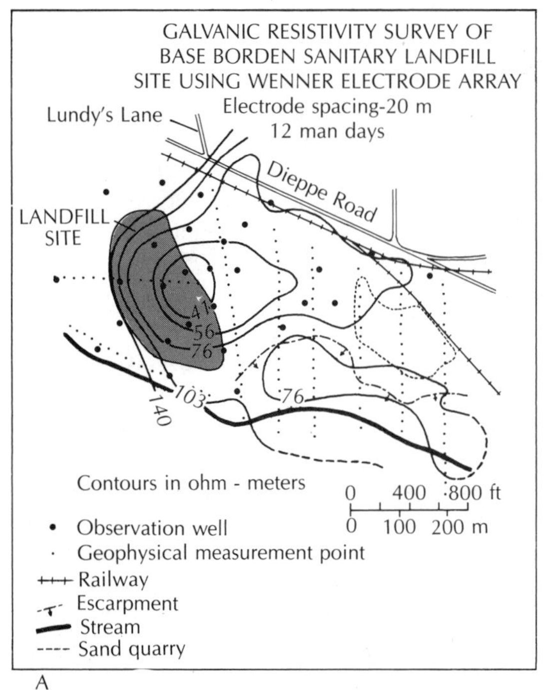


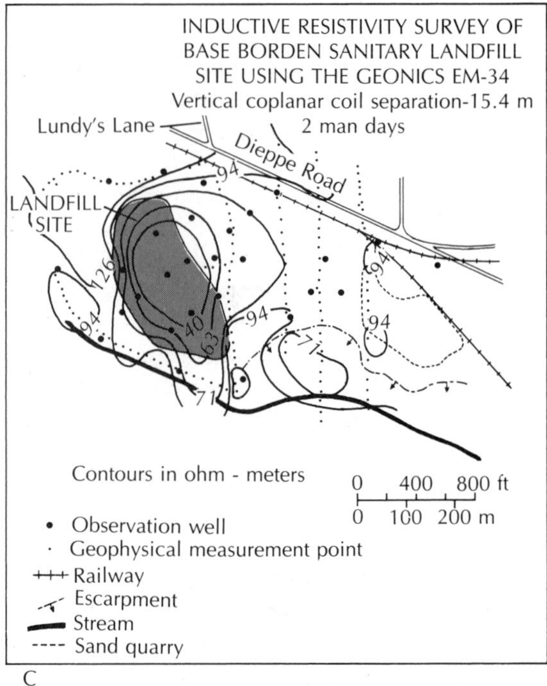


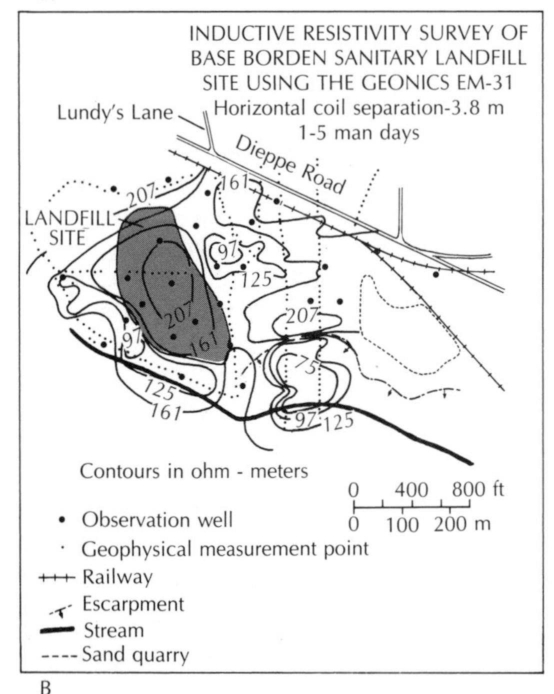


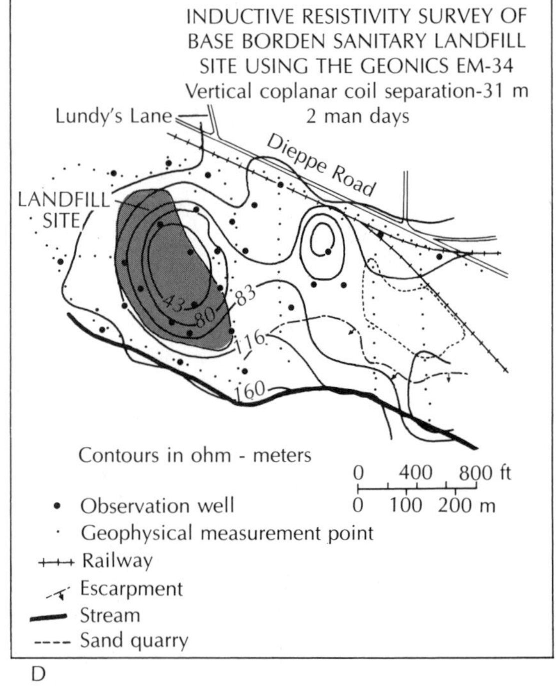


**FIGURE 12.11** Comparison of electrical resistivity and electrical conductivity measurements at a sanitary landfill at Camp Bordon, Ontario. The low-resistivity area represents a plume of contaminated ground water with high total dissolved solids. The plume shows up more clearly on the surveys with greater depth of penetration (Wenner array and EM-34-3). Source: Geonics Ltd. Used with permission.

idated material overlying bedrock. For this purpose, the **seismic refraction method** is superior. The loose material transmits seismic waves more slowly than consolidated bedrock. By studying the arrival times of seismic waves at various distances from the energy source, the depth to bedrock can be determined.

The energy source can be a small explosive charge set in a shallow drill hole. One or two sticks of dynamite is sufficient for depths to bedrock in excess of 100 to 200 feet (30 to 50 meters). Of course, explosives should be handled only by persons trained and licensed to do so. A judgment of how large a charge to use must be made in each case. For shallower work, 15 to 50 feet (5 to 15 meters), a sledgehammer struck on a steel plate lying on the ground may be a sufficient energy source. The seismic wave is detected by geophones placed in the earth in a line extending away from the energy source. A seismograph records the travel time for the wave to go from the energy source to the geophone. The more sophisticated seismographs are multichannel units with a number of geophones attached.

Figure 12.12 illustrates the travel paths of compressive seismic waves traveling through a two-layer earth. The seismic velocity in the lower layer is greater than that in the upper layer. As the energy travels faster in the lower layer, the wave passing through it gets ahead of the wave in the upper layer. At the boundary between the two layers, part of the energy is refracted back upward from the lower-layer boundary to the surface.

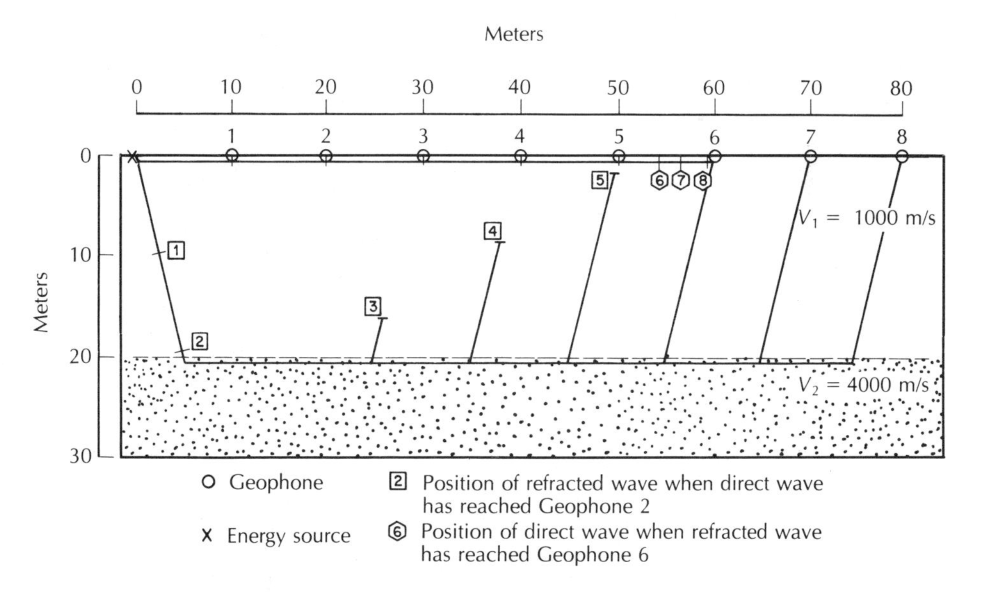


**FIGURE 12.12** Travel paths of a refracted seismic wave and a direct wave. The direct wave will reach the first five geophones first, but for the more distant geophones the first arrival is from a refracted wave. Numbers inside symbols refer to distances traveled by wave paths going toward the indicated geophone.

The angle of refraction of each wave front is called the **critical angle,** $i_c$, and is equal to the arc sin of the ratio of the velocities of the two layers:

$$i_c = \sin^{-1} \frac{V_1}{V_2} \qquad \textbf{(12-6)}$$

Figure 12.13 illustrates a wave front and the path of the refracted energy that travels along the lower-layer boundary. A direct wave in the upper layer is also shown.

**EXAMPLE PROBLEM** Determine the critical angle, $i_c$, when $V_1 = 1000$ meters per second and $V_2 = 4000$ meters per second.

$$i_c = \sin^{-1} V_1/V_2 = \text{arc sin } 0.25 = 14.5°$$

If $V_2$ is less than $V_1$, the wave will be refracted downward and no energy will be directed upward. Thus, the refraction method will show higher-velocity layers but not lower-velocity layers that are overlain by a high-velocity layer.

Energy can travel directly through the upper layer from the source to the geophone. This is the shortest distance, but the waves do not travel as fast as those traveling along the top of the lower layer. The latter must go farther, but they do so with a higher velocity. In Figure 12.12, the positions of waves traveling to each geophone are indicated. Geophones 1 through 5 first receive waves that have traveled through only the upper layer. The sixth and succeeding geophones measure arrival times of refracted waves that have gone through the

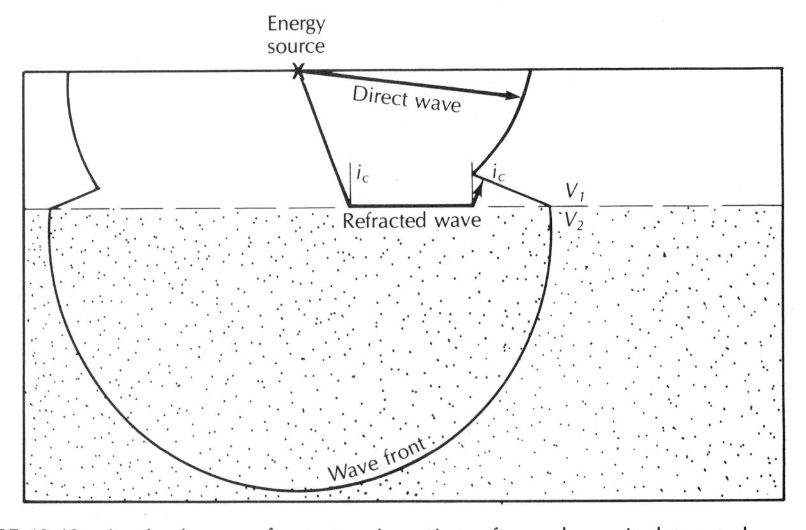


**FIGURE 12.13** A seismic wave front at a given time after a charge is detonated.

high-velocity layer as well. The position of the trailing wave front at each time the leading front reaches each geophone is indicated in the figure.

A graph is made of the arrival time of the first wave to reach the geophone versus the distance from the energy source to the geophone. This is known as a **travel-time** or **time-distance curve.** Figure 12.14 shows the time-distance curve for the shot in Figure 12.12. The reciprocal of the slope of each straight-line segment is the apparent velocity in the layer through which the first arriving wave passed. The slope of the first segment is 10 milliseconds per 10 meters, so that the reciprocal is 10 meters per 10 milliseconds, or 1000 meters per second.

The projection of the second line segment backward to the time-axis ($X = 0$) yields a value known as the **intercept time,** $T_i$. This value can be determined graphically, as shown in Figure 12.14. $T_i$ is 39 milliseconds and $X$ is 52 meters. The depth to the lower layer, $Z$, is found from the equation (24)

$$Z = \frac{T_i}{2} \frac{V_1 V_2}{\sqrt{V_2^2 - V_1^2}} \tag{12-7}$$

The depth to the lower layer can also be found from the equation (24)

$$Z = \frac{X}{2} \sqrt{\frac{V_2 - V_1}{V_1 + V_2}} \tag{12-8}$$

where $X$ is the distance from the shot to the point at which the direct wave and the refracted wave arrive simultaneously. This is shown on Figure 12.14 as the $x$-axis distance where the two line segments cross.

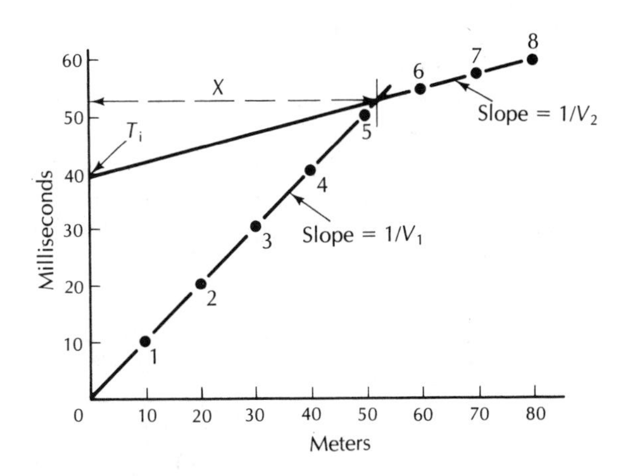


**FIGURE 12.14** Arrival time-distance diagram for a two-layered seismic problem. Numbers refer to geophones in Figure 12.12.

**EXAMPLE PROBLEM** Find the value $Z$ from Figure 12.14.

From the slope of each line segment, $V_1 = 1000$ meters per second and $V_2 = 4000$ meters per second. $T_i$ is 39 milliseconds and $X$ is 52 meters.

$$Z = \frac{T_i}{2} \frac{V_1 V_2}{\sqrt{V_2^2 - V_1^2}}$$

$$= \frac{0.039}{2} \times \frac{1000 \times 4000}{\sqrt{4000^2 - 1000^2}}$$

$$= 20 \text{ m}$$

Also,

$$Z = \frac{X}{2} \sqrt{\frac{V_2 - V_1}{V_1 + V_2}}$$

$$= \frac{52}{2} \sqrt{\frac{4000 - 1000}{4000 + 1000}}$$

$$= 20 \text{ m}$$

---

A more typical case in hydrogeology is a three-layer earth, the top layer being unsaturated, unconsolidated material. In the next layer, below the water table, the unconsolidated deposits are saturated, which yields a higher seismic velocity. The third layer is then bedrock. Under such conditions, the seismic method can be used to find the water table. However, similar velocities are possible from either saturated sand or unsaturated glacial till. Similar seismic refraction patterns could be obtained from a water table in a uniform sand deposit or a layer of unsaturated sand overlying unsaturated glacial till. This illustrates the point that geophysics is best interpreted in light of other data.

The three-layer seismic refraction case with $V_1 < V_2 < V_3$ is shown in Figure 12.15. The first arriving waves show three line segments. The reciprocal of the slope of each line is the seismic velocity of the respective layers. The intercept time for each of the two deeper layers is the projection of the line segment back to the time-axis. Indicated on the figure is the distance, $X_1$, from the shot to the point at which waves from Layers 1 and 2 arrive simultaneously and the distance, $X_2$, to the point at which waves from Layers 2 and 3 arrive simultaneously. The depth, $Z_1$, of Layer 1 is found from the values of $V_I$ and $V_2$ and either $T_{i1}$ or $X_1$ using Equation 12-7 or 12-8. The thickness of the second layer, $Z_2$, is found using (24)

$$Z_2 = \frac{1}{2}\left(T_{i2} - 2Z_1 \frac{\sqrt{V_3^2 - V_1^2}}{V_3 V_1}\right)\left(\frac{V_2 V_3}{\sqrt{V_3^2 - V_2^2}}\right) \quad \textbf{(12-9)}$$

The value of $Z_1$ must be computed before computing the value of $Z_2$.

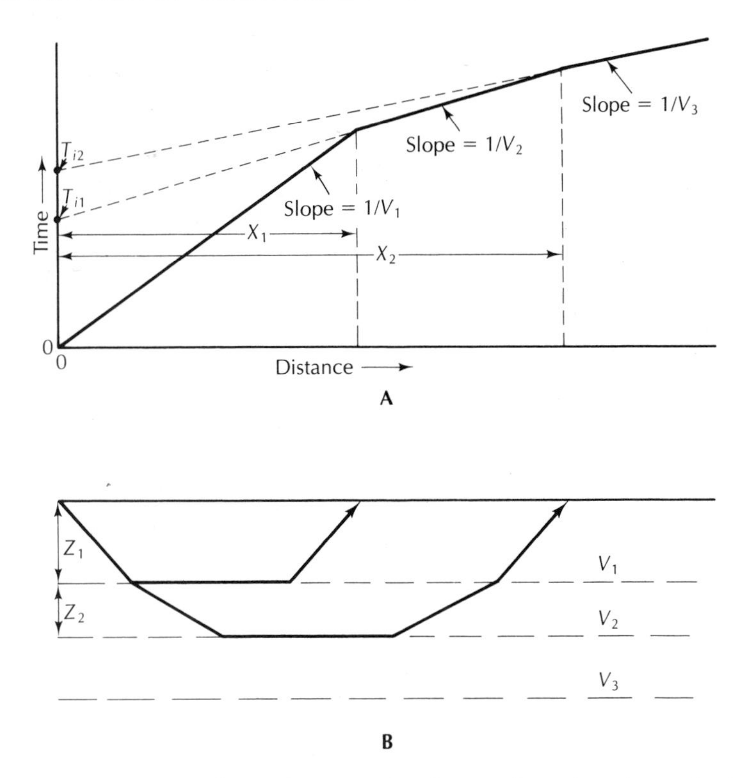


FIGURE 12.15 **A.** Diagram of arrival time versus distance for a three-layered seismic problem. **B.** Wave path for a three-layered seismic problem.

The velocities computed from the reciprocals of the slope are called **apparent velocities.** If the lower layer is horizontal, they represent the actual velocity. However, if the lower layer is sloping, the arrival time for a shot measured downslope will be different from one measured upslope. Seismic lines are routinely run with a shot at either end, so that dipping beds can be determined. Time-distance curves for a dipping stratum are shown in Figure 12.16, with travel times measured from shots at either end of the line. The upper layer is unaffected by the dip of the lower bed, so that the reciprocal of the slope of the first line segment is $V_1$. In order to find the values of $V_2$ and the depth to the bedrock at the updip end of the line, $Z_d$, as well as at the downdip end, $Z_u$, a complex series of computations must be made (24).

The slope of the second line segment of the downdip line is $m_d$, and the slope of the second line segment of the updip line is $m_u$. The value of the angle of refraction, $i_c$, is found from

$$i_c = \frac{1}{2}(\sin^{-1} V_1 m_d + \sin^{-1} V_1 m_u) \qquad \textbf{(12-10)}$$

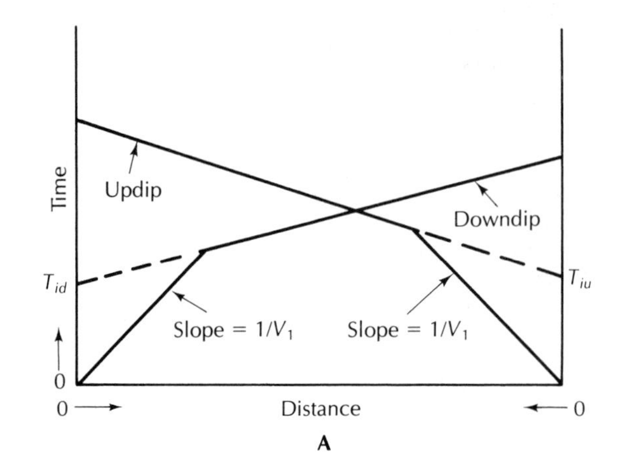


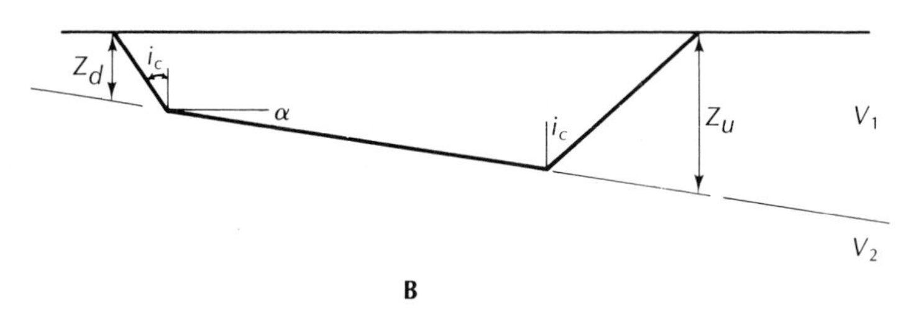


**FIGURE 12.16** **A.** Diagram of arrival time versus distance for a two-layered seismic problem with a sloping lower layer. **B.** Wave path for the preceding problem.

The value of $V_2$ is given by

$$V_2 = \frac{V_1}{\sin i_c} \tag{12-11}$$

The angle of slope of the dipping layer, α, is found from

$$\alpha = \frac{1}{2}(\sin^{-1} V_1 m_d - \sin^{-1} V_1 m_u) \tag{12-12}$$

Finally, the depths to the lower layer at either end of the shot line are found from the expressions

$$Z_u = \frac{V_1 T_{iu}}{\cos \alpha \; 2 \cos i_c} \tag{12-13}$$

$$Z_d = \frac{V_1 T_{id}}{\cos \alpha \; 2 \cos i_c} \tag{12-14}$$

If there are more than two dipping layers, then expressions of greater complexity must be used (25).

**EXAMPLE PROBLEM** A seismic survey yielded data for a dipping two-layer case in which the following values were obtained:

$$V_1 = 1570 \text{ m/sec}$$
$$m_u = 1.67 \times 10^{-4} \text{ sec/m}$$
$$m_d = 1.54 \times 10^{-4} \text{ sec/m}$$
$$T_{id} = 0.046 \text{ sec}$$
$$T_{iu} = 0.050 \text{ sec}$$

Compute $V_2$, $Z_u$, and $Z_d$.

$$\begin{aligned} i_c &= \tfrac{1}{2}(\sin^{-1} V_1 m_d + \sin^{-1} V_1 m_u) \\ &= \tfrac{1}{2}(\sin^{-1} 1570 \times 1.54 \times 10^{-4} + \sin^{-1} 1570 \times 1.67 \times 10^{-4}) \\ &= \tfrac{1}{2}(13.99 + 15.20) \\ &= 14.6^\circ \end{aligned}$$

$$\begin{aligned} V_2 &= V_1/\sin i_c \\ &= 1570/\sin 14.6 \\ &= 6230 \text{ m/sec} \end{aligned}$$

$$\begin{aligned} \alpha &= \tfrac{1}{2}(\sin^{-1} V_1 m_d - \sin^{-1} V_1 m_u) \\ &= \tfrac{1}{2}(\sin^{-1} 1570 \times 1.54 \times 10^{-4} - \sin^{-1} 1570 \times 1.67 \times 10^{-4}) \\ &= \tfrac{1}{2}(13.99 - 15.20) \\ &= -0.6^\circ \end{aligned}$$

$$\begin{aligned} Z_u &= \frac{V_1 T_{iu}}{\cos \alpha \; 2 \cos i_c} \\ &= \frac{1570 \times 0.050}{\cos -0.6 \times 2 \times \cos 14.6} \\ &= 40.6 \text{ m} \end{aligned}$$

$$\begin{aligned} Z_d &= \frac{V_1 T_{id}}{\cos \alpha \; 2 \cos i_c} \\ &= \frac{1570 \times 0.046}{\cos -0.6 \times 2 \times \cos 14.6} \\ &= 37.3 \text{ m} \end{aligned}$$

The cases given in this section are but a few of the many possible cases that might be encountered. Figure 12.17 shows schematic travel-time curves for a number of nonhomogeneous earth models. Should the hydrogeologist suspect that the situation is more than a simple two- or three-layer model with homogeneous beds, an experienced geophysicist should interpret the field data.

The principal application of refraction seismology in hydrogeology has been in the identification of buried refractors such as the top of the water table—which, being saturated, has a greater seismic velocity than the unsaturated equivalent soil unit—and the bedrock surface (26). It has been very useful in the delineation of bedrock valleys buried in glacial drift (27).

### 12.3.4 GROUND-PENETRATING RADAR AND MAGNETOMETER SURVEYS

There are thousands of abandoned waste-disposal sites in the United States alone. For many sites the records of trench locations and drum burial areas are sketchy or nonexistent. If a remedial action plan is to be developed for such a site, it is necessary to know the extent of the buried waste. From a safety standpoint, one needs to know the locations of buried drums of toxic material so that drilling operations for the installation of monitoring wells don't penetrate the drums. There are several methods that are used to locate areas of waste disposal. Two of the more common are **magnetometer surveys** and **ground-penetrating radar** (28–33).

Magnetometer surveys measure the strength of the earth's magnetic field. A proton nuclear magnetic resonance magnetometer is frequently used. This is a hand-held instrument and one person can rapidly perform a survey over a site of a few acres in size. A grid system is set up and measurements are made of the magnetic field at each intersection of the grid. Areas with large amounts of buried metal, such as steel drums, will have magnetic anomalies associated with them. The strength of the anomaly will vary with the amount and depth of the buried metal.

Ground-penetrating radar (GPR) is based on the transmission into the ground of repetitive pulses of electromagnetic waves in the frequencies of 10 to 1000 MHz. The pulses are reflected back to the surface when the radiated energy encounters an interface between two materials of differing dielectric properties. The interfaces that cause the reflections may be due to such things as a change in strata or buried objects. The ground-penetrating radar system has antennae to both send a signal and receive the reflected pulse. The reflected pulse may be recorded on a magnetic tape recorder and generally is also converted into voltage as a function of time and displayed on a graphic recorder. The graphic output from a GPR system looks like a sonar signal. Figure 12.18 shows the graphic output from a ground-penetrating radar survey line. A typical receiving unit will display the output in a number of shades of gray that vary proportionally to the voltage level of the received wave.

The GPR unit is pulled along the ground and a continuous line profile is

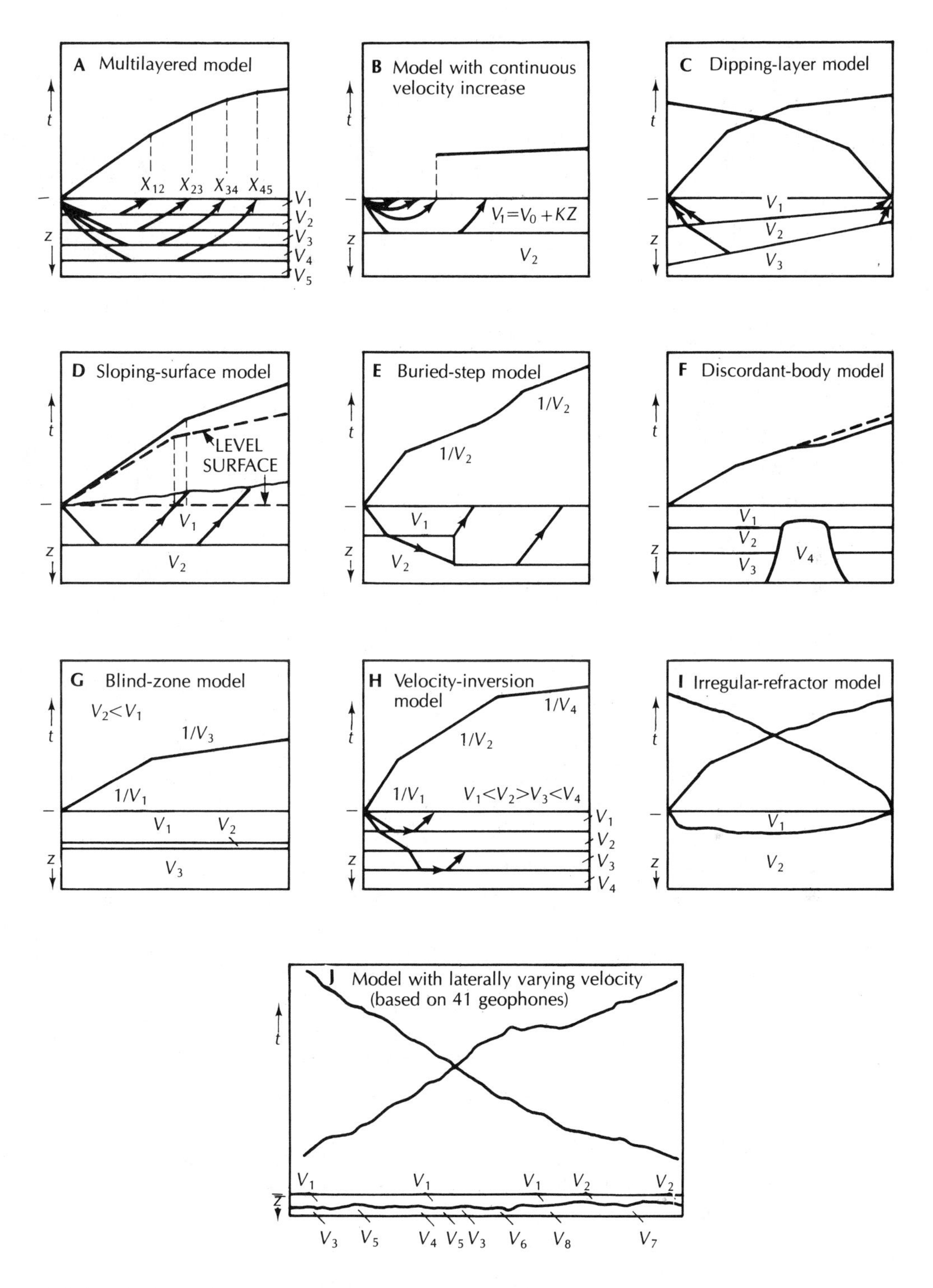


**FIGURE 12.17** Schematic travel-time curves for idealized nonhomogeneous geologic models. Source: G. P. Eaton, "Seismology," in *Techniques of Water-Resources Investigations.* U.S. Geological Survey, 1974, Book 2, Chap. D1, pp. 67–84.

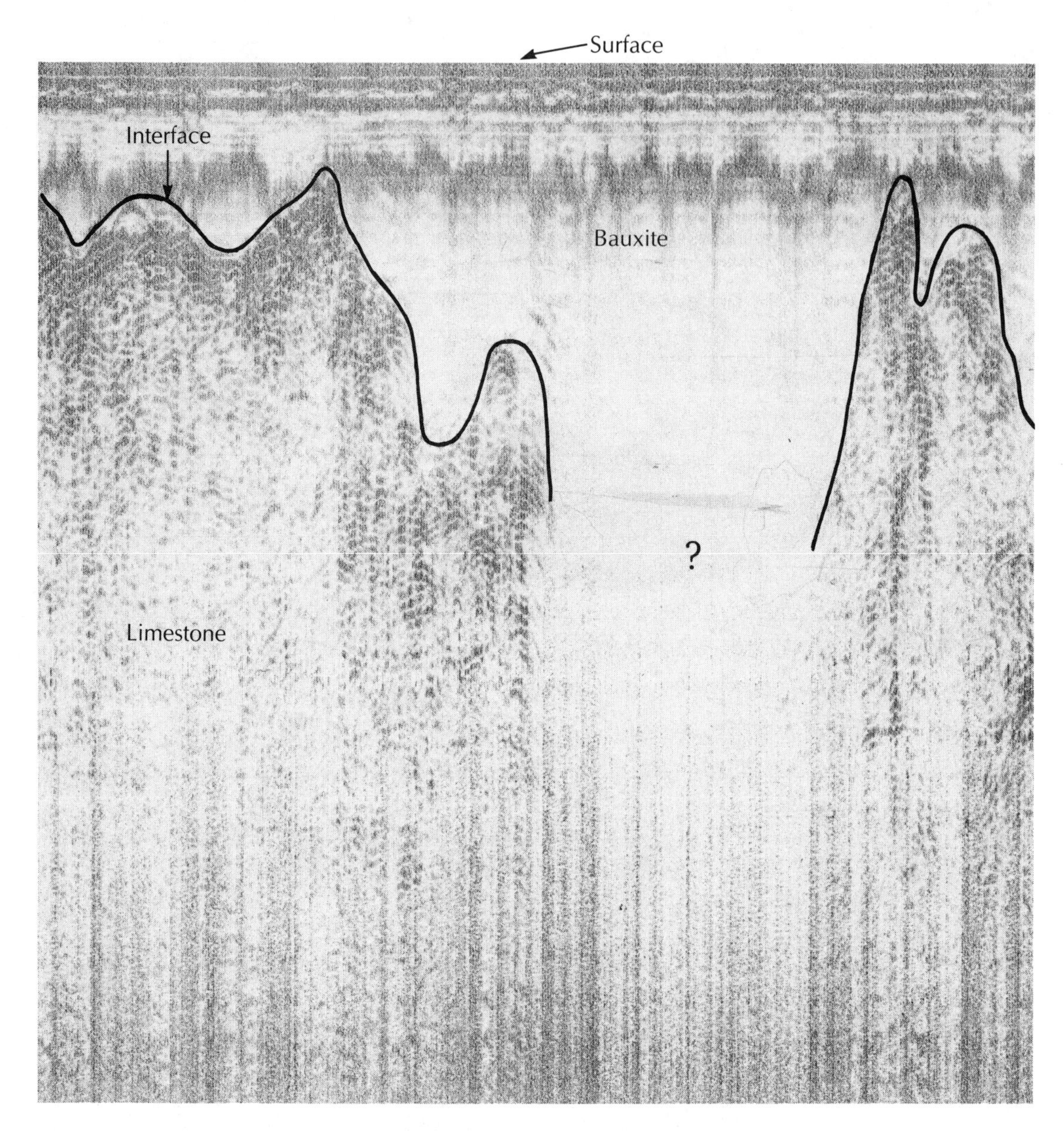


**FIGURE 12.18** Ground-penetrating radar image of a bauxite deposit overlying limestone. No horizontal or vertical scale. Source: Courtesy of Joan Underwood, Donohoe and Associates, Sheboygan, Wisconsin.

generated. Parallel lines are surveyed to cover an area being studied. The depth of penetration of the GPR is a function of the type of geological material and the frequency of the radar being used. The lower frequencies penetrate a given medium to the greatest depths; higher frequencies will not penetrate as deeply, but give greater resolution. Greater resolution increases the ability of the unit to discriminate between interfaces and objects that are closely spaced. For waste-site studies, typical depths of study are 5 to 20 feet (1.5 to 6 meters). GPR is capable of showing the location of a single 55-gallon metal drum buried at depths of 6 to 9 feet (2 to 3 meters) (31). It has been successfully used to show the limits of a buried crystalline mass below a concrete pavement. The great advantage of GPR is that it is able to give a continuous profile of the subsurface, an accomplishment not otherwise possible.

### 12.3.5 GRAVITY AND AEROMAGNETIC METHODS

Measurement of the gravimetric and magnetic fields of the earth are standard geophysical methods used to study the structure and composition of the earth. To the extent that the basic geology influences the hydrogeology, these methods are quite useful in ground-water studies. The collection of data can be relatively simple if ground stations are used. However, the reduction and correction of the data are fairly complex (11, 24, 34). Airborne geomagnetic surveys obviously involve specialized skills and equipment.

Both gravity and magnetics surveys can be used to delineate the area of unconsolidated basin fill or buried stream-channel aquifers. In Figure 12.19, a Cenozoic basin in Antelope Valley, California, is depicted in cross section. The presence of the basin and the area of the deepest part are shown on both an aeromagnetic profile and a gravity profile.

Magnetic anomalies are caused by distortions of the earth's magnetic field created by magnetic materials in the crust. Magnetic anomalies indicate the type of rocks in a very general way. In hydrogeologic studies, magnetic anomalies can be useful in indicating the depth to magnetic basement rocks. Sedimentary rocks are typically nonmagnetic; hence, they do not affect the magnetic field. Some magnetic rocks, such as basalt flows, can be important aquifers. Magnetic surveys might be useful in tracing basalt flows in areas of nonmagnetic rocks.

The mass of the rocks beneath a point on the earth will affect the local value of the acceleration of gravity. To be useful, the measured values must be referenced to a common datum—usually, mean sea level. A free-air correction is made to compensate for elevation differences. To correct for the gravitational attraction of the rock that lies between the gravity station and sea level, a **Bouguer correction** is made. Corrections must also be made for tidal effects, latitude, and terrain. After the measured gravity data are corrected, the result is a **Bouguer anomaly value,** which can be mapped with gravity contours drawn. Such a map could help to define the extent of a buried bedrock valley if there were a density difference between the sediments and the bedrock.

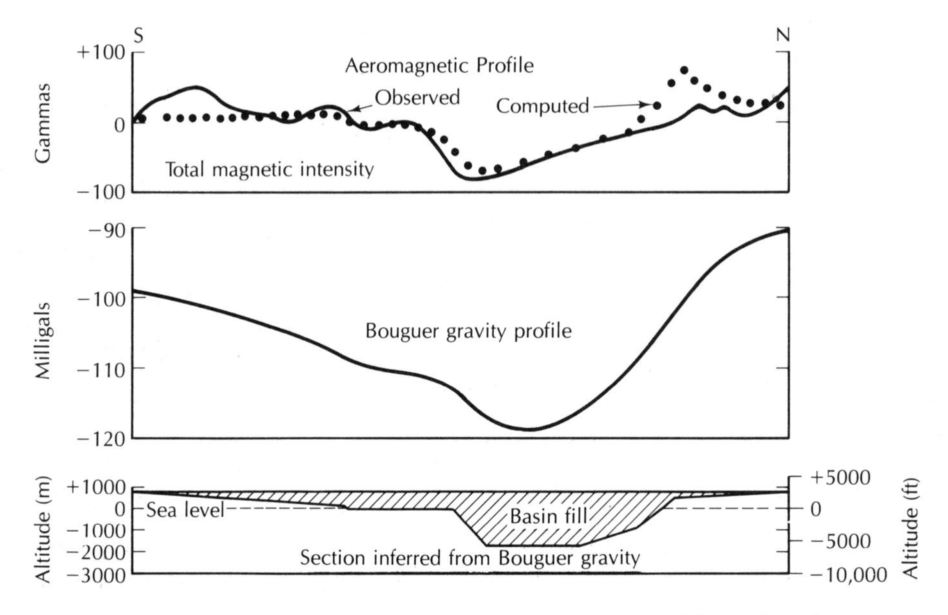


**FIGURE 12.19** Gravity and aeromagnetic profiles across a basin-fill aquifer. The aeromagnetic profile was flown at 490 feet (150 meters). Source: D. R. Mabey "Magnetic Methods," in *Techniques of Water-Resources Investigations,* U.S. Geological Survey, 1974, Book 2, Chap. D1, pp. 85–115.

It should be emphasized that there are many possible earth models that would result in the same gravitational or magnetic anomaly. There is no unique solution to any set of geophysical data, and the person interpreting the data must keep this in mind.

## 12.4 GEOPHYSICAL WELL LOGGING

Direct access to the subsurface is gained wherever there is a well or test boring. When a well is drilled, a record may be made of the geologic formations encountered. The reliability of a **lithologic well log** depends on the method of drilling and sample recovery as well as the knowledge and skills of the person making the log. There are also many existing wells for which there is no available record of the subsurface geology.

Geophysical well logging offers a great deal in the way of practical application to hydrogeology. Borehole geophysical methods were developed primarily in the petroleum industry, and virtually all oil and gas wells are routinely logged when drilled. In the water-well industry, the use of geophysical logging is generally restricted to either research projects or high-capacity municipal and industrial wells. The cost of well logging is not justified by the marginal benefits gained for small-yield domestic wells.

Borehole geophysical data have a number of uses (see Table 12.1). The log of a well can indicate the areas of high porosity and permeability that would produce the most water. Zones of an aquifer with high-salinity water can be identified. If a number of wells are logged over an area, the logs can be used for stratigraphic correlation. The lithology of the rocks penetrated by the wells can be identified, especially if some core samples are available for baseline comparison. Regional ground-water flow patterns might be identified from such characteristics as fluid temperature. Nuclear well-logging techniques can be used in cased wells. This is the only way to get subsurface data under such conditions. Geophysical logs give a permanent record, based on repeatable measurements. Thus, data collected for one purpose are available for other, unanticipated, uses in the future.

Because of the large number of borehole techniques that are applicable to water wells (35–44), a discussion of only the more common methods will be included in this section, with emphasis on qualitative rather then quantitative interpretation of geophysical logs. Generally, a suite of geophyscial logs is made, rather than only a single type. The methods tend to be complementary; one may confirm another. Likewise, certain interpretations are made on the basis of two or more logs. Figure 12.20 consists of six different geophysical logs made on the same borehole, along with a lithologic log. It can readily be seen that the logs deflect with the changes in lithology.

Well logs are usually made as pen-and-ink strip charts. This gives a continuous record, which is most useful. Some instruments give point readings of various values at discrete depths. A probe is lowered into the borehole on a cable. The cable, which contains powerlines from the surface to the probe, supports the weight of the probe and transmits signals from the probe to the recorder at the surface. The probe contains the necessary electronics, energy or nuclear sources, and detectors. Logging information can be obtained as the probe is either lowered or raised.

### 12.4.1 CALIPER LOGS

A **caliper log** is used to measure the diameter of an uncased borehole in bedrock units. It can also be used to find the casing depth. The minimum hole diameter is, of course, the size of the drill bit.* The hole may be enlarged by caving of the formations into the hole or by solution of minerals by the drill water. It also may be enlarged if the drill bit is rotated at a depth while the downward pressure is removed. Another use of caliper logs is to indicate solution-enlarged bedding planes and joints in carbonate aquifers.

---

*When a hole is drilled, the drilling rig turns a pipe (drill stem). The rock at the bottom of the hole is broken up by the rotating drill bit, which cuts it into pieces. Drilling fluids circulating through the hole and drill stem bring the broken rock to the surface.

**TABLE 12.1** Summary of log applications

| Required Information on the Properties of Rocks, Fluid, Wells, or the Ground-Water System | Widely Available Logging Techniques That Might Be Utilized |
|---|---|
| Lithology and stratigraphic correlation of aquifers and associated rocks | Electric, sonic, or caliper logs made in open holes; nuclear logs made in open or cased holes |
| Total porosity or bulk density . . . . . . . | Calibrated sonic logs in open holes, calibrated neutron or gamma-gamma logs in open or cased holes |
| Effective porosity or true resistivity . . . . . | Calibrated long-normal resistivity logs |
| Clay or shale content . . . . . . . . . . . | Gamma logs |
| Permeability . . . . . . . . . . . . . . . . | No direct measurement by logging. May be related to porosity, injectivity, sonic amplitude |
| Secondary permeability—fractures, solution openings | Caliper, sonic, or borehole televiewer or television logs |
| Specific yield of unconfined aquifers . . . . | Calibrated neutron logs |
| Grain size . . . . . . . . . . . . . . . . . | Possible relation to formation factor derived from electric logs |
| Location of water level or saturated zones | Electric, temperature, or fluid conductivity in open hole or inside casing, neutron or gamma-gamma logs in open hole or outside casing |
| Moisture content . . . . . . . . . . . . . | Calibrated neutron logs |
| Infiltration . . . . . . . . . . . . . . . . . | Time-interval neutron logs under special circumstances or radioactive tracers |
| Direction, velocity, and path of ground-water flow | Single-well tracer techniques—point dilution and single-well pulse; multiwell tracer techniques |
| Dispersion, dilution, and movement of waste | Fluid conductivity and temperature logs, gamma logs for some radioactive wastes, fluid sampler |
| Source and movement of water in a well | Injectivity profile; flowmeter or tracer logging during pumping or injection; temperature logs |
| Chemical and physical characteristics of water, including salinity, temperature, density, and viscosity | Calibrated fluid conductivity and temperature in the well; neutron chloride logging outside casing; multi-electrode resistivity |
| Determining construction of existing wells, diameter and position of casing, perforations, screen | Gamma-gamma, caliper, collar, and perforation locator; borehole television |
| Guide to screen setting . . . . . . . . . . | All logs providing data on the lithology, water-bearing characteristics, and correlation and thickness of aquifers |
| Cementing . . . . . . . . . . . . . . . . | Caliper, temperature, gamma-gamma; acoustic for cement bond |
| Casing corrosion . . . . . . . . . . . . . | Under some conditions, caliper or collar locator |
| Casing leaks and (or) plugged screen | Tracer and flowmeter |

Source: W. S. Keys and L. M. MacCary, "Application of Borehole Geophysics to Water-Resources Investigations," in *Techniques of Water-Resources Investigations,* U.S. Geological Survey, 1971, Book 2, Chap. E1.

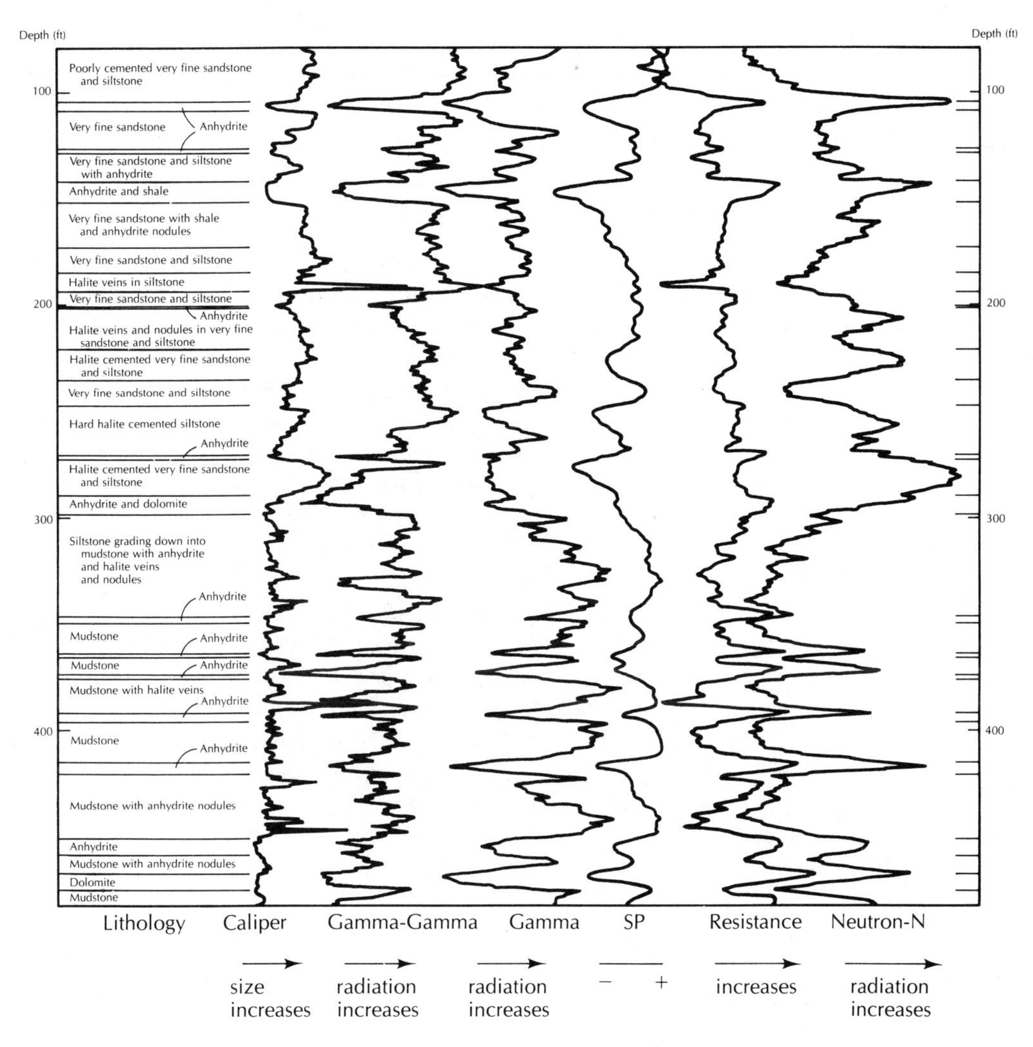


**FIGURE 12.20** The relationship of six different geophysical logs to the lithology of a well in the upper Brazos River Basin, Texas. Source: W. S. Keys and L. M. MacCary, in *Techniques of Water-Resources Investigations,* U.S. Geological Survey, 1971, Book 2, Chap. E1.

### 12.4.2 TEMPERATURE LOGS

A **temperature log** is a continuous vertical record of the temperature of the fluid in the borehole. This may or may not be indicative of the temperature of the fluid in the rocks opposite the borehole fluid. In a recently drilled well, the borehole fluid may be well mixed. After the well has had an opportunity to reach environmental equilibrium, the temperature log can reveal zones of differing temperature in the well. There will be a component of the geothermal gradient present; however, water in different aquifers may be at discrete temperatures, which may be detected on the log. Thermal logging has been used to trace the movement of water previously injected into an aquifer (40). The recharged water had a daily thermal variation of up to 17° C, and the diurnal fluctuation was traced in a series of observation wells (Figure 12.21).

### 12.4.3 SINGLE-POINT RESISTANCE

Electrical resistance can be measured in a borehole by a number of different methods. The simplest case is the **single-point resistance,** in which a single electrode is lowered into the borehole on an insulated cable. The other electrode is at the ground surface. As the electrode is lowered into the borehole, the resistance of the earth between the two electrodes is measured.

The single-point electrode is measuring the resistance of all of the rocks between the electrodes. Most of the variation in resistance is due to the changes in the conductivity of the borehole fluid and to a small volume of rock around the borehole near the downhole electrode. If the borehole fluid is homogeneous, the variation in resistance will be due to the lithologic variations near the borehole.

Lithologies with a high resistance include sand, sand and gravel, sandstone, and lignite. Clay and shale have the lowest resistance. Increasing salinity will cause a decrease in resistance. If the borehole is enlarged (for instance, by a fracture), the resistance will also decrease. If the caliper log indicates a cavity, and the resistance log decreases, the decrease is due to the hole enlargement. Should the borehole be straight, a decrease in resistance might be due to either a shale layer or, perhaps, a sandstone containing a brine. However, other logs can distinguish brine from fresh water and sandstone from shale. Figure 12.22 shows a single-point resistance log.

### 12.4.4 RESISTIVITY

Earth **resistivity** may be measured in a borehole by lowering two current electrodes and measuring the resistivity between two additional electrodes. Resistivity is measured in ohm-meters and is different from resistance, the latter being measured in ohms. Single-point resistance logging measures the total resistance of the earth materials, while resistivity measures a specific property of the rock and the contained pore waters. The trace of a resistivity log is similar to that of a single-point resistance log. However, resistivity logs can be calibrated and used quantitatively.

A number of different electrode configurations are used in resistivity logging. These include the **short-normal, long-normal,** and **lateral configurations,** as shown in Figure 12.23. The three logs have similar traces (Figure 12.24). The short-normal curve indicates the resistivity of the zone close to the borehole. It is in this area that the drilling fluid may have invaded the formations. The long-normal spacing has more spacing between electrodes and thus measures the resistivity farther away from the borehole—presumably, beyond the influence of

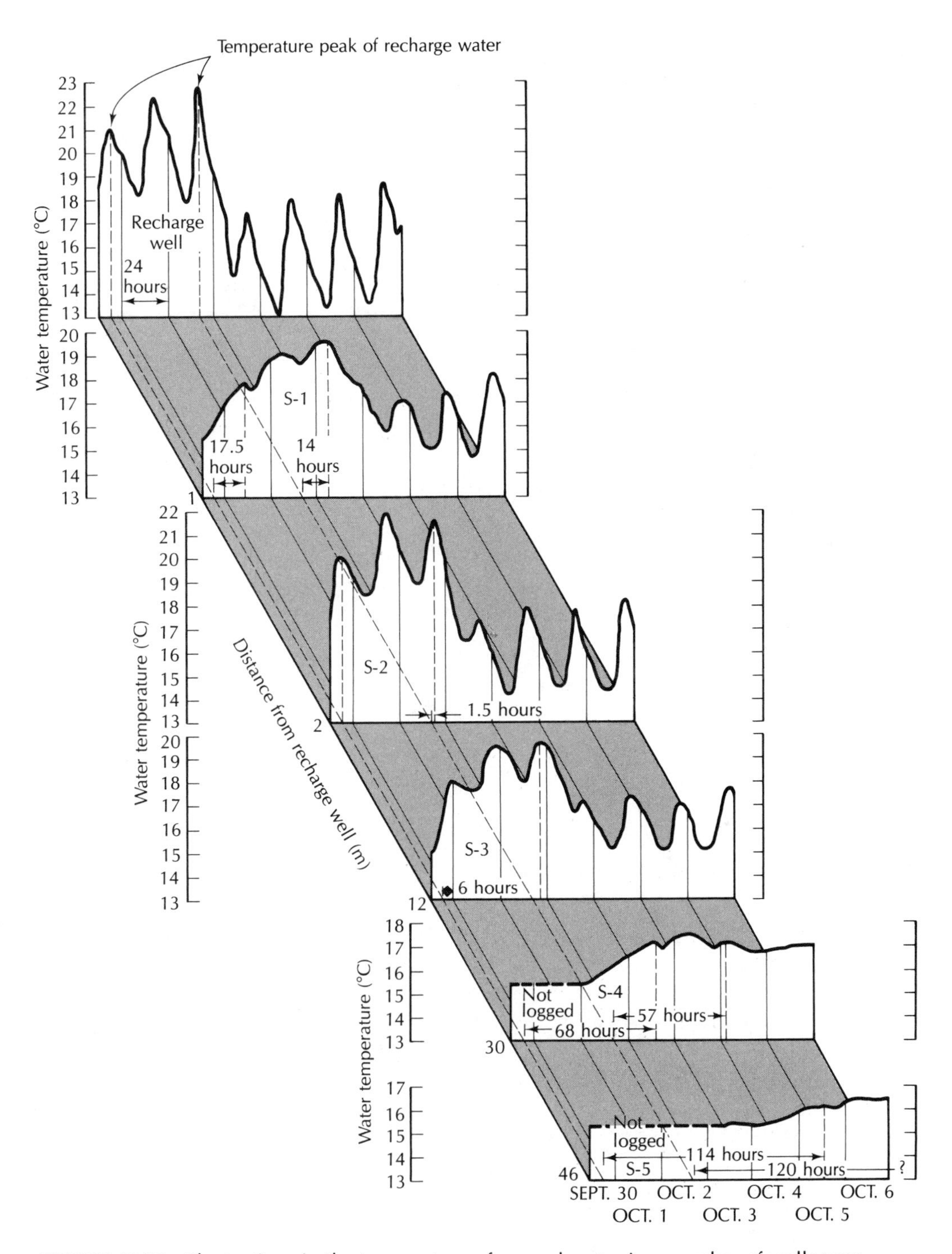


**FIGURE 12.21** Fluctuations in the temperature of ground water in a number of wells near an injection well for artificial recharge. All temperatures are measured at a depth of 161 feet (49 meters). The diurnal variation in the temperature of the recharge water dies out by 98 feet (30 meters). Source: W. S. Keys and R. F. Brown, *Ground Water,* 16 (1978):32–48.

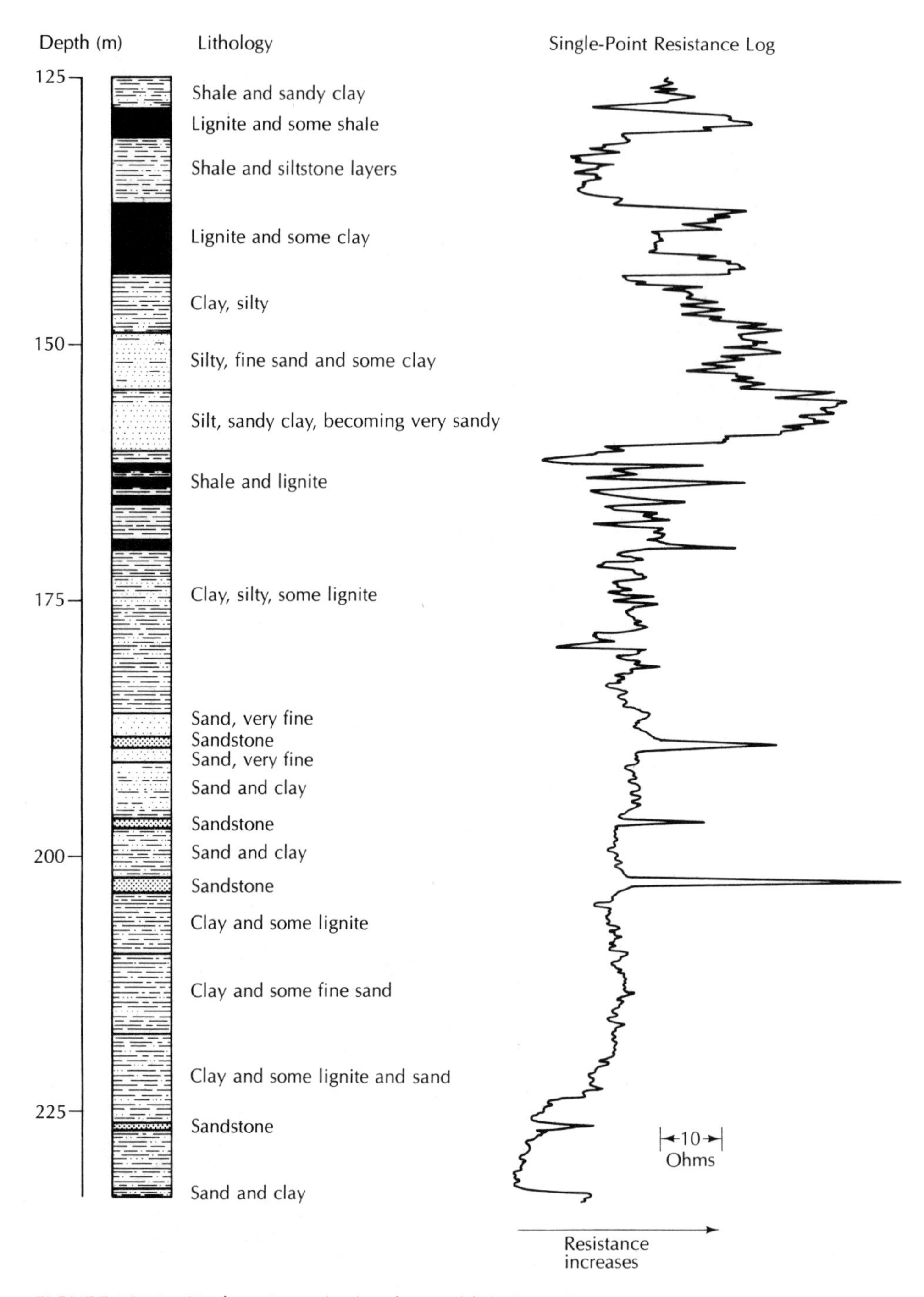


**FIGURE 12.22** Single-point resistance log and lithologic log. Source: W. S. Keys and L. M. MacCary, in *Techniques of Water-Resources Investigations,* U.S. Geological Survey, 1971, Book 2, Chap. E1.

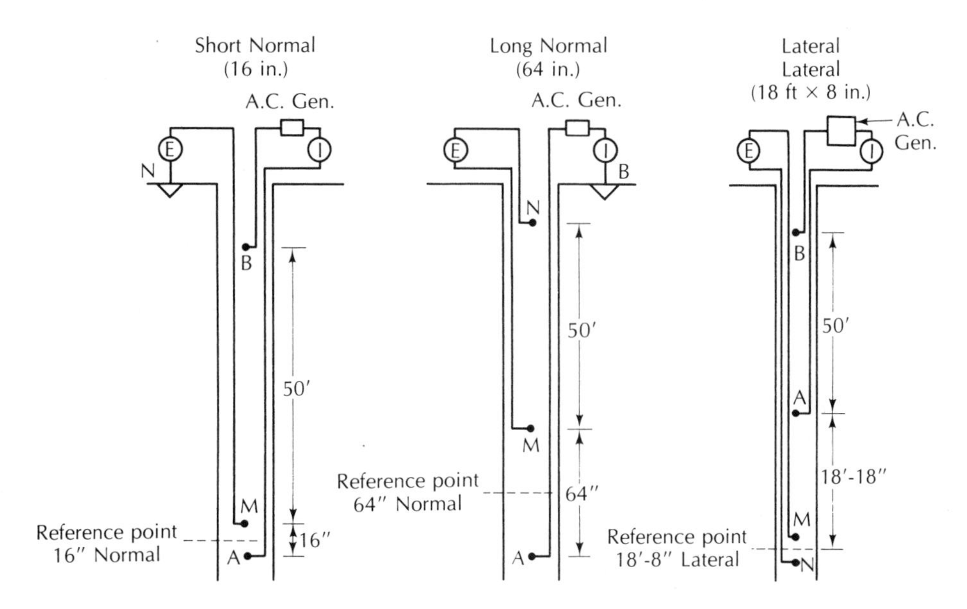


**FIGURE 12.23** Electrode configuration for various resistivity logging devices.

the drilling fluid. Both the short-normal and long-normal resistivity measure a greater radius of influence than the single-point resistance.

Lateral devices have very widely spaced electrodes for measuring zones that are far from the borehole. Because of the wide spacing, lateral devices will not pick out thin beds of different resistivity. For example, the 18-foot × 8-inch lateral log will be best for beds at least 40 feet (12 meters) thick.

### 12.4.5 SPONTANEOUS POTENTIAL

Along with resistance and resistivity, **spontaneous potential** is another form of electrical logging. It is a measure of the natural electrical potential that develops between the formation and the borehole fluids. It is run only on an open hole filled with fluid, as are resistance and resistivity. It can be used below a casing in a partially cased well. The spontaneous-potential (SP) curve can be used for determination of bed thickness, geologic correlation, and also for delineation of permeable rocks. An SP device consists of a surface electrode and a borehole electrode with a voltmeter to measure potential.

One use of the SP curve is to distinguish shale from sandstone lithology. Shale has a positive SP response and sandstone a negative one if the salinity of the formation fluid is greater than that of the borehole fluid.

### 12.4.6 NUCLEAR LOGGING

Some of the most useful logging methods involve the measurement of either natural radioactivity of the rock and fluids or their attenuation of induced

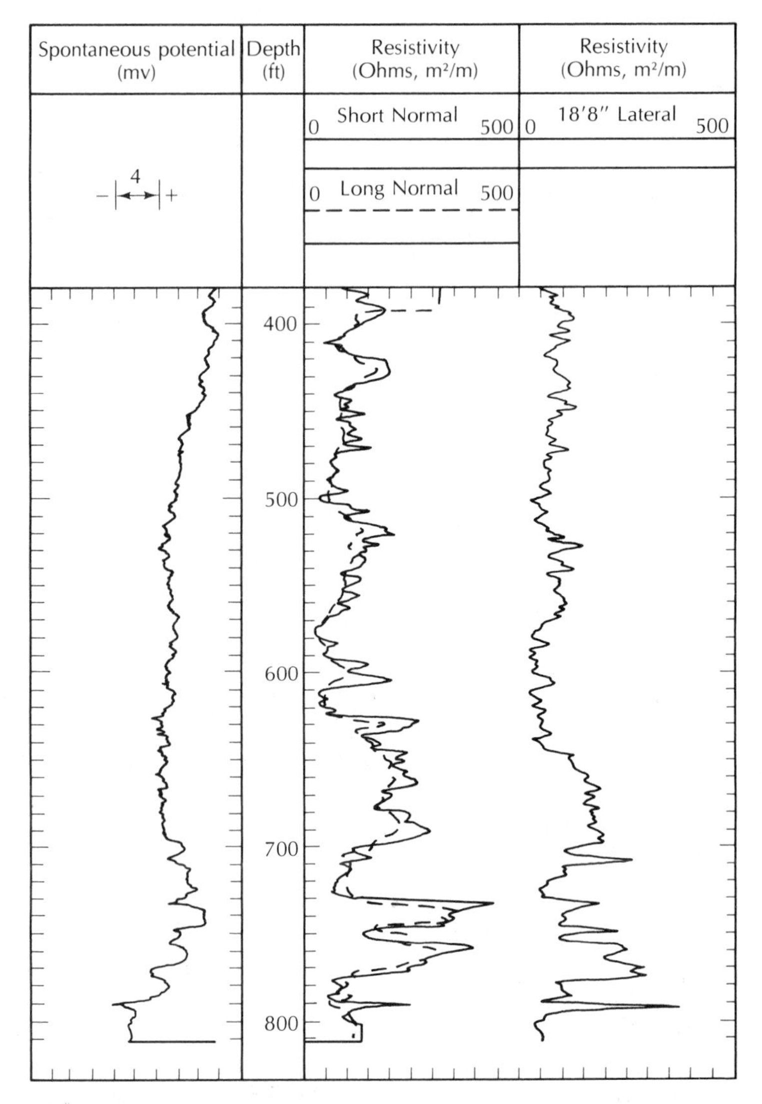


**FIGURE 12.24** Electric logs of a limestone well. The long-normal log is shown as a dashed line. Source: W. S. Keys and L. M. MacCary, in *Techniques of Water-Resources Investigations,* U.S. Geological Survey, 1971, Book 2, Chap. E1.

radiation. **Nuclear logging** can be done in either a cased or an uncased hole, and the logs are not affected by the type of drilling mud. The use of radioactive isotopes involves special safety precautions.

Radioactive decay is a process with a random component; hence, the instantaneous rate of decay fluctuates. Over a long time period, the rate of decay per time period is constant. However, as the time period decreases, the variation in the number of decay events per time period will increase. Nuclear logging records the number of disintegrations over a fixed time period, called the **time constant.** The longer the time constant, the less the likelihood that a variation in radiation intensity is due to random decay and, hence, the more likely the variation reflects different lithology.

Consideration must also be given to the speed at which the probe is moved up or down the hole. If the speed is too great, the probe may pass a thin bed before the time constant has elapsed. Consequently, the selection of the proper time constant and logging speed is very important and depends upon the equipment, logging technique, and lithology (35).

Nuclear logs do not have exact reproducibility, owing to the statistical nature of the decay process. Repeat logging runs are necessary to determine if an observed variation represents a lithologic change or a statistical fluctuation in the decay rate. In Figure 12.25, the first two neutron-gamma logs were made going up and down the hole, respectively. The same peaks are present, but there is variability in the exact radiation count. To the right is a third log of the same hole made with a different radiation source having a longer time constant—10 seconds versus 3 seconds. The right-hand log has a poor ratio of time constant and logging speed. It does not distinguish thin beds, and the positions of the lithologic contacts are incorrect.

The thickness of individual strata can be determined from nuclear logs if there is a change in lithology or porosity from one unit to the next. This is assumed to be equal to the thickness of the anomaly at one-half the maximum amplitude. This method will overestimate the thickness of thin layers. By convention, radiation increases to the right in nuclear logging. In a reversed log, it increases to the left.

There are three nuclear methods that can be used for composite identification (35). The neutron count rate increases with decreasing porosity, while the gamma-gamma count rate decreases. The natural gamma radiation increases with an increasing clay or shale content as well as with increased phosphate and K-feldspars, but has no direct relationship to porosity. A stratum with low natural gamma radiation and a low neutron count (or high gamma-gamma count) could be interpreted as a porous sandstone. A low natural gamma radiation and high neutron count could be a dense quartz sandstone or quartzite (see Figure 12.20).

## NATURAL GAMMA RADIATION

This is the nuclear-logging method most commonly employed in hydrogeology. It is a measure of the natural radiation of rocks as determined by the emission of gamma activity by potassium 40, the uranium 238 decay series, and the thorium 232 decay series. These are constituent materials for some shales and clays with high gamma activity. Certain feldspars and micas are high in $^{40}K$. A natural gamma log shows increasing radiation opposite sedimentary beds that contain potassium-rich shale, or clay or phosphate rock. Thus, a shaly sandstone

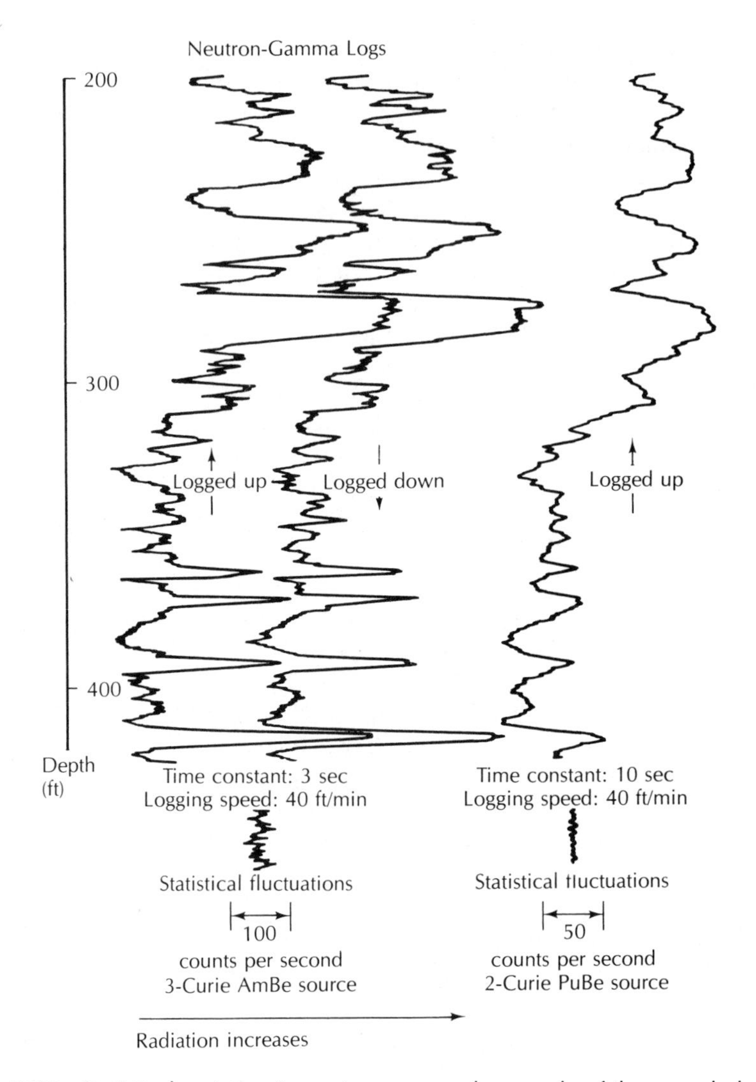


**FIGURE 12.25** Statistical variation in neutron-gamma logs made of the same hole. The log on the right had a poor ratio of time constant and logging speed. Source: W. S. Keys and L. M. MacCary, in *Techniques of Water-Resources Investigations,* U.S. Geological Survey, 1971, Book 2, Chap. E1.

could be distinguished from a clean quartz sandstone. The natural gamma log can be used for lithologic determination—especially of detrital sediments—on the basis of differences in radiation intensity. No calibration of the unit is necessary in this nuclear-logging method. Another advantage is that no radiation sources need be used.

The natural gamma radiation log has a tremendous advantage in that it can be performed on an existing cased well. Part of the radiation penetrates the well casing, and the amount absorbed by the casing is constant. Therefore, the variations in radiation due to lithology show up on the log. The method works equally well in both plastic and steel casing. It may not work inside hollow-stem augers if the thickness of the steel wall of the augers varies at the joints where the augers are joined. In this case the varying steel wall thickness will create a situation where the amount of radiation absorbed by the augers is not constant over the length of the augers and the resulting log is distorted.

## NEUTRON LOGGING

A neutron probe contains a radioactive element, such as PbBe, which is a source of neutrons, and a detector. The emitted neutrons are slowed and scattered by collisions with nuclei of hydrogen atoms. Detectors are available to measure gamma radiation produced by the neutron-hydrogen atom collision, or the number of neutrons present at different energy levels. Thus, a neutron log will be identified as a **neutron-thermal neutron,** a **neutron-epithermal neutron,** or a **neutron-gamma log** on the basis of the method of detection.

Hydrogen is present in the ground primarily in the form of water or hydrocarbons. In almost all rocks of interest to hydrogeologists, there are no hydrocarbons. Therefore, apart from hydrated minerals, water is present as moisture in the pore spaces of the rock. An increase in the amount of water will result in an increase in the number of neutrons that are captured or moderated. As a result, saturated rocks with a high porosity will have a lower neutron count than low-porosity rocks. Above the water table, the neutron-logging equipment can be used to measure the moisture content, but not the porosity. Neutron logging can be used to determine the specific yield of unconfined aquifers (45). It also can distinguish gypsum, with a high proportion of hydrated water, from anhydrite. Both have a very low natural gamma radiation; however, anhydrite has a high neutron count and gypsum has a low count.

## GAMMA-GAMMA RADIATION

In this type of logging, a source of gamma radiation, such as cobalt 60, is lowered into the borehole. Gamma photons are absorbed or scattered by all material with which the cobalt 60 comes in contact. This includes fluid, casing, and rock. The absorption is proportional to the bulk density of the earth material. **Bulk density** is defined as the weight of the rock divided by the total volume, including the porosity. Thus, gamma-gamma radiation increases with decreasing bulk density (increasing porosity). Bulk density can be determined from a calibrated gamma-gamma log. The formation porosity can be determined from the equation

$$\text{Porosity} = \frac{\text{grain density} - \text{bulk density}}{\text{grain density} - \text{fluid density}} \qquad \textbf{(12-15)}$$

Grain density may be determined from drill cuttings or assumed to be 2.65 grams per cubic centimeter for quartz sandstone. Fluid density is 1.000 grams per cubic centimeter for mud-free fresh water. The drilling fluid may contain additives to increase the fluid density.

## 12.5 HYDROGEOLOGIC SITE EVALUATIONS

Many types of construction and similar projects require evaluation of the hydrogeology of a site. This may be a part of a comprehensive study of the engineering geology of a proposed dam or tunnel. The siting of sanitary landfills or areas for toxic- or hazardous-waste disposal or storage almost always requires a hydrogeologic site evaluation. Environmental impact studies often include sections on geology and hydrogeology. Hydrogeological analysis can help ensure the integrity and safety of structures located in areas of high water table, spring discharge, or quicksand potential. Likewise, ground-water pollution generally can be avoided if sanitary landfill sites are selected and engineered on the basis of a thorough hydrogeological investigation. Well-planned and -executed field investigations of the hydrogeology of areas of ground-water contamination are also necessary in order to locate the source of the contamination, define the extent of the plume, and plan for corrective action.

While the detailed design of a hydrogeological site study will be dictated by the specific purpose, there are some elements common to all studies. The first is that they are all relatively expensive. A great expense is in the collection of the basic data. This includes the cost of geophysical surveys, test borings, test wells, pumping tests, permeability lab tests, and so forth. It is frequently the case that data collection is more costly than the evaluation of the data and the formulation of conclusions and recommendations.

The results of a hydrogeological study will generally need to be presented in part on maps. At a minimum, a site-location map will be needed. Standard U.S. Geological Survey topographic maps can be used for site-location maps and perhaps as base maps. A base map should show the important surface features and topography, as well as the locations of all test borings, wells, sampling points, geophysical station locations, and so forth. If a custom base map is needed of an area, it is generally ordered from an aerometric engineering company, which will prepare it from aerial photographs with some surveyed ground elevations. Contour intervals of 1 or 2 feet are desirable for most base maps unless the site has great relief.

A map of the surficial geology can be prepared using standard geologic field techniques. Attention should be given to the nature of bedding planes, fractures, and porosity of rock outcrops. A fracture-trace analysis on air photos can enhance the surficial geology map in areas of fractured bedrock. Detailed mapping of surficial geology can be facilitated if there are soil maps available. Interpretation sheets for the soil-map units indicate the parent material on which the soil formed. They also give a general range of infiltration rate for each soil type.

Surface geophysical surveys may be made as a part of the preliminary site evaluation. Seismic refraction, geoelectric soundings, and geoelectric surveys would be useful for those areas in which the unconsolidated deposits are important. The data from the geophysical surveys can guide the selection of the location of test borings and provide data to correlate between test borings. These borings, or wells, are needed to determine the geologic units present. **Borings** refer to uncased holes drilled in unconsolidated overburden, while **test wells** either have a casing or extend into bedrock. A geologist should examine the drill samples or cuttings, prepare a detailed sedimentological and petrographic description, and make a well log showing the depth and thickness of each stratum.

Slightly disturbed samples of unconsolidated deposits can be collected by driving a thin-walled steel tube (either a Shelby tube or split-spoon sampler) into the bottom of the drilled hole. In sandy soils, this must be done through hollow-stem drill pipe or auger-flights. The slightly disturbed sample can be used for laboratory testing, including permeameter tests for hydraulic conductivity and grain-size analysis. Test borings and wells can also be used for slug and baildown tests for hydraulic conductivity.

**Piezometers** should be installed to determine the configuration of the potentiometric surface. These are small-diameter cased wells with a short well screen. They may be steel or plastic pipe. A minimum of three piezometers is necessary to determine the horizontal direction of ground-water flow if the potentiometric surface is a sloping plane. If it has a more complex shape, then more piezometers are necessary. In order to determine vertical head differences, closely spaced piezometers set at different depths are needed. A set of piezometers at different depths is called a **piezometer nest.** There should be at least one piezometer nest with separate shallow and deep piezometers for a small site, with more for larger sites.

The elevation of the ground surface and the top of the casing of each piezometer should be accurately surveyed. Measurements of the depth to water of each piezometer can then be used to construct a potentiometric map. Successive water-level measurements are necessary, as the elevation and configuration of the potentiometric surface may change with time. Likewise, the amount and direction of vertical flow components may vary.

In addition to the potentiometric maps, the water table may be shown on geologic cross sections. These are prepared on the basis of the test borings and wells. Piezometers are generally installed in a test-boring hole so that both geologic and hydraulic data are collected.

A great deal of hydrogeological judgment is needed in drawing groundwater maps. The number of data points may be limited and several interpretations possible; under such conditions draw the map that makes the most sense hydrogeologically.

Figure 12.26A shows a water-table map where there are some gaining streams and a lake that is in hydrologic contact with the water table. The water-table contours bend around the lake, even though there are wells with a water-table elevation of 775 on either side of the lake. Figure 12.26B shows a water-

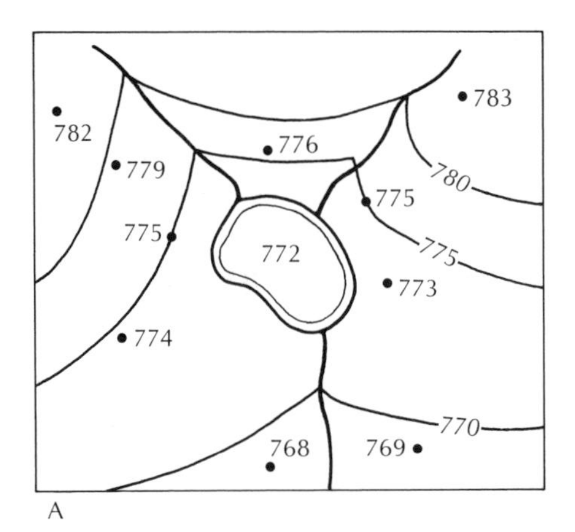


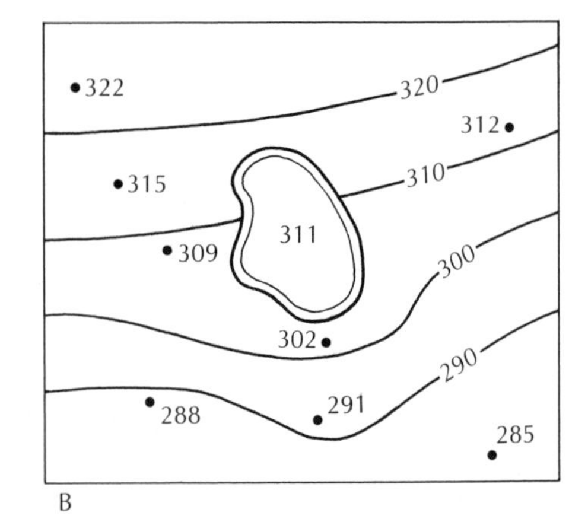


**FIGURE 12.26** Maps showing construction of water-table maps in areas with surface-water bodies. **A.** A water-table lake with two gaining streams draining into it and one gaining stream draining from it. **B.** A perched lake which through outseepage is recharging the water table.

table map with a perched lake. Notice that the surface of the lake is higher than the water table. The water table on the downgradient side of the lake is bowed away from the lake, indicating that outseepage from the lake is recharging the water table.

## 12.6 PROJECT REPORTS

The culmination of a hydrogeological study is usually a report. Some reports, such as a thesis or dissertation, will be written for a highly technical audience. Most project reports will have a readership that will include both technical and nontechnical persons. In these cases, the writer must take care to utilize an approach that can serve the needs of both. This is frequently done by making sure that there are some basic maps and illustrations that convey the general principles of the study. These may be accompanied by an "executive summary," which is written in nontechnical language. This will normally serve the needs of the nontechnical reader (who may well be the client who sponsored the project). In those sections meant for the nontechnical reader, it may be necessary to define terms and explain basic concepts in nontechnical terms.

The report will generally need to have sections that describe the technical aspects of the study in some detail. If the report is going to a reviewing agency, such as a permit granting body, then sufficient background information, basic data, and calculations should be included so that the work can be independently checked and verified. Usually the data and calculations are best put into an appendix so that they don't impede the reader who is not making an independent verification of the work.

The following outline covers the material that would go into a final project report.

1. A title page describing the report, who prepared it, for whom it was prepared, and a date.
2. An executive summary.
3. An introduction stating the purpose of the study, why it was made, and the general conditions under which it was made.
4. The conclusions that can be drawn about the site on the basis of the study.
5. Recommendations to the client or agency regarding the use of the site or the need for additional studies.
6. The body of the report consisting of
   (a) a review of previous work that was done on the site,
   (b) a description of the procedures and methods used in the study,
   (c) the general results of the field study and laboratory analyses,
   (d) an interpretation of the findings, and
   (e) appropriate maps and cross sections.
7. An appendix consisting of
   (a) acknowledgments of help given by others during the study,
   (b) a bibliography, and
   (c) technical data, computations, and other supporting evidence.

## REFERENCES


1. LATTMAN, L. H. "Technique of Mapping Geologic Fracture Traces and Lineaments on Aerial Photographs." *Photogrammetric Engineering,* 24 (1958):568–76.
2. PARIZEK, R. R. "On the Nature and Significance of Fracture Traces and Lineaments in Carbonate and Other Terranes." In *Karst Hydrology and Water Resources,* ed. V. V. Yevjevich. Fort Collins, Colo.: Water Resources Publications, 1976, pp. 3-1–3-62.
3. LATTMAN, L. H., and R. H. MATZKE. "Geological Significance of Fracture Traces." *Photogrammetric Engineering,* 27 (1961):435–38.
4. SETZER, J. "Hydrologic Significance of Tectonic Fractures Detectable on Air Photos." *Ground Water,* 4, no. 4 (1966):23–29.
5. SIDDIQUI, S. H., and R. R. PARIZEK. "Hydrogeologic Factors Influencing Well Yields in Folded and Faulted Carbonate Rocks in Central Pennsylvania." *Water Resources Research,* 7 (1971):1295–1312.
6. LATTMAN, L. H., and R. R. PARIZEK. "Relationship between Fracture Traces and the Occurrence of Ground Water in Carbonate Rocks." *Journal of Hydrology,* 2 (1964):73–91.
7. WOBBER, F. J. "Fracture Traces in Illinois." *Photogrammetric Engineering,* 33 (1967):499–506.
8. MC DONALD, H. R., and D. WANTLAND. "Geophysical Procedures in Ground

Water Study." *Transactions, American Society of Civil Engineers,* 126 (1961):122–35.

9. HEIGOLD, P. C., et al. "Aquifer Transmissivity from Surficial Electrical Methods." *Ground Water,* 17, no. 4 (1979):338–45.
10. BAYS, C. A. "Prospecting for Ground Water—Geophysical Methods." *Journal of American Water Works Association,* 42 (1950):947–56.
11. ZOHDY, A. A. R., G. P. EATON, and D. R. MABEY. "Application of Surface Geophysics to Ground-Water Investigations." In *Techniques of Water-Resources Investigations,* U.S. Geological Survey, 1974, Book 2, Chap. D1.
12. MOONEY, H. M., and W. W. WETZEL. "The Potentials About a Point Electrode and Apparent Resistivity for a Two-, Three-, and Four-Layer Earth." Minneapolis: University of Minnesota Press, 1956, 145 pp., 243 plates.
13. ZOHDY, A. A. R. "Geoelectric and Seismic Refraction Investigations Near San José, California." *Ground Water,* 3, no. 3 (1965):41–48.
14. GILKESON, R. H., and K. CARTWRIGHT. "The Application of Surface Electrical and Shallow Geothermic Methods in Monitoring Network Design." *Ground Water Monitoring Review* 3, no. 3 (1983):30–42.
15. YAZICIGIL, H., and L. V. A. SENDLEIN. "Surface Geophysical Techniques in Ground-Water Monitoring." *Ground Water Monitoring Review,* 2, no. 1 (1982):56–62.
16. STEWART, M., M. LAYTON, and T. LIZANEC. "Application of Surface Resistivity Surveys to Regional Hydrogeologic Reconnaissance." *Ground Water,* 21 (1983):42–48.
17. URISH, D. W. "The Practical Application of Surface Electrical Resistivity to Detection of Ground-Water Pollution." *Ground Water,* 21 (1983):144–52.
18. FRETWELL, J. D., and M. T. STEWART. "Resistivity Study of a Coastal Karst Terrain, Florida. *Ground Water,* 19 (1981):156–62.
19. REED, P. C., K. CARTWRIGHT, and D. OSBY. "Electrical Earth Resistivity Surveys near Brine Holding Ponds in Illinois." Illinois State Geological Survey Environmental Geology Notes 95, 1981, 30 pp.
20. STEWART, M. T. "Evaluation of Electromagnetic Methods for Rapid Mapping of Salt-Water Interfaces in Coastal Aquifers." *Ground Water,* 20 (1982):538–45.
21. SWEENEY, J. J. "Comparison of Electrical Resistivity Methods for Investigation of Ground Water Conditions at a Landfill Site." *Ground Water Monitoring Review,* 4, no. 1 (1984):52–59.
22. STEWART, M. T., and M. C. GAY. "Evaluation of Transient Electromagnetic Soundings for Deep Detection of Conductive Fluids." *Ground Water,* 24 (1986):351–56.
23. GREENHOUSE, J. P., and D. J. SLAINE. "The Use of Reconnaissance Electromagnetic Methods to Map Contaminant Migration." *Ground Water Monitoring Review,* 3, no. 2 (1983):47–59.
24. DOBRIN, M. B. *Introduction to Geophysical Prospecting,* 3rd ed. New York: McGraw-Hill, 1976, 630 pp.
25. MOTA, L. "Determination of Dips and Depths of Geological Layers by the Seismic Refraction Method." *Geophysics,* 19 (1954):242–54.

26. SVERDRUP, K. A. "Shallow Seismic Refraction Survey of Near-Surface Ground Water Flow." *Ground Water Monitoring Review,* 6, no. 1 (1986):80–83.

27. DENNE, J. E., and others. "Remote Sensing and Geophysical Investigations of Glacial Buried Valleys in Northeastern Kansas." *Ground Water,* 22 (1984):56–65.

28. KOERNER, R. M., and others. "Use of NDT Methods to Detect Buried Containers in Saturated Silty Clay Soil." *National Conference on Management of Uncontrolled Hazardous Waste Sites.* Silver Spring, Md.: Hazardous Materials Control Research Institute, 1982, pp. 12–16.

29. EVANS, R. B., R. C. BENSON, and J. RIZZO. "Systematic Hazardous Waste Site Assessments." *National Conference on Management of Uncontrolled Hazardous Waste Sites.* Silver Spring, Md.: Hazardous Materials Control Research Institute, 1982, pp. 17–22.

30. HITCHCOCK, A. S., and H. D. HARMON, JR. "Applications of Geophysical Techniques as a Site Screening Procedure at Hazardous Waste Sites." *Proceedings of the Third National Symposium on Aquifer Restoration and Ground-Water Monitoring.* Dublin, Ohio: National Water Well Association, 1983, pp. 307–12.

31. HORTON, K. A., and others. "The Complementary Nature of Geophysical Techniques for Mapping Chemical Waste Disposal Sites: Impulse Radar and Resistivity." *National Conference on Management of Uncontrolled Hazardous Waste Sites.* Silver Spring, Md.: Hazardous Materials Control Research Institute, 1981, pp. 158–64.

32. KOERNER, R. M., A. E. LORD, JR., and J. J. BOWDERS. "Utilization and Assessment of a Pulsed RF System to Monitor Subsurface Liquids." *National Conference on Management of Uncontrolled Hazardous Waste Sites.* Silver Spring, Md.: Hazardous Materials Control Research Institute, 1981, pp. 165–70.

33. GILKESON, R. H., P. C. HEIGOLD, and D. E. LAYMON. "Practical Application of Theoretical Models to Magnetometer Surveys of Hazardous Waste Disposal Sites—A Case History." *Ground Water Monitoring Review,* 6, no. 1 (1986):54–61.

34. WILSON, M. P., D. N. PETERSON, and T. F. OSTRYE. "Gravity Exploration of a Buried Valley in the Appalachian Plateau." *Ground Water,* 21 (1983):589–96.

35. KEYS, W. S., and L. M. MAC CARY. "Application of Borehole Geophysics to Water-Resources Investigations." In *Techniques of Water-Resources Investigations,* U.S. Geological Survey, 1971, Book 2, Chap. E1.

36. KEYS, W. S. *Borehole Geophysics as Applied to Groundwater.* Canadian Geological Survey Economic Geology Report 26, 1967, pp. 598–614.

37. BALDWIN, A. D., JR., and J. MILLER. "Use of a Gamma Logger to Delineate Glacial and Bedrock Stratigraphy in Southwestern Ohio." *Ground Water,* 17, no. 4 (1979):385–90.

38. BROWN, D. L. "Techniques for Quality-of-Water Interpretations from Calibrated Geophysical Logs, Atlantic Coastal Areas." *Ground Water,* 9, no. 4 (1971):25–38.

39. CROSBY, J. W., III, and J. V. ANDERSON. "Some Applications of Geophysical Well Logging to Basalt Hydrogeology. *Ground Water,* 9, no. 5 (1971):12–20.

40. NORRIS, S. E. "The Use of Gamma Logs in Determining the Character of Unconsolidated Sediments and Well Construction Features." *Ground Water,* 10, 6 (1972):14–21.

41. KEYS, W. S., and R. F. BROWN. "The Use of Temperature Logs to Trace the Movement of Injected Water." *Ground Water,* 16 (1978):32–48.

**42.** MAC CARY, L. M. "Geophysical Logging in Carbonate Aquifers." *Ground Water,* 21 (1983):334–42.

**43.** KEYS, W. S. "Analysis of Geophysical Logs of Water Wells with a Microcomputer." *Ground Water,* 24 (1986):750–60.

**44.** KWADER, T. "The Use of Geophysical Logs for Determining Formation Water Quality." *Ground Water,* 24 (1986):11–15.

**45.** MEYER, W. R. *Use of a Neutron Moisture Probe to Determine the Storage Coefficient of an Unconfined Aquifer.* U.S. Geological Survey Professional Paper 450-E, 1963, pp. 174–76.

# THIRTEEN

# Ground-Water Models

In nature the hydraulic system in an aquifer is in balance; the discharge is equal to the recharge and the water table or other piezometric surface is more or less fixed in position. Discharge by wells is a new discharge superimposed on the previous system. Before a new equilibrium can be established water levels must fall throughout the aquifer to an extent sufficient to reduce the natural discharge or increase the recharge by an amount equal to the amount discharged by the well. Until this new equilibrium is established water must be withdrawn from storage in the aquifer and conversely the new equilibrium cannot be established until an amount of water is withdrawn from storage by the well sufficient to depress the piezometric surface enough to change the recharge or natural discharge the proper amount. The depression of the piezometric surface is called the cone of depression.
*The Significance and Nature of the Cone of Depression in Ground-Water Bodies,*
Charles V. Theis, 1938

## 13.1 INTRODUCTION

There are two areas of hydrogeology where we need to rely upon models of the real hydrogeologic system: to understand why a flow system is behaving in a particular observed manner and to predict how a flow system will behave in the future. The term **model** refers to any representation of a **real system.** In studying a ground-water flow system we develop a **conceptual model.** For example, we might describe a ground-water system as being contained in deposits of glacial sand and gravel overlying a nearly level bedrock surface consisting of Precambrian-age granite. The sand and gravel unit contains a water table near the land surface, and ground water flows from a nearby moraine into a stream. Such a conceptual model is less complex than a real system. No matter how much field work would be performed in describing the above system, our conceptual model could never fully describe all of the minute details of the real system. Nonetheless, conceptual models are necessary for us to be able to understand flow-system behavior.

Conceptual models are static. They describe the present condition of a system. In order to make predictions of future behavior, it is necessary to have some sort of dynamic model that is capable of manipulation. There are many types of dynamic models of ground-water flow. They include **physical scale models, analog models,** and **mathematical models** (1, 2).

A **scale model** is made from the same materials as those of the natural system. For instance, a plastic container may be manufactured to scale and filled with sand or glass beads with a hydraulic conductivity scaled to the actual aquifer material. Dye added to the water helps the observer trace the flow. Pressure is measured in piezometers inserted through the walls. Water can be added to the model aquifer to simulate recharge and can also be pumped from scale-model wells. Sand models have been used for a variety of studies (e.g., 3–5).

The flow of water through porous media is governed by equations similar to those governing the flow of electricity through a conductor. This also is true for the flow of a viscous fluid between two very closely spaced parallel plates. Models can be constructed using electrical circuits or viscous-fluid flow to simulate real or ideal aquifers. These are called **analog models,** since the model is analogous to the actual aquifer. Analog models are typically constructed to model two-dimensional flow. The models can be either horizontal or vertical. If areal flow patterns are being studied, a horizontal model is indicated. In order to study vertical flow, a cross-sectional or vertical model is the one of choice.

**Electrical models** (e.g., 6, 7) are made with a network of resistors scaled to represent the framework of the aquifer, with capacitors to provide for storage. The flow of electrical current in amperes through the model is representative of fluid flow. The voltage in the model corresponds to hydraulic potential, and the volume of water in storage is analogous to coulombs of electricity. Resistivity is inversely proportional to the hydraulic conductivity of the aquifer, while the capacitance network is scaled to aquifer storativity. Measurements of current and voltage made at various points in the model represent the flow of water and hydraulic head in the aquifer. Electrical analog models can be made to represent three-dimensional flow by linking a series of horizontal models together. While electrical analogs can be used to study the flow of water, they are not amenable to studying the flow in unconfined aquifers, where aquifer transmissivity decreases with pumpage as the water table declines. They are also unable to simulate mass transport, dispersion, and diffusion (1).

The **viscous-fluid model** is also known as a **Hele-Shaw model** after H. S. Hele-Shaw, who first used it. It has been applied to many problems in hydrogeology. Hele-Shaw models are especially adapted to the study of immiscible fluids with different densities; for example, as encoutered in salt-water intrusion (8). Two liquids with different densities are used to simulate the fresh and salt water. The injection of wastewater into flow fields can be visually studied using Hele-Shaw models (9). Both horizontal and vertical models have been built. Because of physical constraints, it has not been possible to model three-dimensional flow with a viscous-fluid model.

All analog models, and scale models, as well, have some definite disadvantages. Not the least of these is that the models must be constructed by someone handy at carpentry, plumbing, and wiring. The time and materials cost for large models is substantial. There must be room to construct and house the models, and possibly to store them when not in use. Finally, the models are not very flexible. It is difficult to change the aquifer geometry and hydraulic characteristics built into a model.

Scale models and Hele-Shaw models are advantageous in that the fluid movement is visible through the use of dyes. This is particularly helpful in presentations to those not familiar with subsurface flow. Under some conditions, such as transient flow with closely spaced wells and nonlinear boundaries, the resistance-capacitance electrical analog may be more accurate than digital-computer models (1).

Mathematical models rely upon the solution of the basic equations of ground-water flow, heat flow, and mass transport. The most simple mathematical model of ground-water flow is Darcy's law. To apply Darcy's law we need to have a conceptual model of the aquifer and to develop data on the physical properties of the aquifer system, the potential field, and the fluid properties. Darcy's law is an example of an **analytical model.** In order to solve an analytical model we need to know the initial and boundary conditions of the flow problem. These conditions need to be simple enough so that the flow equation can be solved directly by using calculus. Analytical flow models have been developed to simulate the flow of water to wells and streams (10) as well as for heat and mass transport (11).

Analytical models can be solved rapidly, accurately, and inexpensively with a programmable calculator or small personal computer. A minimal amount of data is needed since all sectors of a parameter are assigned the same value. For example, an average value for hydraulic conductivity would be used for an aquifer. If the problem involved a boundary or multiple wells, image well theory can be applied to the solution (12).

**Numerical models** must be used when there are complex boundary conditions or where the value of parameters varies within the model area (13). Numerical solutions to the flow, heat, and mass transport equation require that they be recast in an algebraic form. These recast equations are numerical approximations and the answers obtained are also approximations. The equations are most commonly in matrix form and they are solved on a digital computer. Numerical models are one of the most important developments in hydrogeology of the last 15 years.

**Stochastic models** of ground-water flow based on statistical theory are being developed. The number of papers on this topic has increased from none in 1975 to more than 125 in 1983 (14). Obviously this approach to understanding subsurface flow is in a state of rapid flux. However, stochastic models, as contrasted with numerical models, have not come into widespread use (15). This is perhaps due to the immature state of development of the models and the theoretical complexity associated with them.

## 13.2 APPLICATIONS OF GROUND-WATER MODELS

Ground-water models have been applied to four general types of problems: ground-water flow, solute transport, heat flow, and aquifer deformation (16–22). All of the models start with the basic equation of ground-water flow (Equation 5-45). This is solved for the head distribution in the aquifer. Solute-transport models add an equation for the changes in chemical concentration in the ground water. Heat-transport equations utilize an equation for transfer of heat in the aquifer. Aquifer deformation models combine the flow equation with other equations that describe the changes in the physical structure of the aquifer with changes in the aquifer head.

Two broad classes of models have appeared: those that deal with flow through porous media and those that deal with flow through fractured media. A sand and gravel or sandstone aquifer is an example of an aquifer best described by use of a porous-medium model. A fractured crystalline rock with widely spaced fractures would be an example of an aquifer best described by use of a fractured-rock model.

Models have been written for the flow of two or more immiscible fluids in an aquifer and for both saturated and unsaturated flow.

Ground-water flow models have been applied to the study of regional steady-state flow in aquifer systems; regional changes in hydraulic head caused by changes in discharge or recharge; changes in head near a well field, dewatering well system, injection well, or infiltration basin; and surface-water–ground-water interactions.

Solute-transport models have been used in studies of salt-water intrusion, leachate movement from landfills and other waste-disposal sites, contamination plumes from seepage ponds, radionuclide movement from radioactive-waste sites, movement of pesticides from agricultural fields, and other types of ground-water pollution.

Heat-transport models have been used in the analysis of thermal impact from high-level radioactive-waste storage, storage of thermal energy in aquifers, and ground-water heat-pump impacts. Land subsidence due to ground-water withdrawals has been studied with the use of deformation models.

## 13.3 DATA REQUIREMENTS FOR MODELS

In order to successfully transform a conceptual model into a mathematical, analog, or scale model it is necessary to have a data base that provides adequate information to apply the requisite equations. All of the models start with a ground-water flow model. For this, one needs to know the physical configuration of the aquifer. This includes the location, areal extent, and thickness of all of the aquifers and confining layers; the locations of the surface water bodies and streams; and the boundary conditions of all aquifers. Hydraulic properties that need to be known include the variation of transmissivity or permeability and storage coefficient of the aquifers, the variation of permeability and specific storage of the confining layers, and the hydraulic connection between the aquifers and surface-water bodies. Hydraulic energy, as indicated by water-table or potentiometric-surface maps, and the amounts of natural-aquifer recharge and natural streamflow are also needed.

In order to model stresses on the natural ground-water flow system the modeler must know the locations, types, and amounts through time of any artificial recharge, such as results from recharge basins and wells or return flow from irrigation, as well as the amounts and locations through time of ground-water withdrawals from wells. Changes in the amounts of water flowing in the streams and changes in the water levels of surface-water bodies should also be known.

Solute-transport models need, in addition to the flow-model data, the following information: distribution of effective porosity, aquifer dispersivity factors, fluid-density variations, and natural concentrations of solutes distributed throughout the ground-water reservoir. The locations and strengths of the sources of contamination must be known as well as retardation factors for the specific solutes with the specific rocks and soils of the area. The flow model is used to compute direction and rate of fluid movement, and the solute-transport equations are "piggybacked" onto the flow model to derive movement and retardation values of contaminants.

A model is initially calibrated by taking the initial estimates of the model parameters and solving the model to see how well it reproduces some known condition of the aquifer. Most models are initially calibrated against the steady-state ground-water heads. A water-table or potentiometric-surface map is required for this type of calibration. As this is almost always known with better accuracy than the distribution of aquifer parameters and/or amount of recharge, the values for the aquifer parameters and recharge are varied until the model closely reproduces the known water-table or potentiometric-surface condition. This process is known as **model calibration.** As the same result can often be obtained be changing two variables simultaneously, it is highly desirable to **verify** a model once it has been calibrated. This is usually done by history matching. A transient response of the model is obtained and compared with a known transient condition in the aquifer. For example, the water levels in the aquifer may have declined over time owing to withdrawal from wells. If the water levels through time and the locations and pumping rates of the wells are known, the model is verified if it reproduces the known water-level changes by inputting the known ground-water withdrawal history. If the known history is not reproduced with a desired degree of accuracy, the model parameters can be changed to recalibrate the model with a new set of model parameters and the verification run repeated. This process will eventually result in a calibrated and verified model.

In verifying against a transient condition, the aquifer storage coefficient is likely to be the parameter that must be adjusted as it will not have been utilized in the calibration against steady-state conditions. Once the model has been verified against a transient event, it should be checked against the steady-state condition to assure that it still is calibrated. **Field verification** can then be performed by actually stressing the aquifer to see if the model correctly predicts the response of the aquifer as it is stressed; for example, by performing a pumping test (23). Most computer models are not subjected to field verification as this is a very time-consuming and expensive step. Transient events for history-matching can include pumping tests in addition to long-term water-level declines. In some cases it may be necessary to use a calibrated model that has not been verified. This can occur if there is no history of water-level changes against which to verify the model. Such a model can certainly be useful, but it should be recognized that any predictions are less likely to be valid than those made with a verified model. The more accurate the data that initially go into a model, and the more detailed the data against which it is verified, the more confidence the modeler can have in the model results.

The process of model calibration and verification frequently requires many changes in the data parameters that make up the model. Scale and analog models might require rebuilding each time a change is made in data values. Numerical models can be easily recomputed with the new data. This is the primary reason that numerical models have apparently almost totally replaced other types of model. The rapid growth in computing power and fall in price of personal computers has put the ability to cheaply and easily solve numerical models on the desks of many, if not most, hydrogeologists. The remainder of this chapter will look at computer models of ground-water flow.

## 13.4 FINITE-DIFFERENCE MODELS

### 13.4.1 FINITE-DIFFERENCE GRIDS

In a conceptual model that is continuous, we presume to know the properties of the aquifer at every point within the boundaries. We can replace the continuous model with a set of discrete points arranged in a grid pattern. Figures 13.1A and 13.1B show two variations of a **finite-difference grid**. Associated with the grid are **node points**, where the equations are solved to obtain unknown values. Also associated with each node are the known values of such parameters as transmissivity and storativity. A **block-centered grid** (Figure 13.1A) is one where the node points fall in the center of the grid; a **mesh-centered grid** (Figure 13.1B) has the node points at the intersections of grid lines. The choice of whether to use a block-centered or a mesh-centered grid depends upon the boundary conditions. A block-centered grid is most useful when a flux is specified across a boundary, and a mesh-centered grid is most convenient for situations where the head is specified at the boundary.

The basic grid is regular, with the rows and columns being normal to each other and the distance in the $x$ direction, $\Delta x$, being equal to the distance in the $y$ direction, $\Delta y$. Often it is convenient to vary the size of the rows and col-

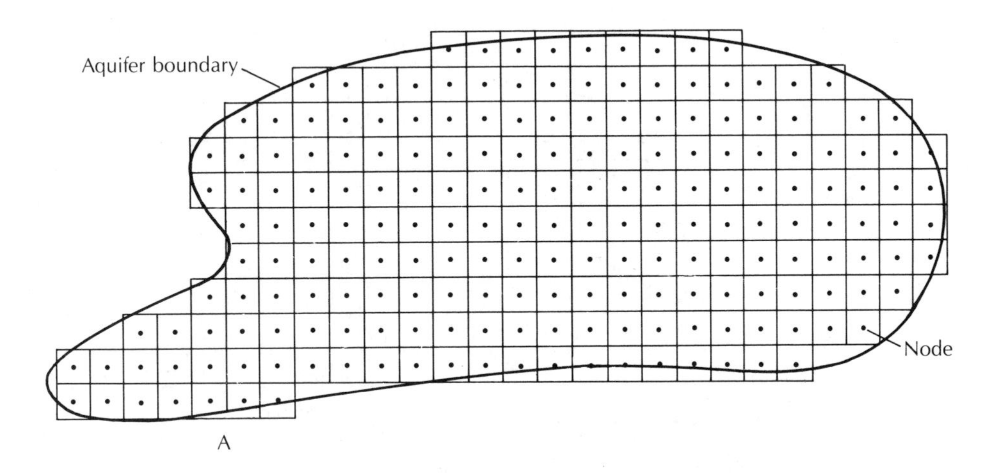


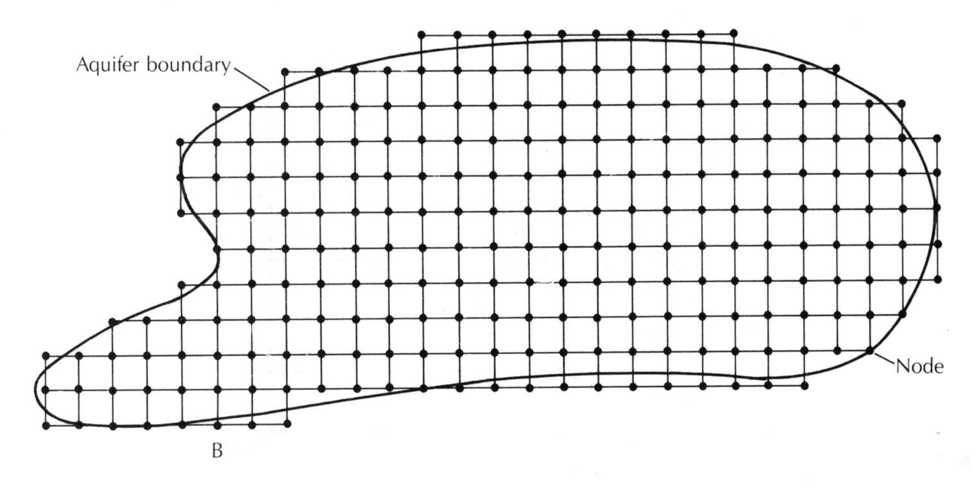


**FIGURE 13.1** Finite-difference grids laid over an aquifer. **A.** Block-centered finite-difference grid. **B.** Mesh-centered finite-difference grid.

umns so that there are more node points in certain parts of the aquifer than others. This might be desirable in the area around a well field, for example, where there are more changes in head expected than in other parts of the aquifer (Figure 13.2). The size of the difference in $\Delta x$ or $\Delta y$ from one column or row to an adjacent one should not change by more than 50 percent in a variable grid spacing.

### 13.4.2 FINITE-DIFFERENCE NOTATION

Special notation is used to describe the positions of the nodes in the finite-difference grids. Figure 13.3A shows a finite-difference grid centered on $(x,y)$. Adjacent points on the grid are located at a distance $\Delta x$ away to the right

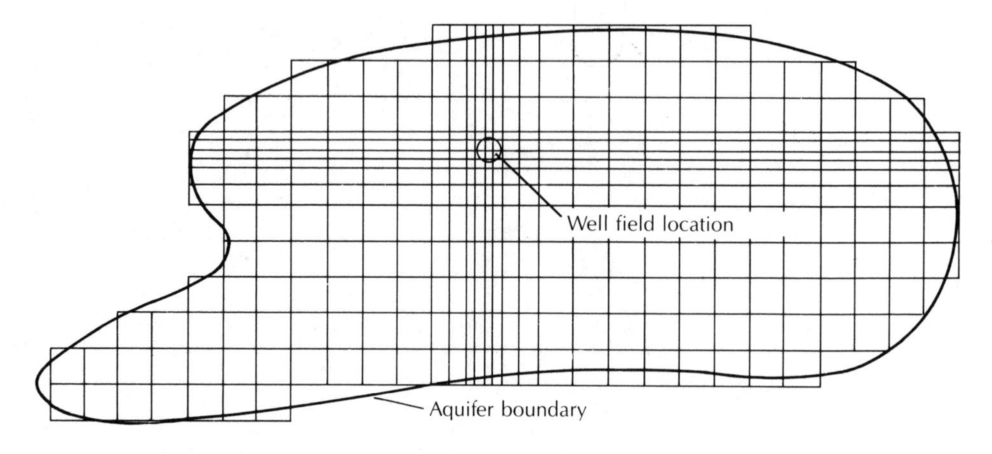


**FIGURE 13.2** Variable spacing finite-difference grid.

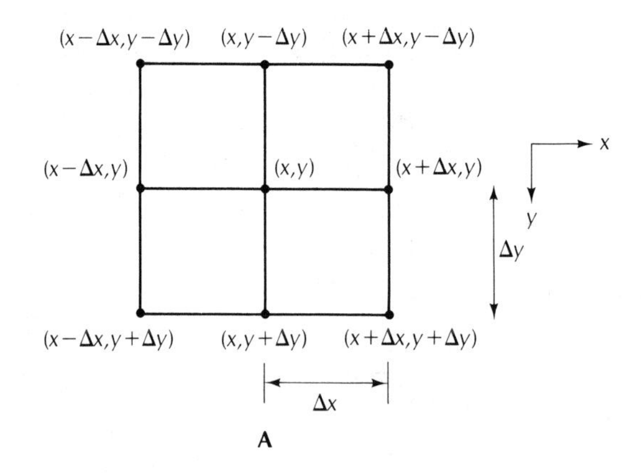


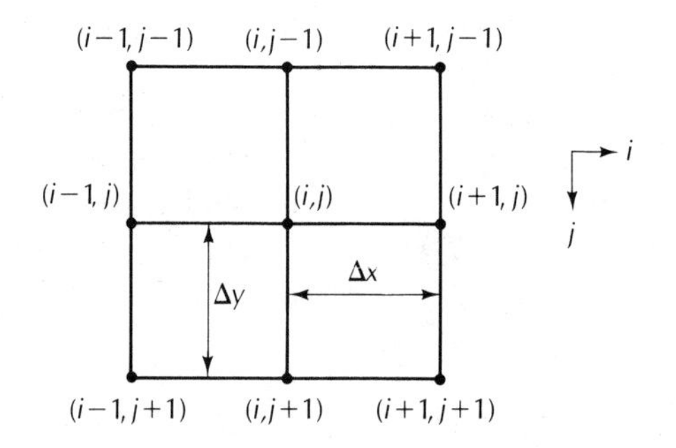


**FIGURE 13.3** **A.** Finite-difference grid. **B.** Computer notation for finite-difference grid.

or left and $\Delta y$ away up or down; $x$ is positive to the right and $y$ is positive downward.

In the computer codes the locations of the nodes are designated with reference to node $i,j$, where $i$ represents the column and $j$ represents the row. The notation for $i$ is positive to the right and for $j$ it is positive downward. Thus the row above the $j$ row is row $j - 1$ and the row below the $j$ row is $j + 1$. The column to the left of column $i$ is column $i - 1$ and the column to the right of column $i$ is $i + 1$.

### 13.4.3 BOUNDARY CONDITIONS

In order to solve the ground-water flow equation we must be able to specify the boundary conditions. There are two basic types of boundary conditions (17). If the head is known at the boundary of the flow region, this is known as a **Dirichlet condition**. If the flux across a boundary to the flow region is known, this is a **Neumann condition**. In some cases the boundary conditions will be mixed, with some portions having known head and some portions having known flux.

As an example of boundary conditions, examine Figure 13.4. This is a sand and gravel aquifer overlying an impermeable basement rock. There are several flow regions present, each leading away from a ground-water divide toward a stream. We shall look at the regional flow in the region bounded by the letters W, X, Y, and Z.

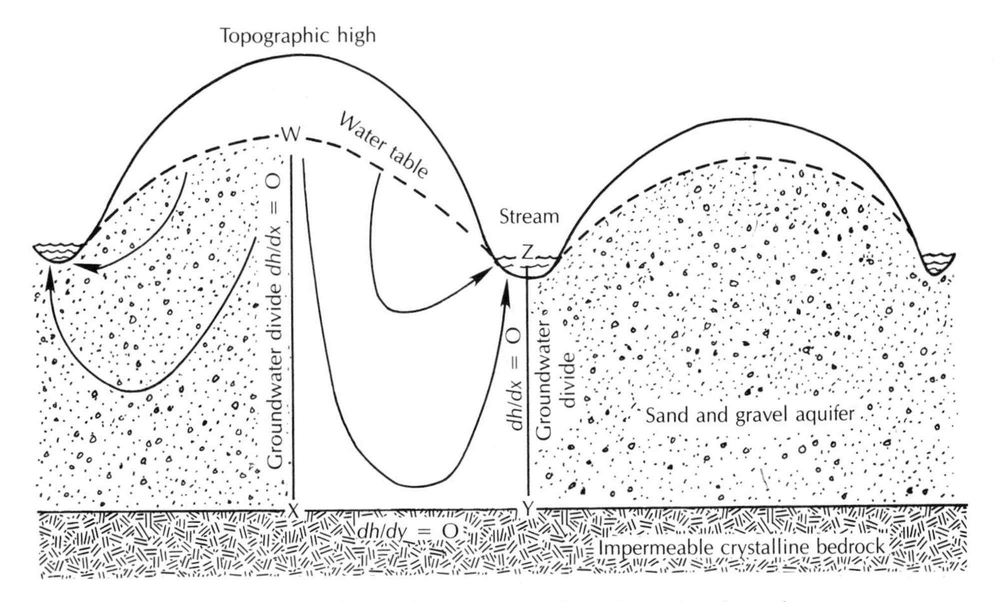


**FIGURE 13.4** Boundary conditions for a cross section of a regional aquifer.

The plane of W to X is a ground-water divide. There is no flow across the divide and the boundary condition along the divide is $dh/dx = 0$. The plane of Z to Y, which underlies the center of the stream, is also a ground-water divide, across which no flow takes place. The boundary condition along this divide is $dh/dx = 0$. No flow occurs from the sand deposit into the impermeable bedrock, so along the X to Y plane there is no flow and the boundary condition is $dh/dy = 0$. These first three boundaries are Neumann boundaries, where we have specified the flow condition.

There are two ways in which we could treat the upper boundary—the W to Z surface—which is the water table. We could specify that the position of the water table is fixed. This would be a Dirichlet boundary. For steady-state conditions—that is, no change of head with time—we could solve this model to give us the amount of recharge needed to maintain the water table in the observed position. The second approach would be to specify the flux across the water table—that is, the amount of recharge. This then becomes a Neumann boundary. This model could be solved to determine the position of the water table under various amounts of recharge. Both of the above models require that the aquifer permeability is known. If we wish to simulate the movement of the water table with time under Neumann conditions, we would also need to know the aquifer storativity.

### 13.4.4 METHODS OF SOLUTION FOR STEADY-STATE CASE FOR SQUARE GRID SPACING

In the absence of recharge to the aquifer, the finite-difference equation for the steady-state case (Laplace equation) is (17)

$$h_{i-1,j} + h_{i+1,j} + h_{i,j-1} + h_{i,j+1} - 4h_{i,j} = 0 \quad \textbf{(13-1)}$$

When solved for $h_{i,j}$, Equation 13-1 becomes (17)

$$h_{i,j} = (1/4)(h_{i-1,j} + h_{i+1,j} + h_{i,j-1} + h_{i,j+1}) \quad \textbf{(13-2)}$$

The value of $h_{i,j}$ is the average of the heads at the four closest nodes in the nodal mesh.

If there is recharge to the aquifer, then the finite-difference equation for the steady-state case (Poisson's equation) is (17)

$$(h_{i-1,j} - 2h_{i,j} + h_{i+1,j})/(\Delta x)^2 + (h_{i,j-1} - 2h_{i,j} + h_{i,j+1})/(\Delta y)^2 = -R/T \quad \textbf{(13-3)}$$

where

$\Delta x$ and $\Delta y$ are the distances between nodes in the $x$ and $y$ directions
$R$ is the recharge
$T$ is the aquifer transmissivity

In a finite-difference mesh there are from tens to hundreds of nodes. Each node requires the solution of a form of Equation 13-2 or 13-3. The method of solution is to make an initial guess of the value of head for each of the nodes in the mesh. For Dirichlet boundary nodes the head will be fixed. For all other boundary nodes and interior nodes the value of the head will not be fixed.

The finite-difference equation is solved by what are known as **iterative methods.** On the basis of the fixed head values, plus the initial guesses, Equation 13.1 is solved for each node on the basis of the values at the surrounding four nodes. A computer code sweeps the solution path through the finite-difference grid so that for each node, other than the first one and the last one, the head values at some of the adjacent nodes will be based on the initial guess, while at the remainder of the adjacent nodes the head value will already have been recomputed by Equation 13.1. Once the head at each node has been recomputed, the difference between the initial guess and the recomputed head is determined. The process is repeated until the maximum difference in head values from one iteration to the next is less than some preset value known as the **convergence criterion.** When the solution has converged, the equation has been solved. The solution is approximate as there is some finite value to the convergence criterion. The smaller that value is, the more iterations, and hence longer period of time, it takes to reach the solution. There is some practical tradeoff between accuracy of the solution and the amount of computer time expended to reach it.

The **Gauss-Seidel** iteration method computes the value of $h_{i,j}$ during an iteration on the basis of heads at two adjacent nodes, which have been computed during the current iteration, and heads at two adjacent nodes, which were computed during the prior iteration.

The Gauss-Seidel equation for the Laplace equation is

$$h_{i,j}^{m+1} = (1/4)(h_{i-1,j}^{m+1} + h_{i,j-1}^{m+1} + h_{i+1,j}^{m} + h_{i,j+1}^{m} \quad \textbf{(13-4)}$$

where the superscript $m$ indicates a value computed during a prior iteration and $m + 1$ indicates a value computed during the current iteration.

The Gauss-Seidel equation for Poisson's equation in a grid where $\Delta x = \Delta y$ is (17)

$$h_{i,j}^{m+1} = 1/4(h_{i-1,j}^{m+1} + h_{i,j-1}^{m+1} + h_{i+1,j}^{m} + h_{i,j+1}^{m} + \Delta x^2 R/T) \quad \textbf{(13-5)}$$

The method of **successive overrelaxation** (SOR) is a variation of the Gauss-Seidel method. The difference in head value computed at a mode by two Gauss-Seidel iterations is known as the residual. During each Gauss-Seidel iteration, the residual shrinks until the convergence criterion is reached. In successive overrelaxation, a factor is added to increase the rate of convergence. The overrelaxation factor is a number between 1 and 2, and an acceptable value is normally determined by trial and error. If we designate the value of $h_{i,j}$ as determined by a Gauss-Seidel iteration as $(h_{i,j}^{m+1})$ and the overrelaxation factor as $f$, the value of $h_{i,j}^{m+1}$ as determined by the SOR method is (17)

$$h_{i,j}^{m+1} = h_{i,j}^{m} + f(h_{i,j}^{m+1} - h_{i,j}^{m}) \quad \textbf{(13-6)}$$

where $(h_{i,j}^{m+1} - h_{i,j}^{m})$ is the residual.

### 13.4.5 METHODS OF SOLUTION FOR THE TRANSIENT CASE

In a transient problem the head is a function of time. This is the type of solution that would be applied to problems such as determining the change in head around a pumping well or the growth of a ground-water mound beneath a recharge basin. Equation 5-47 is the transient-flow equation in two dimensions with a factor for recharge. This equation can be adapted to unsaturated aquifers if the transmissivity varies with time as the saturated thickness of the aquifer changes with changes in head.

The **alternating direction implicit method** of solution is used in a number of published computer codes (18, 19). Equation 5-45 is the two-dimensional flow equation without recharge. The fully implicit finite-difference approximation is (17)

$$h^{n+1}_{i+1,j} + h^{n+1}_{i-1,j} + h^{n+1}_{i,j+1} + h^{n+1}_{i,j-1} - 4h^{n+1}_{i,j} = [(1/\mathrm{T})(Sa^2)][(1/\Delta t)(h^{n+1}_{i,j} - h^{n}_{i,j})] \quad \textbf{(13-7)}$$

where

$S$ is storativity

$T$ is transmissivity

$\Delta t$ is the length of the time step

$a = \Delta x = \Delta y =$ the dimensions of the finite-difference grid

$n$ represents the $n$th time step

The fully implicit finite-difference approximation can be combined with an iterative solution, such as the Gauss-Seidel. In this method the head is first determined along columns and then along rows. For each time step, a solution is obtained for the head at every point on the finite-difference grid by convergence to a residual value less than the convergence criterion. The convergence criterion is checked after each two iterations as the program must go through all the points along columns and then rows for a complete iteration cycle. The program then steps to the next time increment and the process is repeated. In the following equation, the time step counter, $n$, is now shown as a subscript while the iteration step counter, $m$, is shown as a superscript. The heads in column $i$ at time step $n + 1$ may be found from the following equation (17)

$$h^{m+1}_{i,j-1,n+1} + [-4 - (Sa^2)/(T\Delta t)]h^{m+1}_{i,j,n+1} + h^{m+1}_{i,j+1,n+1} = [(-Sa^2)/(T\Delta t)]h_{i,j,n} - h^{m}_{i+1,j,n+1} - h^{m+1}_{i-1,j,n+1} \quad \textbf{(13-8)}$$

For the next iteration ($m + 2$) the solution is oriented along rows. The heads in row $j$ at time step $n + 1$ may be found from the equation (17)

$$h^{m+2}_{i-1,j,n+1} + [-4 - (Sa^2)/(T\Delta t)]h^{m+2}_{i,j,n+1} + h^{m+2}_{i+1,j,n+1} = [(-Sa^2/(T\Delta t)]h_{i,j,n} - h^{m+1}_{i,j+1,n+1} - h^{m+2}_{i,j-1,n+1} \quad \textbf{(13-9)}$$

The solution of a transient problem takes much more time than solution of a steady-state problem as the method is to solve a series of steady-state problems, each separated by a time step. The amount of central processor time in the computer used by the transient type of problem is thus greatly increased.

There are several other methods of solution for transient finite-difference models. These include line-successive overrelaxation and the strongly implicit procedure. Several of the ground water model codes, including the Trescott, Pinder, and Larson model (18) and the McDonald and Harbaugh model (19), offer the user the choice of more than one method of solution. For a particular problem it may be that the user would find one method faster than another. This should be determined during the calibration procedure for each area modeled.

## 13.5 FINITE-ELEMENT MODELS

Finite-element models offer an alternative approach to the numerical modeling of ground-water flow. Rather than use the rectangular network of nodes that is used in the finite-difference method, the aquifer is divided into polygonal cells, typically triangular, but not necessarily so. Figure 13.5 shows the finite-element cells for an aquifer to be modeled. The triangles intersect at nodes that represent the points at which the unknown values, such as heads, will be computed. The value of the head in the interior of each cell is determined by interpolation between the nodal points.

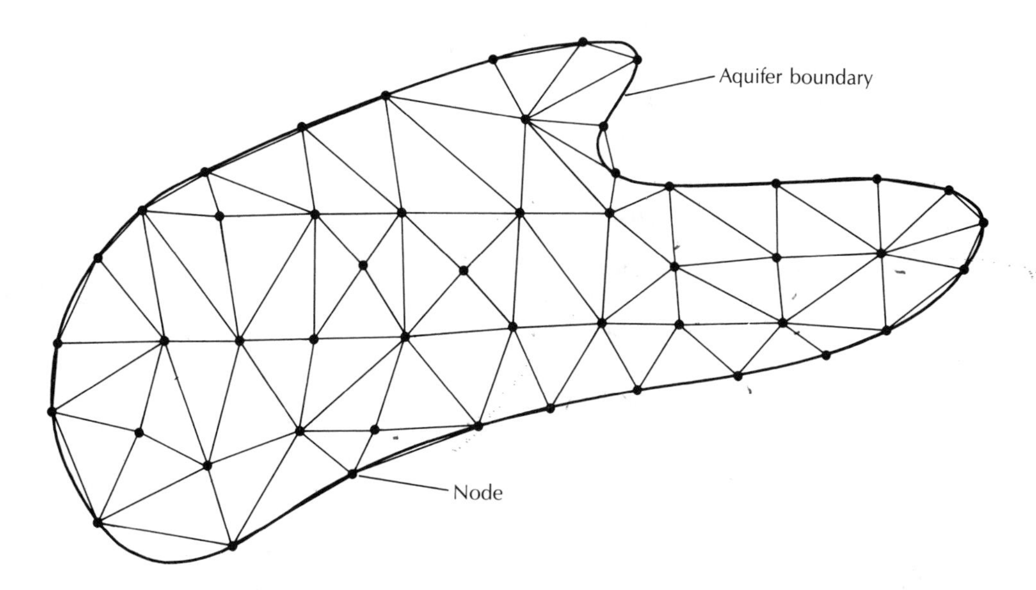


**FIGURE 13.5** Finite-element grid for an aquifer.

The mathematical basis for the finite-element method is much more complex than that of the finite-difference method. Most finite-element solutions are based on Galerkin's method. This method is described elsewhere (17, 20). The mathematics are not intuitively obvious, as are those of the finite-difference method, and will not be pursued further here.

Finite-element models are reported to be somewhat superior to finite-difference models for problems that have a moving boundary, such as a cross-sectional model with a water table that is transient, as well as coupled problems, such as contaminant transport (17). The finite-element model also has the advantage of being much more flexible in terms of mimicking the geometry of an aquifer system than the finite-difference method and requires fewer nodes.

## 13.6 METHOD OF CHARACTERISTICS

This is a useful method for simulating mass transport in ground water. It is usually used in conjunction with a finite-difference model that simulates the flow of the ground water. The finite-difference grid then becomes the coordinate system for the movement of the fluid and the contained solute. The method of characteristics follows the fluid and computes the changing concentrations of the solute. In the two-dimensional problem there are three equations: one each for the $x$-velocity, the $y$-velocity, and the concentration. The solution of each yields what is known as a characteristics curve (16). Points, representing solutes, are placed in various cells of the finite-difference grid. The computer code then moves each of them a distance proportional to the length of the time increment and the ground-water velocity at that point. This simulates convective transport and the remaining factors of the mass-transport equation are solved with finite-difference techniques.

## 13.7 USE OF PUBLISHED MODELS

Most hydrogeologists will never devise a new method of solution for the flow equations or write a new computer code for ground-water modeling. A large number of computer models have already been developed and more will be developed in research institutions. Published models have certain advantages over custom-designed ones. It is much less expensive to apply an already developed model, if an available model will accomplish the desired goal. The model will already have been "debugged" and most often applied to a field situation so the user has a good chance that the model will actually work in the intended application. Finally, if the published model has been widely used and accepted by practicing hydrogeologists, the results of another application will be more readily accepted than if an entirely new model code has been developed. This may be very important if the model results are to be used in a regulatory process or in litigation.

The U.S. Geological Survey has published several computer models. These are well documented and well tested. They are in the public domain and versions are readily available at moderate cost for both mainframe and personal computers.

The **McDonald and Harbaugh model** (19) is an extremely versatile finite-difference ground-water flow model. Documentation is available from the U.S. Geological Survey. It is three-dimensional and the layers can be either water table, confined, or a combination of both. The model simulates recharge, evapotranspiration, areal recharge, flow to wells, flow to drains, and flow through river beds. It is set up as a series of separate modules, which are independent; the user selects only the modules needed for the particular system that is under study. The user may choose between different methods of solution. The code is written in FORTRAN and some knowledge of that language is helpful in formatting the data for input into the model. The model is simple to use and is available on floppy disks for use in an IBM-compatible personal computer. At least 256K of random access memory is needed, but 640K would be better. The solution time is speeded up if the computer has an 8087 math coprocessor (an optional computer chip that is plugged into the motherboard of the computer).

The **Konikow and Bredehoeft model** (21) is a widely used solute-transport model. Documentation is available from the U.S. Geological Survey. It is based on a finite-difference grid and uses the method of characteristics for solution. It is two-dimensional and computes changes of concentration with time caused by advection, hydrodynamic dispersion, and mixing. The model assumes that the solute is nonreactive and that gradients in fluid density, viscosity, and temperature do not affect the flow velocity. This program is also available on floppy disks for use on an IBM-compatible personal computer. A full 640K of memory is required. A preprocessor is available to aid in setting up the program and entering the data.

The **Prickett Lonquist Aquifer Simulation Model,** PLASM (22), is an earlier finite-difference ground-water–flow model, but is still very useful. Documentation is available from the Illinois State Geological Survey. In essence this is a series of related programs that can be used for distinct applications. Much less memory is required than for the U.S. Geological Survey programs as the user inputs only the portion of the model that is needed. (In the U.S. Geological Survey program, the entire program code is in memory and the user then selects the portions that are needed.) Both two- and three-dimensional as well as water-table and confined-aquifer models are included.

**SUTRA** (saturated-unsaturated transport) is a finite-element simulation model for fluid movement and the transport of either dissolved solutes or energy under either saturated or unsaturated conditions. Under the solute-transport mode, the solute may be subjected to adsorption on the porous medium as well as decay or production. In addition, the flow of variable-density fluids, such as saline water or leachate, may be simulated. The heat-transport mode can simulate such things as subsurface heat conduction, geothermal reservoirs, and thermal storage in aquifers. The model may be employed in either a cross-sectional or horizontal configuration.

Model codes and instructions are available from a number of sources. The International Ground Water Modeling Center of the Holcomb Research Institute, Butler University, Indianapolis, Indiana 46208, is a clearinghouse for nonproprietary ground-water models. In addition, a variety of commercial models are marketed through the National Water Well Association, 6375 Riverside Drive, Dublin, Ohio 43017. Documentation and program disks for IBM personal computers for the McDonald and Harbaugh model and the Konikow and Bredehoeft model are sold by Scientific Publications Co., P. O. Box 23041, Washington, D.C. 20026. That supplier also has graphic preprocessors available for the above models. The preprocessors can be used to organize the data and set up the model as well as graphics display programs to exhibit the output. The basic model output is in tabular form. Scientific Publications Co. also has the documentation and computer tapes for SUTRA.

## CASE STUDY: GROUND-WATER MODELING FOR A PLANNED UNDERGROUND MINE

A large massive sulfide ore body was recently discovered in northeastern Wisconsin (24–26). The ore body has recoverable reserves of 67.4 million tons of zinc, copper, and lead ore. It is tabular in shape, with a length of 4200 feet (1280 meters) and a width ranging from 0 to 200 feet (61 meters), and extends from the bedrock surface to a depth of 1800 feet (550 meters) below the bedrock surface. The bedrock surface is overlaid by some 200 feet (60 meters) of glacial drift, primarily permeable outwash.

The bedrock in which the ore body is found is Precambrian-age metasedimentary and metavolcanic rock. The top of the ore body is weathered and more permeable than the surrounding bedrock. The planned mine is to be constructed as an underground working. Because of the large amount of ground water in the ore body and the overlying glacial sediments, mine dewatering will be necessary. The dewatering system is expected to lower the water table in the glacial drift aquifer above the mine. In order to be able to predict how much water will be pumped from the mine, the surface extent and depth of the resulting cone of depression, and the environmental impact of the lowered water table on lakes, streams, and wetlands of the region, a ground-water model of the project was developed.

Extensive test borings in both the glacial drift materials and the bedrock materials were made in order to establish the geologic framework for the model. Numerous piezometers were installed to determine the hydraulic head in the various aquifers. Several lakes are in the area of the expected cone of depression and the water budgets for the lakes and their hydraulic relationship to the water table were studied. Over a period of several years, streamflows were gaged, the heads in piezometers measured, and the stages in lakes measured to obtain baseline hydrologic data. Pumping tests in both glacial-drift aquifers and bedrock wells were performed to measure the hydraulic parameters of the aquifer materials. Figure 13.6 shows the water-table elevations of the glacial-drift aquifer. Some of the

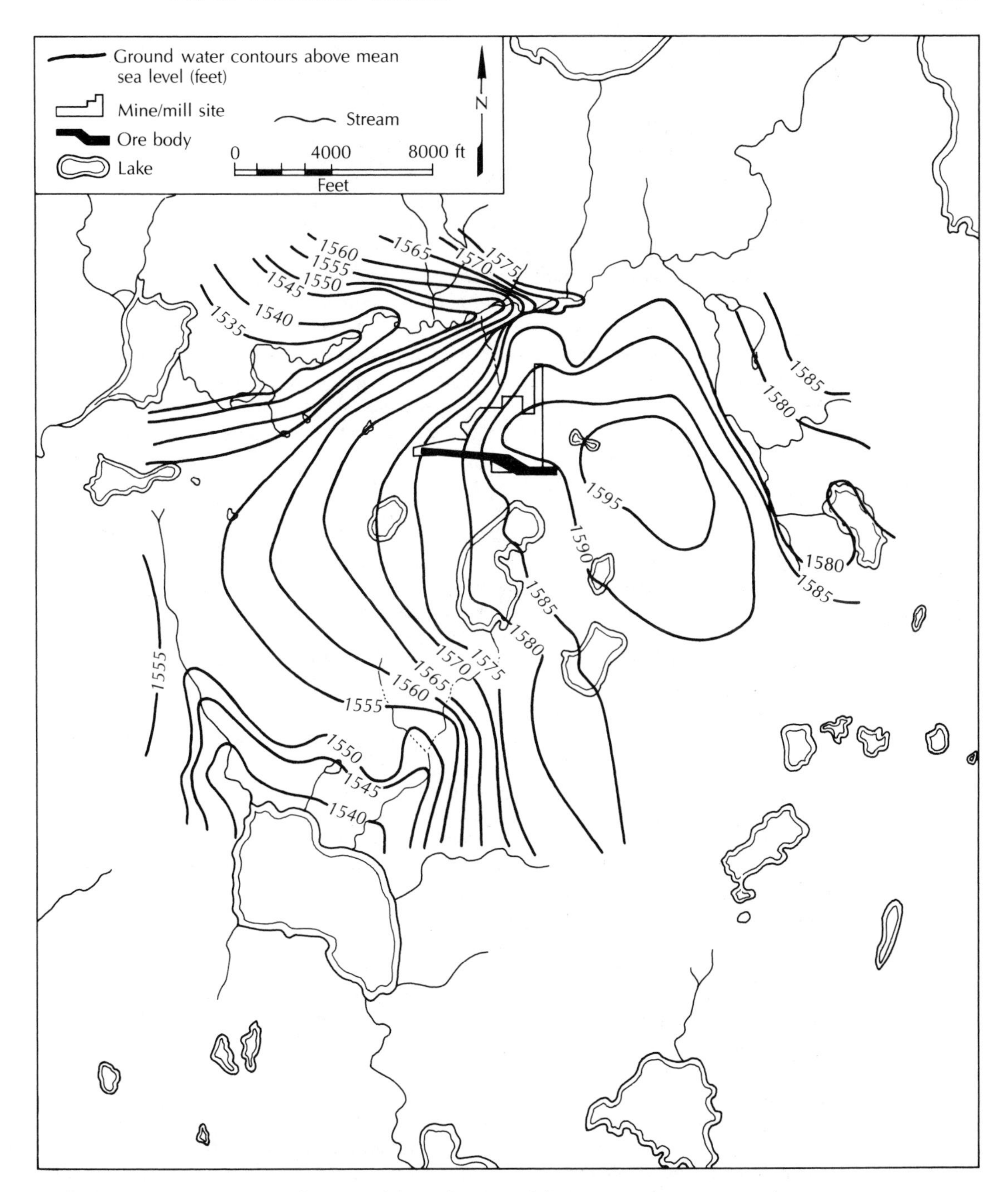


**FIGURE 13.6** Existing ground-water table at the site of the proposed mine in northeastern Wisconsin. Source: Wisconsin Department of Natural Resources, Final Environmental Impact Statement, Exxon Coal and Minerals Co., Zinc-Copper Mine, Crandon, Wisconsin, 1986.

small lakes over the mine area are seepage lakes, which are in contact with the water table, and some are perched above the water table on low-permeability lake-bed sediments. Also shown on this map are the locations of the ore body and the mine/mill facility.

Once the physical boundaries and hydrologic parameters of the area were established, a model grid system was set up for a two-layer, horizontal finite-element model. Figure 13.7 shows the grid system. The boundaries of the model were set at major streams and lakes and were all constant-head boundaries. The model used is proprietary (23).

The model was calibrated against the steady-state water-table position. Steady-state models are calibrated against the ratio of the rate of ground-water recharge to the hydraulic conductivity of the aquifer. Figure 13.8 shows the modeled water-table elevation based on the calibrated model. The amount of recharge was reduced in some places to reflect wetland areas where evapotranspiration is greater than in the upland areas so that there is less recharge. Recharge from the seepage lakes to the water table was also included. The aquifer transmissivity reflected the saturated thickness and the local hydraulic conductivity at each node. The recharge rate and the saturated hydraulic conductivity of the model were varied as the model was calibrated. Figure 13.8 shows the computer-generated water table.

The model was not verified as there were no events that could be used for history-matching. The value of the storage coefficient was obtained from a short-term pumping test. However, a mass-balance analysis of the model indicated that the model-computed values of baseflow to the streams in the area were close to the measured baseflows. It was felt that this was a good indication of the reliability of the model.

A transient form of the model was stressed to simulate the drawdown conditions that would form when the mine becomes fully operational. Mine-dewatering activity was simulated by making the heads in the aquifer above the mine zero. In addition, provisions were made for a potable water-supply well pumping 50 gallons (190 meters) per minute and seepage from a soil absorption field for wastewater at 50 gallons per minute.

Figure 13.9 shows the drawdown from the original water levels after 28 years of mining. This is the end of the planned mining period and represents maximum drawdown. If the cone of depression is superimposed on the initial water table, Figure 13.8, the water table after 28 years of mining can be obtained. This is represented by Figure 13.10. Comparison of Figure 13.10 with Figure 13.8 shows a large cone of depression over the mine as well as many changes in the directions of ground-water flow that are a result of the predicted changes in the configuration of the water table. The model also predicted that reductions in baseflow would occur in many of the streams and that the stage of the seepage lakes would decline.

As neither the recharge rate nor the hydraulic conductivity of the aquifer was known with exact precision, three calibrated versions of the model were developed, each with a different recharge rate. The three recharge rates were 6 inches (15.2 centimeters) per year, 8.5 inches (21.6 centimeters) per year, and 11 inches (27.9 centimeters) per year. The model based on the middle recharge rate of 8.5 inches per year is considered the most credible as it gave the best match of computed baseflows to streams with measured baseflows. The three versions of the model were not necessary in order to determine the steady-state position of the water table, against which they all were calibrated. Obviously, different hydraulic conductivity values were used for each recharge rate. As the recharge rate increases, so does the hydraulic conductivity necessary for calibration.

Greater volumes of water move through the aquifer for the models with higher recharge and greater permeability. This becomes important in predicting the amount of water that must be pumped in order to effect the desired mine dewatering. The maximum

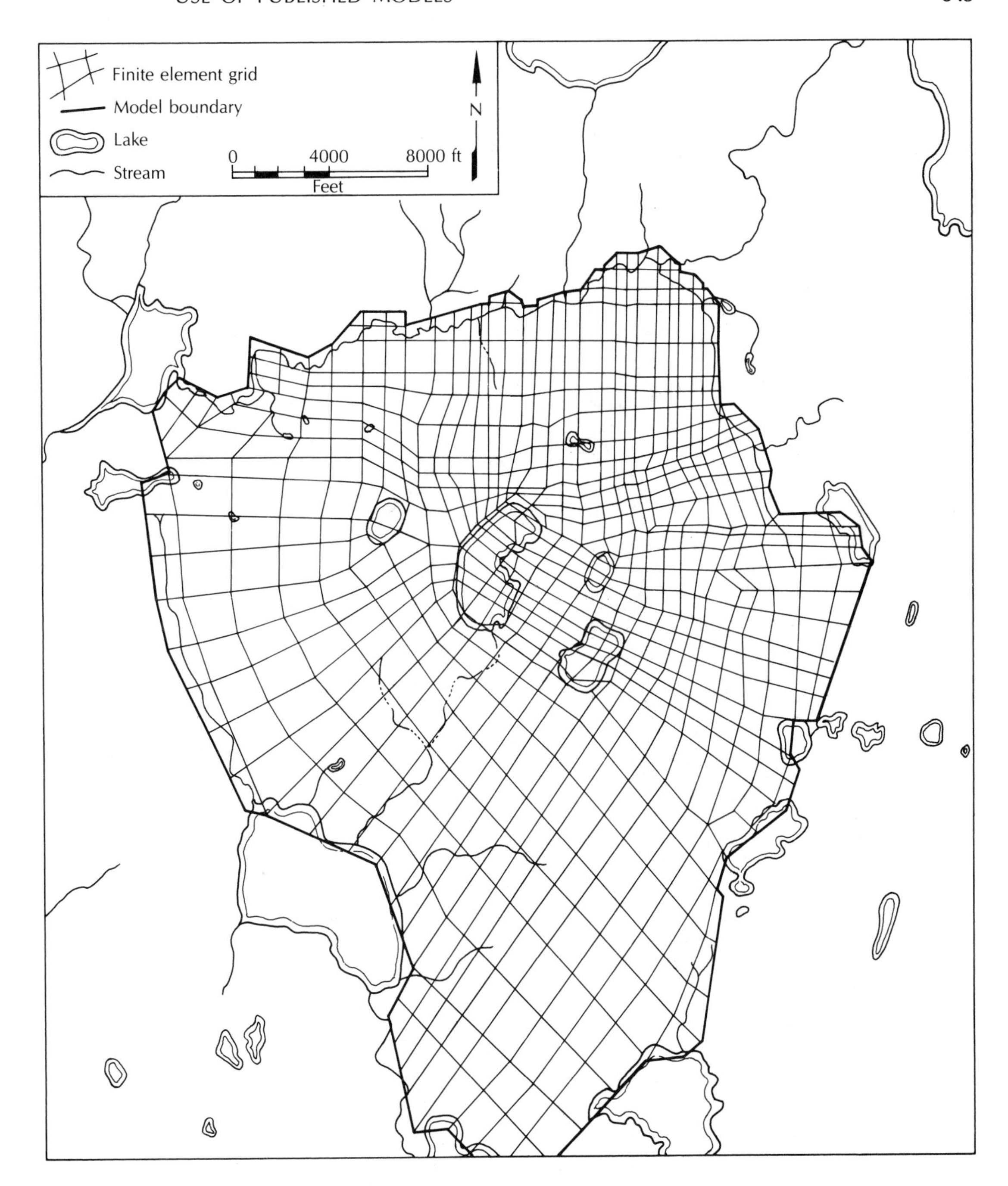


**FIGURE 13.7** Finite-element grid for computer model of ground-water flow at the site of a proposed underground mine in northeastern Wisconsin. Source: Exxon Minerals Company, Preliminary Environmental Report, Crandon Project, 1984.

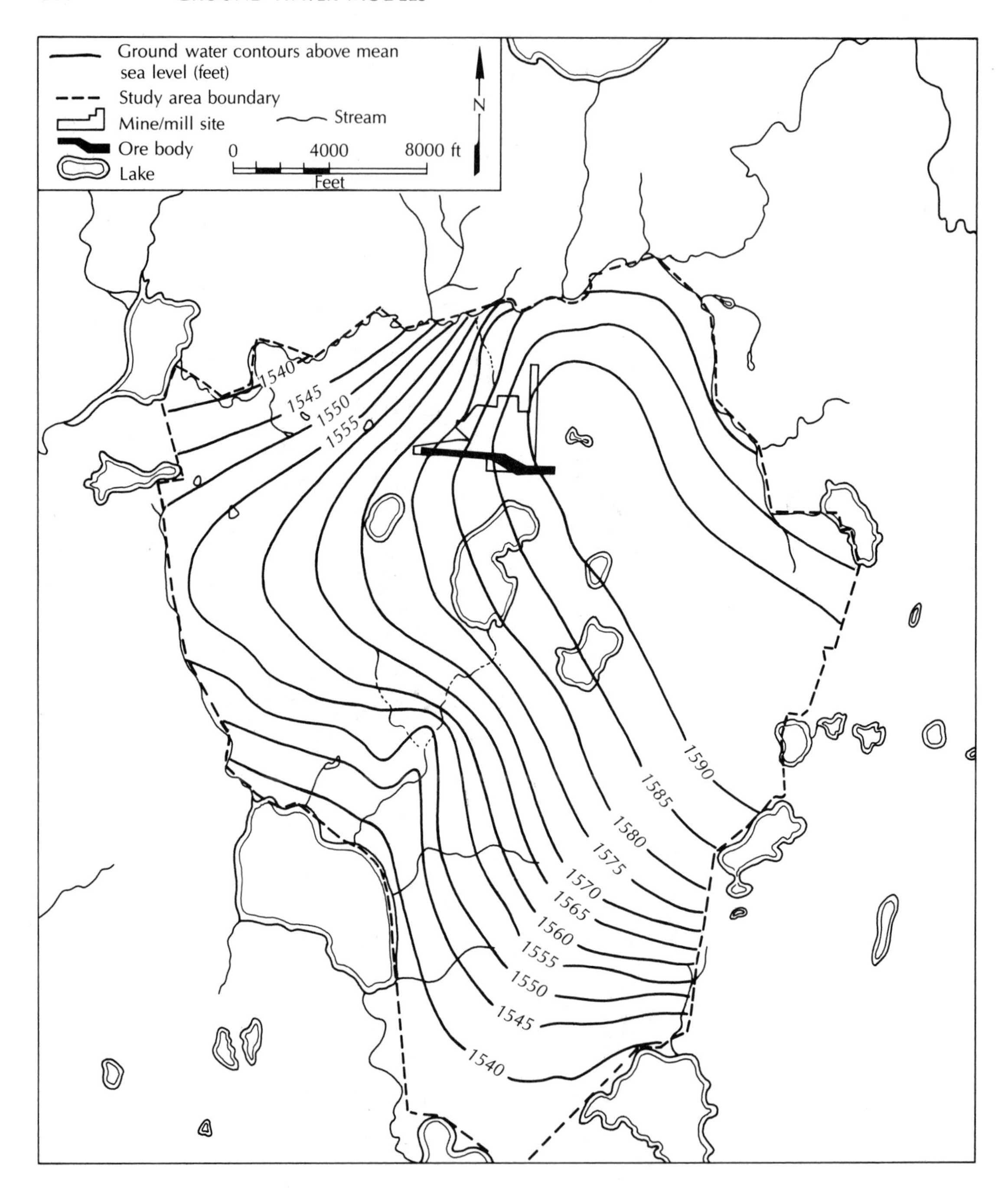


**FIGURE 13.8** Computer model of ground-water table under existing conditions at the site of a proposed underground mine in northeastern Wisconsin. Source: Wisconsin Department of Natural Resources, Final Environmental Impact Statement, Exxon Coal and Minerals Co., Zinc-Copper Mine, Crandon, Wisconsin, 1986.

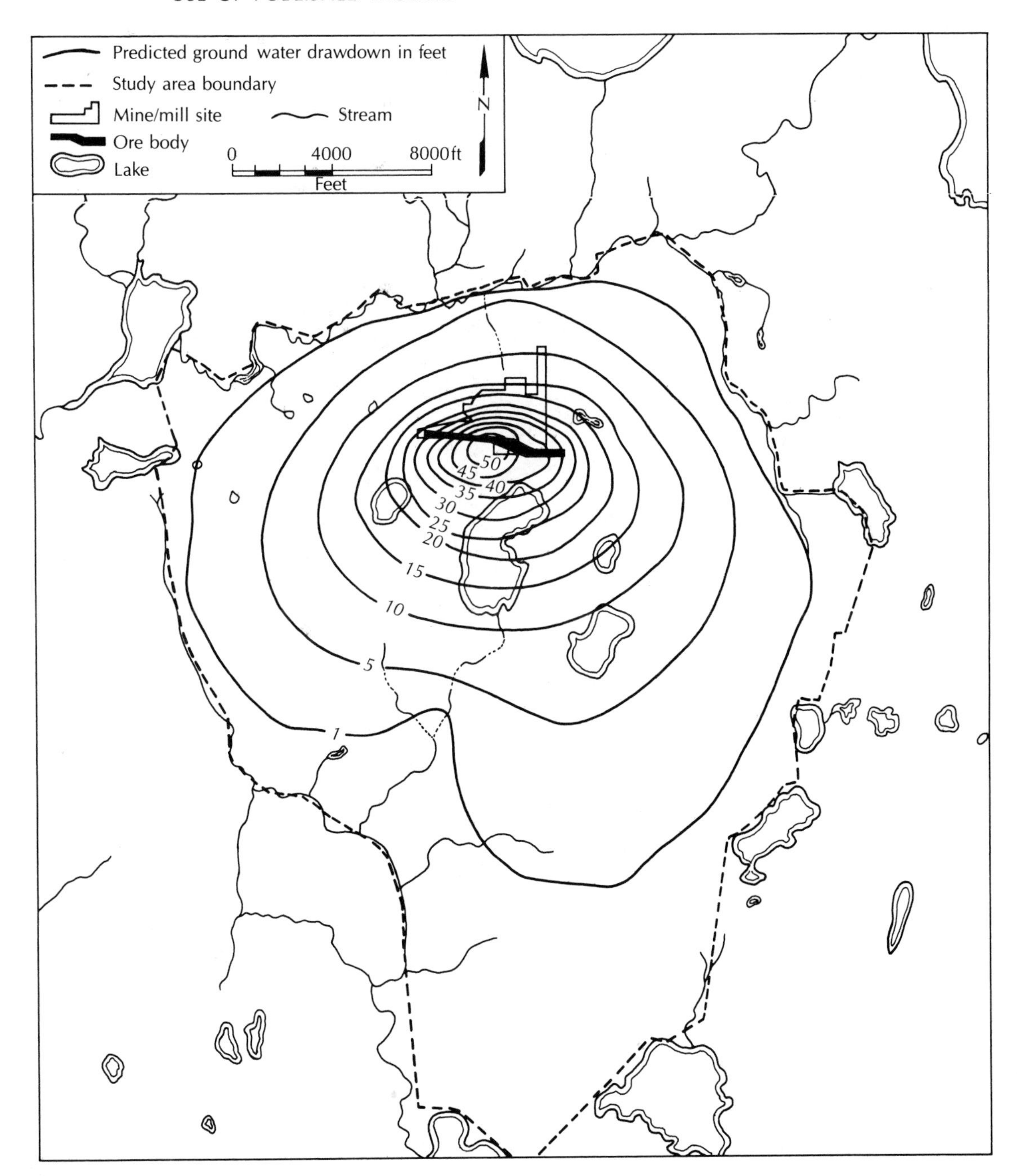


**FIGURE 13.9** Computer model prediction of drawdown of the water table due to mine dewatering after 28 years of mining at the site of a proposed underground mine in northeastern Wisconsin. Source: Wisconsin Department of Natural Resources, Final Environmental Impact Statement, Exxon Coal and Minerals Co., Zinc-Copper Mine, Crandon, Wisconsin, 1986.

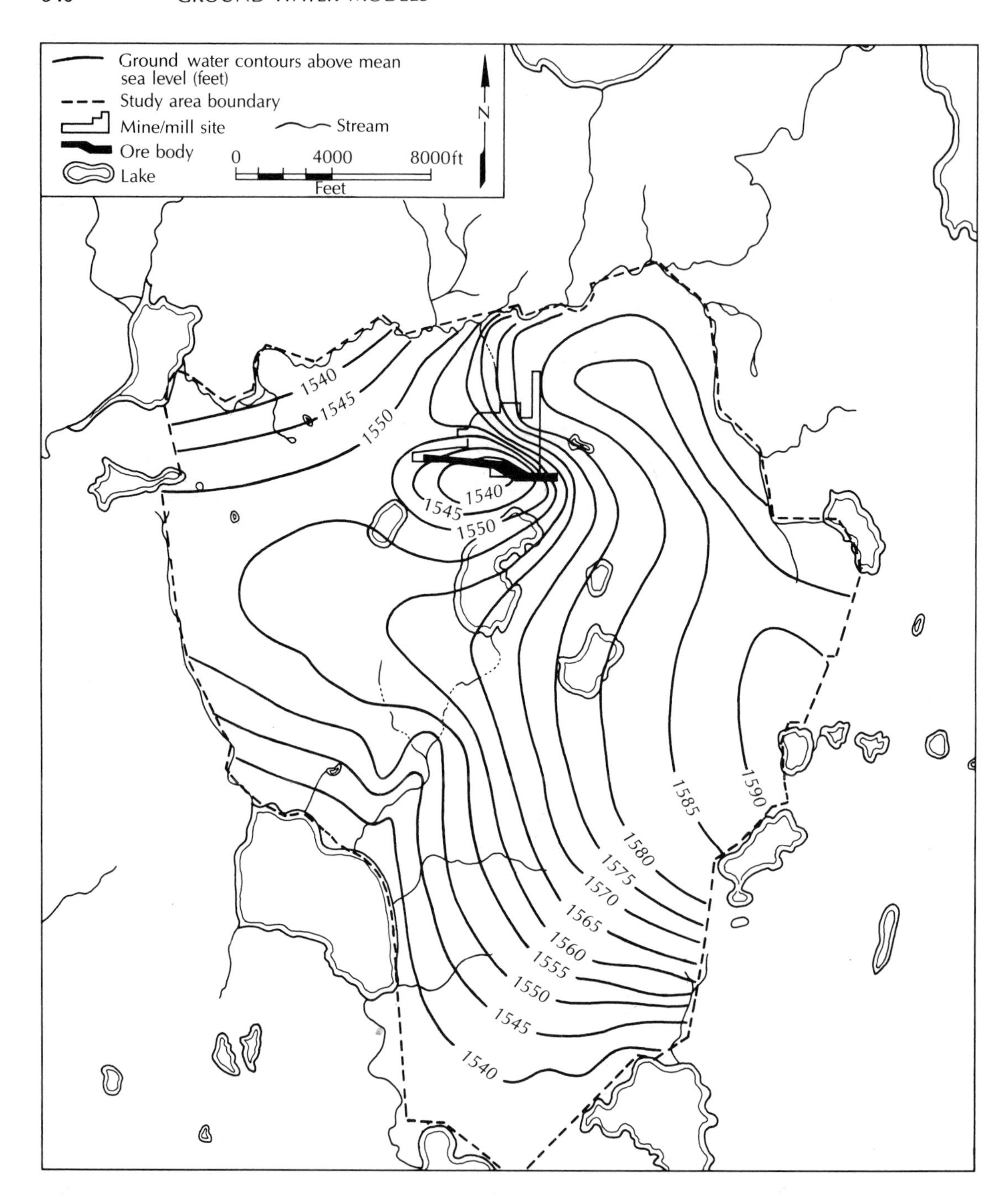


**FIGURE 13.10** Computer model prediction of the configuration of the water table after 28 years of mining at the site of a proposed underground mine in northeastern Wisconsin. Source: Wisconsin Department of Natural Resources, Final Environmental Impact Statement, Exxon Coal and Minerals Co., Zinc-Copper Mine, Crandon, Wisconsin, 1986.

pumping rate for mine dewatering was predicted to be 933 gallons (3531 liters) per minute for the low-recharge model, 1270 gallons (4807 liters) per minute for the middle-recharge model, and 1590 gallons (6018 liters) per minute for the high-recharge model.

It has been proposed that to obviate the impact of the lower regional water table on the seepage lakes, ground water be pumped into the lakes to raise their water levels. This would result in additional seepage into the area of the cone of depression of the mine-dewatering system and add about 200 gallons (757 liters) to the amount of water that must be pumped from the mine. Knowledge of the approximate volume of mine-dewatering pumpage is needed to design the pumping system as well as the treatment system for the mine drainage water.

## REFERENCES


1. PRICKETT, T. A. "Modeling Techniques for Ground Water Evaluation." In *Advances in Hydroscience,* vol. 10, ed., V. T. Chow, New York: Academic Press, 1975, pp. 1–143.
2. PRICKETT, T. A. "Ground Water Computer Models—State of the Art." *Ground Water,* 17 (1979):167–73.
3. KIMBLER, O. K. "Fluid Model Studies of the Storage of Freshwater in Saline Aquifers." *Water Resources Research,* 6 (1970):1522–27.
4. PETER, Y. "Model Tests for a Horizontal Well." *Ground Water,* 8, no. 5 (1970):30–34.
5. JAMES. R. V., and J. RUBIN. "Accounting for Apparatus-Induced Dispersion in Analysis of Miscible Displacement Experiments." *Water Resources Research,* 8 (1972):717–21.
6. ANDERSON, T. W. *Electrical-Analog Analysis of the Hydrologic System, Tucson Basin, Southeastern Arizona.* U.S. Geological Survey Water-Supply Paper 1939-C, 1972.
7. SPIEKER, A. M. *Effect of Increased Pumping of Ground Water in the Fairfield-New Baltimore Area, Ohio—A Prediction by Analog-Model Study.* U.S. Geological Survey Professional Paper 605-C, 1968.
8. COLLINS, M. A., L. W. GELHAR, and J. L. WILSON, III. "Hele-Shaw Model of Long Island Aquifer System." *Journal of Hydraulics Division, Proceedings of the American Society of Civil Engineers,* 98, HY9 (1972):1701–14.
9. PETERSON, F. L., J. A. WILLIAMS, and S. W. WHEATCRAFT. "Waste Injection into a Two-Phase Flow Field: Sand Box and Hele-Shaw Model Study." *Ground Water,* 16 (1978):410–16.
10. WALTON, W. C. "Analytical Ground Water Modeling with Programmable Calculators and Hand-Held Computers." In *Ground Water Hydraulics,* ed. J. Rosenshein and G. D. Bennett. American Geophysical Union Monograph 9, 1984, pp. 298–312.
11. JAVANDEL, I., C. DOUGHTY, and C.-F. TSANG. *Ground Water Transport: Handbook of Mathematical Models.* American Geophysical Union, Water Resources Monograph 10, 1984, 228 pp.
12. WALTON, W. C. "Progress in Analytical Groundwater Modeling." *Journal of Hydrology,* 43 (1979):149–59.

**13.** VAN DER HEIJDE, P. and others. *Ground Water Management: The Use of Numerical Models,* 2nd ed. American Geophysical Union, Water Resources Monograph 5, 1985, 180 pp.

**14.** DAGEN, G. "Statistical Theory of Groundwater Flow and Transport: Pore to Laboratory, Laboratory to Formation and Formation to Regional Scale." *Water Resources Research* 22, no. 9 (Supplement) (1986):120S–34S.

**15.** GELHAR, L. W. "Stochastic Subsurface Hydrology from Theory to Applications." *Water Resources Research* 22, no. 9 (Supplement) (1986):135S.

**16.** MERCER, J. W., and C. R. FAUST. *Ground-Water Modeling*. Dublin, Ohio: National Water Well Association, 1981, 60 pp.

**17.** WANG, H. F., and M. P. ANDERSON. *Introduction to Groundwater Modeling—Finite Difference and Finite Element Methods*. San Francisco: W. H. Freeman, 1982, 237 pp.

**18.** TRESCOTT, P. C., G. F. PINDER, and S. P. LARSON. *Finite Difference Model for Aquifer Simulation in Two Dimensions with Results of Numerical Experiments*. U.S. Geological Survey Techniques of Water-Resources Investigations, Book 7, Chap. C1, 1976, 116 pp.

**19.** MC DONALD, M. G., and A. W. HARBAUGH. *A Modular Three-Dimensional Finite-Difference Ground-Water Flow Model*. U.S. Geological Survey, 1984, 528 pp.

**20.** PINDER, G. F., and W. G. GRAY. *Finite Element Simulation in Surface and Subsurface Hydrology*. New York: Academic Press, 1977, 295 pp.

**21.** KONIKOW, L. F., and J. D. BREDEHOEFT. *Computer Model of Two-Dimensional Solute Transport and Dispersion in Ground Water*. U.S. Geological Survey, Techniques of Water-Resources Investigations, Book 7, Chap. C2, 1978, 90 pp.

**22.** PRICKETT, T. A., and C. G. LONQUIST. *Selected Digital Computer Techniques for Groundwater Resources Evaluation*. Illinois State Water Survey Bulletin 55, 1971, 55 pp.

**23.** ANDERSON, M. "Field Verification of Ground Water Models." In *Evaluation of Pesticides in Ground Water,* ed. W. Y. Garner and others. American Chemical Society Symposium Series, 315 (1986):396–412.

**24.** D'APPOLONIA CONSULTING ENGINEERS, INC. User's Manual. GEOFLOW Ground Water Model, 1980.

**25.** EXXON MINERALS COMPANY. Environmental Impact Report, Crandon Project, Volume X, Appendia 4.1A, 1982, 70 pp., tables and figures.

**26.** WISCONSIN DEPARTMENT OF NATURAL RESOURCES. Final Environmental Impact Statement, Exxon Coals and Minerals Co. Zinc-Copper Mine, Crandon, Wisconsin, 1986, 446 pp.

# Appendices

1. Values of the function $W(u)$ for various values of $u$
2. Values of the function $F(\eta,\mu)$ for various values of $\eta$ and $\mu$
3. Values of the functions $W(u,r/B)$ and $W(u_A,r/B)$ for various values of $u$ or $u_A$
4. Values of the function $H(\mu,\beta)$
5. Values of the functions $K_0(x)$ and $\exp(x)K_0(x)$
6. Values of the functions $W(u_A, \Gamma)$, and $W(u_B, \Gamma)$ for water-table aquifers
7. Table for length conversion
8. Table for area conversion
9. Table for volume conversion
10. Table for time conversion
11. Solubility products for selected minerals and compounds
12. Atomic weights and numbers of naturally occurring elements
13. Table of values of erf $(x)$ and erfc $(x)$
14. Absolute density and absolute viscosity of water

**APPENDIX 1.** Values of the function *W(u)* for various values of *u*

| *u* | *W(u)* | *u* | *W(u)* | *u* | *W(u)* | *u* | *W(u)* |
|---|---|---|---|---|---|---|---|
| $1 \times 10^{-10}$ | 22.45 | $7 \times 10^{-8}$ | 15.90 | $4 \times 10^{-5}$ | 9.55 | $1 \times 10^{-2}$ | 4.04 |
| 2 | 21.76 | 8 | 15.76 | 5 | 9.33 | 2 | 3.35 |
| 3 | 21.35 | 9 | 15.65 | 6 | 9.14 | 3 | 2.96 |
| 4 | 21.06 | $1 \times 10^{-7}$ | 15.54 | 7 | 8.99 | 4 | 2.68 |
| 5 | 20.84 | 2 | 14.85 | 8 | 8.86 | 5 | 2.47 |
| 6 | 20.66 | 3 | 14.44 | 9 | 8.74 | 6 | 2.30 |
| 7 | 20.50 | 4 | 14.15 | $1 \times 10^{-4}$ | 8.63 | 7 | 2.15 |
| 8 | 20.37 | 5 | 13.93 | 2 | 7.94 | 8 | 2.03 |
| 9 | 20.25 | 6 | 13.75 | 3 | 7.53 | 9 | 1.92 |
| $1 \times 10^{-9}$ | 20.15 | 7 | 13.60 | 4 | 7.25 | $1 \times 10^{-1}$ | 1.823 |
| 2 | 19.45 | 8 | 13.46 | 5 | 7.02 | 2 | 1.223 |
| 3 | 19.05 | 9 | 13.34 | 6 | 6.84 | 3 | 0.906 |
| 4 | 18.76 | $1 \times 10^{-6}$ | 13.24 | 7 | 6.69 | 4 | 0.702 |
| 5 | 18.54 | 2 | 12.55 | 8 | 6.55 | 5 | 0.560 |
| 6 | 18.35 | 3 | 12.14 | 9 | 6.44 | 6 | 0.454 |
| 7 | 18.20 | 4 | 11.85 | $1 \times 10^{-3}$ | 6.33 | 7 | 0.374 |
| 8 | 18.07 | 5 | 11.63 | 2 | 5.64 | 8 | 0.311 |
| 9 | 17.95 | 6 | 11.45 | 3 | 5.23 | 9 | 0.260 |
| $1 \times 10^{-8}$ | 17.84 | 7 | 11.29 | 4 | 4.95 | $1 \times 10^{0}$ | 0.219 |
| 2 | 17.15 | 8 | 11.16 | 5 | 4.73 | 2 | 0.049 |
| 3 | 16.74 | 9 | 11.04 | 6 | 4.54 | 3 | 0.013 |
| 4 | 16.46 | $1 \times 10^{-5}$ | 10.94 | 7 | 4.39 | 4 | 0.004 |
| 5 | 16.23 | 2 | 10.24 | 8 | 4.26 | 5 | 0.001 |
| 6 | 16.05 | 3 | 9.84 | 9 | 4.14 | | |

Source: Adapted from L. K. Wenzel, *Methods for Determining Permeability of Water-Bearing Materials with Special Reference to Discharging Well Methods.* U.S. Geological Survey Water-Supply Paper 887, 1942.

**APPENDIX 2** Values of the function $F(\eta,\mu)$ for various values of $\eta$ and $\mu$

| $\eta = Tt/r_c^2$ | $\mu = 10^{-6}$ | $\mu = 10^{-7}$ | $\mu = 10^{-8}$ | $\mu = 10^{-9}$ | $\mu = 10^{-10}$ |
|---|---|---|---|---|---|
| 0.001 | 0.9994 | 0.9996 | 0.9996 | 0.9997 | 0.9997 |
| 0.002 | 0.9989 | 0.9992 | 0.9993 | 0.9994 | 0.9995 |
| 0.004 | 0.9980 | 0.9985 | 0.9987 | 0.9989 | 0.9991 |
| 0.006 | 0.9972 | 0.9978 | 0.9982 | 0.9984 | 0.9986 |
| 0.008 | 0.9964 | 0.9971 | 0.9976 | 0.9980 | 0.9982 |
| 0.01 | 0.9956 | 0.9965 | 0.9971 | 0.9975 | 0.9978 |
| 0.02 | 0.9919 | 0.9934 | 0.9944 | 0.9952 | 0.9958 |
| 0.04 | 0.9848 | 0.9875 | 0.9894 | 0.9908 | 0.9919 |
| 0.06 | 0.9782 | 0.9819 | 0.9846 | 0.9866 | 0.9881 |
| 0.08 | 0.9718 | 0.9765 | 0.9799 | 0.9824 | 0.9844 |
| 0.1 | 0.9655 | 0.9712 | 0.9753 | 0.9784 | 0.9807 |
| 0.2 | 0.9361 | 0.9459 | 0.9532 | 0.9587 | 0.9631 |
| 0.4 | 0.8828 | 0.8995 | 0.9122 | 0.9220 | 0.9298 |
| 0.6 | 0.8345 | 0.8569 | 0.8741 | 0.8875 | 0.8984 |
| 0.8 | 0.7901 | 0.8173 | 0.8383 | 0.8550 | 0.8686 |
| 1.0 | 0.7489 | 0.7801 | 0.8045 | 0.8240 | 0.8401 |
| 2.0 | 0.5800 | 0.6235 | 0.6591 | 0.6889 | 0.7139 |
| 3.0 | 0.4554 | 0.5033 | 0.5442 | 0.5792 | 0.6096 |
| 4.0 | 0.3613 | 0.4093 | 0.4517 | 0.4891 | 0.5222 |
| 5.0 | 0.2893 | 0.3351 | 0.3768 | 0.4146 | 0.4487 |
| 6.0 | 0.2337 | 0.2759 | 0.3157 | 0.3525 | 0.3865 |
| 7.0 | 0.1903 | 0.2285 | 0.2655 | 0.3007 | 0.3337 |
| 8.0 | 0.1562 | 0.1903 | 0.2243 | 0.2573 | 0.2888 |
| 9.0 | 0.1292 | 0.1594 | 0.1902 | 0.2208 | 0.2505 |
| 10.0 | 0.1078 | 0.1343 | 0.1620 | 0.1900 | 0.2178 |
| 20.0 | 0.02720 | 0.03343 | 0.04129 | 0.05071 | 0.06149 |
| 30.0 | 0.01286 | 0.01448 | 0.01667 | 0.01956 | 0.02320 |
| 40.0 | 0.008337 | 0.008898 | 0.009637 | 0.01062 | 0.01190 |
| 50.0 | 0.006209 | 0.006470 | 0.006789 | 0.007192 | 0.007709 |
| 60.0 | 0.004961 | 0.005111 | 0.005283 | 0.005487 | 0.005735 |
| 80.0 | 0.003547 | 0.003617 | 0.003691 | 0.003773 | 0.003863 |
| 100.0 | 0.002763 | 0.002803 | 0.002845 | 0.002890 | 0.002938 |
| 200.0 | 0.001313 | 0.001322 | 0.001330 | 0.001339 | 0.001348 |

Source: After I. S. Papadopulos, J. D. Bredehoeft, and H. H. Cooper, Jr., "On the Analysis of 'Slug Test' Data," *Water Resources Research*, 9 (1973):1087–89.

Note: For slug tests on wells of finite diameter, $H/H_0 = F(\eta,\mu)$.

**APPENDIX 3** Values of the functions $W(u,r/B)$ for various values of $u$

| $u$ \ $r/B$ | 0.002 | 0.004 | 0.006 | 0.008 | 0.01 | 0.02 | 0.04 | 0.06 | 0.08 | 0.1 | 0.2 | 0.4 | 0.6 | 0.8 | 1 | 2 | 4 | 6 | 8 |
|---|---|---|---|---|---|---|---|---|---|---|---|---|---|---|---|---|---|---|---|
| 0 | 12.7 | 11.3 | 10.5 | 9.89 | 9.44 | 8.06 | 6.67 | 5.87 | 5.29 | 4.85 | 3.51 | 2.23 | 1.55 | 1.13 | 0.842 | 0.228 | 0.0223 | 0.0025 | 0.0003 |
| 0.000002 | 12.1 | 11.2 | 10.5 | 9.89 | 9.44 | | | | | | | | | | | | | | |
| 0.000004 | 11.6 | 11.1 | 10.4 | 9.88 | 9.44 | | | | | | | | | | | | | | |
| 0.000006 | 11.3 | 10.9 | 10.4 | 9.87 | 9.44 | | | | | | | | | | | | | | |
| 0.000008 | 11.0 | 10.7 | 10.3 | 9.84 | 9.43 | | | | | | | | | | | | | | |
| 0.00001 | 10.8 | 10.6 | 10.2 | 9.80 | 9.42 | 8.06 | | | | | | | | | | | | | |
| 0.00002 | 10.2 | 10.1 | 9.84 | 9.58 | 9.30 | 8.06 | | | | | | | | | | | | | |
| 0.00004 | 9.52 | 9.45 | 9.34 | 9.19 | 9.01 | 8.03 | 6.67 | | | | | | | | | | | | |
| 0.00006 | 9.13 | 9.08 | 9.00 | 8.89 | 8.77 | 7.98 | 6.67 | | | | | | | | | | | | |
| 0.00008 | 8.84 | 8.81 | 8.75 | 8.67 | 8.57 | 7.91 | 6.67 | | | | | | | | | | | | |
| 0.0001 | 8.62 | 8.59 | 8.55 | 8.48 | 8.40 | 7.84 | 6.67 | 5.87 | 5.29 | | | | | | | | | | |
| 0.0002 | 7.94 | 7.92 | 7.90 | 7.86 | 7.82 | 7.50 | 6.62 | 5.86 | 5.29 | | | | | | | | | | |
| 0.0004 | 7.24 | 7.24 | 7.22 | 7.21 | 7.19 | 7.01 | 6.45 | 5.83 | 5.29 | 4.85 | | | | | | | | | |
| 0.0006 | 6.84 | 6.84 | 6.83 | 6.82 | 6.80 | 6.68 | 6.27 | 5.77 | 5.27 | 4.85 | | | | | | | | | |
| 0.0008 | 6.55 | 6.55 | 6.54 | 6.53 | 6.52 | 6.43 | 6.11 | 5.69 | 5.25 | 4.84 | | | | | | | | | |
| 0.001 | 6.33 | 6.33 | 6.32 | 6.32 | 6.31 | 6.23 | 5.97 | 5.61 | 5.21 | 4.83 | 3.51 | | | | | | | | |
| 0.002 | 5.64 | 5.64 | 5.63 | 5.63 | 5.63 | 5.59 | 5.45 | 5.24 | 4.98 | 4.71 | 3.50 | | | | | | | | |
| 0.004 | 4.95 | 4.95 | 4.95 | 4.94 | 4.94 | 4.92 | 4.85 | 4.74 | 4.59 | 4.42 | 3.48 | 2.23 | | | | | | | |
| 0.006 | 4.54 | | | | 4.54 | 4.53 | 4.48 | 4.41 | 4.30 | 4.18 | 3.43 | 2.23 | | | | | | | |
| 0.008 | 4.26 | | | | 4.26 | 4.25 | 4.21 | 4.15 | 4.08 | 3.98 | 3.36 | 2.23 | | | | | | | |
| 0.01 | 4.04 | | | | 4.04 | 4.03 | 4.00 | 3.95 | 3.89 | 3.81 | 3.29 | 2.23 | 1.55 | 1.13 | | | | | |
| 0.02 | 3.35 | | | | 3.35 | 3.35 | 3.34 | 3.31 | 3.28 | 3.24 | 2.95 | 2.18 | 1.55 | 1.13 | | | | | |
| 0.04 | 2.68 | | | | 2.68 | 2.68 | 2.67 | 2.66 | 2.65 | 2.63 | 2.48 | 2.02 | 1.52 | 1.13 | 0.842 | | | | |
| 0.06 | 2.30 | | | | 2.30 | 2.29 | 2.29 | 2.28 | 2.27 | 2.26 | 2.17 | 1.85 | 1.46 | 1.11 | 0.839 | | | | |
| 0.08 | 2.03 | | | | | 2.03 | 2.02 | 2.02 | 2.01 | 2.00 | 1.94 | 1.69 | 1.39 | 1.08 | 0.832 | | | | |
| 0.1 | 1.82 | | | | | | 1.82 | 1.82 | 1.81 | 1.80 | 1.75 | 1.56 | 1.31 | 1.05 | 0.819 | 0.228 | | | |
| 0.2 | 1.22 | | | | | | 1.22 | 1.22 | 1.22 | 1.22 | 1.19 | 1.11 | 0.996 | 0.857 | 0.715 | 0.227 | | | |
| 0.4 | 0.702 | | | | | | 0.702 | 0.702 | 0.701 | 0.700 | 0.693 | 0.665 | 0.621 | 0.565 | 0.502 | 0.210 | | | |
| 0.6 | 0.454 | | | | | | 0.454 | 0.454 | 0.454 | 0.453 | 0.450 | 0.436 | 0.415 | 0.387 | 0.354 | 0.177 | 0.0222 | | |
| 0.8 | 0.311 | | | | | | 0.311 | 0.310 | 0.310 | 0.310 | 0.308 | 0.301 | 0.289 | 0.273 | 0.254 | 0.144 | 0.0218 | | |
| 1 | 0.219 | | | | | | | | | 0.219 | 0.218 | 0.213 | 0.206 | 0.197 | 0.185 | 0.114 | 0.0207 | 0.0025 | |
| 2 | 0.049 | | | | | | | | | | 0.049 | 0.048 | 0.047 | 0.046 | 0.044 | 0.034 | 0.011 | 0.0021 | 0.0003 |
| 4 | 0.0038 | | | | | | | | | | | 0.0038 | 0.0037 | 0.0037 | 0.0036 | 0.0031 | 0.0016 | 0.0006 | 0.0002 |
| 6 | 0.0004 | | | | | | | | | | | | | | 0.0004 | 0.0003 | 0.0002 | 0.0001 | 0 |
| 8 | 0 | | | | | | | | | | | | | | | | | | 0 |

Source: After M. S. Hantush, "Analysis of Data from Pumping Tests in Leaky Aquifers," *Transactions, American Geophysical Union*, 37 (1956):702–14.

**APPENDIX 4** Values of the function $H(\mu,\beta)$

| $\mu$ \ $\beta$ | 0.001 | 0.005 | 0.01 | 0.05 | 0.10 | 0.20 | 0.50 | 1.0 | 2.0 | 5.0 | 10.0 | 20.0 |
|---|---|---|---|---|---|---|---|---|---|---|---|---|
| 0.000001 | 11.9842 | 10.5908 | 9.9259 | 8.3395 | 7.6497 | 6.9590 | 6.0463 | 5.3575 | 4.6721 | 3.7756 | 3.1110 | 2.4671 |
| 0.000005 | 10.8958 | 9.7174 | 9.0866 | 7.5284 | 6.8427 | 6.1548 | 5.2459 | 4.5617 | 3.8836 | 3.0055 | 2.3661 | 1.7633 |
| 0.00001 | 10.3739 | 9.3203 | 8.7142 | 7.1771 | 6.4944 | 5.8085 | 4.9024 | 4.2212 | 3.5481 | 2.6822 | 2.0590 | 1.4816 |
| 0.00005 | 9.0422 | 8.3171 | 7.8031 | 6.3523 | 5.6821 | 5.0045 | 4.1090 | 3.4394 | 2.7848 | 1.9622 | 1.3943 | 0.8994 |
| 0.0001 | 8.4258 | 7.8386 | 7.3803 | 5.9906 | 5.3297 | 4.6581 | 3.7700 | 3.1082 | 2.4658 | 1.6704 | 1.1359 | 0.6878 |
| 0.0005 | 6.9273 | 6.6024 | 6.2934 | 5.1223 | 4.4996 | 3.8527 | 2.9933 | 2.3601 | 1.7604 | 1.0564 | 0.6252 | 0.3089 |
| 0.001 | 6.2624 | 6.0193 | 5.7727 | 4.7290 | 4.1337 | 3.5045 | 2.6650 | 2.0506 | 1.4776 | 0.8271 | 0.4513 | 0.1976 |
| 0.005 | 4.6951 | 4.5786 | 4.4474 | 3.7415 | 3.2483 | 2.6891 | 1.9250 | 1.3767 | 0.8915 | 0.4001 | 0.1677 | 0.0493 |
| 0.01 | 4.0163 | 3.9334 | 3.8374 | 3.2752 | 2.8443 | 2.3325 | 1.6193 | 1.1122 | 0.6775 | 0.2670 | 0.0955 | 0.0221 |
| 0.05 | 2.4590 | 2.4243 | 2.3826 | 2.1007 | 1.8401 | 1.4872 | 0.9540 | 0.5812 | 0.2923 | 0.0755 | 0.0160 | 0.00164 |
| 0.1 | 1.8172 | 1.7949 | 1.7677 | 1.5768 | 1.3893 | 1.1207 | 0.6947 | 0.3970 | 0.1789 | 0.0359 | 0.00552 | 0.00034 |
| 0.5 | 0.5584 | 0.5530 | 0.5463 | 0.4969 | 0.4436 | 0.3591 | 0.2083 | 0.1006 | 0.0325 | 0.00288 | 0.00015 | |
| 1.0 | 0.2189 | 0.2169 | 0.2144 | 0.1961 | 0.1758 | 0.1427 | 0.0812 | 0.0365 | 0.00993 | 0.00055 | 0.00002 | |
| 5.0 | 0.00115 | 0.00114 | 0.00112 | 0.00104 | 0.00093 | 0.00076 | 0.00042 | 0.00017 | 0.00003 | | | |

Source: Condensed from M. S. Hantush, "Modification of the Theory of Leaky Aquifers," *Journal of Geophysical Research, 65 (1960):3713–25.*

**APPENDIX 5** Values of the functions $K_0(x)$ and $\exp(x)K_0(x)$

| $x$ | $K_0(x)$ | $\exp(x)K_0(x)$ | $x$ | $K_0(x)$ | $\exp(x)K_0(x)$ |
|---|---|---|---|---|---|
| 0.01 | 4.72 | 4.77 | 0.35 | 1.23 | 1.75 |
| 0.015 | 4.32 | 4.38 | 0.40 | 1.11 | 1.66 |
| 0.02 | 4.03 | 4.11 | 0.45 | 1.01 | 1.59 |
| 0.025 | 3.81 | 3.91 | 0.50 | 0.92 | 1.52 |
| 0.03 | 3.62 | 3.73 | 0.55 | 0.85 | 1.47 |
| 0.035 | 3.47 | 3.59 | 0.60 | 0.78 | 1.42 |
| 0.04 | 3.34 | 3.47 | 0.65 | 0.72 | 1.37 |
| 0.045 | 3.22 | 3.37 | 0.70 | 0.66 | 1.33 |
| 0.05 | 3.11 | 3.27 | 0.75 | 0.61 | 1.29 |
| 0.055 | 3.02 | 3.19 | 0.80 | 0.57 | 1.26 |
| 0.06 | 2.93 | 3.11 | 0.85 | 0.52 | 1.23 |
| 0.065 | 2.85 | 3.05 | 0.90 | 0.49 | 1.20 |
| 0.07 | 2.78 | 2.98 | 0.95 | 0.45 | 1.17 |
| 0.075 | 2.71 | 2.92 | 1.0 | 0.42 | 1.14 |
| 0.08 | 2.65 | 2.87 | 1.5 | 0.21 | 0.96 |
| 0.085 | 2.59 | 2.82 | 2.0 | 0.11 | 0.84 |
| 0.09 | 2.53 | 2.77 | 2.5 | 0.062 | 0.760 |
| 0.095 | 2.48 | 2.72 | 3.0 | 0.035 | 0.698 |
| 0.10 | 2.43 | 2.68 | 3.5 | 0.020 | 0.649 |
| 0.15 | 2.03 | 2.36 | 4.0 | 0.011 | 0.609 |
| 0.20 | 1.75 | 2.14 | 4.5 | 0.006 | 0.576 |
| 0.25 | 1.54 | 1.98 | 5.0 | 0.004 | 0.548 |
| 0.30 | 1.37 | 1.85 | | | |

Source: Adapted from M. S. Hantush, "Analysis of Data From Pumping Tests in Leaky Aquifers," *Transactions, American Geophysical Union,* 37 (1956):702–14.

**APPENDIX 6A** Values of the function $W(u_A,\Gamma)$ for water-table aquifers

| $1/u_A$ | $\Gamma = 0.001$ | $\Gamma = 0.01$ | $\Gamma = 0.06$ | $\Gamma = 0.2$ | $\Gamma = 0.6$ | $\Gamma = 1.0$ | $\Gamma = 2.0$ | $\Gamma = 4.0$ | $\Gamma = 6.0$ |
|---|---|---|---|---|---|---|---|---|---|
| $4.0 \times 10^{-1}$ | $2.48 \times 10^{-2}$ | $2.41 \times 10^{-2}$ | $2.30 \times 10^{-2}$ | $2.14 \times 10^{-2}$ | $1.88 \times 10^{-2}$ | $1.70 \times 10^{-2}$ | $1.38 \times 10^{-2}$ | $9.33 \times 10^{-3}$ | $6.39 \times 10^{-3}$ |
| $8.0 \times 10^{-1}$ | $1.45 \times 10^{-1}$ | $1.40 \times 10^{-1}$ | $1.31 \times 10^{-1}$ | $1.19 \times 10^{-1}$ | $9.88 \times 10^{-2}$ | $8.49 \times 10^{-2}$ | $6.03 \times 10^{-2}$ | $3.17 \times 10^{-2}$ | $1.74 \times 10^{-2}$ |
| $1.4 \times 10^{0}$ | $3.58 \times 10^{-1}$ | $3.45 \times 10^{-1}$ | $3.18 \times 10^{-1}$ | $2.79 \times 10^{-1}$ | $2.17 \times 10^{-1}$ | $1.75 \times 10^{-1}$ | $1.07 \times 10^{-1}$ | $4.45 \times 10^{-2}$ | $2.10 \times 10^{-2}$ |
| $2.4 \times 10^{0}$ | $6.62 \times 10^{-1}$ | $6.33 \times 10^{-1}$ | $5.70 \times 10^{-1}$ | $4.83 \times 10^{-1}$ | $3.43 \times 10^{-1}$ | $2.56 \times 10^{-1}$ | $1.33 \times 10^{-1}$ | $4.76 \times 10^{-2}$ | $2.14 \times 10^{-2}$ |
| $4.0 \times 10^{0}$ | $1.02 \times 10^{0}$ | $9.63 \times 10^{-1}$ | $8.49 \times 10^{-1}$ | $6.88 \times 10^{-1}$ | $4.38 \times 10^{-1}$ | $3.00 \times 10^{-1}$ | $1.40 \times 10^{-1}$ | $4.78 \times 10^{-2}$ | $2.15 \times 10^{-2}$ |
| $8.0 \times 10^{0}$ | $1.57 \times 10^{0}$ | $1.46 \times 10^{0}$ | $1.23 \times 10^{0}$ | $9.18 \times 10^{-1}$ | $4.97 \times 10^{-1}$ | $3.17 \times 10^{-1}$ | $1.41 \times 10^{-1}$ | | |
| $1.4 \times 10^{1}$ | $2.05 \times 10^{0}$ | $1.88 \times 10^{0}$ | $1.51 \times 10^{0}$ | $1.03 \times 10^{0}$ | $5.07 \times 10^{-1}$ | | | | |
| $2.4 \times 10^{1}$ | $2.52 \times 10^{0}$ | $2.27 \times 10^{0}$ | $1.73 \times 10^{0}$ | $1.07 \times 10^{0}$ | | | | | |
| $4.0 \times 10^{1}$ | $2.97 \times 10^{0}$ | $2.61 \times 10^{0}$ | $1.85 \times 10^{0}$ | $1.08 \times 10^{0}$ | | | | | |
| $8.0 \times 10^{1}$ | $3.56 \times 10^{0}$ | $3.00 \times 10^{0}$ | $1.92 \times 10^{0}$ | | | | | | |
| $1.4 \times 10^{2}$ | $4.01 \times 10^{0}$ | $3.23 \times 10^{0}$ | $1.93 \times 10^{0}$ | | | | | | |
| $2.4 \times 10^{2}$ | $4.42 \times 10^{0}$ | $3.37 \times 10^{0}$ | $1.94 \times 10^{0}$ | | | | | | |
| $4.0 \times 10^{2}$ | $4.77 \times 10^{0}$ | $3.43 \times 10^{0}$ | | | | | | | |
| $8.0 \times 10^{2}$ | $5.16 \times 10^{0}$ | $3.45 \times 10^{0}$ | | | | | | | |
| $1.4 \times 10^{3}$ | $5.40 \times 10^{0}$ | $3.46 \times 10^{0}$ | | | | | | | |
| $2.4 \times 10^{3}$ | $5.54 \times 10^{0}$ | | | | | | | | |
| $4.0 \times 10^{3}$ | $5.59 \times 10^{0}$ | | | | | | | | |
| $8.0 \times 10^{3}$ | $5.62 \times 10^{0}$ | | | | | | | | |
| $1.4 \times 10^{4}$ | $5.62 \times 10^{0}$ | $3.46 \times 10^{0}$ | $1.94 \times 10^{0}$ | $1.08 \times 10^{0}$ | $5.07 \times 10^{-1}$ | $3.17 \times 10^{-1}$ | $1.41 \times 10^{-1}$ | $4.78 \times 10^{-2}$ | $2.15 \times 10^{-2}$ |

**APPENDIX 6B** Values of the function $W(u_B,\Gamma)$ for water-table aquifers

| $1/u_B$ | $\Gamma = 0.001$ | $\Gamma = 0.01$ | $\Gamma = 0.06$ | $\Gamma = 0.2$ | $\Gamma = 0.6$ | $\Gamma = 1.0$ | $\Gamma = 2.0$ | $\Gamma = 4.0$ | $\Gamma = 6.0$ |
|---|---|---|---|---|---|---|---|---|---|
| $4.0 \times 10^{-4}$ | $5.62 \times 10^{0}$ | $3.46 \times 10^{0}$ | $1.94 \times 10^{0}$ | $1.09 \times 10^{0}$ | $5.08 \times 10^{-1}$ | $3.18 \times 10^{-1}$ | $1.42 \times 10^{-1}$ | $4.79 \times 10^{-2}$ | $2.15 \times 10^{-2}$ |
| $8.0 \times 10^{-4}$ | | | | | | | | $4.80 \times 10^{-2}$ | $2.16 \times 10^{-2}$ |
| $1.4 \times 10^{-3}$ | | | | | | | | $4.81 \times 10^{-2}$ | $2.17 \times 10^{-2}$ |
| $2.4 \times 10^{-3}$ | | | | | | | | $4.84 \times 10^{-2}$ | $2.19 \times 10^{-2}$ |
| $4.0 \times 10^{-3}$ | | | | | $5.08 \times 10^{-1}$ | $3.18 \times 10^{-1}$ | $1.42 \times 10^{-1}$ | $4.88 \times 10^{-2}$ | $2.21 \times 10^{-2}$ |
| $8.0 \times 10^{-3}$ | | | | | $5.09 \times 10^{-1}$ | $3.19 \times 10^{-1}$ | $1.43 \times 10^{-1}$ | $4.96 \times 10^{-2}$ | $2.28 \times 10^{-2}$ |
| $1.4 \times 10^{-2}$ | | | | | $5.10 \times 10^{-1}$ | $3.21 \times 10^{-1}$ | $1.45 \times 10^{-1}$ | $5.09 \times 10^{-2}$ | $2.39 \times 10^{-2}$ |
| $2.4 \times 10^{-2}$ | | | | | $5.12 \times 10^{-1}$ | $3.23 \times 10^{-1}$ | $1.47 \times 10^{-1}$ | $5.32 \times 10^{-2}$ | $2.57 \times 10^{-2}$ |
| $4.0 \times 10^{-2}$ | | | | | $5.16 \times 10^{-1}$ | $3.27 \times 10^{-1}$ | $1.52 \times 10^{-1}$ | $5.68 \times 10^{-2}$ | $2.86 \times 10^{-2}$ |
| $8.0 \times 10^{-2}$ | | | | $1.09 \times 10^{0}$ | $5.24 \times 10^{-1}$ | $3.37 \times 10^{-1}$ | $1.62 \times 10^{-1}$ | $6.61 \times 10^{-2}$ | $3.62 \times 10^{-2}$ |
| $1.4 \times 10^{-1}$ | | | $1.94 \times 10^{0}$ | $1.10 \times 10^{0}$ | $5.37 \times 10^{-1}$ | $3.50 \times 10^{-1}$ | $1.78 \times 10^{-1}$ | $8.06 \times 10^{-2}$ | $4.86 \times 10^{-2}$ |
| $2.4 \times 10^{-1}$ | | | $1.95 \times 10^{0}$ | $1.11 \times 10^{0}$ | $5.57 \times 10^{-1}$ | $3.74 \times 10^{-1}$ | $2.05 \times 10^{-1}$ | $1.06 \times 10^{-1}$ | $7.14 \times 10^{-2}$ |
| $4.0 \times 10^{-1}$ | | | $1.96 \times 10^{0}$ | $1.13 \times 10^{0}$ | $5.89 \times 10^{-1}$ | $4.12 \times 10^{-1}$ | $2.48 \times 10^{-1}$ | $1.49 \times 10^{-1}$ | $1.13 \times 10^{-1}$ |
| $8.0 \times 10^{-1}$ | $5.62 \times 10^{0}$ | $3.46 \times 10^{0}$ | $1.98 \times 10^{0}$ | $1.18 \times 10^{0}$ | $6.67 \times 10^{-1}$ | $5.06 \times 10^{-1}$ | $3.57 \times 10^{-1}$ | $2.66 \times 10^{-1}$ | $2.31 \times 10^{-1}$ |
| $1.4 \times 10^{0}$ | $5.63 \times 10^{0}$ | $3.47 \times 10^{0}$ | $2.01 \times 10^{0}$ | $1.24 \times 10^{0}$ | $7.80 \times 10^{-1}$ | $6.42 \times 10^{-1}$ | $5.17 \times 10^{-1}$ | $4.45 \times 10^{-1}$ | $4.19 \times 10^{-1}$ |
| $2.4 \times 10^{0}$ | $5.63 \times 10^{0}$ | $3.49 \times 10^{0}$ | $2.06 \times 10^{0}$ | $1.35 \times 10^{0}$ | $9.54 \times 10^{-1}$ | $8.50 \times 10^{-1}$ | $7.63 \times 10^{-1}$ | $7.18 \times 10^{-1}$ | $7.03 \times 10^{-1}$ |
| $4.0 \times 10^{0}$ | $5.63 \times 10^{0}$ | $3.51 \times 10^{0}$ | $2.13 \times 10^{0}$ | $1.50 \times 10^{0}$ | $1.20 \times 10^{0}$ | $1.13 \times 10^{0}$ | $1.08 \times 10^{0}$ | $1.06 \times 10^{0}$ | $1.05 \times 10^{0}$ |
| $8.0 \times 10^{0}$ | $5.64 \times 10^{0}$ | $3.56 \times 10^{0}$ | $2.31 \times 10^{0}$ | $1.85 \times 10^{0}$ | $1.68 \times 10^{0}$ | $1.65 \times 10^{0}$ | $1.63 \times 10^{0}$ | $1.63 \times 10^{0}$ | $1.63 \times 10^{0}$ |
| $1.4 \times 10^{1}$ | $5.65 \times 10^{0}$ | $3.63 \times 10^{0}$ | $2.55 \times 10^{0}$ | $2.23 \times 10^{0}$ | $2.15 \times 10^{0}$ | $2.14 \times 10^{0}$ | $2.14 \times 10^{0}$ | $2.14 \times 10^{0}$ | $2.14 \times 10^{0}$ |
| $2.4 \times 10^{1}$ | $5.67 \times 10^{0}$ | $3.74 \times 10^{0}$ | $2.86 \times 10^{0}$ | $2.68 \times 10^{0}$ | $2.65 \times 10^{0}$ | $2.65 \times 10^{0}$ | $2.64 \times 10^{0}$ | $2.64 \times 10^{0}$ | $2.64 \times 10^{0}$ |
| $4.0 \times 10^{1}$ | $5.70 \times 10^{0}$ | $3.90 \times 10^{0}$ | $3.24 \times 10^{0}$ | $3.15 \times 10^{0}$ | $3.14 \times 10^{0}$ | $3.14 \times 10^{0}$ | $3.14 \times 10^{0}$ | $3.14 \times 10^{0}$ | $3.14 \times 10^{0}$ |
| $8.0 \times 10^{1}$ | $5.76 \times 10^{0}$ | $4.22 \times 10^{0}$ | $3.85 \times 10^{0}$ | $3.82 \times 10^{0}$ | $3.82 \times 10^{0}$ | $3.82 \times 10^{0}$ | $3.82 \times 10^{0}$ | $3.82 \times 10^{0}$ | $3.82 \times 10^{0}$ |
| $1.4 \times 10^{2}$ | $5.85 \times 10^{0}$ | $4.58 \times 10^{0}$ | $4.38 \times 10^{0}$ | $4.37 \times 10^{0}$ | $4.37 \times 10^{0}$ | $4.37 \times 10^{0}$ | $4.37 \times 10^{0}$ | $4.37 \times 10^{0}$ | $4.37 \times 10^{0}$ |
| $2.4 \times 10^{2}$ | $5.99 \times 10^{0}$ | $5.00 \times 10^{0}$ | $4.91 \times 10^{0}$ | $4.91 \times 10^{0}$ | $4.91 \times 10^{0}$ | $4.91 \times 10^{0}$ | $4.91 \times 10^{0}$ | $4.91 \times 10^{0}$ | $4.91 \times 10^{0}$ |
| $4.0 \times 10^{2}$ | $6.16 \times 10^{0}$ | $5.46 \times 10^{0}$ | $5.42 \times 10^{0}$ | $5.42 \times 10^{0}$ | $5.42 \times 10^{0}$ | $5.42 \times 10^{0}$ | $5.42 \times 10^{0}$ | $5.42 \times 10^{0}$ | $5.42 \times 10^{0}$ |
| $8.0 \times 10^{2}$ | $6.47 \times 10^{0}$ | $6.11 \times 10^{0}$ | $6.11 \times 10^{0}$ | $6.11 \times 10^{0}$ | $6.11 \times 10^{0}$ | $6.11 \times 10^{0}$ | $6.11 \times 10^{0}$ | $6.11 \times 10^{0}$ | $6.11 \times 10^{0}$ |
| $1.4 \times 10^{3}$ | $6.67 \times 10^{0}$ | $6.67 \times 10^{0}$ | $6.67 \times 10^{0}$ | $6.67 \times 10^{0}$ | $6.67 \times 10^{0}$ | $6.67 \times 10^{0}$ | $6.67 \times 10^{0}$ | $6.67 \times 10^{0}$ | $6.67 \times 10^{0}$ |
| $2.4 \times 10^{3}$ | $7.21 \times 10^{0}$ | $7.21 \times 10^{0}$ | $7.21 \times 10^{0}$ | $7.21 \times 10^{0}$ | $7.21 \times 10^{0}$ | $7.21 \times 10^{0}$ | $7.21 \times 10^{0}$ | $7.21 \times 10^{0}$ | $7.21 \times 10^{0}$ |
| $4.0 \times 10^{3}$ | $7.72 \times 10^{0}$ | $7.72 \times 10^{0}$ | $7.72 \times 10^{0}$ | $7.72 \times 10^{0}$ | $7.72 \times 10^{0}$ | $7.72 \times 10^{0}$ | $7.72 \times 10^{0}$ | $7.72 \times 10^{0}$ | $7.72 \times 10^{0}$ |
| $8.0 \times 10^{3}$ | $8.41 \times 10^{0}$ | $8.41 \times 10^{0}$ | $8.41 \times 10^{0}$ | $8.41 \times 10^{0}$ | $8.41 \times 10^{0}$ | $8.41 \times 10^{0}$ | $8.41 \times 10^{0}$ | $8.41 \times 10^{0}$ | $8.41 \times 10^{0}$ |
| $1.4 \times 10^{4}$ | $8.97 \times 10^{0}$ | $8.97 \times 10^{0}$ | $8.97 \times 10^{0}$ | $8.97 \times 10^{0}$ | $8.97 \times 10^{0}$ | $8.97 \times 10^{0}$ | $8.97 \times 10^{0}$ | $8.97 \times 10^{0}$ | $8.97 \times 10^{0}$ |
| $2.4 \times 10^{4}$ | $9.51 \times 10^{0}$ | $9.51 \times 10^{0}$ | $9.51 \times 10^{0}$ | $9.51 \times 10^{0}$ | $9.51 \times 10^{0}$ | $9.51 \times 10^{0}$ | $9.51 \times 10^{0}$ | $9.51 \times 10^{0}$ | $9.51 \times 10^{0}$ |
| $4.0 \times 10^{4}$ | $1.94 \times 10^{1}$ | $1.94 \times 10^{1}$ | $1.94 \times 10^{1}$ | $1.94 \times 10^{1}$ | $1.94 \times 10^{1}$ | $1.94 \times 10^{1}$ | $1.94 \times 10^{1}$ | $1.94 \times 10^{1}$ | $1.94 \times 10^{1}$ |

Source: Adapted from S. P. Neuman, *Water Resources Research,* 11 (1975):329–42.

**APPENDIX 7** Table for length conversion

| Unit | mm | cm | m | km | in | ft | yd | mi |
|---|---|---|---|---|---|---|---|---|
| 1 millimeter | 1 | 0.1 | 0.001 | $10^{-6}$ | 0.0397 | 0.00328 | 0.00109 | $6.21 \times 10^{-7}$ |
| 1 centimeter | 10 | 1 | 0.01 | 0.0001 | 0.3937 | 0.0328 | 0.0109 | $6.21 \times 10^{-6}$ |
| 1 meter | 1000 | 100 | 1 | 0.001 | 39.37 | 3.281 | 1.094 | $6.21 \times 10^{-4}$ |
| 1 kilometer | $10^6$ | $10^5$ | 1000 | 1 | 39,370 | 3281 | 1093.6 | 0.621 |
| 1 inch | 25.4 | 2.54 | 0.0254 | $2.54 \times 10^{-5}$ | 1 | 0.0833 | 0.0278 | $1.58 \times 10^{-5}$ |
| 1 foot | 304.8 | 30.48 | 0.3048 | $3.05 \times 10^{-4}$ | 12 | 1 | 0.333 | $1.89 \times 10^{-4}$ |
| 1 yard | 914.4 | 91.44 | 0.9144 | $9.14 \times 10^{-4}$ | 36 | 3 | 1 | $5.68 \times 10^{-4}$ |
| 1 mile | $1.61 \times 10^6$ | $1.01 \times 10^5$ | $1.61 \times 10^3$ | 1.6093 | 63,360 | 5280 | 1760 | 1 |

**APPENDIX 8** Table for area conversion

| Unit | $cm^2$ | $m^2$ | $km^2$ | ha | $in^2$ | $ft^2$ | $yd^2$ | $mi^2$ | ac |
|---|---|---|---|---|---|---|---|---|---|
| 1 sq. centimeter | 1 | 0.0001 | $10^{-10}$ | $10^{-8}$ | 0.155 | $1.08 \times 10^{-3}$ | $1.2 \times 10^{-4}$ | $3.86 \times 10^{-11}$ | $2.47 \times 10^{-8}$ |
| 1 sq. meter | $10^4$ | 1 | $10^{-6}$ | $10^{-4}$ | 1550 | 10.76 | 1.196 | $3.86 \times 10^{-7}$ | $2.47 \times 10^{-4}$ |
| 1 sq. kilometer | $10^{10}$ | $10^6$ | 1 | 100 | $1.55 \times 10^9$ | $1.076 \times 10^7$ | $1.196 \times 10^6$ | 0.3861 | 247.1 |
| 1 hectare | $10^8$ | $10^4$ | 0.01 | 1 | $1.55 \times 10^7$ | $1.076 \times 10^5$ | $1.196 \times 10^4$ | $3.861 \times 10^{-3}$ | 2.471 |
| 1 sq. inch | 6.452 | $6.45 \times 10^{-4}$ | $6.45 \times 10^{10}$ | $6.45 \times 10^{-8}$ | 1 | $6.94 \times 10^{-3}$ | $7.7 \times 10^{-4}$ | $2.49 \times 10^{-10}$ | $1.574 \times 10^7$ |
| 1 sq. foot | 929 | 0.0929 | $9.29 \times 10^{-8}$ | $9.29 \times 10^{-6}$ | 144 | 1 | 0.111 | $3.587 \times 10^{-8}$ | $2.3 \times 10^{-5}$ |
| 1 sq. yard | 8361 | 0.8361 | $8.36 \times 10^{-7}$ | $8.36 \times 10^{-5}$ | 1296 | 9 | 1 | $3.23 \times 10^{-7}$ | $2.07 \times 10^{-4}$ |
| 1 sq. mile | $2.59 \times 10^{10}$ | $2.59 \times 10^6$ | 2.59 | 259 | $4.01 \times 10^9$ | $2.79 \times 10^7$ | $3.098 \times 10^6$ | 1 | 640 |
| 1 acre | $4.04 \times 10^7$ | 4047 | $4.047 \times 10^{-3}$ | 0.4047 | $6.27 \times 10^6$ | 43,560 | 4840 | $1.562 \times 10^{-3}$ | 1 |

**APPENDIX 9** Table for volume conversion

| Unit | mL | liters | $m^3$ | $in^3$ | $ft^3$ | gal | ac-ft | million gal |
|---|---|---|---|---|---|---|---|---|
| 1 milliliter | 1 | 0.001 | $10^{-6}$ | 0.06102 | $3.53 \times 10^{-5}$ | $2.64 \times 10^{4}$ | $8.1 \times 10^{-10}$ | $2.64 \times 10^{-10}$ |
| 1 liter | $10^3$ | 1 | 0.001 | 61.02 | 0.0353 | 0.264 | $8.1 \times 10^{-7}$ | $2.64 \times 10^{-7}$ |
| 1 cu. meter | $10^6$ | 1000 | 1 | 61,023 | 35.31 | 264.17 | $8.1 \times 10^{-4}$ | $2.64 \times 10^{-4}$ |
| 1 cu. inch | 16.39 | $1.64 \times 10^{-2}$ | $1.64 \times 10^{-5}$ | 1 | $5.79 \times 10^{-4}$ | $4.33 \times 10^{-3}$ | $1.218 \times 10^{-8}$ | $4.329 \times 10^{-9}$ |
| 1 cu. foot | 28,317 | 28.317 | 0.02832 | 1728 | 1 | 7.48 | $2.296 \times 10^{-5}$ | $7.48 \times 10^{6}$ |
| 1 U.S. gallon | 3785.4 | 3.785 | $3.78 \times 10^{-3}$ | 231 | 0.134 | 1 | $3.069 \times 10^{-6}$ | $10^6$ |
| 1 acre-foot | $1.233 \times 10^{9}$ | $1.233 \times 10^{6}$ | 1233.5 | $75.27 \times 10^{6}$ | 43,560 | $3.26 \times 10^{5}$ | 1 | 0.3260 |
| 1 million gallons | $3.785 \times 10^{9}$ | $3.785 \times 10^{6}$ | 3785 | $2.31 \times 10^{8}$ | $1.338 \times 10^{5}$ | $10^6$ | 3.0684 | 1 |

**APPENDIX 10** Table for time conversion

| Unit | sec | min | hours | days | years |
|---|---|---|---|---|---|
| 1 second | 1 | $1.67 \times 10^{-2}$ | $2.77 \times 10^{-4}$ | $1.157 \times 10^{-5}$ | $3.17 \times 10^{-8}$ |
| 1 minute | 60 | 1 | $1.67 \times 10^{-2}$ | $6.94 \times 10^{-4}$ | $1.90 \times 10^{-6}$ |
| 1 hour | 360 | 60 | 1 | $4.17 \times 10^{-2}$ | $1.14 \times 10^{-4}$ |
| 1 day | $8.64 \times 10^{4}$ | 1440 | 24 | 1 | $2.74 \times 10^{-3}$ |
| 1 year | $3.15 \times 10^{7}$ | $5.256 \times 10^{5}$ | 8760 | 365 | 1 |

**APPENDIX 11** Solubility products for selected minerals and compounds

| Compound | Solubility product | Mineral name |
|---|---|---|
| Chlorides | | |
| $CuCl$ | $10^{-6.7}$ | |
| $PbCl_2$ | $10^{-4.8}$ | |
| $Hg_2Cl_2$ | $10^{-17.9}$ | |
| $AgCl$ | $10^{-9.7}$ | |
| Fluorides | | |
| $BaF_2$ | $10^{-5.8}$ | |
| $CaF_2$ | $10^{-10.4}$ | Fluorite |
| $MgF_2$ | $10^{-8.2}$ | Sellaite |
| $PbF_2$ | $10^{-7.5}$ | |
| $SrF_2$ | $10^{-8.5}$ | |
| Sulfates | | |
| $BaSO_4$ | $10^{-10.0}$ | Barite |
| $CaSO_4$ | $10^{-4.5}$ | Anhydrite |
| $CaSO_4 \cdot 2H_2O$ | $10^{-4.6}$ | Gypsum |
| $PbSO_4$ | $10^{-7.8}$ | Anglesite |
| $Ag_2SO_4$ | $10^{-4.8}$ | |
| $SrSO_4$ | $10^{-6.5}$ | Celestite |
| Sulfides | | |
| $Cu_2S$ | $10^{-48.5}$ | |
| $CuS$ | $10^{-36.1}$ | |
| $FeS$ | $10^{-18.1}$ | |
| $PbS$ | $10^{-27.5}$ | Galena |
| $HgS$ | $10^{-53.3}$ | Cinnebar |
| $ZnS$ | $10^{-22.5}$ | Wurtzite |
| $ZnS$ | $10^{-24.7}$ | Sphalerite |
| Carbonates | | |
| $BaCO_3$ | $10^{-8.3}$ | Witherite |
| $CdCO_3$ | $10^{-13.7}$ | |
| $CaCO_3$ | $10^{-8.35}$ | Calcite |
| $CaCO_3$ | $10^{-8.22}$ | Aragonite |
| $CoCO_3$ | $10^{-10.0}$ | |
| $FeCO_3$ | $10^{-10.7}$ | Siderite |
| $PbCO_3$ | $10^{-13.1}$ | |
| $MgCO_3$ | $10^{-7.5}$ | Magnesite |
| $MnCO_3$ | $10^{-9.3}$ | Rhodochrosite |
| Phosphates | | |
| $AlPO_4 \cdot 2H_2O$ | $10^{-22.1}$ | Variscite |
| $CaHPO_4 \cdot 2H_2O$ | $10^{-6.6}$ | |
| $Ca_3(PO_4)_2$ | $10^{-28.7}$ | |
| $Cu_3(PO_4)_2$ | $10^{-36.9}$ | |
| $FePO_4$ | $10^{-21.6}$ | |
| $FePO_4 \cdot 2H_2O$ | $10^{-26.4}$ | |

Source: K. B. Krauskopf, *Introduction to Geochemistry,* 2nd ed. New York: McGraw-Hill, 1979.

**APPENDIX 12** Atomic weights and numbers of naturally occurring elements

| Element | Symbol | Atomic number | Atomic weight |
|---|---|---|---|
| Actinium | Ac | 89 | 227.03 |
| Aluminum | Al | 13 | 26.98 |
| Antimony | Sb | 51 | 121.75 |
| Argon | Ar | 18 | 39.95 |
| Arsenic | As | 33 | 74.92 |
| Barium | Ba | 56 | 137.33 |
| Beryllium | Be | 4 | 9.01 |
| Bismuth | Bi | 83 | 208.98 |
| Boron | B | 5 | 10.81 |
| Bromine | Br | 35 | 79.90 |
| Cadmium | Cd | 48 | 112.41 |
| Calcium | Ca | 20 | 40.08 |
| Carbon | C | 6 | 12.01 |
| Cerium | Ce | 58 | 140.12 |
| Cesium | Cs | 55 | 132.91 |
| Chlorine | Cl | 17 | 35.45 |
| Chromium | Cr | 24 | 52.00 |
| Cobalt | Co | 27 | 58.93 |
| Copper | Cu | 29 | 63.55 |
| Dysprosium | Dy | 66 | 162.50 |
| Erbium | Er | 68 | 167.26 |
| Europium | Eu | 63 | 151.96 |
| Fluorine | F | 9 | 19.00 |
| Gadolinium | Gd | 64 | 157.25 |
| Gallium | Ga | 31 | 69.72 |
| Germanium | Ge | 32 | 72.59 |
| Gold | Au | 79 | 196.97 |
| Hafnium | Hf | 72 | 178.49 |
| Helium | He | 2 | 4.003 |
| Holmium | Ho | 67 | 164.93 |
| Hydrogen | H | 1 | 1.008 |
| Indium | In | 49 | 114.82 |
| Iodine | I | 53 | 126.90 |
| Iridium | Ir | 77 | 192.22 |
| Iron | Fe | 26 | 55.85 |
| Krypton | Kr | 36 | 83.80 |
| Lanthanum | La | 57 | 138.91 |
| Lead | Pb | 82 | 207.19 |
| Lithium | Li | 3 | 6.94 |
| Lutetium | Lu | 71 | 174.97 |
| Magnesium | Mg | 12 | 24.31 |
| Manganese | Mn | 25 | 54.94 |
| Mercury | Hg | 80 | 200.59 |
| Molybdenum | Mo | 42 | 95.94 |

**APPENDIX 12** Continued

| Element | Symbol | Atomic number | Atomic weight |
|---|---|---|---|
| Neodymium | Nd | 60 | 144.24 |
| Neon | Ne | 10 | 20.18 |
| Nickel | Ni | 28 | 58.70 |
| Niobium | Nb | 41 | 92.91 |
| Nitrogen | N | 7 | 14.01 |
| Osmium | Os | 76 | 190.2 |
| Oxygen | O | 8 | 16.00 |
| Palladium | Pd | 46 | 106.4 |
| Phosphorus | P | 15 | 30.97 |
| Platinum | Pt | 78 | 195.09 |
| Polonium | Po | 84 | 209 |
| Potassium | K | 19 | 39.10 |
| Praseodymium | Pr | 59 | 140.91 |
| Protactinium | Pa | 91 | 231.04 |
| Radium | Ra | 88 | 226.03 |
| Radon | Rn | 86 | 222 |
| Rhenium | Re | 75 | 186.21 |
| Rhodium | Rh | 45 | 102.91 |
| Rubidium | Rb | 37 | 85.47 |
| Ruthenium | Ru | 44 | 101.07 |
| Samarium | Sm | 62 | 150.35 |
| Scandium | Sc | 21 | 44.96 |
| Selenium | Se | 34 | 78.96 |
| Silicon | Si | 14 | 28.09 |
| Silver | Ag | 47 | 107.87 |
| Sodium | Na | 11 | 22.99 |
| Strontium | Sr | 38 | 87.62 |
| Sulfur | S | 16 | 32.06 |
| Tantalum | Ta | 73 | 180.95 |
| Tellurium | Te | 52 | 127.60 |
| Terbium | Tb | 65 | 158.93 |
| Thallium | Tl | 81 | 204.37 |
| Thorium | Th | 90 | 232.04 |
| Thulium | Tm | 69 | 168.93 |
| Tin | Sn | 50 | 118.69 |
| Titanium | Ti | 22 | 47.90 |
| Tungsten | W | 74 | 183.85 |
| Uranium | U | 92 | 238.03 |
| Vanadium | V | 23 | 50.94 |
| Xenon | Xe | 54 | 131.30 |
| Ytterbium | Yb | 70 | 173.04 |
| Yttrium | Y | 39 | 88.91 |
| Zinc | Zn | 30 | 65.38 |
| Zirconium | Zr | 40 | 91.22 |

Source: K. B. Krauskopf, *Introduction to Geochemistry,* 2nd ed. New York: McGraw Hill, 1979.

**APPENDIX 13** Values of the error of *x* [erf *(x)*] and the complementary error function of *x* [erfc *(x)*]. Note that erfc *(x)* = 1 − erf *(x)*.

| x | erf *(x)* | erfc *(x)* |
|---|---|---|
| 0 | 0 | 1.0 |
| 0.05 | 0.056372 | 0.943628 |
| 0.1 | 0.112463 | 0.887537 |
| 0.15 | 0.167996 | 0.832004 |
| 0.2 | 0.222703 | 0.777297 |
| 0.25 | 0.276326 | 0.723674 |
| 0.3 | 0.328627 | 0.671373 |
| 0.35 | 0.379382 | 0.620618 |
| 0.4 | 0.428392 | 0.571608 |
| 0.45 | 0.475482 | 0.524518 |
| 0.5 | 0.520500 | 0.479500 |
| 0.55 | 0.563323 | 0.436677 |
| 0.6 | 0.603856 | 0.396144 |
| 0.65 | 0.642029 | 0.357971 |
| 0.7 | 0.677801 | 0.322199 |
| 0.75 | 0.711156 | 0.288844 |
| 0.8 | 0.742101 | 0.257899 |
| 0.85 | 0.770668 | 0.229332 |
| 0.9 | 0.796908 | 0.203092 |
| 0.95 | 0.820891 | 0.179109 |
| 1.0 | 0.842701 | 0.157299 |
| 1.1 | 0.880205 | 0.119795 |
| 1.2 | 0.910314 | 0.089686 |
| 1.3 | 0.934008 | 0.065992 |
| 1.4 | 0.952285 | 0.047715 |
| 1.5 | 0.966105 | 0.033895 |
| 1.6 | 0.976348 | 0.023652 |
| 1.7 | 0.983790 | 0.016210 |
| 1.8 | 0.989091 | 0.010909 |
| 1.9 | 0.992790 | 0.007210 |
| 2.0 | 0.995322 | 0.004678 |
| 2.1 | 0.997021 | 0.002979 |
| 2.2 | 0.998137 | 0.001863 |
| 2.3 | 0.998857 | 0.001143 |
| 2.4 | 0.999311 | 0.000689 |
| 2.5 | 0.999593 | 0.000407 |
| 2.6 | 0.999764 | 0.000236 |
| 2.7 | 0.999866 | 0.000134 |
| 2.8 | 0.999925 | 0.000075 |
| 2.9 | 0.999959 | 0.000041 |
| 3.0 | 1.999978 | 0.000022 |
| ∞ | 1.00000 | 0.00000 |

**APPENDIX 14.** Absolute density and absolute viscosity of water

| Temperature (°C) | Density ($g/cm^3$) | Viscosity (poise) |
|---|---|---|
| 0 | 0.999841 | 0.017921 |
| 1 | 0.999900 | 0.017313 |
| 2 | 0.999941 | 0.016728 |
| 3 | 0.999965 | 0.016191 |
| 4 | 0.999973 | 0.015674 |
| 5 | 0.999965 | 0.015188 |
| 6 | 0.999941 | 0.014728 |
| 7 | 0.999902 | 0.014284 |
| 8 | 0.999849 | 0.013860 |
| 9 | 0.999781 | 0.013462 |
| 10 | 0.999700 | 0.013077 |
| 11 | 0.999605 | 0.012713 |
| 12 | 0.999498 | 0.012363 |
| 13 | 0.999377 | 0.012028 |
| 14 | 0.999244 | 0.011709 |
| 15 | 0.999099 | 0.011404 |
| 16 | 0.998943 | 0.011111 |
| 17 | 0.998774 | 0.010828 |
| 18 | 0.998595 | 0.010559 |
| 19 | 0.998405 | 0.010299 |
| 20 | 0.998203 | 0.010050 |
| 21 | 0.997992 | 0.009810 |
| 22 | 0.997770 | 0.009579 |
| 23 | 0.997538 | 0.009358 |
| 24 | 0.997296 | 0.009142 |
| 25 | 0.997044 | 0.008937 |
| 26 | 0.996783 | 0.008737 |
| 27 | 0.996512 | 0.008545 |
| 28 | 0.996232 | 0.008360 |
| 29 | 0.995944 | 0.008180 |
| 30 | 0.995646 | 0.008007 |
| 35 | 0.994029 | 0.007225 |
| 40 | 0.992214 | 0.006560 |
| 45 | 0.990212 | 0.005988 |
| 50 | 0.988047 | 0.005494 |

Source: *Handbook of Chemistry and Physics* (Cleveland, Ohio: CRC Publishing Company, 1976).

# Glossary

**Adiabatic expansion** The process that occurs when an air mass rises and expands without exchanging heat with its surroundings.

**Adsorption** The attraction and adhesion of a layer of ions from an aqueous solution to the solid mineral surfaces with which it is in contact.

**Advection** The process by which solutes are transported by the motion of flowing ground water.

**Aliquot** One of a number of equal-sized portions of a water sample that is being analyzed.

**Alluvium** Sediments deposited by flowing rivers. Depending upon the location in the floodplain of the river, different-sized sediments are deposited.

**American Rule** A ground-water doctrine that holds that an overlying property owner has the right to use only a reasonable amount of ground water.

**Anisotropy** The condition under which one or more of the hydraulic properties of an aquifer vary according to the direction of flow.

**Antecedent moisture** The soil moisture present before a particular precipitation event.

**Aquiclude** A low-permeability unit that forms either the upper or lower boundary of a ground-water flow system.

**Aquifer** Rock or sediment in a formation, group of formations, or part of a formation which is saturated and sufficiently permeable to transmit economic quantities of water to wells and springs.

**Aquifer, confined** An aquifer that is overlain by a confining bed. The confining bed has a significantly lower hydraulic conductivity than the aquifer.

**Aquifer, perched** A region in the unsaturated zone where the soil may be locally saturated because it overlies a low-permeability unit.

**Aquifer, semiconfined** An aquifer confined by a low-permeability layer that permits water to slowly flow through it. During pumping of the aquifer, recharge to the aquifer can occur across the confining layer. Also known as a leaky artesian or leaky confined aquifer.

**Aquifer test** *See* pumping test.

**Aquifer, unconfined** An aquifer in which there are no confining beds between the zone of saturation and the surface. There will be a water table in an unconfined aquifer. Water-table aquifer is a synonym.

**Aquifuge** An absolutely impermeable unit that will neither store nor transmit water.

**Aquitard** A low-permeability unit that can store ground water and also transmit it slowly from one aquifer to another.

**Artificial recharge** The process by which water can be injected or added to an aquifer. Dug basins, drilled wells, or simply the spread of water across the land surface are all means of artificial recharge.

**Average linear velocity** *See* seepage velocity.

**Bail-down test** A type of slug test performed by using a bailer to remove a volume of water from a small-diameter well.

**Bailer** A device used to withdraw a water sample from a small-diameter well or piezometer. A bailer typically is a piece of pipe attached to a wire and having a check valve in the bottom.

**Barrier boundary** An aquifer-system boundary represented by a rock mass that is not a source of water.

**Baseflow** That part of stream discharge from ground water seeping into the stream.

**Baseflow recession** The declining rate of discharge of a stream fed only by baseflow for an extended period. Typically, a baseflow recession will be exponential.

**Baseflow-recession hydrograph** A hydrograph that shows a baseflow-recession curve.

**Bladder pump** A positive-displacement pumping device that uses pulses of gas to push a water-quality sample toward the surface.

**Borehole geochemical probe** A water-quality monitoring device that is lowered into a well on a cable and that can make a direct reading of such parameters as pH, *Eh,* temperature, and specific conductivity.

**Borehole geophysics** The general field of geophysics developed around the lowering of various probes into a well.

**Boring** A hole advanced into the ground by means of a drilling rig.

**Boussinesq equation** The general equation for two-dimensional unconfined transient flow.

**Caliper log** A borehole log of the diameter of an uncased well.

**Capillary forces** The forces acting on soil moisture in the unsaturated zone, attributable to molecular attraction between soil particles and water.

**Capillary fringe** The zone immediately above the water table, where water is drawn upward by capillary attraction.

**Casing** *See* well casing.

**Cation exchange capacity** The ability of a particular rock or soil to absorb cations.

**Cementation** The process by which some of the voids in a sediment are filled with precipitated materials, such as silica, calcite, and iron oxide, and which is a part of diagenesis.

**Chemical activity** The molal concentration of an ion multiplied by a factor known as the activity coefficient.

**Clastic dike** Intrusion of sediment forced into fractures in rock or sediments.

**Cleat** The vertical planes of fracture that are found in coal.

**Collection lysimeter** A device installed in the unsaturated zone to collect a water-quality sample by having the water drain downward by gravity into a collection pit.

**Combining weight** *See* equivalent weight.

**Common-ion effect** The decrease in the solubility of a salt dissolved in water already containing some of the ions of the salt.

**Condensation** The process that occurs when an air mass is saturated and water droplets form around nuclei or on surfaces.

**Confining bed** A body of material of low hydraulic conductivity that is stratigraphically adjacent to one or more aquifers. It may lie above or below the aquifer.

**Connate water** Interstitial water that was not buried with a rock but which has been out of contact with the atmosphere for an appreciable part of a geologic period.

**Contact spring** A spring that forms at a lithologic contact where a more permeable unit overlies a less permeable unit.

**Contaminant** *See* pollutant.

**Current meter** A device that is lowered into a stream in order to record the rate at which the current is moving.

**Darcian velocity** *See* specific discharge.

**Darcy's law** An equation that can be used to compute the quantity of water flowing through an aquifer.

**Debye-Hückel equation** A means of computing the activity coefficient for an ionic species.

**Density** The mass or quantity of a substance per unit volume. Units are kilograms per cubic meter or grams per cubic centimeter.

**Depression spring** A spring formed when the water table reaches a land surface because of a change in topography.

**Depression storage** Water from precipitation that collects in puddles at the land surface.

**Dew point** The temperature of a given air mass at which condensation will begin.

**Diagenesis** The chemical and physical changes occurring in sediments before consolidation or while in the environment of deposition.

**Diffusion** The process by which both ionic and molecular species dissolved in water move from areas of higher concentration to areas of lower concentration.

**Digital computer model** A model of ground-water flow in which the aquifer is described by numerical equations, with specified values for boundary conditions, that are solved on a digital computer.

**Dipole array** A particular arrangement of electrodes used to measure surface electrical resistivity.

**Direct precipitation** Water that falls directly into a lake or stream without passing through any land phase of the runoff cycle.

**Dirichlet condition** A boundary condition for a ground-water computer model where the head is known at the boundary of the flow field.

**Discharge** The volume of water flowing in a stream or through an aquifer past a specific point in a given period of time.

**Discharge area** An area in which there are upward components of hydraulic head in the aquifer. Ground water is flowing toward the surface in a discharge area and may escape as a spring, seep, or baseflow or by evaporation and transpiration.

**Discharge velocity** *See* specific discharge.

**Dispersion** The phenomenon by which a solute in flowing ground water is mixed with uncontaminated water and becomes reduced in concentration. Dispersion is caused by both differences in the velocity that the water travels at the pore level and differences in the rate at which water travels through different strata in the flow path.

**Distribution coefficient** The slope of a linear Freundlich isotherm.

**Drainage basin** The land area from which surface runoff drains into a stream system.

**Drainage divide** A boundary line along a topographically high area that separates two adjacent drainage basins.

**Drawdown** A lowering of the water table of an unconfined aquifer or the potentiometric surface of a confined aquifer caused by pumping of ground water from wells.

**Dupuit assumptions** Assumptions for flow in an unconfined aquifer that (1) the hydraulic gradient is equal to the slope of the water table, (2) the streamlines are horizontal, and (3) the equipotential lines are vertical.

**Dupuit equation** An equation for the volume of water flowing in an unconfined aquifer; based upon the Dupuit assumptions.

**Duration curve** A graph showing the percentage of time that the given flows of a stream will be equaled or exceeded. It is based upon a statistical study of historic streamflow records.

**Dynamic equilibrium** A condition in which the amount of recharge to an aquifer equals the amount of natural discharge.

**Effective grain size** The grain size corresponding to the 10 percent finer by weight line on the grain-size distribution curve.

**Effective pore fraction** The ratio of the porosity available for fluid flow to the total porosity of a rock or sediment.

**Effective porosity** *See* porosity, effective.

**Electrical resistance model** An analog model of ground-water flow based upon the flow of electricity through a circuit containing resistors and capacitors.

**Electrical sounding** An earth-resistivity survey made at the same location by putting the electrodes progressively farther apart. It shows the change of apparent resistivity with depth.

**Electromagnetic conductivity** A method of measuring the induced electrical field in the earth to determine the ability of the earth to conduct electricity. Electromagnetic conductivity is the inverse of electrical resistivity. Also known as electric conductivity and terrain conductivity.

**English Rule** A ground-water doctrine that holds that property owners have the right of absolute ownership of the ground water beneath their land.

**Equilibrium constant** The number defining the conditions of equilibrium for a particular reversible chemical reaction.

**Equipotential line** A line in a two-dimensional ground-water flow field such that the total hydraulic head is the same for all points along the line.

**Equipotential surface** A surface in a three-dimensional ground-water flow field such that the total hydraulic head is the same everywhere on the surface.

**Equivalent weight** The formula weight of a dissolved ionic species divided by the electrical charge. Also known as combining weight.

**Eutrophication** The process of accelerated aging of a surface-water body; caused by excess nutrients and sediments being brought into the lake.

**Evaporation** The process by which water passes from the liquid to the vapor state.

**Evapotranspiration** The sum of evaporation plus transpiration.

**Evapotranspiration, actual** The evapotranspiration that actually occurs under given climatic and soil-moisture conditions.

**Evapotranspiration, potential** The evapotranspiration that would occur under given climatic conditions if there were unlimited soil moisture.

**Fault spring** A spring created by the movement of two rock units on a fault.

**Field blank** A water-quality sample where highly purified water is run through the field-sampling procedure and sent to the laboratory to detect if any contamination of the samples is occurring during the sampling process.

**Field capacity** The maximum amount of water that the unsaturated zone of a soil can hold against the pull of gravity. The field capacity is dependent on the length of time the soil has been undergoing gravity drainage.

**Finite-difference model** A particular kind of a digital computer model based upon a rectangular grid that sets the boundaries of the model and the nodes where the model will be solved.

**Finite-element model** A digital ground-water–flow model where the aquifer is divided into a mesh formed of a number of polygonal cells.

**Flow net** The set of intersecting equipotential lines and flowlines representing two-dimensional steady flow through porous media.

**Flow, steady** The flow that occurs when, at any point in the flow field, the magnitude and direction of the specific discharge are constant in time.

**Flow, unsteady** The flow that occurs when, at any point in the flow field, the magnitude or direction of the specific discharge changes with time. Also called transient flow or nonsteady flow.

**Fluid potential** The mechanical energy per unit mass of fluid at any given point in space and time.

**Force potential** The sum of the kinetic energy, elevation energy, and pressure at a point in an aquifer. It is equal to the hydraulic head times the acceleration of gravity.

**Fossil water** Interstitial water that was buried at the same time as the original sediment.

**Fracture spring** A spring created by fracturing or jointing of the rock.

**Fracture trace** The surface representation of a fracture zone. It may be a characteristic line of vegetation or linear soil-moisture pattern or a topographic sag.

**Free energy** A measure of the thermodynamic driving energy of a chemical reaction. Also known as Gibbs free energy or Gibbs function.

**Freundlich isotherm** An empirical equation that describes the amount of solute adsorbed onto a soil surface.

**Gamma-gamma radiation log** A borehole log in which a source of gamma radiation as well as a detector are lowered into the borehole. This log measures bulk density of the formation and fluids.

**Gamma log** *See* natural gamma radiation log.

**Gauss-Seidel** A particular type of method for solving for the head in a finite-difference ground-water model.

**Ghyben-Herzberg principle** An equation that relates the depth of a salt-water interface in a coastal aquifer to the height of the fresh-water table above sea level.

**Glacial-lacustrine sediments** Silt and clay deposits formed in the quiet waters of lakes that received meltwater from glaciers.

**Glacial outwash** Well-sorted sand, or sand and gravel, deposited by the meltwater from a glacier.

**Glacial till** A glacial deposit composed of mostly unsorted sand, silt, clay, and boulders and laid down directly by the melting ice.

**Gouge** Soft, ground-up rock formed between the moving surfaces of a geological fault.

**Ground-penetrating radar** A surface geophysical technique based upon the transmission of repetitive pulses of electromagnetic waves into the ground. Some of the radiated energy is reflected back to the surface and the reflected signal is captured and processed.

**Ground water** The water contained in interconnected pores located below the water table in an unconfined aquifer or located in a confined aquifer.

**Ground-water basin** A rather vague designation pertaining to a ground-water reservoir that is more or less separate from neighboring ground-water reservoirs. A ground-water basin could be separated from adjacent basins by geologic boundaries or by hydrologic boundaries.

**Ground water, confined** The water contained in a confined aquifer. Pore-water pressure is greater than atmospheric at the top of the confined aquifer.

**Ground-water flow** The movement of water through openings in sediment and rock; occurs in the zone of saturation.

**Ground-water mining** The practice of withdrawing ground water at rates in excess of the natural recharge.

**Ground water, perched** The water in an isolated, saturated zone located in the zone of aeration. It is the result of the presence of a layer of material of low hydraulic conductivity, called a perching bed. Perched ground water will have a perched water table.

**Ground water, unconfined** The water in an aquifer where there is a water table.

**Grout curtain** An underground wall designed to stop ground-water flow; can be created by injecting grout into the ground, which subsequently hardens to become impermeable.

**Hantush-Jacob formula** An equation to describe the change in hydraulic head with time during pumping of a leaky confined aquifer.

**Hardness** A measure of the amount of calcium, magnesium, and iron dissolved in the water.

**Hazen method** An empirical equation that can be used to approximate the hydraulic conductivity of a sediment on the basis of the effective grain size.

**Head, total** The sum of the elevation head, the pressure head, and the velocity head at a given point in an aquifer.

**Hele-Shaw model** An analog model of ground-water flow based upon the movement of a viscous fluid between two closely spaced, parallel plates.

**Heterogeneous** Pertaining to a substance having different characteristics in different locations. A synonym is nonuniform.

**Hollow-stem auger** A particular kind of a drilling device whereby a hole is rapidly advanced into sediments. Sampling and installation of the equipment can take place through the hollow center of the auger.

**Homogeneous** Pertaining to a substance having identical characteristics everywhere. A synonym is uniform.

**Horizontal profiling** A method of making an earth-resistivity survey by measuring the apparent resistivity using the same electrode spacings at different grid points around an area.

**Humidity, absolute** The amount of moisture in the air as expressed by the number of grams of water per cubic meter of air.

**Humidity, relative** Percent ratio of the absolute humidity to the saturation humidity for an air mass.

**Humidity, saturation** The maximum amount of moisture that can be contained by an air mass at a given temperature.

**Hvorslev method** A procedure for performing a slug test in a piezometer that partially penetrates a water-table aquifer.

**Hydraulic conductivity** A coefficient of proportionality describing the rate at which water can move through a permeable medium. The density and kinematic viscosity of the water must be considered in determining hydraulic conductivity.

**Hydraulic diffusivity** A property of an aquifer or confining bed defined as the ratio of the transmissivity to the storativity.

**Hydraulic gradient** The change in total head with a change in distance in a given direction. The direction is that which yields a maximum rate of decrease in head.

**Hydraulic head** *See* head, total.

**Hydrochemical facies** Bodies of water with separate but distinct chemical compositions contained in an aquifer.

**Hydrodynamic dispersion** The process by which ground water containing a solute is diluted with uncontaminated ground water as it moves through an aquifer.

**Hydrogeology** The study of the interrelationships of geologic materials and processes with water, especially ground water.

**Hydrograph** A graph that shows some property of ground water or surface water as a function of time.

**Hydrologic equation** An expression of the law of mass conservation for purposes of water budgets. It may be stated as inflow equals outflow plus or minus changes in storage.

**Hydrology** The study of the occurrence, distribution, and chemistry of all waters of the earth.

**Hydrophyte** A type of plant that grows with the root system submerged in standing water.

**Hydrostratigraphic unit** A formation, part of a formation, or group of formations in which there are similar hydrologic characteristics allowing for grouping into aquifers or confining layers.

**Hygroscopic water** Water that clings to the surfaces of mineral particles in the zone of aeration.

**Ideal gas** A gas having a volume that varies inversely with pressure at a constant temperature and that also expands by 1/273 of its volume at 0° C for each degree rise in temperature at constant pressure.

**Image well** An imaginary well that can be used to simulate the effect of a hydrologic barrier, such as a recharge boundary or a barrier boundary, on the hydraulics of a pumping or recharge well.

**Infiltration** The flow of water downward from the land surface into and through the upper soil layers.

**Infiltration capacity** The maximum rate at which infiltration can occur under specific conditions of soil moisture. For a given soil, the infiltration capacity is a function of the water content.

**Injection well** A well drilled and constructed in such a manner that water can be pumped into an aquifer in order to recharge it.

**Interception** The process by which precipitation is captured on the surfaces of vegetation before it reaches the land surface.

**Interception loss** Rainfall that evaporates from standing vegetation.

**Interflow** The lateral movement of water in the unsaturated zone during and immediately after a precipitation event. The water moving as interflow discharges directly into a stream or lake.

**Intermediate zone** That part of the unsaturated zone below the root zone and above the capillary fringe.

**Intrinsic permeability** Pertaining to the relative ease with which a porous medium can transmit a liquid under a hydraulic or potential gradient. It is a property of the porous medium and is independent of the nature of the liquid or the potential field.

**Ion exchange** A process by which an ion in a mineral lattice is replaced by another ion that was present in an aqueous solution.

**Isocon** A line drawn on a map to indicate equal concentrations of a solute in ground water.

**Isohyetal line** A line drawn on a map, all points along which receive equal amounts of precipitation.

**Isotropy** The condition in which hydraulic properties of the aquifer are equal in all directions.

**Jacob straight-line method** A graphical method using semilogarithmic paper and the Theis equation for evaluating the results of a pumping test.

**Juvenile water** Water entering the hydrologic cycle for the first time.

**Karst** The type of geologic terrane underlain by carbonate rocks where significant solution of the rock has occurred due to flowing ground water.

**Kemmerer sampler** A sampling device that can be lowered either into a deep well or into a lake in order to retrieve a water sample from a particular depth in the well or the lake.

**Kinematic viscosity** The ratio of dynamic viscosity to mass density. It is obtained by dividing dynamic viscosity by the fluid density. Units of kinematic viscosity are square meters per second.

**Laminar flow** That type of flow in which the fluid particles follow paths that are smooth, straight, and parallel to the channel walls. In laminar flow, the viscosity of the fluid damps out turbulent motion. *Compare with* Turbulent flow.

**Langmuir isotherm** An empirical equation that describes the amount of solute adsorbed onto a soil surface.

**Land pan** A device used to measure free-water evaporation.

**Laplace equation** The partial differential equation governing steady-state flow of ground water.

**Law of mass action** The law stating that for a reversible chemical reaction the rate of reaction is proportional to the concentrations of the reactants.

**Leachate** Water that contains a high amount of dissolved solids and is created by liquid seeping from a landfill.

**Leachate collection system** A system installed in conjunction with a liner to capture the leachate that may be generated from a landfill so that it may be taken away and treated.

**Leaky confining layer** A low-permeability layer that can transmit water at sufficient rates to furnish some recharge to a well pumping from an underlying aquifer. Also called aquitard.

**Lineament** A natural linear surface longer than a mile (1500 meters).

**Liner** A low-permeability material, such as clay or plastic sheeting, that is put beneath a landfill in order to capture any leachate generated so as to help to prevent ground-water contamination.

**Lithologic log** A record of the lithology of the rock and soil encountered in a borehole from the surface to the bottom. Also known as a well log.

**Lysimeter** A field device containing a soil column and vegetation; used for measuring actual evapotranspiration.

**Magmatic water** Water associated with a magma.

**Magnetometer** A geophysical device that can be used to locate items that disrupt the earth's localized magnetic field; can be used for finding buried steel.

**Manning equation** An equation that can be used to compute the average velocity of flow in an open channel.

**Maximum contaminant level** The highest concentration of a solute permissible in a public water supply as specified in the National Interim Primary Drinking Water Standards for the United States.

**Maximum contaminant level goal** A nonenforceable health goal for solutes in drinking water; set at a level to prevent known or anticipated adverse effects with an adequate margin of safety.

**Micrograms per liter** A measure of the amount of dissolved solids in a solution in terms of micrograms of solute per liter of solution.

**Milliequivalents per liter** A measure of the concentration of a solute in solution; obtained by dividing the concentration in milligrams per liter by equivalent weight of the ion.

**Milligrams per liter** A measure of the amount of dissolved solids in a solute in terms of milligrams of solute per liter of solution.

**Model calibration** The process by which the independent variables of a digital computer model are varied in order to calibrate a dependent variable such as a head against a known value such as a water-table map.

**Model field verification** The process by which a digital computer model that has been calibrated and verified is tested to see if it can predict the field response of an aquifer to some transient condition.

**Model verification** The process by which a digital computer model that has been calibrated against a steady-state condition is tested to see if it can generate a transient response, such as the decline in the water table with pumping, that matches the known history of the aquifer.

**Moisture potential** The tension on the pore water in the unsaturated zone due to the attraction of the soil–water interface.

**Molality** A measure of chemical concentration. A one-molal solution has one mole of solute dissolved in 1000 grams of water. One mole of a compound is its formula weight in grams.

**Molarity** A measure of chemical concentration. A one-molar solution has one mole of solute dissolved in one liter of solution.

**Mutual-prescription doctrine** A ground-water doctrine stating that in the event of an overdraft of a ground-water basin, the available ground water will be apportioned among all the users in amounts proportional to their individual pumping rates.

**Natural gamma radiation log** A borehole log that measures the natural gamma radiation emitted by the formation rocks. It can be used to delineate subsurface rock types.

**Neumann condition** The boundary condition for a ground-water–flow model where a flux across the boundary of the flow region is known.

**Neutron log** A borehole log obtained by lowering a radioactive element, which is a source of neutrons, and a neutron detector into the well. The neutron log measures the amount of water present; hence, the porosity of the formation.

**Nonequilibrium type curve** A plot on logarithmic paper of the well function $W(u)$ as a function of $u$.

**Observation well** A nonpumping well used to observe the elevation of the water table or the potentiometric surface. An observation well is generally of larger diameter than a piezometer and typically is screened or slotted throughout the thickness of the aquifer.

**Overland flow** The flow of water over a land surface due to direct precipitation. Overland flow generally occurs when the precipitation rate exceeds the infiltration capacity of the soil and depression storage is full. Also called Horton overland flow.

**Packer test** An aquifer test performed in an open borehole; the segment of the borehole to be tested is sealed off from the rest of the borehole by inflating seals, called packers, both above and below the segment.

**Permafrost** Perenially frozen ground, occurring wherever the temperature remains at or below 0° C for two or more years in a row.

**Permeameter** A laboratory device used to measure the intrinsic permeability and hydraulic conductivity of a soil or rock sample.

**Phreatic cave** A cave that forms below the water table.

**Phreatophyte** A type of plant that typically has a high rate of transpiration by virtue of a taproot extending to the water table.

**Phreatic water** Water in the zone of saturation.

**Piezometer** A nonpumping well, generally of small diameter, that is used to measure the elevation of the water table or potentiometric surface. A piezometer generally has a short well screen through which water can enter.

**Piezometer nest** A set of two or more piezometers set close to each other but screened to different depths.

**Polar coordinates** The means by which the position of a point in a two-dimensional plane is described; based upon the radial distance from the origin to the given point and the angle between a horizontal line passing through the origin and a line extending from the origin to the given point.

**Pollutant** Any solute or cause of change in physical properties that renders water unfit for a given use.

**Pore space** The volume between mineral grains in a porous medium.

**Porosity** The ratio of the volume of void spaces in a rock or sediment to the total volume of the rock or sediment.

**Porosity, effective** The volume of the void spaces through which water or other fluids can travel in a rock or sediment divided by the total volume of the rock or sediment.

**Porosity, primary** The porosity that represents the original pore openings when a rock or sediment formed.

**Porosity, secondary** The porosity that has been caused by fractures or weathering in a rock or sediment after it has been formed.

**Potentiometric map** A contour map of the potentiometric surface of a particular hydrogeologic unit.

**Potentiometric surface** A surface that represents the level to which water will rise in tightly cased wells. If the head varies significantly with depth in the aquifer, then there may be more than one potentiometric surface. The water table is a particular potentiometric surface for an unconfined aquifer.

**Prior-appropriation doctrine** A doctrine stating that the right to use water is separate from other property rights and that the first person to withdraw and use the water holds the senior right. The doctrine has been applied to both ground and surface water.

**Public trust doctrine** A legal theory holding that certain lands and waters in the public domain are held in trust for use by the entire populace. It is especially applicable to navigable waters.

**Pumping cone** The area around a discharging well where the hydraulic head in the aquifer has been lowered by pumping. Also called cone of depression.

**Pumping test** A test made by pumping a well for a period of time and observing the change in hydraulic head in the aquifer. A pumping test may be used to determine the capacity of the well and the hydraulic characteristics of the aquifer. Also called aquifer test

**Quantification limit** The lower limit to the range in which the concentration of a solute can be determined by a particular analytical instrument.

**Radial flow** The flow of water in an aquifer toward a vertically oriented well.

**Rating curve** A graph of the discharge of a river at a particular point as a function of the elevation of the water surface.

**Recharge area** An area in which there are downward components of hydraulic head in the aquifer. Infiltration moves downward into the deeper parts of an aquifer in a recharge area.

**Recharge basin** A basin or pit excavated to provide a means of allowing water to soak into the ground at rates exceeding those that would occur naturally.

**Recharge boundary** An aquifer system boundary that adds water to the aquifer. Streams and lakes are typically recharge boundaries.

**Recharge well** A well specifically designed so that water can be pumped into an aquifer in order to recharge the ground-water reservoir.

**Recovery** The rate at which the water level in a well rises after the pump has been shut off. It is the inverse of drawdown.

**Regolith** The upper part of the earth's surface that has been altered by weathering processes. It includes both soil and weathered bedrock.

**Resistivity log** A borehole log made by lowering two current electrodes into the borehole and measuring the resistivity between two additional electrodes. It measures the electrical resistivity of the formation and contained fluids near the probe.

**Retardation** A general term for the many processes that act to remove the solutes in ground water; for many solutes the solute front will travel more slowly than the rate of the advecting ground water.

**Reverse type curve** A plot on logarithmic paper of the well function $W(u)$ as a function of $1/(u)$.

**Reynolds number** A number, defined by an equation, that can be used to determine whether flow will be laminar or turbulent.

**Riparian doctrine** A doctrine that holds that the property owner adjacent to a surface-water body has first right to withdraw and use the water.

**Rock, igneous** A rock formed by the cooling and crystallization of a molten rock mass called magma.

**Rock, metamorphic** A rock formed by the application of heat and pressure to preexisting rocks.

**Rock, plutonic** An igneous rock formed when magma cools and crystallizes within the earth.

**Rock, sedimentary** A rock formed from sediments through a process known as diagenesis or formed by chemical precipitation in water.

**Rock, volcanic** An igneous rock formed when molten rock called lava cools on the earth's surface.

**Root zone** The zone from the land surface to the depth penetrated by plant roots. The root zone may contain part or all of the unsaturated zone, depending upon the depth of the roots and the thickness of the unsaturated zone.

**Runoff** The total amount of water flowing in a stream. It includes overland flow, return flow, interflow, and baseflow.

**Safe yield** The amount of naturally occurring ground-water that can be economically and legally withdrawn from an aquifer on a sustained basis without impairing the native ground-water quality or creating an undesirable effect such as environmental damage. It cannot exceed the increase in recharge or leakage from adjacent strata plus the reduction in discharge, which is due to the decline in head caused by pumping.

**Saline-water encroachment** The movement, as a result of human activity, of saline ground water into an aquifer formerly occupied by fresh water. Passive saline-water encroachment occurs at a slow rate owing to a general lowering of the fresh-water potentiometric surface. Active saline-water encroachment proceeds at a more rapid rate owing to the lowering of the fresh-water potentiometric surface below sea level.

**Salt-water encroachment** *See* saline-water encroachment.

**Sand model** A scale model of an aquifer; built using a porous medium to demonstrate ground-water flow.

**Sanitary landfill** The disposal of solids and, in some instances, semisolid and liquid wastes by burying the material to shallow depths, usually in unconsolidated materials.

**Saprolite** A soft, earthy, decomposed rock, typically clay-rich, formed in place by chemical weathering of igneous and metamorphic rocks.

**Saturated zone** The zone in which the voids in the rock or soil are filled with water at a pressure greater than atmospheric. The water table is the top of the saturated zone in an unconfined aquifer.

**Saturation ratio** The ratio of the volume of contained water in a soil to the volume of the voids of the soil.

**Schlumberger array** A particular arrangement of electrodes used to measure surface electrical resistivity.

**Screen** *See* well screen.

**Sediment** An assemblage of individual mineral grains that were deposited by some geologic agent such as water, wind, ice, or gravity.

**Seepage velocity** The actual rate of movement of fluid particles through porous media.

**Seismic refraction** A method of determining subsurface geophysical properties by measuring the length of time it takes for artificially generated seismic waves to pass through the ground.

**Shelby tube** A sampling device that is pushed into an unconsolidated aquifer ahead of the drill bit. Typically, the Shelby tube is pushed by hydraulic means.

**Single-point resistance log** A borehole log made by lowering a single electrode into the well with the other electrode at the ground surface. It measures the overall electrical resistivity of the formation and drilling fluid between the surface and the probe.

**Sinkhole spring** A spring created by ground water flowing from a sinkhole in karst terrane.

**Slug test** An aquifer test made either by pouring a small instantaneous charge of water into a well or by withdrawing a slug of water from the well. A synonym for this test, when a slug of water is removed from the well, is a bail-down test.

**Slurry wall** An underground wall designed to stop ground-water flow; constructed by digging a trench and backfilling it with a slurry rich in bentonite clay.

**Soil liquefaction** A process that occurs when saturated sediments are shaken by an earthquake. The soil can lose its strength and cause the collapse of structures with foundations in the sediment.

**Soil moisture** The water contained in the unsaturated zone.

**Solubility product** The equilibrium constant that describes a solution of a slightly soluble salt in water.

**Specific capacity** An expression of the productivity of a well, obtained by dividing the rate of discharge of water from the well by the drawdown of the water level in the well. Specific capacity should be described on the basis of the number of hours of pumping prior to the time the drawdown measurement is made. It will generally decrease with time as the drawdown increases.

**Specific discharge** An apparent velocity calculated from Darcy's law; represents the flow rate at which water would flow in an aquifer if the aquifer were an open conduit.

**Specific electrical conductance** The ability of water to transmit an electrical current. It is related to the concentration and charge of ions present in the water.

**Specific retention** The ratio of the volume of water the rock or sediment will retain against the pull of gravity to the total volume of the rock or sediment.

**Specific weight** The weight of a substance per unit volume. The units are newtons per cubic meter.

**Specific yield** The ratio of the volume of water a rock or soil will yield by gravity drainage to the volume of the rock or soil. Gravity drainage may take many months to occur.

**Spiked sample** A water sample to which a known quantity of a solute has been added so that the accuracy of the laboratory in analyzing the sample can be determined.

**Split-spoon sample** A sample of unconsolidated material taken by driving a sampling device ahead of the drill bit in a boring. The split-spoon sampler is typically advanced by the repetitive dropping of a weight.

**Spontaneous potential log** A borehole log made by measuring the natural electrical potential that develops between the formation and the borehole fluids.

**Stagnation point** A place in a ground-water flow field at which the ground water is not moving. The magnitude of vectors of hydraulic head at the point are equal but opposite in direction.

**Stem flow** The process by which rainwater drips and flows down the stems and branches of plants.

**Stiff pattern** A graphical means of presenting the chemical analysis of the major cations and anions of a water sample.

**Storage, specific** The amount of water released from or taken into storage per unit volume of a porous medium per unit change in head.

**Storativity** The volume of water an aquifer releases from or takes into storage per unit surface area of the aquifer per unit change in head. It is equal to the product of specific storage and aquifer thickness. In an unconfined aquifer, the storativity is equivalent to the specific yield. Also called storage coefficient.

**Storm hydrograph** A graph of the discharge of a stream over the time period when, in addition to direct precipitation, overland flow, interflow, and return flow are adding to the flow of the stream. The storm hydrograph will peak owing to the addition of these flow elements.

**Stream, gaining** A stream or reach of a stream, the flow of which is being increased by inflow of ground water. Also known as an effluent stream.

**Stream, losing** A stream or reach of a stream that is losing water by seepage into the ground. Also known as an influent stream.

**Successive overrelaxation method** A particular type of method for solving for the head in a finite-difference ground-water model.

**Suction lysimeter** A device for withdrawing pore water samples from the unsaturated zone by applying tension to a porous ceramic cup.

**Swallow hole** A vertical shaft in a karst terrane leading from a surface stream into an underground cavern.

**Tensiometer** A device used to measure the soil-moisture tension in the unsaturated zone.

**Tension** The condition under which pore water exists at a pressure less than atmospheric.

**Theis equation** An equation for the flow of ground water in a fully confined aquifer.

**Theissen method** A process used to determine the effective uniform depth of precipitation over a drainage basin with a nonuniform distribution of rain gages.

**Throughflow** The lateral movement of water in an unsaturated zone during and immediately after a precipitation event. The water from throughflow seeps out at the base of slopes and then flows across the ground surface as return flow, ultimately reaching a stream or lake.

**Time of concentration** The time it takes for water to flow from the most distant part of the drainage basin to the measuring point.

**Tortuosity** The actual length of a ground-water–flow path, which is sinuous in form, divided by the straight-line distance between the ends of the flow path.

**Transmissivity** The rate at which water of a prevailing density and viscosity is transmitted through a unit width of an aquifer or confining bed under a unit hydraulic gradient. It is a function of properties of the liquid, the porous media, and the thickness of the porous media.

**Transpiration** The process by which plants give off water vapor through their leaves.

**Trilinear diagram** A method of graphically plotting the chemical composition of the major anions and cations of a water sample.

**Turbidity** Cloudiness in water due to suspended and colloidal organic and inorganic material.

**Turbulent flow** That type of flow in which the fluid particles move along very irregular paths. Momentum can be exchanged between one portion of the fluid and another. Compare with Laminar flow.

**Uniformity coefficient** The ratio of the grain size that is 60 percent finer by weight to the grain size that is 10 percent finer by weight on the grain-size distribution curve. It is a measure of how well or poorly sorted sediment is.

**Unsaturated zone** The zone between the land surface and the water table. It includes the root zone, intermediate zone, and capillary fringe. The pore spaces contain water at less than atmospheric pressure, as well as air and other gases. Saturated bodies, such as perched ground water, may exist in the unsaturated zone. Also called zone of aeration and vadose zone.

**Vadose cave** A cave that occurs above the water table.

**Vadose water** Water in the zone of aeration.

**Vadose zone** *See* unsaturated zone.

**Viscosity** The property of a fluid describing its resistance to flow. Units of viscosity are newton-seconds per meter squared or pascal-seconds. Viscosity is also known as dynamic viscosity.

**Volatile organic compound (VOC)** An organic compound that is characterized by being highly mobile in ground water and which is readily volatilized into the atmosphere.

**Water budget** An evaluation of all the sources of supply and the corresponding discharges with respect to an aquifer or a drainage basin.

**Water content** The weight of contained water in a soil divided by the total weight of the soil mass.

**Water equivalent** The depth of water obtained by melting a given thickness of snow.

**Water quality criteria** Values for dissolved substances in water based upon their toxicological and ecological impacts.

**Water table** The surface in an unconfined aquifer or confining bed at which the pore water pressure is atmospheric. It can be measured by installing shallow wells extending a few feet into the zone of saturation and then measuring the water level in those wells.

**Water-table cave** A cave that forms at the approximate position of the water table.

**Water-table map** A specific type of potentiometric-surface map for an unconfined aquifer; shows lines of equal elevation of the water table.

**Weir** A device placed across a stream and used to measure the discharge by having the water flow over a specifically designed spillway.

**Well casing** A solid piece of pipe, typically steel or PVC plastic, used to keep a well open in either unconsolidated materials or unstable rock.

**Well development** The process whereby a well is pumped or surged to remove any fine material that may be blocking the well screen or the aquifer outside the well screen.

**Well, fully penetrating** A well drilled to the bottom of an aquifer, constructed in such a way that it withdraws water from the entire thickness of the aquifer.

**Well function** An infinite-series term that appears in the Theis equation of ground-water flow.

**Well interference** The result of two or more pumping wells, the drawdown cones of which intercept. At a given location, the total well interference is the sum of the drawdowns due to each individual well.

**Well log** *See* lithologic log.

**Well, partially penetrating** A well constructed in such a way that it draws water directly from a fractional part of the total thickness of the aquifer. The fractional part may be located at the top or the bottom or anywhere in between in the aquifer.

**Well screen** A tubular device with either slots, holes, gauze, or continuous-wire wrap; used at the end of a well casing to complete a well. The water enters the well through the well screen.

**Wenner array** A particular arrangement of electrodes used to measure surface electrical resistivity.

**Wilting point** The soil-moisture content below which plants are unable to withdraw soil moisture.

**Winters Doctrine** A United States doctrine holding that when Indian reservations were established, the federal government also reserved the water rights necessary to make the land productive.

**Xerophyte** A desert plant capable of existing by virtue of a shallow and extensive root system in an area of minimal water.

**Zone of aeration** *See* unsaturated zone.

# Index